Inside Rhinoceros® 4

Ron K. C. Cheng

THOMSON
DELMAR LEARNING™

Australia Canada Mexico Singapore Spain United Kingdom United States

Inside Rhinoceros 4®

Ron K. C. Cheng

Vice President, Technology and Trades ABU:
David Garza

Director of Learning Solutions:
Sandy Clark

Managing Editor:
Larry Main

Senior Acquisitions Editor:
James Gish

Senior Product Manager:
John Fisher

Marketing Director:
Deborah S. Yarnell

Channel Manager:
Kevin Rivenburg

Marketing Coordinator:
Mark Pierro

Production Director:
Patty Stephen

Production Manager:
Stacy Masucci

Production Technology Analyst:
Thomas Stover

Senior Content Project Manager:
Elizabeth C. Hough

Art Director:
Benj Gleeksman

Editorial Assistant:
Sarah Timm

Cover Images:
Getty Images

Printed in Canada
1 2 3 4 5 XX 11 10 09 08 07

For more information contact
Thomson Delmar Learning
Executive Woods, 5 Maxwell Drive,
PO Box 8007, Clifton Park, NY
12065-8007

Or find us on the World Wide Web at
www.delmarlearning.com

Library of Congress Cataloging-in-Publication Data
Cheng, Ron.
Inside Rhinoceros 4 / Ron K.C. Cheng.
p. cm.
ISBN-13: 978-1-4180-2101-6
ISBN-10: 1-4180-2101-6
1. Computer graphics. 2. Three-dimensional display systems. 3. Rhino (Computer file) I. Title.
T385.C4746 2007 006.6--dc22

2007028313
ISBN-10: 1-4180-2101-6
ISBN-13: 978-14180-2101-6

About the Author

Cheng leads the Engineering Design and Communication Unit of the Industrial Center of The Hong Kong Polytechnic University. His main areas of interest are industrial and engineering product development and realization, as well as application of computer-aided tools in product design. He is actively involved in development and realization of industrial design projects, engineering development projects, integrated learning projects, and manufacturing projects. Including this book, Cheng has written 17 books related to computer-aided design applications, including AutoCAD, Mechanical Desktop, Autodesk Inventor, Pro/Desktop, and SmartSketch. Among them, the Autodesk Inventor book has been translated into Russian.

Acknowledgments

Grateful acknowledgment is made to Jerry Hambley, Bill Adamoski, John Novak, and Bruce Weirich for their reviews of content and helpful suggestions. I would also like to thank the Delmar team for their efforts in association with this project, in particular, Acquisitions Editor Jim Gish, Senior Product Manager John Fisher, Project Manager Gina Dishman, and production house GEX Publishing Services.

Contents

Introduction

Rhinoceros, also known as Rhino, is a 3D surface modeling application. It is continually gaining popularity in the industry, including industrial design, marine design, jewelry design, CAD/CAM, rapid prototyping, reverse engineering, graphic design, and multimedia. It is based on the popular NURBS (Non-Uniform Rational B-Spline) mathematics, which enables the construction of free form organic surfaces that are compatible with most other computer models used in the industry.

To cope with rapid prototyping and some animation and visualization applications that use faceted approximation instead of NURBS surfaces to represent 3D free-form objects, Rhino also provides a set of polygon mesh tools.

Making NURBS surfaces usually requires a set of curve frameworks. Therefore, Rhino has a set of curve tools. To enable modification and improvising surface models, Rhino incorporates a comprehensive set of editing, transformation, and analysis tools. Apart from that, it has tools for outputting 2D engineering drawings and rendered images and animations.

Audience and Prerequisites

This book is intended for students and practitioners in the following industry: industrial design, marine design, jewelry design, CAD/CAM, rapid prototyping, reverse engineering, graphic design, and multimedia. There is no prerequisite needed for several reasons, because the book is written in a non-linear manner, enabling novice to learn the application in a short period of time, and allowing experienced users to discover more advanced use of Rhino in design.

Philosophy and Approach

By having a balanced emphasis on theory, concepts, and tutorials, this book attempts to bridge the theoretical and software-oriented approaches to modeling in the computer, because theory is relatively useless without hands-on experience and vice versa.

To enable both novice and more advanced users to learn about Rhino in a non-linear way, this book has two major components: a set of hands-on tutorials and the main chapters. The set of tutorials is placed at the beginning of the book, allowing all readers to have hands-on experience on using Rhino to construct some real world objects. In order to provide a clear picture about what can be produced by using Rhino, the main chapters are written logically with concepts explained together with simple tutorials focusing on individual elements.

Content

This book is written to Rhinoceros Release 4, addressing the new tools provided as well as tools that are already provided in previous releases. In order to meet the needs of novice as well as more advanced users, the book in written in a logical but non-linear manner.

At the beginning of the book, there is a set of four easy to follow, hands-on case studies. In the first case study, guides are provided to construct the chassis and mechanical components of the famous bubble car. The second case study completes the bubble car by making the free-form body work. Case study three relates to a set of small objects to highlight the use of Rhino in various ways. Finally, in case study four, way to assemble a set of components is suggested. (Case studies 3 and 4 can be found in the companion CD-ROM.)

There are 14 main chapters.

Chapter 1 provides an overview about digital modeling, explains the concepts of surface modeling, and introduces Rhinoceros as a digital modeling tool, as well as an examination of the Rhino user interface, including its key functional components. Chapter 2 familiarizes you with the concepts of construction plane, basic geometry construction and modification methods, layer management, and surface display methods.

Chapter 3 provides you with a clear understanding on the kind of surfaces that you can construct by using Rhino and the framework of points and curves required for building such surfaces. Chapters 4 through 6 delineate the ways to construct and manipulate points and curves for making surfaces. After learning curves, in Chapter 7 you learn more detailed surface manipulation methods. Chapter 8 covers the manner in which Rhino represents solids in the computer and various ways of solid construction. In Chapter 9, you will learn the ways to construct and manipulate polygon meshes, which is significantly used by rapid prototyping.

Chapter 10 explains various methods to transform curves, surfaces, and polygon meshes. Chapter 11 deals with analyzing curves, surfaces, and polygon meshes. In Chapter 12, you learn about grouping drawing objects, construction of data blocks, and use of work sessions. Chapter 13 addresses producing 2D engineering drawings from 3D models, as well as import and export of files. Chapter 14 explains how rendered images and animations can be produced from 3D models.

Finally, Appendix A outlines vaious digital model concepts, Appendix details the ways construction planes are manipulated, and Appendix C introduces the use of digitizer. (Appendices A to C can be found in the companion CD-ROM.)

Features and Conventions

This edition includes a companion CD-ROM at the back of the book (see "About the Companion CD-ROM" at the end of this introduction). Exercises in the book are supported by content found on the companion CD-ROM. Italic font in regular text is used to distinguish certain command names, code elements, file names, directory and path names, user input, and similar items. Italic is also used to highlight terms and for emphasis.

About the Companion CD-ROM

The companion CD-ROM found at the back of this book contains all Rhino files used in conjunciton with exercise, Case Studies 3 and 4, Appendices A to C, as well as a trial version of Rhinoceros Release 4.

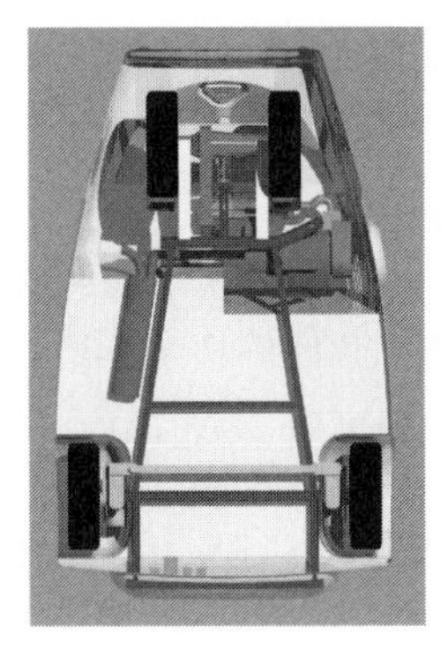

Case Studies

We will begin with a set of case studies, which are intended to let you appreciate how to use Rhinoceros as a tool in computer modeling. After these case studies, there are the main chapters, delineating various computer modeling concepts and Rhinoceros commands in a logical way. It is hoped that you will find these case studies interesting and not too difficult to follow. However, if you are a novice in surface modeling or using Rhinoceros and find these case studies too difficult, it is suggested that you should first go to the main chapters and then come back to the case studies later. Figure C–1 shows the rendered views of the bubble car that we will construct in Case Studies 1 and 2.

Figure C–1. Bubble car case study

NURBS Curves, NURBS Surfaces, Polygon Meshes, and Solids

As will be explained in detail in the main chapters of this book, Rhinoceros is a 3D modeling tool and we can use it to construct three major kinds of geometric objects: NURBS curves, NURBS surfaces, and polygon meshes.

Among them, we use curves as framework on which surfaces are constructed. As for NURBS surfaces and polygon meshes, they are two distinct ways of composing a 3D surface model, with NURBS surfaces being used for making accurate models and polygon meshes used for approximated representation of 3D objects.

In Rhinoceros terms, by joining two or more contiguous NURBS surfaces together, a polysurface is formed. If a single NURBS surface, such as a sphere or ellipsoid, or a polysurface forms a closed loop enclosing a volume without any gap or opening, a solid is implied.

Design Development

Prior working on the case studies, let us have a quick review on the two-stage iterative design development process.

Top-Down Thinking Process

With an idea or concept in mind, you start deconstructing the 3D object into discrete surface elements by identifying and matching various portions of the object with various types of surfaces. Unless the individual surfaces are primitive surfaces, like box, sphere, or ellipsoid, you need to think about the shapes and locations of curves and points required to build the surface, as well as the surface construction commands to be applied on the curves and points. This is a top-down thinking process.

Bottom-Up Construction Process

After this thinking process, you create the curves and points and apply appropriate surface construction commands. By putting the surfaces together properly in 3D space, you obtain the 3D object. This is the bottom-up construction process. If the surfaces you construct do not conform to your concept, you think about the curves and points again.

Detailing

With practice you gain the experience that makes this process more efficient. Logically, you construct points and curves, and from these construct surfaces and solids. In detailing your design, you add surface features. To construct surface features that conform in shape to existing surfaces, you create points and curves from existing objects.

Surface Modeling Case Studies

The prime goal of surface modeling is to construct a set of surfaces and put them together to represent an object depicting a design. Constructing surfaces inevitably requires curves and points. In comparison to surface construction, curve construction is a tedious job.

Because the location and shape of the curves have a direct impact on the shape of surfaces constructed from them, the particulars of most curves in these case studies are given to you. While working on these case studies, you should try to relate the 3D curves to the 3D surfaces. It is hoped that you reverse the process, seeing the curves when a surface is given.

CASE STUDY 1

Bubble Car Chassis

Introduction

This case study, together with the next one, concerns making a 1:10 scale model of the famous bubble car, which was first designed and produced by an Italian refrigerator manufacturer over fifty years ago. The design was later franchised to a number of car manufacturers, who built the car in huge quantities with a variety of configurations. The computer model that we are going to construct, as shown in Figures C1–1 and C1–2, is based on an earlier model.

Figure C1–1. Images of the finished Rhinoceros bubble car model rendered using Rhino renderer (left) and Penguin renderer (right)

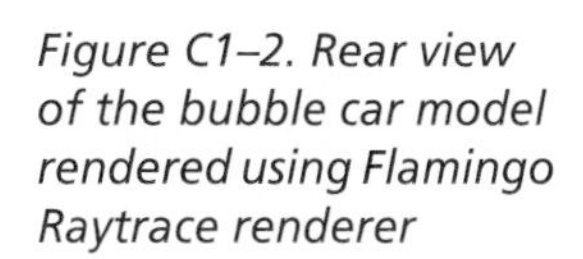

Figure C1–2. Rear view of the bubble car model rendered using Flamingo Raytrace renderer

The bubble car is chosen as a Rhinoceros computer modeling case study because the car has a simple engineering structure and a stylish body. In addition, you can find a significant amount of information about the car by searching over the Internet, using the keyword "isetta." For the sake of simplicity, construction of the model will be carried out in two unique case study sessions. In the first session, we will work on the engineering components. Making the engineering components can help you understand the process of constructing geometric shapes of regular pattern. In the second session, we will concentrate on body panel parts. Building the body panels lets you experience how free-form surfaces can be constructed. The components that we will construct here are shown in Figure C1–3. The main body panels, consisting of an egg-like shell (Figure C1–4) and a refrigerator-like front door (Figure C1–5), will be addressed in the next session.

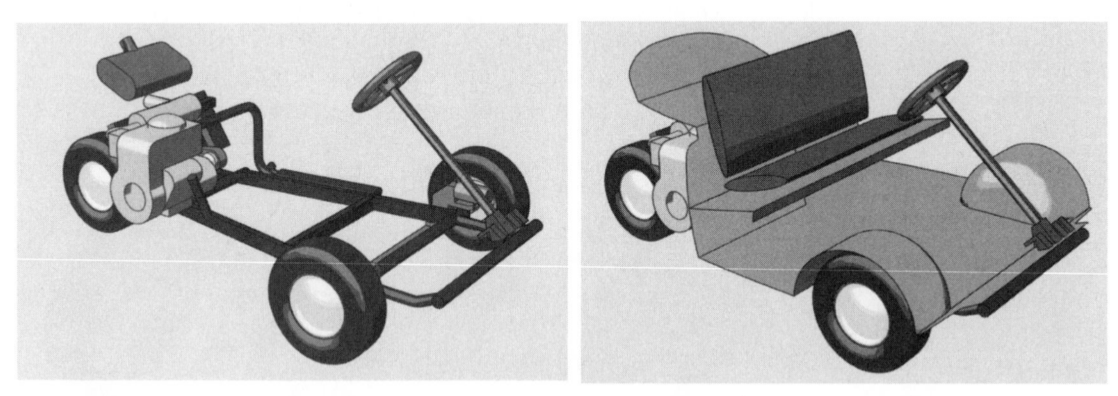

Figure C1–3. Bubble car mechanical components

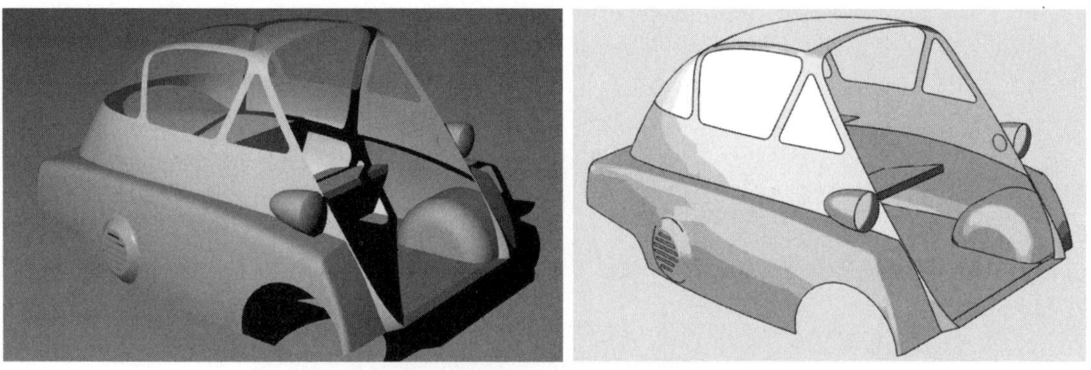

Figure C1–4. Bubble car body panels rendered in Rhino renderer (left) and Penguin renderer (right)

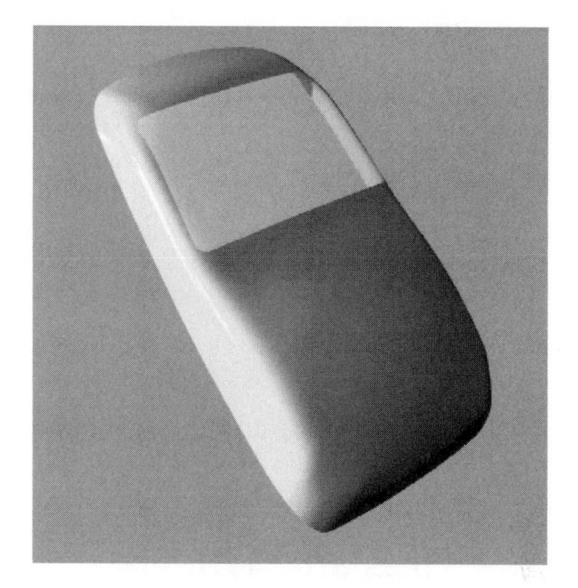

Figure C1–5. Bubble car's front door panel

Constructing the Mechanical Components

Figure C1–6 shows the bottom views of the components we will make. They include the wheels, tires, rear axle, chassis, rear suspension, engine, exhaust system, intake system, gas tank, front suspension, drive controls, floor panels, and seat.

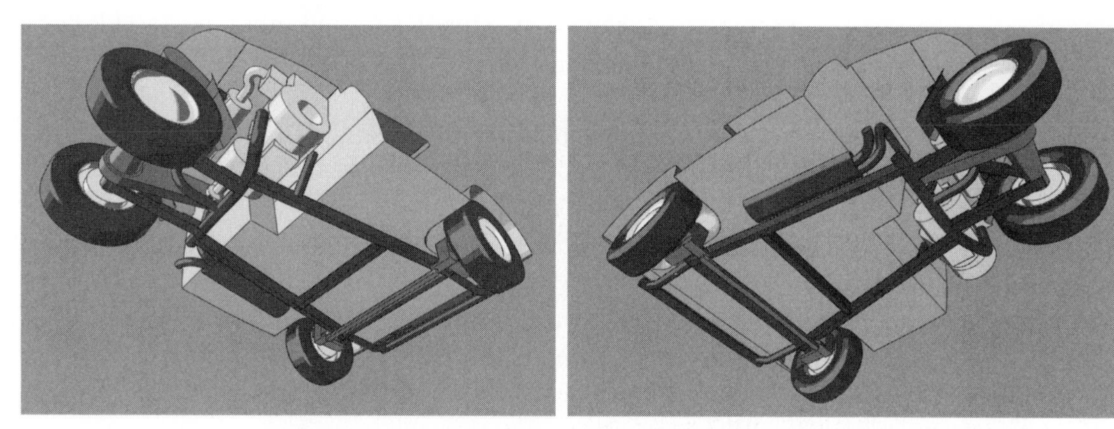

Figure C1–6. Bottom views of the mechanical components of the bubble car model

Template File

To build the mechanical components, we need NURBS curves as wire-frame skeletons, upon which we will build the NURBS surfaces, polysurfaces, and solids. Because curve building is a tedious task, we have constructed all the required curves and placed them in various layers of a file saved in the CD accompanying this book. As the computer model is a 1:10 scale model of the real car, we have simplified most of the component parts appropriately. Now open the template file.

1. Select File > Open and select the file Bubble Car 1.3dm from the Case Study 1 folder on the companion CD.

In the template file, the grid lines are turned on and grid extents have been set to 200 mm. However, grid lines are omitted in the illustrations shown here to improve clarity.

Tires and Wheels

As shown in Figure C1–6, the Top viewport is maximized and there are several curves residing on three separate layers. Curves X, Y, and Z serve to indicate the rear track, wheelbase, and front track, respectively. They are used as a reference for positioning the tires and wheels. As for curves A and D, they depict the cross section of the tire and wheel. We will revolve them to construct a set of revolved surfaces. We will continue with the following steps:

2. Select Surface > Revolve, or click on the Revolve/Rail Revolve button on the Surface toolbar.
3. Select curve A, shown in Figure C1–7, and press the ENTER key.
4. Use the Osnap toolbar to set OSNAP to End, and then select endpoints B and C.
5. Select the FullCircle option on the command area. A polysurface depicting the wheel of the bubble car is constructed.
6. Set current layer to Tire.
7. Select Surface > Revolve, or click on the Revolve/Rail Revolve button on the Surface toolbar.
8. Select curve D, shown in Figure C1–7, and press the ENTER key.
9. Select endpoints B and C, and select the FullCircle option on the command area. A closed polysurface representing the tire is constructed.
10. Turn off the Curves Wheel and Curves Tire layers.

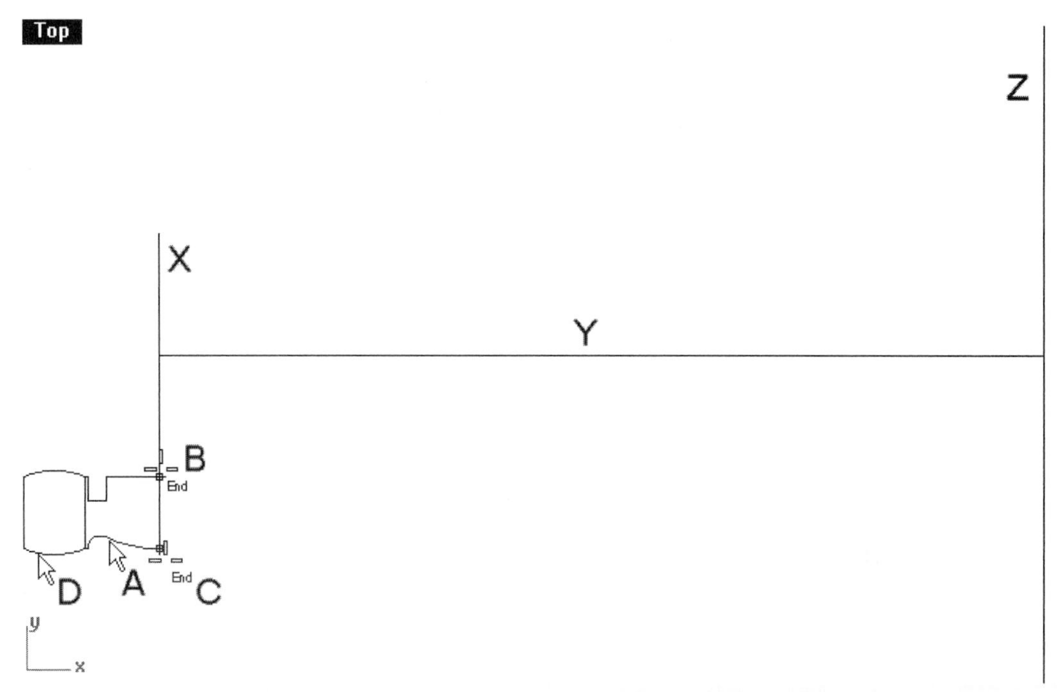

Figure C1–7. Curves for making the tire and wheel as well as curves indicating the wheelbase and track

Because curves A and D are sets of curve elements joined together, the outcome is sets of surfaces joined together. In other words, they are polysurfaces. As each of these polysurfaces encloses a volume without any opening, they are regarded as solids in Rhinoceros terms.

The edges of the polysurface representing the tire are to be rounded by filleting. Basically, there are two different ways to construct a filleted edge: using the fillet command from the Surface menu or the fillet commands in the Solid menu. Because the surfaces are already joined as a polysurface, we will use the solid filleting command to round off two edges, as follows:

11 Select Solid > Fillet Edge > Fillet Edge, or click on the Variable Radius Fillet/Variable Radius Blend button on the Solid Tools toolbar.

12 Select the Current Radius option on the command area and set it to 2 mm.

13 Select edges A and B, indicated in Figure C1–8, and press the ENTER key twice. The edges are filleted.

We will now copy the tire and wheel from the rear axis to the front axis of the car, using the front and rear track lines as references.

14 Select Transform > Copy, or click on the Copy button on the Transform toolbar.

15 Select the tire B and wheel C (Figure C1–8), and press the ENTER key.

16 Select endpoint D and endpoint E, and press the ENTER key. The tire and wheel are copied.

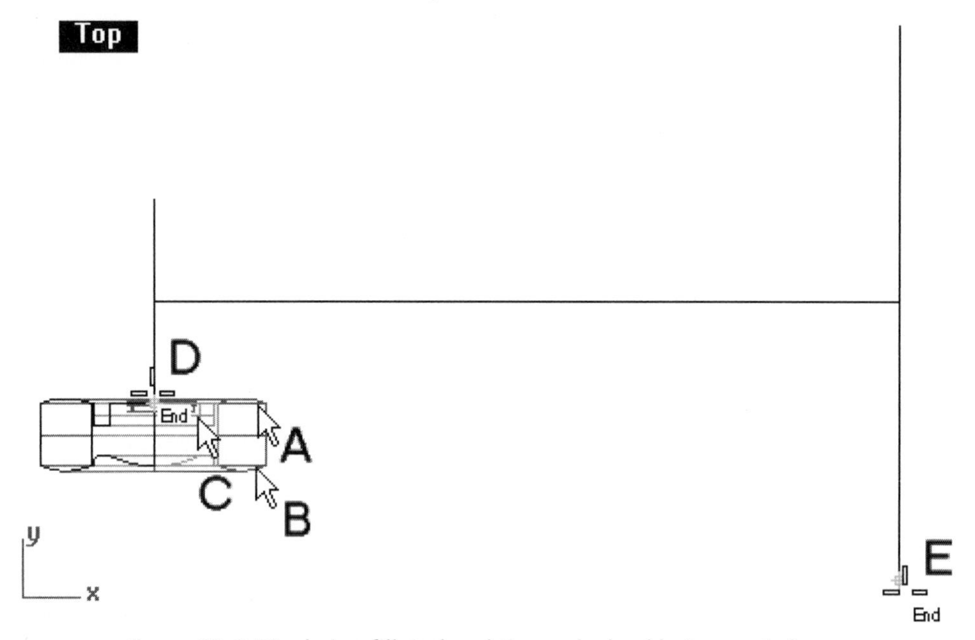

Figure C1–8. Tire being filleted and tire and wheel being copied

As the car is symmetric left and right, we will make a mirror copy of the tires and wheels, as follows.

17 Select Transform > Mirror, or click on the Mirror/Mirror on 3 point plane button on the Transform toolbar.

18 Select A, B, C, and D, shown in Figure C1–9, and press the ENTER key.

19 Click on endpoints E and F, as shown in Figure C1–9.

The tires and wheels are complete. For your reference, Figure C1–9 also shows the perspective rendered view of the tires and wheels.

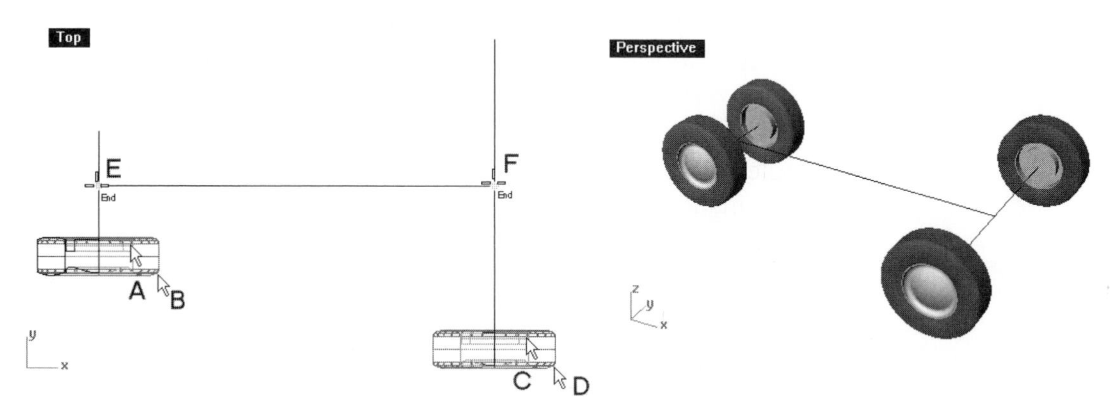

Figure C1–9. Tires and wheels being mirrored (left) and tires and wheels mirrored (right)

Rear Axle

In our model, the rear axle, as shown in Figure C1–10, will be simplified and constructed by first making two closed polysurfaces and then combining them together. To construct these polysurfaces, we need to extrude one curve and revolve another curve.

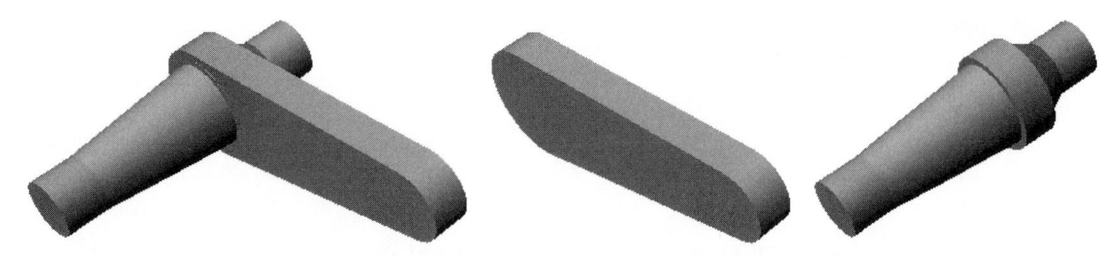

Figure C1–10. Rear axle (left) and two polysurfaces (middle and right)

Now we will continue with the following steps.

20 Turn on the Curves Rear Axle layer and set the current layer to Rear Axle.

21 Select Solid > Extrude Planar Curve > Straight, or click on the Extrude Closed Planar Curve button on the Solid toolbar. (Using this command will extrude the curve as well as make two planar surfaces, thus producing a closed polysurface—a solid.)

22 Select curve A, indicated in Figure C1–11, and press the ENTER key.

23 Click on endpoint B. A closed polysurface is constructed.

24 Select Surface > Revolve, or click on the Revolve button on the Revolve/Rail Revolve button on the Surface toolbar.

25 Select curve C, shown in Figure C1–11, and press the ENTER key.

26 Select endpoints D and E, and select the FullCircle option on the command area. A polysurface is constructed.

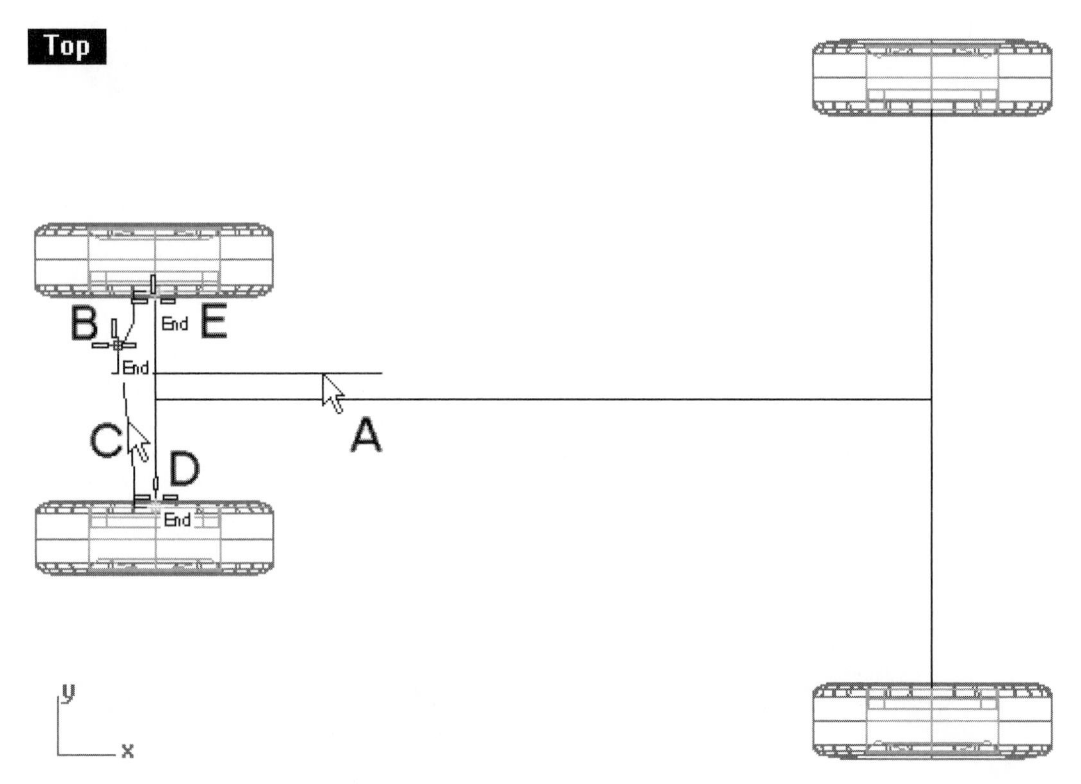

Figure C1–11. A curve being extruded and another curve being revolved

Now we have two closed polysurfaces (Rhinoceros solids). To combine them, we will use the Boolean Union operation, as follows:

27 Turn off Layer Curves Rear Axle.

28 Select Solid > Union, or click on the Boolean Union button on the Solid Tools toolbar.

29 Select polysurfaces A and B, shown in Figure C1–12, and press the ENTER key. The polysurfaces are combined.

The rear axle is complete.

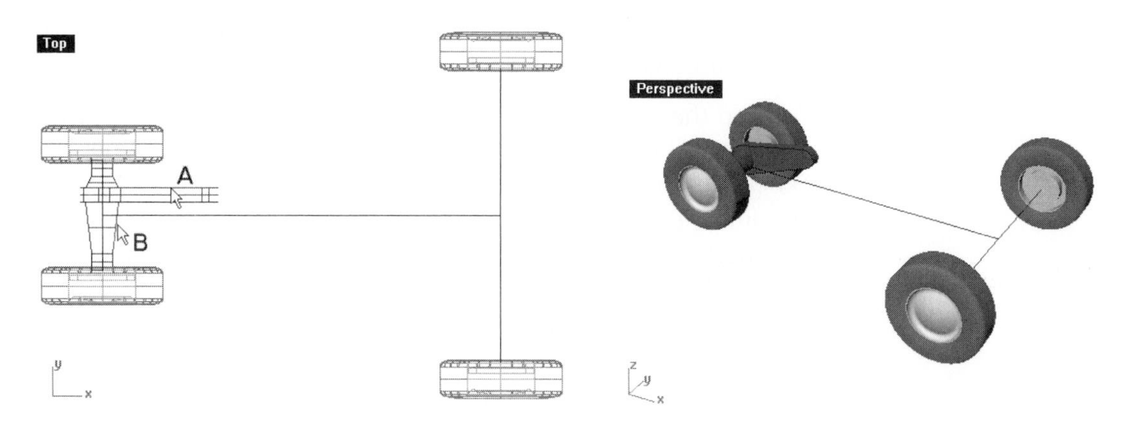

Figure C1–12. Polysurfaces being combined using Boolean Union (left) and perspective view of the model (right)

Chassis

Here, the chassis of the bubble car includes not only the tubular framework, but also the mud guards. For the sake of simplicity, the hollowness of the frame and the mounting bolts and nuts are disregarded. For your reference, Figure C1–13 shows the chassis and mud guards as well as the chassis, mud guards, and components already constructed.

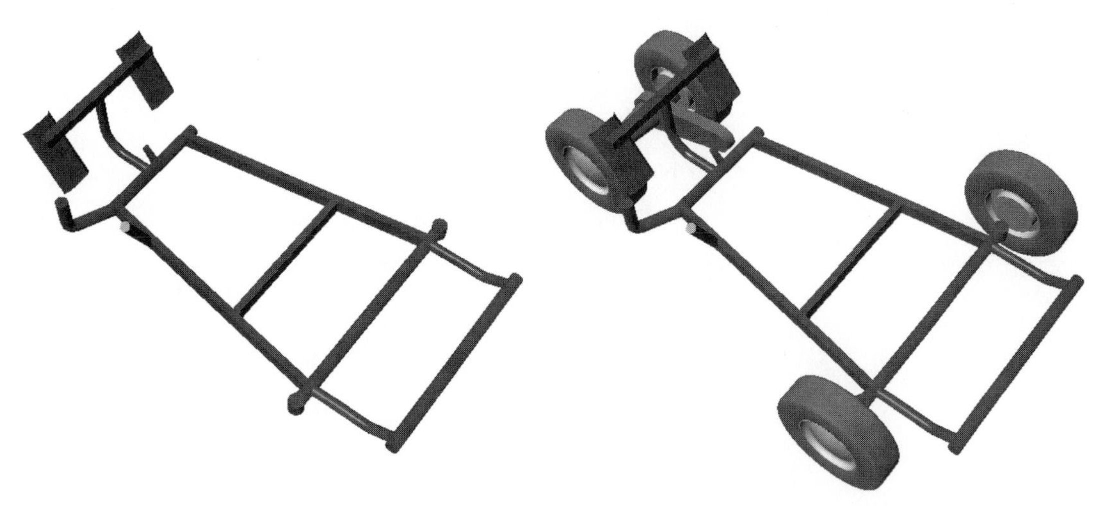

Figure C1–13. Chassis of the bubble car (left) and chassis and other components already constructed (right)

So far, we have been only extruding and revolving curves. To make the chassis, we will perform a sweeping operation as well. Again, curves for making these objects are already constructed and reside on a layer. Now we will continue with the following steps:

30 Turn on the Curves Chassis layer, set the current layer to Chassis, and turn off the layers Curves Wheelbase, Curves Rear Axle, Tire, Wheel, and Rear Axle.

31 Maximize the Perspective viewport by double-clicking the Top viewport's label and then double-clicking the Perspective viewport's label.

32 Select Surface > Sweep 1 Rail, or click on the Sweep 1 Rail button on the Surface toolbar.

33 Select curve A (Figure C1–14) as the rail, select circle B as the section, and press the ENTER key. Because the location where we click on the rail curve has an effect on the direction of the resulting swept surface, when selecting the rail curve A, we have to click near circle B.

34 Click on the OK button of the Sweep 1 Rail Options dialog box. A swept surface is constructed.

35 Repeat the command three more times, using curve C as the rail and circle D as the section, using curve E as the rail and circle F as the section, and then using curve G as the rail and circle H as the section.

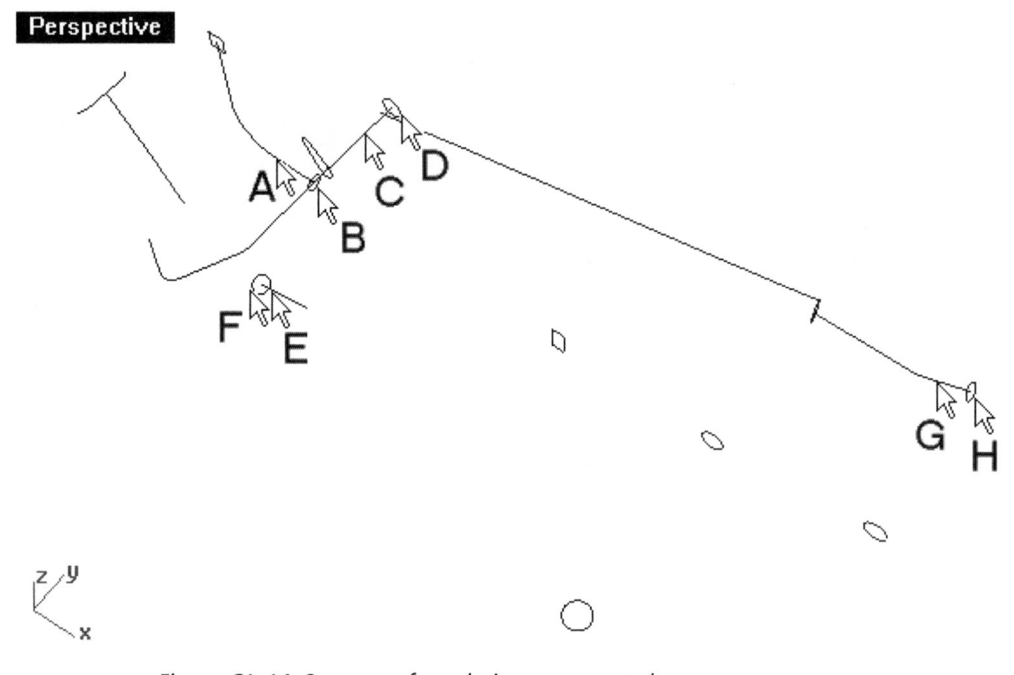

Figure C1–14. Swept surfaces being constructed

Four swept surfaces are constructed. If you take a closer look, you will find that both ends of the swept surfaces are opened. In order to use the Boolean Union command to combine these surfaces with other polysurfaces that we will construct later on, we will cap their planar holes, causing the objects to become closed polysurfaces. In Rhinoceros terms, they will become solids. We will continue with the following steps.

36 Select Solid > Cap Planar Holes, or click on the Cap Planar Holes button on the Solid Tools toolbar.

37 Select surfaces A, B, C, and D, shown in Figure C1–15, and press the ENTER key. The swept surfaces are capped and become closed polysurfaces.

Let us construct a few more polysurfaces, as follows:

38 Select Solid > Extrude Planar Curve > Along Curve, or click on the Extrude Curve Along Curve/Extrude Curve Along Sub Curve button on the Extrude Solid toolbar.

39 Select curve E (Figure C1–15) and press the ENTER key. You may have to zoom in more closely to select curve E.

40 Select curve F as the path curve. Because the location where we click on the path curve has an effect on the direction of the resulting extruded surface, when selecting the rail curve F, we have to click near curve E. An extruded solid is constructed.

41 Repeat the command, using curve G as the curve to extrude and curve H as the path curve. Because curve G is an open curve with two sharp corners, the resulting object is an open polysurface consisting of three surfaces joined together. This polysurface will represent the mud guard of the bubble car.

42 Select Solid > Extrude Planar Curve > Straight, or click on the Extrude Closed Planar Curve button on the Solid toolbar.

43 Select curve J and press the ENTER key.

44 Click on the Both Sides option on the command area, if Both Sides = No. Otherwise, proceed to next step.

45 Type 30 at the command area. The curve is extruded 30 mm in both directions. In other words, the total distance of extrusion is 60 mm.

46 Repeat the command to extrude curve K a distance of 1 mm in both directions. Similar to the previous step, the total extrusion distance is two times the specified distance, 2 mm.

47 Repeat the command to extrude curve L a distance of 33 mm in both directions. (Total distance is 66 mm.)

48 Repeat the command to extrude curve M a distance of 46 mm in both directions. (Total distance is 92 mm.)

49 Repeat the command to extrude curve N a distance of 3 mm in both directions. (Total distance is 6 mm.)

50 Repeat the command to extrude curve P a distance of 42 mm in both directions. (Total distance is 84 mm.)

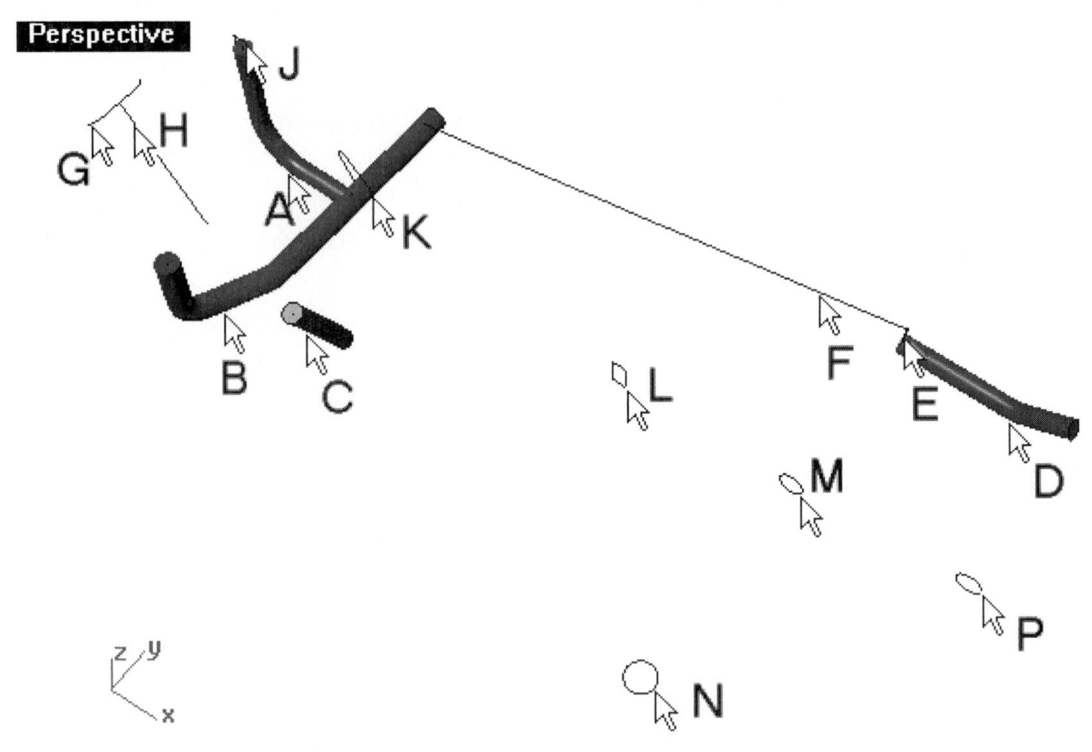

Figure C1–15. Swept surfaces being capped and curves being extruded

To complete the components of the chassis, we will construct three mirrored objects, as follows.

51 Turn off the Curves Chassis layer and turn on the Curves Wheelbase layer.

52 Select Transform > Mirror, or click on the Mirror/Mirror on 3 point plane button on the Transform toolbar.

53 Select polysurfaces A, B, C, and D, indicated in Figure C1–16, and press the ENTER key.

54 Select endpoints E and F. The polysurfaces are mirrored.

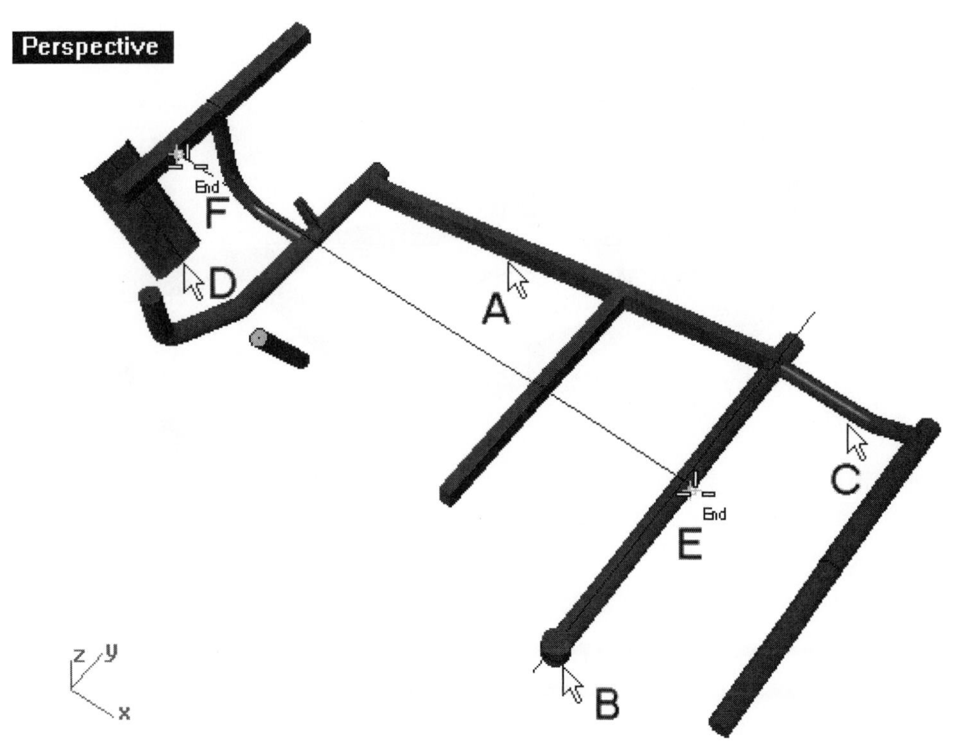

Figure C1–16. Polysurfaces being mirrored

To complete the chassis, we will combine the chassis elements by using the Boolean Union operation. To clean up the overlapping surfaces of a Boolean joined object, we will merge these faces. Now we will continue with the following steps.

55 Select Solid > Union, or click on the Boolean Union button on the Solid Tools toolbar.

56 Select all the polysurface objects except A and B, shown in Figure C1–17, and press the ENTER key. The selected polysurfaces are combined.

57 Select Solid > Solid Edit Tools > Faces > Merge All Faces, or click on the Merge two coplanar faces/Merge all coplanar faces button on the Solid Tools toolbar.

58 Select C (this is the combined chassis) and press the ENTER key. Coplanar faces are merged.

The chassis of the bubble car is complete.

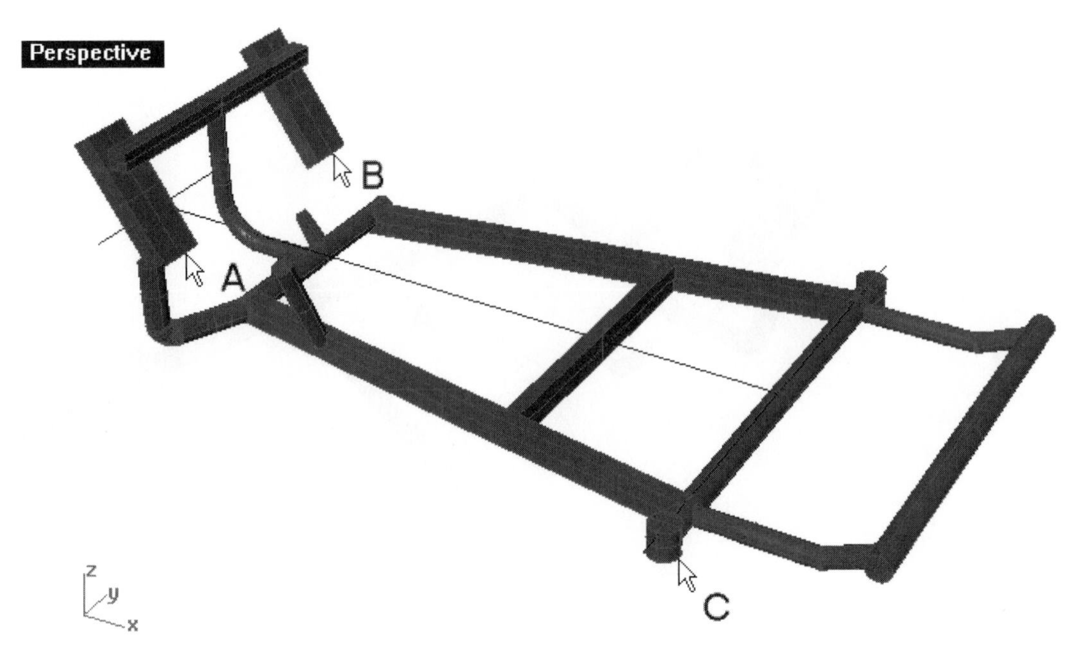

Figure C1–17. Chassis elements being combined and faces being merged

Rear Suspension

The rear suspension of the car consists of two sets of leaf springs and shock absorbers. To depict these objects, we will use the curves already built to construct a revolved polysurface and an extruded polysurface, as follows:

59 Turn on the Curve Rear Suspension layer, set the current layer to Rear Suspension, and turn off the Chassis and Curves Wheelbase layer.

60 Select Solid > Extrude Planar Curve > Straight, or click on the Extrude Closed Planar Curve button on the Solid toolbar.

61 Select curve A (Figure C1–18) and press the ENTER key.

62 Click on the Both Sides option on the command area to set Both Sides = No.

63 Type 5 at the command area. The curve is extruded 5 mm in one direction.

64 Select Surface > Revolve, or click on the Revolve button on the Revolve/Rail Revolve button on the Surface toolbar.

65 Select curve B (Figure C1–18) and press the ENTER key.

66 Select endpoints C and D, and select the FullCircle option on the command area. A polysurface is constructed

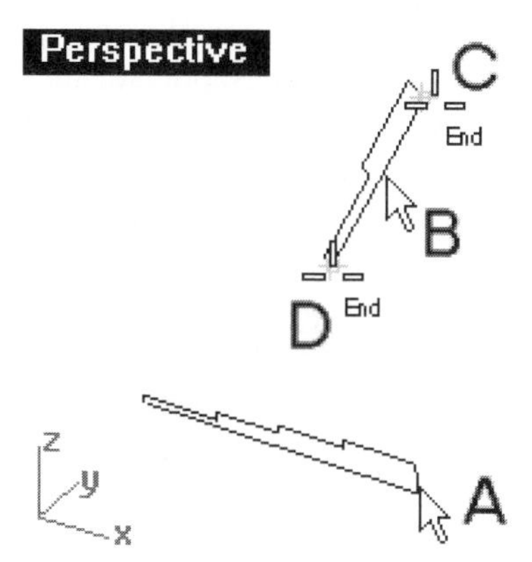

Figure C1–18. Leaf spring and shock absorber being constructed

Now we will mirror the polysurfaces representing the leaf spring and shock absorber, as follows:

67 Turn off the Curve Rear Suspension layer and turn on Curves Wheelbase.

68 Select Transform > Mirror, or click on the Mirror/Mirror on 3 point plane button on the Transform toolbar.

69 Select polysurfaces A and B, shown in Figure C1–19, and press the ENTER key.

70 Select endpoints C and D. The polysurfaces are mirrored

For your reference, all objects constructed to this point are also shown in Figure C1–19 (right).

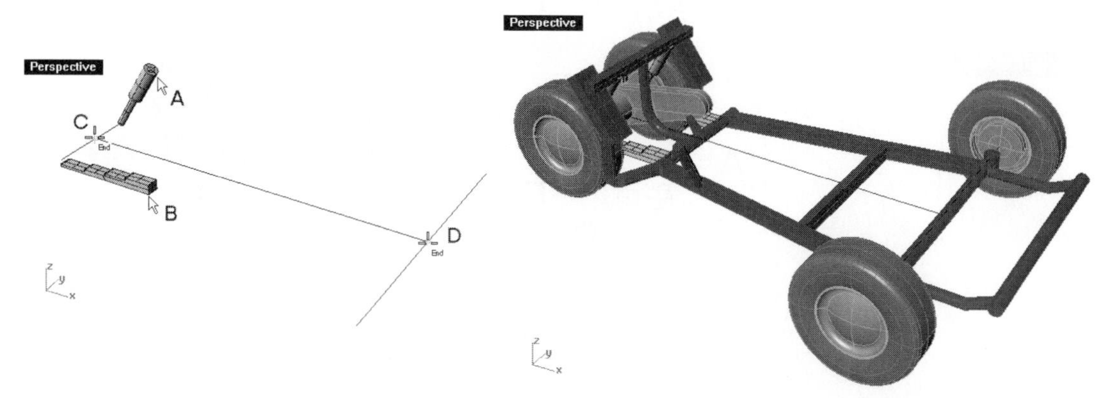

Figure C1–19. Leaf spring and shock absorber being mirrored (left) and all components constructed so far (right)

Engine Unit

The engine unit of the bubble car model will be simplified and represented by two polysurfaces produced by making a set of extruded and revolved surfaces/polysurfaces combined together and then filleted. Figure C1–20 shows the engine unit added to the model.

Figure C1–20. Engine unit added to the model

Now we will continue with the following steps to revolve two curves:

71 Turn on the Curves Engine layer, set the current layer to Engine, and turn off the layers Curves Wheelbase and Rear Suspension.

72 Referencing Figure C1–21, revolve curves A and Dseparately around axes BC and EF to construct two polysurfaces.

Because revolving curve A produces two separate objects, we have to join them for performing Boolean operations in later stages.

73 Select Edit > Join, or click on the Join button on the Main 1 toolbar.

74 Select G and H, and press the ENTER key.

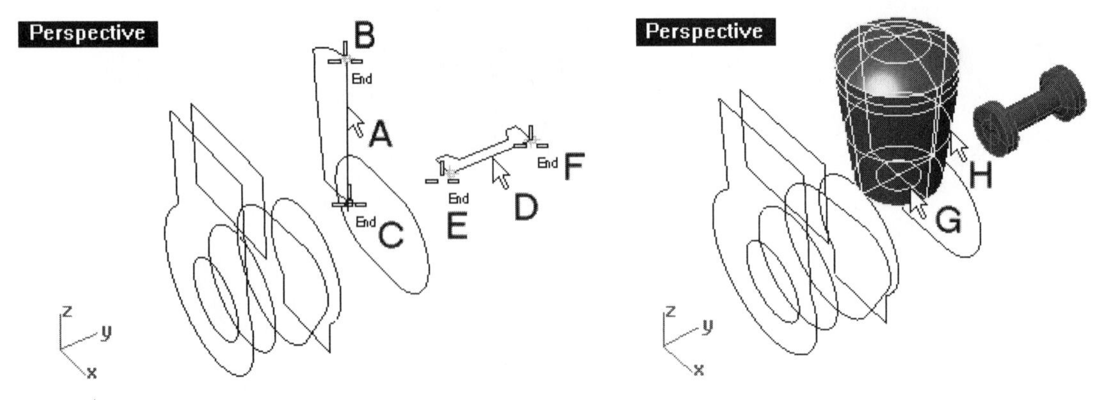

Figure C1–21. Curves being revolved (left) and revolved polysurfaces (right)

Now we will continue with the following steps to construct four closed polysurfaces.

75 Extrude curves A, B, C, and D (Figure C1–22) separately to construct four closed polysurfaces. Extrusion distance for A is 16 mm, B is 15 mm, C is 18 mm, and D is 10 mm.

76 Combine the polysurfaces E, F, G, H, J, and K to form a single polysurface by using Boolean Union.

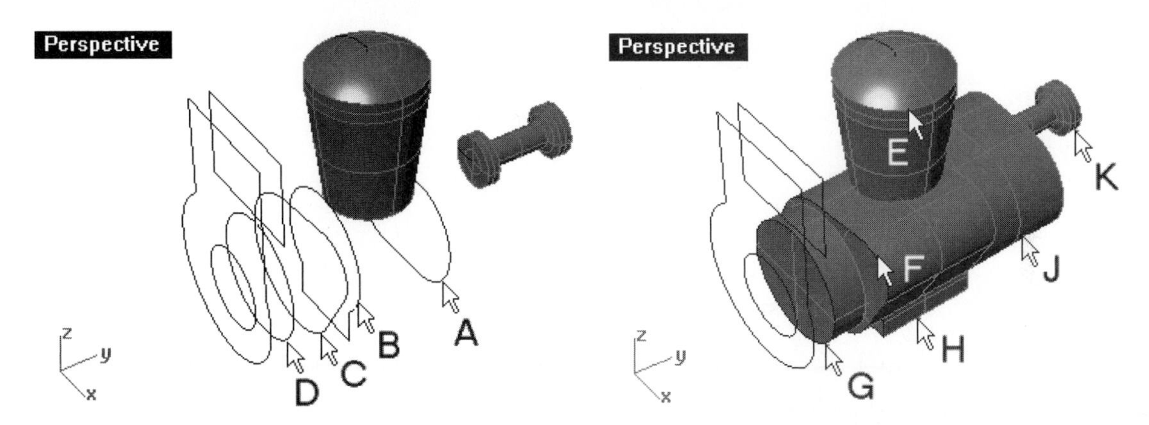

Figure C1–22. Curves being extruded (left) and polysurfaces being combined (right)

The polysurface representing the engine's main body is complete. We will continue with the following steps to construct two polysurfaces and combine them into one:

77 Extrude curve A (Figure C1–23) a distance of 33 mm to build a closed polysurface.

78 Repeat the command to extrude curve B and C together for a distance of 8 mm respectively to another closed polysurface. As can be seen, extruding two curves (one inside another) produces a hollow object.

79 Use Boolean Union to combine polysurfaces D and E to form a single polysurface.

80 Select Solid > Solid Edit Tools > Faces > Merge All Faces, and click on the combined polysurface D to remove unwanted overlapped faces.

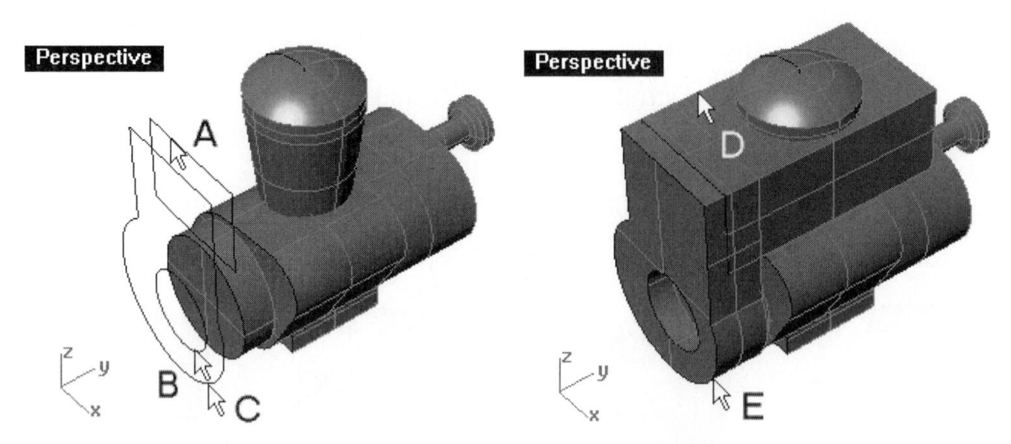

Figure C1–23. Curves being extruded (left) and two polysurfaces being combined (right)

To complete the engine, we will fillet an edge and change the color property of a polysurface, as follows:

81 Select Solid > Fillet Edge > Fillet Edge, or click on Variable Radius Fillet/Variable Radius Blend button on the Solid Tools toolbar.

82 Select the Current Radius option on the command area.

83 Change the radius to 5 mm.

84 Select edge A (Figure C1–24) and press the ENTER key twice. The edge is filleted.

85 Select Edit > Object Properties, or click on the Object Properties/ Hide Properties Window button on the Standard toolbar.

86 Select the filleted polysurface B, shown in Figure C1–24, and press the ENTER key.

87 In the Properties dialog box, change the display color to Magenta.

The engine is complete.

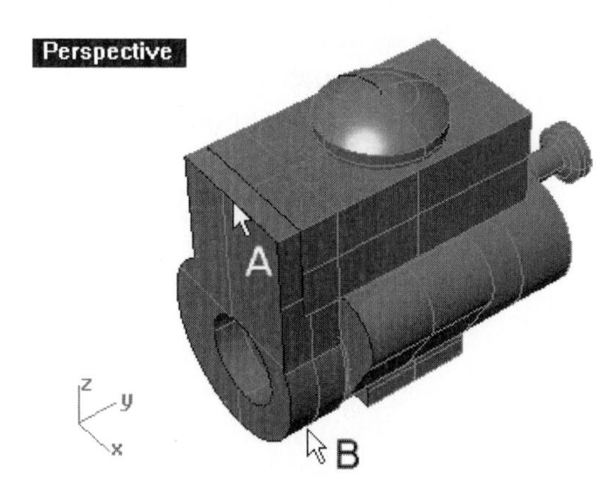

Figure C1–24. Edge being filleted

Exhaust System

The exhaust system of the bubble car is represented by two capped swept surfaces and an extruded polysurface. Figure C1–25 shows the exhaust system incorporated in the model.

Figure C1–25. Exhaust system included in the model

Now we will continue with the following steps.

88 Turn on the Curves Exhaust layer, set the current layer to Exhaust, and turn off the layers Curves Engine and Engine.

89 Construct a Sweep 1 Rail surface, using curve A (Figure C1–26) as rail and circle B as cross section.

90 Construct another Sweep 1 Rail surface, using curve C as rail and circle D as cross section.

91 Extrude planar ellipse E along curve F to construct a solid.

92 Cap the planar holes of swept surfaces G and H to form two solids.

93 Combine polysurfaces G, H, and J to form a single polysurface by using Boolean Union.

The exhaust system is complete.

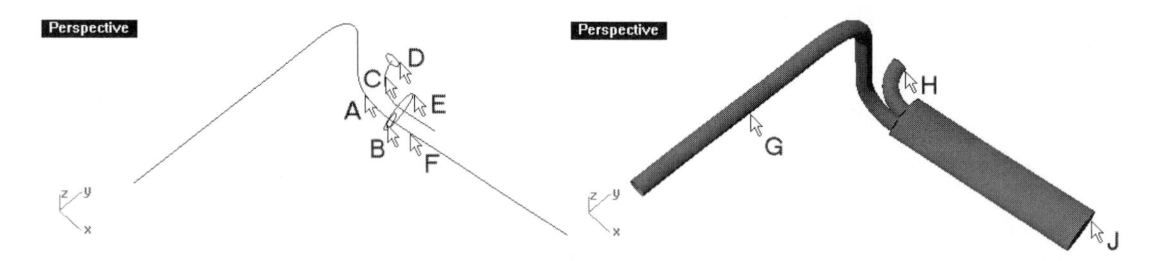

Figure C1–26. Curves being used to construct polysurfaces (left) and swept surfaces being capped and combined with extruded polysurface (right)

Intake System

The polysurface that we will construct to represent the intake system consists of four solid elements, which will be constructed by revolving, sweeping, and extruding. The intake system, together with other objects already constructed, are shown in Figure C1–27.

Figure C1–27. Intake system included in the model

Now we will continue with the following steps.

94 Turn on layer Curves Intake, set current layer to Intake, and turn off layers Curves Exhaust and Exhaust.

95 Revolve curve A (Figure C1–28) around endpoints B and C to construct a revolved polysurface.

96 Construct a Sweep 1 Rail surface with curve D as rail and circle E as cross section.

97 Extrude planar curve F a distance of 36 mm to construct a solid.

98 Extrude planar curve G a distance of 10 mm to construct another solid.

99 Cap the planar holes of the sweep surface H to convert it into a solid.

100 Combine solids H, J, K, and L to form a single solid by using Boolean Union.

The intake system is complete.

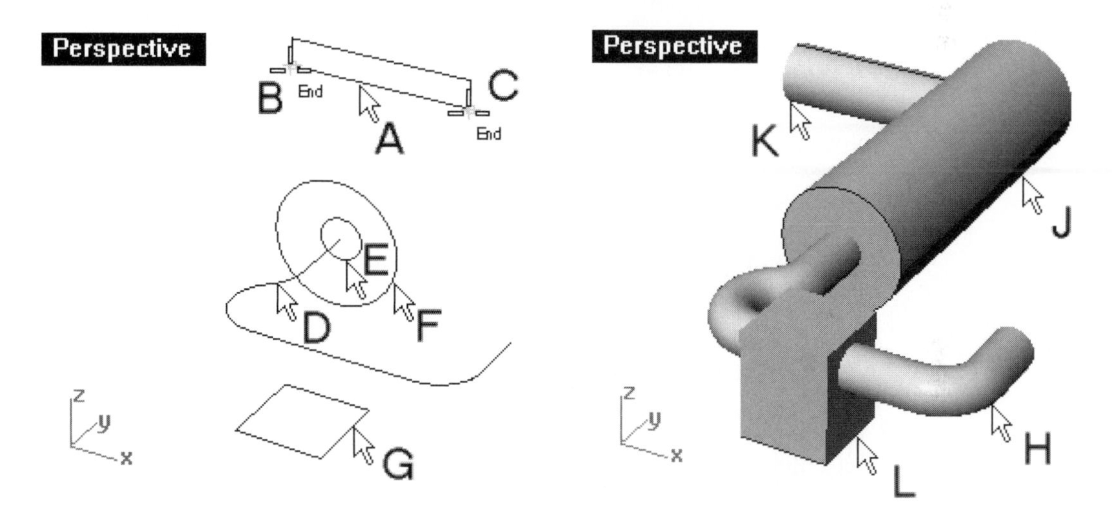

Figure C1–28. Curves for making four solids (left) and four solids being combined (right)

Fuel Tank

The polysurface representing the fuel tank is quite simple. We only need to construct a revolved solid and an extruded solid and then combine them to form a single solid. Figure C1–29 shows the fuel tank added to the model.

Figure C1–29. Fuel tank added to the model

Now we will continue with the following steps.

101 Turn on the Curves Fuel layer, set the current layer to Fuel, and turn off the layers Curves Intake and Intake.

102 Revolve curve A (Figure C1–30) around endpoints B and C to construct a solid.

103 Extrude curve D a distance of 19 mm in both directions. The total extrusion distance is 38 mm.

104 Combine solids E and F by using Boolean Union.

The fuel tank is complete.

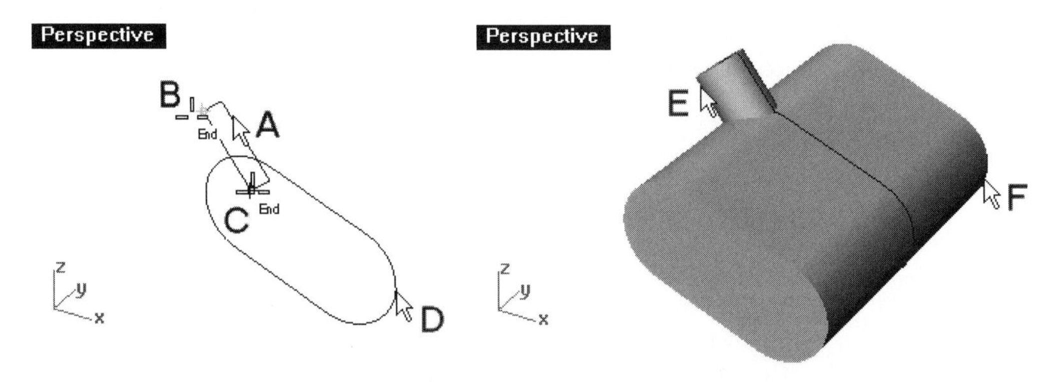

Figure C1–30. Curves for making the fuel tank (left) and solids being combined (right)

Front Suspension

The objects representing the front suspension are a set of extruded polysurfaces. We will use the curves already constructed as framework. After making the extruded solids, we will combine them by using Boolean Union and round off some corners by filleting. Figure C1–31 shows the front suspension added to the model.

Figure C1–31. Bottom view showing front suspension added to the model

Now we will continue with the following steps:

105 Turn on the Curves Front Suspension layer, set the current layer to Front Suspension, and turn off the layers Curves Fuel and Fuel.

106 Construct six solid objects by extruding curves A, B, C, D, E, and F (Figure C1–32). Extrusion distance for A is 3 mm, B is 5 mm, C is 15 mm, D is 6 mm, E is 2 mm, and F is 2 mm.

107 Turn off the Curves Front Suspension layer.

108 Round off edges G, H, J, and K with a radius of 2 mm.

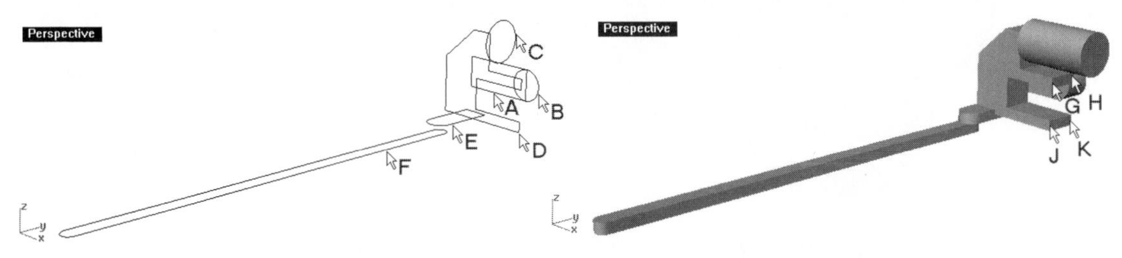

Figure C1–32. Curves being extruded (left) and edges being filleted (right)

109 Combine solids A, B, C, D, and E (Figure C1–33) by using Boolean Union.

110 Turn on the Curves Wheelbase layer.

111 Mirror the combined solid around endpoints F and G.

The front suspension is complete.

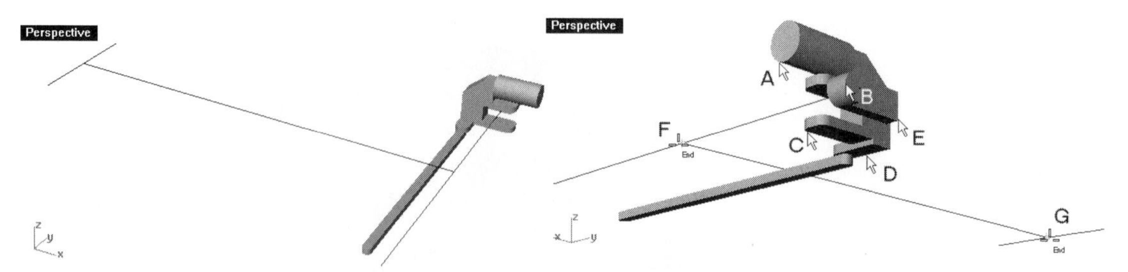

Figure C1–33. Two views showing solids being combined and mirrored

Drive and Steering

Figure C1–34 shows the drive and steering unit added to the model.

Figure C1–34. Drive and steering added to the model

Now we will continue with the following steps:

112 Turn on the Curves Drive layer, set the current layer to Drive, and turn off the layers Front Suspension and Curves Wheelbase.

113 Construct a Sweep 1 Rail surface, using curve A (Figure C1–35) as rail and curve B as cross section.

114 Construct another Sweep 1 Rail surface, using curve C as rail and curve D as cross section.

115 Construct a solid by extruding curve E (the smaller circle) along path curve F.

116 Construct a solid by extruding curve G a distance of 4 mm in one direction.

117 Construct a solid by extruding curve H a distance of 3 mm in both directions (total 6 mm).

118 Construct a solid by extruding curve J a distance of 2 mm in both directions.

119 Cap the planar holes of swept surface K to construct a solid.

120 Select Solid > Sphere > Center, Radius, or click on the Sphere: Center, Radius button on the Solid toolbar.

121 Select endpoint L.

122 Type 4 at the command area to specify the diameter. A solid sphere is constructed.

123 Select Transform > Copy, or click on the Copy button on the Transform toolbar.

124 Select solid M and press the ENTER key.

125 Click on any location in the viewport to specify the point to copy from.

126 Type r14 < 90 at the command area. The select object is copied 14 mm in 90 degree direction from the reference point.

127 Type r6 < 270. Another copy of the selected object is constructed.

128 Press the ENTER key to terminate the command.

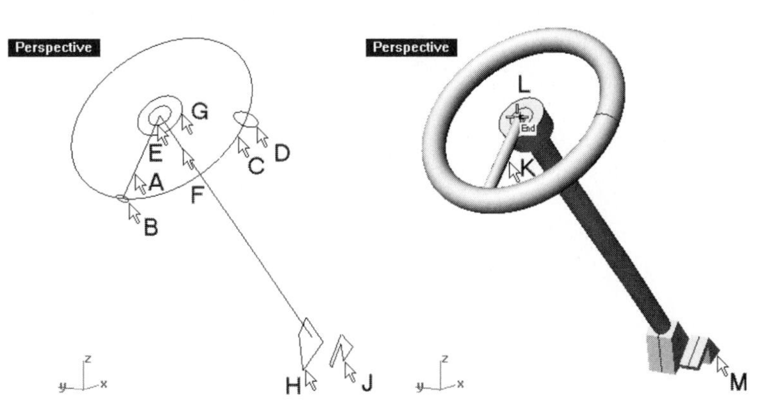

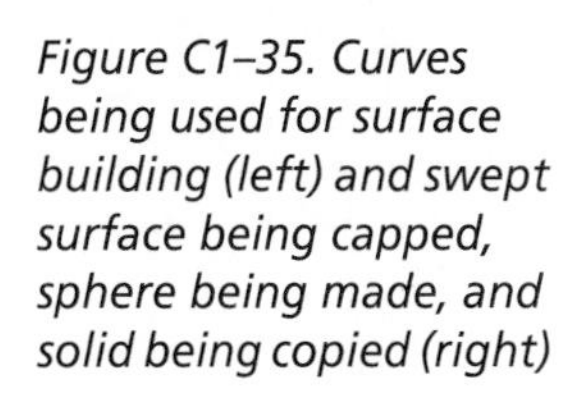

Figure C1–35. Curves being used for surface building (left) and swept surface being capped, sphere being made, and solid being copied (right)

Now we will construct an array, as follows:

129 Right-click on the Perspective viewport's label and select Wireframe to set the display to wireframe.

130 Select Transform > Array > Along Curve, or click on the Array Along Curve button on the Array toolbar.

131 Select polysurface A (Figure C1–36) and press the ENTER key.

132 Select curve B.

133 In the Array Along Curve Options dialog box, set the number of items to 3 and click on the OK button. The polysurface is arrayed.

134 Combine all the solids, except C, D, and E, by using Boolean Union.

The drive and steering unit is complete.

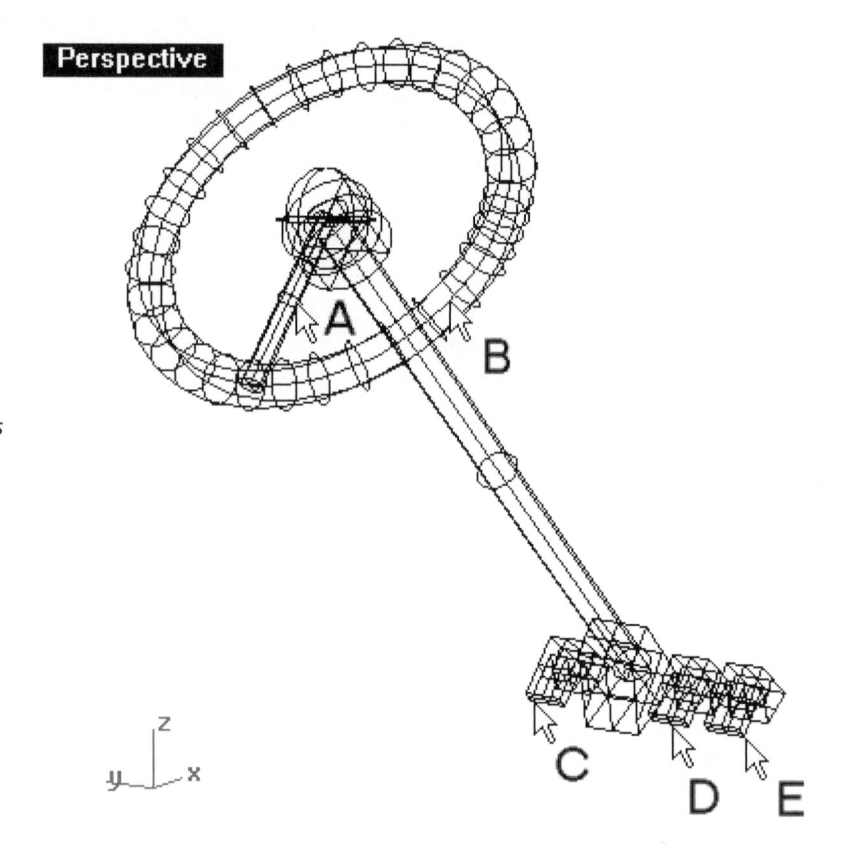

Figure C1–36. A solid being arrayed and solids being combined

Floor Panel and Seat

To finish this case study session, we will add floor panels and a bench seat. Because the outer boundary of the floor panels needs to fit with the external body panels, we will construct a floor panel a bit wider than required and trim it in the next case study. Figure C1–37 shows the floor panel (trimmed by the external body panels) and the bench seat.

Figure C1–37. Bench seat and trimmed floor panels added to the model

Now we will continue with the following steps to construct the floor panels.

135 Turn on the Curve Floor layer, set the current layer to Floor Panel, and turn off the layers Curves Drive and Drive.

136 Construct a solid by extruding curve A (Figure C1–38) a distance of 100 mm in one direction.

137 Construct another solid by extruding curve B a distance of 60 mm in one direction.

138 Construct a fillet edge of 11 mm radius at edge C.

139 Select Edit > Explode.

140 Select solids D and G. They are decomposed into individual surfaces.

141 Delete faces D, E, F, and G.

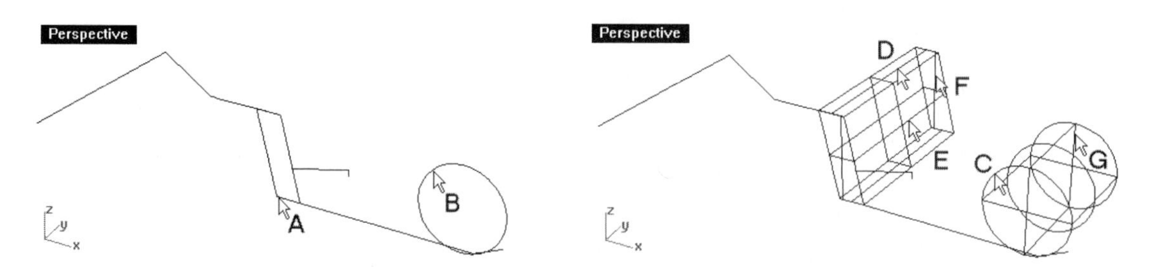

Figure C1–38. Curves for the floor panels being extruded (left) and polysurfaces being exploded and surfaces being deleted (right)

142 Set the display to Rendered.

143 Turn on the Curve Wheelbase layer.

144 Mirror surface A, B, and C around endpoints D and E.

145 Extrude curve F (Figure C1–39) along path G.

146 Extrude curves H, J, and K separately a distance of 100 mm in both directions (total distance 200 mm).

147 Turn off the Curves Wheelbase and Curves Floor layers.

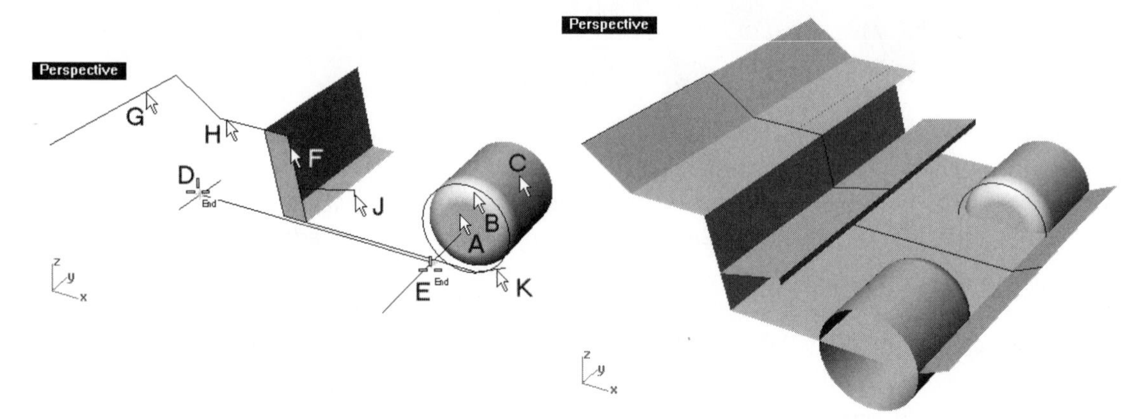

Figure C1–39. Surfaces being constructed and mirrored (left) and surface constructed and mirrored (right)

148 Turn off the Curves Wheelbase and Curves Floor layers.

149 Referencing Figure C1–40, trim away the unwanted parts of the surfaces.

150 Join the trimmed surfaces together to form a polysurface.

The floor panels are complete.

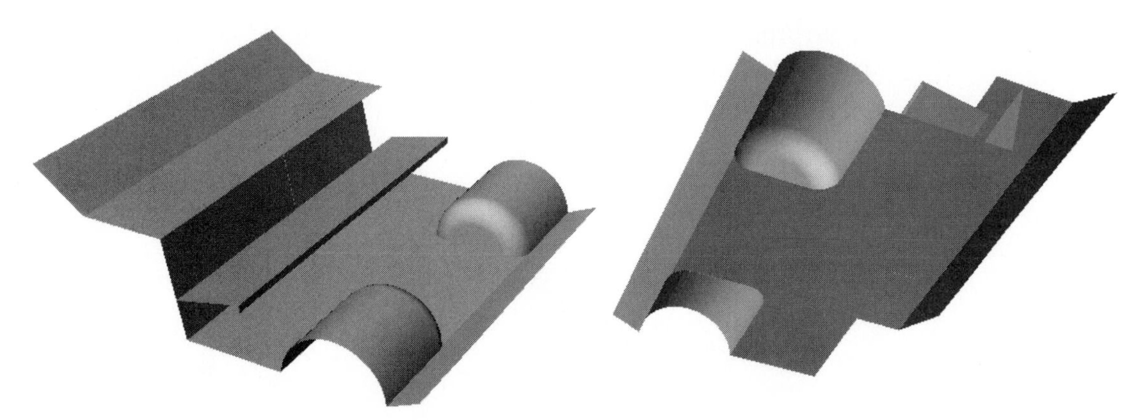

Figure C1–40. Two views showing floor panel surfaces trimmed and joined

To complete the model, we will work on the seat, as follows:

151 Turn on the Curves Seat layer, set the current layer to Seat, and turn off the Floor layer.

152 Construct two solids by extruding curves A and B (Figure C1–41) a distance of 49 mm in both directions.

153 Turn off the Curves Seat layer and turn on the layers Wheel, Tire, Rear Axle, Chassis, Rear Suspension, Engine, Exhaust, Intake, Fuel, Front Suspension, Drive, and Floor panel.

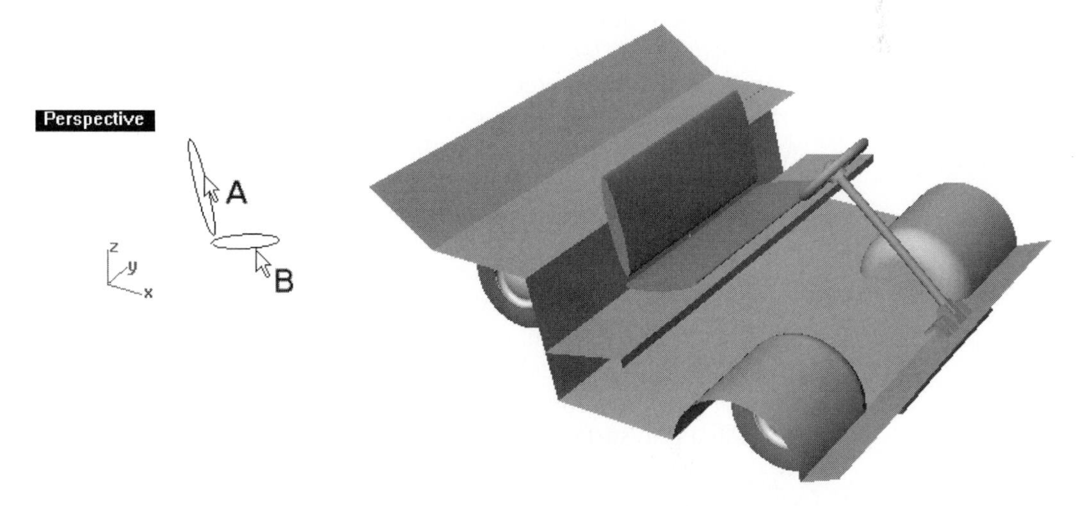

Figure C1–41. Curves for the seat being extruded (left) and bench seat constructed (right)

Apart from the exterior body panels, the model car is complete. In the next xase study, we will continue with making the body panels and trim away unwanted part of the floor panels, as shown in Figure C1–42.

Figure C1–42. Floor panels trimmed by the car's exterior body

Summary

In this case study, we have experienced using the curves as framework to construct surfaces, polysurfaces, and solids of regular geometric shapes. To let you appreciate how to use Rhinoceros as a tool to construct 3D surfaces and polysurfaces quickly, we have done the tedious work of curve construction and saved these curves for 3D surface/polysurfaces building in a template file. If you were to construct similar objects from scratch, you have to think about the object's cross section profile and location and then construct them using the curves tools.

Practically, Rhinoceros polysurfaces and solids are derived from NURBS surfaces by joining a set of contiguous surfaces together. If a polysurface or a single surface, such as a sphere or an ellipsoid, encloses a volume without any opening, a solid is formed. Contrary to joining, we can explode a polysurface to become individual surfaces.

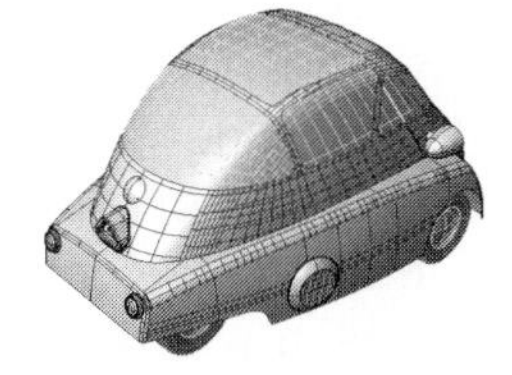

Case Study 2 Bubble Car Body

Introduction

This is the second part of the bubble car modeling case study. In this study, we will work on the bubble car's bodywork, which mainly concerns building free-form surfaces and making necessary modifications to the boundaries of the surfaces. Making free-form surfaces inevitably requires a set of curves to depict various cross-sections of the surface. Unlike modeling the internal structure of the car—in which we needed only to extrude a curve along a straight line, revolve a curve about an axis, or sweep a single curve along a path—we have to use various surface-making techniques. Obviously, having a thorough understanding of what types of surface profiles can be used, how the surfaces can be made, and what type of curves are required for making such surfaces is essential. Figure C2–1 shows two preliminary sketches of the bubble car body, depicting the car's general profile and silhouette.

Figure C2–1. Sketches showing the front and rear views of the bubble car

The car's bodywork can be divided into two major components: main body and skirt. We will first make the main body, which consists of an egg-shape body shell, a front-opening door, and a number of window sections. To make these components, we need to construct a number of contiguous surfaces. To ensure proper surface profile continuity between the surfaces, we will treat them holistically when designing the curves. To help construct the curves, we have prepared some sketches based on photos from various sources. Some of the sketches are shown in Figure C2–2. With these and other sketches, a wireframe of curves is produced and a set of contiguous surfaces can then be made, as shown in Figure C2–3.

Figure C2–2. Sketches made while thinking about various cross sections of the main body

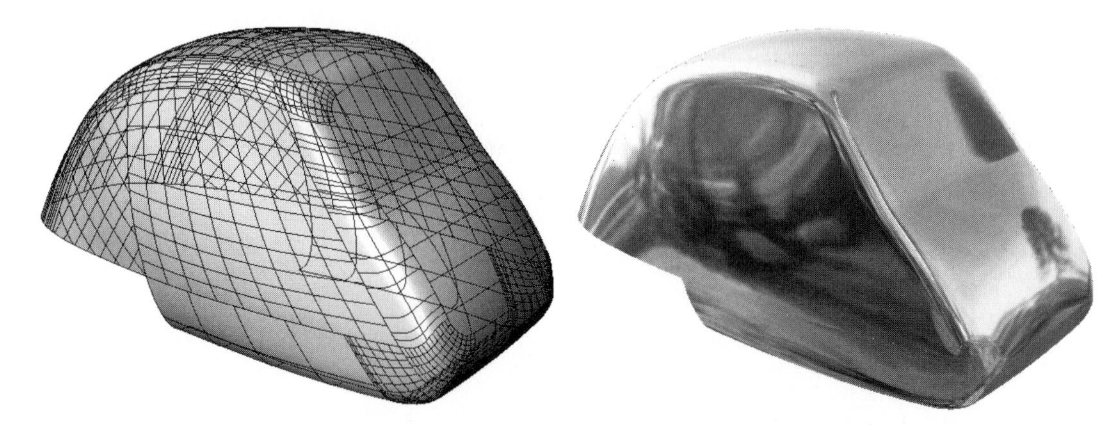

Figure C2–3. Surfaces (left) and rendered image (right) representing the bubble car's main body panels

Upon completion of the main body, we will proceed to make the skirt by producing five intersecting surfaces, trimming away the unwanted portions, and rounding off the edges. Figure C2–4 shows the preliminary sketches, and Figure C2–5 shows the surfaces and trimmed surfaces. To finish the bodywork, we will cut the window openings, trim away the unwanted portions of the surfaces, and add lights and other accessories, as shown in Figure C2–6.

Figure C2–4. Sketches showing four intersected surfaces (left) and surfaces trimmed (right)

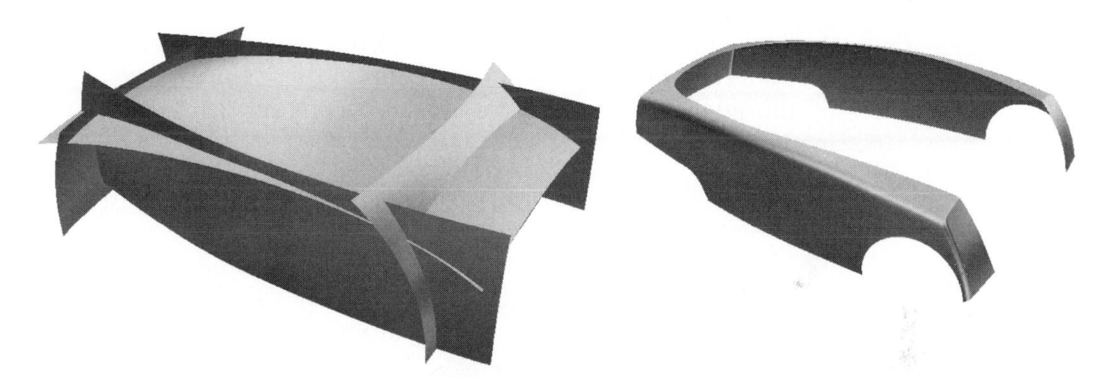

Figure C2–5. Surfaces for making the skirt of the bubble car (left) and skirt surfaces trimmed (right)

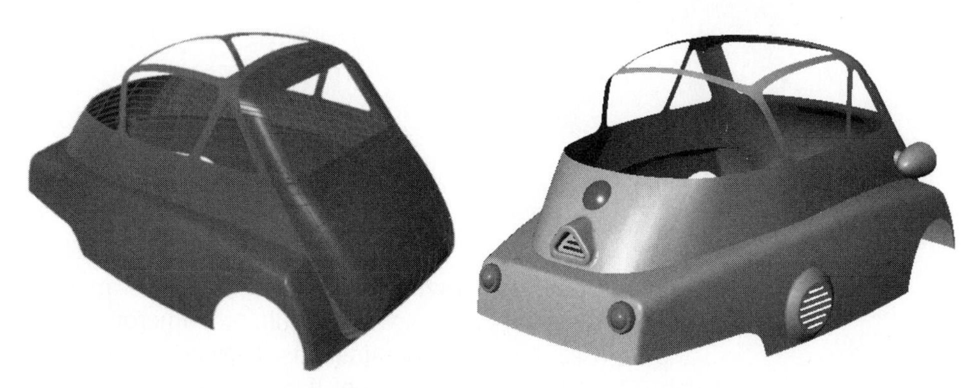

Figure C2–6. Main body, door, and skirt (left) and car body with accessories (right)

Bodywork Construction

Using the curves provided, we will first make the bubble car's main body. Then we will proceed to make the door panel, skirt, and window sections. Finally, we will complete the model by adding lights and other small fittings.

Template File

Although this is the second part of the bubble car case study, we can skip the previous case study and work directly on this section because we have prepared a template in which all the work for Case Study 1 is saved. In addition, the template file also has all the curves required to build the bodywork. Now open the template file.

1. Select File > Open and select the file Bubble Car 2.3dm from the Case Study 2 folder on the companion CD.

As shown in Figure C2–7 (left), the work that we have done in the previous case study is saved; we will proceed from this point. We will now turn off those layers containing the previous work and set the current working layer, as follows.

2. Select Edit > Layers > One Layer On, or click on the One Layer On button on the Layer toolbar.
3. In the Layer to leave on dialog box, click on the Curves Body Panel and the OK button.

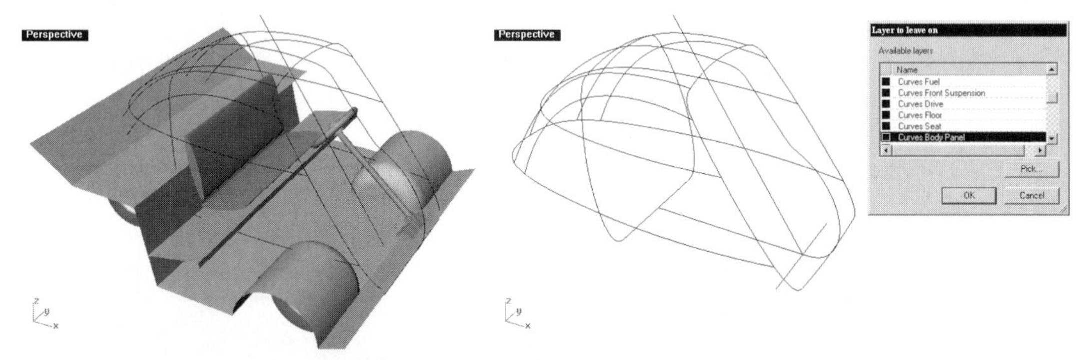

Figure C2–7. Template file (left) and curves for making the body panel (right)

The curves that are shown in Figure C2–7 (right) for making the main body panel of the bubble car are the result of numerous trials and errors on various possibilities and combinations. They are smooth, continuous curves with G2 continuity, and some of them are split at their intersections with other curves in order to build surfaces with smooth G2 continuity.

Main Body

We will now construct the surfaces for making the main body panel of the bubble car, as follows:

4 Set the current layer to Body Panel.

5 Referencing Figure C2–8, rotate the display.

6 Select Surface > Curve Network, or click on the Surface from network of curves button on the Surface toolbar.

7 Select curves A, B, C, D, E, F, and G (Figure C2–8) and press the ENTER key.

8 In the Surface from Curve Network dialog box, set the tolerances for edge curves and interior curves to 0.991, click on the Position buttons for all four surface edges, and click on the OK button. A curve network surface is constructed.

By checking the Position boxes in the dialog box, this curve network surface's edges will coincide exactly with the input curves A, C, D, and G.

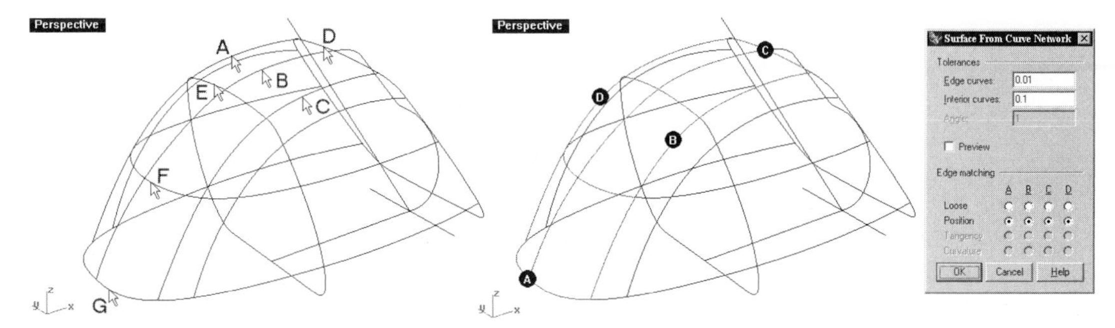

Figure C2–8. Curves being selected (left) and curve network surface being constructed (right)

Now we will construct the second surface. Because surface continuity is very important, we will use an existing edge as one of the input curves. Let us perform the following steps:

9 Select Surface > Curve Network, or click on the Surface from network of curves button on the Surface toolbar.

10 Select surface edge A, curves B, C, D, E, and F (Figure C2–9) and press the ENTER key.

(***NOTE:*** *The effects of selecting curve A and surface edge A are different. Using surface edge A can provide a smooth continuity between the existing surface and the surface to be constructed.)*

11 In the Surface from Curve Network dialog box, click on the Curvature check box for the edge connecting with the previous surface and click on the OK button. Another curve network surface is constructed.

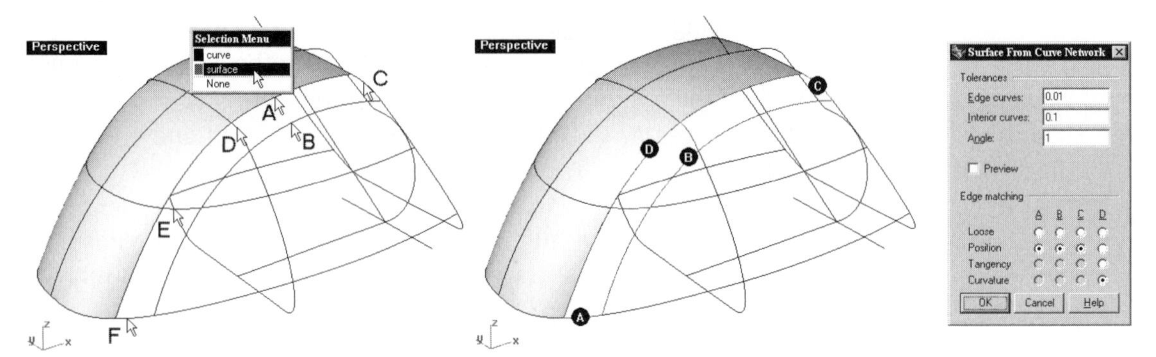

Figure C2–9. Surface edge and curves being selected (left) and curve network surface being constructed (right)

Because the next and subsequent surfaces to be constructed will use a segment of an existing surface edge as the input curve, we will first split an edge of a surface into three segments, as follows:

12 Select Surface > Edge Tools > Split Edge, or click on the Split Edge/Merge Edge button on the Edge Tools toolbar.

13 Select surface edge C (Figure C2–10) and endpoints P and Q, and press the ENTER key. *(Note: It is assumed that END osnap is activated.) The surface edge C is split into three segments, at points P and Q.)*

Now we will construct another network curve surface, as follows:

14 Select Surface > Curve Network, or click on the Surface from network of curves button on the Surface toolbar.

15 Select curves A and B, surface edge segment C, and curves D and E (Figure C2–10), and press the ENTER key. (The sequence of selection is important because the first two selected curves will be used as the rails.)

16 Click on the OK button. A network curve surface is constructed.

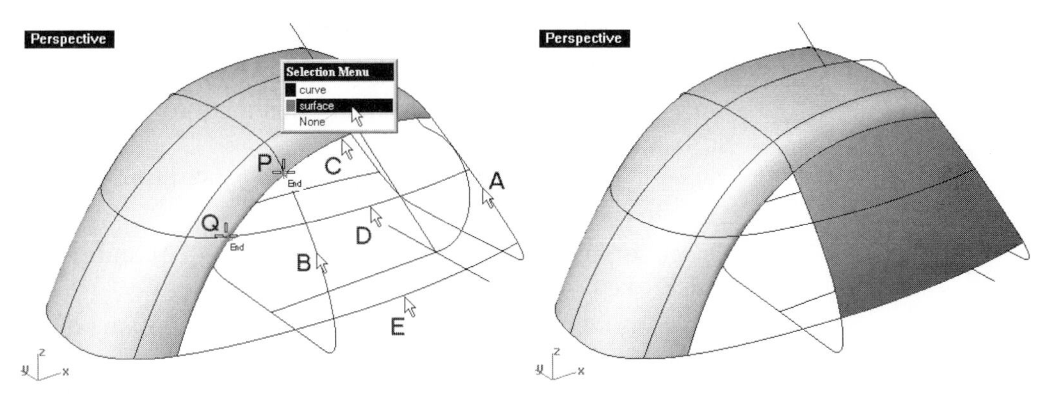

Figure C2–10. Surface edge split and curves and edge being selected (left) and sweep 2 rails surface constructed (right)

We will now split one of the edges of the surface we just constructed and then construct another sweep 2 rails surface, as follows:

17 Split surface edge D at endpoint X (Figure C2–11).

18 Construct a sweep 2 rails surface, using surface edge segments C and D as rails, and curves A and curve B as cross sections, as shown in Figure C2–11. Remember to click on the Curvature buttons on the Sweep 2 Rails Options dialog box to ensure a G2 continuity between the surfaces to be constructed with the existing surface edges.

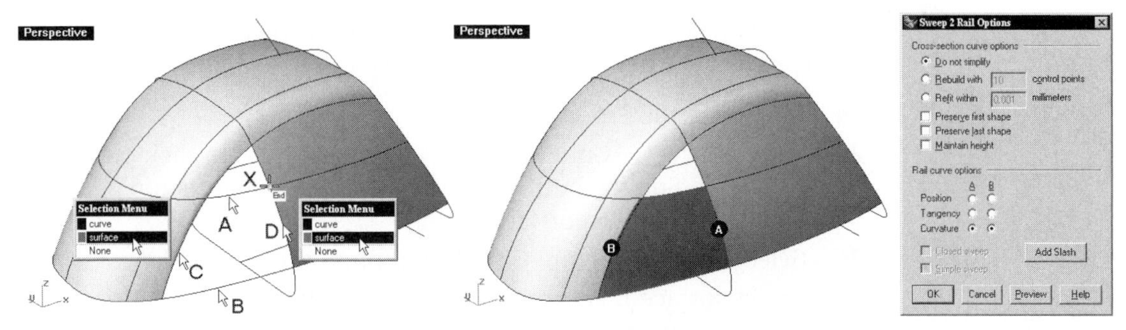

Figure C2–11. Surface edge split and curves and surface edges selected (left) and sweep 2 rails surface being constructed (right)

To fill up the triangular opening among the surfaces, we will construct a patch surface, as follows:

19 Turn off layer Curves Body Panel. This enables you to select surface edges more easily.

20 Select Surface > Patch, or click on the Patch button on the Surface toolbar.

21 Select surface edges A, B, and C (Figure C2–12) and press the ENTER key.

22 Click on the OK button on the Patch Surface Options dialog box. A patch surface is constructed.

23 Set current layer to Curves Body Panel. We will use the curves to make other surfaces.

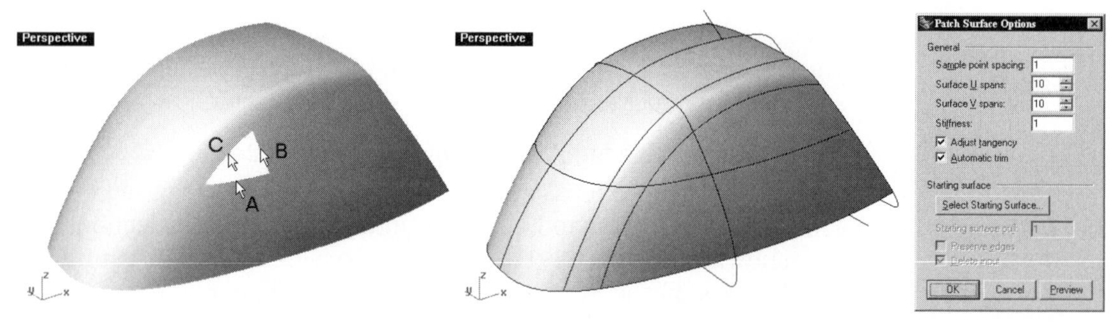

Figure C2–12. Surface edges selected (left) and patch surface constructed (right)

Now we will construct two curves for surface construction by projecting a curve onto a surface and blending two curves, as follows:

24 Maximize the Front viewport.

25 Select Curve > Curve from Objects > Project, or click on the Project to Surface button on the Curve from Objects toolbar. Projection is viewport dependent.

26 Select curve A (Figure C2–13) and press the ENTER key.

27 Select surface B and press the ENTER key. A curve is projected.

28 Maximize the Perspective viewport and change its display to wireframe.

29 Turn off the Body Panel layer.

30 Select Curve > Blend Curves, or click on the Blend curves/Blend perpendicular to two curves button on the Curve Tools toolbar.

31 Select curves C and D (Figure C2–13). A blend curve is constructed.

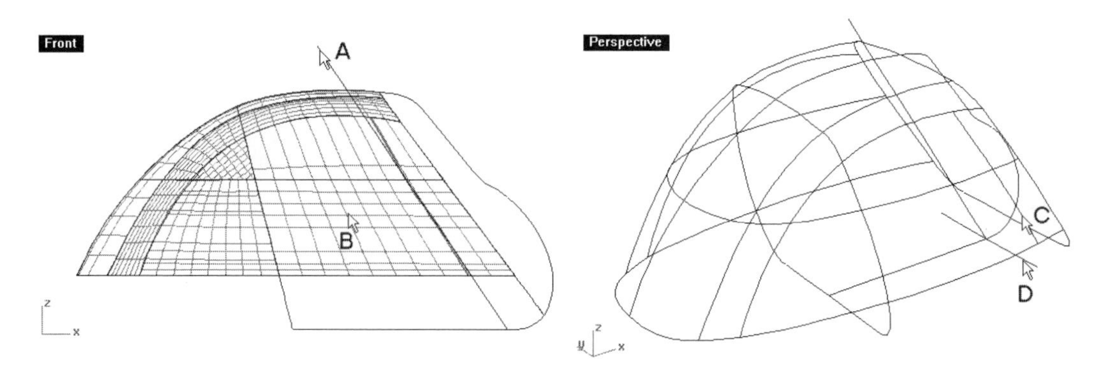

Figure C2–13. Curve being projected (left) and blend curve being constructed (right)

Now we will construct three sweep surfaces, as follows:

32 Set current layer to Body Panel, and set the display to Rendered.

33 Referencing Figure C2–14, split surface edge A at endpoint P.

34 Construct a sweep 1 rail surface, using surface edge segment A as rail and curves B and C as cross-sections.

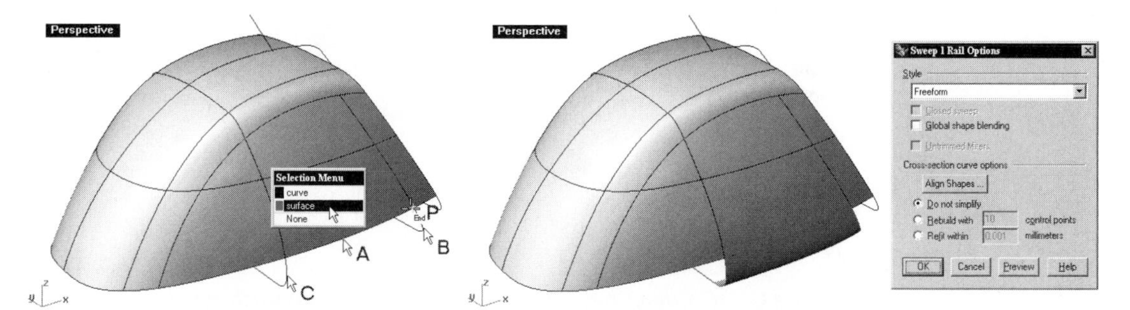

Figure C2–14. Surface edge split and surface edge and curves being selected (left) and sweep surface being constructed (right)

35 Rotate the display with reference to Figure C2–15.

36 Construct a sweep 2 rails surface, using curves A and B as rails and curve C as cross-section.

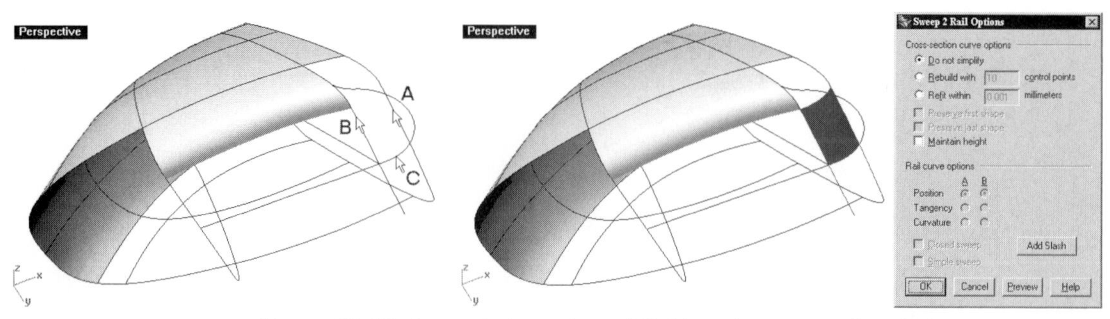

Figure C2–15. Curves being selected (left) and sweep surface being constructed (right)

37 Construct a sweep 2 rails surface, using surface edges A and B as rail and surface edge C and curve D (Figure C2–16) as cross-sections. Remember to click on the Curvature buttons on the Sweep 2 Rails options dialog box to obtain a G2 continuity.

Surface E is not needed, so we will delete it.

38 Select surface E (Figure C2–16) and press the DEL key.

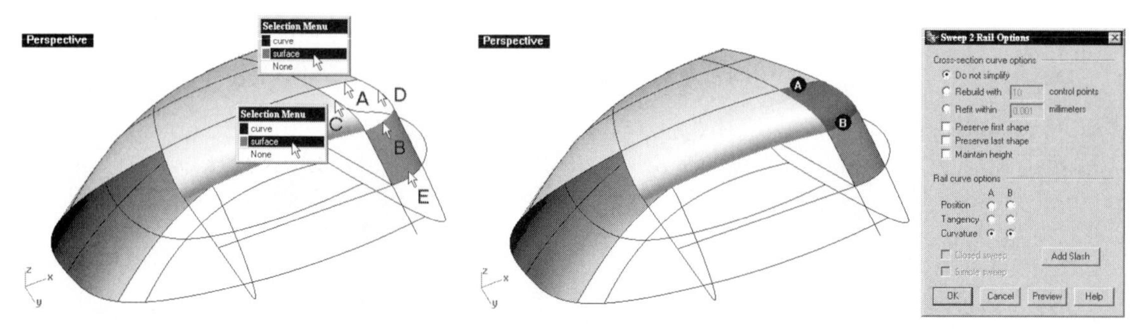

Figure C2–16. Curves being selected (left) and sweep surface being constructed (right)

Now we will mirror a set of surfaces, as follows:

39 Referencing Figure C2–17, rotate the display.

40 Select Transform > Mirror, or click on the Mirror/Mirror on 3 point plane button on the Transform toolbar.

41 Select surfaces A, B, C, D, E, and F (Figure C2–17) and press the ENTER key.

42 Type w0,0 at the command area. This specifies the origin point with reference to world construction plane.

43 Type w1,0 at the command area. This point, together with the previous point, specifies a plane about which selected surfaces are mirrored. The surfaces are mirrored.

44 Rotate the display to see the result of mirroring.

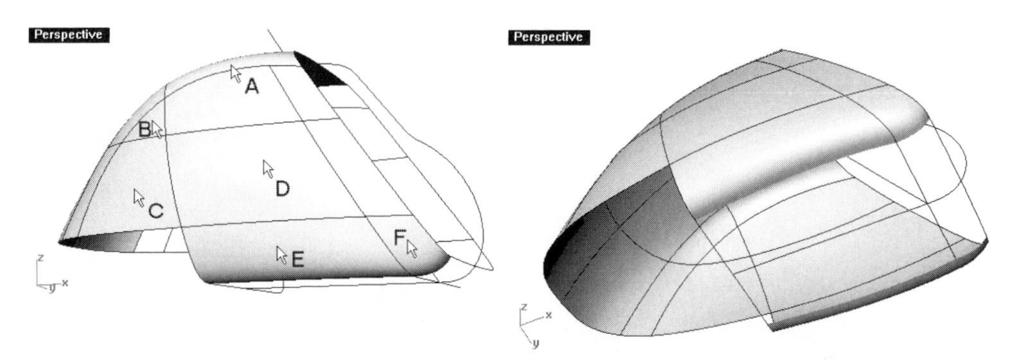

Figure C2–17. Surfaces being selected (left) and surfaces mirrored and display rotated (right)

Now we will construct the final piece of surface for the car's body main panel.

45 Construct a sweep 2 rails surface, using curves A and B (Figure C2–18) as rails and surface edge C, curve D, and surface edge E as cross-sections.

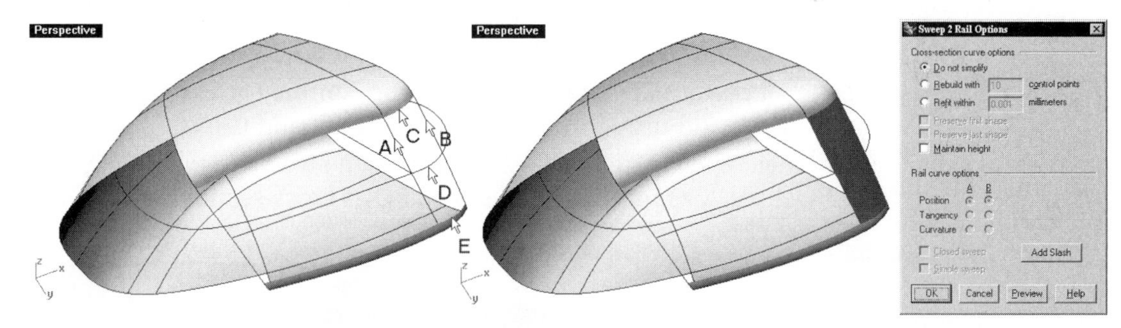

Figure C2–18. Curves and surface edges being selected (left) and sweep surface being constructed (right)

To complete the body panel, we will join the surfaces together by joining edges of all the contiguous surfaces.

46 Select Surface > Edge Tools > Join 2 Naked Edges, or click on the Join 2 Naked Edges button on the Edge Tools toolbar.

47 Select edge A (Figure C2–19) twice.

48 In the Edge Joining dialog box, click on the OK button. The surfaces are joined. *(Note: The tolerance value shown here may be different from yours.)*

49 Repeat the command to join all the contiguous edges.

(Note that edge joining simply makes the edges appear joined.)

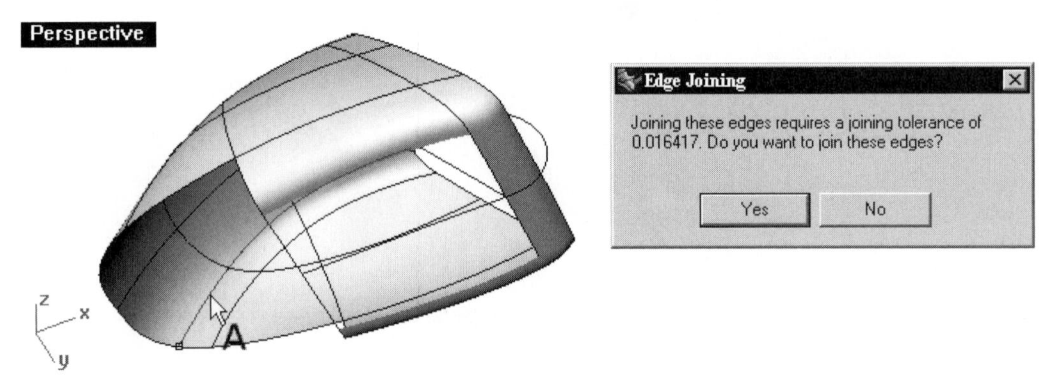

Figure C2–19. Edges of two contiguous surfaces being joined

Frontal Door

To make the door, we will first make two contiguous surfaces, merge them into a single surface, trim the merged surface, split the edge of the trimmed surface, and construct a set of blend surfaces. Now let us continue with the following steps to construct a sweep 1 rail surface and a curve network surface:

50 Set current layer to Door Panel, turn on the Curves Door Panel layer, and turn off the layers Body Panel and Curves Body Panel.

51 Referencing Figure C2–20, construct a sweep 1 rail surface, using curve A as rail and curves B and C as cross-sections.

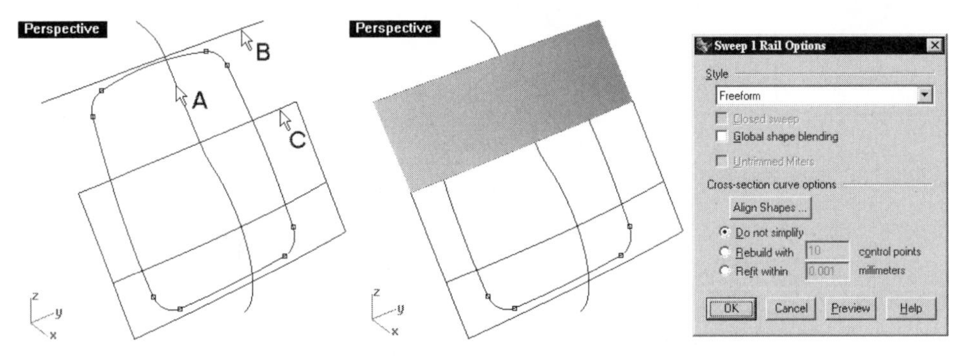

Figure C2–20. Curves selected (left) and sweep one rail surface being constructed (right)

52 Referencing Figure C2–21, use surface edge A and curves B, C, D, E, and F to construct a curve network surface. A curvature continuity (G2) is needed for the edge, contiguous with the previous surface.

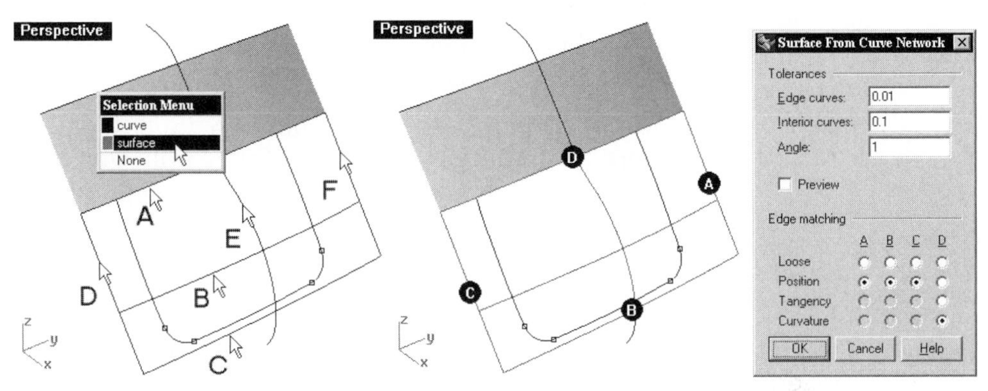

Figure C2–21. Curves selected (left) and curve network surface being constructed

We will now merge the contiguous surfaces into one single surface. In addition, we will set the construction plane orientation for subsequent operations.

53 Select Surface > Edit Tools > Merge, or click on the Merge Surfaces/Divide Along Creases button on the Surface Tools toolbar.

54 Select surfaces A and B (Figure C2–22) and press the ENTER key. The surfaces are merged into one.

55 Select View > Set CPlane > 3 Points, or click on the Set CPlane by 3 Points button on the Set CPlane toolbar.

56 Select endpoints C and D, and press the ENTER key. The construction plane is set.

To help you see the effect of changing the construction plane orientation, grid lines are shown in Figure C2–22 (right). However, for the sake of clarity in illustration, grid lines are not shown in other figures.

Figure C2–22. Surfaces being merged and CPlane being set (left) and surface merged and CPlane set (right)

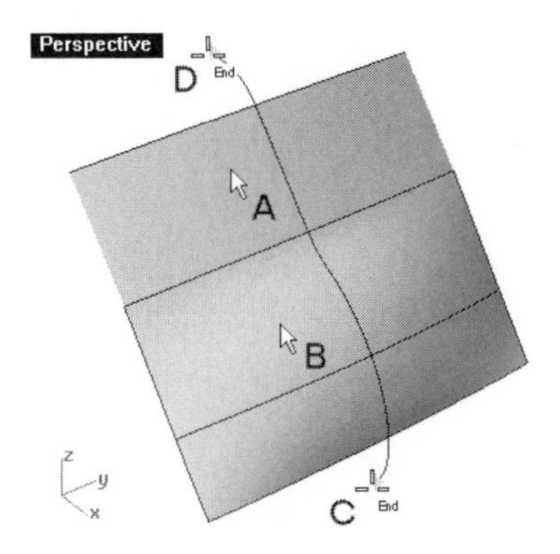

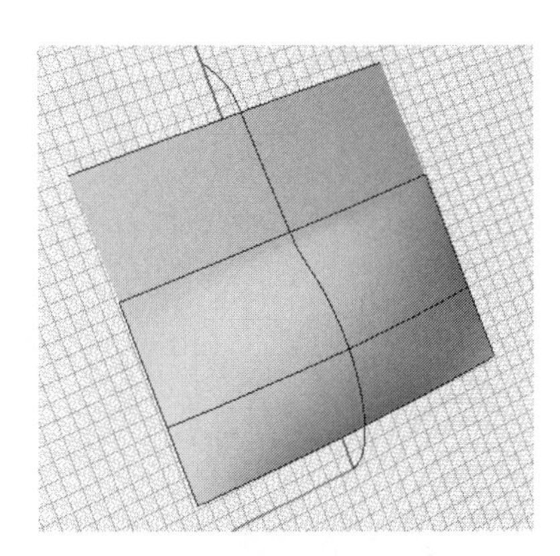

Now we will use a curve to trim the merged surface. Prior to trimming, we will project the endpoints of the trimming curve to the merged surface. The projected points will be used for subsequent splitting of the trimmed surface's edge.

57 Set the display to wireframe.

58 Set the current layer to Curves Body Panel.

59 Select Curve > Curve from Objects > Project, or click on the Project to Surface button on the Curve from Objects toolbar.

60 Select points A, B, C, D, E, F, G, and H (Figure C2–23) and press the ENTER key.

61 Select surface J. The selected points are projected onto the surface.

62 Select Edit > Trim, or click on the Trim button on the Main1 toolbar.

63 Select curve K (Figure C2–23) and press the ENTER key.

64 Select location L and press the ENTER key. The surface is trimmed.

Note that projection and trimming directions are construction-plane dependent. They work in a direction perpendicular to the active construction plane.

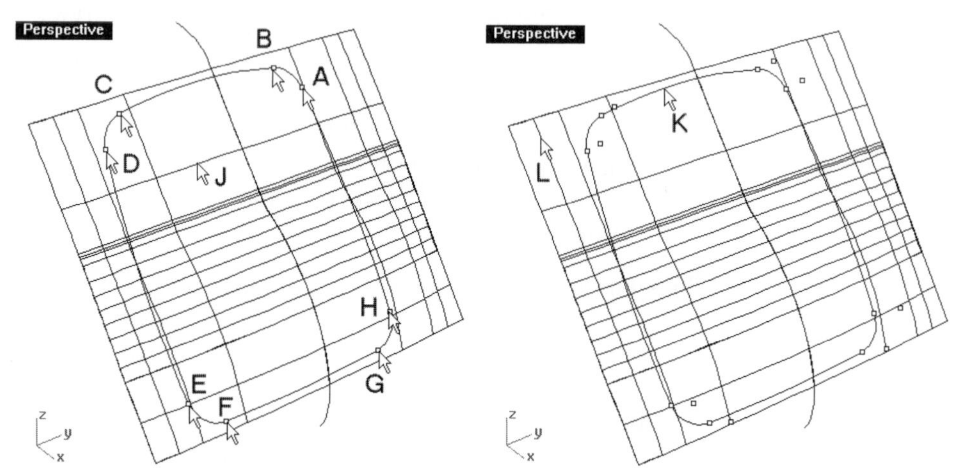

Figure C2–23. Endpoints being projected (left) and surface being trimmed (right)

Now we will use the projected points to split the trimmed surface's edge into eight segments.

65 Set OSNAP to Point.

66 Split surface edge A at points B, C, D, E, F, G, H, and J.

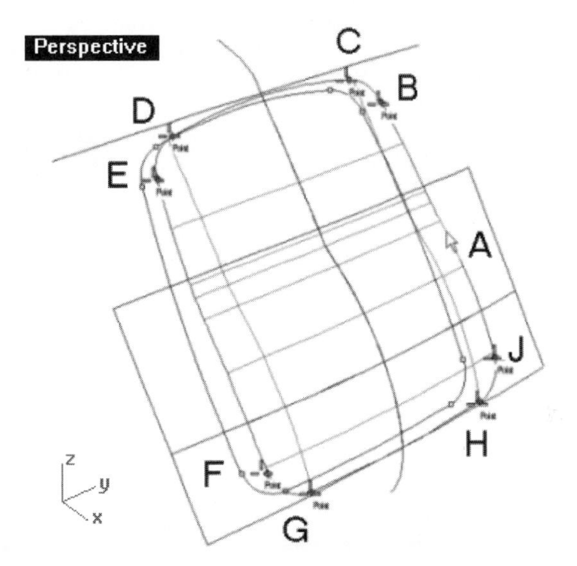

Figure C2–24. Points being projected (left) and surface edge being split (right)

We will now use the split edge segments, together with the surface edges of the main body panel, to construct blend surfaces and a curve network surface, as follows:

67 Set the current layer to Door Panel, turn off the Curves Door Panel layer, and turn on the Body Panel layer.

68 Referencing Figure C2–25, rotate the display and set the display to Shaded.

69 Select Surface > Blend Surface, or click on the Blend Surface button on the Surface Tools toolbar.

70 Select edges A and B. It is necessary to select both edges near the right or left side, because the selected location has a direct impact on the final outcome.

71 In the Adjust Blend Bulge dialog box, click on the OK button. A blend surface is constructed.

72 Repeat the command to construct two more blend surfaces between surface edges C and D and surface edges E and F.

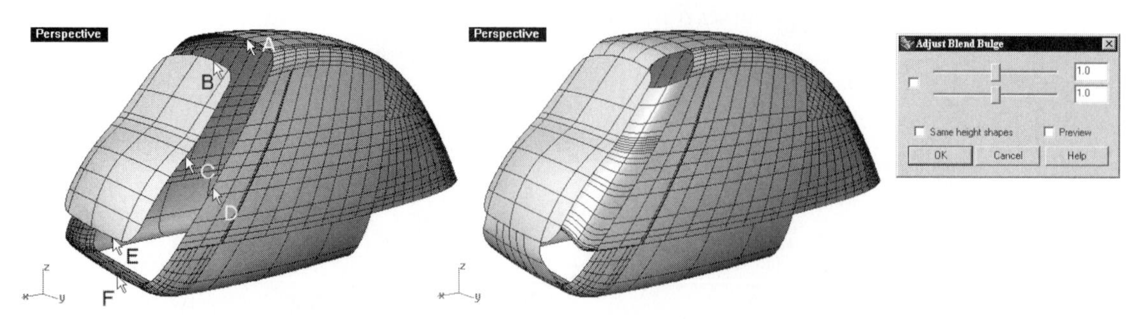

Figure C2–25. Surface edges being selected (left) and blend surfaces constructed (right)

73 Referencing Figure C2–26, construct two curve network surfaces. The first surface uses surface edges A, B, C, and D as input curves, and the second surface uses surface edges E, F, G, and H. Edge matching must be curvature.

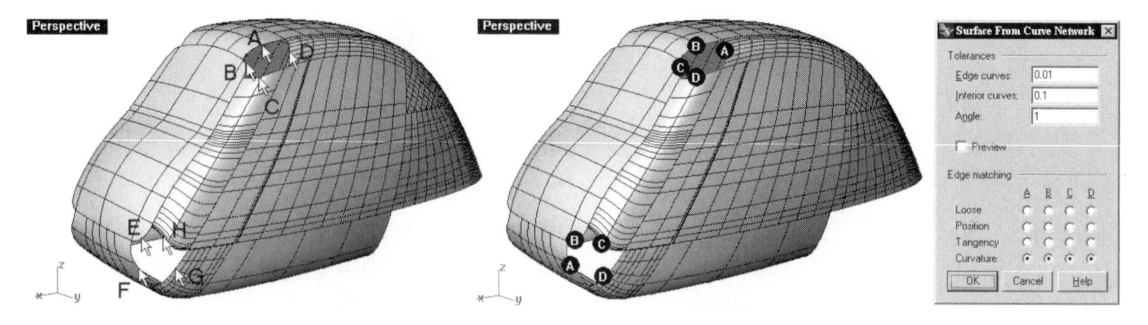

Figure C2–26. Surface edges being selected (left) and curve network surfaces constructed (right)

Now we will mirror three surfaces and join the surfaces into a single polysurface.

74 Select Transform > Mirror, or click on the Mirror/Mirror on 3 point plane button on the Transform toolbar.

75 Select surfaces A, B, and C (Figure C2–27) and press the ENTER key.

76 Type w0,0 at the command area.

77 Type w1,0 at the command area. The surfaces are mirrored.

78 Turn off the Body Panel layer.

79 Select Edit > Join, or click on the Join button on the Main 1 toolbar.

80 Select all the door panel surfaces, and press the ENTER key.

The door panel is complete.

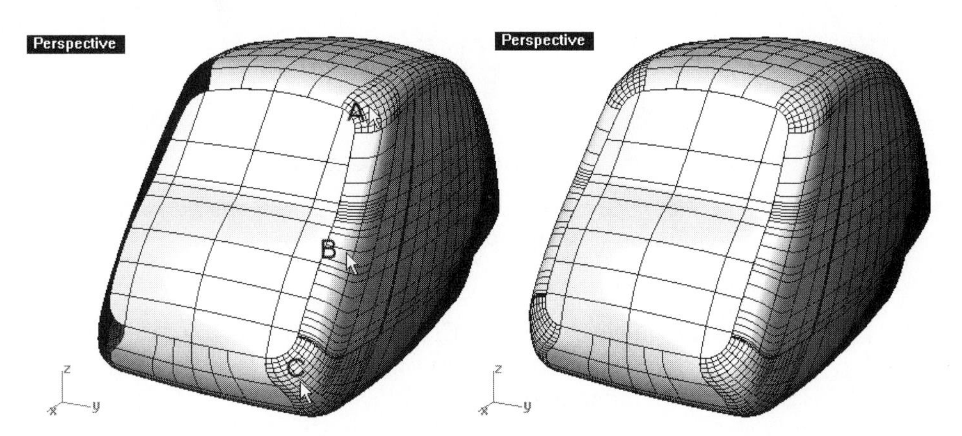

Figure C2–27. Surfaces selected (left) and surfaces mirrored (right)

Windows

We will make the window sections by splitting the body panel and the door panel.

81 Turn on the Curves Door Glass and Door Glass layers.

82 Set the display to wireframe.

83 Select, Edit > Split, or click on the Split button on the Main 2 toolbar.

84 Select polysurface A, shown in Figure C2–28, and press the ENTER key.

85 Select curve B and press the ENTER key. The polysurface representing the door panel is split into two sections.

86 Select Edit > Layers > Change Object Layer, or click on the Change Object Layer button on the Layers toolbar.

87 Select polysurface C and press the ENTER key.

88 In the Layer for Objects dialog box, select Door Glass and click on the OK button.

89 Shade the display. The door window is complete.

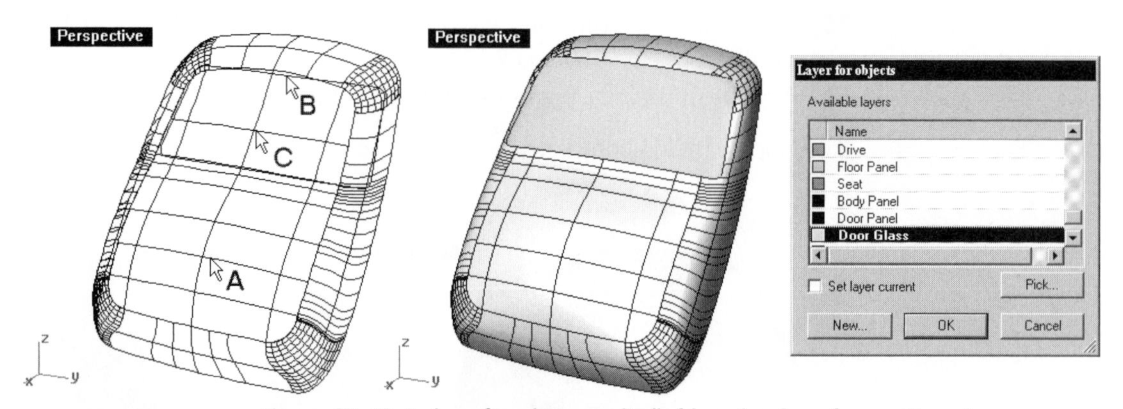

Figure C2–28. Polysurface being split (left) and polysurface split and one polysurface's layer being changed

90 Select View > CPlane > World Top, or click on the Set CPlane World Top button on the Set CPlane toolbar. The construction plane is reset.

We will continue to split the polysurface representing the body panel into window and sun roof elements, as follows:

91 Set the current layer to Body Panel; turn on the Curves Body Glass and Curves Sun Roof, Body Glass, and Sun Roof layers; and turn off the Door Panel, Door Glass, and Curves Door Glass layers.

92 Maximize the Front viewport.

93 Select, Edit > Split, or click on the Split button on the Main 2 toolbar.

94 Select polysurface A (Figure C2–29) and press the ENTER key.

95 Select curves A, B, C, and D, and press the ENTER key. The polysurface representing the body panel is split into seven polysurfaces.

96 Referencing Figure C2–29, change polysurfaces F to layer Sun Roof, and change all polysurfaces except A and F to layer Body Glass.

97 Maximize the Perspective viewport and rotate it around to see the change.

The window and sun roof sections are complete.

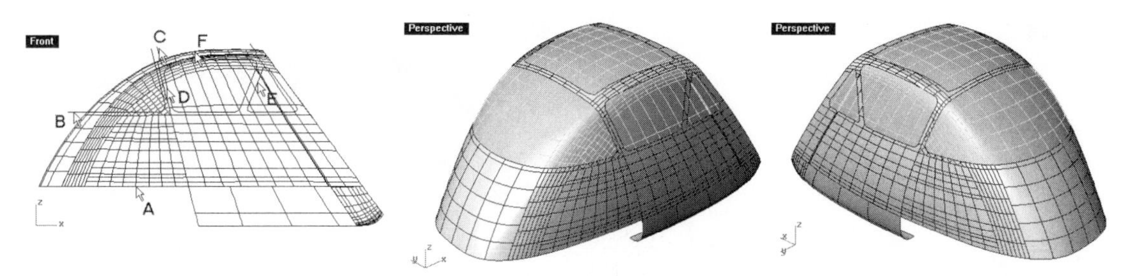

Figure C2–29. From left to right: polysurface being split, right rear view of the body, and left rear view of the body

Skirts

We will make the skirt by using five intersecting surfaces, trimming the surfaces to remove unwanted portions, and rounding off the edges. Now we will continue with the following steps to construct five surfaces.

98 Set the current layer to Skirt, turn on the Curves Skirt layer, and turn off the layers Body Panel, Body Glass, Sun Roof, Curves Sun Roof, and Curves Body Glass.

99 Construct a sweep 1 rail surface, using curve A as rail and curves B, C, and D (Figure C2–30) as cross-sections.

100 Construct another sweep 1 rail surface, using curve E as rail and curves F and G as cross-sections.

101 Select Surface > Loft, or click on the Loft button on the Surface toolbar.

102 Select curves H, J, and K, and press the ENTER key. When selecting the curves, select either their upper ends or lower ends. Otherwise, you may have to click on the Align Curve button on the Loft Options dialog box and make necessary adjustments.

103 Click on the OK button.

104 Extrude curve L a distance of 80 mm in both directions (total 160 mm).

105 Mirror sweep surface M around coordinates w0,0 and w1,0.

The surfaces for making the main body of the skirt are complete.

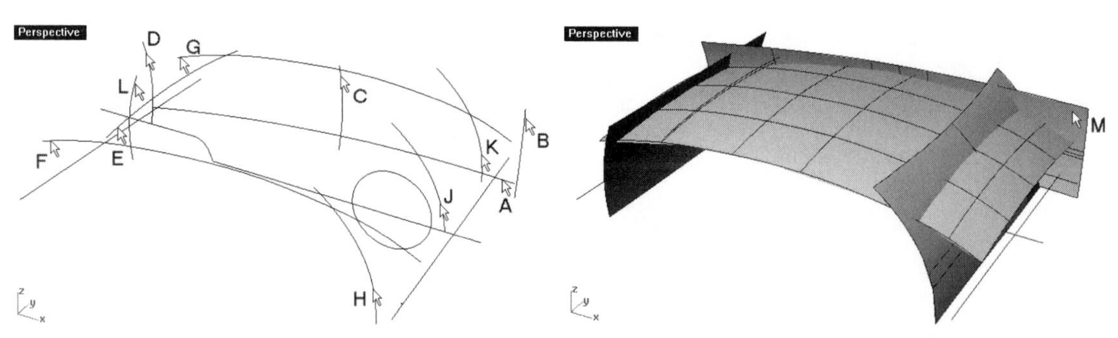

Figure C2–30. Five surfaces being constructed (left) and surfaces being mirrored (right)

Now we will trim the surfaces and join them together, as follows:

106 Maximize the Top viewport, and shade the display.

107 Use curves A and B (Figure C2–31) to trim away the outer portions of the surfaces at C, D, E, F, G, and H.

108 Maximize the Perspective viewport.

109 Using the trim command, select surfaces J, K, and L as cutting objects to trim away portions J, K, L, and M (Figure C2–31).

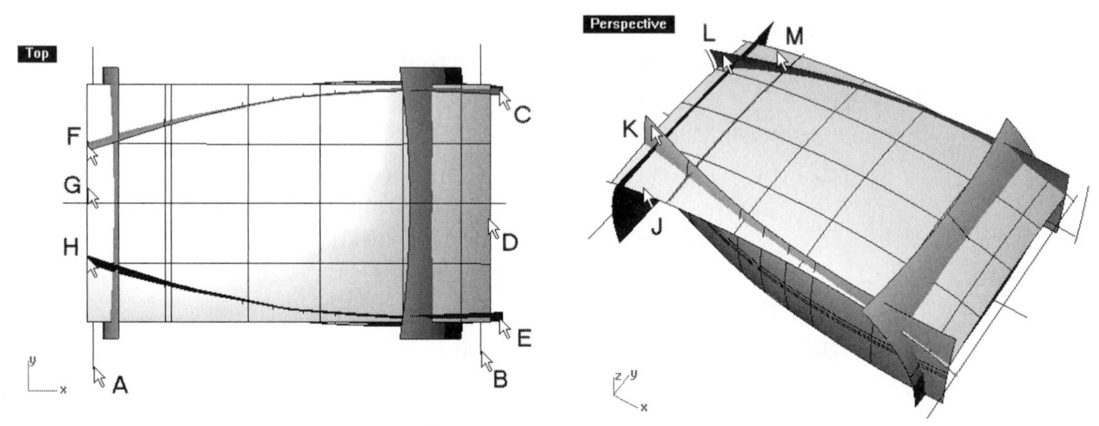

Figure C2–31. Surfaces being trimmed

110 Maximize the Front Viewport.

111 Referencing Figure C2–32, use curves A and B to trim away the lower portions of all the surfaces.

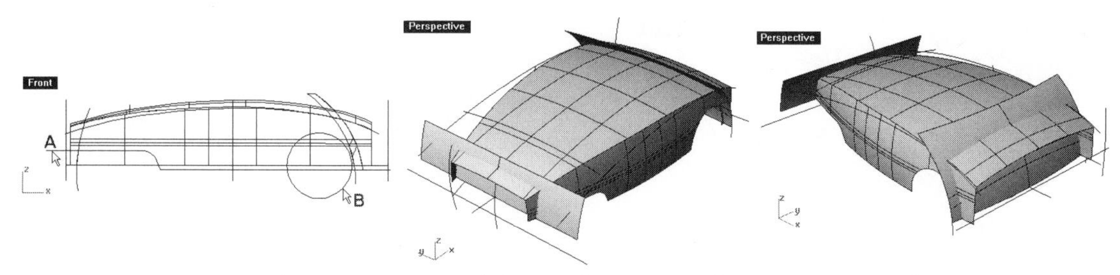

Figure C2–32. From left to right: Cutting objects, rear view of trimmed surfaces, a trimmed surfaces

112 Turn off layer Curves Skirt.

113 Referencing Figure C2–33, trim the surfaces accordingly.

114 Join all the surfaces into a single polysurface.

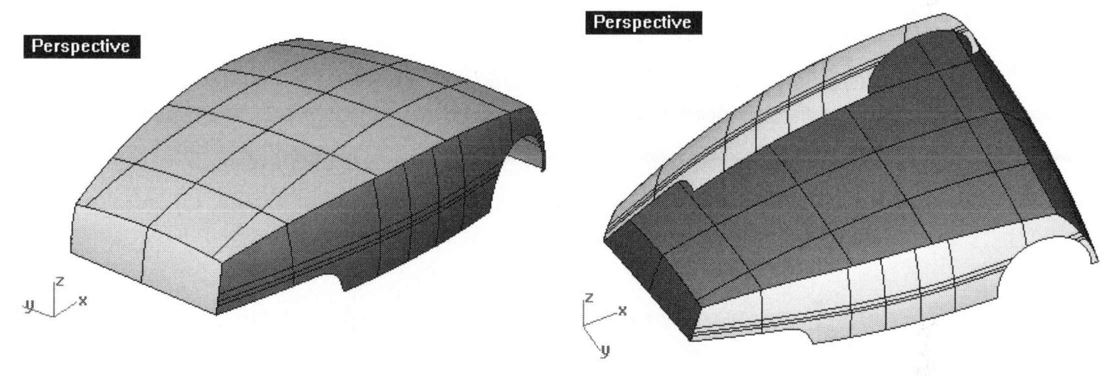

Figure C2–33. Surfaces trimmed, joined, and curves turned off

115 Select Solid > Fillet Edge > Fillet Edge, or click on the Variable Radius Fillet/Variable Radius Blend button on the Solid Tools toolbar.

116 Select the Current Radius option on the command area, and change it to 6 mm.

117 Select edges A and B (Figure C2–34) and press the ENTER key twice. Two edges are filleted.

118 Repeat the command to construct fillets of radius 3 mm on the remaining edges, as shown in Figure C2–34.

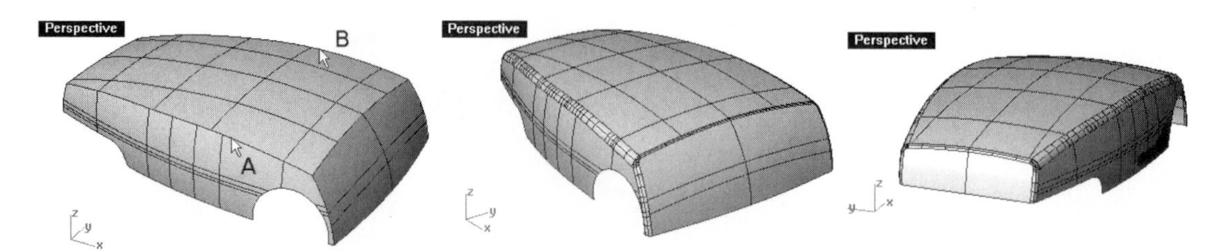

Figure C2–34. From left to right: two edges being filleted, frontal filleted, and rear portion filleted

Trimming Body, Skirt, and Floor Panels

We will now finish the main bodywork by trimming away unwanted portions of the main body, the skirt, and the floor panel that we constructed in the previous case study.

119 Turn on the Body Panel layer.

120 Referencing Figure C2–35, use the two polysurfaces as cutting objects to trim away the unwanted portions.

If you encounter any problem in trimming the unwanted portions, you can explode the skirt's polysurface into individual surfaces and try again. After trimming, you have to join the surfaces back into a single polysurface after trimming.

121 Join the two polysurfaces into a single polysurface by first selecting the main body and then the skirt.

(***NOTE:*** *Selection sequence determines in which layer the joined polysurface will reside. Here, because we select the body panel first, the joined polysurface will reside in the Body Panel layer.*)

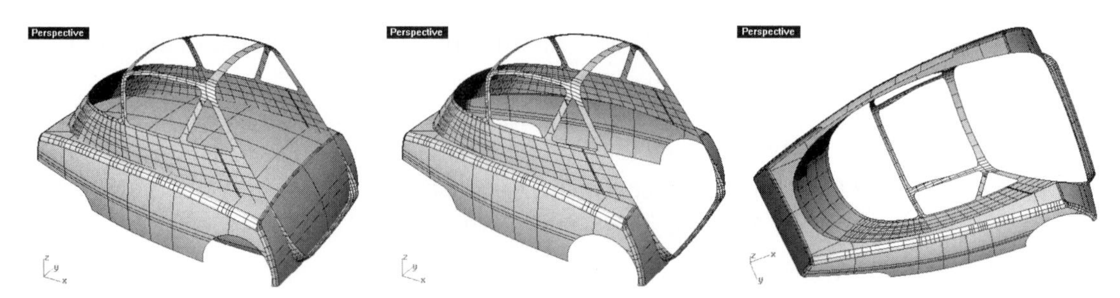

Figure C2–35. From left to right: Two polysurfaces and two views of the trimmed polysurfaces

Because a cutting object in a trimming operation needs to have a trimming boundary larger than the objects to be trimmed, we will cap the polysurface, as follows:

122 Select Solid > Cap Planar Holes, or click on the Cap Planar Holes button on the Solid Tools toolbar.

123 Select polysurface A (Figure C2–36) and press the ENTER key. A planar hole is constructed and joined to the polysurface.

If you encounter any problem in constructing a planar surface by capping, you can try using the PLANARSRF command (by selecting Surface > Planar Curves) and selecting the edges where the body panel meets the door panel.

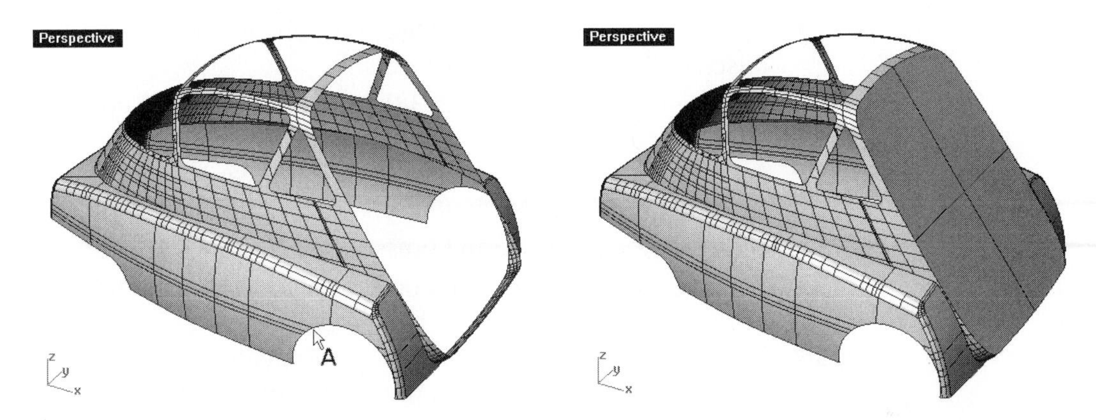

Figure C2–36. Polysurface selected (left) and polysurface capped (right)

Now we will use the capped body and skirt polysurface to trim away the unwanted portions of the floor panels. After trimming, we will remove the capped planar surface.

124 Turn on layer Floor Panel.

125 Referencing Figure C2–37, use polysurface A as a cutting object and trim away portions B, C, and D of the floor panels.

126 Select Solid > Extract Surface, or click on the Extract Surface button on the Solid Tools toolbar.

127 Select element E and press the ENTER key. The selected surface element is extracted from the polysurface.

128 Select surface E and press the ENTER key. The extracted surface is deleted.

The car body panels are complete.

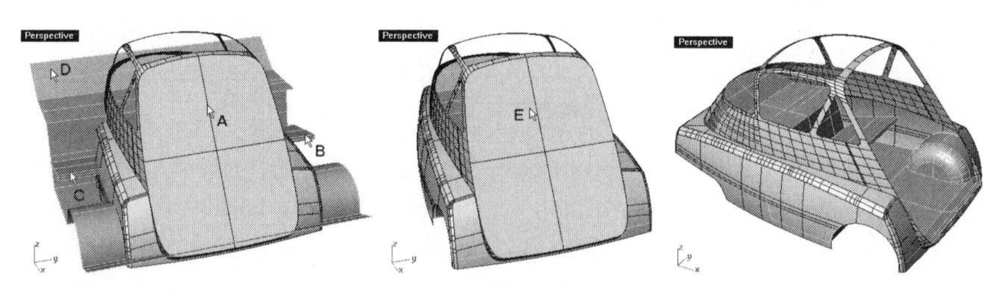

Figure C2–37. From left to right: Polysurfaces being trimmed, surface being extracted, and extracted surface deleted

Lights, Gas Cap, and Engine Vents

We will now finish the car model by adding lights, gas cap, and engine vents. We will first construct a revolved polysurface, an extruded solid, and a sphere, as follows:

129 Set current layer to Engine Misc, turn on layer Curves Engine Misc, and turn off all other layers.

130 Revolve curve A (Figure C2–38) 360 degrees around endpoints B and C.

131 Select Solid > Extrude Planar Curve > Tapered, or click on the Extrude Curve Tapered button on the Extrude Solid toolbar.

132 Select curves D and E, and press the ENTER key.

133 Select the DraftAngle option on the command area, and set it to -10 degrees.

134 Type 15 at the command area. The selected curves are extruded a distance of 15 mm, with a draft angle of -10 degrees.

135 Construct a sphere with center at point F (Use Point Osnap) and a radius of 10 mm.

Now we have three objects: a sphere, a revolved polysurface, and an extruded polysurface. The sphere will be used for making the gas cap, the revolved polysurface for making the engine's intake vent, and the extruded polysurface for making the engine's exhaust vent.

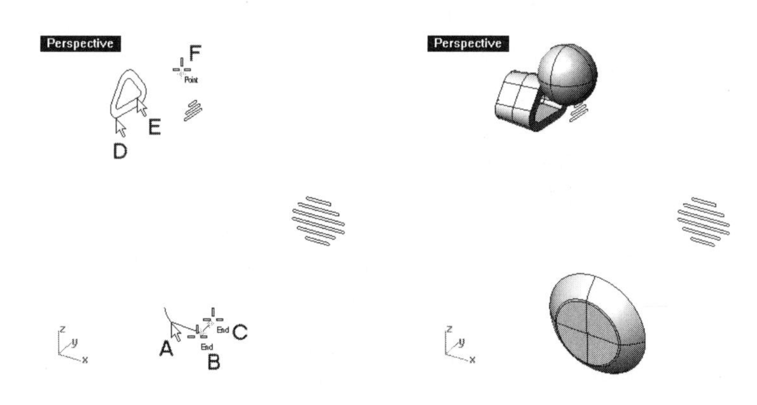

Figure C2–38. Curves and point selected (left) and revolved surface, extruded solid, and sphere constructed (right)

We will now cut some holes in the revolved polysurface in the Front viewport, as follows:

136 Maximize the Front viewport and set its display to Wireframe.

137 Referencing Figure C2–39, use curve A to trim polysurface B.

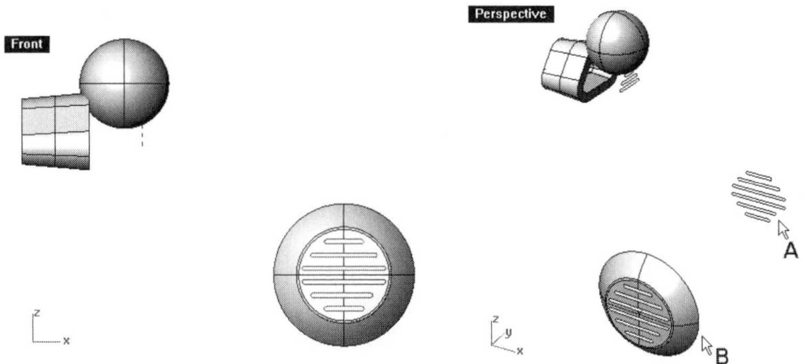

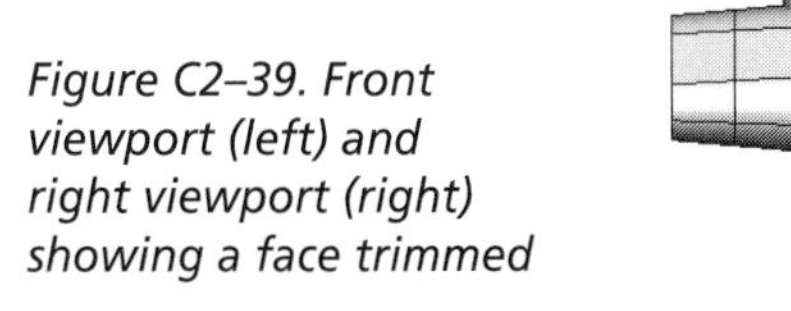
Figure C2–39. Front viewport (left) and right viewport (right) showing a face trimmed

We will rotate the revolved polysurface and fillet the edges of the extruded polysurface, as follows:

138 Select Transform > Rotate, or click on the Rotate 2D/Rotate 3D button on the Transform toolbar.

139 Select polysurface A (Figure C2–40) and press the ENTER key.

140 Select endpoint B as the base point. This is the center location of the circular face.

141 Type -10 at the command area. The polysurface is rotated.

142 Use the solid filleting command to fillet edges C and D with a radius of 1 mm.

Again, many commands, including the Rotate command, are construction-plane dependent. Therefore, we must not forget to reset the construction plane of the perspective viewport to World Top. Otherwise, the effect of rotation is very different.

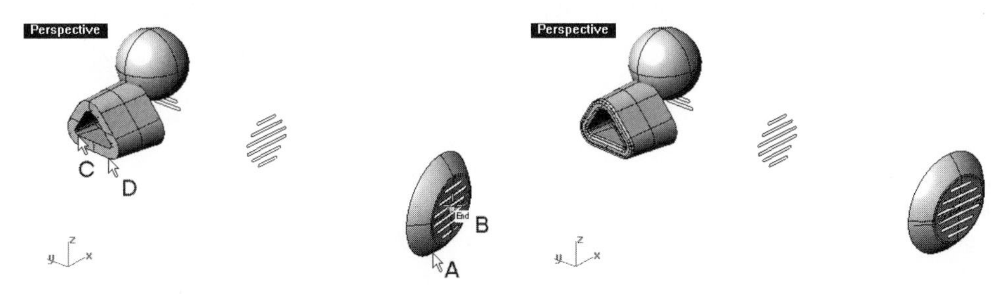

Figure C2–40. Polysurface being rotated and edges being filleted (left) and polysurface rotated and edges filleted (right)

We will now construct the front and rear lights of the car, as follows:

143 Set the current layer to Lights, and turn on the Curves Lights layer.

144 Referencing Figure C2–41, rotate the display.

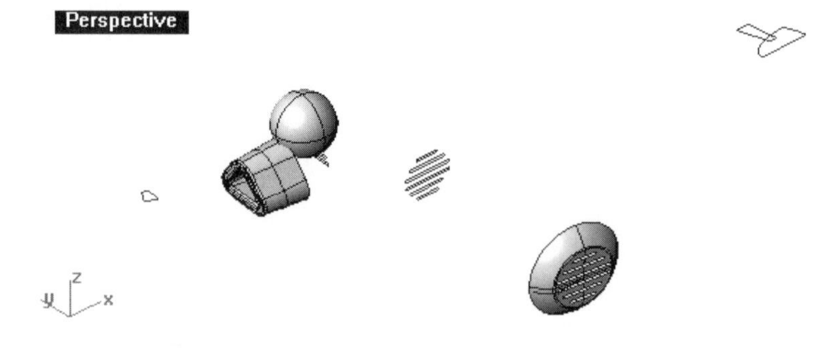

Figure C2–41. Curves for constructing lights turned on and display rotated

145 Referencing Figure C2–42, zoom in as necessary and construct two revolved polysurfaces by revolving curve A around endpoint B and C and curve D around endpoints E and F.

146 Combine polysurfaces G and H by Boolean Union operation.

147 Delete surface J.

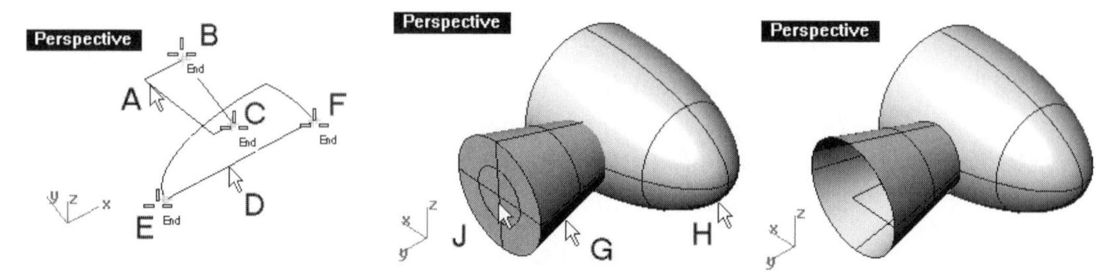

Figure C2–42. From left to right: curves being rotated, polysurface being united, and surface deleted

148 Referencing Figure C2–43, rotate curve A around endpoints B and C.

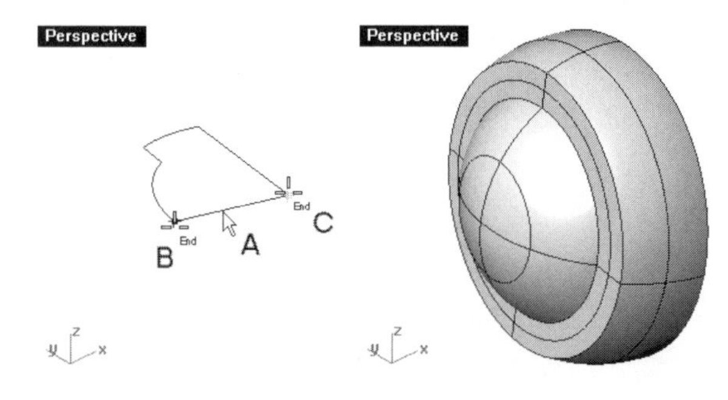

Figure C2–43. Curves for the rear light (left) and rear light constructed (right)

The objects for making the lights, gas cap, and engine vents are ready. We will perform some trimming operations to remove some unwanted portions, as follows:

149 Turn on the Body Panel and Skirt layer.

150 Use Boolean Union operation to combine polysurfaces A and B (Figure C2–44). Because these two polysurfaces reside on different layers, we have to select A and then B to have the combined polysurface placed in the Body Panel layer.

151 Construct a solid fillet of 1 mm radius at edge C.

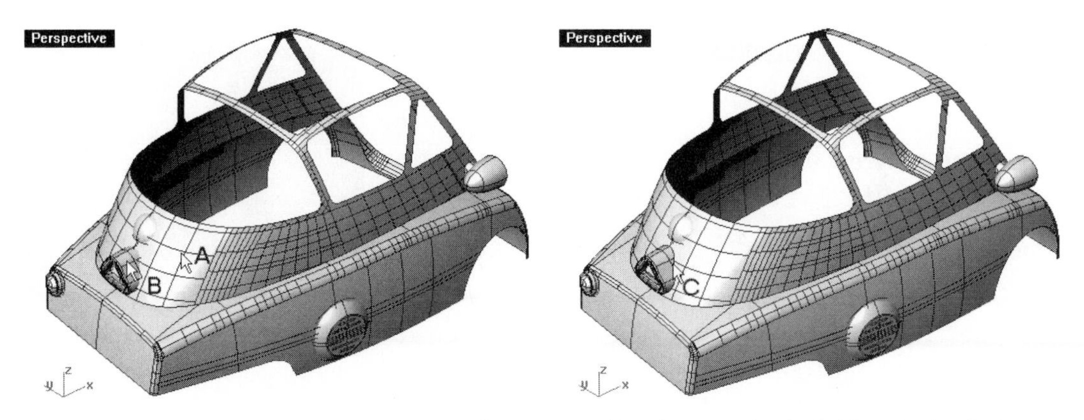

Figure C2–44. Rear vent being joined to the main body (left) and edge being filleted (right)

152 Set the display to Wireframe.

153 Referencing Figure C2–45, zoom in as necessary.

154 Use polysurface A to trim surface B and polysurface C.

155 Use polysurfaces D and E to trim away portions D and E.

156 Use polysurface F to trim away portion G.

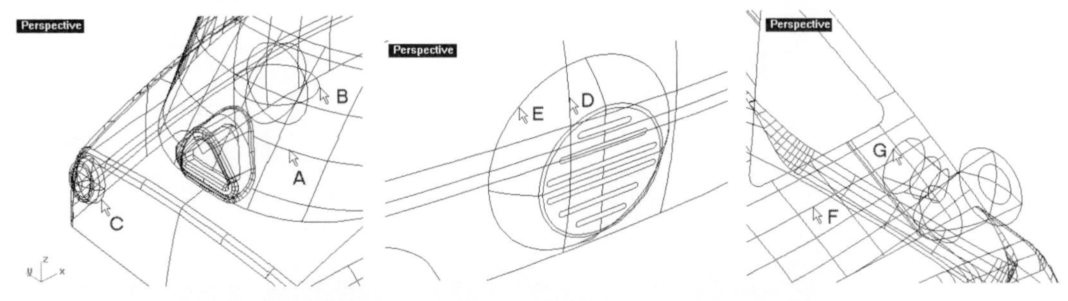

Figure C2–45. From left to right: trimming the light and sphere, trimming the body and engine vent, and trimming the light

We will construct a fillet edge and cut several holes, as follows:

157 Select Surface > Fillet Surfaces, or click on the Fillet Surface button the Surface Tools toolbar.

158 Select the Radius option on the command area.

159 Type 1 at the command area to change the fillet radius to 1 mm.

160 Select polysurfaces A and B. A fillet surface is constructed.

161 Join polysurfaces A and B and the fillet surface.

162 Set the display to Rendered.

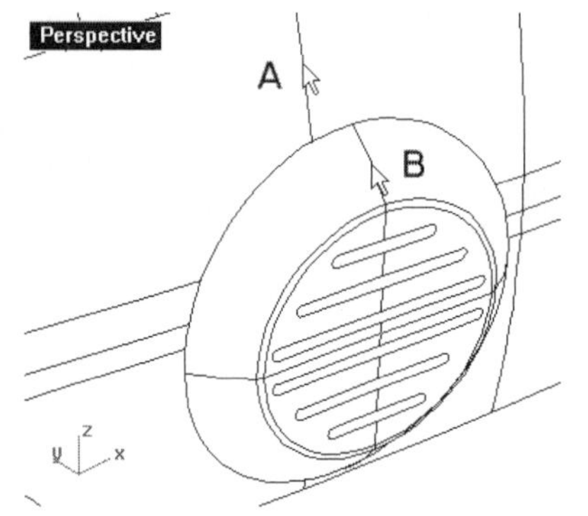

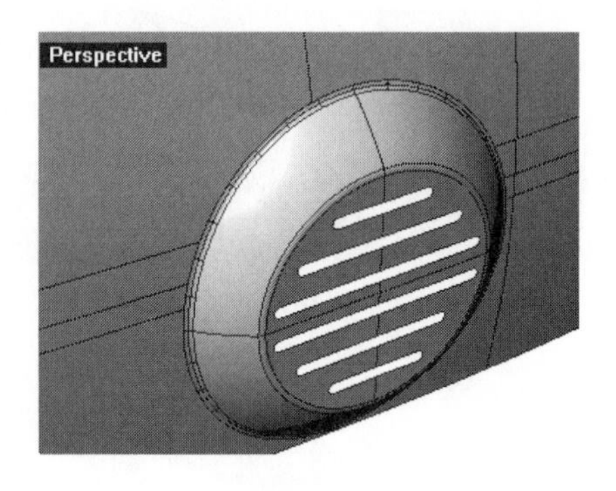

Figure C2–46. Surface being selected (left) and rendered display of the filleted edge

163 Maximize the Right viewport.

164 Referencing Figure C2–47, use curves A, B, and C to cut three holes on the main body.

165 Turn off the Curves Engine Misc and Curves Lights layers.

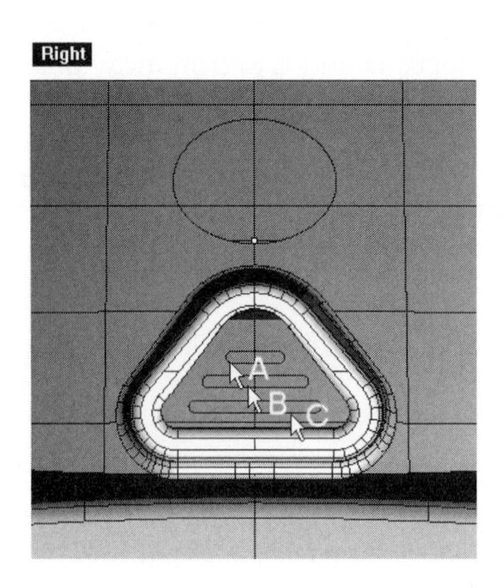

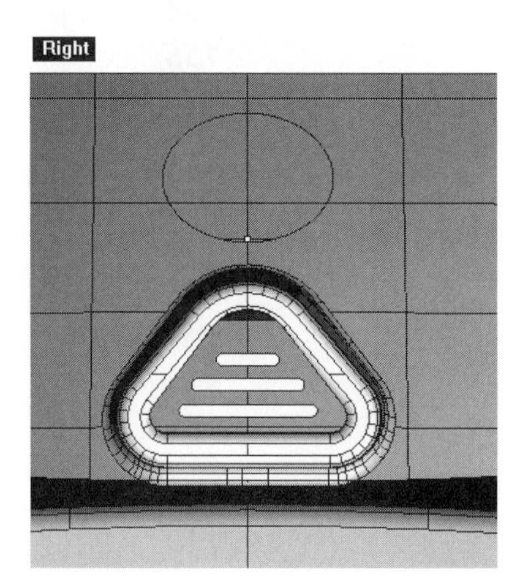

Figure C2–47. Curves selected (left) and polysurface trimmed (right)

We will complete the model by mirroring the lights, as follows:

166 Maximize the Perspective viewport, and set it to shaded display.

167 Mirror lights A and B (Figure C2–48) around coordinates w0,0 and w1,0.

168 Turn on all the layers except the curves layers.

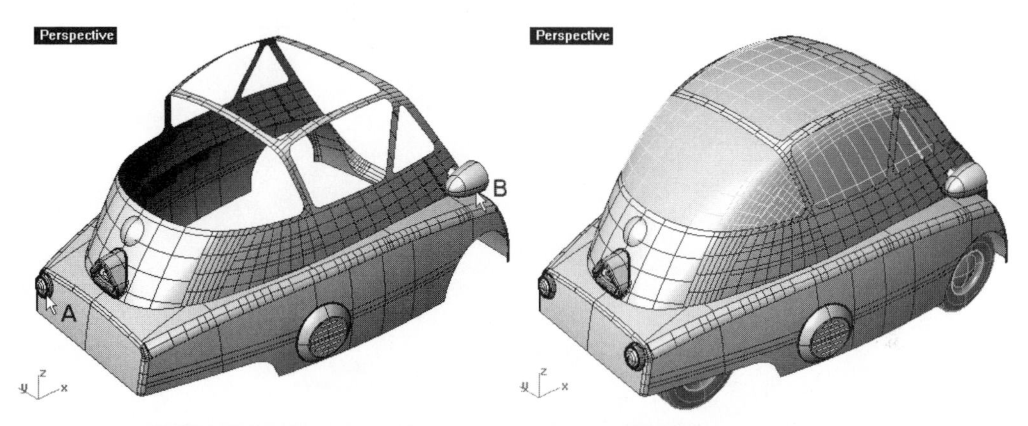

Figure C2–48. Lights being mirrrored (left) and lights mirrored (right)

The model is complete. Rendered images of the bubble car are shown in Figures C2–49 and C2–50.

Figure C2–49. Render images of the finished bubble car

Figure C2–50. Bottom view

Summary

To construct a model of a 3D free-form object in the computer, we use a set of free-form surfaces. Before making the surfaces, we first study and analyze the 3D object to determine what types of surfaces are needed. We consider various types of primitive surfaces, basic free-form surfaces, and derived surfaces. Among the surfaces, free-form surfaces are those most commonly used. All free-form surfaces have one thing in common: They need to be constructed from smooth curves and/or point objects.

It is natural to start thinking about the surfaces but not the points and curves while we are designing and making a surface model. However, because the computer constructs basic free-form surfaces from defined curves and/or point objects, the first task we need to tackle in making free-form surfaces is to think about what types of curves and/or points are needed and how they can be constructed. After making the curves and/or points, we then let the computer generate the required surfaces.

CHAPTER 1

Rhinoceros Functions and User Interface

Introduction

This chapter introduces Rhinoceros 4's functions, user interface, help system, and file saving methods.

Objectives

After studying this chapter, you should be able to

- ❒ Describe the key functions of Rhinoceros in relation to digital modeling
- ❒ Use Rhinoceros user interface and help system and save files

Overview

Rhinoceros (also known as Rhino) is a 3D digital modeling application that enables you to construct computer models in terms of NURBS (non-uniform rational B-spline) surfaces, polysurfaces (sets of contiguous NURBS surfaces joined together), solids ('water-tight' surfaces and polysurfaces), and polygon meshes (an approximation of an object by using a set of small contiguous planar faces). It also enables you to produce photorealistic rendered images from surfaces, polysurfaces, solids, and polygon meshes. To facilitate downstream computerized operations and reuse of existing digital models constructed using some other computer application, you can export Rhino models to various file formats and import various file formats into Rhino. To enable human interpretation, you can construct 2D engineering drawings. You may refer to Appendix A to learn more about various digital modeling concepts.

What is Rhinoceros?

In essence, Rhino is a very flexible and user-friendly 3D surface modeling tool, enabling you to construct NURBS surfaces as well as polygonal meshes for making 3D models of free-form objects. To facilitate NURBS surface construction, it provides a comprehensive set of tools for making and manipulating NURBS curves and point objects.

Constructing the model of an object usually concerns the making of two or more surfaces with different surface patterns. For easy handling of surface objects, you may join two or more contiguous surfaces sharing common edges to form a polysurface. Among the many ways to represent a solid in the computer; Rhino's solid is a surface or a polysurface enclosing a volume without any gaps, openings, or intersections among the individual surfaces. To obtain special form and shape effects from surfaces that are already constructed, you can use Rhino's transformation tools.

To help improvise your design, Rhino provides a set of analysis tools. To facilitate the management of models of products or systems with a number of components, Rhino enables you to construct block definition and block instances. To cope with other upstream and downstream computerized operation, Rhino enables you to import and export various file formats. To facilitate human interpretation of design, Rhino enables you to output 2D engineering drawing. In terms of rendering and animation, you can use Rhino's basic rendering tool as well as other plug-in tools, such Flamingo, Bongo, and Penguin.

Rhino's Points and Curves Tools

Although wireframe models by themselves have limited utility in design and manufacture, curves and points are required in many surface and solid construction operations. Therefore, you need to learn how to construct curves and points for the purpose of making surfaces and solids. Using Rhino, you can easily construct points and various types of 3D curves. Using the points and curves as framework, you construct various kinds of free-form surfaces. You will learn about points and curves in Chapters 4 through 6. Figure 1–1 shows a free-form object and the curves for making it.

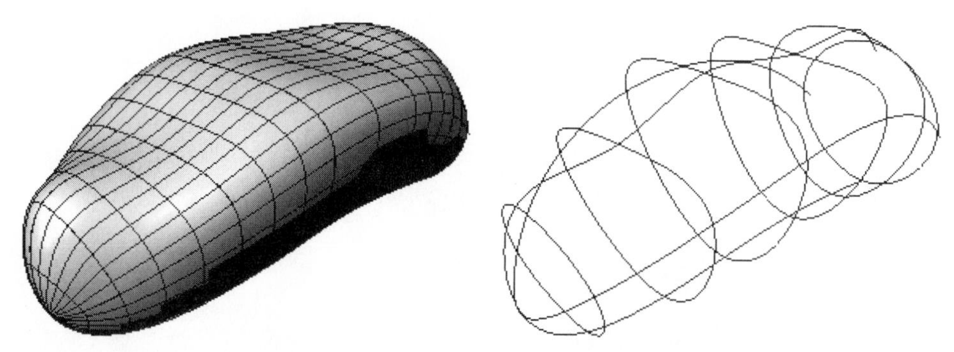

Figure 1–1. NURBS curves (right) for making free-form surface (left)

Rhino's NURBS Surfaces and Polygon Meshes Tools

Naturally, the prime objective of using Rhino as a tool in digital modeling is to construct free-form objects. As mentioned earlier, there are two basic ways to represent a surface in the computer: using a NURBS (Non-Uniform Rational B-Spline) surface to exactly represent the surface or using a polygon mesh to approximate the surface. Using Rhino, you can construct both NURBS surfaces and polygon meshes. You will learn about NURBS surfaces in Chapters 3 and 7, and about polygonal meshes in Chapter 9. Figure 1–2 shows a NURBS surface, and Figure 1–3 shows a polygonal mesh of a mobile phone casing.

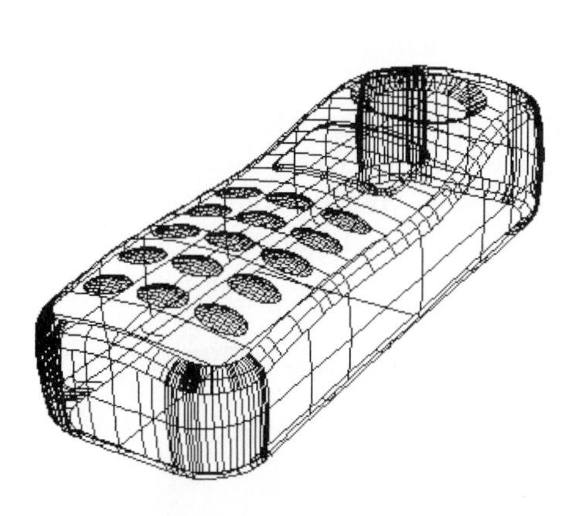

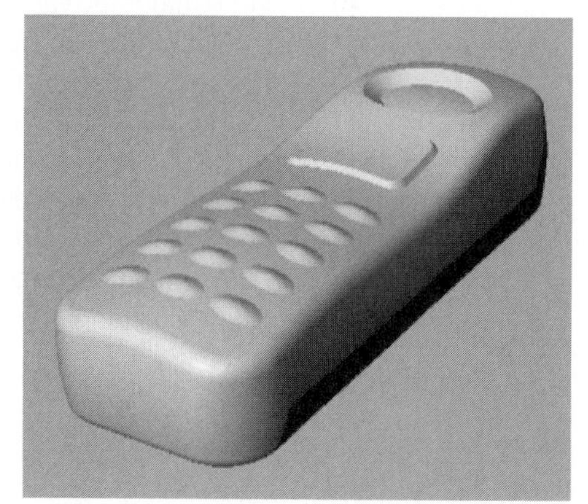

Figure 1–2. NURBS surface model of a mobile phone (left) and its rendered image (right)

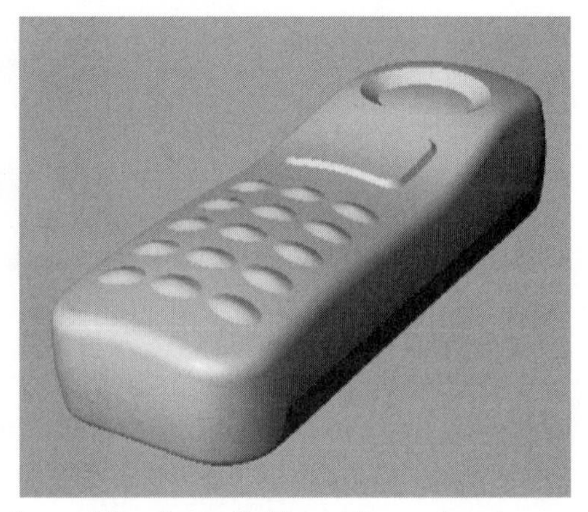

Figure 1–3. NURBS surface model of a mobile phone (left) and its rendered image (right)

Render Meshes

You may notice that the rendered images for both (NURBS surfaces and polygon meshes) are more or less the same. This is the case because Rhino will first generate a set of meshes from the NURBS surface prior rendering. The mesh that is produced is called render meshes. To fine tune the mesh, you may change the settings in the Mesh tab of the Rhinos Options dialog box accessible from the Tools pull-down menu.

Rhino's Polysurface and Solid Tools

One major advantage of using surface modeling tool rather than solid modeling tool in free-form modeling is its flexibility, in terms of constructing individual surfaces to represent unique facets of an object, because individual surfaces are independent of each other in the database of the file, and there is no relationship between them. To help handle a set of contiguous surfaces collectively, you can join them together at their connecting edges to form a polysurface. Figure 1–4 shows a set of contiguous surfaces joined together to form the model of a model car body.

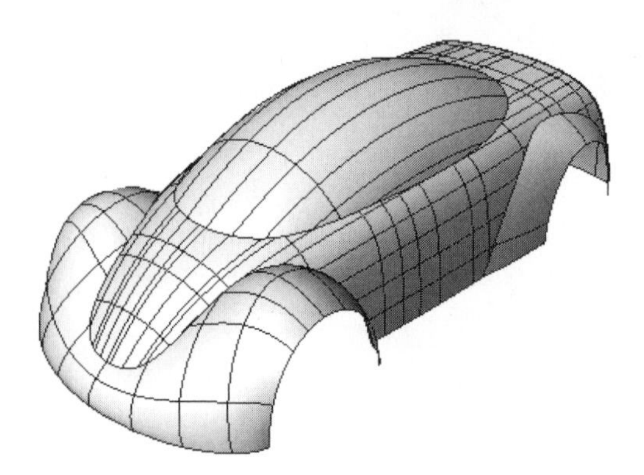

Figure 1–4. Polysurface representing a model car body

In Rhino, if a polysurface or any single surface (such as a sphere or an ellipsoid having their edges collapsed to a point) encloses a watertight volume (without any gaps, openings, or intersections among the individual surfaces), a solid is formed. You can construct Rhino solids in two basic ways: directly, using the Rhino solid modeling tools, or by converting a set of contiguous NURBS surfaces into a solid volume by joining them. You will learn more about Rhino's solid modeling in Chapter 8. Figure 1–5 shows a Rhino solid, which consists of a set of NURBS surfaces joined together without any gaps, openings, or intersection among the surfaces.

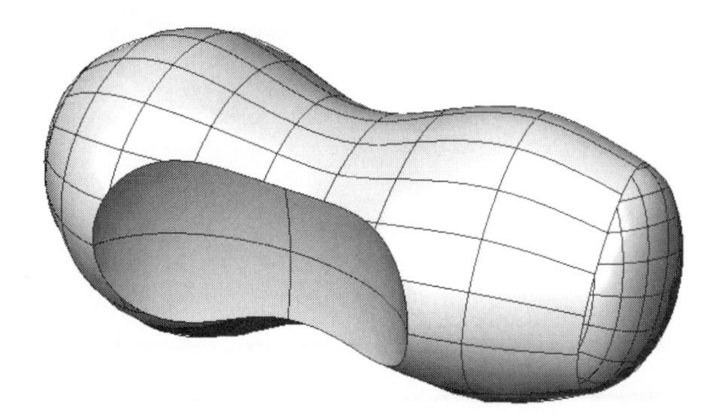

Figure 1–5. Rhino solid

Rhino's Transformation and Analysis Tools

To change the shape of your design by manipulating surfaces that are already constructed, you use the transformation tools. Figure 1–6 shows how a complex shape can be obtained by transforming an object of simple shape. To analyze objects in order to improvise the design, you use the analysis tools. You will learn about transformation in Chapter 10 and analysis in Chapter 11. Figure 1–7 shows the use of zebra lines to help visualize the smoothness of a polysurface.

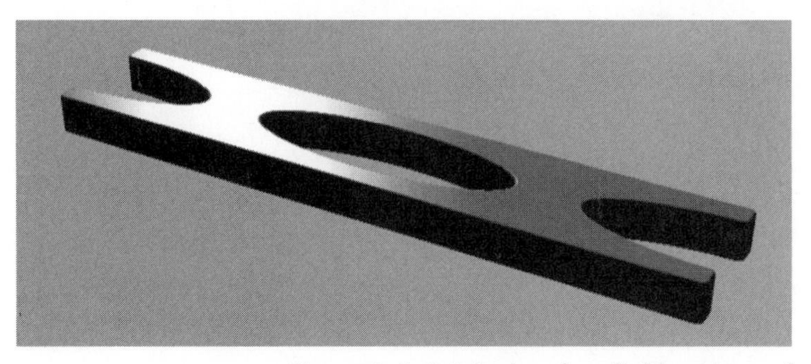

Figure 1–6. Original surface (left) and transformed surface (right)

Figure 1–7. Analyzing the smoothness of a surface model

Assembly Tools

To facilitate evaluation of how various component parts of a product or system can or should be put together, you assemble individual components in an assembly model.

Using Rhinos, an assembly can be simulated by manipulating and inserting individual files representing various components into a single file, using groups and blocks that you will learn about in Chapter 12. Figure 1–8 shows the assembly of a set of components.

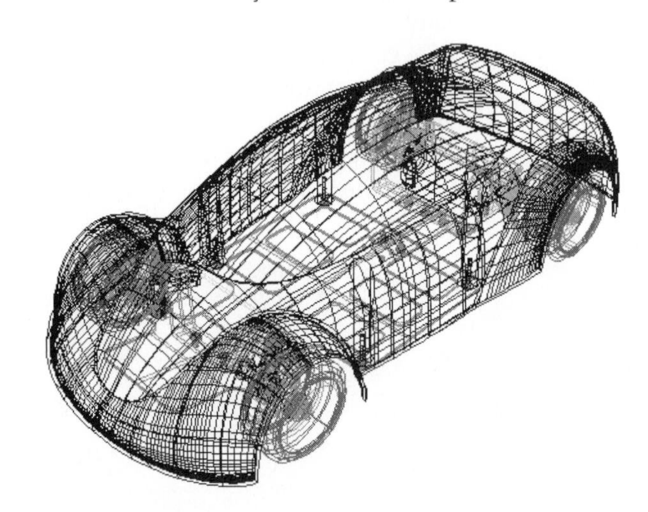

Figure 1–8. An assembly of components

Data Exchange and Engineering Drawing Tools

To reuse digital data constructed by using other applications and to facilitate downstream computerised processes, you exchange data in various formats. To represent a 3D object in a 2D drawing sheet, you use an orthographic engineering drawing. If you already have a 3D digital model, you use the computer to generate orthographic views of the model. Using Rhino, you can open files saved in other data formats, insert files saved in other data formats into a Rhino file, export Rhino file to other data formats, and generate a 2D drawing from the digital model of the 3D object and add appropriate dimensions and annotations to the drawing. You will learn about data exchange and generation of 2D drawings from digital models in Chapter 13. Figure 1–9 shows a 2D engineering drawing generated from the model shown in Figure 1–8.

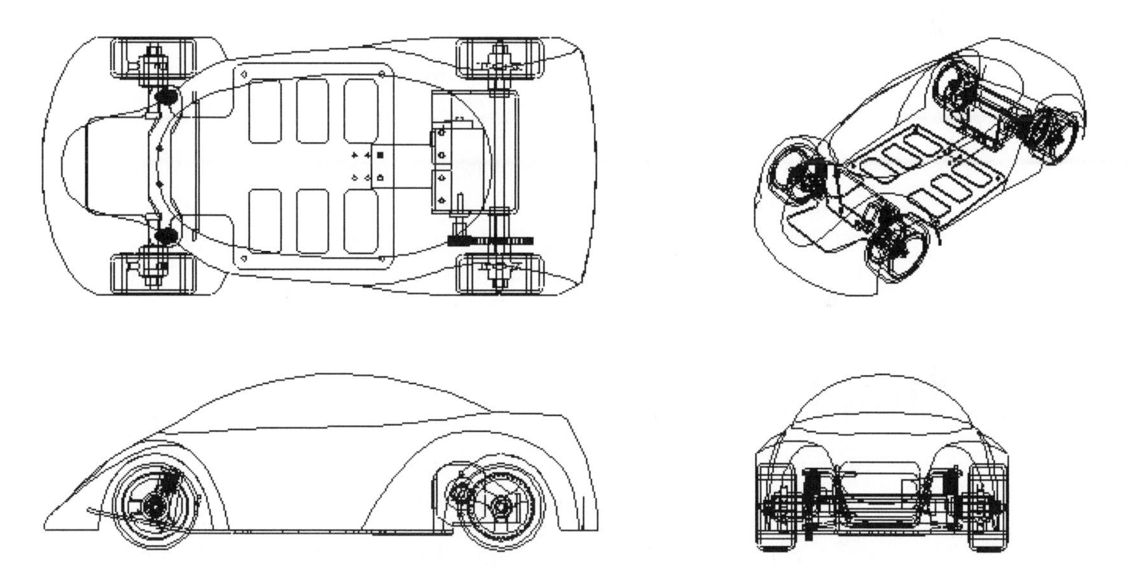

Figure 1–9. 2D drawing generated from 3D surface model

Rendering and Animation Tools

Rendering is a way to produce photorealistic images from 3D surfaces, polysurfaces, and polygon meshes. To add reality to the images, you have to include material and lighting information. Figure 1–10 shows the two rendered images of the model shown in Figure 1–8.

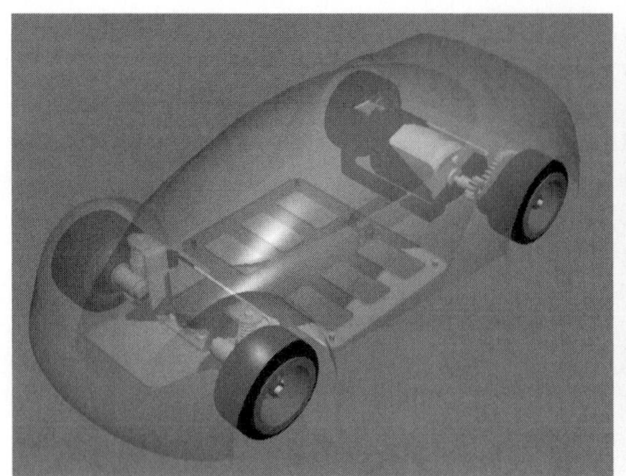

Figure 1–10. Rhino rendering (left) and Penguin' cartoon style rendering (right)

To produce a rendered image in Rhino, you can use the basic renderer, the Tree Frog renderer, or the Flamingo Raytrace/Photometric renderer. To obtain a cartoon-like or hand-sketching-like image, you can use the Penguin render. To further realize the 3D model in terms of photorealistic animation, you can use the Bongo animator. Note that you need to install Flamingo before you can use the Flamingo Raytrace/Photometric renderer, install Penguin before you can use the Penguin render, and install Bongo before you can produce photorealistic animations. You will learn about rendering and animation in Chapter 14.

Rhino's User Interface

If you are a novice, then have a bit of hands-on examination of Rhinoceros 4's user interface. Perform the following step.

1. Start Rhino by selecting the Rhinoceros 4 icon from your desktop. The Rhino Startup Template dialog box (Figure 1–11.), together with the user interface will display (Figure 1–12).

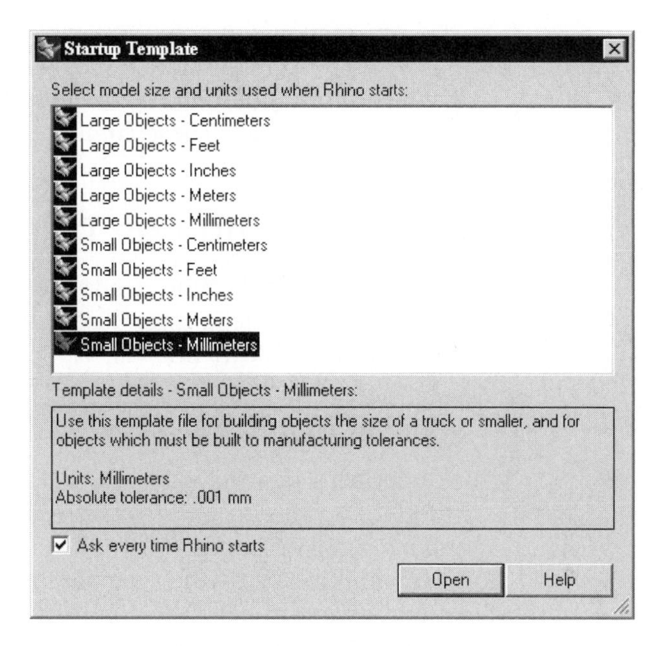

Figure 1–11. Startup Template dialog box

The Startup Template dialog box serves to provide you with a choice of template file upon startup. You should select one of the templates and click on the Open button to proceed. If you do not want this dialog box to show up next time you start Rhino, clear the Ask every time Rhino starts check box.

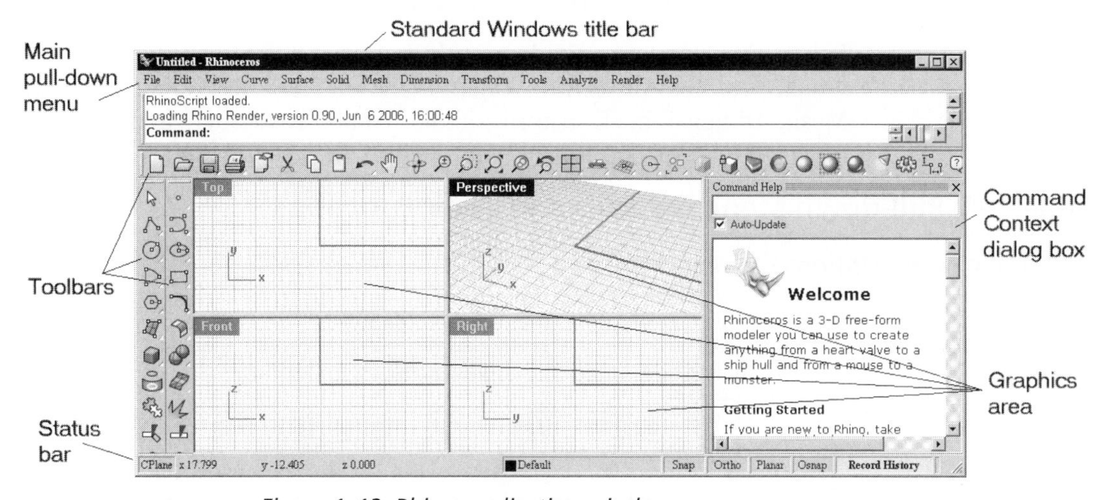

Figure 1–12. Rhino application window

In the Application window, you will find five major areas: standard Windows title bar, main pull-down menu, command area, graphics area, and status bar. In addition, there are a number of toolbars and a command help dialog box. These components are described in the sections that follow.

Standard Windows Title Bar

At the top of the Application window there is the standard Windows title bar. This title bar functions no differently and contains nothing different than the basic Windows title bar.

Command Help Dialog Box

The Command Help dialog box, by default, is activated the first time you start Rhino. It is an instant help system, providing help messages to any command that you activate. If you want to close the command help dialog box, simply click on the x sign at the upper right corner of the dialog box. If you want to open the command help dialog box after closing it, you may call it out by typing the command name "CommandHelp" at the command area (to be explained later in this chapter).

2 Click on the X mark at the upper right corner of the Command Help dialog box to close it. (If you prefer, you may leave it on all the time. We will discuss the use of Command Help dialog box later.)

Main Pull-down Menu

Below the standard Windows title bar is the main pull-down menu, which contains thirteen options: File, Edit, View, Curve, Surface, Solid, Mesh, Dimension, Transform, Tools, Analyze, Render, and Help. The functions of these options are outlined in Table 1–1.

Table 1–1: Pull-down Menu Options and Their Functions

Pull-down Menu Options	Function
File	For working on files and templates
Edit	For editing points, curves, surfaces, and solids
View	For manipulating display settings and establishing construction planes
Curve	For constructing and manipulating points and curves
Surface	For constructing and manipulating NURBS surfaces

Pull-down Menu Options	Function
Solid	For constructing and manipulating solids and polysurfaces
Mesh	For constructing and manipulating polygon meshes
Dimension	For constructing 2D drawings and adding annotations
Transform	For transforming objects you have constructed
Tools	Provides various types of useful tools
Analyze	Helps analyze objects you have constructed
Render	For shading and rendering
Help	Provides useful help information

One way to perform commands and operations is to select options from pull-down menu. To appreciate how to use the pull-down menu, continue with the following steps to construct a cone and a sphere.

3 Select Solid > Cone. (Select Cone from the Solid pull-down menu.)

4 Click on locations A and B (Figure 1–13) in the Top viewport (to indicate the base center and to specify the radius of the cone) and then location C in the Front viewport (to indicate the vertex of the cone). (Note: For the purpose of this tutorial, exact location is unimportant.) A cone is constructed.

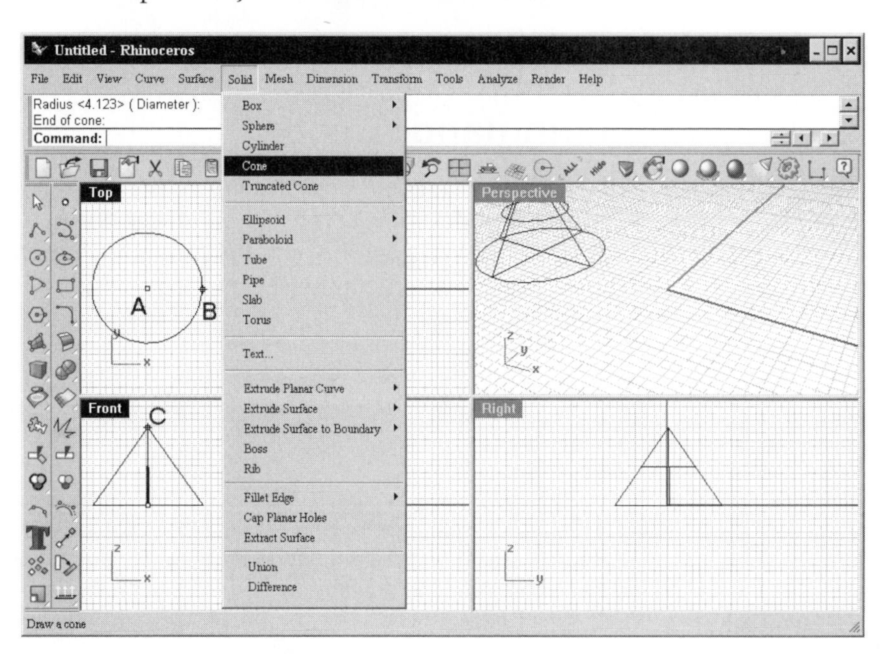

Figure 1–13. Cone command activated from the Solids pull-down menu

5 Select Solids > Sphere > Center, Radius. The Center, Radius command from the Sphere cascading menu of the Solids pull-down menu is used to construct a sphere by specifying its center and its radius.

6 Click on two locations A and B indicated in Figure 1–14 to specify the center and a point on the sphere. A sphere is constructed.

*(**NOTE:** A cone is a polysurface consisting of a slant surface and a flat surface joined together. A sphere is a single surface with edges collapsed to a single point. Because they are regarded as solids in Rhino, these commands are grouped under the Solid menu.)*

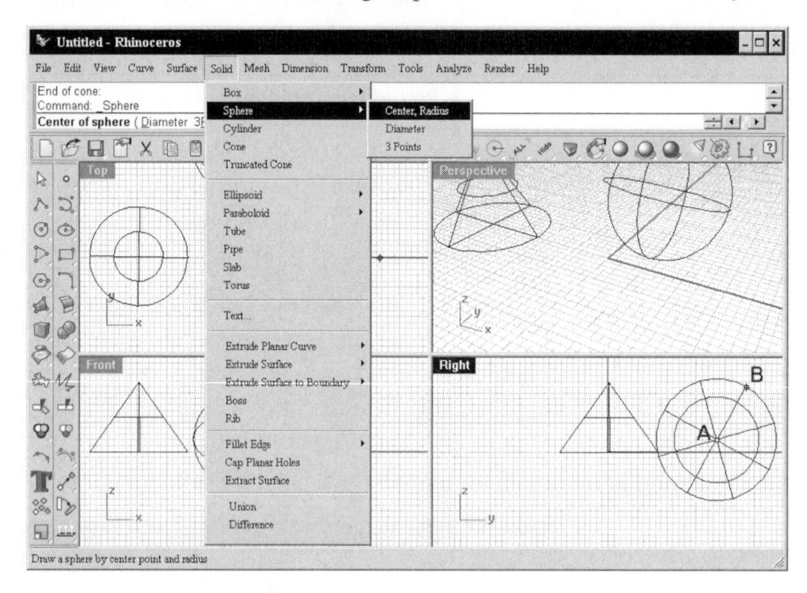

Figure 1–14. Sphere cascading menu for constructing a center-radius sphere

Command Area

Below the main pull-down menu is the command area (Figure 1–15), which provides a place for textual interaction. By default, the command area is docked below the pull-down menu. However, you may select and drag it to anywhere on the screen.

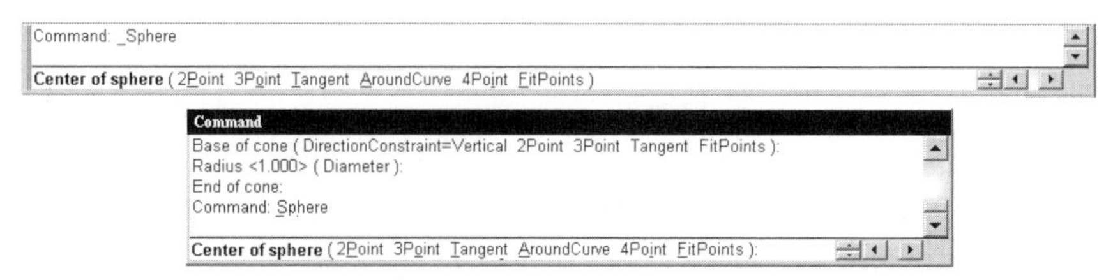

Figure 1–15. Command area docked below the pull-down menu (above) and floating command area (below)

Case Sensitivity

Command names are NOT case sensitive; and therefore, you can use any combination of small and capital letters to specify a command name. Here in the command area, you run a command by typing the command name or alias of the command and then pressing the ENTER key or the space bar. After a command is run, further prompts or instructions will appear in this area or in any associated pop-up dialog boxes.

Command List

If you are not too sure about the spelling of a command name, you may simply type the first letter and then move the cursor to the command area to display the list of command beginning with the letter that you just typed.

7 Type A at command area and move the cursor to the command area. A list of command beginning with the letter A is displayed. You may click on a command to select it. (See Figure 1–16.)

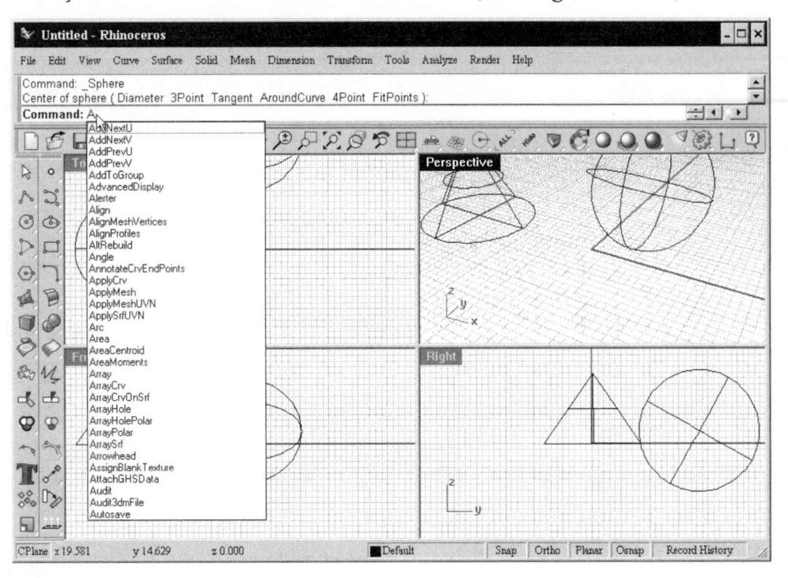

Figure 1–16. List of commands displayed

Auto-Complete Feature

An "auto-complete" function is incorporated here at the command line. That is, if you type a command here, you do not have to type the full name. For example, after you type AR, the system will automatically complete the typing to display the word ARC. Continue with the following steps to learn more about the command area.

8 Type R at command area, together with the letter A that you typed previously, you have two letters AR typed.

Full name of the command "arc" is displayed. (This is the auto-complete feature.)

9 Press the ENTER key. The command is executed. (See Figure 1–17.)

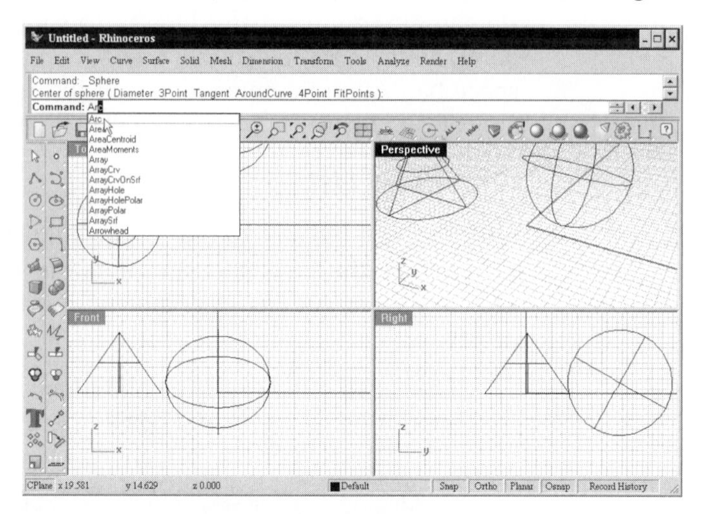

Figure 1–17. Auto-complete feature

Clickable Options in the Command Area

Options in the command area are clickable, which means that you can use the mouse to click on an option as well as typing the keyword of the options. Continue with the following steps.

10 Use the mouse to click-on the Deformation option in the command area. The deformable option is selected. (A deformable arc is a degree 3 curve in the shape of an arc. You will learn more about degree of polynomial in Chapter 4.)

11 Click on locations A, B, and C (Figure 1–18). An arc is constructed.

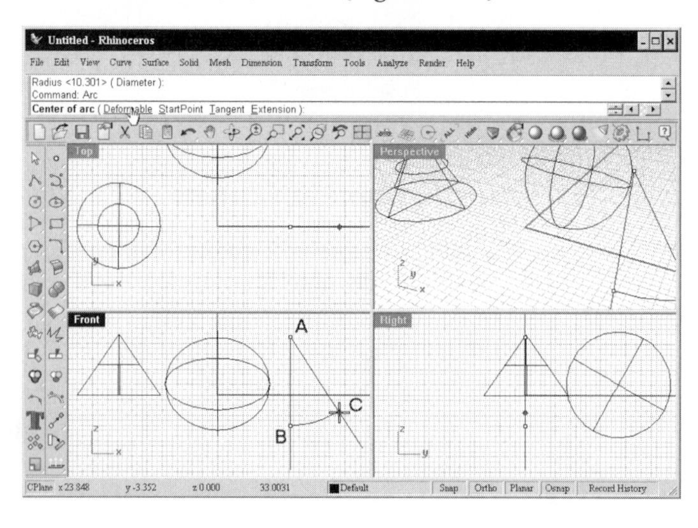

Figure 1–18. Clickable options

Displaying Recently Used Commands

By right-clicking in the command area, a pop-up menu showing all recently used commands is displayed, providing a quick access to commands that are used. Continue with the following steps:

12 Right-click on the command area. A pop-up menu appears, displaying all the recently used commands. (See Figure 1–19.)

13 Click on Cone from the pop-up menu. The recently used command, cone, is executed again.

14 Press the Esc key to terminate the command.

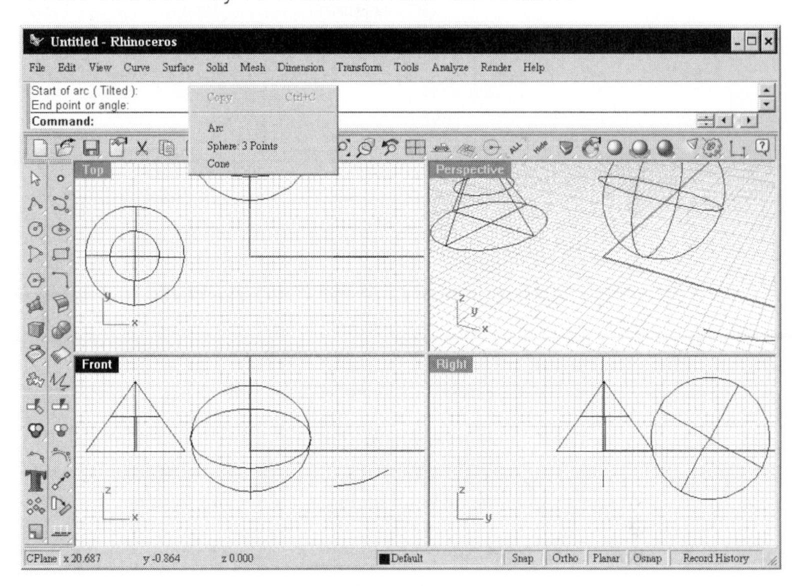

Figure 1–19. Pop-up menu

Displaying Command History

Apart from displaying the command name, you may display the command line history in a dialog box. History can be saved in a text file for reference or for construction of a script file (see scripting in the next paragraph). Continue with the following steps:

15 Select Tools > Commands > Command History or click on Command History from the Tools toolbar. The Command History dialog box is displayed. (See Figure 1–20.)

In the Command History dialog box, there are three buttons. Clicking the Copy All button will copy the command line history to the Windows clipboard for subsequent pasting to another file. Clicking on the Save As button will save the command line to a text file.

16 Click on the Close button. The Command History dialog box is closed.

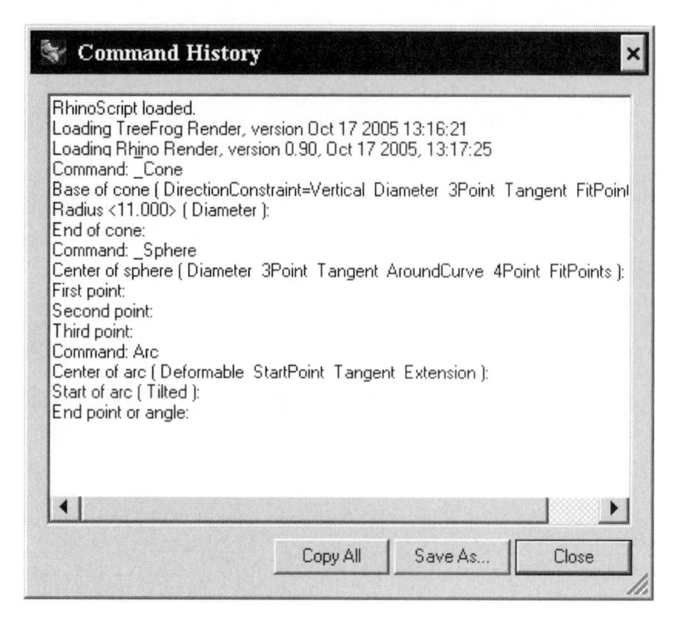

Figure 1–20. Command History dialog box

Complete List of Commands

Another way of knowing Rhino command names is to list them or save them to a text file, as follows:

17 Type CommandList at the command area to display the Print Window dialog box as shown in Figure 1–21.

18 Click on the Save As button.

19 Specify a file name. All Rhino commands are exported to the text file.

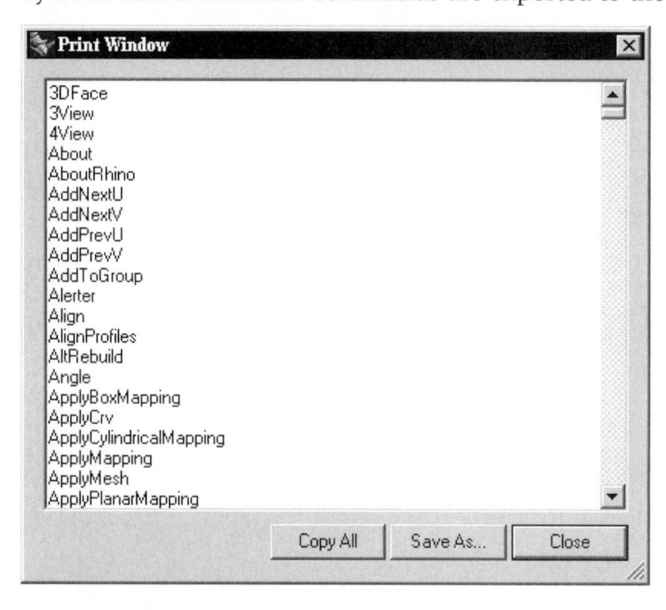

Figure 1–21. Print Window dialog box

Command Scripting

All Rhino commands are scriptable. It means that if you write down the command sequence in a text file with an extension of .txt, you can then run the script file to have all the command sequence executed automatically. Simply speaking, a script file consists of a number of lines that you will type at the command area to execute an action. In the script file, you may use a [Space] or [Return] to execute a command. If a command calls for a dialog box, you can prefix the command name with the character -. At the beginning of the script file, you may use the exclamation mark (!) and a space to terminate any previous commands. Continue with the following steps.

20 Select Tools > Commands > Read from File, or click on Read Command File from the Tools toolbar.

21 In the Open Text File dialog box, select the file "Rhino Script 1" from the Chapter 1 folder of the CD accompanying this book and click on the Open button. A circle is constructed by the script file. (See Figure 1–22.)

To view the script file, you may use any text editor.

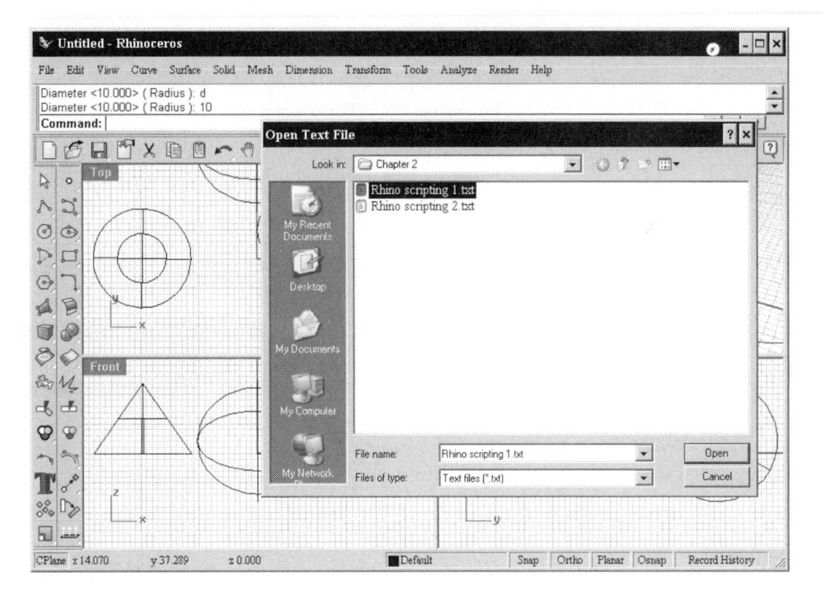

Figure 1–22. Opening the script file and a circle constructed by the script file

Script Editor

To test run a sequence of commands prior writing a script, you may use the Macro Editor, as follows:

22 Select Tools > Commands > Macro Editor, or click on the Open Macro Editor button on the Utilities toolbar.

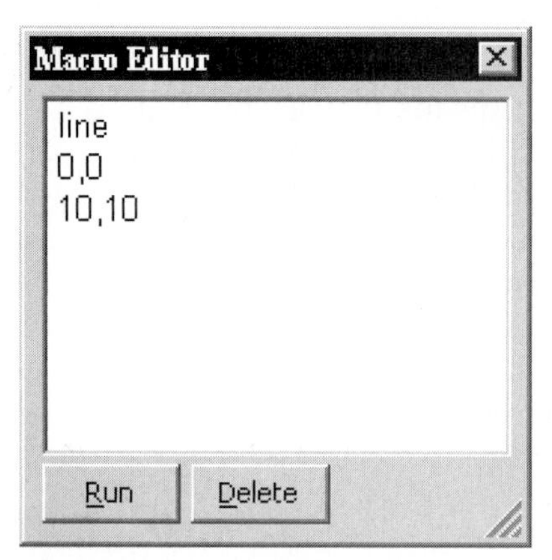

Figure 1–23. Macro Editor and a line constructed

23 In the Macro Editor dialog box shown in Figure 1–23, type the following lines:

line
0,0
10,10

24 Click on the Top viewport to make it the current viewport.

25 Click on the Run button, a line is constructed. If the sequence of command lines is correct, you may copy it from the Macro Editor and paste it to a text file, which is subsequently saved as a script file.

Command Aliases

To speed up calling a command, you may use command aliases that are already defined by default or define you own command aliases. To view the command aliases available, you may open the Options dialog box. Continue with the following steps.

26 Select Tools > Options to display the Options dialog box.

27 In the Options dialog box shown in Figure 1–24, select Rhino Options > Aliases from the left pane.

28 In Aliases tab, there are a list of aliases and four buttons, enabling you to import and export aliases to a text file, define a new alias, and to delete an existing alias.

29 Click on the Cancel button to close the Options dialog box without changing anything.

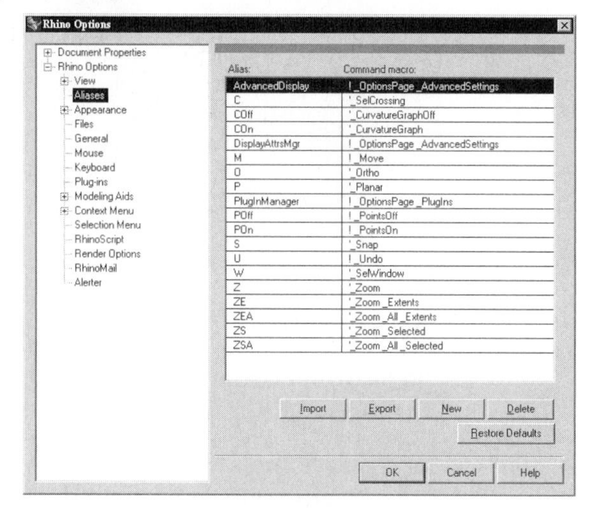

Figure 1–24. Aliases tab of the Options dialog box

Command Alerter

By making appropriate settings in the Alerter page of the Options dialog box, you may set the system to perform a specified operation if a command takes longer time than expected, as follows:

30 Click on the Alerter Utility button on the Utilities toolbar, or type Alerter at the command area.

31 In the Alerter tab of the Rhino Options dialog box (Figure 1–25), check the Enable command alerting.

32 In the Commands to watch area, input the command names to be watches.

33 If you want to play a sound file, check on the Play a sound file button and specify a sound file.

34 If you want to perform an operation, check on the Run Rhino commands button and specify commands to run.

35 Click on the OK button to close the dialog box.

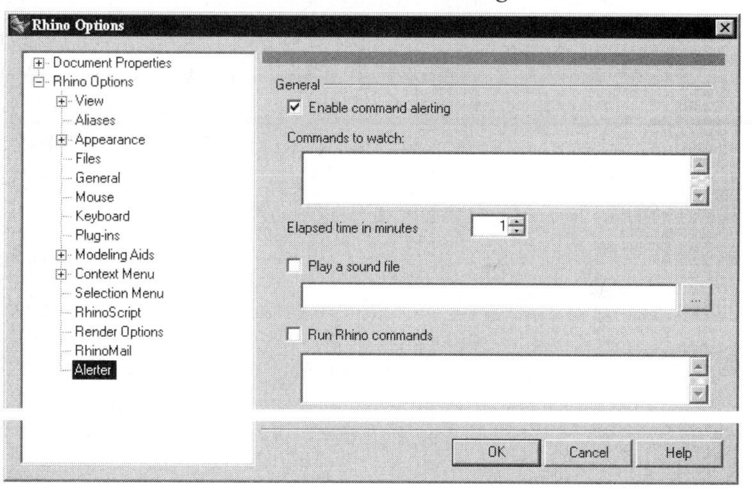

Figure 1–25. Alerter page

Graphics Area

Down below the command area is the graphics area, where you construct your model. This area can be divided into a number of viewports, which can be docked or floating. The default file that you are working on has a default four-viewport configuration (i.e., Top, Front, Right, and Perspective viewports).

You can double-click the viewport's label (at the upper left corner of the viewport) to maximize it. To return to the previous viewport configuration, you double-click it again. Continue with the following steps:

36 Double-click the label of the Perspective viewport. The viewport is maximized. (See Figure 1–26.)

37 Double-click the label again. The previous viewport configuration is restored.

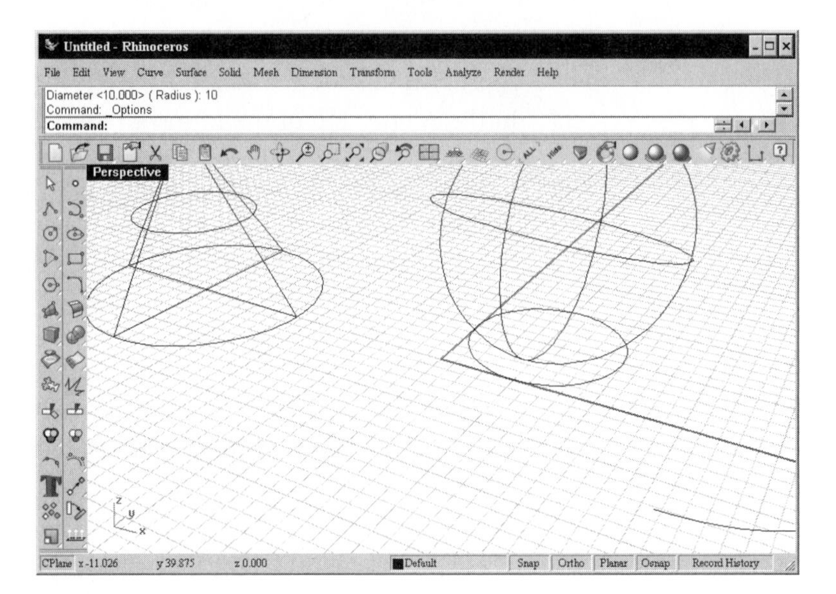

Figure 1–26. Perspective viewport maximized

Viewport Configuration

Although the number of viewports and their orientation are determined by the viewport setting of the template file you use to start a new file, you may configure it and the number of viewports is unlimited. To use a template, you start a new file by selecting New from the File pull-down menu. This accesses the Open Template File dialog box (shown in Figure 1–27), in which you select a template. Perform the following steps:

1 Select File > New.

2 Do not save your file. In the dialog box that pops up, click on the No button.

(***NOTE:*** *If you are using the Evaluation version, the number of times you can perform a save is limited. To be able to see the result of your work, do not save your files unless you are working on the case studies. Opening a new file, starting a new file, or exiting Rhino automatically closes your current file.)*

3 In the Open Template File dialog box (Figure 1–27), select the file “Small Objects”, Millimeters and click on the Open button. A new file is started.

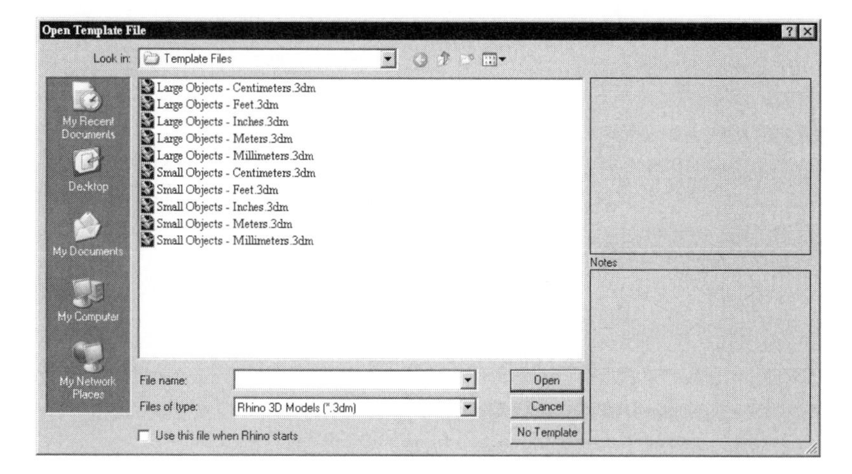

Figure 1–27. Open Template File dialog box

4 Select View > Viewport Layout > 3 Viewports, or 3 Viewport/3 Default Viewports button on the Viewport Layout toolbar. The display is set to a 3 viewport display. (See Figure 1–28.)

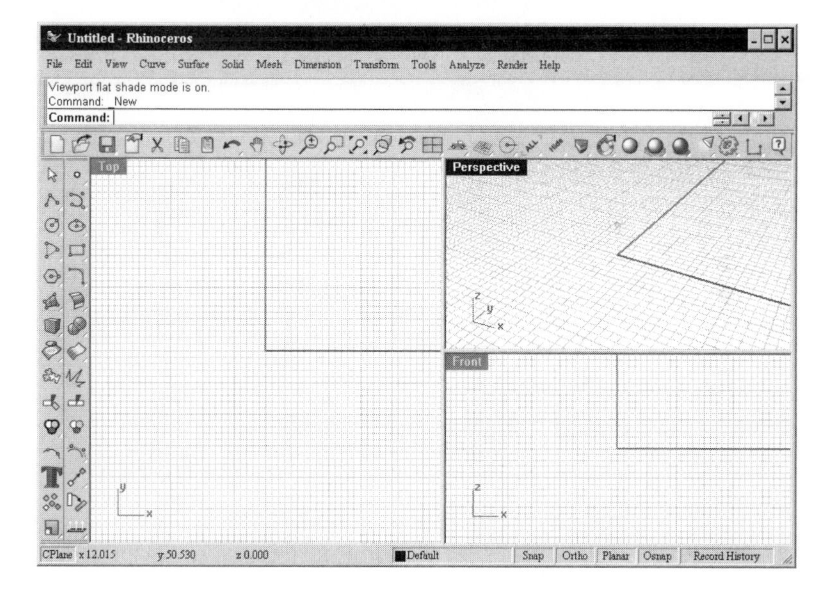

Figure 1–28. Three-viewport configuration

Continue with the following steps to manipulate viewports:

5 Click on the Perspective viewport to set it as the current viewport.

6 Select View > Viewport Layout > Split Vertical. The selected viewport is split vertically into two viewports. (See Figure 1–29.)

7 Select View > Viewport Layout > New Viewport. A new viewport is constructed. (See Figure 1–30.)

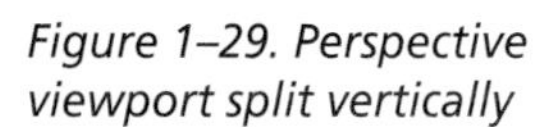

Figure 1–29. Perspective viewport split vertically

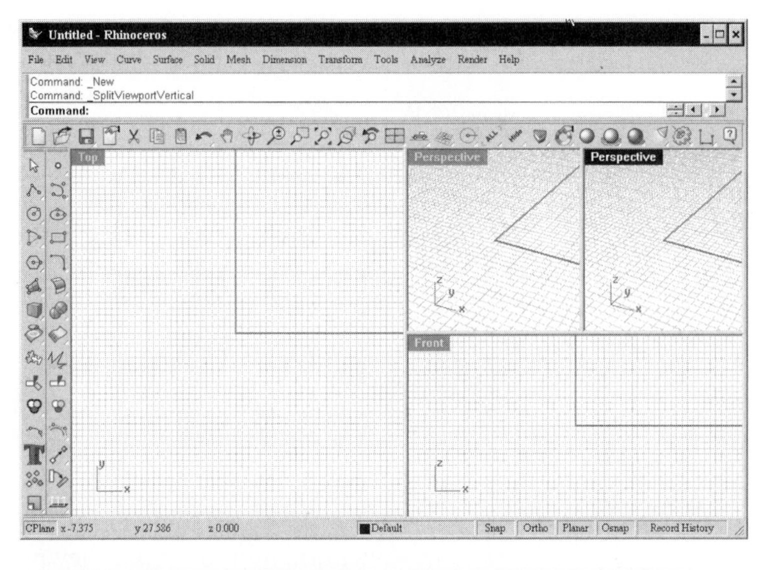

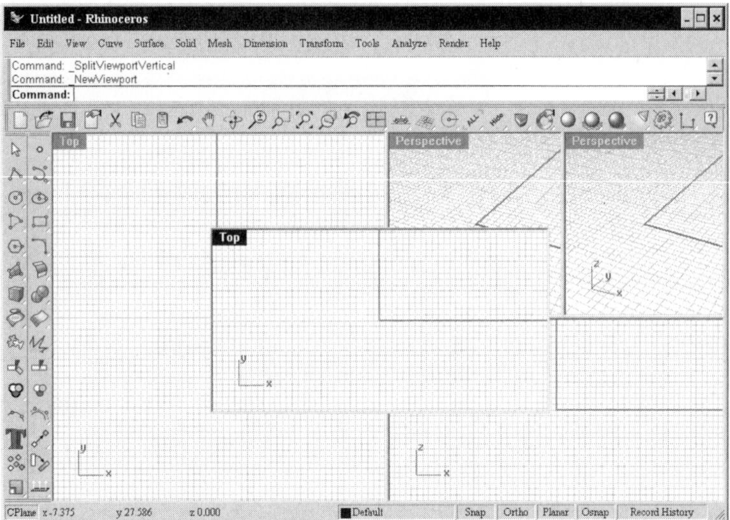

Figure 1–30. New viewport constructed

8 Select View > Viewport Layout > 4 Viewports. The graphics area is set to 4-viewport configuration.

Viewport's Gradient Background

A viewport's background can be set to gradient color, as follows:

9 Type -GradientView at the command area.

10 If the View option is All, select it from the command area to change it to Active.

11 Select the State option to change it to On, the active viewport has a gradient color. (See Figure 1–31.)

12 Type Gradientview at the command area. The viewport's background is reset.

This command has two versions: Adding the "-" hyphen sign adds command line options.

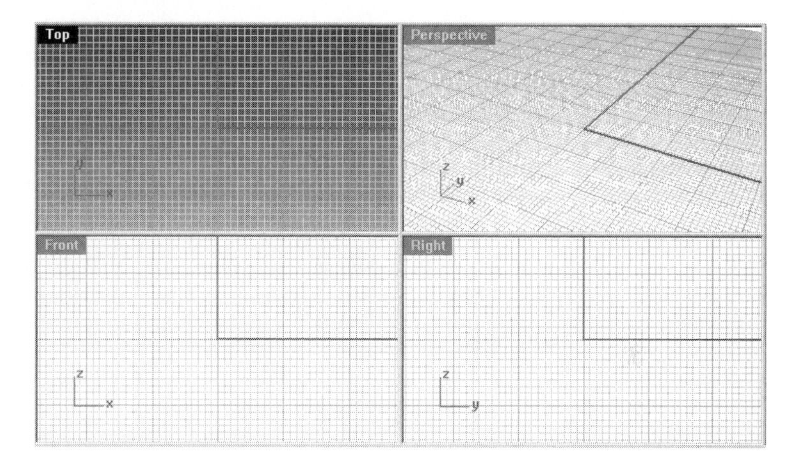

Figure 1–31. Viewport's background changed

Viewport Properties

Viewport properties concern how the viewport's display. You may set its properties as follows:

13 Select View > Viewport Properties or right-click the viewport's label and select Viewport Properties to bring up the Viewport Properties dialog box. See Figure 1–32.

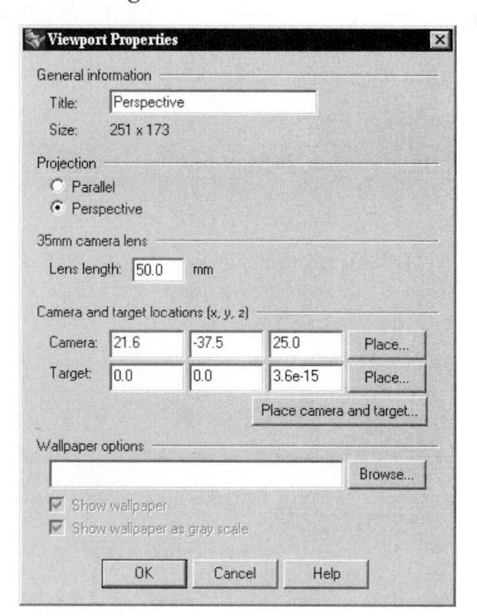

Figure 1–32. Viewport Properties dialog box

As shown in Figure 1–32, the Viewport Properties dialog box has five areas, as follows:

- The General Information area tells the size of the viewport in terms of pixel size and enables you to set the viewport's name.
- The Projection area sets the viewport's display method: parallel projection or perspective projection.
- If the project is set to perspective, you can set the camera's lens length in the third area.
- By setting the camera and target locations in the fourth area, the direction of viewing is determined.
- To enhance the display, the fifth area allows you to include and display a wallpaper in the viewport.

14 Try out various buttons to experience how to set the viewport's properties. When you are done, click on the OK button to close the dialog box.

If you simply want to learn about the properties of a viewport, continue with the following steps:

15 Do not select any object. You may press the ESC key once or twice to stop any command running and clearing any selection of objects.

16 Select Edit > Object Properties, or click on the Object Properties button on the Standard toolbar. The Properties dialog box is display, showing the properties of the active viewport. (See Figure 1–33.)

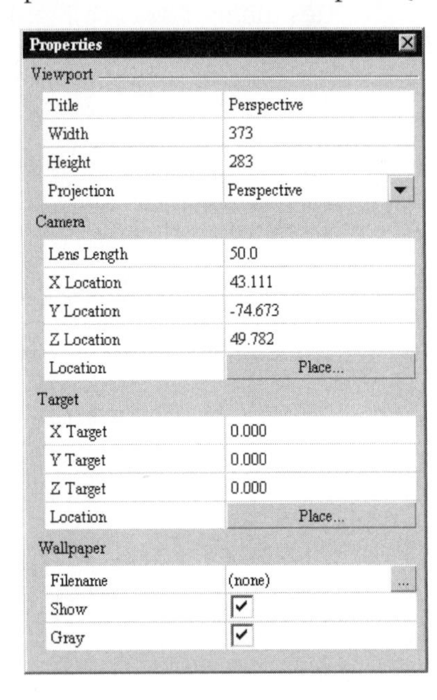

Figure 1–33. Viewport properties

Floating and Docked Viewports

The viewports that you constructed are called docked viewports because they are restricted within the graphics window. Apart from these viewports, you can construct floating viewports that can float outside the main Rhino window. The advantage of having floating viewport is that, if you are using two display units together with extended display mode, you can display Rhino's application in one display unit and a floating viewport in another display unit.

Floating viewports and docked viewports are interchangeable; you can dock a floating viewport or make a docked viewport floating. Continue with the following steps.

17 Select New Floating Viewport button on the Viewport Layout toolbar or type NewFloatingViewport at the command area.

18 Press the ENTER key to accept the default Perspective viewport. A new floating viewport showing the perspective view is constructed. (See Figure 1–34.)

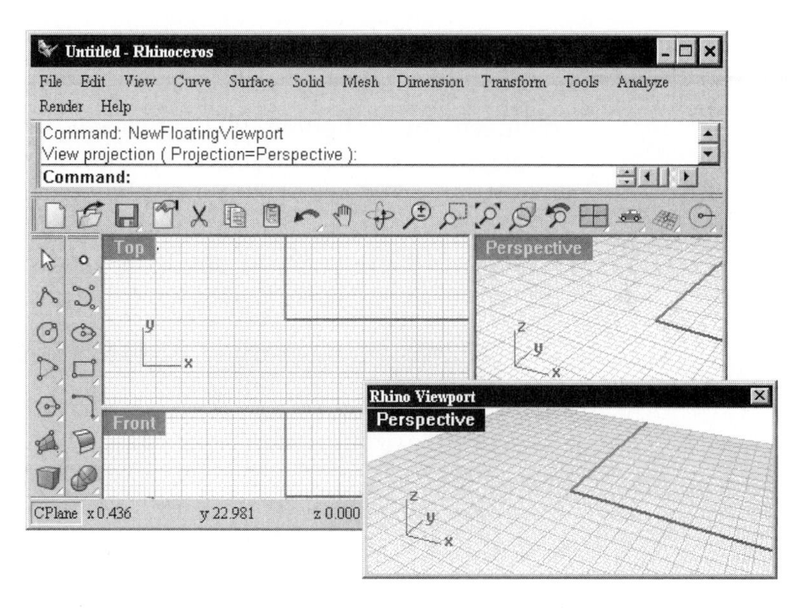

Figure 1–34. New floating viewport constructed

19 Select Toggle floating viewport state on the Viewport Layout toolbar, or type ToggleFloatingViewport at the command area. The new floating viewport, being the active viewport, is docked.

20 Click on the Top viewport to make it current.

21 Select Toggle floating viewport state on the Viewport Layout toolbar, or type ToggleFloatingViewport at the command area. The Top viewport becomes a floating viewport. (See Figure 1–35.)

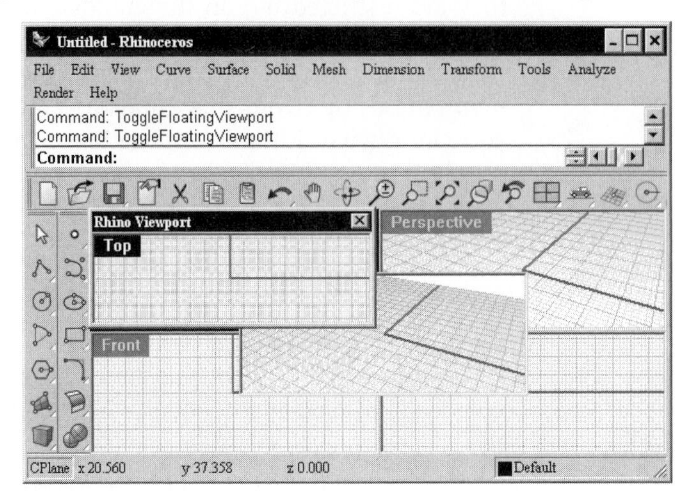

Figure 1–35. New floating viewport docked and Top viewport made floating

Viewport Tab

You may display a set of viewport tab alongside the viewports, at the right, left, top, or bottom. By right-clicking on one of the tabs, a menu is displayed, as shown in Figure 1–36. Viewport tab can be manipulated by selecting the Viewport tab Controls button on the Viewport Layout toolbar or typing ViewportTab at the command area. To simply toggle turning on or off the viewport tab, right-click on the Toggle viewport tabs button on the Viewport Layout toolbar.

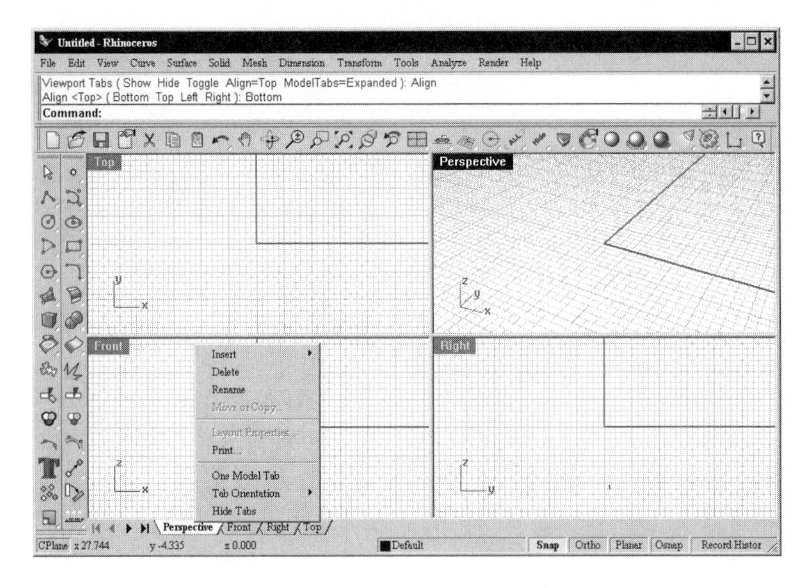

Figure 1–36. Viewport tab and its right-click menu displayed

Viewport Synchronization

A way to help visualize objects' size in various viewports, you synchronize them such that their display scales are the same for all the viewports, as follows:

1. Select File > Open and select the file Synchron.3dm from the Chapter 1 folder on the companion CD-ROM.
2. Right-click on the Zoom Extent All Views button on the Standard toolbar. All the viewports are zoomed to their extent. However, their zoom scales are different. (See Figure 1–37.)

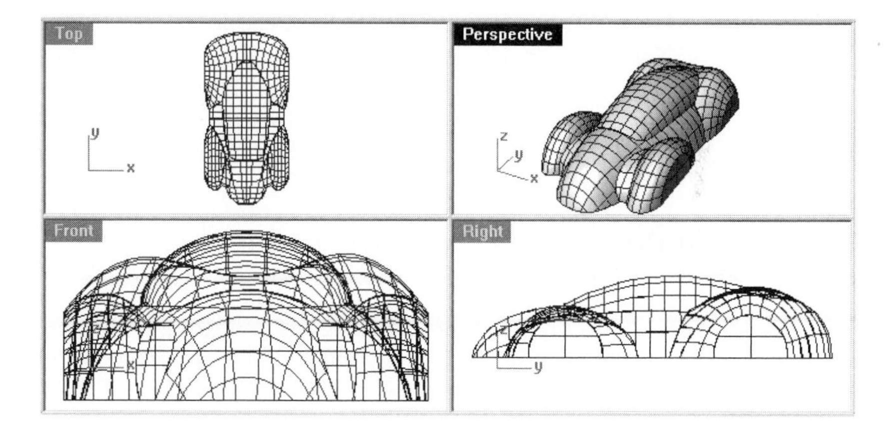

Figure 1–37. Viewports zoomed to extend

3. Click on the Perspective viewport to make it the current viewport.
4. Select View > Set Camera > Synchronize, or click on the Synchronize View button on the Viewport Layout toolbar. The viewports are synchronized. All the viewports' zoom scales are synchronized to the zoom scale of the perspective viewport, as shown in Figure 1–38.

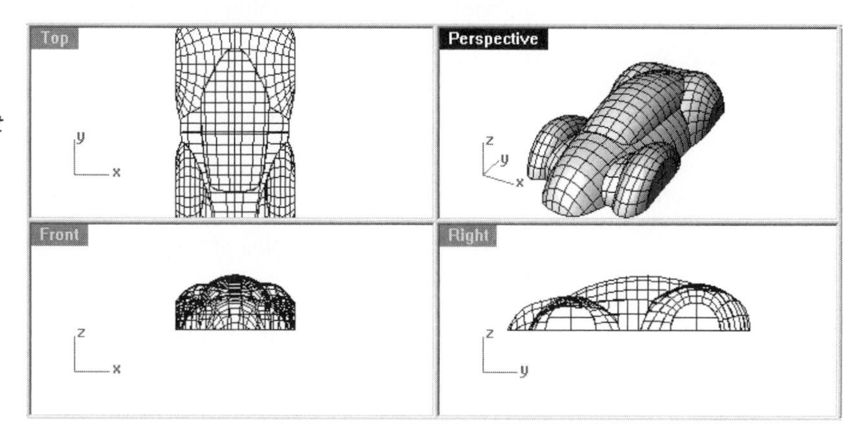

Figure 1–38. Viewports synchronized to the front viewport

Status Bar

At the bottom of the Application window is the status bar (Figure 1–39).

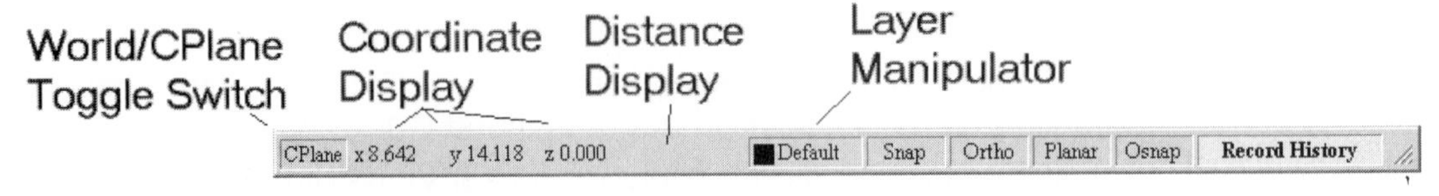

Figure 1–39. Status bar

The status bar contains eight panes, showing the location of the cursor marker, the layer manipulator, and five buttons controlling the state of the drawing aids (Snap, Ortho, Planar, and Osnap) and history recording. By right-clicking these buttons, you can bring out a context menu. The functions of which are outlined in Table 1–2. (Details of them will be explained later in this chapter.)

Table 1–2: Status Bar Panes and Their Functions

Pane	Function
World/CPlane toggle switch	Toggles the coordinate display to show either World coordinates or construction plane coordinates
Coordinate Display	Shows the coordinates of the mouse pointer
Distance	Optionally displays the distance from the mouse pointer to the last picked point
Layer Manipulator	Shows the current layer and, when clicked on, displays the layer manager
Snap	Toggles on/off constraining the cursor movement to the specified snap intervals
Ortho	Toggles on/off constraining the cursor movement to be orthogonal or some other preset direction
Planar	Toggles on/off constraining the cursor movement to be parallel to the current construction plane from the last point
Osnap	Toggles on/off the display of the Osnap dialog box (Osnap stands for "object snap")
Record History	Toggles on/off History recording

Toolbars

Buttons on toolbars represent commands in a graphical way. To run a command, you select a button on a toolbar.

Docking and Floating Toolbars

Like most Windows applications, Rhino toolbars can be docked at the top, bottom, or sides of the application window; and can also be made floating anywhere on the screen. You can drag a floating toolbar to the top, bottom, or sides of the window to dock it; and you can drag a docked toolbar to change it to floating.

Locking Toolbars

By checking Tools > Lock the Toolbars, all docked and floating toolbars will remain locked or floating. When this feature is turned on, dragging a floating toolbar over the docking area will not dock it.

Displaying Toolbars

Because there are many commands and toolbars, displaying all of them would take up the entire screen display. Therefore, only the Standard, Main1, and Main2 toolbars are displayed by default. To find out which toolbars are available, and to display them on the screen, you use the Toolbar command by typing the command name at the command area or by selecting Tools > Toolbar Layout > Edit from the main pull-down menu. In the Toolbars dialog box, shown in Figure 1–40, check the box next to a toolbar to display that toolbar.

1. Start a new file, using the Small Objects, Millimeters template file.
2. Select Tools > Toolbar Layout.
3. In the Toolbars dialog box, click on the Line box. The Line toolbar is displayed.
4. Click on the X mark at the upper right corner of the Toolbars dialog box to close it.
5. Click on the X mark at the upper right corner of the Line toolbar to close it also.

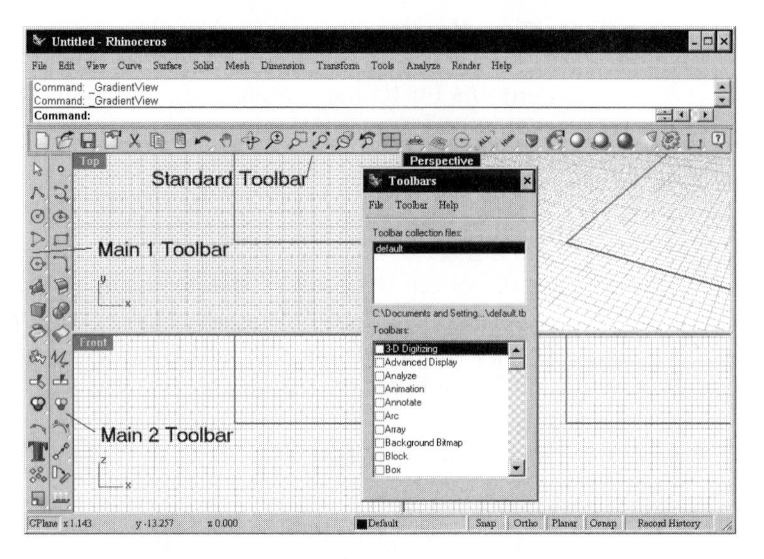

Figure 1–40. Standard, Main1 and Main2 toolbars and the Toolbars dialog box

Another way of displaying toolbars not shown on the screen is to right-click on the blank space of a docked toolbar to display the complete list of toolbars, as shown in Figure 1–41, and click on the check boxes. To lock the docked toolbars, click on the Lock toolbars check box at the bottom of the list.

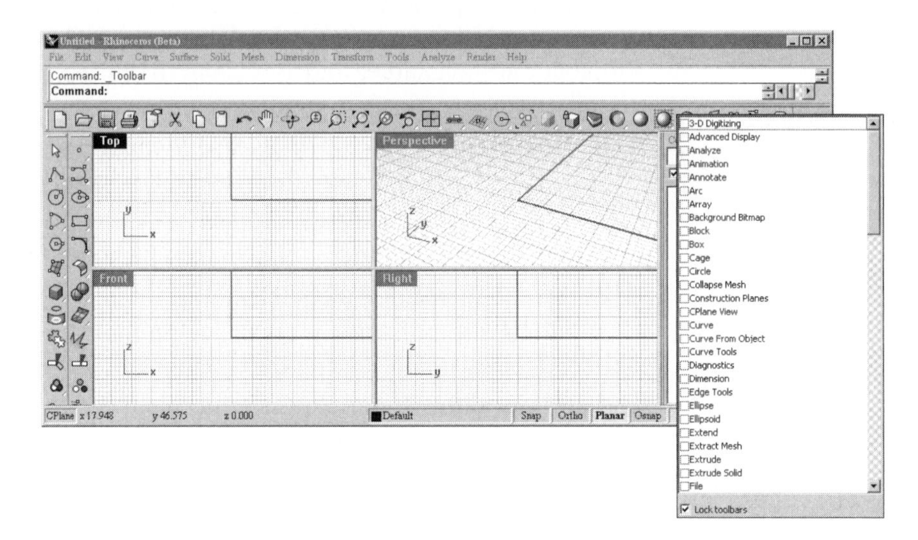

Figure 1–41. Right-clicking blank space of a docked toolbat

Toolbar Fly-Outs

Some buttons on the toolbar have a small triangular mark at their lower right corner. Clicking on this mark will bring out a fly-out toolbar. Selecting and dragging the header of the fly-out will cause it to stay on the screen.

6 With reference to Figure 1–42, click on the small triangular mark at the lower right corner of the Edit Layer button of the Standard Toolbar. The Layer toolbar is brought out.

7 Select the Layer toolbar's header and drag it away from the button. The Layer toolbar is displayed.

8 Click on the X mark at the upper right corner of the Layer toolbar. The toolbar is closed.

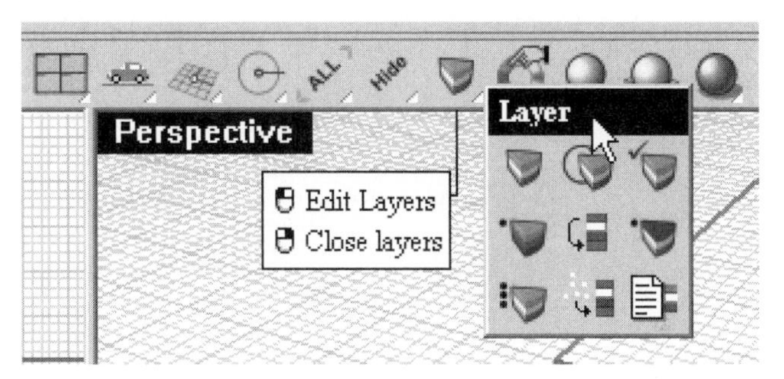

Figure 1–42. Bringing out a fly-out toolbar

Making Your Own Toolbar

Apart from the toolbars provided, you may make your own toolbar as follows:

9 Select Tools > Toolbar Layout.

10 In the Toolbars dialog box, select Toolbar > New. The Toolbar Properties dialog box is displayed. (See Figure 1–43.)

11 Specify a toolbar name or accept the default name.

12 Click on the OK button. A new toolbar is constructed.

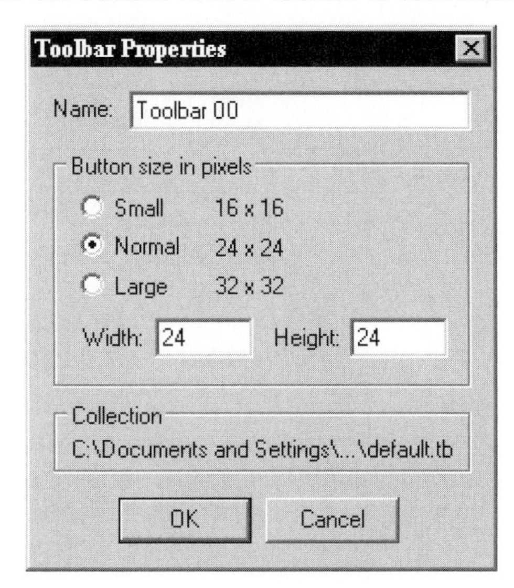

Figure 1–43. Toolbar Properties dialog box

On returning to the Toolbars dialog box, there should be a new entry in the toolbar list and the new entry should be selected.

13 Select the new toolbar from the toolbar list of the Toolbar dialog box, if it is not already selected.

14 In the Toolbars dialog box, select Toolbar > Add Button. A new button is added to the new toolbar.

Together with the default button given, there should be two buttons on this new toolbar. Continue with the following steps to configure the buttons.

15 Move the cursor over one of the buttons, hold down the SHIFT key, and right-click. The Edit Toolbar dialog box is displayed.

16 With reference to Figure 1–44, fill in the following entries in the dialog box.

Left Tooltips: Command Help
Right Tooltips: What is it?
Button text: Context/What
Left mouse button command: CommandHelp
Right mouse button command: What

17 Click on the Show bitmap and text and the Float to top buttons.

18 In the Linked toolbar pull down list, select Utilities.

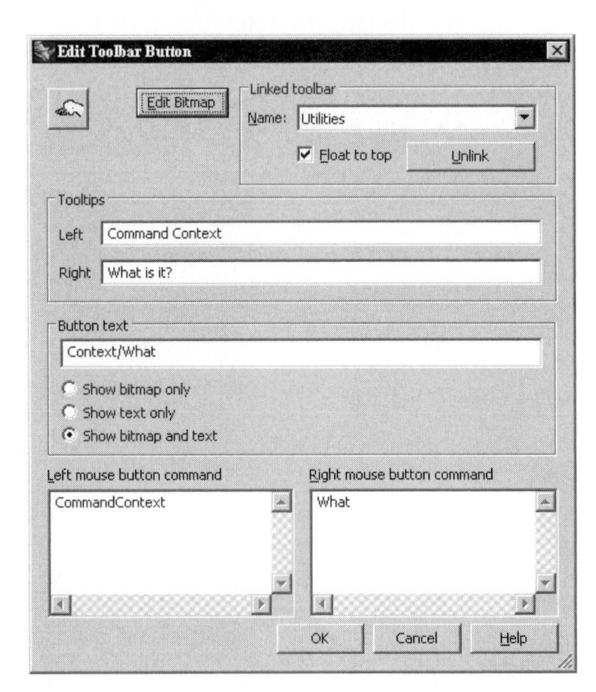

Figure 1–44. Edit Toolbar Button dialog box

19 Click on the Edit Bitmap button.

20 In the Edit Bitmap dialog box shown in Figure 1–45, construct a bitmap and click on the OK button.

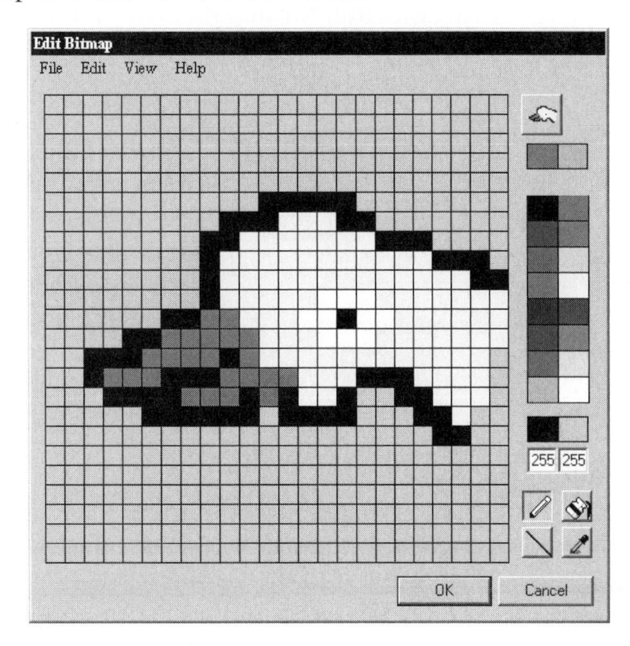

Figure 1–45. Edit Bitmap dialog box

21 On returning to the Edit Toolbar dialog box, click on the OK button.

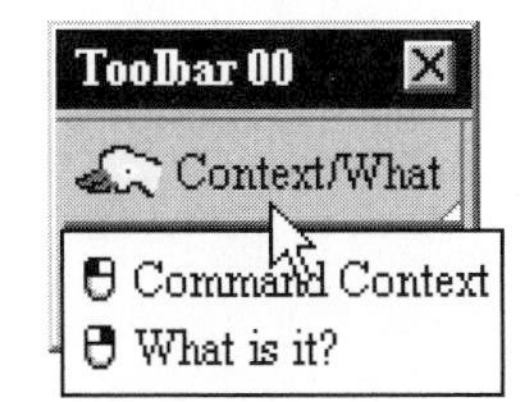

Figure 1–46. New toolbar

A toolbar with two buttons are constructed. One of the buttons is blank and the other button is referenced to two commands, CommandHelp and What. You may run these commands by left-clicking and right-clicking respectively. At the lower right corner of the button, there is a small triangular mark, clicking on it will bring out the Utilities toolbar. (See Figure 1–46.)

22 Do not save your file.

Command Interaction

To summarize, there are three ways to run a Rhino command: Make selections from the main pull-down menu, type a command at the command area, and select a button on the toolbar

Mouse Left-click and Right-click

Normally, your pointing device (mouse) has two buttons (left and right). You use the left button to select an item from the main pull-down menu, a button on a toolbar, or a check box in a dialog box, as well as to specify a location on the graphics area. Depending on where you place your cursor, right-clicking has different effects. Some toolbars have two commands sharing a single button. For example, the commands Zoom Extents and Zoom Extents All Views, shown in Figure 1–47, share a single button on the Standard toolbar. Left-clicking activates the Zoom Extents command, and right-clicking starts the Zoom Extents All Viewports command.

Figure 1–47. Zoom Extents and Zoom Extents All Viewports commands

Several commands are accessed by placing the cursor over the graphics area and right-clicking in one of three ways. Simply right-clicking repeats the last command. Right-clicking and holding down the mouse button for a while (or holding down the Shift key) accesses the Pan command. Right-clicking and holding down the Ctrl key accesses the Zoom command.

Middle Mouse Button

If your mouse has the third (middle) button, pressing it will access a menu or toolbar based on the settings in the Mouse tab (shown in Figure 1–48) of the Options dialog box (accessible by selecting Tools > Options). If the third button can be slided forward and backward, doing so will zoom in and out on the display.

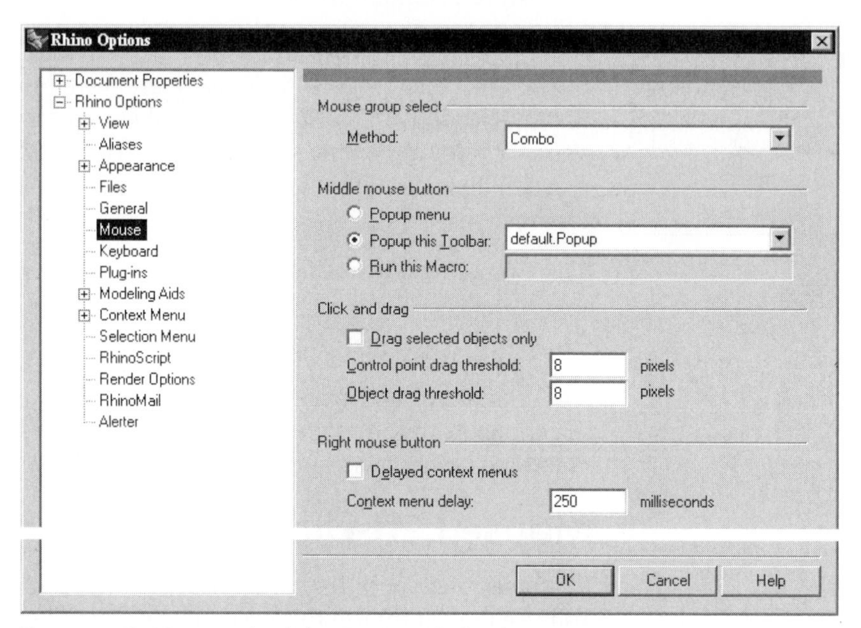

Figure 1–48. Mouse tab of the Options dialog box, pop-up menu, and pop-up toolbar

Display Control

The 3D space within each viewport is unlimited but the physical size of the viewport is limited. Therefore, you may need to zoom in, zoom out, pan, and rotate the viewport, as follows:

1. Select File > Open and select the file Display.3dm from the Chapter 1 folder on the companion CD-ROM.
2. Click on the Zoom Dynamic button on the Standard toolbar.
3. Click on anywhere on the graphics area, hold down the left button, and drag up and down. The display is zoomed in and out.
4. Click on the Pan button on the Standard toolbar.
5. Click on anywhere on the graphics area, hold down the left button, and drag left and right. The display is paned to the left and to the right.
6. Click on the Rotate View button on the Standard toolbar.
7. Click on anywhere on the graphics area, hold down the left button, and drag around. The display is rotated.
8. Right-click on the Perspective viewport and select Set View > Perspective. The view is set to perspective.

Apart from using the pull-down menu or the appropriate toolbar button, you may perform the followings:

- Hold down the Control and Shift keys, right-click, and drag to rotate the display
- Hold down the Shift key, right-click, and drag to pan the display
- Hold down the Control key, right-click, and drag, or spin the mouse wheel to zoom in and out

Figure 1–49 shows the various kinds of icons while performing various display manipulation tasks.

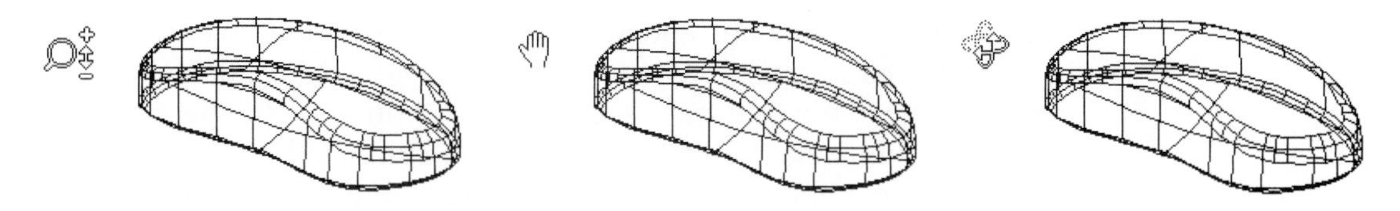

Figure 1–49. From left to right: Zooming in and out, panning around, and rotating

Panning in the Perspective Viewport

Panning in the perspective is carried out by moving the location of the imaginary camera. Continue with the following steps:

9 Double-click on the Perspective viewport's label to maximize the viewport.

10 Select View > Set Camera > Rotate Camera.

11 Click on the viewport and drag the mouse. The perspective viewport is rotated. This is equivalent to panning in a parallel projection viewports.

Zooming in the Perspective Viewport

Zooming in the perspective viewport is done by adjusting the lens length of the camera, as follows:

12 Select View > Set Camera > Adjust Lens Length.

13 Click on the viewport and drag the mouse. As the lens length changes, the viewport appears to be zoomed in and out.

Dollying in the Perspective Viewport

Dollying is to change the lens length of the camera and at the same time move the camera nearer or farther away from the object, keeping the displayed size of the object in the viewport more or less constant. Continue with the following steps.

14 Select View > Set Camera > Adjust Lens Length and Dolly.

15 Click on the viewport and drag the mouse. Because both the lens change in length and the location of the camera is moving nearer or farther away from the objects, the perspective effect is changed but the object remains more or less the same size in the viewport.

Transparent Zoom and Pan

To zooming in and out and panning during drafting and design, the zoom and pan commands can work transparently inside other commands. The term transparent means that you can, in the middle of another command, run the zoom and pan command and, upon finishing the zoom and pan command, continue with the original command.

Undo View Change

The Undo command does not undo view display changes. To undo or redo viewport changes, select View > Undo View Change or select View > Redo View Change.

Full Screen Display

Contrary to displaying all the major areas, you can set the display to full screen, showing only the graphics area and occupying the entire screen display. This can be useful in a presentation.

- ❒ To set full screen display, type FullScreen at the command area. (A hyphenated version of the command adds command line options.)
- ❒ To return to normal display mode, press the ESC key.

Display Mode

To help visualize surface and polygon mesh objects in the viewport, you may set the display to various display modes: wireframe, shaded, rendered, ghosted, and x-ray. Perform the following steps:

1. Select File > Open and select the file DisplayMode.3dm from the Chapter 1 folder on the companion CD-ROM.
2. Type SetDisplayMode at the command area, select the Mode option, and then select one of the five display modes, or right-click on the active viewport's label and select Wireframe, Shaded, Rendered, Ghosted, or X-ray. The display is set.

Figure 1–50 shows five different settings. Because the image here is not large enough, you may not find the subtle difference between Ghosted and X-ray display.

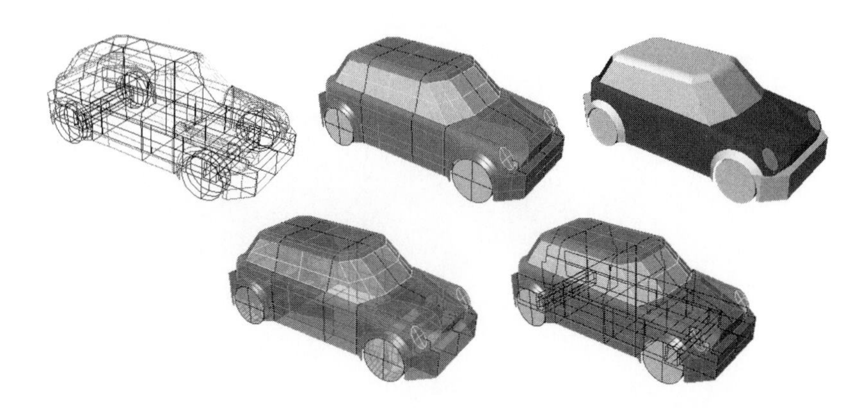

Figure 1–50. From left to right and from top to bottom: wireframe, shaded, rendered, ghosted, and x-ray display modes

Viewport Zoom Aspect Ratio

To help visualize the effect of scaling an object un-uniformly, you may set the viewport's horizontal and vertical zooming scale differently, as follows:

3 Maximize the Front viewport and set the display to shaded.

4 Select Tools > Options.

5 In the Rhino Options dialog box, select Rhino Options > Appearance

6 Suppose we would like to change the aspect ratio for the shaded display, click on the + sign prefixing the Shaded folder under the Advanced Settings folder and then click on Other Settings.

7 Set the horizontal scale to 1.5, as shown in Figure 1–51 and click on the OK button. The horizontal display scale of the Shaded viewport is changed. (See Figure 1–52.)

8 Open the Advanced setting tab of the Options dialog box and reset the horizontal scale to 1.

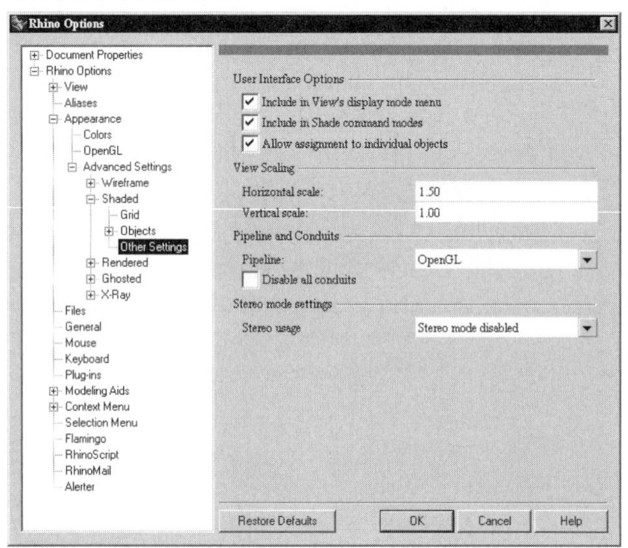

Figure 1–51. Advanced settings

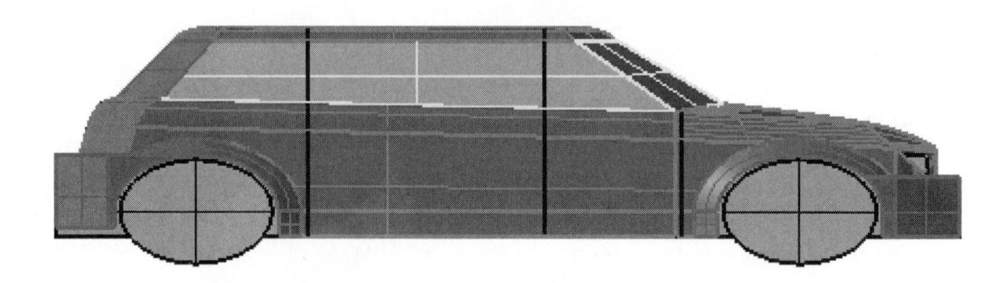

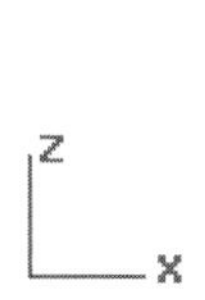

Figure 1–52. Viewports' horizontal and vertical aspect ratio changed

Turn Table Display

You may rotate the viewport continuously as if it is a turn table, as follows:

9. Maximize the Perspective viewport and set its display to shaded.
10. Select Set Camera > Turntable, or click on Turn table/Turn table; One cycle button on the View toolbar.
11. The perspective viewport, being the active viewport, will rotate continuously.
12. You may set the rotation speed and the number of revolution in the Turntable dialog box. (See Figure 1–53.)
13. To stop rotating, press the ESC key.
14. If you want to rotate only one revolution, set the number of revolution to 1 in the Turntable dialog box after the turn table is activated or right-click on the Turn table/Turn table: one cycle button.
15. Do not save your file.

Figure 1–53. Turn table dialog box

Utilities and Help System

The help system contains several categories: Command Help dialog box, Help Topics, Frequently Asked Questions, Learn Rhino, Help on the Web, Command List, Feature Overview, and Technical Support. The sections that follow describe Help Topics and Frequently Asked Questions.

Web Browser

You can open a stay on top web browser window that you can read while working in Rhino by selecting Tools > Web Browser, or clicking on the Web Browser button on the HTML toolbar.

Hyperlink

To incorporate additional information to selected objects, you may attach a hyperlink. Naturally, unwanted hyperlink can be removed. If a hyperlink is already attached to an object, you may open its hyperlink, view its hyperlink, and select it by specifying the hyperlink.

- You may perform these operations by selecting an item from the Tools > Hyperlink cascading menu.

Calculators

There are two calculators available in Rhino, the ordinary calculator and the reverse polish notation calculator. They can be accessed by selecting Tools > Calculator and selecting Tools > RPN Calculator. Both calculators provide a dialog box for inputting and they output the results to the command line. (See Figure 1–54.)

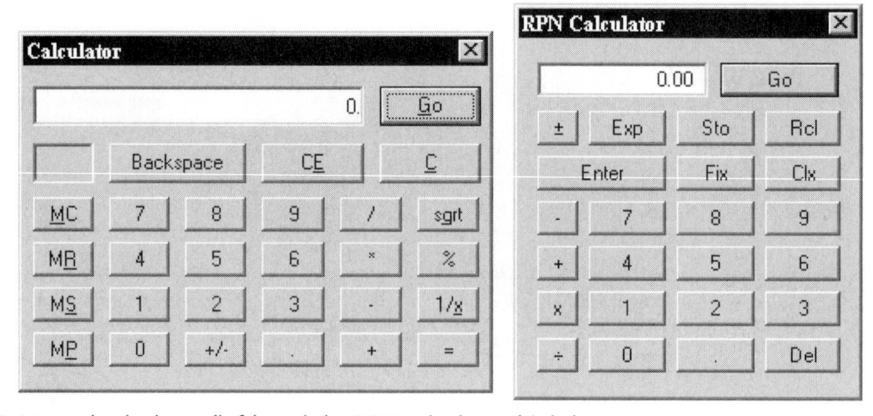

Figure 1–54. Normal calculator (left) and the RPN calculator (right)

Command Help

You may type a command name at the uppermost column of the Command Help dialog box. Similar to the command area, auto-complete feature is also available here and, therefore, you do not have to type the full name of the command. After typing the command name, appropriate help message will be provided. If the Auto-update box in the dialog box is checked, the help for the current command will display.

1. If the Command Help dialog box is not displayed, click on the Command Context Help button on the Help toolbar or type CommandHelp at the command area.

2 Type Line at the command area and press the ENTER key. Help topic on Line command is displayed in the Command Help dialog box. (See Figure 1–55.)

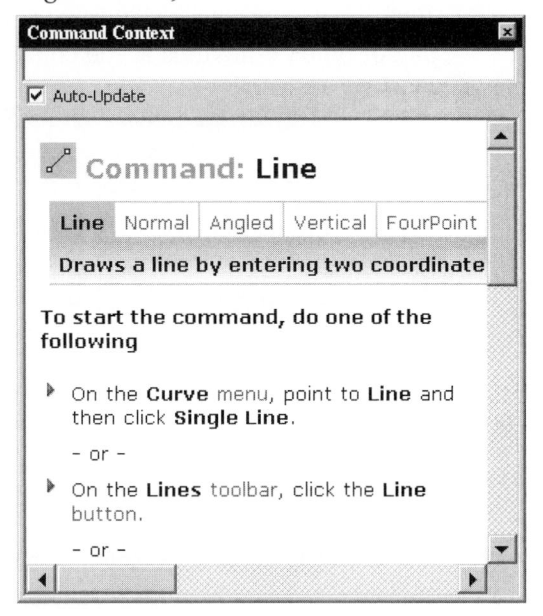

Figure 1–55. Command Help dialog box showing help for the Line command

Help Topics

There are several ways you can display the Help dialog box, by pressing the F1 key, selecting Help Topics from either the standard toolbar, the Help toolbar, or Help pull-down menu, you gain access to the comprehensive Help dialog box, shown in Figure 1–56. Here you will find all the information you need.

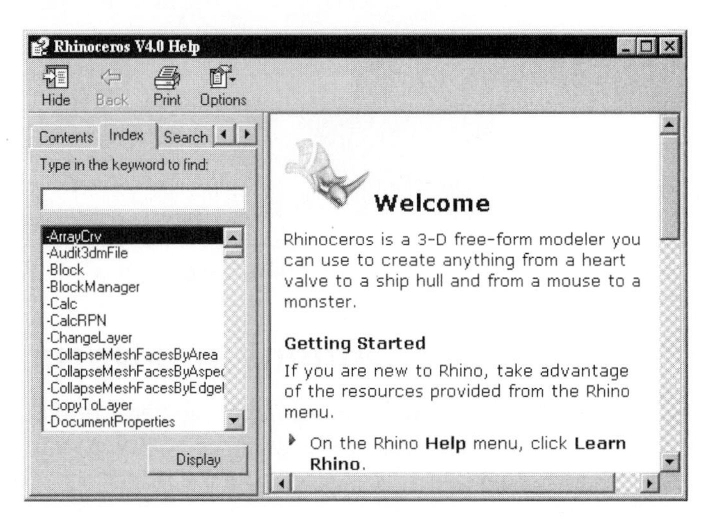

Figure 1–56. Help dialog box

Display All Commands

If you wish to view a complete list of Rhino commands, you may right-click on the Help Topics/Display all Commands button on the Help toolbar, or type Commands at the command area. See Figure 1–57.

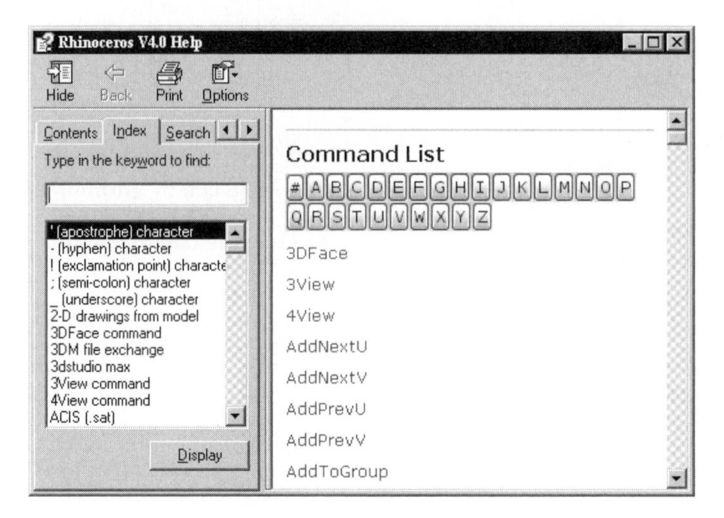

Figure 1–57. Help dialog box displaying the command list

Frequently Asked Questions

Selecting Frequently Asked Questions from the Help pull-down menu brings you to the web site *www.rhino3d.com/support/faq/*. This site includes the Frequently Asked Questions section, shown in Figure 1–58.

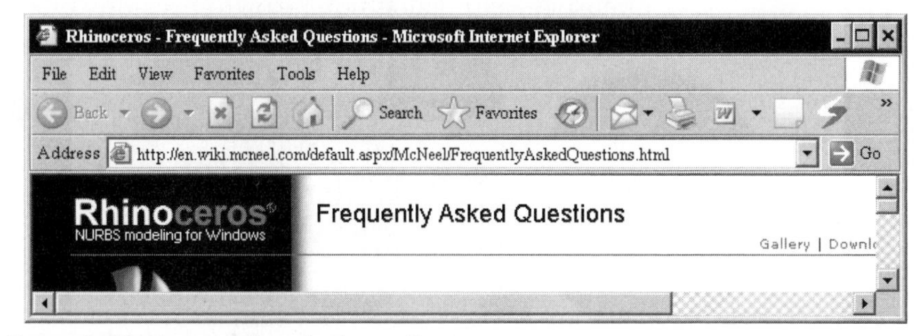

Figure 1–58. Frequently asked questions

Rhino Options and Document Properties

By now you should have a general understanding of Rhino's user interface and various drawing aids. Before you proceed to the other chapters to learn how to construct various types of objects, spend some time learning the meaning of various settings in the Rhino Options and Document Properties dialog boxes, which share the same dialog box. To access this dialog box, select File > Properties (or Tools) > Options. Some of these options are discussed in the sections that follow.

Shortcut Keys

Using the Keyboard tab, you can assign shortcut keys to the commands you use most frequently.

Files Location

To set the template file and auto-save file location, you use the Files tab.

Appearance

In the Appearance tab, you set the color and appearance of the user interface.

Units

Before you start constructing a model, you should check the units of measurement so that the model you construct is compatible with any upstream and downstream operations. In the Units tab, you set the units of measurement and the tolerance of the model. In addition, you set the display tolerance.

Export and Import Options

Options that you saved in a Rhino file can be exported to an option file and then imported to another Rhino file, thus saving a lot of time to set options repeatedly.

- To export options set in a file to an option file, select Tools > Export Options, or type ExportOptions at the command area. In the SaveAs dialog box, specify a file name, and click on the OK button.
- To import options set in another file, select Tools > Import Options, or type ImportOptions at the command area. In the Import Options dialog box, select an option file, click on the option items that you want to import, and click on the OK button.

File Saving Methods

This section discusses way to add note, save methods, render mesh extraction, view capture, and file sending via email.

Add Notes to File

You can include textual information in a Notes window that is saved in the file, as follows:

1. Select File > Open and select the file Notes.3dm from the Chapter 1 folder on the companion CD-ROM.
2. Select File > Notes, or click on the Notes button on the Files toolbar.
3. With reference to Figure 1–59, add textual information in the Notes window.
4. Click on the X mark of the Notes window to close it (You may leave the Window there and continue with your work.)
5. Do not save the file.

Figure 1–59. Notes window

Save Methods

A Rhino file can be saved via five operations: Save, Save Small, Incremental Save, Save As, and Save as Template. Save file format will be discussed in the next section.

Save

- By selecting File > Save, you save your file together with a preview and associated rendering meshes.

A NURBS surface is an accurate representation of a smooth surface. However, producing a rendered image of such surfaces typically requires the existence of a set of polygon meshes representing the surfaces. When you invoke the Render command to construct a rendered image, the computer automatically constructs a set of polygon meshes and uses the meshes for rendering purposes. (You will learn rendering in the Chapter 14.) Normally, these polygon meshes are preserved when you save your file. The next time you render the object (if it is not modified), the computer will use the preserved meshes (saved in the file) for rendering. As a result, rendering time is reduced because the program does not have to construct the polygon meshes.

Save Small

❒ By selecting File > Save Small, you save your file without the preview and rendering meshes.

The inclusion of polygon mesh data, however, makes file sizes larger. To minimize the storage requirement for polygon mesh data, you can use the Save Small command, which saves NURBS surfaces but not the polygon meshes required for construction of a rendered image. However, as a result, the next time you open the file rendering time will be longer because the computer will have to reconstruct the set of polygon meshes. Apart from that, save small also deletes the preview image. If you want to keep the preview image but do not want to save the rendered meshes, type ClearAllMeshes at the command area to clear the meshes and then save normally.

If render meshes are needed for any purpose, you may extract them from selected objects.

❒ To extract render meshes, type ExtractRenderMesh at the command area or right-click on the RenderMesh Settings/Extract Render Mesh button on the Render toolbar and select objects from which you want to extract render meshes.

Incremental Save

❒ By selecting File > Incremental Save, you save your model in a set of versions so that you can experiment with changes and retrieve previous versions.

If the file has not been saved before, the IncrementalSave command will prompt for a file name. Running this command again will save the file incrementally with a file name suffixed by three numbers, starting from 001.

Save As

❒ By selecting File > Save As, you save your model with a different file name.

The first time you save a file, using the Save or SaveAs command will require you to specify a file name; and the file will be saved accordingly. If the file is already saved previously, using the Save command will simply save the file without any prompt. However, using the SaveAs command will prompt for a file name. After specifying a file name, Rhino will save the data from the current file to a new file, close the current file without saving, and let you work on the new file.

Save As Template

- By selecting File > Save as Template, the model is saved as a template.

If the objects constructed in a file are to be used repeatedly as a starting point for many other files, you may consider saving the file as a template in the template folder.

Send File via Email

You can send email with the current Rhino file as attachment by selecting File > Send. Naturally, you need to configure your email profile.

Summary

Rhinoceros is a 3D digital modeling tool. You use it to construct points, curves, NURBS surfaces, polysurfaces, solids, and polygon meshes. In addition, you output photorealistic renderings, 2D drawings, and file formats of various types for downstream computerized operations. On the other hand, you can reuse upstream digital models by opening various file formats.

The Rhino user interface contains five major areas: standard Windows title bar, main pull-down menu, command area, graphics area, and status bar. In addition, there are a number of toolbars and a command help dialog box. You perform commands via the main pull-down menu, command area, or toolbar. The graphics area is where you construct digital models. To help visualizing 3D objects in different directions, there are several viewports. You can also zoom, pan, and rotate the display. In addition, there are a number of utility tools, such as web browser, hyperlink, and calculator.

Review Questions

1. What kinds of curve and surface continuity can be achieved in Rhino?
2. What types of objects can you construct by using Rhino?
3. Explain the methods by which an object is shaded in the viewport.

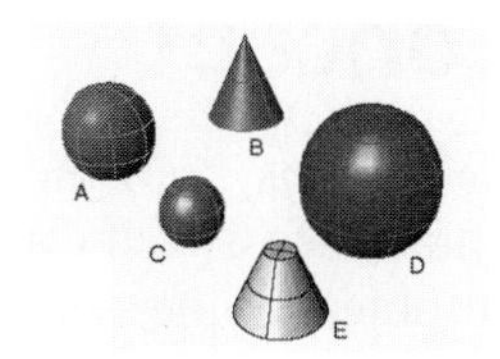

CHAPTER 2

Rhinoceros Basic Operating Methods

Introduction

This chapter serves to familiarize you with Rhino's basic operation methods.

Objectives

After studying this chapter, you should be able to:

- ❒ Demonstrate an understanding of the concepts of construction plane
- ❒ Input precise coordinates and use various drawing aids in geometric object construction
- ❒ Manipulate geometric objects that are already constructed
- ❒ Explain the concepts of history management
- ❒ Manage layers
- ❒ Set the display of objects

Overview

Working in a computer-aided design system involves two categories of tasks: you construct objects and manipulate them. To construct objects accurately, you need to know what is meant by a construction plane, how precise coordinates can be input, and how various tools can be used in geometric object construction. To be able to manipulate objects that are already constructed, you need to have an understanding of various object translation tools, Rhino's history management concept, layer management methods, and various display settings.

Appreciating the Construction Plane Concept

To construct geometric objects, apart from using the options from the pull-down menu, toolbar, or command area, you must also specify a location or a number of locations for the geometric object to be constructed. To specify a location using the pointing device, you pick a point in one of the viewports. Naturally, you may key in a set of coordinates at the command area to precisely specify a point. Coordinate systems will be explained in the next section.

When you pick a point in one of the viewports, you have selected a point on an imaginary construction plane corresponding to the selected viewport. If you pick a point in the Top viewport, you have selected a point on a construction plane parallel to the Top viewport and passing through the origin. (The Perspective viewport has the same construction plane.) Similarly, if you pick a point in the Front viewport, you have selected a point on a construction plane parallel to the Front viewport and passing through the origin. You can construct objects on more than one construction plane. For example, to construct two circles on two different construction planes, you would perform the following steps:

1. Start a new file. Use the Small Objects, Millimeters template.
2. From the main pull-down menu, select Curve > Circle > Center, Radius. Alternatively, select the Circle: Center, Radius option from the Circle toolbar, or type Circle at the command area.
3. Move the cursor over the Top viewport. Indicating that the viewport is active, the label of the viewport (top left-hand corner) should be highlighted and the labels in the other viewports should be grayed out, as shown in Figure 2–1.
4. Pick a point to specify the center location.
5. Pick to specify a point on the circumference. A circle is constructed on a construction plane parallel to the Top viewport.

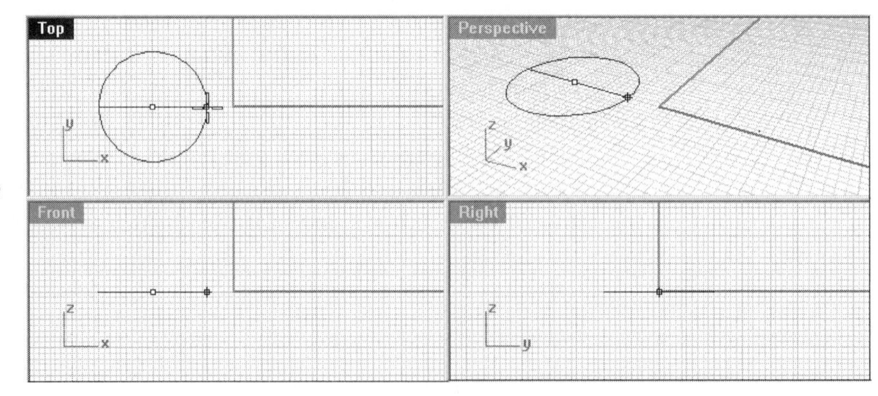

Figure 2–1. Circle being constructed on the construction plane corresponding to the Top viewport

6 Press the ENTER key to repeat the Circle command.
7 Move the cursor over the Front viewport.

Now the label of the Front viewport is highlighted. You are working on a construction plane parallel to the Front viewport.

8 Pick a point specifying the center location.
9 Pick to specify a point on the circumference. A circle is constructed on a construction plane parallel to the Front viewport, as shown in Figure 2–2.

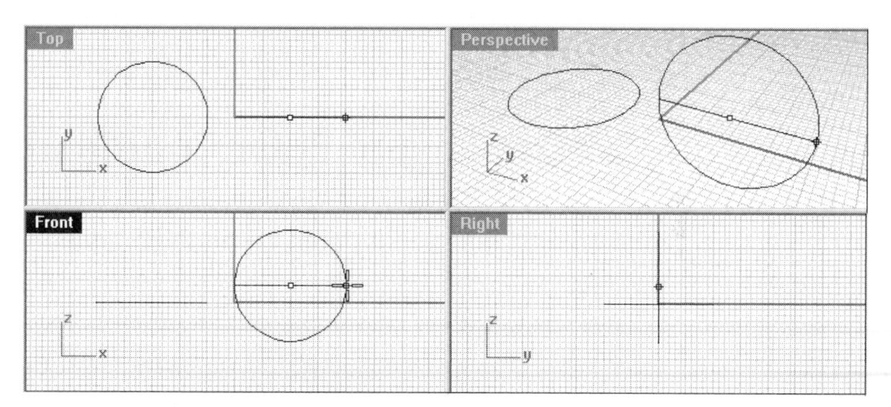

Figure 2–2. Circle being constructed on a construction plane (the front construction plane) parallel to the Front viewport

Ways to Manipulate the Construction Plane

To reiterate, the orientation of the objects you construct depends on the active construction plane, and the active construction plane depends on which viewport you select. In a four-viewport display, there are three construction planes. With the exception of the Perspective viewport, which has the same construction plane as the Top viewport, each viewport has a construction plane parallel to itself. In addition to the default construction planes, you can construct new construction planes using one of the options in the Set CPlane cascading menu of the View pull-down menu. Details about setting up construction planes can be foundare provided in Appendix B.

Using Coordinates Input

To specify a point precisely, you input its coordinates. There are two coordinate systems: the construction plane coordinate system (corresponding to the active viewport) and World coordinate system (independent of the active viewport). The construction plane coordinate system corresponds to the construction plane.

Coordinate Axes

In each viewport, a red line and a green line depict, respectively, the X and Y axes of the construction plane in the viewport. (The color of these lines, and other color settings, are configurable via the Color tab of the Rhino Options dialog box, accessed by selecting Tools > Options from the main pull-down menu.) The Z direction is perpendicular to the construction plane.

In addition to the red and green lines, there is a World Axes icon in the lower left-hand corner of the respective viewport. The World Axes icon is in the shape of a tripod. The lines on the tripod depict the absolute X, Y, and Z axes of the World coordinate system.

Cartesian and Polar Coordinate Systems

To specify a location using the command area, you indicate the nature of the Cartesian coordinate system or polar coordinate system you want to use via the nature of the coordinate input.

For the Cartesian coordinate system options, you specify X and Y values or X, Y, and Z values (separated by a comma or commas, as shown in Table 2–1). If you specify only the X and Y values, the Z value is assumed to be zero.

For the polar coordinate system options, you specify a distance value and an angular value, separated by a less-than (<) sign.

For either coordinate system, if you want to specify a point relative to the last selected point, you prefix the coordinate with the letter r or the @ sign. If you want to use a World coordinate, you prefix the coordinate with the letter w. Coordinate systems that operate under Rhino, and what each system specifies, are summarized in Table 2–1.

Table 2–1 Coordinate Systems and Their Functions

Coordinate System	Example	Specifies
Construction plane Cartesian system	2,3	A point 2 units in the X direction and 3 units in the Y direction from the origin (Z is zero)
	2,3,4	A point 2 units in the X direction, 3 units in the Y direction, and 4 units in the Z direction from the origin
Construction plane polar system	2<45	A point 2 units at an angle of 45 degrees from the origin

Coordinate System	Example	Specifies
Relative construction plane Cartesian system	R2,3 or @2,3	A point 2 units in the X direction and 3 units in the Y direction from the last reference point
Relative construction plane polar system	R4<60 or @4<60	A point 4 units at an angle of 60 degrees from the last reference point
World Cartesian system	W3,5	A point 3 units in the absolute X direction and 5 units in the absolute Y direction from the absolute origin, regardless of the location of the current construction plane
World relative Cartesian system	wr3,6	A point 3 units in the absolute X direction and 6 units in the absolute Y direction from the last reference point, regardless of the location of the current construction plane
World polar system	W4<30	A point 4 units at an angle of 30 degrees on the absolute XY plane from the absolute origin, regardless of the location of the current construction plane
World relative polar system	wr5<45	A point 5 units at an angle of 45 degrees on the absolute XY plane from a reference point, regardless of the location of the current construction plane

To appreciate how precise coordinates can be input to construct Rhino objects, perform the following steps:

1. Start a new file. Use the Small Objects, Millimeters template. (Do not save the previous file.)
2. Double-click on the Top viewport title to maximize the viewport.

Double-clicking again on the maximized Top viewport title will set the graphics area to the previous four-viewport configuration.

3. Select Curve > Line > Single Line, or click on the Line button on the Line toolbar.
4. Type 10,10 at the command area to specify the first endpoint of the line segment, shown as A in Figure 2–3. (This is a construction plane Cartesian coordinate, indicating that the endpoint is 10 units from the origin in the construction plane's X direction and 10 units from the origin in the construction plane's Y direction.)

5. Type 20 < 170 at the command area to specify the second endpoint of the line segment, shown as B in Figure 2–3. (This is a construction plane polar coordinate, indicating that the endpoint is 20 units away from the origin and at an angle of 170 degrees from the construction plane's origin.)
6. Press the ENTER key to repeat the last command, select Curve > Line > Single Line, or click on the Line button on the Line toolbar.
7. Type r10,0 at the command area to specify the first endpoint, shown as C in Figure 2–3. (This is a relative construction plane Cartesian coordinate, indicating that it is 10 units from the last input point, shown as C in Figure 2–3, in the construction plane's X direction and 0 units from the last input point in the construction plane's Y direction.)
8. Type r20 < 300 at the command area to specify the second endpoint, shown as D in Figure 2–3. (This is a relative construction plane polar coordinate, indicating that it is 20 units from the last input point and at an angle of 300 degrees from the last input point.)

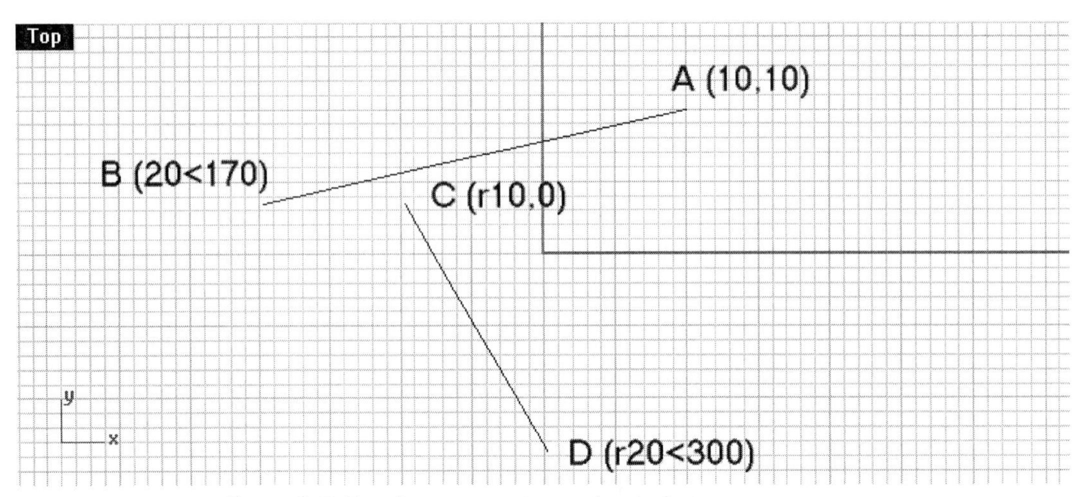

Figure 2–3. Two line segments constructed

Angle Constraint

If you want to construct a line at a certain angle, but have no definite idea on its length, you may first select a point to specify the start point of the line and then type < 60 in the command area. The endpoint of the line will be constrained at an angle of 60 degrees. Continue with the following steps.

9 Select Curve > Line > Single Line, or click on the Line button on the Line toolbar.
10 Click on point A in the Top viewport. (Exact location is unimportant.)
11 Type < 60 at the command area to specify an angle constraint.
12 While the cursor is being constrained at an angle of 60 degrees, click on B, shown in Figure 2–4. A line at an angle of 60 degrees is constructed.
13 Do not save your file.

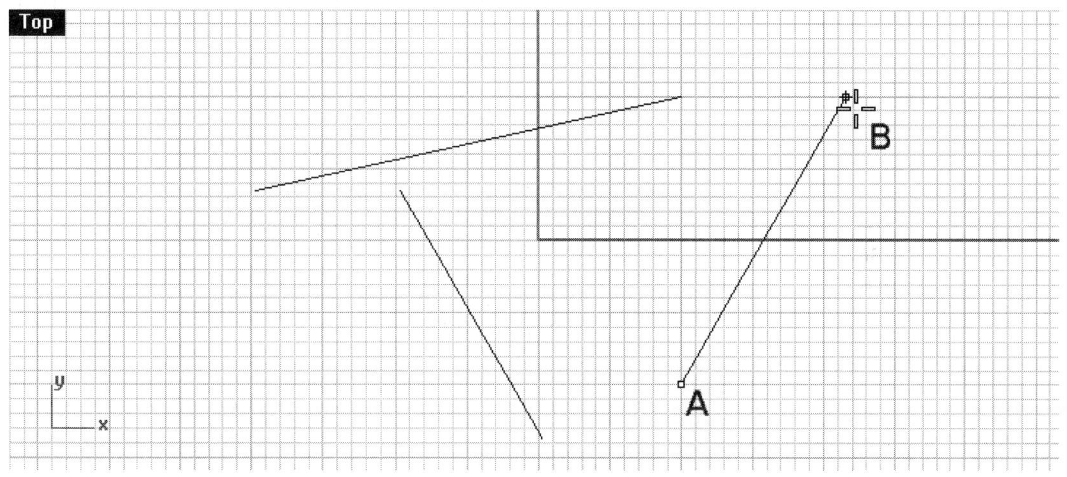

Figure 2–4. A line constrained at an angle of 60 degrees constructed

Using Drawing Aids

To specify locations for the construction of geometric objects, you use various drawing aids such as grid meshes, snapping to grids, planar mode, elevator mode, ortho mode, and object snap.

Setting Grid Mesh

A grid mesh of known spacing on the screen gives you a sense of the actual size of the current viewport. Perform the following steps.

1 Start a new file. Use the "Small Objects, Millimeters" template. (Do not save the previous file.)
2 Select View > Grid Options.
3 Change Grid extents to 150 millimeters. (See Figure 2–5.)
4 Click on the OK button. Grid spacing is set.

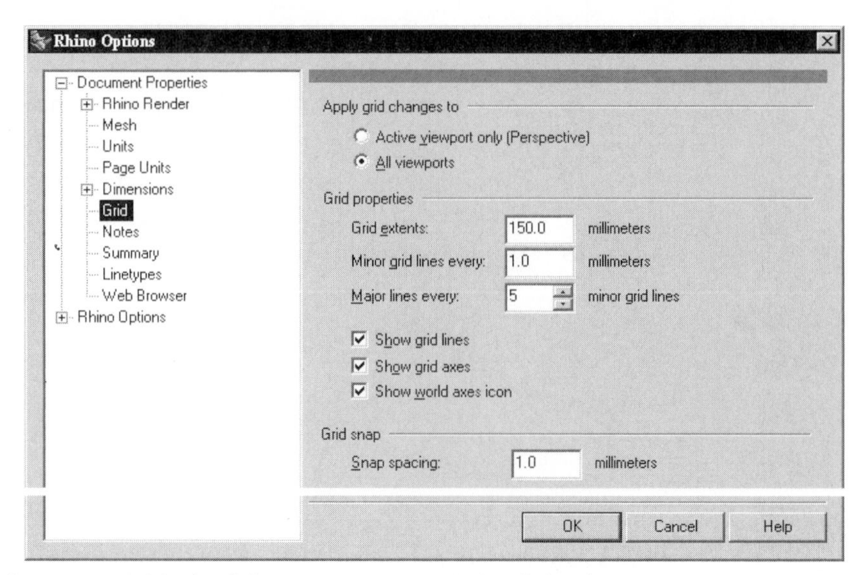

Figure 2–5. Grid tab of the Document Properties dialog box

In the Grid tab of the Document Properties dialog box, you control the display of grid lines and set their spacing. In addition, you control the display of the World Axes icon. To quickly turn on or off the grid mesh in a viewport, press the F7 key.

An alternative way to set grid settings is to type Grid at the command area and make appropriate settings.

Snapping to Grid

The grid display shown in the viewport is for visual reference only. Without the use of further aids, it is virtually impossible to select these points precisely using the pointing device. To restrict the movement of the cursor so that it will stop only at the grid intervals, you use the Snap option, or select or deselect the Snap button on the status bar. Snap interval can be set in the Grid tab of the Options dialog box. Continue with the following steps:

5 Click on the Snap button of the Status bar. Snap is turned on.

6 Move the cursor around in the graphics area. You will find that the cursor is restricted to grid locations.

7 Click on the Snap button of the Status bar. Snap is turned off.

Using Planar Mode

As we have said, there are three pre-set construction planes in a default four-viewport Rhino file. By using the cursor to pick a point in the viewport, you are restricted to picking a location from one of these planes. To restrict the current construction plane to that of the last selected point, you can use planar mode (established via the Planar option), or you can select or deselect the Planar button on the status bar. Continue with the following steps:

8. Check the Planar button from the Status bar, if it is not already checked.
9. Select Curve > Polyline > Polyline.
10. Click on location A of the Front viewport, as indicated in Figure 2–6.
11. Move the cursor to the Top viewport and continue to click on locations B, C, and D (Figure 2–6).
12. Press the ENTER key. A planar polyline is constructed.
13. Click on the Planar button from the Status bar to deselect Planar mode.

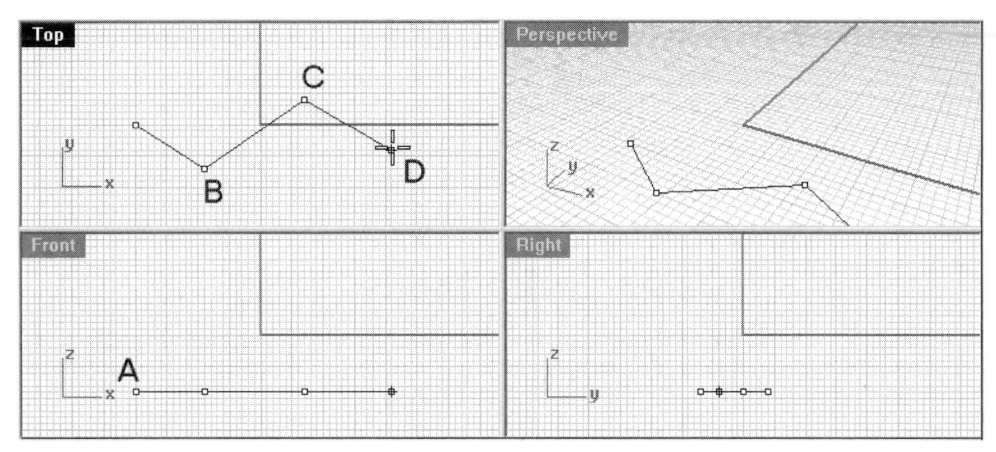

Figure 2–6. Planar polyline constructed

Using Elevator Mode

To help construct 3D points by first picking the X- and Y- coordinates of a point from a construction plane and then picking a point from an adjacent construction plane to depict the Z coordinate, you use the elevator mode, which is activated by holding down the Ctrl Key while picking a point from the construction plane. Continue with the following steps:

14. Select Curve > Line > Single Line.
15. Hold down the CONTROL key and click on location A, shown in Figure 2–7. This specifies the X- and Y-coordinates relative to the Front viewport.
16. Move the cursor to location B in the Top viewport, and click on location B. This indicates the Z-coordinate relative to the Front viewport's construction plane.
17. With the CONTROL key still held down, click on location C in the Front viewport and then location D in the Top viewport. This specifies the endpoint of the line.

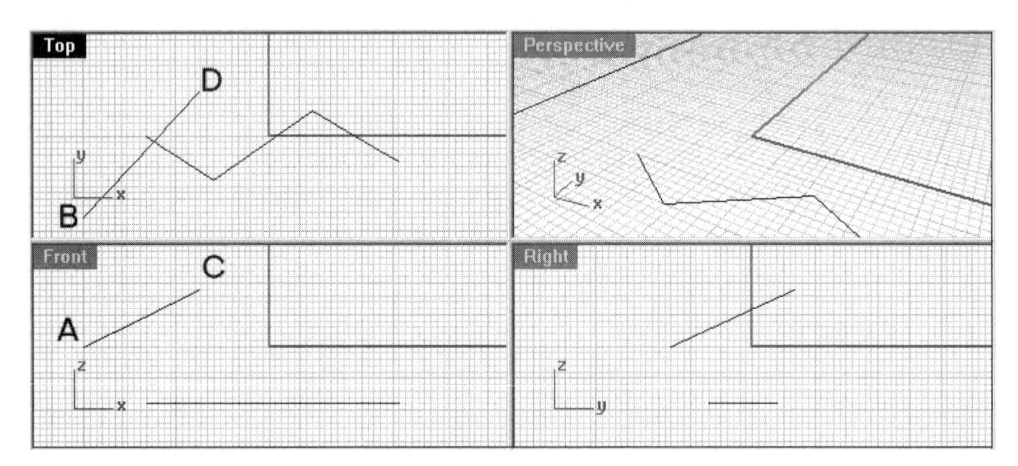

Figure 2–7. Elevator mode being used to construct a line

Using Ortho Mode

To restrict cursor movement in a specified angular direction, you select or deselect Ortho on the status bar. Continue with the following steps:

1. Start a new file. Use the Small Objects, Millimeters template. (Do not save the previous file.)
2. Double-click on the Top viewport title to maximize the viewport.
3. Click on the Ortho button from the Status bar to turn on Ortho mode.
4. Select Curve > Line > Single Line.
5. Click on location A, shown in Figure 2–8, to specify the start point.
6. Move the cursor around; you will find that it is restricted to horizontal or vertical movements.

7. Click on location B, shown in Figure 2–8, to specify the endpoint of the line. A horizontal line is constructed.

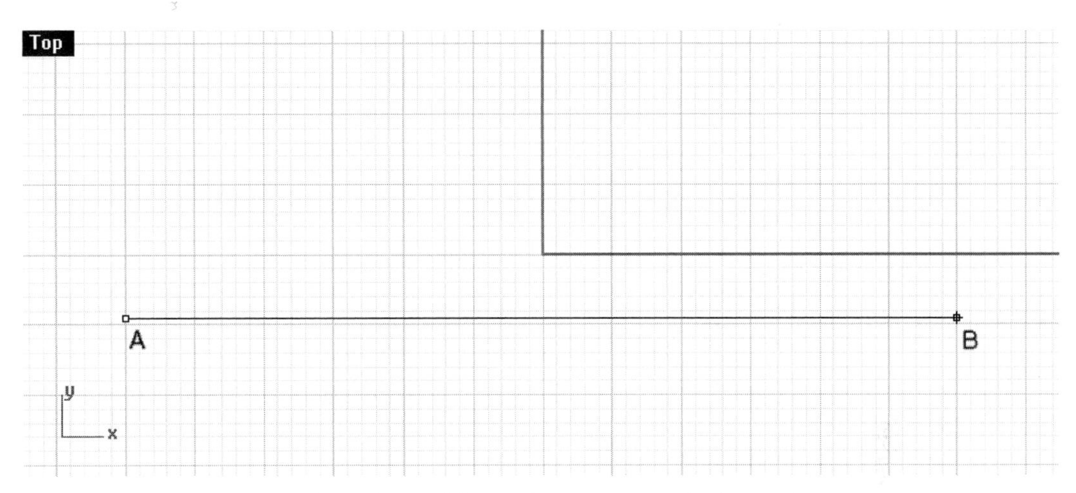

Figure 2–8. Horizontal line being constructed using ortho mode

Setting Ortho Angle

By default, the ortho angle is 90 degrees. In other words, the second selected point is restricted horizontally or vertically. However, you may set the ortho angle to any value other than 90 degrees by typing Orthoangle at the command area while constructing a line or a curve, or by setting the ortho angle in the modeling aids tab of the Options dialog box.

Continue with the following steps to set the ortho angle at the command area.

8. Select Curve > Line > Single Line.
9. Click on location A, indicated in Figure 2–9, to specify the start point.
10. Type orthoangle at the command area.
11. Type 40 to specify an ortho angle of 40 degrees.
12. Click on location B, shown in Figure 2–9, to specify the start point. A line at an angle of 40 degrees is constructed.

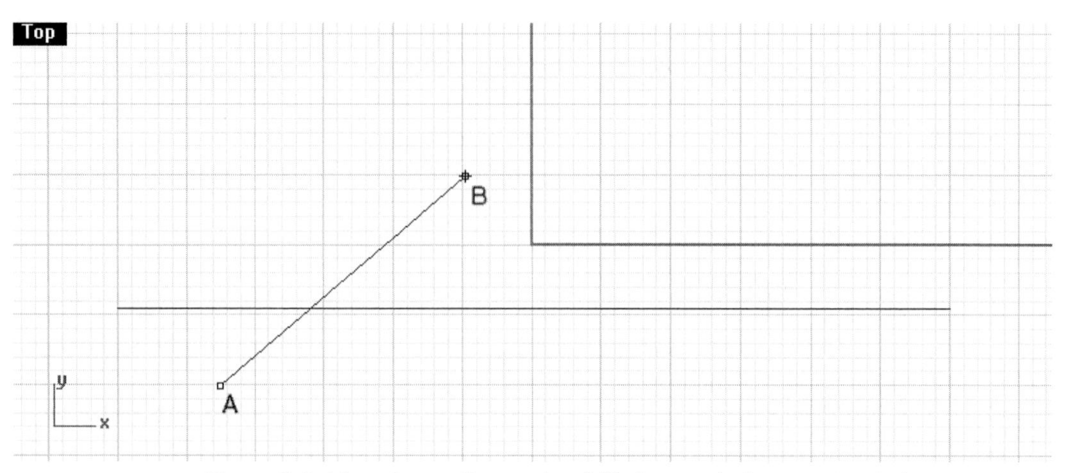

Figure 2–9. Line at an ortho angle of 40 degrees being constructed

To use the Options dialog box to reset the ortho angle to 90 degrees, perform the following steps:

13 Select Tools > Options.

14 On the left pane of the Options dialog box, select Rhino Options > Modeling Aids.

15 Set Ortho Snap every 90 degrees.

16 Click the OK button to close the dialog box. Ortho angle is set to 90 degrees.

17 Click on the Ortho button on the Status bar to turn off ortho mode.

Using Object Snap (Osnap) Tools

While constructing geometric objects, you can use object snap (Osnap) tools to help locate the cursor in relationship to selected features of existing geometric objects. Feature aspects you snap to are end, near, point, midpoint, center, intersection, perpendicular, tangent, quadrant, and knot.

There are two ways to set object snap. You can temporarily set object snap mode (i.e., for each snap to a feature) by specifying the snap mode before you select an object. In Rhino terms, this is one-shot object snap. Alternatively, you can set persistent object snap by selecting Osnap on the status bar to use the Snap option from the Osnap dialog box. Continue with the following steps to use one-shot object snap in geometry construction.

18 Select Curve > Circle > Center, Radius.

19 Select Tools > Object Snap > Intersection.

20 Move the cursor to location A, shown in Figure 2–10, until you see the Int symbol displaying at the cursor, indicating that the intersection between two line segments is found.

21 While Int is still displaying, click on the graphics area.

22 Type mid at the command area to specify a midpoint object snap.

23 Move the cursor to location B, shown in Figure 2–10, until you see mid displaying at the cursor.

24 While mid is still displaying, click on the graphics area. A circle with its center at the intersection of two lines and passing through the midpoint of a line is constructed.

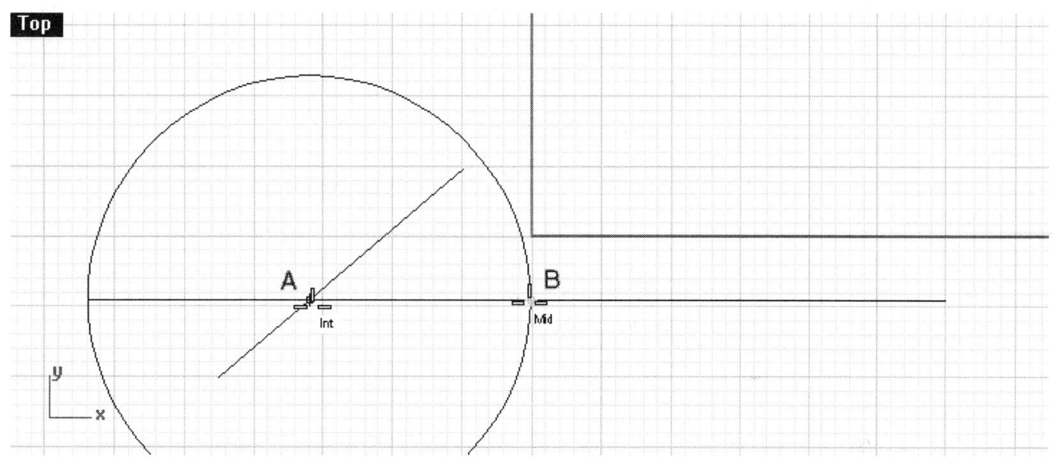

Figure 2–10. Circle being constructed

Osnap Dialog Box

If a certain object snap mode is required a number of times, it is more convenient to use persistent object snap via the Osnap dialog box. Continue with the following steps:

25 Check the Osnap button from the Status bar to display the Osnap dialog box, if it is not already displayed.

As shown in Figure 2–11 (left), there are 10 persistent object snap modes: End, Near, Point, Mid (midpoint), Cen (center point), Int (Intersection), Perp (perpendicular), Tan (tangent), Quad (quadrant), and Knot. In addition, there are three buttons: Project, STrack, and Disable.

Checking the Project button will enable projection mode, checking the STrack button will enable smart tracking, and checking the Disable button will disable object snap mode specified in the dialog box. Right-clicking the Disable button clears all the persistent object snaps.

If you hold down the Control key, the Osnap dialog box will display its second page of icons, as shown in Figure 2–11 (right).

Osnap
End Near Point Mid
Cen Int Perp Tan
Quad Knot Project STrack
Disable

Osnap
From PerpFrc TanFro: AlongL
AlongP Betwee: OnCrv OnSrf
OnPSrf POnCrv POnSrf POnPS
Disable

Figure 2–11. Normal Osnap dialog box (left) and dialog box with Control key held down (right)

Now continue with the following steps:

26 Check the End box of the Osnap dialog box to specify a persistent endpoint object snap.

27 Select Curve > Line > Single Line.

28 Click on endpoints A and B, shown in Figure 2–12, to specify the start and end points of the line.

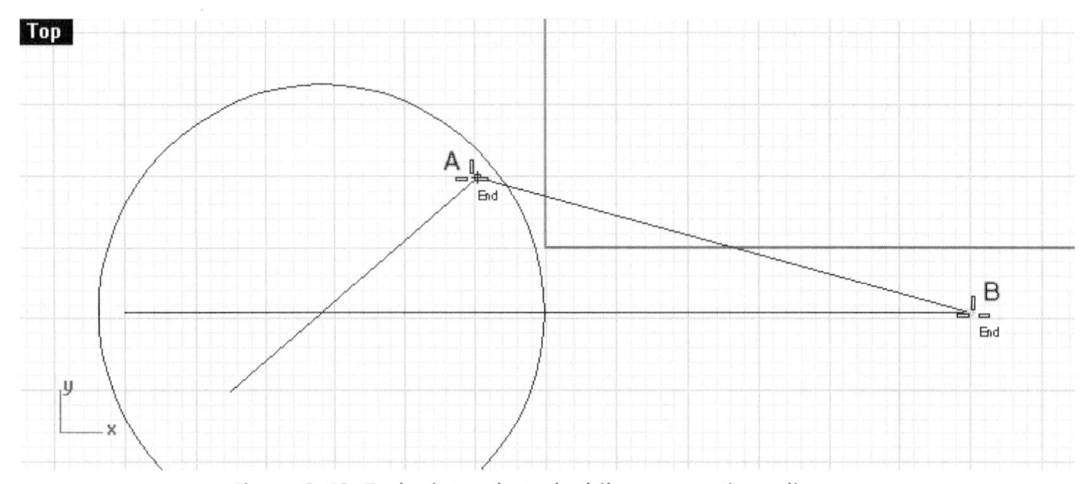

Figure 2–12. Endpoints selected while constructing a line

Enabling/Disabling Osnap by using the ALT key

When Rhino is asking for a point, you can press the ALT key to temporarily enable Osnap check boxes if Osnap is disabled. You can press the ALT key to temporarily disable Osnap check boxes if Osnap is enabled.

Osnap: Along Line

Along Line is a snapping tool. It enables you to select two points to establish a tracking line, along which you can select any point. Perform the following steps.

1. Open the file DrawingAids1.3dm from the Chapter 2 folder of the CD accompanying this book. (Do not save the previous file.)
2. Click on the Osnap button from the Status line.
3. In the Osnap dialog box, click on End and Point, and clear all other buttons.
4. Select Curve > Line > Single Line.
5. Select point A, indicated in Figure 2–13, to specify the start point of the line.
6. Select Tools > Osnap > Along Line.
7. Select endpoints B and C, indicated in Figure 2–13, to specify the start and end of the tracking line.
8. Click on point D. Endpoint E is constructed along the tracking line and closest to point D.

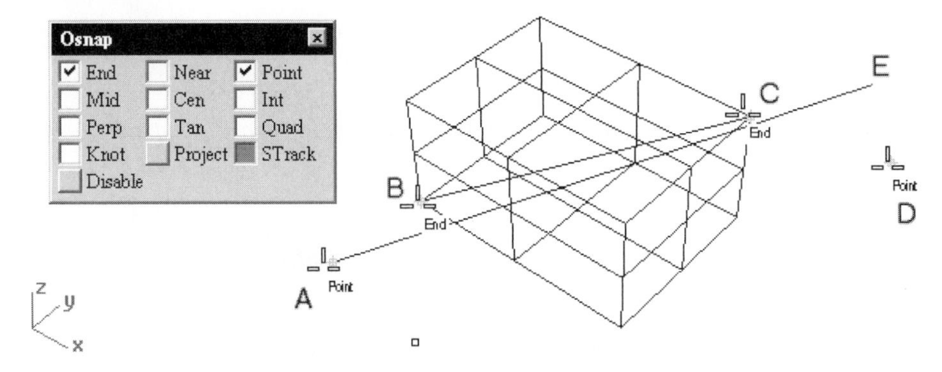

Figure 2–13. Line's endpoint E constructed along tracking line BC

Osnap: Along parallel

Along Parallel is a snapping tool. It establishes a tracking line passing through a point and parallel to a reference line. You can select a location along this parallel line. Continue with the following steps:

9 Select Curve > Line > Single Line.

10 Select point A, shown in Figure 2–14, to specify the start point of the line.

11 Select Tools > Osnap > Along Parallel.

12 Select endpoints B and C, shown in Figure 2–14, to specify the start and end of the reference line.

13 Select endpoint D to specify the start point of the parallel tracking line.

14 Click on point E. Endpoint F is constructed along a tracking line passing through point D, parallel to reference line BC, and closest to point E.

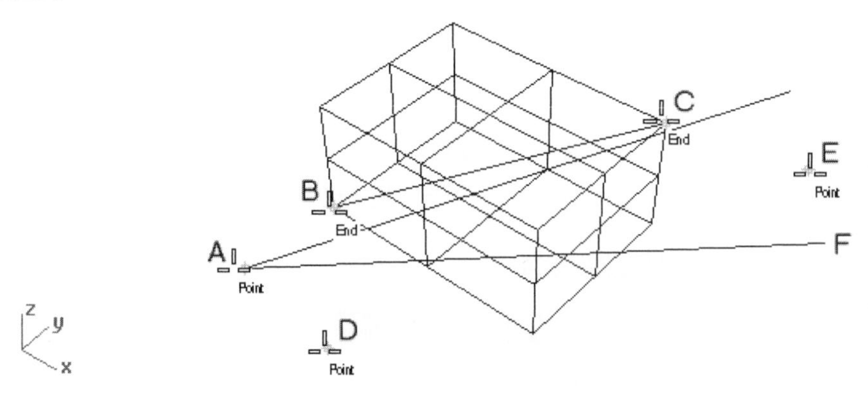

Figure 2–14. A line's endpoint being constructed along a line parallel to a track line

Osnap: Between Two Selected Points

Between is a snapping tool that enables you to specify a point midway between two selected points. Continue with the following steps:

15 Select Curve > Line > Single Line.

16 Select point A, shown in Figure 2–15.

17 Select Tools > Osnap > Between.

18 Select points A and B, shown in Figure 2–15. The endpoint of the line is constructed midway between two selected points.

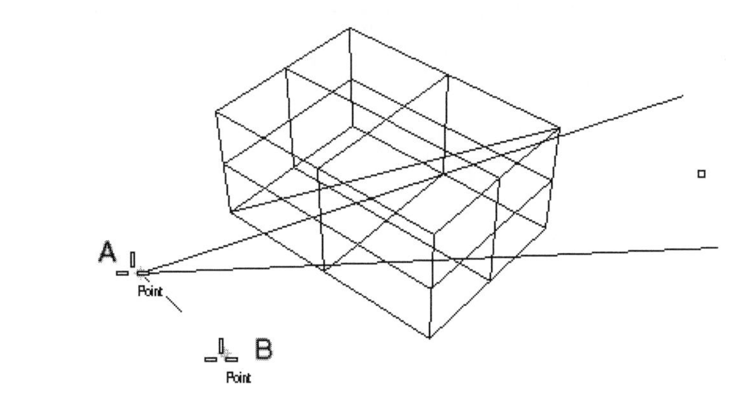

Figure 2–15. Midway between two selected points

Osnap: From, Perpendicular From, and Tangent From

These three snapping tools enable you to specify a point a distance from a selected point, perpendicular from a point, and tangent from a curve. Perform the following steps:

1. Open the file DrawingAids2.3dm from the Chapter 2 folder of the companion CD. (Do not save the previous file.)
2. Check the End and Point buttons, and clear all other buttons in the Osnap dialog box.
3. Select Curve > Line > Single Line.
4. Select Tools > Object Snap > From.
5. Select endpoint A, shown in Figure 2–16.
6. Type 6 at the command area to specify the distance from the selected point.
7. Select point B, shown in Figure 2–16. The starting point of the line is specified. It is 6 units from the endpoint A and closest to point B.
8. Select Tools > Osnap > Perpendicular From.
9. Select curve C, shown in Figure 2–16.
10. Click on point D, indicated in Figure 2–16. A tracking line is defined. It is perpendicular to curve C and closest to point D.
11. Click on point E. The line's endpoint is defined. It resides on the defined tracking line and is closest to point E.

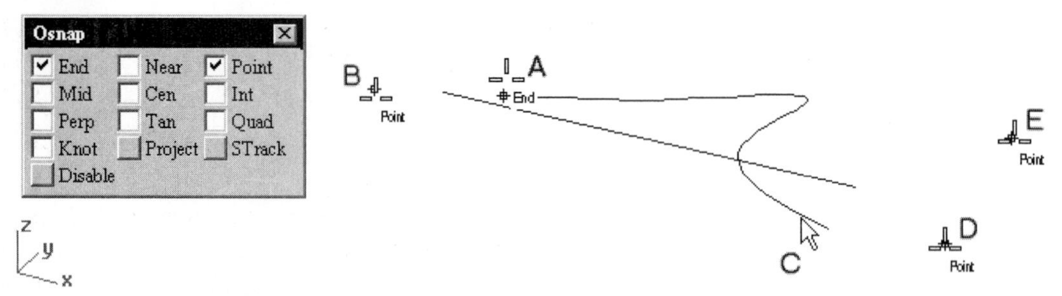

Figure 2–16. A line being constructed by using from and perpendicular from object snaps

12 Select Curve > Line > Single Line.
13 Select Tools > Osnap > Tangent From.
14 Select curve A, shown in Figure 2–17.
15 Select point B, shown in Figure 2–17. A tracking line tangent to curve A and closest to point B is defined.
16 Click on point B again. The start point of the line is defined. It is on the tracking line and is closes to point B.
17 Select endpoint C, (Figure 2–17). A line is constructed.

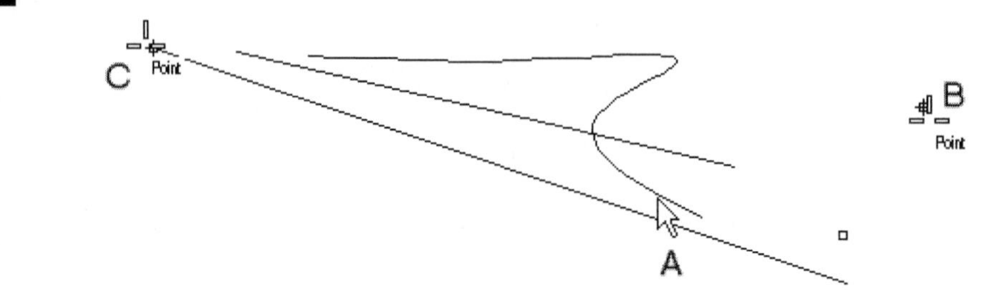

Figure 2–17. A line being constructed by using the tangent from object snap

Osnap: On Curve and On Surface

These two snapping tool enable you to snap to a point on a curve and on a surface. Perform the following steps:

1 Open the file DrawingAids3.3dm from the Chapter 2 folder of the companion CD. (Do not save the previous file.)
2 Select Curve > Line > Single Line.

3. Select Tools > Osnap > On Curve.
4. Select curve A, indicated in Figure 2–18.
5. Click on endpoint B, shown in Figure 2–18. The line's start point is defined. It is along curve A and closest to endpoint B.
6. Select Tools > Osnap > On Surface.
7. Select surface C, shown in Figure 2–18.
8. Click on point D(Figure 2–18).The line's second point is defined. It lies on surface C and is closest to point D.

Perspective

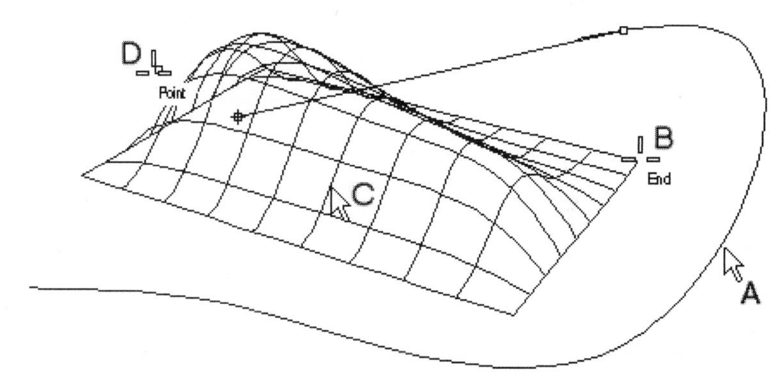

Figure 2–18. A line constructed with its endpoints on a curve and a surface

*(**NOTE:** Lines on the surfaces are called isoparametric curves. You can use the intersection snapping tool to snap to the intersections of these curves.)*

Osnap: Project Mode

If the project button on the Osnap dialog box is selected, the resulting snap point is projected onto the active construction plane. Perform the following steps:

1. Open the file DrawingAids4.3dm from the Chapter 2 folder of the companion CD. (Do not save the previous file.)
2. Click on the Osnap button from the Status bar.
3. Clear all the buttons except Mid and Project.
4. Select Curve > Line > Single Line.
5. Click on location A of the Top viewport, shown in Figure 2–19. The midpoint of the curve that is projected onto the current construction plane is used as the starting point of the line.

6 Click on the Project button of the Osnap dialog box to clear it.
7 Click on location A of the Top viewport again. The midpoint of the curve is used as the endpoint of the line. A line is constructed.

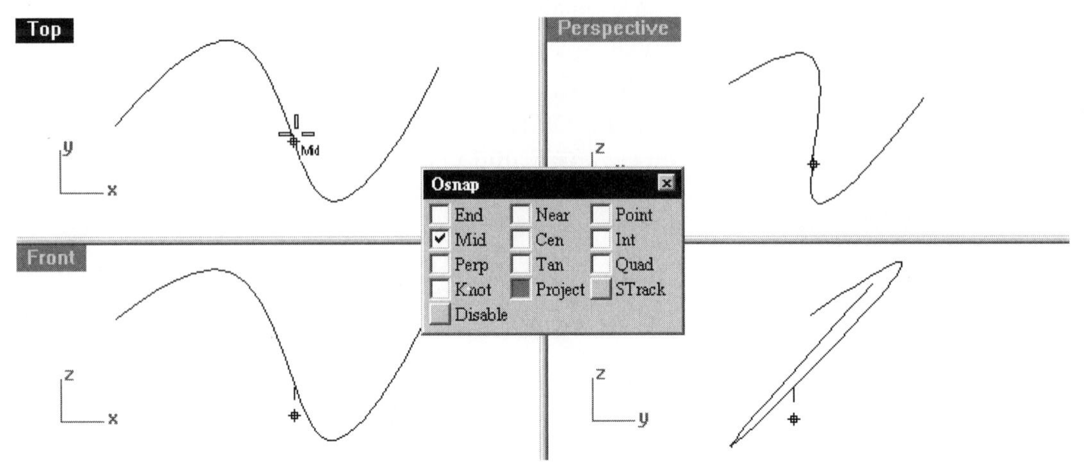

Figure 2–19. A line being constructed

Right-Click Menu on the Toolbar's Snap or Grid Button

Right-clicking on the status bar will bring out a context menu from which you may make various grid and snap settings.

Smart Tracking

Smart Tracking is a drawing aid, providing a set of temporary reference lines and points on the viewport to help indicate implicit relationships among various 3D points, geometry in space, and coordinate axes' direction.

To quickly turn on/off Smart Tracking, you can check or clear the STrack button on the Osnap dialog box and select Tools > Object Snap > Smart Track.

With Smart Tracking turned on, temporary infinite lines, called tracking lines, will appear on the screen after a point or a geometric object is encountered. With two or more nonparallel tracking lines, you get temporary intersection points called smart points. You can snap to these smart points as if they are real point objects. Perform the following steps:

1. Open the file DrawingAids5.3dm from the Chapter 2 folder of the companion CD . (Do not save the previous file.)
2. Click on the Osnap button from the Status bar.
3. Clear all the buttons except End and STrack.
4. Select Curve > Line > Single Line.
5. Move the cursor to endpoint A, shown in Figure 2–20, but do not click on it.
6. Move the cursor to endpoint B, shown in Figure 2–20, but do not click on it.
7. Move the cursor vertically upward until two tracking lines and an intersection (C) between the two tracking lines appear.
8. While the tracking lines are still displayed, click on the intersection mark. The start point of the line is specified.

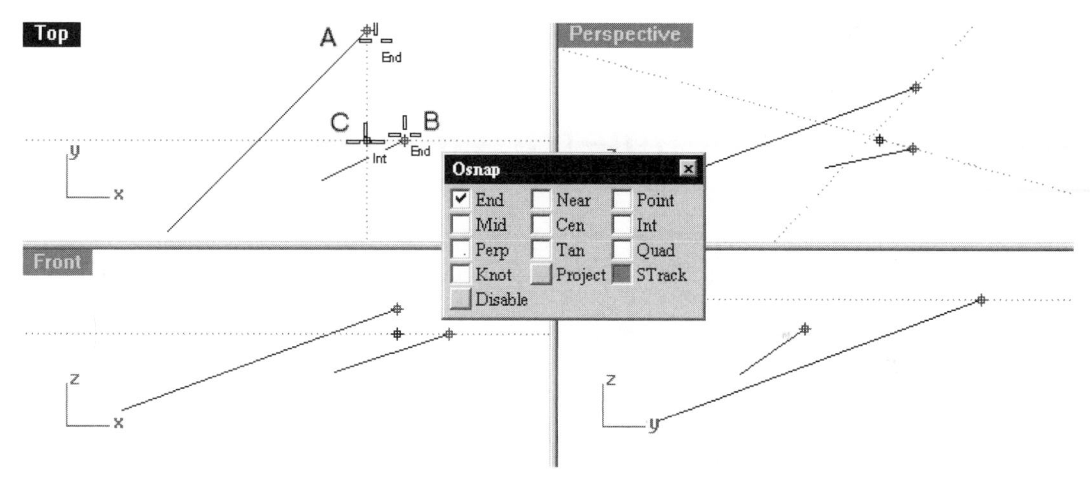

Figure 2–20. Smart tracking tool being used to define the start point of a line

9. Move the cursor to endpoint D, shown in Figure 2–21, but do not click on it.
10. Move the cursor to endpoint E, shown in Figure 2–21, but do not click on it.
11. Move the cursor vertically upward until two tracking lines and an intersection (F) between the two tracking lines appear.
12. While the tracking lines are still displayed, click on the intersection mark. The endpoint of the line is specified.

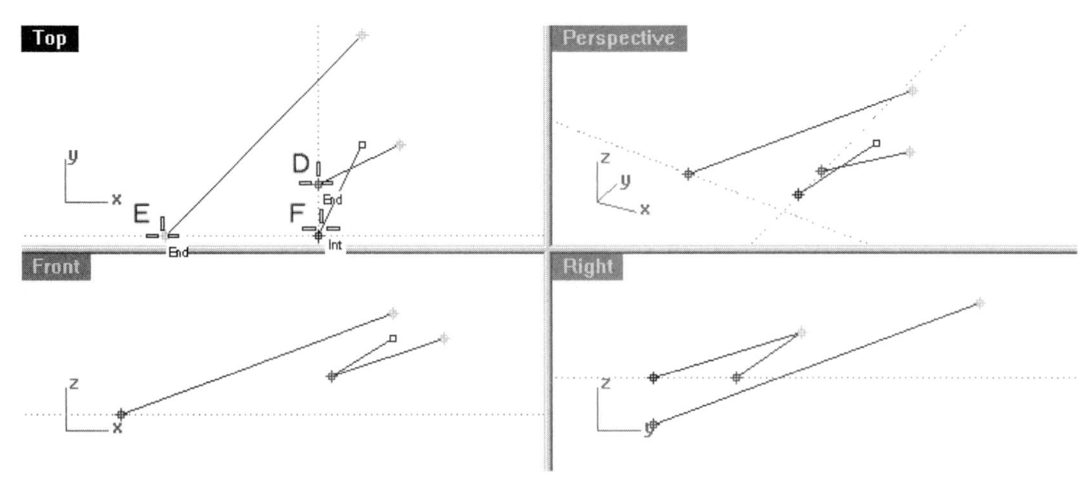

Figure 2–21. Smart tracking tool being used to define the endpoint of the line

Manipulating Geometric Objects

To manipulate geometric objects, you have to selected them and apply a command. The sequence of first selecting the objects and then applying a command and vice versa is unimportant.

Drag Mode

One way to manipulate objects is to select and drag them. By default, objects are dragged along a plane parallel to the construction plane. However, you may set drag mode to world' xy plane, a plane parallel to the view plane, or along the UVN of a surface. U and V are directions along the U and V of a surface, and N is the normal direction of a surface.

Geometric Object Selection Methods

Obviously, the simplest way to select an object is to move the cursor over the object and click. If you want to select multiple objects, you can hold down the SHIFT key or the CONTROL key. However, the SHIFT key favors adding object to the selection set and the CONTROL key favors removing objects from the selection set.

To select multiple objects, you can drag a rectangular zone. If you drag from right to left, you describe a crossing zone, with objects partly or fully within the zone being selected. If you drag from left to right, you describe a window zone. Only objects fully inside the zone are selected. If more than one object is located at where you drag, a pop-up selection menu will display. You can click one of the objects or None if none of the highlighted objects are correct.

Now perform the following steps:

1 Open the file Selection.3dm from the Chapter 2 folder of the companion CD . (Do not save the previous file.)

2 Hold down the SHIFT key and select A, B, and C (Figure 2–22). Three circles are selected.

3 Hold down the CONTROL key and select B and C again. Two selected circles are deselected.

4 Drag from D to E. A crossing zone is defined, and three circles are selected.

5 Drag from F to G. A window zone is defined, and only one circle is selected.

6 Click on H. A pop-up selection menu is displayed. *(Note: There are two identical circles located at the same position.)*

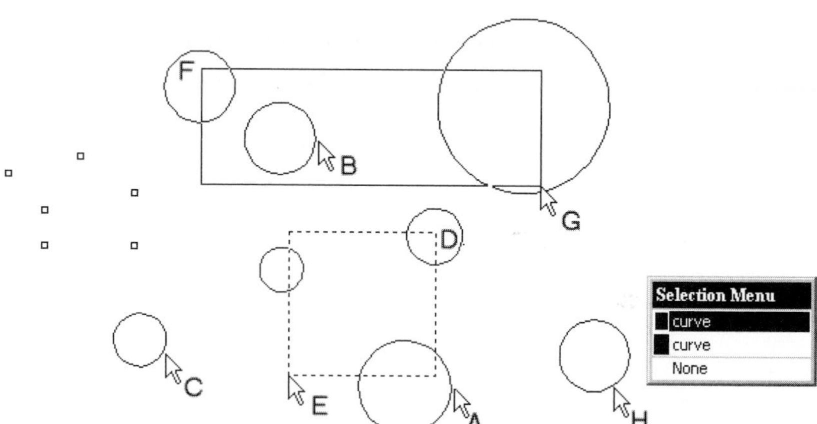

Figure 2–22. Object selection

Selection by Lasso

This command selects point objects by sketching an irregular shape around them. Continue with the following steps:

7 Select Edit > Control Points > Select Control Points > Lasso, or type Lasso at the command area.

8 Referencing Figure 2–23, drag an irregular sketch. A set of points is selected.

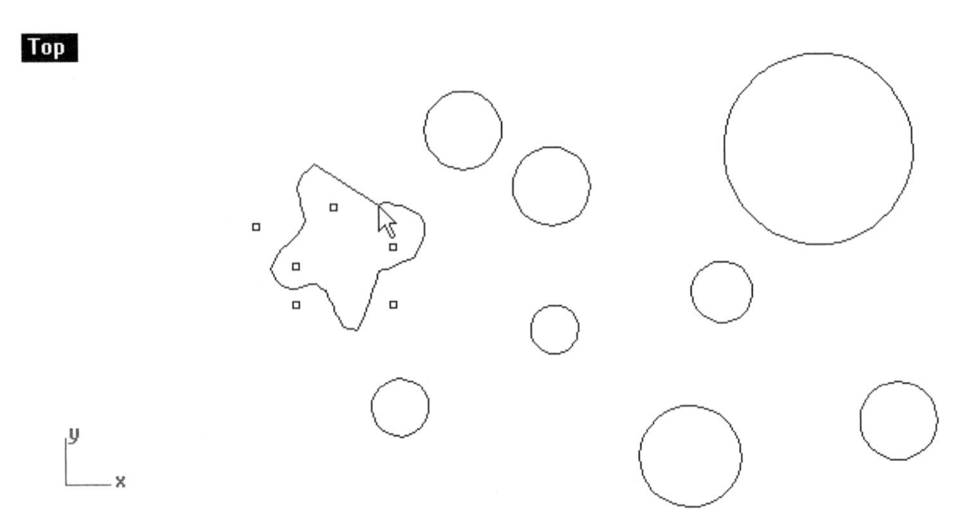

Figure 2–23. Selecting a set of points

After executing the Lasso command and before you make the first click, you may change viewport and perform various zoom and pan operations.

Moving Objects

To relocate one or more geometric objects, you move them. Perform the following steps:

1. Select File > Open and select the file Movecopy.3dm from the Chapter 2 folder on the companion CD.
2. Select Transform > Move, or click on the Move button on the Transform toolbar.
3. Select surface A, shown in Figure 2–24, and press the ENTER key to terminate the selection process.
4. Click anywhere on the graphics area to specify the point to move from.
5. Type r20 < 0 at the command area to specify a point to move to. This is a relative construction plane coordinate, indicating that the second point is 20 units from the last input point and at an angle of 0 degrees from it. As a result, the circle is moved a distance of 20 units in the 0 degree direction.

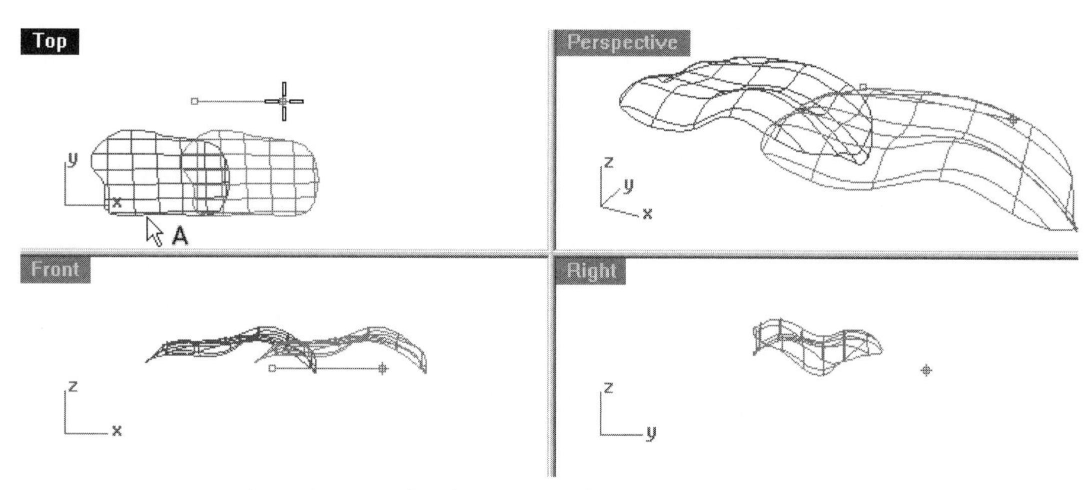

Figure 2–24. Surface being moved

Moving by Dragging

In addition to using the Move command to move objects from one location to another location, you can select objects and drag them to a new location. However, when you use the dragging method, you must have prior understanding of the dragging direction, which can be set via the Edit menu.

To set dragging mode, select Edit > View Based Drag Mode. As can be seen at the command area, there are five options, providing four dragging modes, as follows:

- CPlane dragging mode: Dragging is parallel to the CPlane (construction plane) setting of the active viewport.
- World dragging mode: Dragging is parallel to the world coordinate system's XY plane.
- View dragging mode: Dragging is parallel to the active viewport.
- UVN dragging mode: Dragging is along the U, V, and N directions. (U and V are two orthogonal directions on the surface, which are parallel to the isocurves. N is the normal direction.) To drag along the normal direction, hold down the Control key.

(Note that the Next option sets the dragging mode to the next in the list. In other words, if the current dragging mode is CPlane, selecting the Next option will change the dragging mode to World.)

To remind you about any drag mode setting other than the default CPlane option, the cursor icons are slightly different, as shown in Figure 2–25.

Figure 2–25. Cursor icons (left to right): CPlane, World, View, and UVN

Copying Objects

To duplicate geometric objects, you copy them. Perform the following steps:

6. Select Transform > Copy, or click on the Copy button on the Transform toolbar.
7. Select surface A, shown in Figure 2–26, and press the ENTER key to terminate the selection process.
8. Click anywhere on the graphics area to specify the point to copy from.
9. Type r20 < 180 at the command area to specify the point to copy to.
10. Type r40 < 180 at the command area to specify the point to copy to.
11. Press the ENTER key to terminate the command. The selected circle is copied twice.

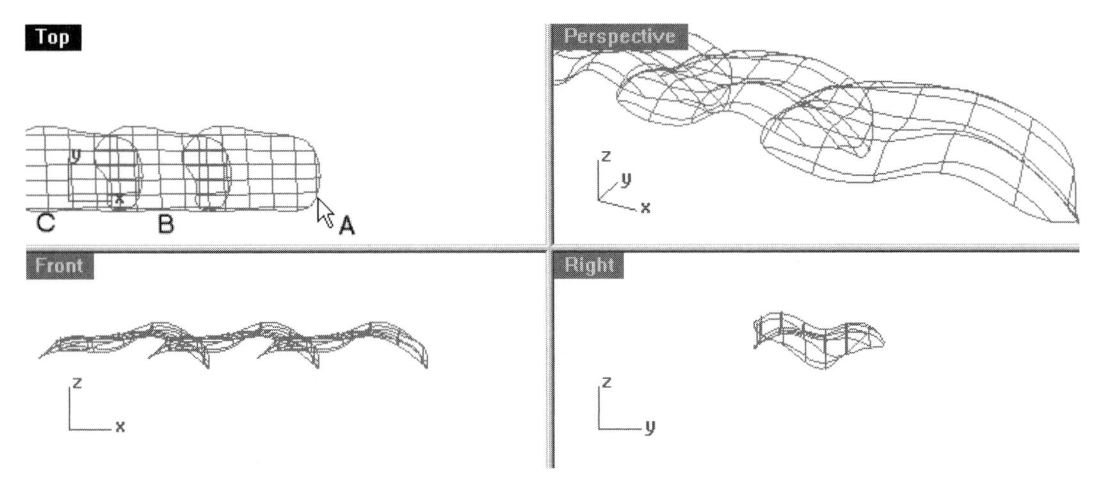

Figure 2–26. Surface copied twice

Holding Down the ALT key While Dragging

By holding down the ALT key while dragging an object, the object will be copied instead of moved. The ALT key also suspends the auto-close function of polyline, curve, and interpolated curve. In addition, it can force a window/crossing selection. Normally, window/crossing selection needs to have the first selection point picked on empty space. Furthermore, it suspends locked-object snap.

Delete

To erase unwanted an geometric object, you delete it. Perform the following steps:

12 Hold down the SHIFT key, and select objects B and C, indicated in Figure 2–26.

13 Press the DELETE key. The selected circles are deleted.

14 Do not save your file.

Rotating Objects

To change the orientation of a geometric object, you rotate it 2D or 3D. 2D rotation causes the selected objects to be rotated around a point on the construction plane. 3D rotation, on the other hand, rotates the selected objects around an axis defined in 3D space.

Perform the following steps:

1 Select File > Open and select the file Rotate.3dm from the Chapter 2 folder on the companion CD.

2 Select Transform > Rotate, or click on the Rotate 2-D/Rotate 3-D button on the Transform toolbar.

3 Select ellipsoid A, shown in Figure 2–27, and press the ENTER key.

4 Type 10,10 at the command area to specify the center of rotation.

5 Type 45 to specify the angle of rotation. The ellipsoid is rotated an angle of 45 degrees around a point located at 10,10.

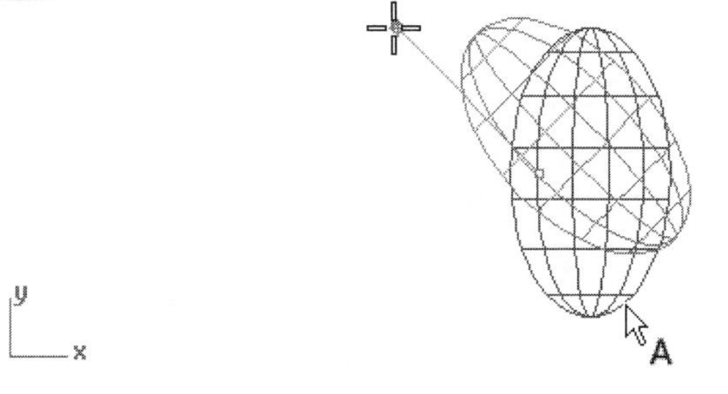

Figure 2–27. Ellipsoid being rotated

6. Double-click on the Top viewport title to return the display to a four-viewport configuration.
7. Select Transform > Rotate 3-D, or right-click on the Rotate 2-D/Rotate 3-D button on the Transform toolbar.
8. Select the ellipsoid and press the ENTER key.
9. Type w10,10 at the command area to specify the start-of-rotation axis.
10. Type w20,20,10 at the command area to specify the end-of-rotation axis.
11. Type 45 at the command area to specify the angle of rotation. The ellipsoid is rotated 45 degrees around an axis formed by the start and end points of the axis. (See Figure 2–28.)
12. Do not save your file.

The direction of rotation follows a right-hand rule in which the thumb indicates the direction of the axis and the fingers indicate the direction of rotation. In other words, the direction of rotation will change to the opposite direction if the sequence of input of the axis endpoints is swapped.

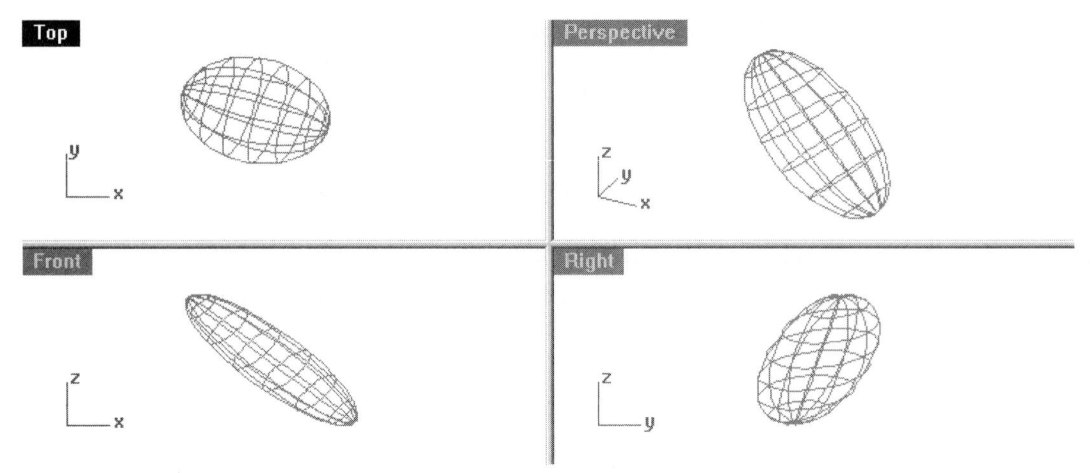

Figure 2–28. Ellipsoid rotated in 3D

Scaling Objects

To change the shape and size of a selected geometric object, you may scale it 3D, 2D, 1D, or nonuniformly. Perform the following steps:

1. Select File > Open and select the file Scale.3dm from the Chapter 2 folder on the companion CD.
2. Select Transform > Scale > Scale 3-D, or click on the Scale 3-D/Scale 2-D button of the Transform toolbar.
3. Select sphere A, shown in Figure 2–29, and press the ENTER key.

4 Type 0,0 at the command area to specify the origin point.
5 Type 0.5 at the command area to specify the scale factor. The sphere is scaled uniformly in 3D.

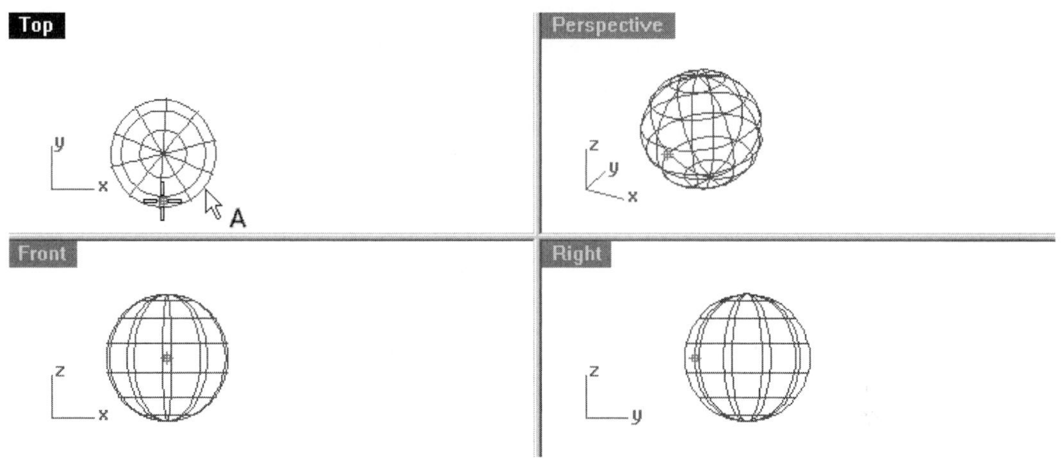

Figure 2–29. Sphere being scaled uniformly in 3D

Scaling 2D changes the size of the geometric object in two directions only. Continue with the following steps.

6 Select Transform > Scale > Scale 2-D, or right-click the Scale 3-D/Scale 2-D button of the Transform toolbar.
7 Select object A, shown in Figure 2–30, and press the ENTER key.
8 Type 0,0 at the command area to specify the origin point.
9 Type 2 at the command area to specify the scale factor. The sphere is scaled in 2D, with the third axis unchanged.

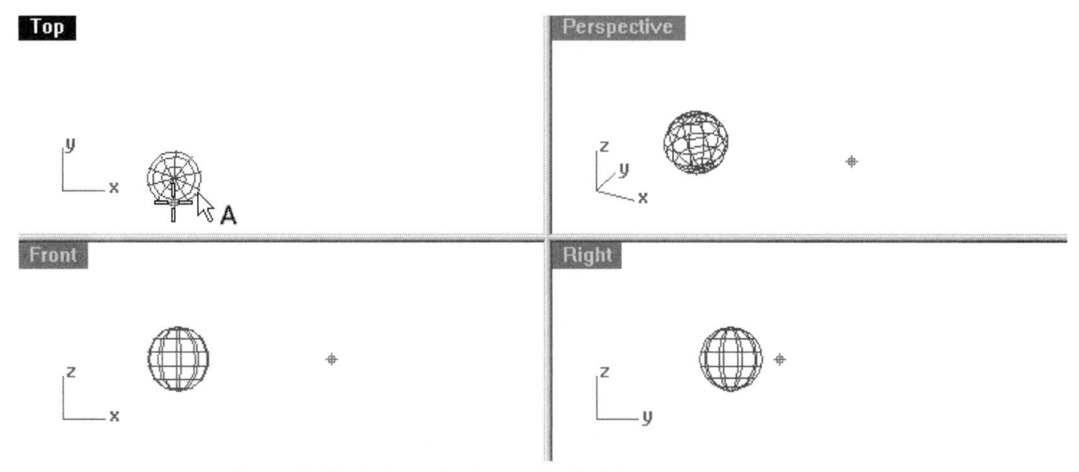

Figure 2–30. Sphere being scaled in 2D

Scaling 1D stretches a geometric object in a specified direction. Continue with the following steps:

10 Select Transform > Scale > Scale 1–D, or click on the Scale 1–D button of the Scale toolbar.

11 Select object A, shown in Figure 2–31, and press the ENTER key.

12 Type 0,0 at the command area to specify the origin point.

13 Type r20 < 0 to specify the first reference point.

14 Type r40 < 0 to specify the second reference point. The sphere is stretched in1D.

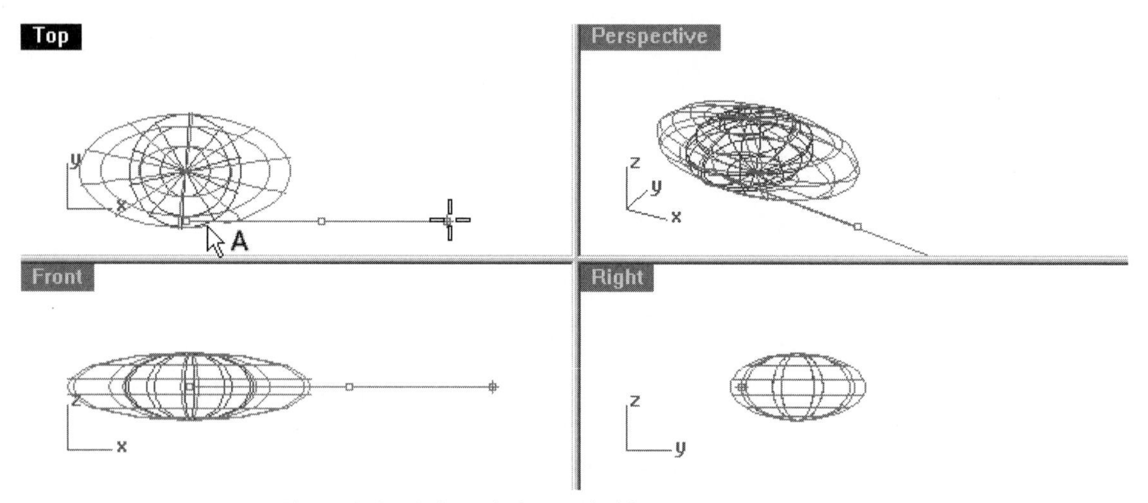

Figure 2–31. Sphere being scaled in 1D

15 Select Transform > Scale > Non-Uniform Scale, or click on the Non-Uniform Scale button of the Scale toolbar.

16 Select object A, shown in Figure 2–32, and press the ENTER key.

17 Type 0,0 at the command area to specify the origin point.

18 Type 0.4 at the command area to specify the X-axis scale.

19 Type 0.2 at the command area to specify the Y-axis scale.

20 Type 2 at the command area to specify the Z-axis scale. The sphere is scaled nonuniformly. (See Figure 2–33.)

21 Do not save your file.

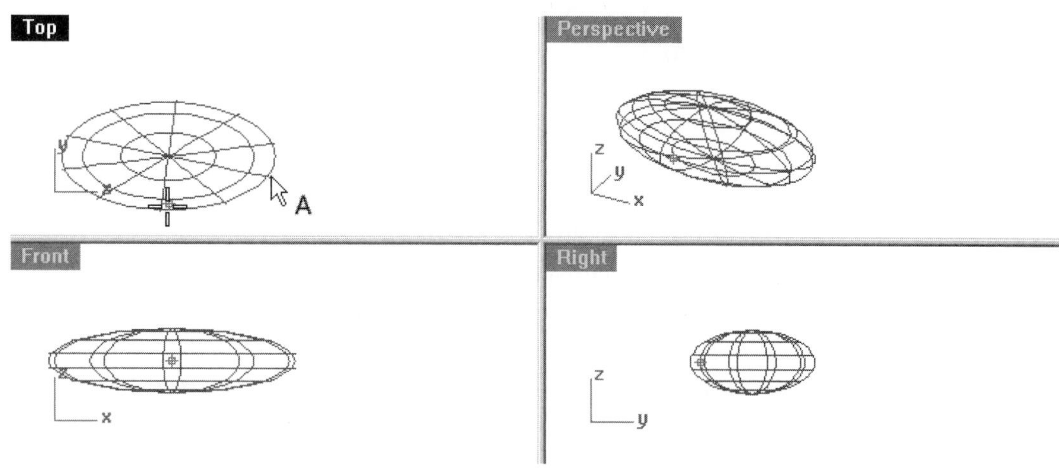

Figure 2–32. Sphere being scaled nonuniformly

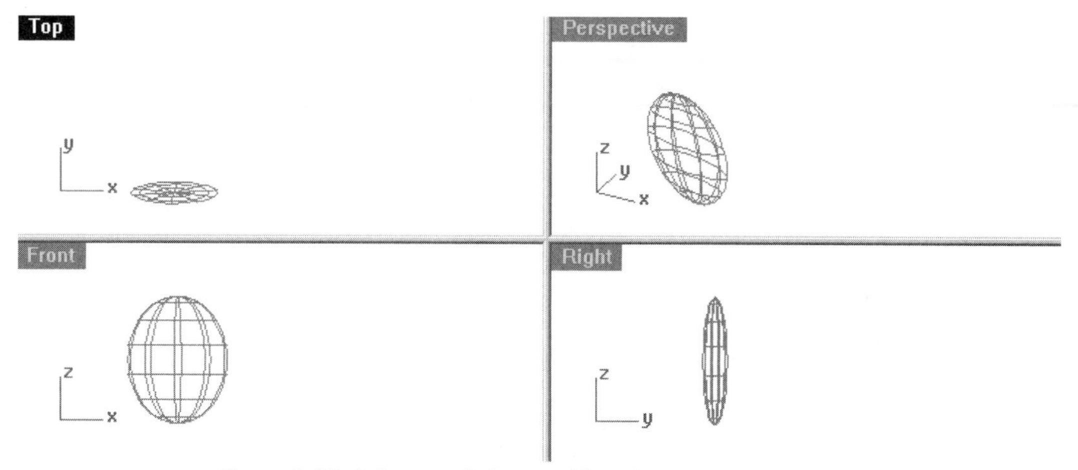

Figure 2–33. Sphere scaled nonuniformly

Mirroring Objects

To mirror objects, you need to define a mirror plane, which can be specified using two methods. In the first method, the mirror plane is perpendicular to the active construction plane and you need to select only two points. In the second method, the mirror plane is 3D, and you need to specify three points to define the plane. To construct a mirror copy of selected objects, perform the following steps:

1. Select File > Open and select the file Mirror.3dm from the Chapter 2 folder on the companion CD.
2. Click on the Osnap button on the Status bar to display the Osnap dialog box, if it is not already displayed.
3. Check the End box in the Osnap dialog box.
4. Select Transform > Mirror, or click on the Mirror button on the Transform toolbar.
5. Select surface A, shown in Figure 2–34, and press the ENTER key.
6. Select points B and C, shown in Figure 2–34. The selected curve is mirrored.

(***NOTE:*** *If the Copy option is set to Yes, the source object is retained. Otherwise, it is deleted after being mirrored.)*

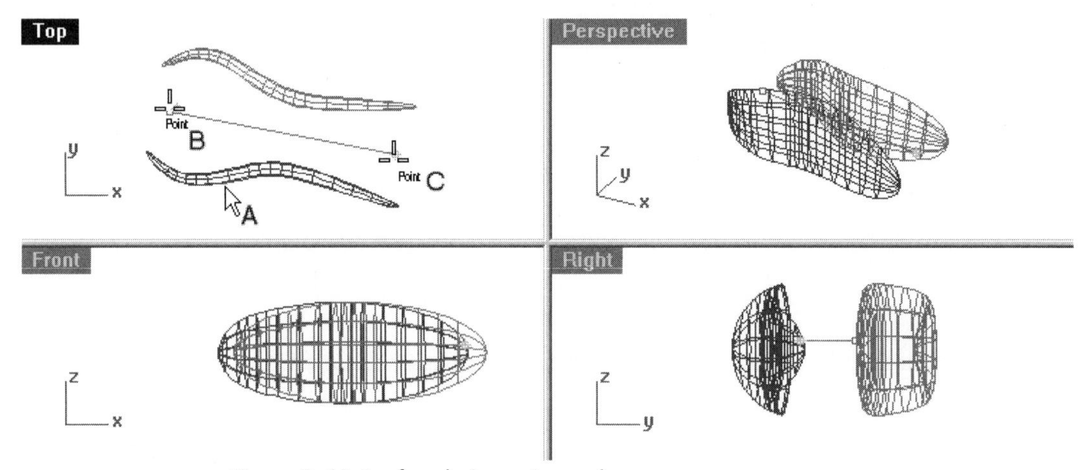

Figure 2–34. Surface being mirrored

Undo and Redo

If you make any mistakes while working, you can use the undo command to revert to the previous state of your work. To reapply the undone commands, you use the redo command.

Number of Undos

Theoretically, you can perform undo as many times as you wish. However, there is a memory limitation because the system has to remember your work in order for the undo command to function properly. To set the minimum number of undos and the maximum memory used for storing undo information, you use the General tab of the Rhino Options dialog box.

When the memory allocated for storing undo information is used up, the undo information will be removed to hold new undo information.

Clearing Undo

To enable the redo command, recently undone work is stored in the memory. One way to free memory space is to clear the undo buffer using the ClearUndo command.

History Manager

History Manager is a modeling aid that helps keep track of the changes made to the source objects from which another object is constructed or derived. If the History Manager is turned on and an object is constructed from another object, the two objects will have a parent-child relationship. When you change the parent, the child will change accordingly. If it is not necessary to keep track of the changes, you may consider turning off the History Manager. If a file is saved with the History Manager turned on, history information will be saved as well. Because the History Manager remembers the relationship between parent and child objects, file size may become very large. Naturally, you may purge history to reduce file size.

To manage history, you can right-click on the Record History button of the status bar, as shown in Figure 2–35. The History Manager has four options: Record, Update, Lock, and Broken History Warning.

- To turn on the History Manager, click on Always Record History.
- To set the children to update automatically as the parent changes, click on Update Children.
- To lock the children so that they will not change even if the parent is modified, click on Lock Children.
- To display the Warning dialog box if history is broken due to an operation, click on History Break Warning.

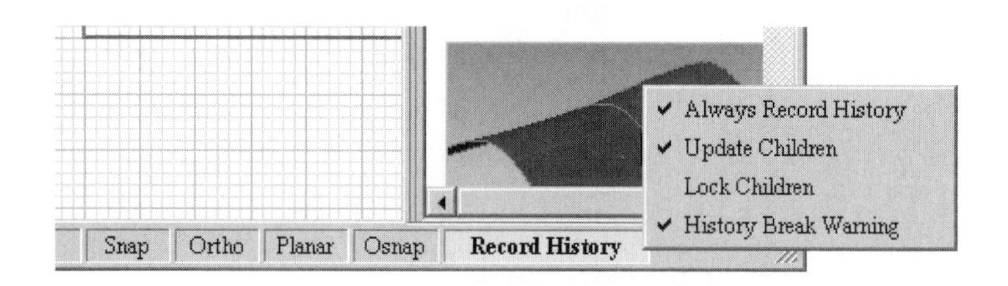

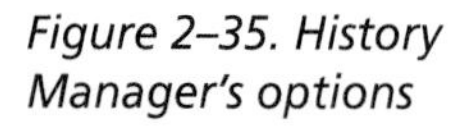
Figure 2–35. History Manager's options

To appreciate how the History Manager works, perform the following steps:

1. Select File > Open and select the file HistoryManager.3dm from the Chapter 2 folder on the companion CD.
2. Click on the History Settings button on the History toolbar, or type History on the command area.
3. If "Record = No," select it to change it to "Yes." Otherwise, proceed to the next step. History data will be recorded.
4. If "Update = No," select it to change it to "Yes." Otherwise, proceed to the next step. Any change made to a source object will cause an automatic update in the repeated objects.
5. Press the ENTER key to exit the command.
6. Select Transform > Copy, or click on the Copy button on the Transform toolbar.
7. Select curve A, surface B, and polygon mesh C, indicated in Figure 2–36, and press the ENTER key.
8. Set object snap mode to Point.
9. Select points D, E, and F and press the ENTER key. Point D is the point to copy from and points E and F are the points to copy to.

The curve, surface, and polygon mesh are copied. To appreciate how the History Manager works, continue with the following steps.

10. Select Transform > Scale > Scale 3-D, or click on the Scale 3-D button on the Transform toolbar.
11. Select curve A, surface B, and polygon mesh C, indicated in Figure 2–36, and press the ENTER key.
12. Select point D, shown in Figure 2–36, to specify the base point.
13. Type 0.5 at the command area to specify the scale factor. Because history update is turned on, the copied objects are also scaled.
14. Click on the History Settings button on the History toolbar, or type History on the command area.
15. Select the Update option on the command area to change it to "No."

16 Select Transform > Scale > Scale 3-D, or click on the Scale 3-D button on the Transform toolbar.

17 Select curve A, surface B, and polygon mesh C, shown in Figure 2–36, and press the ENTER key.

18 Select point D, shown in Figure 2–36, to specify the base point.

19 Type 2 at the command area to specify the scale factor. Because history update is recorded and turned off, the copied objects do not change at all.

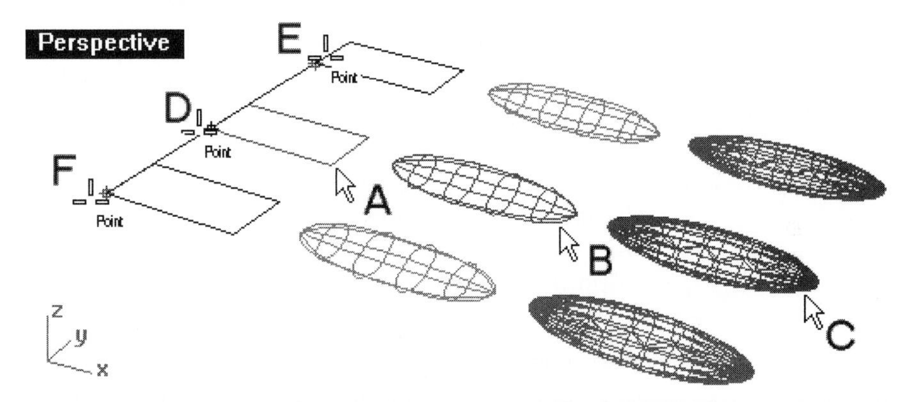

Figure 2–36. Curve, surface, and polygon mesh being copied

20 To discover the update situation, type HistoryReport on the command area.

21 Select object A, shown in Figure 2–37. The History Report is displayed.

22 Do not save your file.

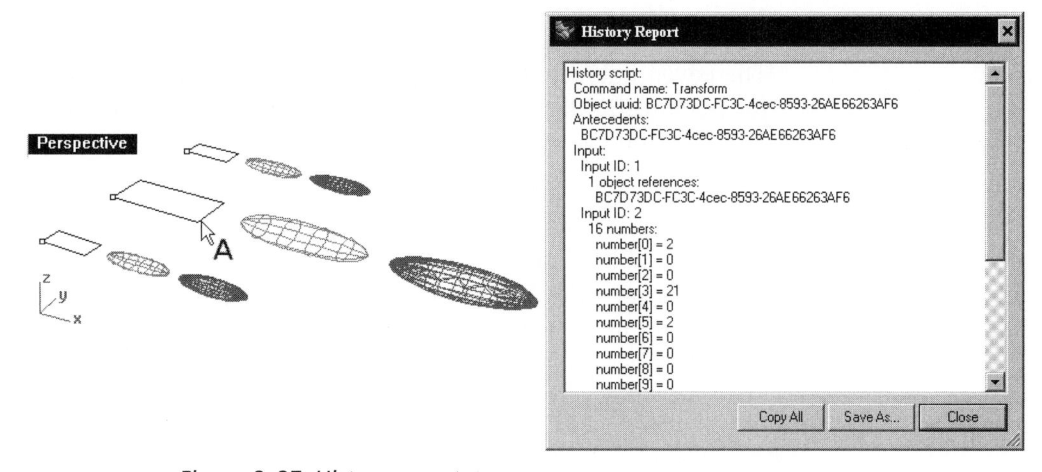

Figure 2–37. History report

Layer Manipulation

The term layer originates from manual drafting. It refers to overlay of clear transparent sheets. Layering is a management mechanism in which sets of objects are drawn on different transparent sheets. By removing or overlaying the sheets, you control which set or sets of objects are shown on a drawing.

Managing Objects in Layers

In computer-aided design, layers are not physical sheets; they are conceptual layers. You construct layers in a file and place objects on different layers. By turning layers on or off, you control the display of the objects on the screen. You can also lock a layer so that objects placed on the layer can be seen and snapped to but cannot be selected or manipulated (such as moving or erasing). In a multilayer Rhino file setup, you can move objects from one layer to another. To try editing a layer, perform the following steps:

1. Select File > Open and select the file Layer.3dm from the Chapter 2 folder on the companion CD.
2. From the main pull-down menu, select Edit > Layers > Edit Layers, or select Edit Layers from the Standard toolbar. Alternatively, right-click on the Layer (pane) option on the status bar.
3. In the Layers dialog box, shown in Figure 2–38, you can add new layers, delete existing layers, set visibility of layers, lock objects on layers, and define the color and material properties of objects placed on a layer as desired.

Layer Dialog Box Column Title

The Layer dialog box has a number of columns. You may right-click over the columns to display a list of column titles. Those with checks are visible. Continue with the following steps to hide some columns.

4. Right-click over the columns of the Layer dialog box. A list of column titles is displayed. By clearing the check mark of a title, the column is hidden.
5. Another way to control column display is to select the columns from the Tools menu of the Layers dialog box.

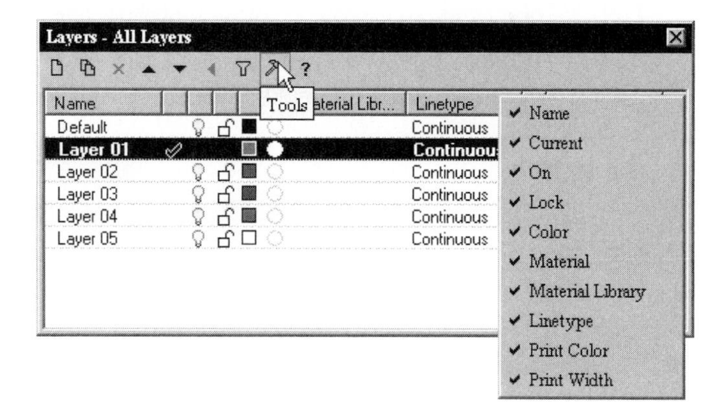

Figure 2–38. Layers dialog box displayed

Setting Color and Material

Continue with the following steps to set the color and material of a layer.

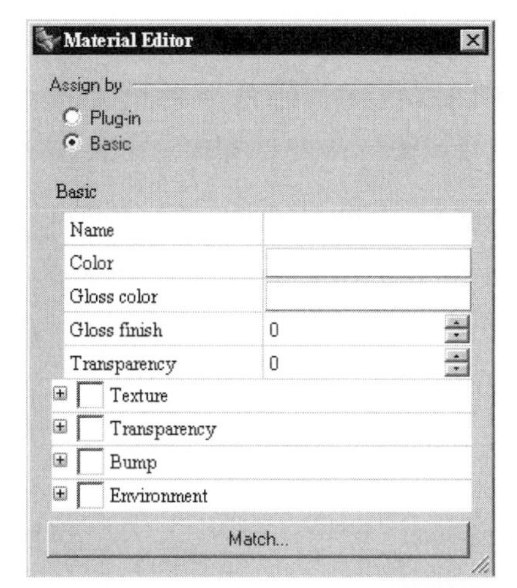

Figure 2–39. Material Properties dialog box

6 If you wish to set the color of objects on a layer, select a layer and then click in the Color column of the Select Color dialog box.

7 In the Select Color dialog box, specify a color by selecting a color from the color swatch or by inputting hue, saturation, and value (RGB, or red, green, and blue) values in these respective fields.

8 Specify the materialof the object (its visual appearance as rendered) by selecting the layer and then the Material column of the Edit Layer dialog box. This accesses the Material Properties dialog box, shown in Figure 2–39. In the Material Properties dialog box, establish the settings as desired. You will learn more about material assignment in Chapter 14.

Changing an Object's Layer

Changing an object's layer refers to moving an object from one layer to another layer. Continue with the following steps.

9 Click on the X button at the upper-right corner of the Layers dialog box to close it. *(Note that you may leave the dialog box on all the time.)*

10 Select Edit > Layers > Change Object Layer, or click on the Change Object Layer/Match Object Layer button on the Layer toolbar.

11 Select sphere A, shown in Figure 2–40, and press the ENTER key.

12 In the Layer for Objects dialog box, click on Layer02 and the OK button. The selected object is moved to Layer02.

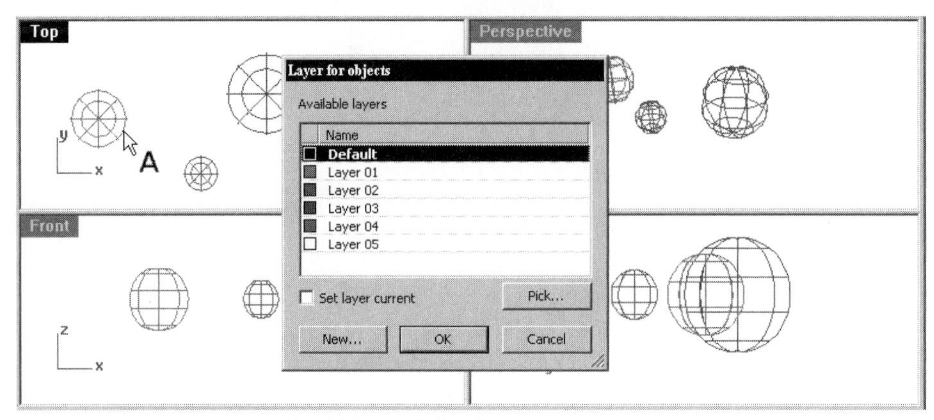

Figure 2–40. Layers for Objects dialog box

Copying Selected Objects from One Layer to Another Layer

You can copy selected objects from one layer to another layer, as follows:

13 Select Edit > Layers > Copy Objects to Layer, or click on the Copy Objects to Layer button on the Layer toolbar.

14 Select sphere A, shown in Figure 2–41, and press the ENTER key.

15 In the Copy Objects to New Layer dialog box, select Layer 01 and click on the OK button. The selected object is copied to a new layer.

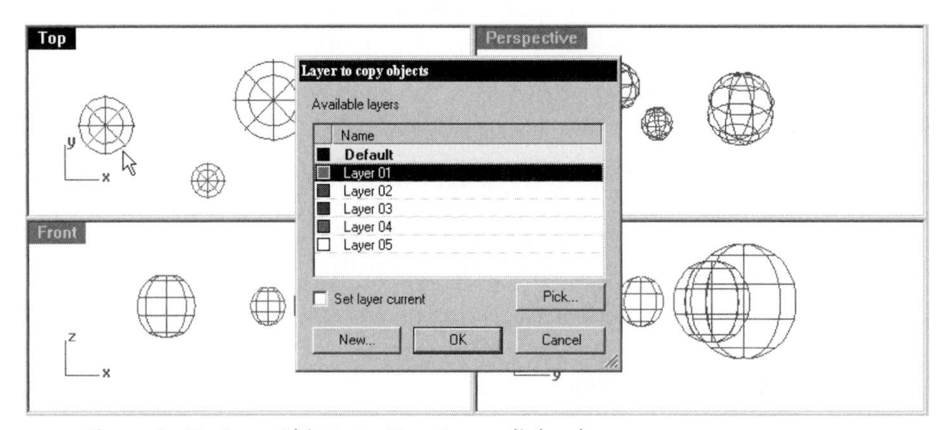

Figure 2–41. Copy Objects to New Layer dialog box

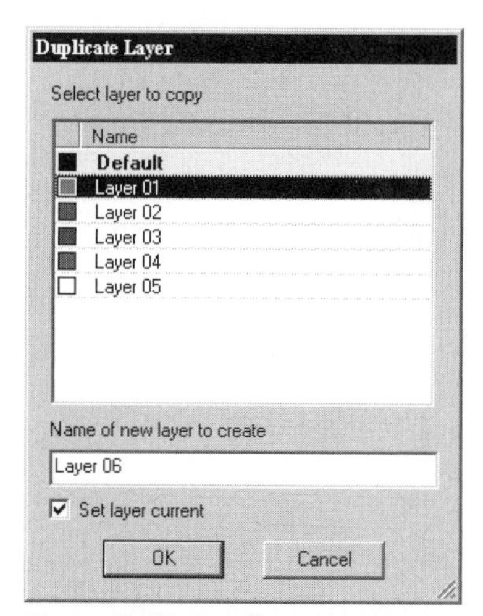

Figure 2–42. Duplicate Layer dialog box

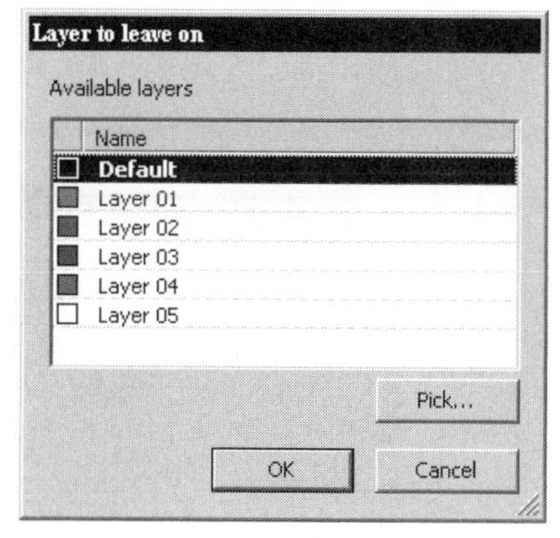

Figure 2–43. Layer to leave on dialog box

Duplicating Objects from One Layer to Another Layer

Instead of copying individual selected objects from one layer to another layer, you can copy all the objects residing in one layer to another layer. Continue with the following steps.

16 Select Edit > Layers > Duplicate Layer, or click on the Duplicate Layer button on the Layer toolbar.

17 In the Duplicate Layer dialog box, select Layer 01, type Layer 06 in the Name of new layer to create box, and click on the OK button.

All the objects in Layer 01 are copied to a new layer, as shown in Figure 2–42.

Ways to Turn On/Off Layers

To turn off all layers except one, continue with the following steps.

18 Select Edit > Layers > One Layer On, or click on the One Layer On button on the Layer toolbar.

19 In the Layer to leave on dialog box, select Layer 06 and click on the OK button.

All layers, except Layer 06, are turned off, as shown in Figure 2–43.

To turn on all layers, continue with the following step.

20 Select Edit > Layers > All Layers On, or click on the All Layers On button on the Layer toolbar.

All the layers are turned on. To turn off a layer on which an object is residing, continue with the following steps.

21 Select Edit > Layers > One Layer Off, or click on the One Layer Off button on the Layer toolbar.

22 Select an object (any object). The object's layer is turned off. Naturally, the selected object becomes invisible.

23 Do not save your file.

Layer Group

You can put one or more layers under another selected layer to form a layer group. Layers that are grouped under another layer are called sublayers. Apart from being manipulated in a normal way individually, a sublayer can also be manipulated collectively by manipulating the layer that governs the sublayer. Perform the following steps:

1. Select File > Open and select the file LayerGroup.3dm from the Chapter 2 folder on the companion CD.
2. Select Edit > Layers > Edit Layers, or click on the Edit Layers button on the Layers toolbar.
3. Hold down the Control key, select Layer 02 and Layer 03 from the Layer dialog box, and drag the layers to Layer 01.
4. Release the mouse button. The selected layers (Layer 02 and Layer 03) are grouped under Layer 01. They now become the sublayers of Layer 01. (See Figure 2–44.)

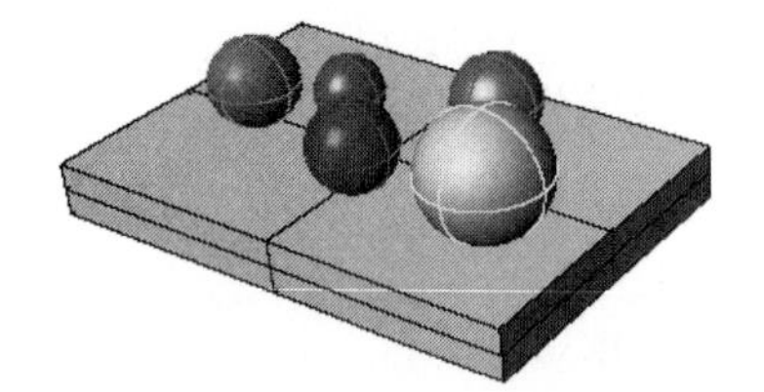

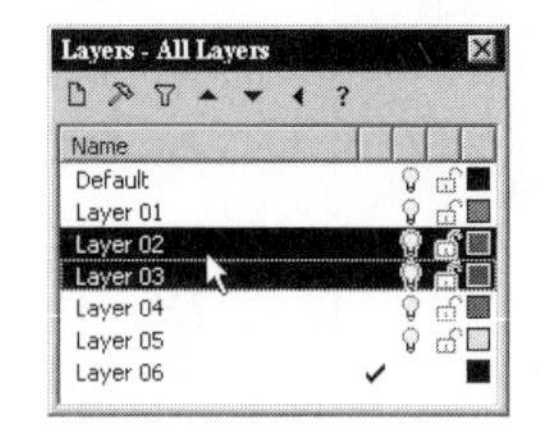

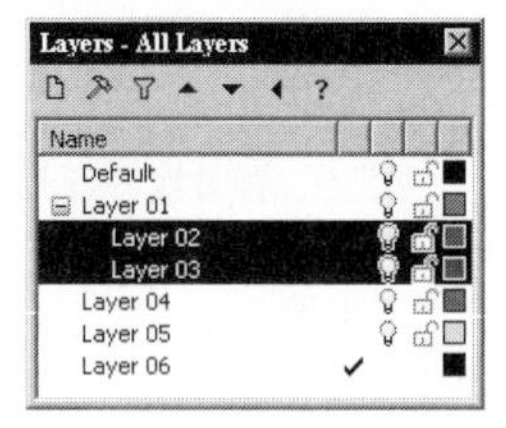

Figure 2–44. From left to right: models, selecting layers in the Layers dialog box, and layers grouped

5. To appreciate how sublayers can be manipulated collectively, turn off Layer 01. Layer 02 and Layer 03, being grouped under Layer 01, are also turned off.
6. Turn on Layer 01. Layer 02 and Layer 03 are also turned on. (See Figure 2–45.)

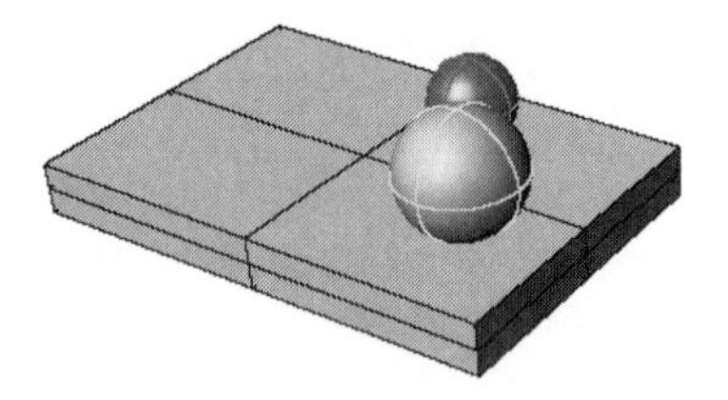

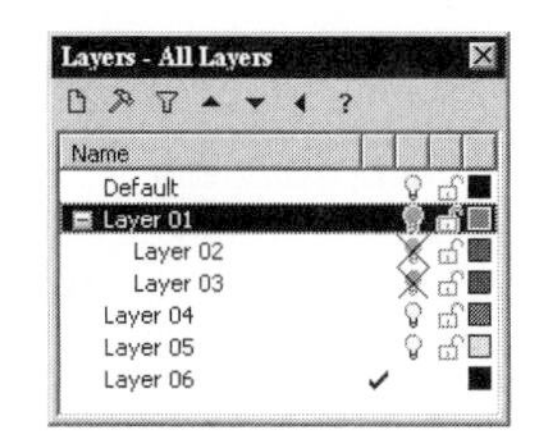

Figure 2–45. Layer group turned off

Sublayers can be ungrouped, as follows:

7. Drag Layer 02 from its location to a location below Layer 06. Layer 02 is ungrouped. (See Figure 2–46.)

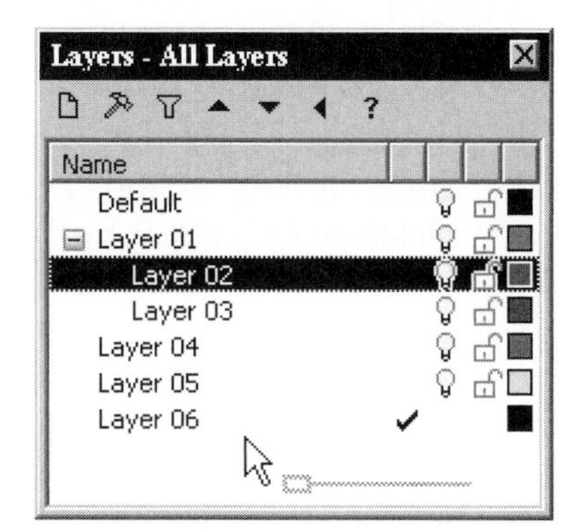

Figure 2–46. Layer being ungrouped

8. Select Layer 02, right-click, and select New Sublayer. A new sublayer is constructed. (See Figure 2–47.)
9. Do not save your file.

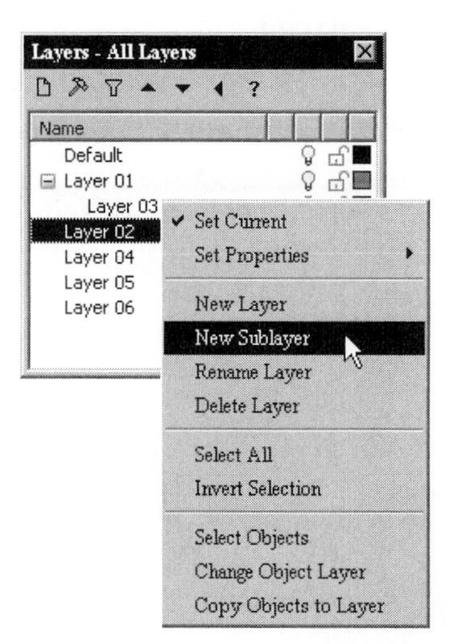

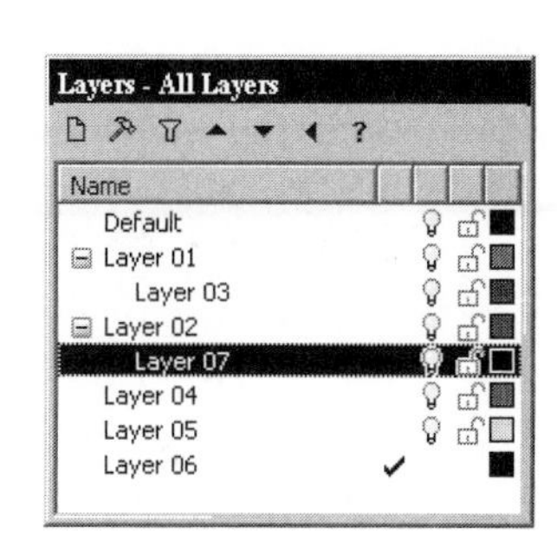

Figure 2–47. New sublayer being constructed (left) and sublayer constructed (right)

Visibility and Locking

Other than turning off a layer to make objects residing on the layer invisible, you can hide selected objects so that they are not displayed, even if the layer on which they reside is still turned on. You can unhide a hidden object to make it visible again. You can also lock an object, whereby it is visible but cannot be modified. Perform the following steps:

1. Select File > Open and select the file Visibility.3dm from the Chapter 2 folder on the companion CD.

2 Select Edit > Visibility > Hide, or click on the Hide Objects/ Show Objects button on the Visibility toolbar.
3 Select objects A, B, C, D, and E, indicated in Figure 2–48, and press the ENTER key. The selected objects are hidden.
4 Select Edit > Visibility > Show Selected, or click on the Show Selected Objects button on the Visibility toolbar. In the display, the visible objects are hidden temporarily and the hidden objects are shown.
5 Select hidden objects A and B, and press the ENTER key. The selected hidden objects are unhidden.
6 Select Edit > Visibility > Swap Hidden and Visible, or click on the Swap Hidden and Visible Objects button on the Visibility toolbar. The visible objects are hidden, and the hidden objects are unhidden.
7 Select Edit > Visibility > Show, or right-click the Hide Objects/ Show Objects button on the Visibility toolbar. All the hidden objects are displayed.
8 Select Edit > Visibility > Lock, or click on the Lock Objects/ Unlock Objects button on the Visibility toolbar.
9 Select objects A, B, and C, shown in Figure 2–48, and press the ENTER key. The objects are locked. A locked object is still displayed but cannot be manipulated.
10 Do not save your file.

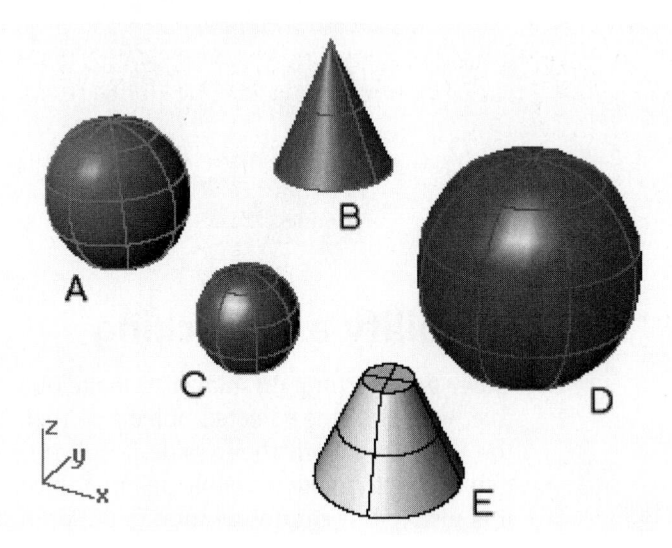

Figure 2–48. Visibility

Layer State Manager

The manner in which one layer is turned on/off, made invisible/invisible, and locked/unlocked can be managed by using the Layer State Manager, as follows:

1. Select File > Open and select the file Visibility.3dm from the Chapter 2 folder on the companion CD.
2. Select Edit > Layers > Edit Layers, or click on the Edit Layers button on the Layer toolbar.
3. Select Edit > Layers > Layer State Manager, or click on the Layer State Manager button on the Layer toolbar.
4. Click on the Save button on the Layer State Manager dialog box.
5. In the Save Layer State dialog box, click on the OK button. The current layer state is saved. *(Note: You may specify another layer state name.)*
6. Click on the Close button of the Layer State Manager to close it.
7. Turn off Layer02, Layer03, and Layer04.
8. Repeat steps 3 through 5 to save the current layer state.
9. Select Layer state 01 in the Saved Layer States list.
10. Click on the Restore button on the Layer State Manager dialog box. The saved layer state is restored. (See Figure 2–49.)
11. Do not save your file.

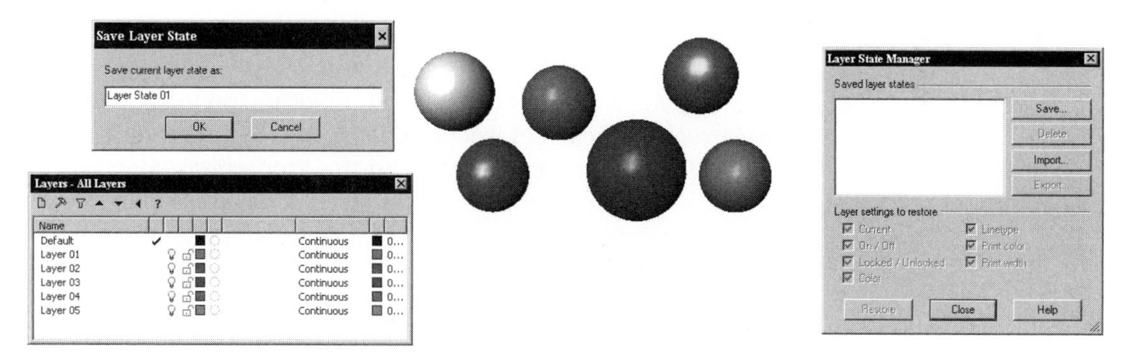

Figure 2–49. Layer State Manager

Object Display Control

This section introduces ways to control an object's display in the computer's display unit.

Isocurve Display for NURBS Surfaces

In reality, there are no curves or lines on a 3D free-form smooth surface. Therefore, you should find only edges on the surface's boundaries. However, boundary edges alone do not provide sufficient information to depict the profile and silhouette of the surface. Hence, a set of isocurves in two orthogonal directions, color shading, or a set of isocurves together with color shading is used to better illustrate a free-form object in the computer display. To avoid confusion with the X- and Y-axes of the coordinate system, the isocurve directions are called U and V. Figure 2–50 shows how isocurves are used to help visualize a surface.

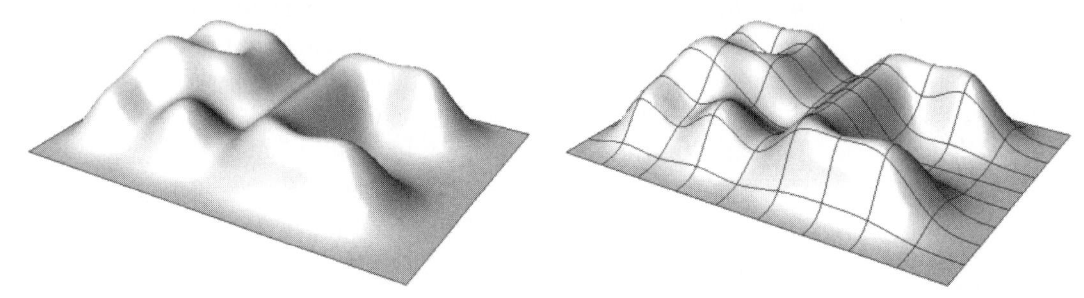

Figure 2–50. A free-form smooth surface (left) and the same surface with U and V isocurves (right)

To select a surface, you select one of the isocurves or the boundary of the surface. Although the isocurves are not physical curves, you may still use object snap tools, such as intersection, to locate the intersections of U and V isocurves. You can set isocurve density by selecting Edit > Properties and changing the density value in the Properties dialog box. Isocurves are also known as isoparametric curves.

Isocurve Density

For a very simple surface, such as a planar surface, one or two isocurves are adequate to provide enough information on the curvature of the surface. For more complex surfaces, you need more isocurves. Although there are more isocurves to better represent the profile of the surface, selection of individual objects from a bunch of objects with high isocurve density may become difficult. Perform the following steps:

1. Select File > Open and select the file Isocurves.3dm from the Chapter 2 folder on the companion CD.
2. Select Edit > Object Properties, or click on the Object Properties button on the Properties toolbar.

3 Select surface A, shown in Figure 2–51, and press the ENTER key.

4 In the Properties dialog box, change the isocurve density to 3. The isocurve density of the surface is changed, as shown in Figure 2–51 (right).

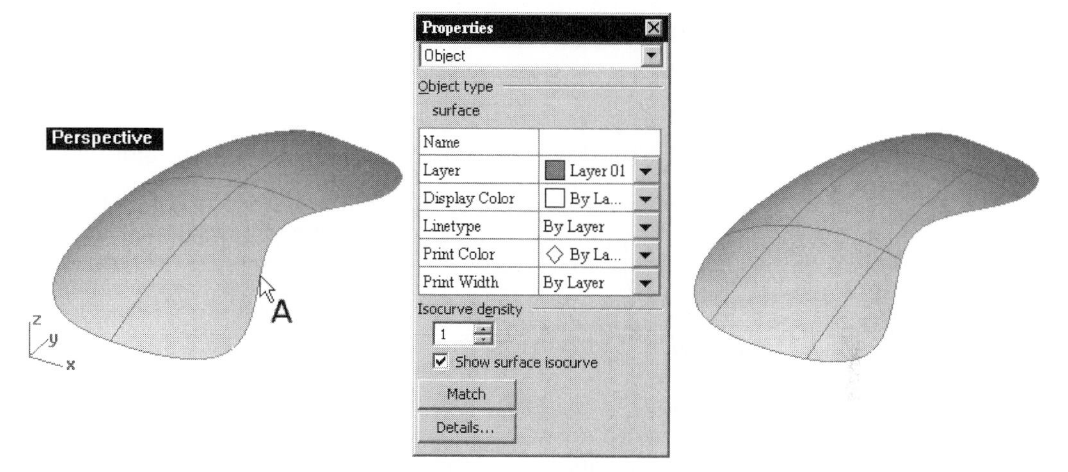

Figure 2–51. From left to right: Original surface, Properties dialog box, and isocurve density changed

Geometric Object's Color

By default, objects are displayed in the viewport in a color assigned to the layer on which the objects reside. In wireframe mode, they are displayed with isocurves. To change the color of an object, you change the layer's color or the object's color.

The default color of a Rhino object is determined by the setting Bylayer. This means that the color of the object is determined by the color assigned to the layer in which the object resides. You can change an object's color by either changing the layer's color or changing the object's color property.

- To change the color assignment of a layer, you use the Layer dialog box by selecting Edit > Layers > Edit Layers.
- To change an object's color property, you use the Properties dialog box by selecting Edit > Object Properties or by clicking on the Object Properties button on the Properties toolbar and then selecting the object.

Color Setting Methods

Color setting can be done in three ways in the Select Color dialog box, shown in Figure 2–52.

- The first way is to select a color from the pre-established color list.
- The second way is to use the HSB (hue, saturation, and brightness) color system.
- The third way is to use the RGB (red, green, blue) color system.

In a collaborative design environment in which a project is handled by a team, specifying color values (HSB or RGB) can accurately describe the color.

HSB Color System

In the Select Color dialog box, the HSB settings are labeled Hue, Sat, and Val. The hue represents a color ranging from red through yellow, green, and blue. Saturation describes the intensity of the hue. Brightness concerns the color's value or luminance. To select a color by using the HSB color system, you first select a color from the hue color range; you then set the saturation (intensity) and the brightness (luminance).

RGB Color System

Specifying the RGB value instructs the computer to project a color mix of red, green, and blue to each individual pixel (picture element) of the monitor. Continue with the following steps to specify color via this method.

5. In the Properties dialog box, select Other from the Color pull-down box. (If you already closed the Properties dialog box, open it again by selecting Edit > Object Properties, or by clicking on the Object Properties button on the Properties toolbar and then selecting the surface.)
6. In the Select Color dialog box, shown in Figure 2–52, select a color in the circular ring in the color swatch to set the hue. You may regard the hue as the basic color. For example, you click on the green zone to select a green color.
7. In the square box, select a location to set the saturation and value. If you select a location near the left-hand edge of the box, you obtain a fully saturated color. If you select a location near the upper edge of the box, you obtain a higher brightness value.

8 Click on the OK button. The object's wireframe color is set.
9 Close the Properties dialog box.
10 Do not save the file.

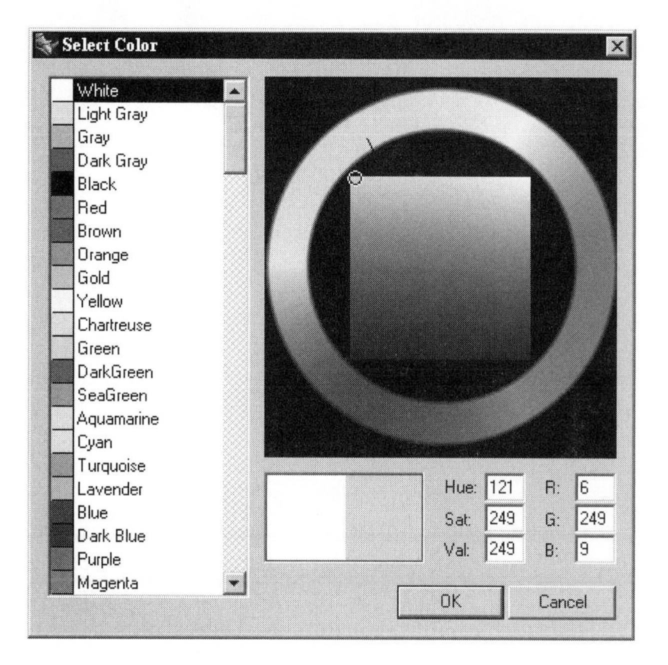

Figure 2–52. Select Color dialog box

Shaded Display Modes

Viewport display can be set to wireframe, shaded, rendered, ghosted, and x-ray. Perform the following steps to explore other viewport display methods:

1 Select File > Open and select the file Shade.3dm from the Chapter 2 folder on the companion CD.
2 Click on the Shaded viewport/Wireframe viewport button on the Shade toolbar. The display is shaded.
3 Click on the X-ray viewport button on the Shade toolbar.
4 Click on the Ghosted viewport button on the Shade toolbar.
5 Click on the Toggle Flat Shade Mode button on the Shade toolbar.

Various types of shaded mode are shown in Figure 2–53.

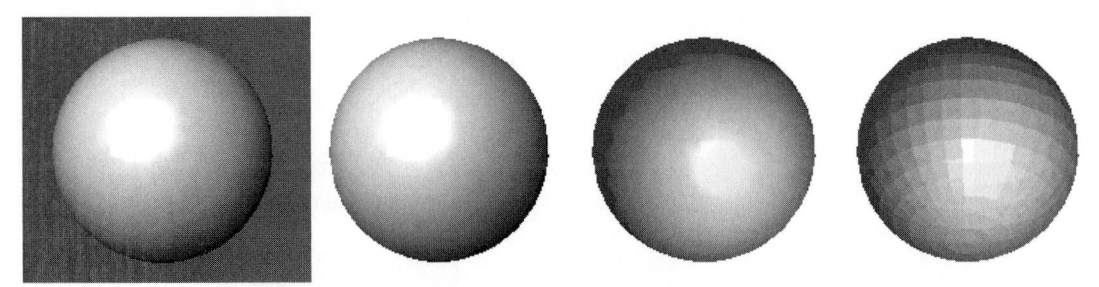

Figure 2–53. From left to right: Shade, X-ray viewport, Ghosted viewport, and flat shade

View Capture

A Rhino active viewport can be captured as a bitmap file to the Windows clipboard for subsequent pasting to other applications. Continue with the following steps if you want to capture the viewport.

6 Select View > Capture > To Clipboard, or right-click on the Capture Viewport to File/Capture Viewport to Clipboard button on the Shade toolbar. The current viewport is captured and the captured image is placed in the Window clipboard. You can now paste it into another application hat supports copy and paste.
7 Select View > Capture > To File, or click on the Capture Viewport to File/Capture Viewport to Clipboard button on the Shade toolbar.
8 In the Save Bitmap dialog box, specify a file name, select a file type, and click on the OK button.
9 Do not save your file.

Chapter Summary

When you use the mouse to select a point in one of the default viewports in the graphics area, you select a point on a construction plane corresponding to the selected viewport. In addition to these three default construction planes, you can set up construction planes in various ways described in Appendix B.

To input a point at the command area, you use either the construction plane coordinate system or the World coordinate system. To help construct objects in the graphics area, you can use various drawing aids. In particular, the Smart Tracking tool is very useful in terms of tracking geometric objects.

To manipulate objects that are already constructed, you have to select them by using various selection tools. Basic object manipulation methods include moving, copying, rotating, scaling, and mirroring. (More advanced manipulation tools will be discussed in Chapter 10.) To help track changes made to objects, you can turn on the History Manager.

As a surface model becomes more complicated, there will be many geometric objects. To organize these objects, you put them into layers. Objects you construct are placed on the current layer.

A smooth surface does not have any thickness or profile curves on it. In essence, you see only its boundaries and its silhouette. To enhance visualization, isocurves are placed on the surface. To contrast an object from its surrounding background, you shade the viewport. Shading is a means of enhancing the visual representation of a 3D object by applying a color shade to the surface of the object.

Review Questions

1. Explain the concepts of the construction plane.
2. Describe the ways to input precise coordinate systems.
3. List the drawing aids available.
4. Give a brief account of the History Manager.
5. Outline how objects can be organized by using layers.
6. What is meant by isocurves?
7. In what ways can a surface be displayed?

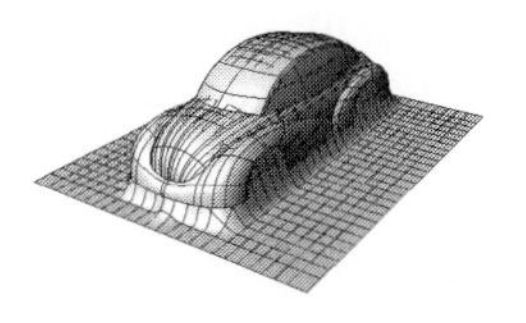

CHAPTER 3

Rhinoceros NURBS Surfaces

Introduction

For easy reference, we will divide Rhino's surfaces into four categories. This chapter delineates how these surfaces are constructed.

Objectives

After studying this chapter you should be able to:

- ❐ Describe four major categories of Rhino surfaces
- ❐ Apply Rhino's tools to construct these surfaces
- ❐ Overview

Surface modeling involves a top-down thinking process to consider how an object can be decomposed into a number of surfaces and how these surfaces are constructed individually. After completing this analysis, you proceed to the bottom-up process of making the individual surfaces and eventually composing a model from the set of surfaces produced. To be able to start the top-down thinking process and subsequently perform the bottom-up process of building the surface model, you need to know what Rhino surfaces are available and how they can be built. To help you gain a thorough understanding in a logical way, we will divide Rhino's surface modeling commands into four main categories and discuss each one. After you gain more knowledge on curves and points construction in Chapters 4 through 6, you will further enhance your NURBS surface modeling skill later in Chapters 7 and 8.

Rhino's NURBS Surface

In terms of free-form surface design and manufacture, NURBS surfaces are more significant because surfaces can be represented in a more accurate way. Therefore, we will first focus on NURBS surfaces in the early chapters of this book and discover more about polygon meshes in Chapter 9. For the sake of easy reference, we will divide Rhino's surfaces into four categories:

- The first category concerns the four general kinds of surfaces that you may find in most computer-aided design applications.
- The second category covers surfaces that are more specific to Rhinoceros.
- The third category is the planar surfaces.
- The fourth category covers surfaces derived from existing surfaces.

Concerns

There are three concerns while constructing surfaces for making a model:

1. The use of curves and surface edges as framework,
2. Continuity between contiguous surface, and
3. Application of the History Manager.

Curves and Surface Edges

In building surfaces requiring a framework of curves, you can use existing surface edges as well as curves.

Continuity Between Contiguous Surfaces

If surface edges are used in the construction of surfaces, continuity between the new surface to be built and the existing surface is sometimes taken into account. As mentioned in Chapter 1, whenever two or more contiguous curves/surfaces are concerned, continuity at the joint has to be considered. Rhino provides five types of continuity: G0 (positional), G1 (tangency), G2 (curvature), G3 (smooth change of curvature), and G4 (constant rate of change of curvature). Among these types, G4 continuity provides such a smooth joint between contiguous curves/surfaces that it is hardly visible.

History Manager

Quit a lot of surface construction commands explained in this chapter can be governed by the History Manager. As explained previously, the History Manager remembers changes made to the source objects and causes corresponding changes to the objects derived from the source. Therefore, if a surface construction command is governed by the History Manager and the History Manager is turned on, changes made to the curves that are used to construct the surface during the working session will cause the surface to modify automatically. However, it must be emphasized that the History Manager only takes effect in the current session and before the surface is further modified by other means.

Common Free-Form Surfaces

The first surface category covers surfaces that are common to most computer-aided design applications. There are four major kinds of free-form surfaces: extruded, revolved, swept, and lofted. In common, these surfaces are all constructed from a framework defining their cross-section profiles.

Extruded Surfaces

Basically, an extruded surface has a cross-section of uniform shape. The surface profile is constructed by translating a curve in a straight line or along a curve. Using Rhino's surface modeling tool, you can extrude a curve in six ways: Straight, Tapered, To Point, Ribbon, Along Curve, and Fin.

Extrude Straight

This option extrudes a curve along a straight line direction. If the curve is a planar curve, the direction of extrusion is perpendicular to the plane of the curve. If the curve is 3D, the direction of extrusion is perpendicular to the active construction plane. Perform the following steps:

1. Select File > Open and select the file Extrude01.3dm from the Chapter 3 folder on the companion CD.
2. Select Surface > Extrude Curve > Straight, or click on the Extrude Straight button on the Extrude toolbar.
3. Select curve A, indicated in Figure 3–1, and press the ENTER key. *(Note: Curve A is a planar curve.)*
4. Click on location B, shown in Figure 3–1. The planar curve is extruded in a straight line perpendicular to the plane of the curve.

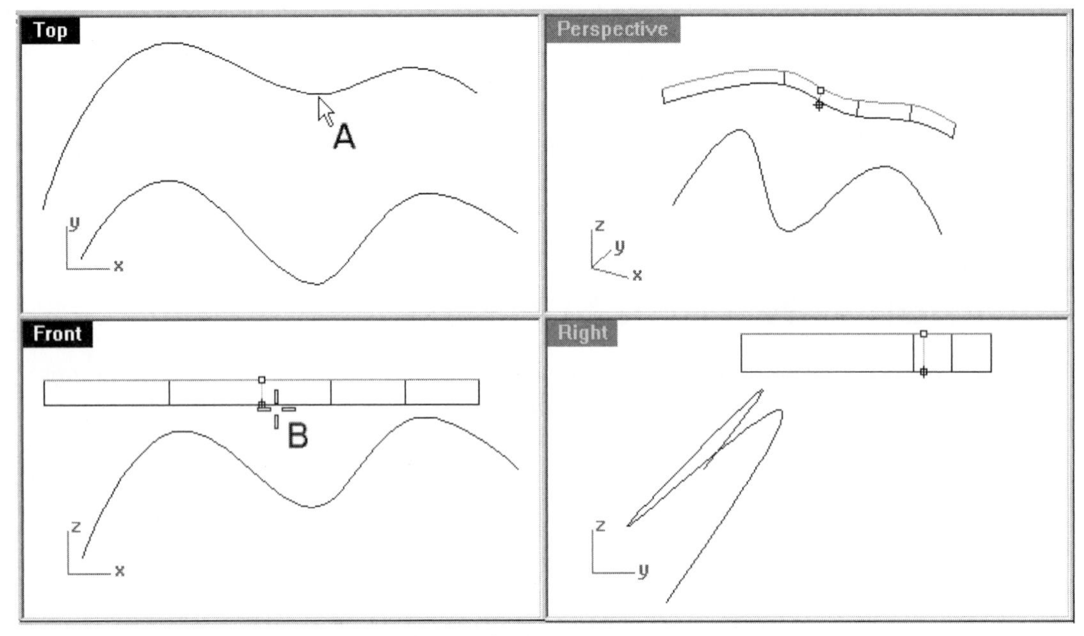

Figure 3–1. Extruding a planar curve

Continue with the following steps to extrude a 3D curve in a direction perpendicular to the active construction plane.

5. Select Surface > Extrude > Straight, or click on the Extrude Straight button on the Extrude toolbar.
6. Select curve A, indicated in Figure 3–2, and press the ENTER key. *(Note: This is a 3D curve.)*
7. Click on location B, shown in Figure 3–2. The 3D curve is extruded in a direction perpendicular to the construction plane from which the curve is selected. In other words, if you select the curve in a different viewport, the outcome will be different.
8. Do not save your file.

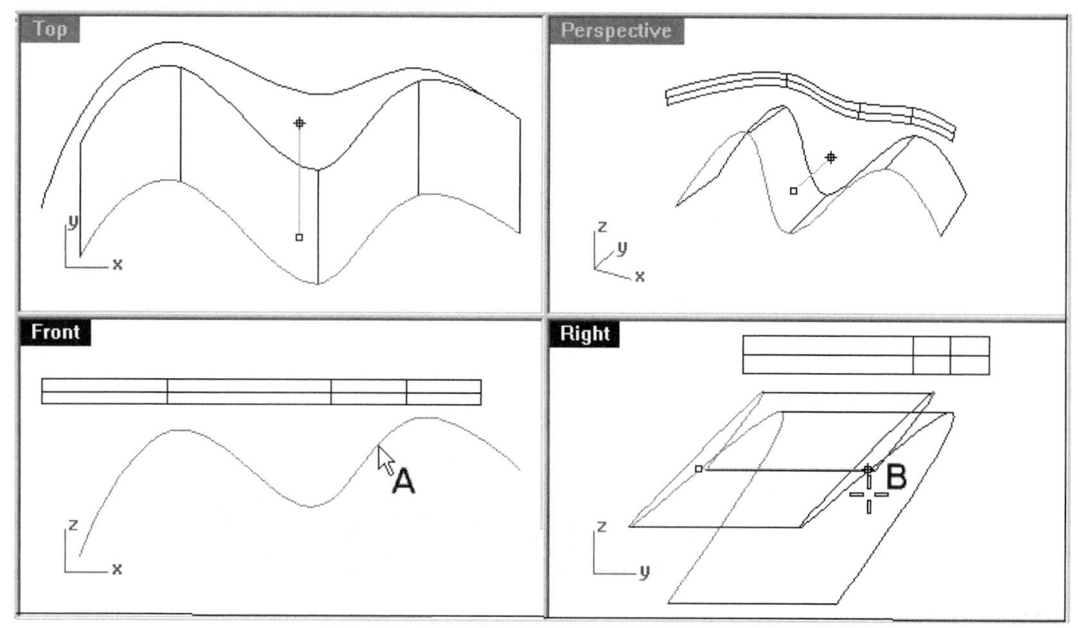

Figure 3–2. Extruding a 3D curve in a direction perpendicular to the active construction plane

Straight Extrusion with a Taper Angle

To uniformly scale up or down the cross-section profile while extruding the defining curve, you apply a taper angle. Perform the following steps:

1. Select File > Open and select the file Extrude02.3dm from the Chapter 3 folder on the companion CD.
2. Select Surface > Extrude Curve > Tapered, or click on the Extrude Tapered button on the Extrude toolbar.
3. Select curve A, shown in Figure 3–3 (Top viewport), and press the ENTER key.
4. Select the DraftAngle option on the command area.
5. Type 10 to set the draft angle (taper angle) to 10 degrees.
6. Select location B, shown in Figure 3–3. A tapered extruded surface is constructed.
7. Do not save the file.

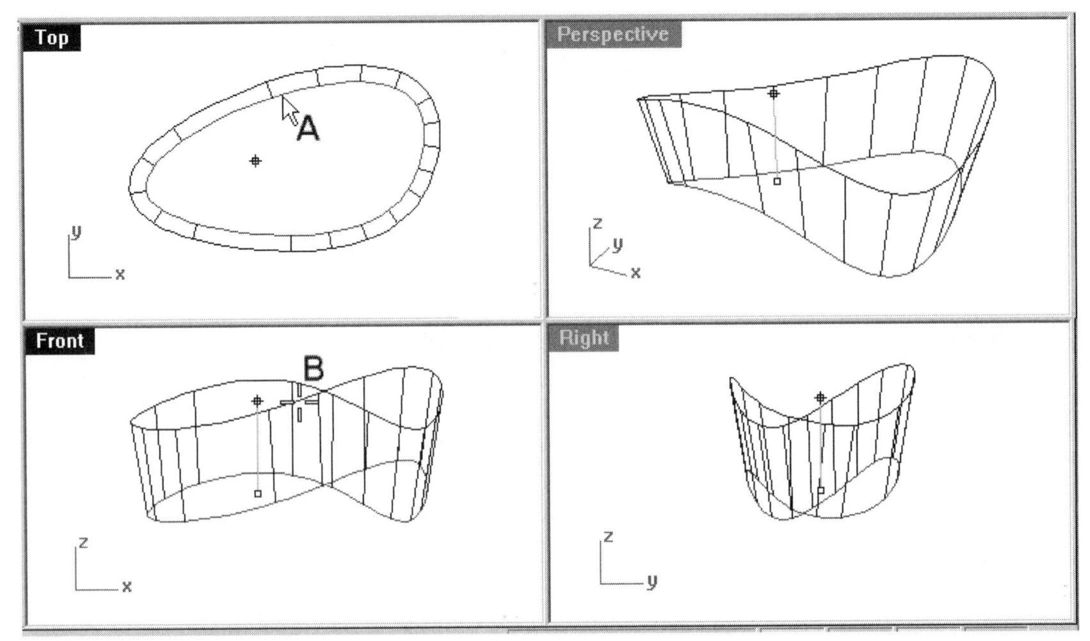

Figure 3–3. Tapered extruded surface being constructed

Extruding to a Point

An extreme case of extruding a curve with a tapered angle is to extrude it to a point. In essence, the surface profile uniformly diminishes to a single point. Perform the following steps:

1. Select File > Open and select the file Extrude03.3dm from the Chapter 3 folder on the companion CD.
2. Select Surface > Extrude Curve > To Point, or click on the Extrude to Point button on the Extrude toolbar.
3. Select curve A, shown in Figure 3–4, and press the ENTER key.
4. Select location B, shown in Figure 3–4. (Exact location is unimportant for the purpose of this tutorial.) A curve is extruded to a point.
5. Do not save the file.

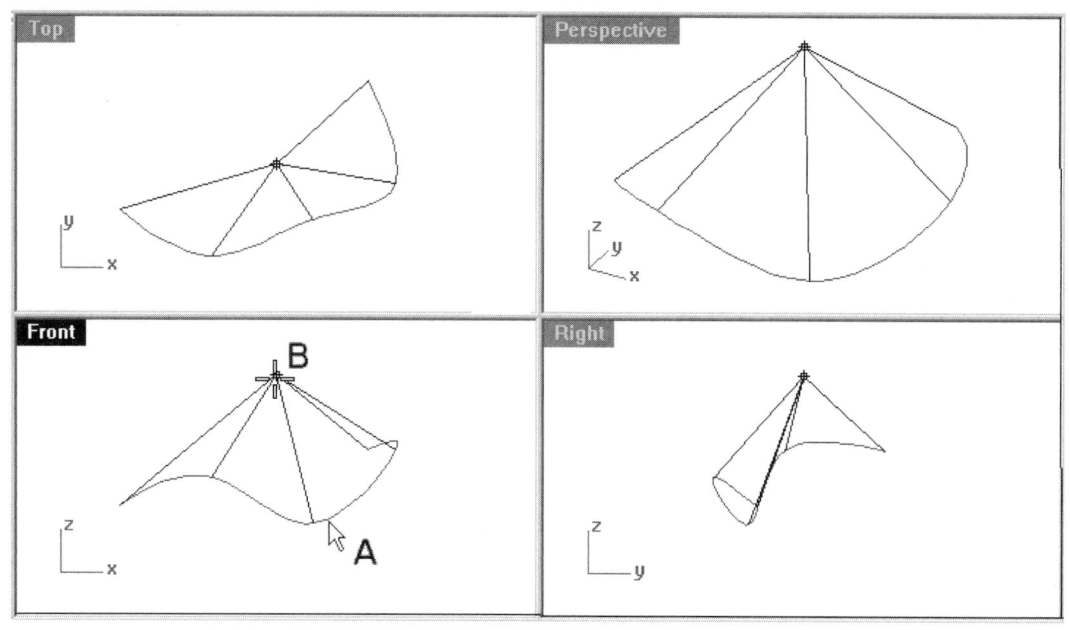

Figure 3–4. Curve being extruded to a point

Extruding Along a Curve or a Sub-Curve

Instead of extruding the curve in a straight line, you use a curve or a portion of a curve (sub-curve) as the path of extrusion. While extruding, the cross-section remains constant and parallel Perform the following steps:

1. Select File > Open and select the file Extrude04.3dm from the Chapter 3 folder on the companion CD.
2. Select Surface > Extrude Curve > Along Curve, or click on the Extrude Along Curve/Extrude Along Sub Curve button on the Extrude toolbar.
3. Select curve A, shown in Figure 3–5, and press the ENTER key.
4. Click on curve B, shown in Figure 3–5. A curve is extruded along a curve.

(***NOTE:*** *The direction of extrusion depends on which end of the path curve you select.)*

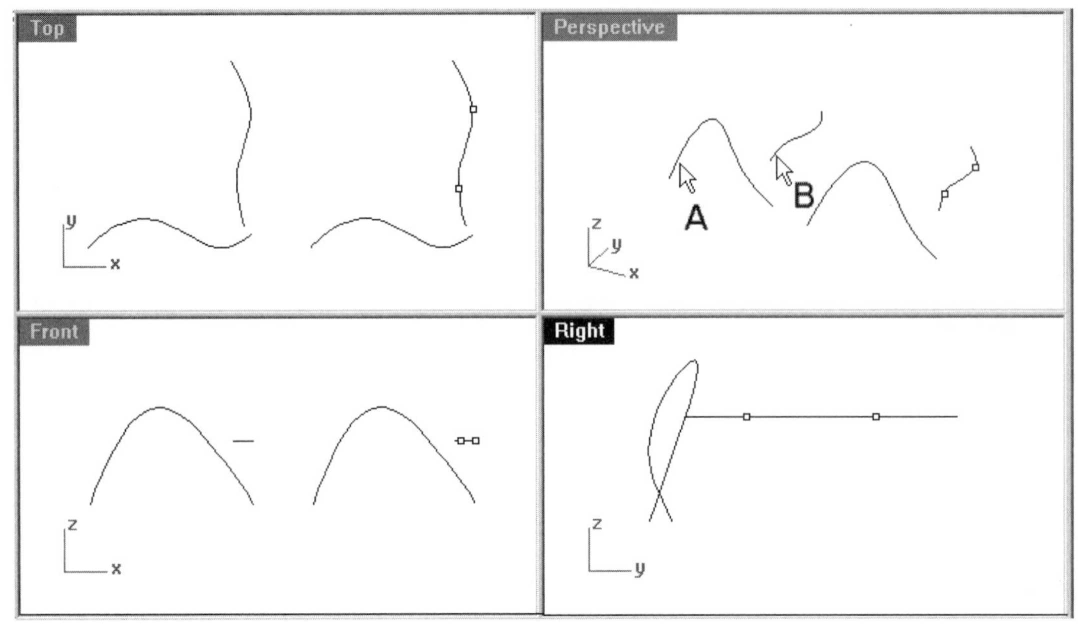

Figure 3–5. Curve being extruded along a path curve

5. Set object snap mode to Point.
6. Select Surface > Extrude Curve > Along Curve, or right-click on the Extrude Along Curve/Extrude Along Sub Curve button on the Extrude toolbar.
7. Select curve A (Figure 3–6) and press the ENTER key.

Skip steps 8 and 9 if you run this command from the toolbar.

8. Select the Mode option on the command area.
9. Select the AlongSubCurve option on the command area.
10. Select curve B, shown in Figure 3–6.
11. Select location C along curve B, shown in Figure 3–6.
12. Select location D along curve B (Figure 3–6). A curve is extruded along a path defined by two points along a path curve. (See Figure 3–7.)
13. Do not save your file.

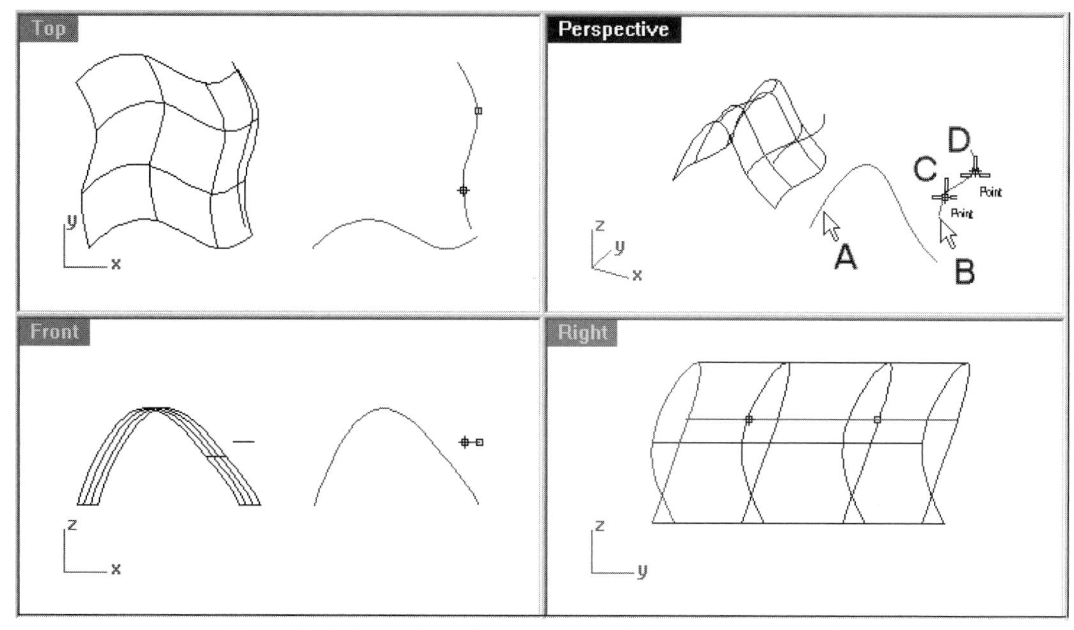

Figure 3–6. A portion of a path curve being used to extrude a curve

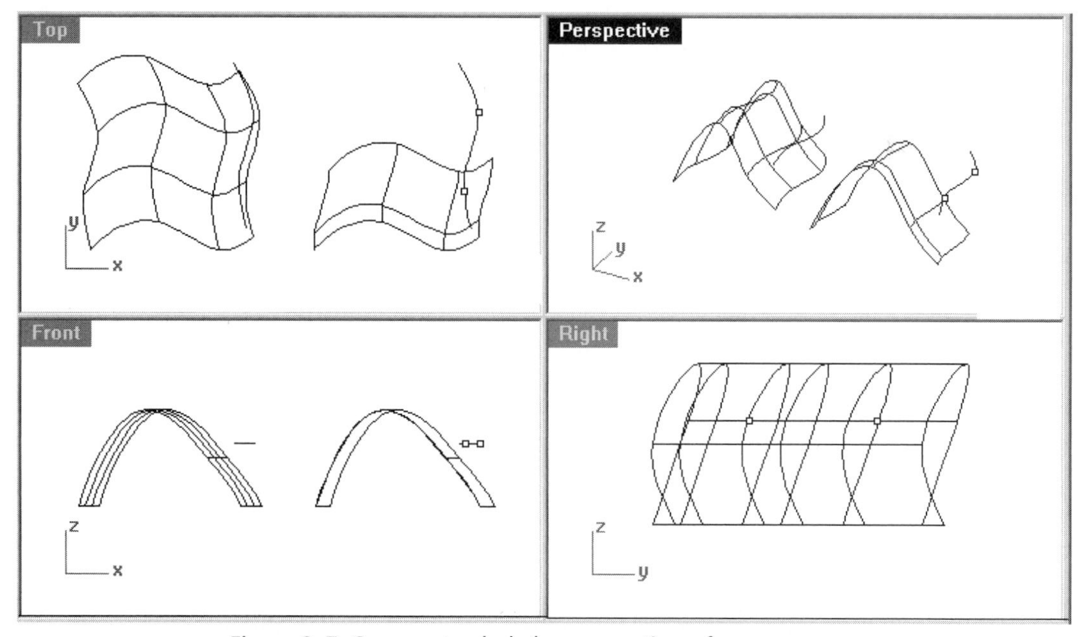

Figure 3–7. Curve extruded along a portion of a curve

Extruding a Ribbon

This is a special kind of extruded surface. It is called a ribbon because the resulting surface resembles a ribbon. In essence, the operation includes offsetting the original curve and then filling the gap between the original curve and the offset curve with a surface. Perform the following steps:

1. Select File > Open and select the file Extrude05.3dm from the Chapter 3 folder on the companion CD.
2. Select Surface > Extrude Curve > Ribbon, or click on the Ribbon button on the Extrude toolbar.
3. Select curve A, shown in Figure 3–8.
4. Select the Distance option on the command area.
5. Type 4 to assign the ribbon's width.
6. Click on location B (Figure 3–8 in the Top viewport). A ribbon surface is constructed.

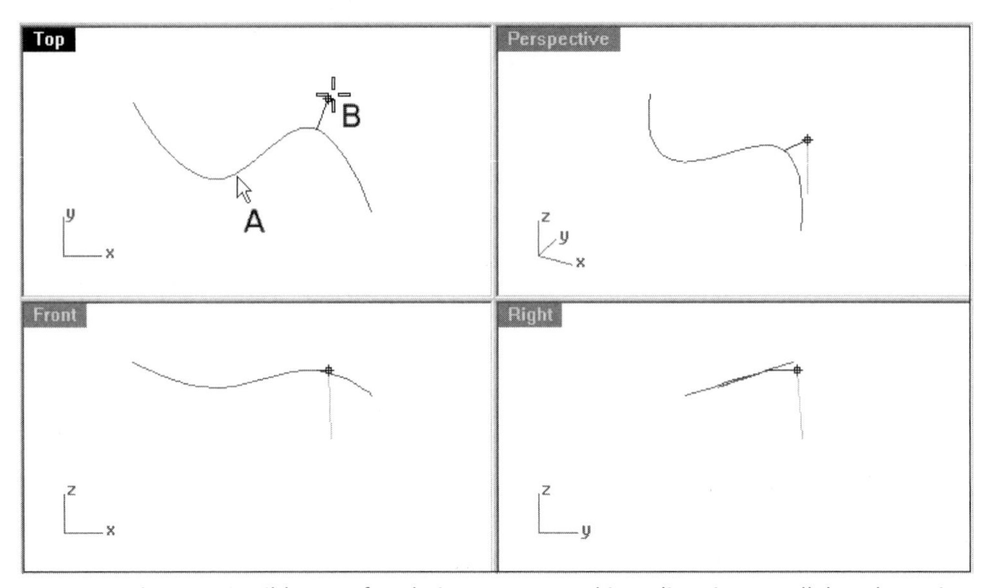

Figure 3–8. Ribbon surface being constructed in a direction parallel to the active construction plane

7. Select Surface > Extrude > Ribbon, or click on the Ribbon button on the Extrude toolbar.
8. Select curve A, shown in Figure 3–9, in the Front viewport.
9. Because there are two edges of the ribbon surface you just constructed and a curve is located in the same position, you need to select one of them from the pop-up menu that opens. Select Curve from the pop-up menu.

10 Select location C, shown in Figure 3–9, in the Front viewport. The second ribbon surface is constructed. (See Figure 3–10.)

11 Do not save your file.

As you can see, the outcome depends on the active construction plane.

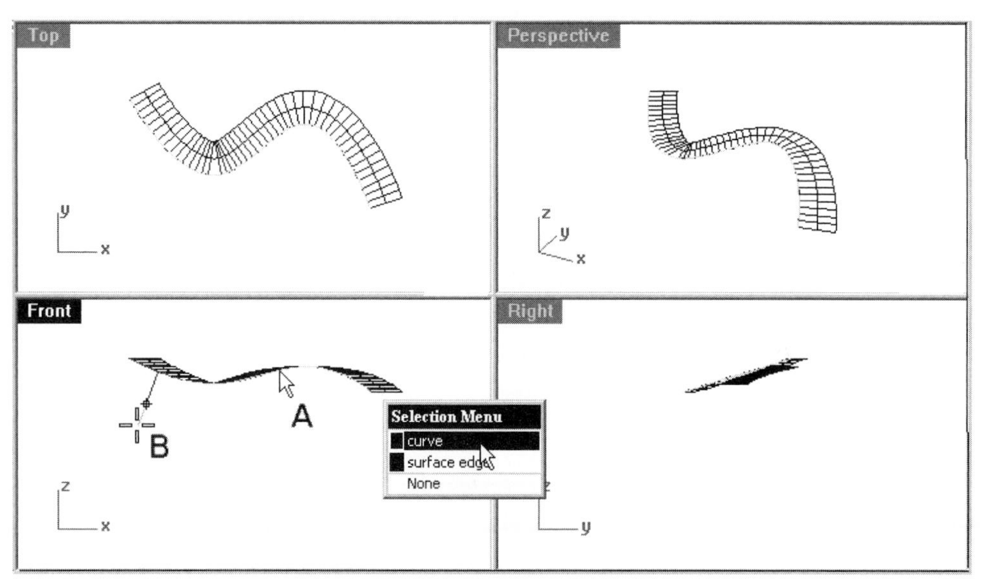

Figure 3–9. Second ribbon surface being constructed

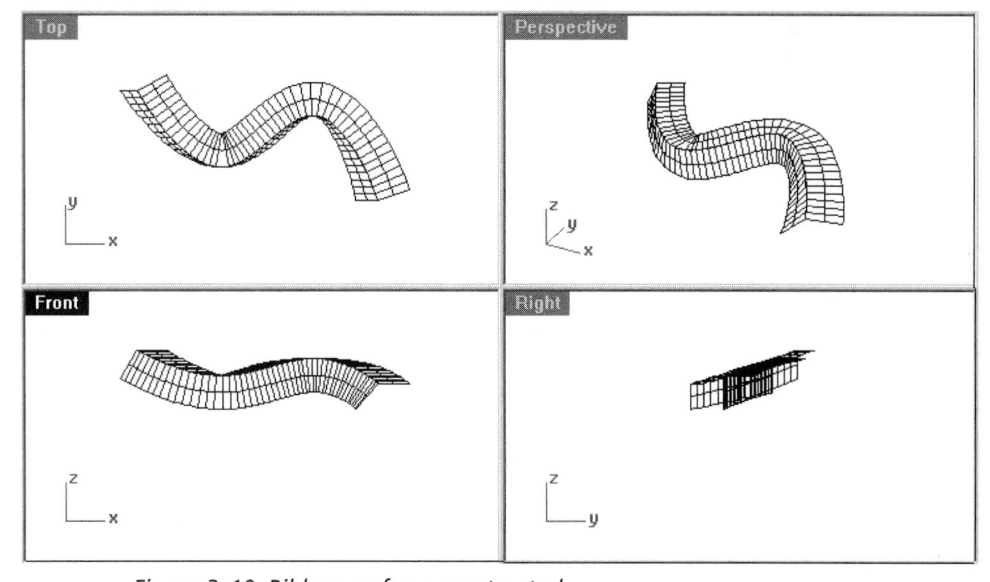

Figure 3–10. Ribbon surfaces constructed

Extruding Normal to a Surface (Making a Fin)

The Fin command produces a surface that resembles the shape of a fish fin. It extrudes a curve in a direction perpendicular to a selected surface. To construct a fin, perform the following steps:

1. Select File > Open and select the file Extrude06.3dm from the Chapter 3 folder on the companion CD.
2. Select Surface > Extrude Curve > Normal to Surface, or click on the Extrude Curve normal to surface button on the Surface Tools toolbar.
3. Select curve A, shown in Figure 3–11.
4. Select surface B, shown in Figure 3–11.
5. Click on location C, shown in Figure 3–11. A fin surface is constructed.
6. Do not save your file.

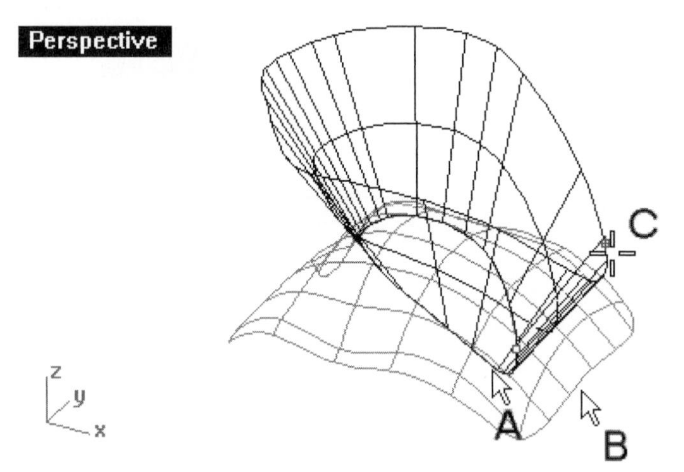

Figure 3–11. Fin surface being constructed

History Manager

Among the six ways of constructing an extruded surface, Straight, Tapered, To Point, and Along Curve are governed by the History Manager. In other words, if the History Manager is turned on prior to making these surfaces, subsequent modification of the curves will cause the surface to change accordingly.

Revolved Surfaces

A revolved surface is constructed by revolving a curve around an axis. Using Rhino's tool, you can revolve a curve in two ways: revolve the curve around an axis or revolve the curve along a rail. Both commands are governed by the History command, if it is turned on.

Revolving Around an Axis

A revolved surface has a uniform cross-section. It is constructed by revolving a curve around an axis. Perform the following steps:

1. Select File > Open and select the file Revolve01.3dm from the Chapter 3 folder on the companion CD.
2. Check the Point box in the Osnap dialog box. (Point objects will be snapped onto automatically.)
3. Select Surface > Revolve, or click on the Revolve/Rail Revolve button on the Surface toolbar.
4. Select curve A, shown in Figure 3–12, and press the ENTER key.
5. Select points B and C, shown in Figure 3–12.
6. Click on location D (Figure 3–12) to specify the start point of revolution.
7. Drag the cursor in a clockwise direction, and click on location E, shown in Figure 3–13. A revolved surface is constructed.

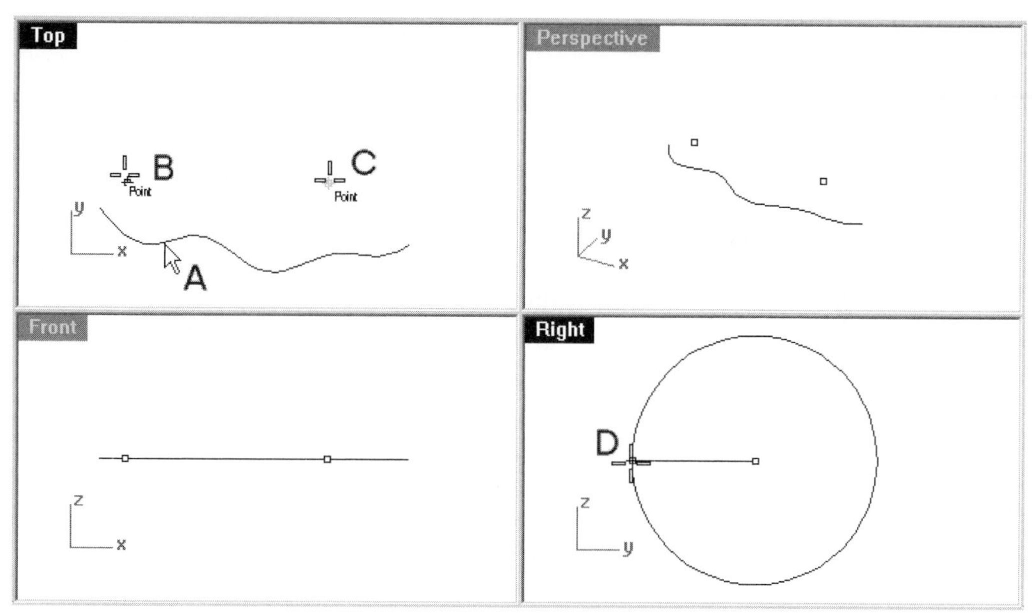

Figure 3–12. Revolved surface being constructed

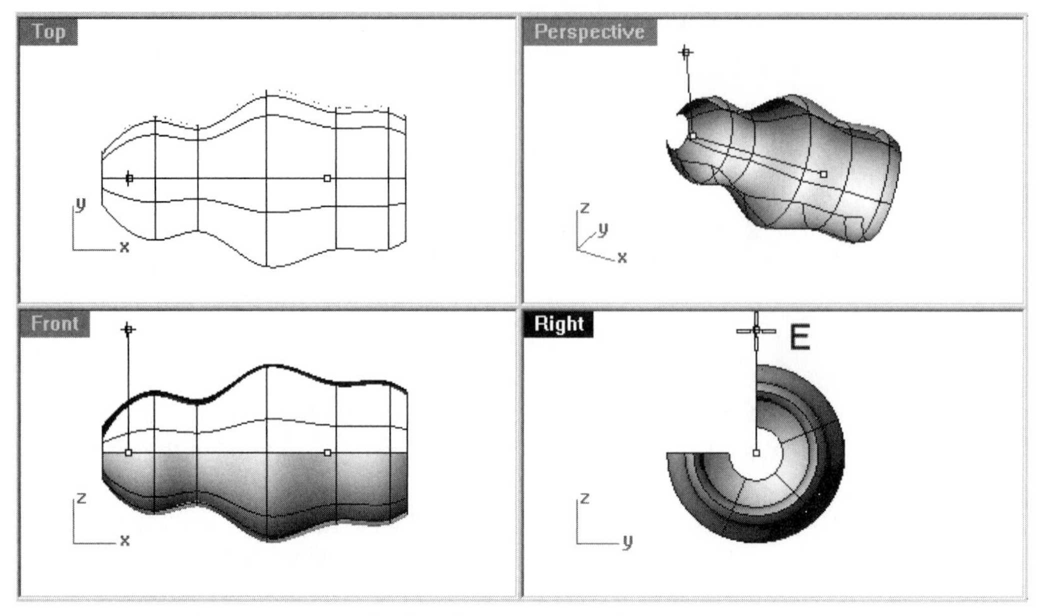

Figure 3–13. Revolved surface constructed

Revolved Surface and History Manager

The History Manager, if it is turned on, can help track changes. You may modify the curve used to make the revolve surface and the surface will update automatically. You will learn more about curve manipulation, including modifying the shape of a curve, in the next chapter. To appreciate how the History Manager helps track changes, continue with the following steps.

8 With reference to Figure 3–14 (left), select the curve that is used to construct the revolve surface. Because there are a curve and a surface edge, a pop-up menu will display, allowing you to choose the curve or the surface edge.

9 Drag it to a new position as shown in Figure 3–14 (center). The surface is modified, as shown in Figure 3–14 (right). This change is caused by a change in location between the curve and the revolve axis.

10 Do not save the file.

Note that the History Manager will work fine as long as no further work is done on the surface. For example, if you move the surface, you may obtain unexpected result.

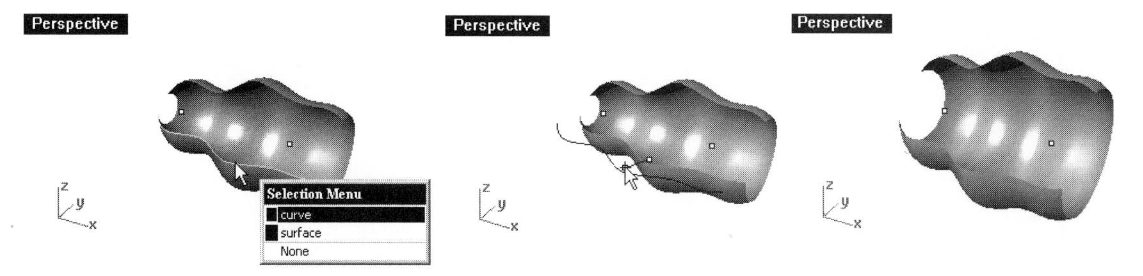

Figure 3–14. From left to right: curve selected, curve dragged, and surface modified by the History Manager

Rail Revolve

To impose control over the surface cross-section while it is being revolved around an axis, you can add a guide rail. The angle of revolution of the surface will depend on the shape of the guide rail curve and its relative location to the revolve axis. Simply speaking, if the guide rail is a closed loop curve, the rail revolve surface will revolve 360 degrees. Perform the following steps:

1. Select File > Open and select the file Revolve02.3dm from the Chapter 3 folder on the companion CD.
2. Select Surface > Rail Revolve, or right-click on the Revolve/Rail Revolve button on the Surface toolbar.
3. Select curve A, shown in Figure 3–15, as the profile curve.
4. Select curve B, shown in Figure 3–15, as the rail curve.
5. Select points C and D, shown in Figure 3–15, to define the axis of revolution. A revolved surface is constructed, as shown in Figure 3–16.

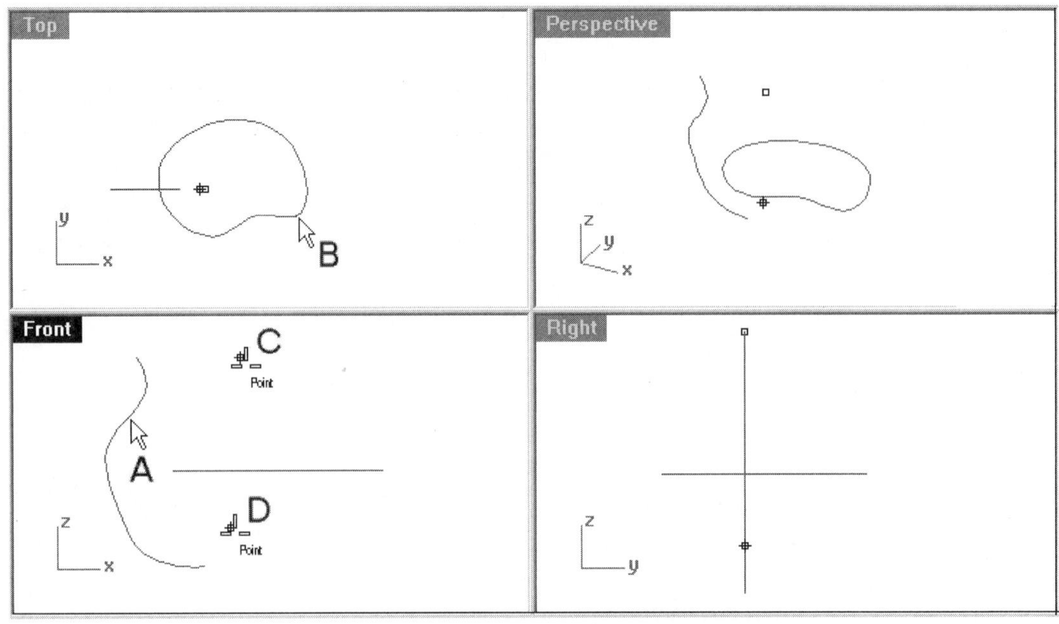

Figure 3–15. Rail revolved surface being constructed

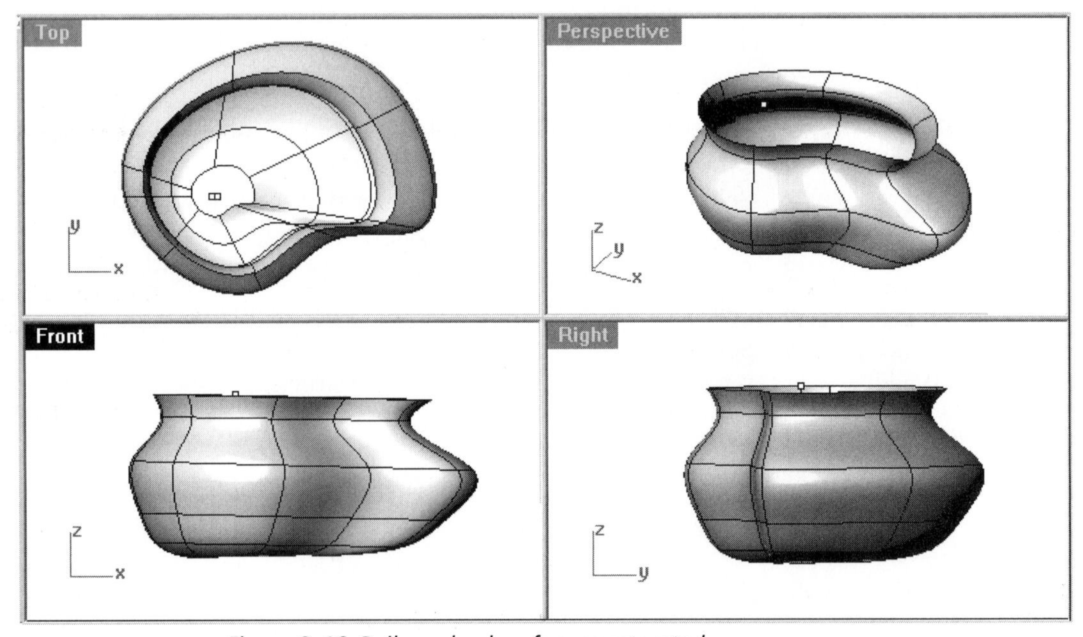

Figure 3–16. Rail revolved surface constructed

History Manager and Rail Revolve Surface

Here, two curves are involved; changes to either or both of them will cause corresponding changes to the rail revolve surface. Continue with the following steps to drag the rail to a new location.

6. Referencing Figure 3–17 (left), select the rail curve and drag it to a new location. The surface is modified, as shown in Figure 3–17 (right). The change in surface is caused by the change of the rail in relation to the revolve axis.
7. Do not save the file.

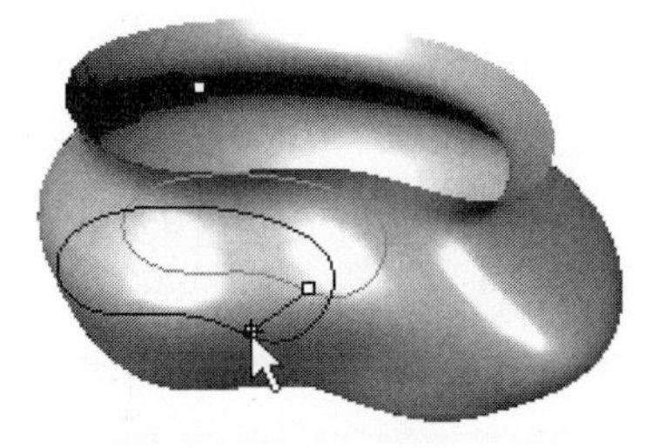

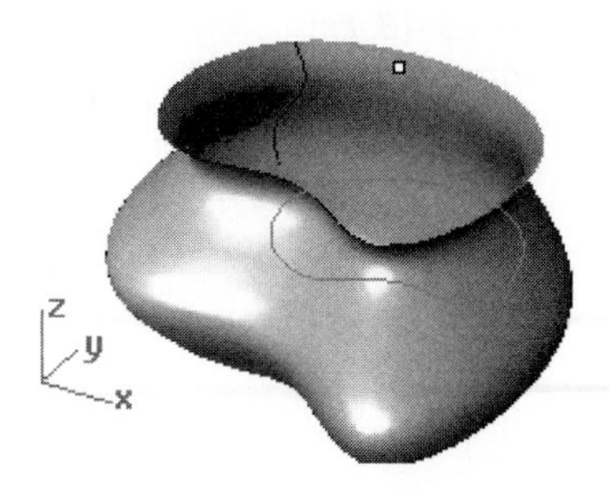

Figure 3–17. Rail curve being dragged to new location (left) and surface modified (right)

Swept Surfaces

A swept surface is constructed by a sweeping process in which one or more cross-section curves sweep along one or two guide rails.

Sweep 1 Rail Surface

A sweep 1 rail surface is constructed by first selecting a rail curve and then selecting one or more cross-section curves. The rail curve can be a single curve or a number of curves in a chain. If one section curve is selected, the surface profile will run along the entire length of the rail. If more than one section curveis selected, the surface file will interpolate from the first through the last section curve.

Comparing Extruding Along a Curve and Sweeping a Curve Along One Rail

Naturally, the simplest swept surface is constructed by sweeping a cross-section curve along a rail curve. However, you must not confuse sweeping a single curve along a single rail with extruding a curve along a curve. As shown in Figure 3–18, the cross-section profiles of an extruded along curve surface is always parallel to the original curve. As for sweeping a single profile along a single curve, the cross-section maintains a constant angle between the normal of the rail curve.

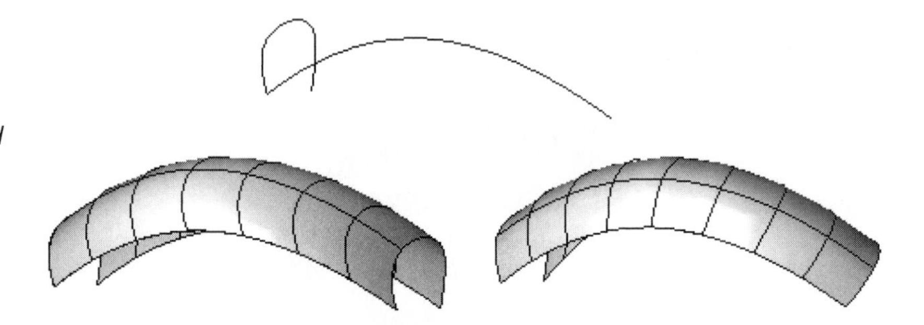

Figure 3–18. Extruding along a curve (left) and sweeping a section along a rail (right)

Perform the following steps.

1. Select File > Open and select the file Sweep101.3dm from the Chapter 3 folder on the companion CD.
2. Select Surface > Sweep 1 Rail, or click on the Sweep 1 Rail button on the Surface toolbar.
3. Select curve A, shown in Figure 3–19, as the rail curve, select curve B, indicated in Figure 3–19, as the cross-section curve, and press the ENTER key.
4. In the Sweep 1 Rail Options dialog box, accept the default and then click on the OK button. A surface is constructed, as shown in Figure 3–20.

Figure 3–19. A rail and a section selected and Sweep 1 Rail Options dialog box

5 Referencing Figure 3–20, repeat the command with curve A as the rail and curves B and C as the sections.

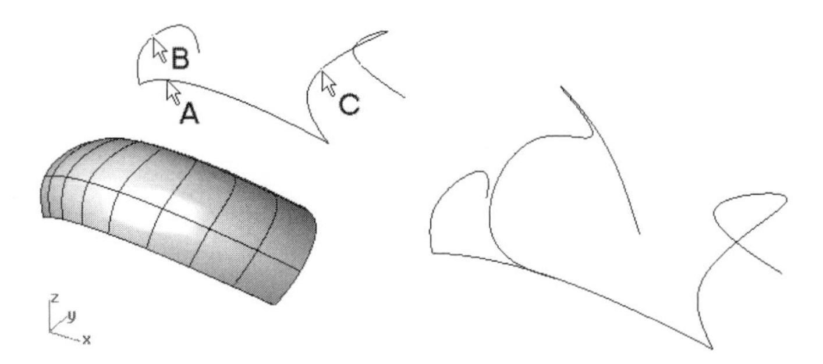

Figure 3–20. A rail and two sections selected

6 Using curve A, shown in Figure 3–21, and curves B, C, and D as the sections, construct another swept surface.

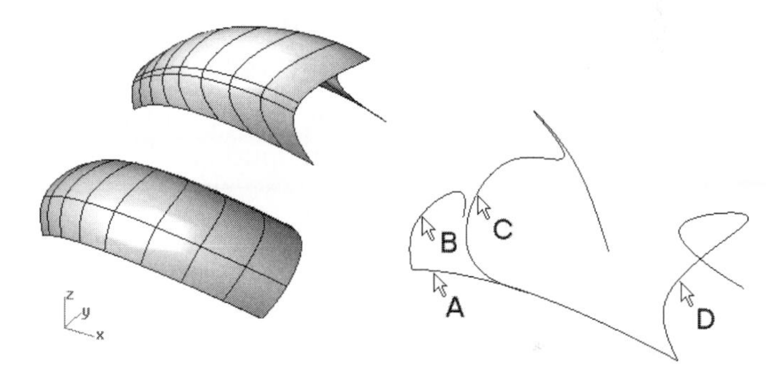

Figure 3–21. A rail and three sections selected

7 Three swept surfaces are constructed. (See Figure 3–22.) Do not save your file.

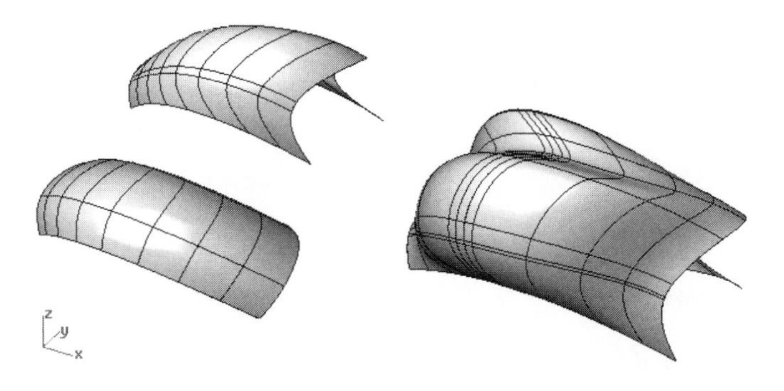

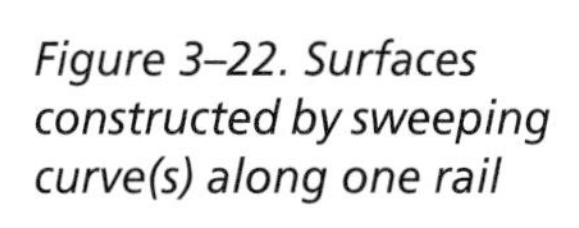

Figure 3–22. Surfaces constructed by sweeping curve(s) along one rail

When two or more cross-sections are used, the surface profile will transit gradually from the first section to the second section and then to the next cross-section.

Surface Continuity in Sweeping

If the rail curve used in making a sweep 1 rail surface is the edge of another surface, the swept surface that is constructed can be made to twist with the surface edge, and continuity between the swept surface and the existing surface can be enabled. Perform the following steps:

1. Select File > Open and select the file Sweep102.3dm from the Chapter 3 folder on the companion CD.
2. Select Surface > Sweep 1 Rail, or click on the Sweep 1 Rail button on the Surface toolbar.
3. Select edge A, shown in Figure 3–23, as the rail curve, select curves B and C, shown in Figure 3–23, as the cross-section curves, and press the ENTER key.
4. In the Style pull-down list of the Sweep 1 Rail Options dialog box (Figure 3–23), select Align with Surface, and click on the OK button. A swept surface that aligns with the surface whose edge is used as sweeping rail is constructed.

For comparison, the swept surface constructed from the same set of curves with freeform option is also shown in Figure 3–24.

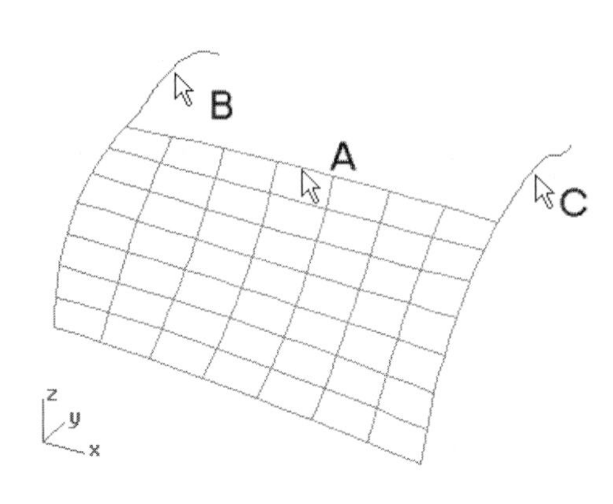

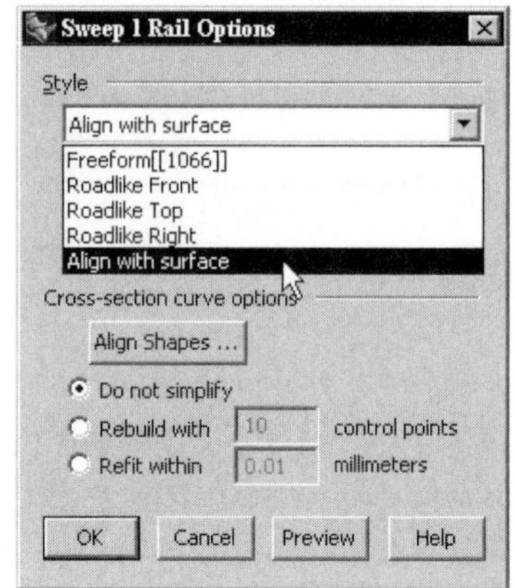

Figure 3–23. Surface edge being selected as rail curve in swept surface construction and Align with surface option selecte

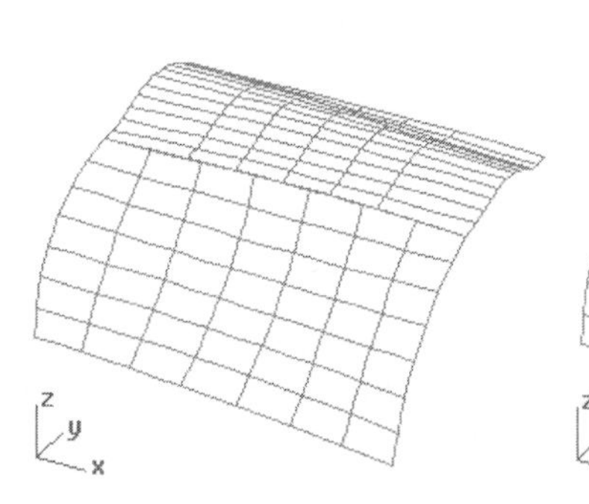

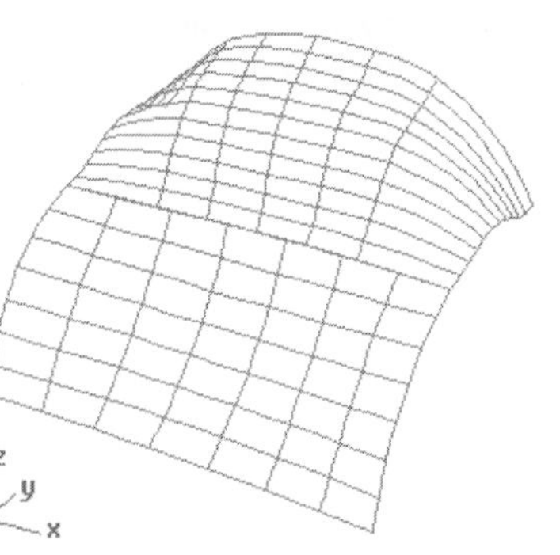

Figure 3–24. Swept surface with freeform option (left) and align with surface option (right)

Edge Chaining for Sweeping Rail

If a rail is made up of a number of curves and/or surface edges, you do not have to first join the curves to form a single rail. Instead, you can use edge chaining option to select multiple curves and/or edges. At the command area, select the ChainEdges option on the command area and then select contiguous rail curves.

1. Select File > Open and select the file Sweep103.3dm from the Chapter 3 folder on the companion CD.
2. Select Surface > Sweep 1 Rail, or click on the Sweep 1 Rail button on the Surface toolbar.
3. Select the ChainEdges option on the command area.
4. Select the ChainContinuity option on the command area.
5. Select the Curvature option on the command area.
6. Select surface edges A, B, and C, indicated in Figure 3–25, and press the ENTER key.
7. Select curves D and E and press the ENTER key.
8. In the Sweep 1 Rail Options dialog box, select Align with surface from the Style pull-down list box, check the Global shape blending box, and click on the OK button. A sweep surface is constructed, as shown in Figure 3–26.

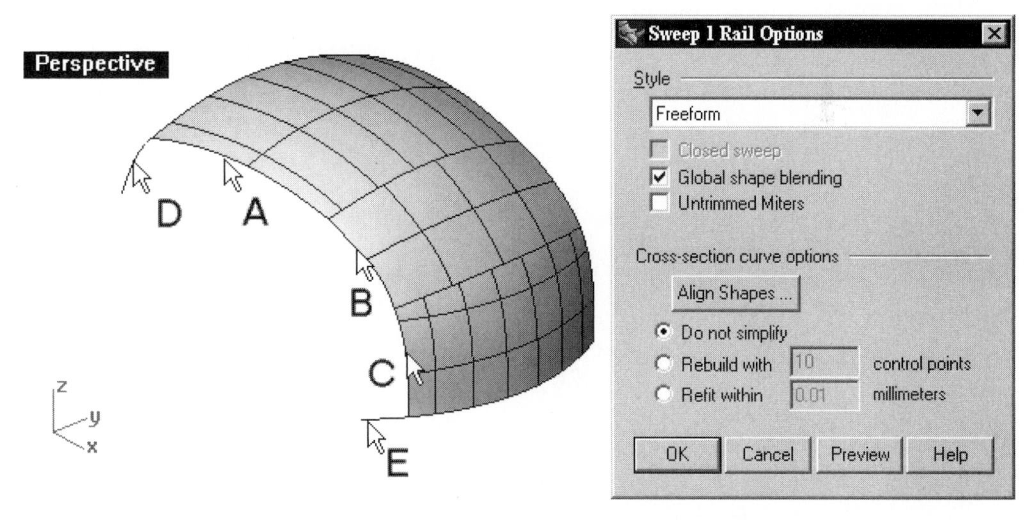

Figure 3–25. Chain edges with two cross-sections selected and Sweep 1 Rail Options dialog box

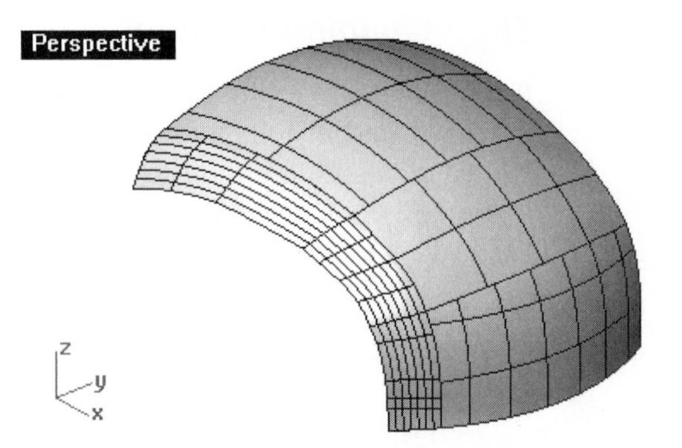

Figure 3–26. Sweep surface constructed

Shape Alignment in Sweeping

In multiple cross-section sweep 1 rail surface construction, you can reverse the direction of selected cross-section curves. Perform the following steps.

1. Select File > Open and select the file Sweep104.3dm from the Chapter 3 folder on the companion CD.
2. Select Surface > Sweep 1 Rail, or click on the Sweep 1 Rail button on the Surface toolbar.
3. Select surface edge A and curves B and C, shown in Figure 3–27, and press the ENTER key.
4. Click on the Align Shapes button on the Sweep 1 Rail Options dialog box.

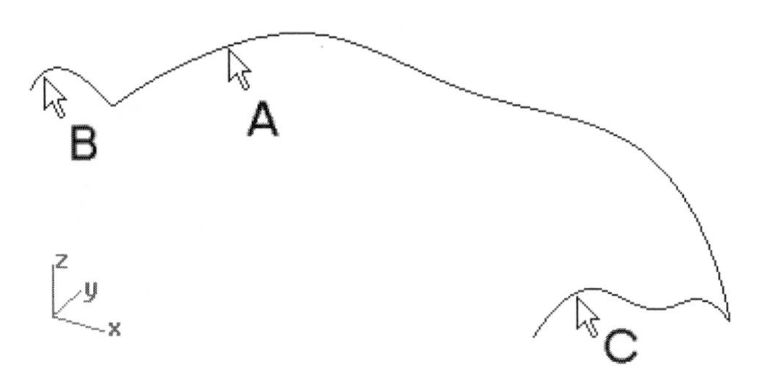

Figure 3–27. Curves selected for making sweep 1 rail surface

5. Select endpoint A, shown in Figure 3–28. The cross-section curve is reversed.
6. Press the ENTER key, and then click on the OK button. A surface is constructed.
7. Do not save the file.

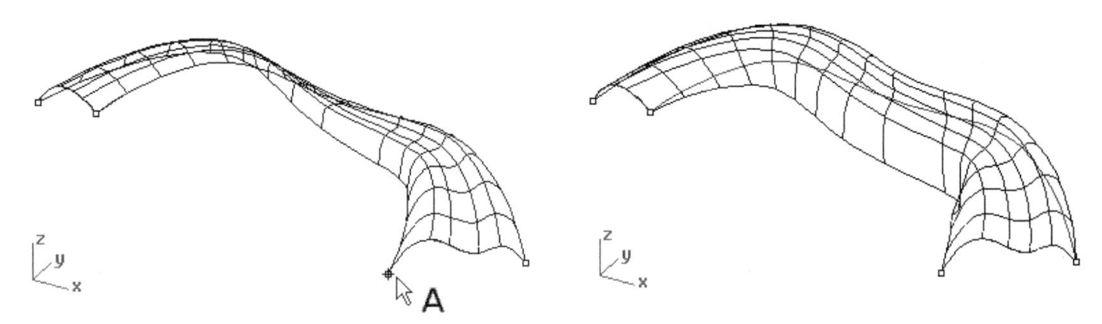

Figure 3–28. Cross-section curve being selected (left) and reversed (right)

Sweeping to a Point

Apart from interpolating among the cross-section curves, you can sweep a surface to a point. Perform the following steps:

1. Select File > Open and select the file Sweep105.3dm from the Chapter 3 folder on the companion CD.
2. Click on Osnap button on the Status bar, and set object snap mode to Point.
3. Select Surface > Sweep 1 Rail, or click on the Sweep 1 Rail button on the Surface toolbar.
4. Select curves A and B, shown in Figure 3–29.
5. Select the Point option on the command area.
6. Select point C, shown in Figure 3–29, and press the ENTER key.
7. Click on the OK button in the Sweep 1 Rail Options dialog box. A surface is constructed. (See Figure 3–29.)
8. Do not save the file.

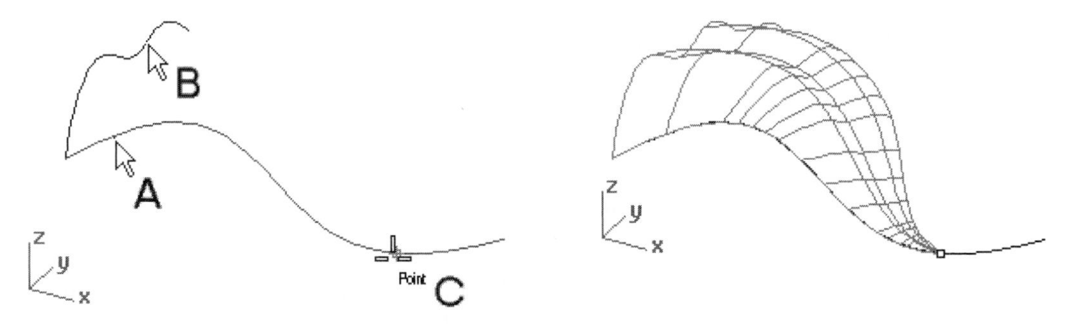

Figure 3–29. Curves and point location selected (left) and Swept to a point surface constructed (right)

Sweep 2 Rails Surface

To add further control to the surface profile while sweeping, you use two guiding rails instead of one guiding rail. In addition to interpolating from one cross-section curve to another cross-section curve, the cross-section translates in accordance with the distance between the rails and the locations of the rails. Using Rhino's sweep 2 rails command, you select two rails and then one or more cross-section curves to construct a sweep 2 rails surface. Moreover, you can incorporate additional cross-section alignments to control how the surface is constructed. Similar to sweep 1 rail surface, you can use a series of curves and/or surface edges as sweeping rails. Perform the following steps:

1. Select File > Open and select the file Sweep201.3dm from the Chapter 3 folder on the companion CD.
2. Select Surface > Sweep 2 Rails, or click on the Sweep 2 Rails button on the Surface toolbar.
3. Select curves A, B, and C, shown in Figure 3–30, and press the ENTER key. (The first two selected curves are rails and subsequently selected curves are sections.)

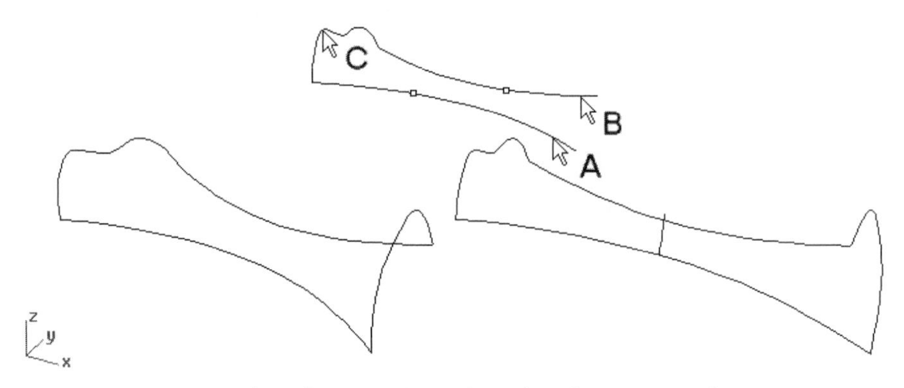

Figure 3–30. Two rails and one section selected and Sweep 2 Rails Options dialog box

Adding Slash to a Sweep 2 Rails Surface

Slashes are additional cross-section alignments. To incorporate additional slash along the rails, continue with the following steps.

4. In the Sweep 2 Rails Options dialog box, click on the Add Slash button. (See Figure 3–31.)
5. Select points A and B, shown in Figure 3–31. *(Note: the effect of adding a slash.)*
6. Press the ENTER key.

7 Click on the OK button. A swept surface with additional slash is constructed.

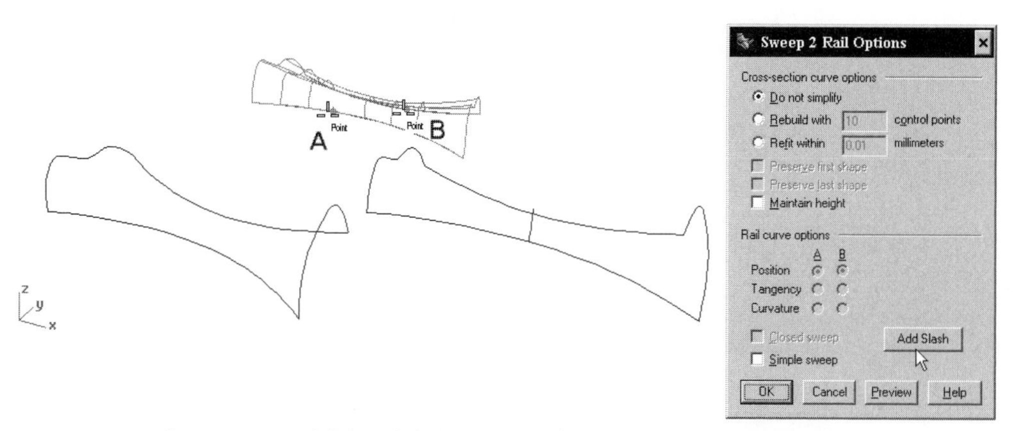

Figure 3–31. Additional slash being defined and Sweep 2 Rails Options dialog box

Maintaining Height in a Sweep 2 Rails Surface

As a cross-section curve is swept along two rails, its overall cross-section is scaled in accordance with the distance between the rails. In order words, the height of cross-sections as they interpolate along the rail curves depends on the distance between the rails. To maintain the height during interpolation, you can use the Maintain Height option. Continue with the following steps.

8 Repeat the command.

9 Select curves A, B, C, and D, shown in Figure 3–32, and press the ENTER key.

10 Check the Maintain height button and the OK button in the Sweep 2 Rails Options dialog box. A swept surface with height adjustment is constructed. Note the effect of having height adjustment on the final shape of the surface.

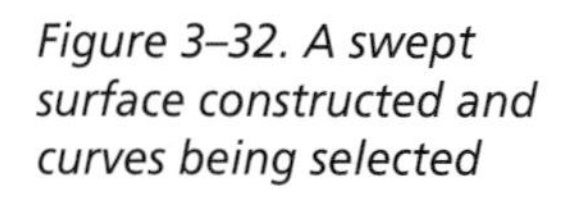

Figure 3–32. A swept surface constructed and curves being selected

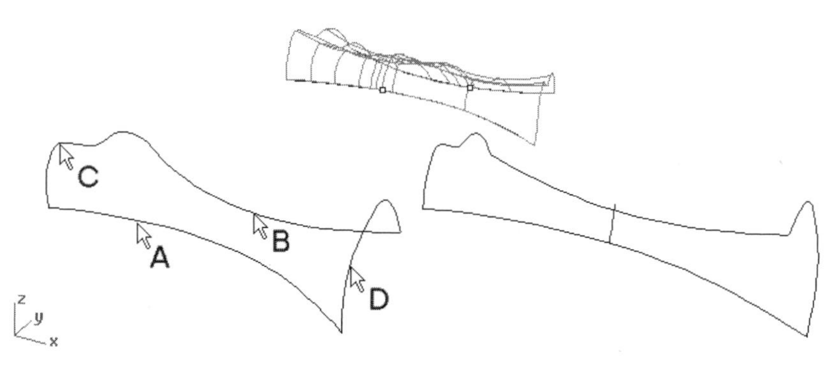

11 Repeat the command.

12 Select curves A, B, C, D, and E, shown in Figure 3–33, and press the ENTER key.

13 Check the Maintain height button, and click on the Preview button to discover the difference between having and not having the Preview button checked.

14 Click on the OK button. The third swept surface is constructed. (See Figure 3–34.)

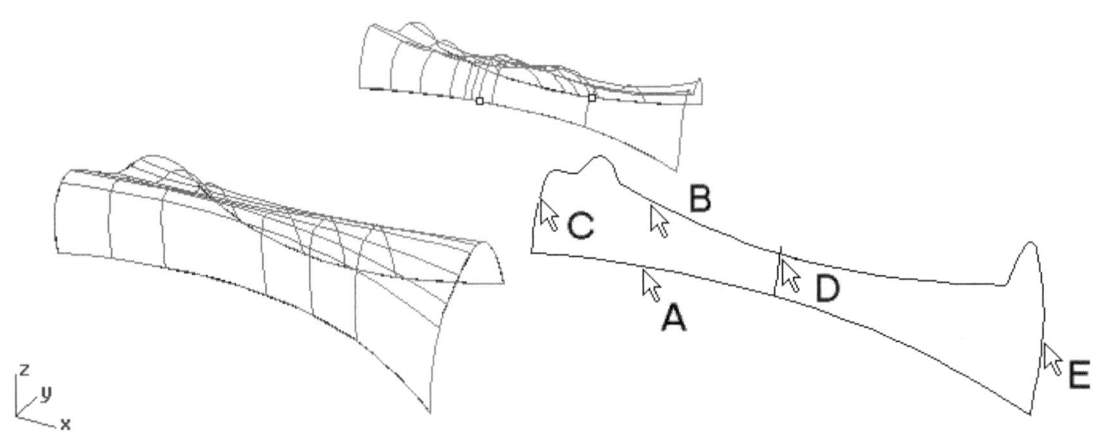

Figure 3–33. Second swept surface constructed and curves being selected for the third swept surface

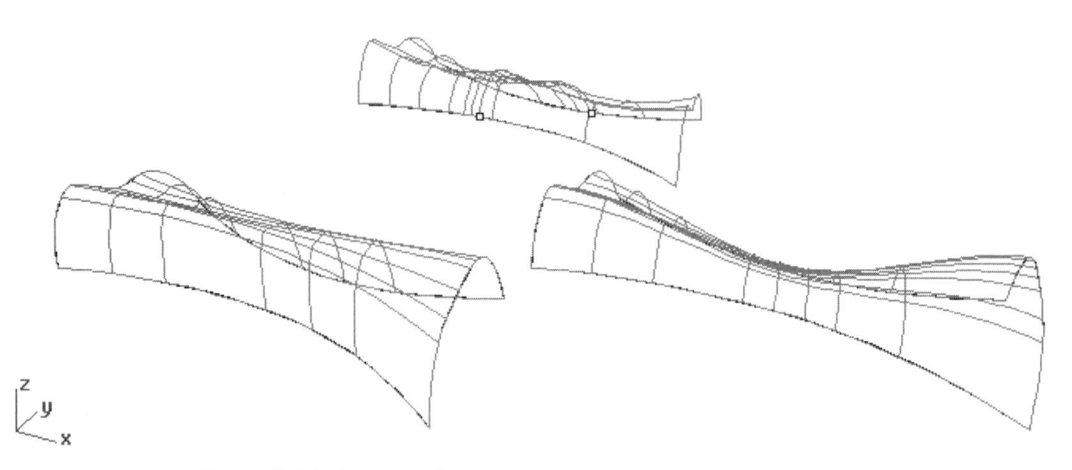

Figure 3–34. Swept surfaces constructed

Edge Continuity Between Edge Rails and Sweep 2 Rails Surface

If the rails for making a sweep 2 rails surface are surface edges, edge continuity between the swept surface and the existing surfaces can be adjusted. Perform the following steps:

1. Select File > Open and select the file Sweep202.3dm from the Chapter 3 folder on the companion CD.
2. Select Surface > Sweep 2 Rails, or click on the Sweep 2 Rails button on the Surface toolbar.
3. Select curves A, B, C, D, and E, shown in Figure 3–35, and press the ENTER key.
4. Check the Curvature buttons in the Sweep 2 Rails dialog box, and click on the OK button.
5. A swept surface is constructed. Do not save your file.

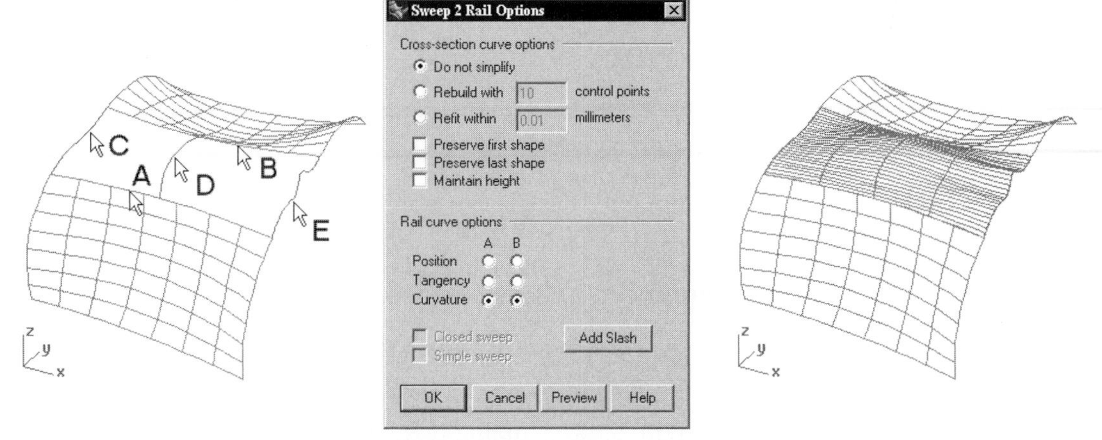

Figure 3–35. From left to right: surface edges as rails, Sweep 2 Rails Options dialog box, Sweep 2 rails surface constructed with surface edge matching

Lofted Surfaces

Lofted surfaces are constructed by interpolating one or two sets of cross-section curves. Using Rhino's tool, you can construct this type of surface in three ways: Loft, Curve Network, and Edge Curve. All three types of surfaces are governed by the History Manager, if it is turned on.

Loft Surface

Rhino's loft command uses only one set of cross-section curves. It produces a surface profile that interpolates from the first cross-section curve to the second, and to the next curve. Perform the following steps:

1. Select File > Open and select the file Loft01.3dm from the Chapter 3 folder on the companion CD.
2. Select Surface > Loft, or click on the Loft button on the Surface toolbar.
3. Select curves A, B, and C, shown in Figure 3–36, and press the ENTER key.

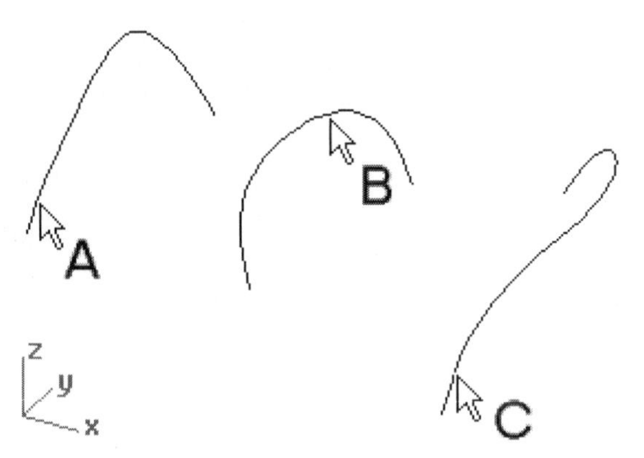

Figure 3–36. Curves being selected

Curve Alignment in Loft Surface Construction

Due to the way the curves are selected, the preview shows a bow tie shape surface, which is obviously not what we want.

4. In the Loft Options dialog box, click on the Align Curves button and then click on endpoint A, shown in Figure 3–37. The curve's direction is reversed and is aligned with other curves.

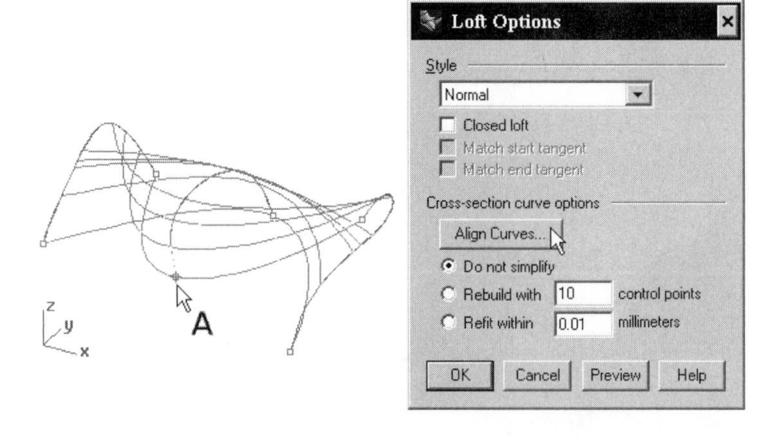

Figure 3–37. Cross-section curve being aligned

Style in Loft Surface

To cope with downstream operations such as sheet metal fabrication, you may need to have a developable surface. A developable surface is a surface that can be developed into a flat sheet. Continue with the following steps:

5. In the Style pull-down list box, there are straight, developable, uniform sections. Select developable and click on the preview button; a developable surface is displayed. There is slight difference between straight sections and developable in that the former produces a lofted surface with straight sections between contiguous cross-section curves, and the latter produces a surface that can be developed into a flat sheet.
6. Select Uniform and then click on the OK button. A uniform loft surface is constructed. Do not save your file. (See Figure 3–38.)

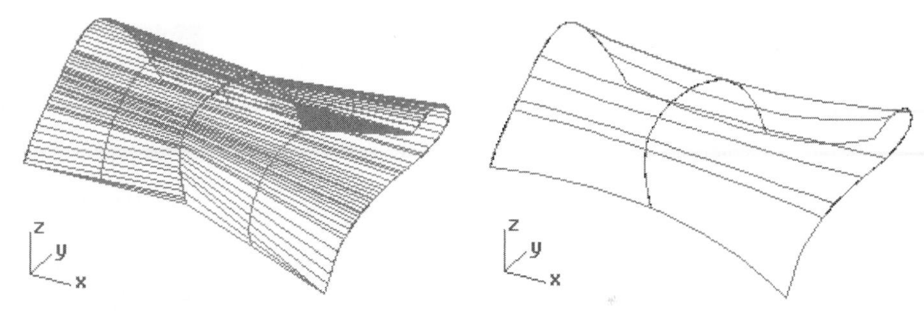

Figure 3–38. Developable surface (left) and uniform surface (right)

Tangency between Loft Surface and Edge Curves

To appreciate how surface edges can be used in making a lofted surface, perform the following steps:

1. Select File > Open and select the file Loft02.3dm from the Chapter 3 folder on the companion CD.
2. Select Surface > Loft, or click on the Loft button on the Surface toolbar.
3. Referencing Figure 3–39, select edge A, click on Surface edge from the pop-up menu, select curve B, and select edge C.
4. Press the ENTER key.
5. In the Loft Options dialog box, check the Match Start Tangent box and the Match End Tangent box, and click on the OK button. A lofted surface with tangent matching is constructed. (See Figure 3–40.)
6. Do not save your file.

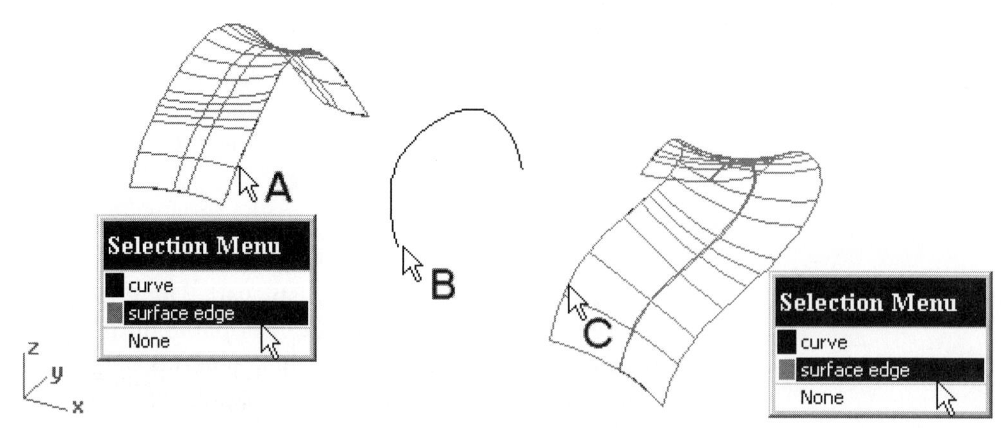

Figure 3–39. Surface edges selected as loft sections

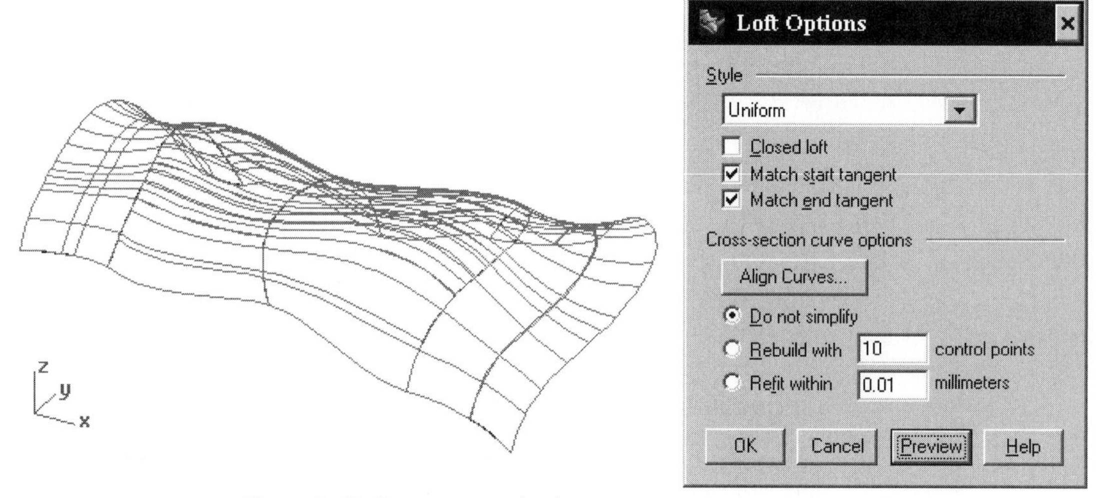

Figure 3–40. Tangent matched

Loft to a Point

Using the point option, you can have a lofted surface terminated at a point. Perform the following steps:

1. Select File > Open and select the file Loft03.3dm from the Chapter 3 folder on the companion CD.
2. Set object snap mode to Point.
3. Select Surface > Loft, or click on the Loft button on the Surface toolbar.

4. Select curves A and B, shown in Figure 3–41.
5. Select the Point option on the command area.
6. Select Point C, shown in Figure 3–41, and press the ENTER key.
7. Press the ENTER key in response to the prompt to adjust curve seams.
8. In the Loft Options dialog box, click on the OK button. A lofted surface terminating at a point is constructed.
9. Do not save your file.

In this tutorial, the curves for making the lofted surface are closed loop curves. Therefore, there is a prompt to tell you to adjust the curve seam. As for the point object, it is provided here for your convenience. You can specify a point by clicking anywhere or specifying a set of coordinates.

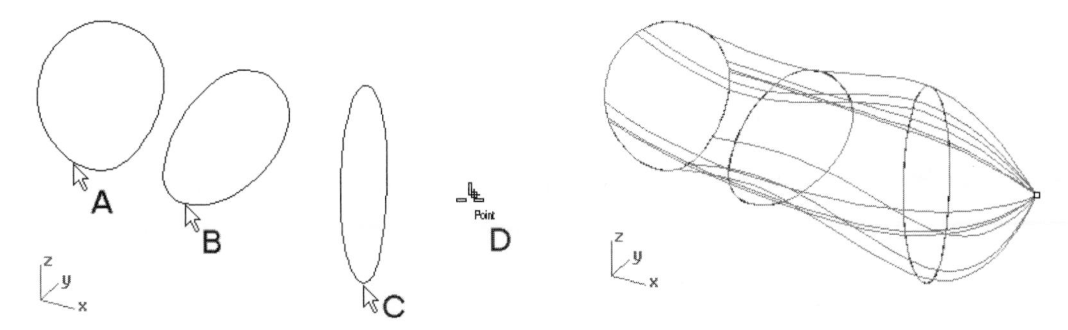

Figure 3–41. Curves and point selected (left) and lofted surface terminating at a point is constructed (right)

Curve Network Surface

Rhino's curve network surface can be regarded as a kind of lofted surface. It is constructed from two sets of cross-section curves in two orthogonal directions. The surface profile interpolates in two directions, defined by two sets of curves in two orthogonal directions.

In essence, valid curves will be automatically sorted into two sets of orthogonal curves, regardless of the sequence of selection. However, if one of the curves is invalid or ambiguously defined, the system will prompt you to select the curve sets manually. Then you should select one set of curves, press the ENTER key, continue to select the second set of orthogonal curves, and press the ENTER key again. Perform the following steps:

1. Select File > Open and select the file CurveNetwork01.3dm from the Chapter 3 folder on the companion CD.

2. Select Surface > Curve Network, or click on the Surface from Network of Curves button on the Surface toolbar.
3. Select all curves and press the ENTER key. After selecting the curves, the Surface From Curve Network dialog box displays. Here, you specify edge tolerance and continuity of matching edges, as follows.
4. In the Surface From Curve Network dialog box, shown in Figure 3–42, accept the default and then click on the OK button. A surface is constructed from a curve network.
5. Do not save your file.

Note in the Surface From Curve Network dialog box that there are options for you to decide how the edge of the surface is to match with the input curves.

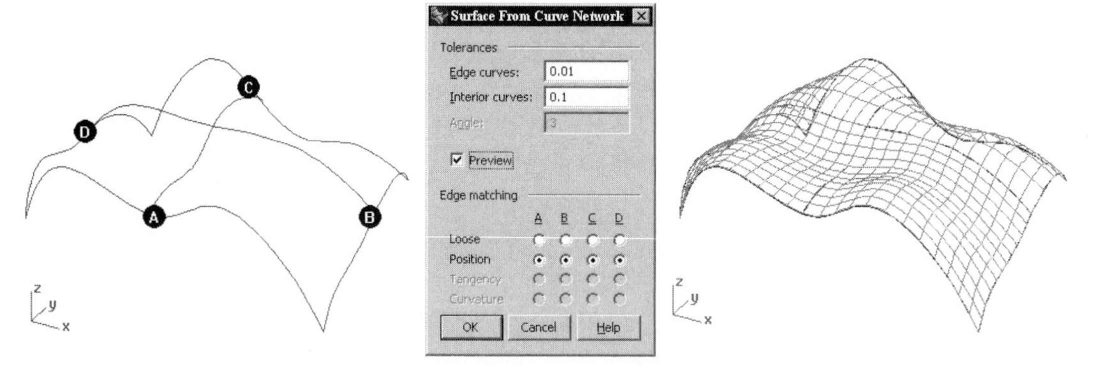

Figure 3–42. From left to right: curves for from curve network surface, Surface from curve network dialog box, and surface constructed

Manual Sorting of U and V Curves in Network Curve Surface Construction

Perform the following steps to understand what would happen if the system fails to sort the selected curves automatically.

1. Select File > Open and select the file CurveNetwork02.3dm from the Chapter 3 folder on the companion CD.
2. Select Surface > Curve Network, or click on the Surface from Network of Curves button on the Surface toolbar.
3. Select all curves and press the ENTER key.

Because a curve that does not conform to the requirement for making a network curve surface is put here intentionally, the NetworkSrf Sorting Problem dialog box displays, as shown in Figure 3–43 (left).

4 Click the Yes button.
5 Select curves A, B, C, and D (Figure 3–43) and press the ENTER key. (These are the curves in the first direction.)
6 Select curves E, F, G, and H (Figure 3–43) and press the ENTER key. (These are the curves in the second direction.)
7 Click the OK button from the Surface From Curve Network dialog box. A surface is constructed.
8 Do not save your file.

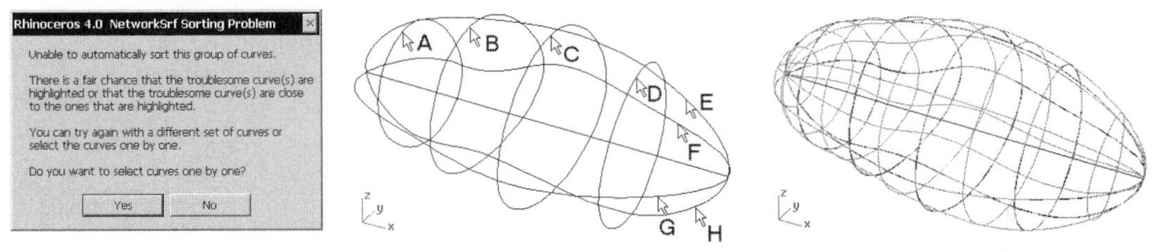

Figure 3–43. From left to right: NetworkSrf Sorting Problem dialog box, curves being selected sequentially, and surface constructed

Edge Matching in Curve Network Surface Construction

If surface edges are used in making curve network surfaces, edge matching can be set. Perform the following steps.

1 Select File > Open and select the file CurveNetwork03.3dm from the Chapter 3 folder on the companion CD.
2 Select Surface > Curve Network, or click on the Surface from Network of Curves button on the Surface toolbar.
3 Referencing Figure 3–44, select surface edges A and D and curves B, C, and E. Then press the ENTER key.

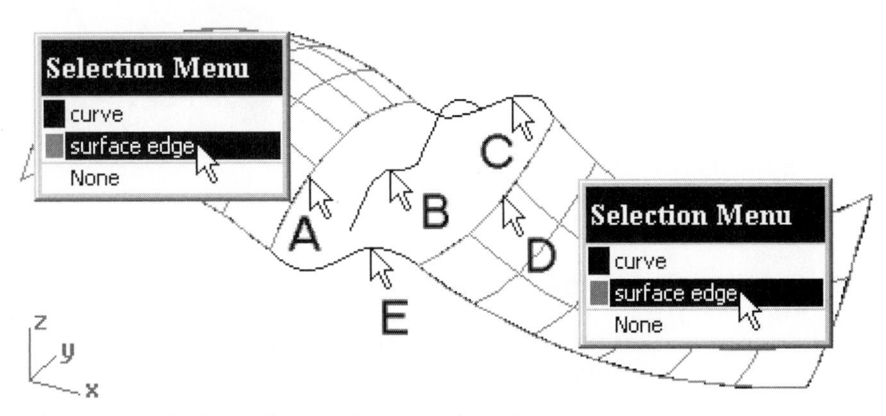

Figure 3–44. Surface edges and curves selected

4. In the Surface From Curve Network dialog box, set the angle value to 0. This is the angle tolerance between the tangent directions of adjacent edges. (See Figure 3–45.)
5. Check the Curvature boxes for edges B and D.
6. Click on the OK button. A surface is constructed. (See Figure 3–46.)
7. Do not save the file.

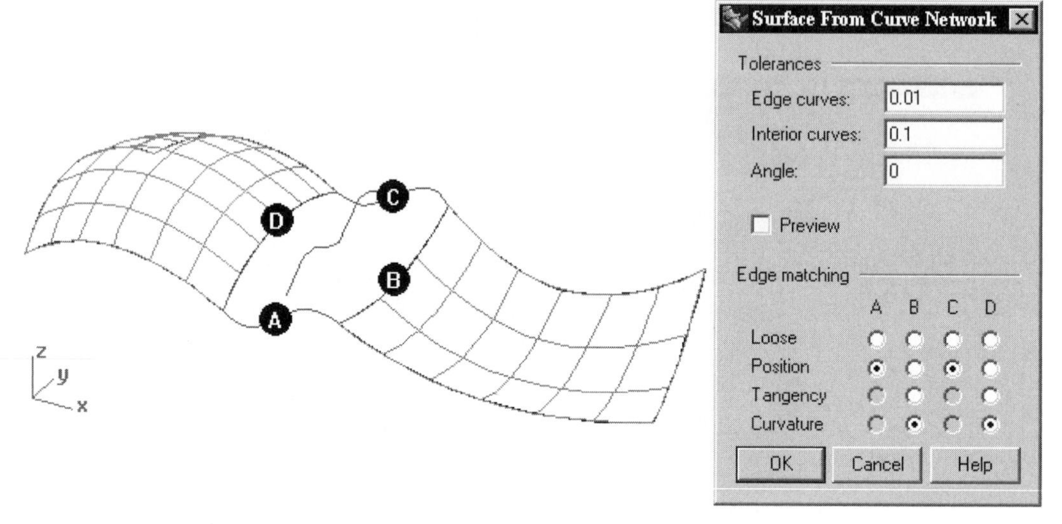

Figure 3–45. Curvature continuity at edges B and D

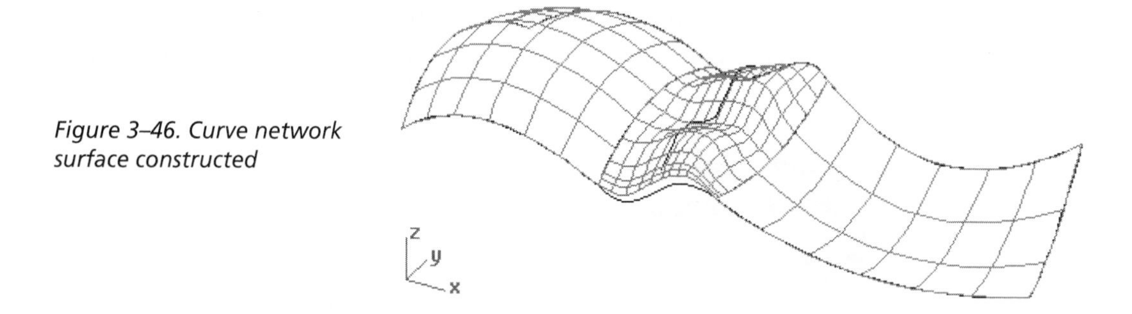

Figure 3–46. Curve network surface constructed

Edge Curves Surface

You may also regard Rhino's edge curves surface as a special kind of lofted surface. It is constructed by specifying two, three, or four edges of the surface. Perform the following steps to construct surfaces from edge curves.

1. Select File > Open and select the file EdgeCurves.3dm from the Chapter 3 folder on the companion CD.
2. Select Surface > Edge Curves, or click on the Surface from 2, 3, or 4 Edge Curves button on the Surface toolbar.
3. Select curves A and B, shown in Figure 3–47, and press the ENTER key. A surface is constructed from two edge curves.
4. Select Surface > Edge Curves, or click on the Surface from 2, 3, or 4 Edge Curves button on the Surface toolbar.
5. Select curves C, D, and E, shown in Figure 3–47, and press the ENTER key. A surface is constructed from three edge curves.
6. Select Surface > Edge Curves, or click on the Surface from 2, 3, or 4 Edge Curves button on the Surface toolbar.
7. Select curves F, G, H, and J, shown in Figure 3–47.

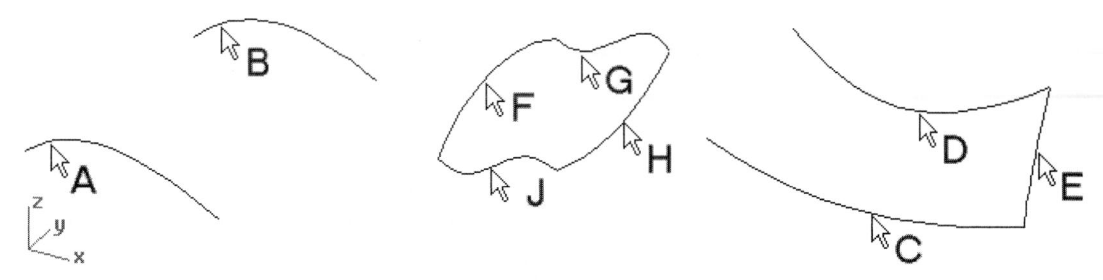

Figure 3–47. Curves being selected

8. Do not save your file.

Three surfaces constructed are shown in Figure 3–48.

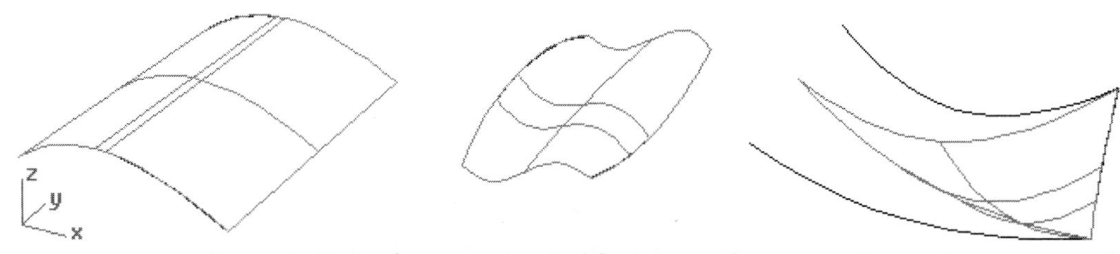

Figure 3–48. Surfaces constructed from two edge curves, three edge curves, and four edge curves

Other Kinds of Rhino Free-form Surfaces

Apart from the four basic kinds of free-form surfaces delineated above, you can use Rhino tools to construct the following types of free-form surfaces: Patch, Corner Points, Point Grid, Drape, and Heightfield from Image.

Patch Surface

A patch surface provides a very flexible way to construct a surface, in particular, by filling holes in a surface model. The input data can be curves and/or points, creating a patch surface from a closed-loop curve, from point objects, from a closed-loop curve and point objects, from an open-loop curve and point objects, and from a number of curves. One major application of patch surface is to "patch" openings in existing surface models. Perform the following steps:

1. Select File > Open and select the file Patch.3dm from the Chapter 3 folder on the companion CD.
2. Set the current layer to Layer01.
3. Select Surface > Patch, or click on the Patch button on the Surface toolbar.
4. Select surface edge curve A (Figure 3–49) and press the ENTER key.
5. In the Patch Surface Options dialog box, click on boxes Adjust Tangency and Automatic Trim, if they are not already checked.
6. Click on the OK button. A patch is constructed from the edge of a close-loop surface.
7. Repeat the command.
8. Select surface edge B and curve C (Figure 3–49), and press the ENTER key.
9. Click on the OK button of the Patch Surface dialog box. A patch surface is constructed from a close-loop surface edge and a curve. Note that the curve has to be reasonably close to the edge for the surface to be valid.
10. Repeat the command.
11. Select surface edge D and points E, F, and G (Figure 3–49), and press the ENTER key.
12. Click on the OK button of the Patch Surface dialog box. A patch surface is constructed from a close-loop surface edge and three point objects. Again, the point objects have to be reasonably close to the surface edge.
13. Repeat the command.

14 Select curve H (there is a curve and surface edge here) and point objects J, K, and L, and press the ENTER key.

15 Click the OK button of the Patch Surface dialog box. A patch surface is constructed from a curve and three point objects. *(Note: the difference between using a curve and a surface edge, as shown in Figure 3–50.)*

16 Do not save your file.

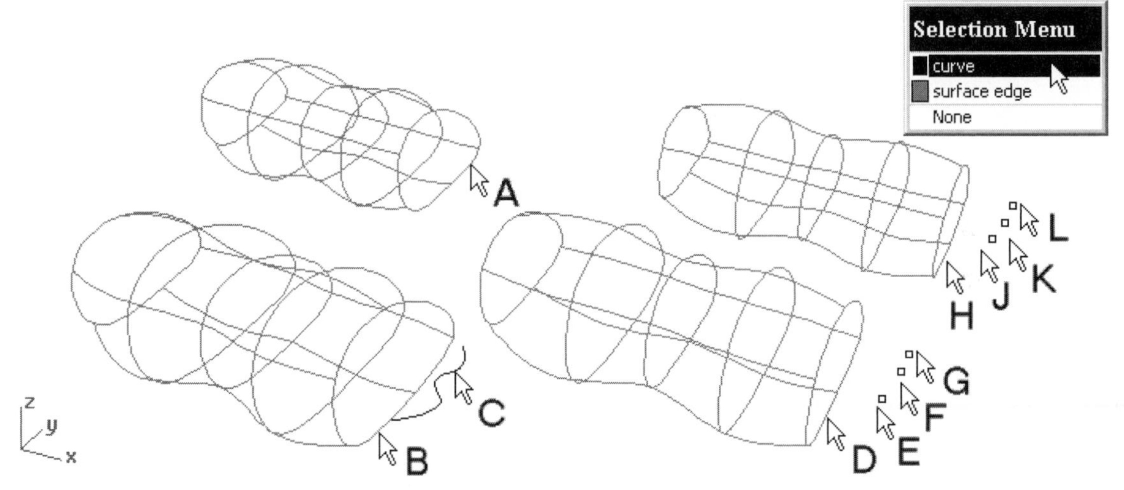

Figure 3–49. Surface edges, point objects, and curves being selected

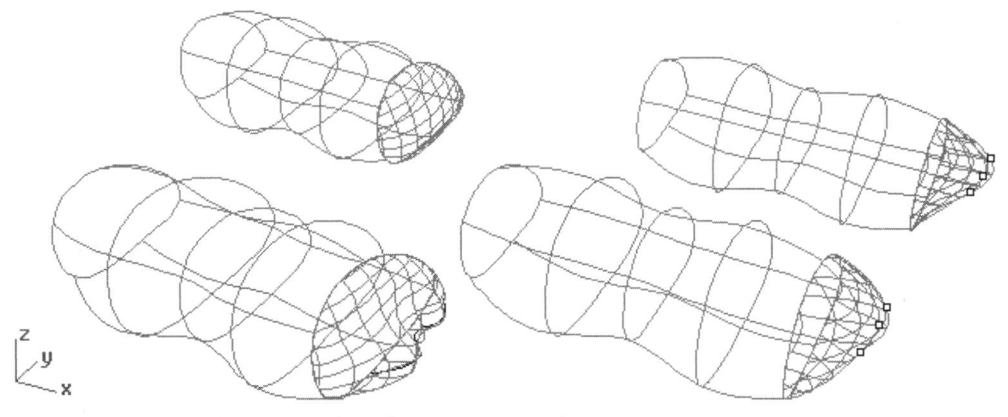

Figure 3–50. Patch surfaces constructed

Corner Points Surface

A corner points surface is a surface defined by three or four corner points. If three input points are used, a triangular planar surface is produced. If four input points are used, a quadrilateral surface is constructed. Perform the following steps:

1. Select File > Open and select the file CornerPoints.3dm from the Chapter 3 folder on the companion CD.
2. Check the Point button on the Osnap dialog box.
3. Select Surface > Corner Points, or click on the Surface from 3 or 4 Corner Points button on the Surface toolbar.
4. Select points A, B, and C, shown in Figure 3–51, and press the ENTER key. A corner point surface from three points is constructed.
5. Select Surface > Corner Points, or click on the Surface from 3 or 4 Corner Points button on the Surface toolbar.
6. Select points D, E, F, and G, shown in Figure 3–51. A corner surface from four points is constructed.
7. Do not save your file.

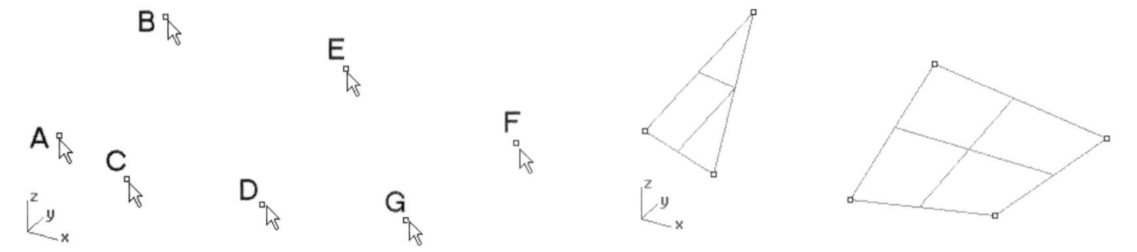

Figure 3–51. Point objects (left) and corner points surfaces constructed (right)

Point Grid Surface

A point grid surface is derived from a matrix of points arranged in rows and columns. Surface constructed using this method may use these points as interpolation points or control points. Using them as interpolation points, the surface profile will lie on the points. Using them as control points, they become the control points of the surface. If you use them as control points, you need to specify the degree of polynomial as well. The concept of polynomial degree and control points will be explained in Chapter 4. Perform the following steps:

1. Select File > Open and select the file PointGrid.3dm from the Chapter 3 folder on the companion CD.
2. Check the Point box on the Osnap dialog box.

3 Select Surface > Point Grid, or click on the Surface From Point Grid/Surface From Control Point Grid button on the Surface toolbar.
4 Type 4 at the command area twice to specify the number of points in rows and columns.
5 Select points A1, A2, A3, A4, B1, B2, B3, B4, C1, C2, C3, C4, D1, D2, D3, and D4, shown in Figure 3–52.
6 Right-click on the Surface From Point Grid/Surface From Control Point Grid button on the Surface toolbar.
7 Press the ENTER key twice to accept the number of rows and columns, which should be 4 for both.
8 Select points E1, E2, E3, E4, F1, F2, F3, F4, G1, G2, G3, G4, H1, H2, H3, and H4, shown in Figure 3–52.
9 Do not save your file.

Surfaces constructed are shown in Figure 3–53.

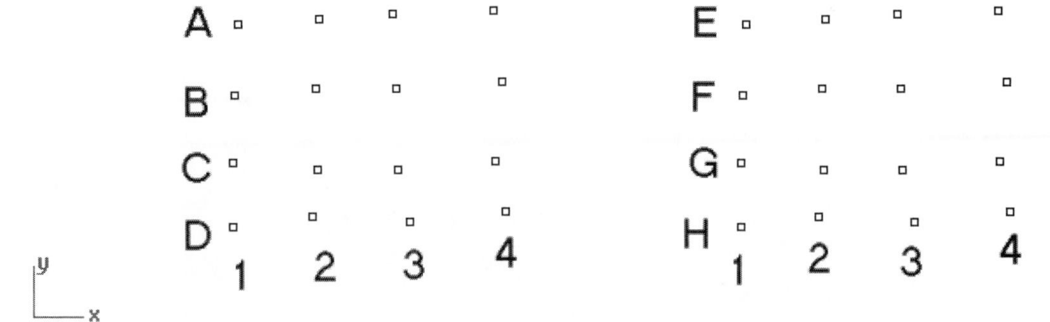

Figure 3–52. Point and surfaces being constructed

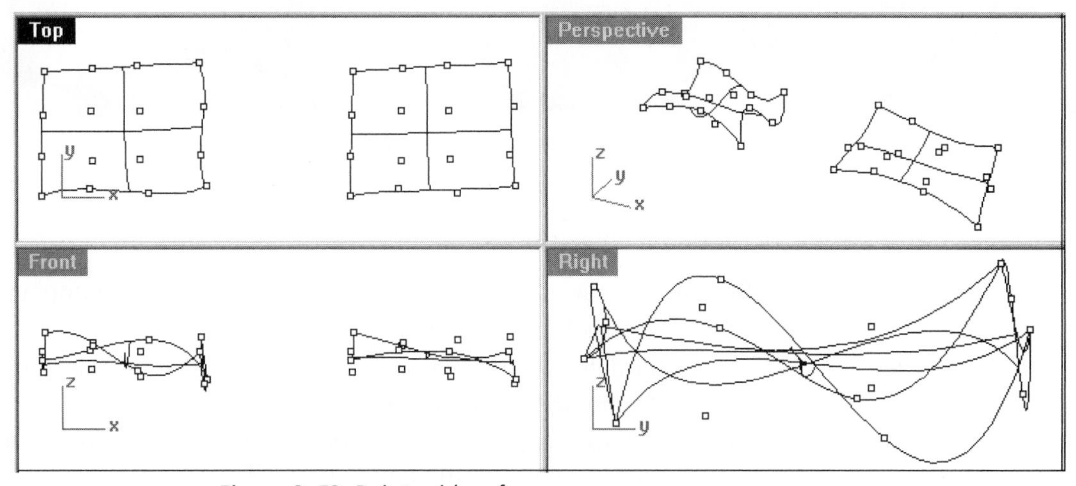

Figure 3–53. Point grid surface

Heightfield from Image Surface

A heightfield from image surface is an interesting way to construct a surface. It uses the color values of a bitmap image to define a matrix of points. Similar to a point grid surface, you can use these points as interpolation points on the surface or control points for the surface. To construct a heightfield from image surface, you select an image, indicate the location of the image on the construction plane, and then specify the number of sample points in two orthogonal directions and the height value of the sample points.

Basically, the bitmap can be color or black and white. However, it is more appropriate to use a black-and-white image because only the brightness value is considered when a surface is derived. If you use a black-and-white image, you can better perceive the outcome before making the surface. Naturally, you need a bitmap to make a surface of this type. Figure 3–54 shows digital black-and-white images being taken via digital camera.

Figure 3–54. Digital black-and-white images being taken

Perform the following steps:

1. Start a new file. Use the “Small Objects, Millimeters” template.
2. Maximize the Top viewport.
3. Select Surface > Heightfield from Image, or click on the Heightfield from Image button on the Surface toolbar.
4. In the Select Bitmap dialog box, select the image file Car.tga from the Chapter 3 folder on the companion CD.

5 Pick two points A and B in the Top viewport to indicate the size of the surface to be derived from the bitmap, as shown in Figure 3–55.
6 In the Heightfield dialog box, set the number of sample points to 200 times 200, set Height to 0.5, select Interpolate through samples, and then click on the OK button. A surface is constructed from the bitmap image, as shown in Figure 3–56. The surface is complete.
7 Do not save your file.

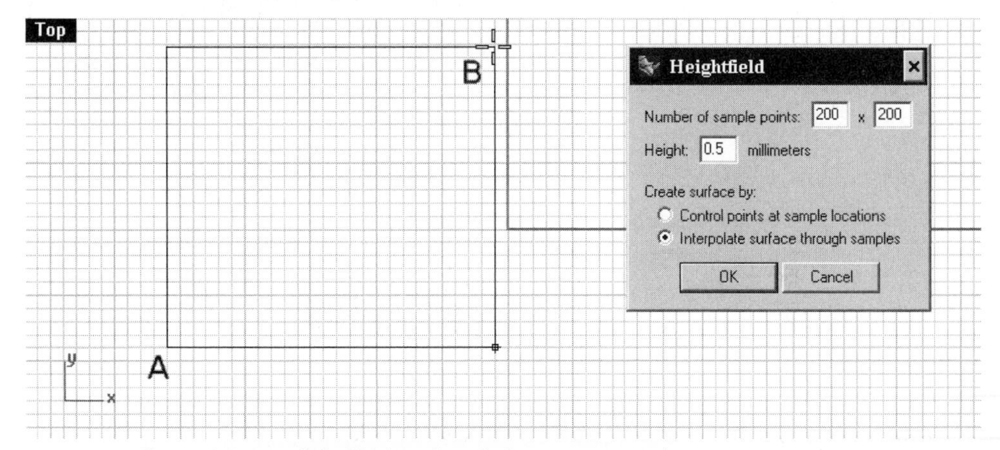

Figure 3–55. Heightfield surface being constructed

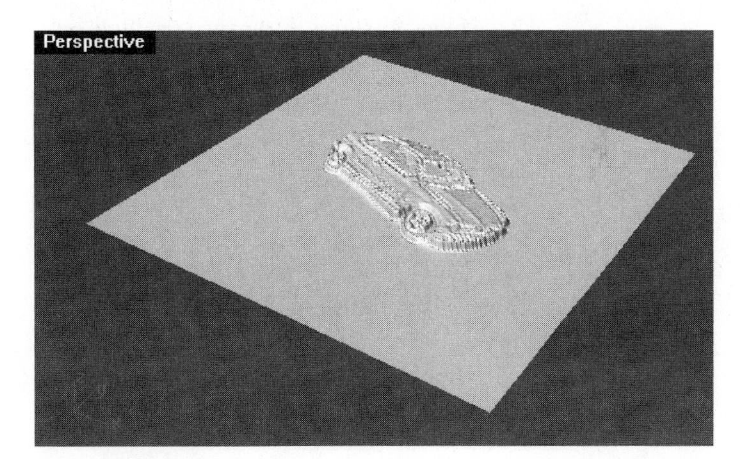

Figure 3–56. Heightfield surface

Drape Surface

Making a draped surface is analogous to wrapping a rectangular piece of plastic sheeting on a set of 3D objects in a way similar to vacuum forming. Vacuum forming is a type of plastics forming process in which plastic sheeting is heated, wrapped onto a 3D object (the mold), and a vacuum is applied to deform the sheet. Figure 3–57 shows a vacuum-form machine.

Figure 3–57. Vacuum-forming machine

Perform the following steps:

1. Select File > Open and select the file Drape.3dm from the Chapter 3 folder on the companion CD.
2. Select Surface > Drape, or click on the Drape Surface over Objects button on the Surface toolbar.
3. Select locations A and B, shown in Figure 3–58.
4. A draped surface is constructed, as shown in Figure 3–59.
5. Do not save your file.

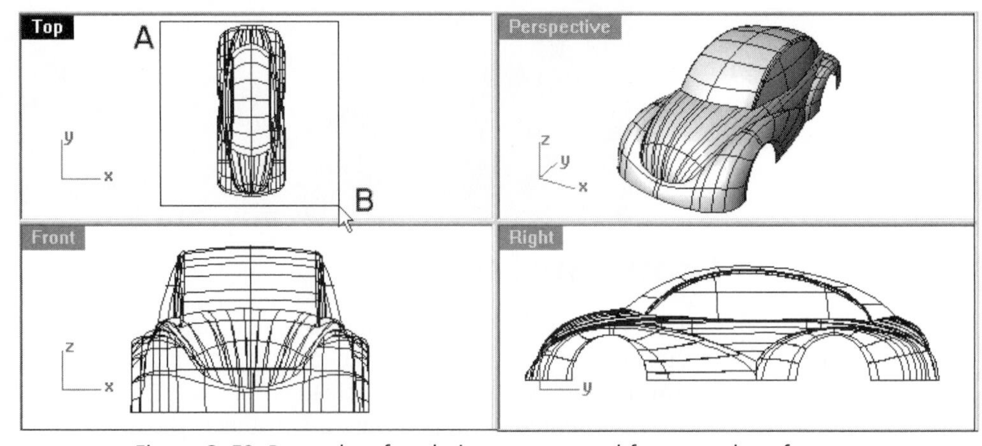

Figure 3–58. Draped surface being constructed from a polysurface

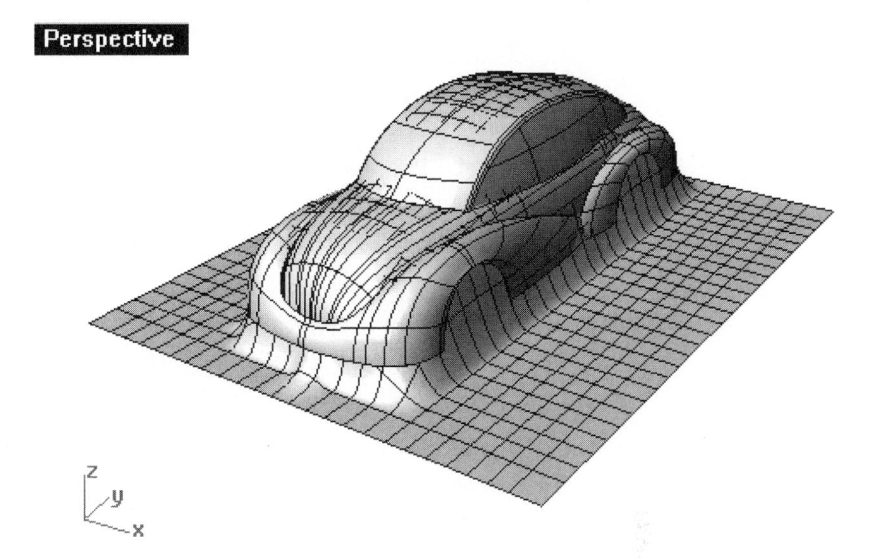

Figure 3–59. Draped surface constructed

Planar Surfaces

A planar surface is a flat surface which is rectangular in shape. However, you may also construct a planar surface with irregular boundary edges.

Rectangular Planar Surfaces

Using Rhino's tool, you can construct planar rectangular surfaces in five ways: Corner-to-Corner, 3 Points, Vertical, Through Points, and Cutting Plane.

Corner-to-Corner Rectangular Surface

A corner-to-corner surface is a rectangular planar surface constructed by specifying two diagonal points of the surface. Perform the following steps:

1. Select File > Open and select the file Rectangle01.3dm from the Chapter 3 folder on the companion CD.
2. Check the Point box on the Osnap dialog box.
3. Select Surface > Plane > Corner to Corner, or click on the Plane: Corner to Corner button on the Plane toolbar.
4. Select Deformable option on the command area. (Deformable refers to how the surface can be deformed by manipulating its control points. In essence, it concerns the degree of polynomial of

the surface produced. Normally, a planar rectangular surface is degree 2. A deformable rectangular has a degree of 3 or above. You will learn more about polynomial degree in the next chapter.)

5 Select points A and B, shown in Figure 3–60. A rectangular planar surface is constructed.

3 Points Rectangular Surface

A 3 points surface is a rectangular planar surface constructed by specifying two endpoints of an edge and a point along the opposite edge of the surface. Continue with the following steps.

6 Select Surface > Plane > 3 Points, or click on the Rectangular Plane: 3 Points on the Plane toolbar.

7 Select points C, D, and E, shown in Figure 3–60. A rectangular planar surface is constructed.

Vertical Rectangular Surface

A vertical surface is a rectangular planar surface constructed perpendicular to a construction plane on which you specify two endpoints of an edge. Perform the following steps.

8 Select Surface > Plane > Vertical, or click on the Vertical Plane button on the Plane toolbar.

9 Select points F, G, and H, shown in Figure 3–60. A vertical planar surface is constructed.

10 Do not save your file.

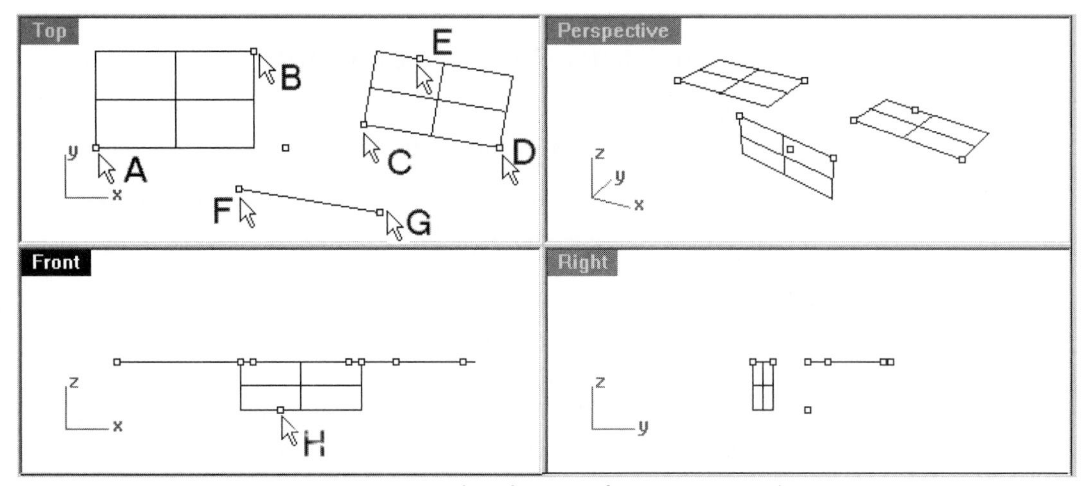

Figure 3–60. Rectangular planar surfaces constructed

Through Points Rectangular Surface

A through points surface is a planar rectangular surface constructed by interpolating among a set of points. Perform the following steps:

1. Select File > Open and select the file Rectangle02.3dm from the Chapter 3 folder on the companion CD.
2. Select Surface > Plane > Through Points, or click on the Fit Plane Through Points button on the Plane toolbar.
3. Select all the point objects, and press the ENTER key. A planar surface is fitted through the points. (See Figure 3–61.)
4. Do not save the file.

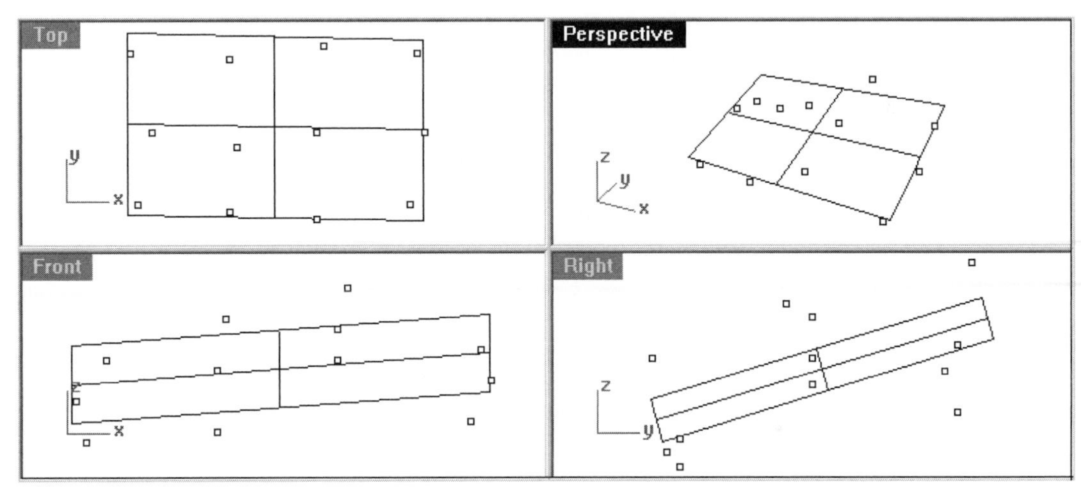

Figure 3–61. Planar surface fitted through a set of points

Cutting Plane Rectangular Surface

A cutting plane rectangle is a planar rectangular surface constructed by defining a section plane across a set of surfaces. Perform the following steps:

1. Select File > Open and select the file Rectangle03.3dm from the Chapter 3 folder on the companion CD.
2. Select Surface > Plane > Cutting Plane, or click on the Cutting Plane button on the Plane toolbar.
3. Select surface A and curves B and C, shown in Figure 3–62, and press the ENTER key.
4. Select locations D and E, F and G, and H and J. (Exact location is unimportant for the purpose of this tutorial.)

5. Press the ENTER key. Three rectangular planar surfaces that cut through the selected objects are constructed.
6. Do not save your file.

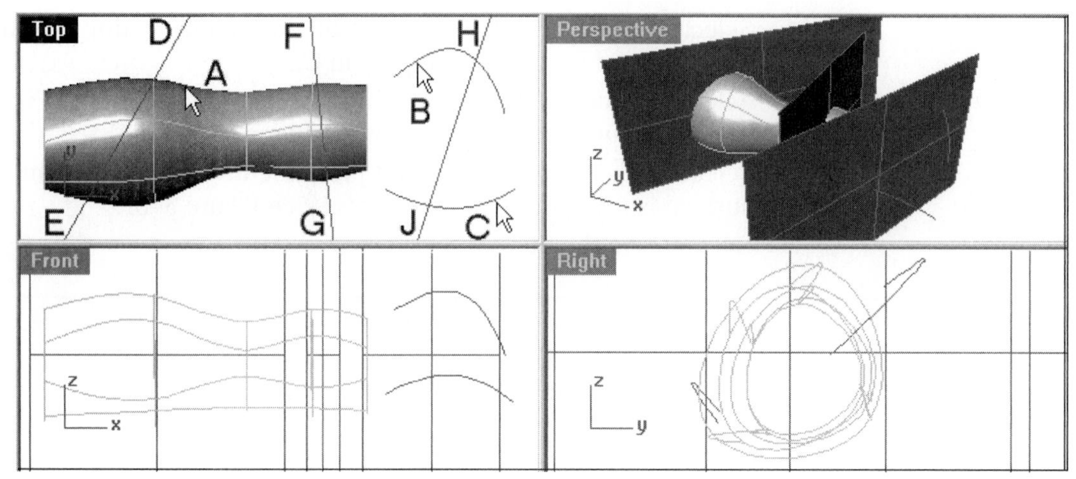

Figure 3–62. Cutting plane being constructed

Picture Frame Rectangular Surface

A picture frame rectangular surface is a rectangular surface with a specified bitmap image attached as texture. You can view the surface's mapped texture by setting the display to Rendered. Now perform the following steps:

1. Select File > Open and select the file PictureFrame.3dm from the Chapter 3 folder on the companion CD.
2. Click on the Picture Frame button on the Plane toolbar, or type PictureFrame at the command area.
3. Select the image file "Bubble Car.tif" from the Chapter 3 folder of the accompanying CD.
4. Click on point A, shown in Figure 3–63, and then point B. Note that the second selected point serves only to specify the width and orientation of the rectangular surface, because the height of the rectangle is governed by the bitmap's aspect ratio.
5. Do not save your file.

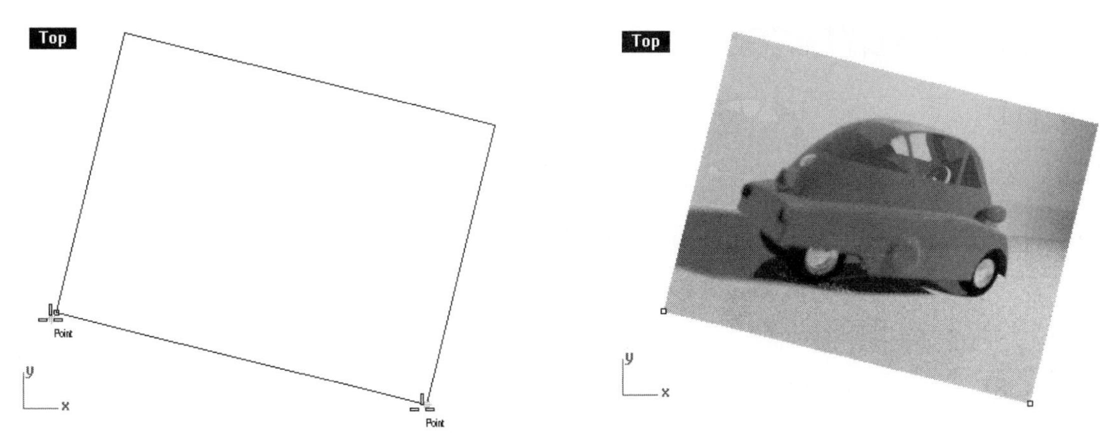

Figure 3–63. Points being selected (left) and picture frame constructed (right)

Planar Surface with Irregular Boundary Edges

A planar surface with irregular boundary edges is constructed from one or more planar curves. In essence, they are trimmed surfaces. You will learn more about trimming a surface in Chapter 7. Perform the following steps:

1. Select File > Open and select the file PlanarTrim.3dm from the Chapter 3 folder on the companion CD.
2. Select Surface > Planar Curve, or click on the Surface from Planar Curve button on the Surface toolbar.
3. Select curves A, B, and C, shown in Figure 3–64, and press the ENTER key. Two trimmed planar surfaces are constructed.
4. Do not save the file.

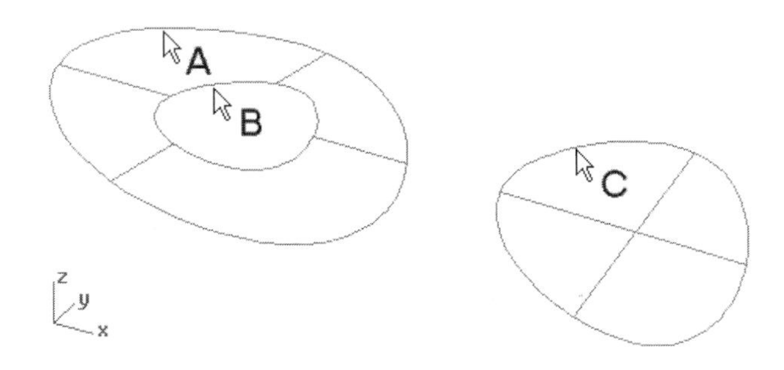

Figure 3–64. Trimmed planar surface being constructed (from a planar closed-loop curve)

Derived Surfaces

You can derive two kinds of surfaces from existing surfaces. You can obtain a new surface by offsetting an existing surface, and you can obtain a 2D flat pattern of an existing 3D surface.

Offset Surfaces

There are two kinds of offset surfaces: uniform offset surface and variable offset surface.

Uniform Offset Surface

To construct a surface that has resemblance in shape to an existing surface and at a uniform distance from that surface, you construct an offset surface. Every point on the offset surface is equal in distance from the original surface. An offset surface is used frequently in product styling. Perform the following steps to construct an offset surface:

1. Select File > Open and select the file OffsetSurface.3dm from the Chapter 3 folder on the companion CD.
2. Select Surface > Offset Surface, or click on the Offset Surface button on the Surface Tools toolbar.
3. Select surface A, shown in Figure 3–65,. (If you do not find the tracking lines indicating the normal direction of the surface, set the color of the Tracking Lines option on the Appearance > Color tab of the Rhino Options dialog box to black.)

Among the five options available, the FlipAll option enables you to flip the offset direction and the Solid option enables you to construct a solid object. (Rhino solid modeling method will be explained in the next chapter.)

4. If the arrow direction is not the same as that, shown in Figure 3–65, select the FlipAll option from the command area. *(Note: You can also click on the surface to flip the normal.)*
5. Type 4 to specify the offset distance. An offset surface is constructed.

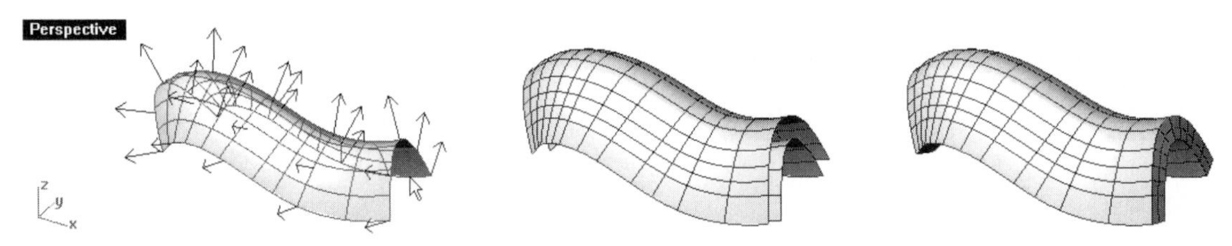

Figure 3–65. From left to right: Surface being offset, offset surface constructed, and solid constructed

Variable Offset Surface

To meet certain aesthetic requirements, you may wish to have an offset surface with variable offsetting distances. Continue with the following steps to construct a variable offset surface.

6. Undo the last command, or open the OffsetSurface.3dm from the Chapter 3 folder on the companion CD again.
7. Select Surface > Variable Offset Surface, or click on the Variable Offset of surface button on the Surface Tools toolbar.
8. Select surface A, shown in Figure 3–66.
9. Click on points B, C, D, and E one by one and drag to new locations.
10. Press the ENTER key.
11. Do not save your file.

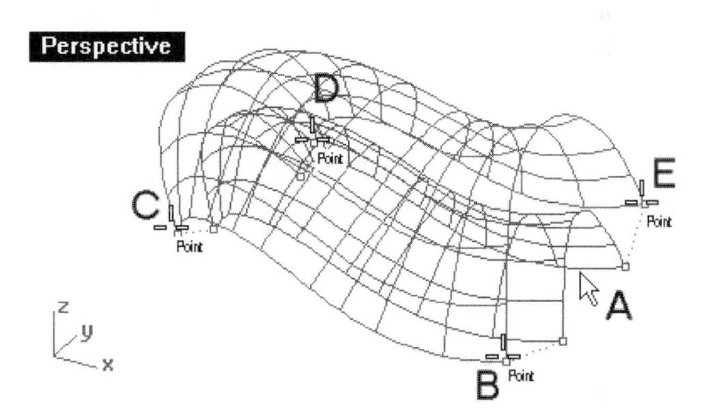

Figure 3–66. Variable distance offset surface being constructed

2D Flat Pattern

There are two ways to obtain a 2D flat pattern from an existing 3D surface: by unrolling and smashing.

Unrolling Developable Surfaces

Basically, only single-curve surface can be unrolled into a 2D surface accurately. Unrolling is useful in the sheet metal working industry. Before you make a sheet metal object, you need a development (flat) pattern of the object as a 2D sheet. You then roll or fold the 2D sheet to create the 3D object. An object constructed this way can be unrolled into a flat sheet. A cylindrical surface, for example, can be unrolled to a rectangle. In essence, a developable surface has to be a single-curve surface. Perform the following steps to unroll a developable surface:

1. Select File > Open and select the file Develop.3dm from the Chapter 3 folder on the companion CD.
2. Select Surface > Unroll Developable Srf, or click on the Unroll Developable Surface/Flatten Surface button on the Surface Tools toolbar.
3. Select surface A and then press the ENTER key. The surface is unrolled, as shown in Figure 3–67.
4. Do not save your file.

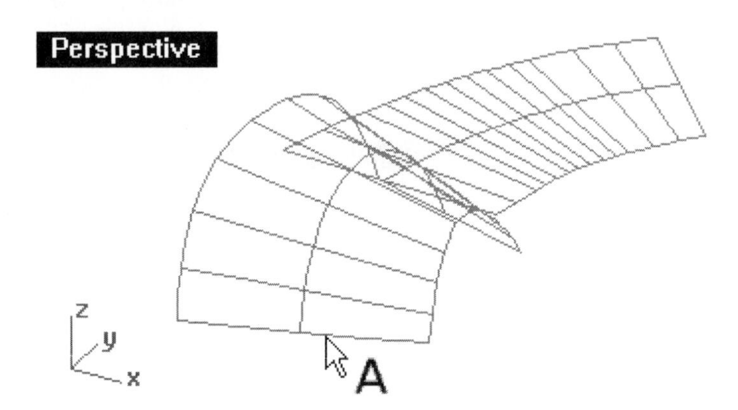

Figure 3–67. Surface being unrolled and unrolled surface

Smashing

If a surface is not readily developable, you may smash it to obtain an approximated unrolled surface. Perform the following steps:

1. Select File > Open and select the file Smash.3dm from the Chapter 3 folder on the companion CD.
2. Select Surface > Smash, or click on the Smash button on the Surface Tools toolbar.
3. If Label = No, select it to change it to Yes.
4. Select surface A, shown in Figure 3–68.
5. Select curve B and press the ENTER key. A smashed surface is constructed.

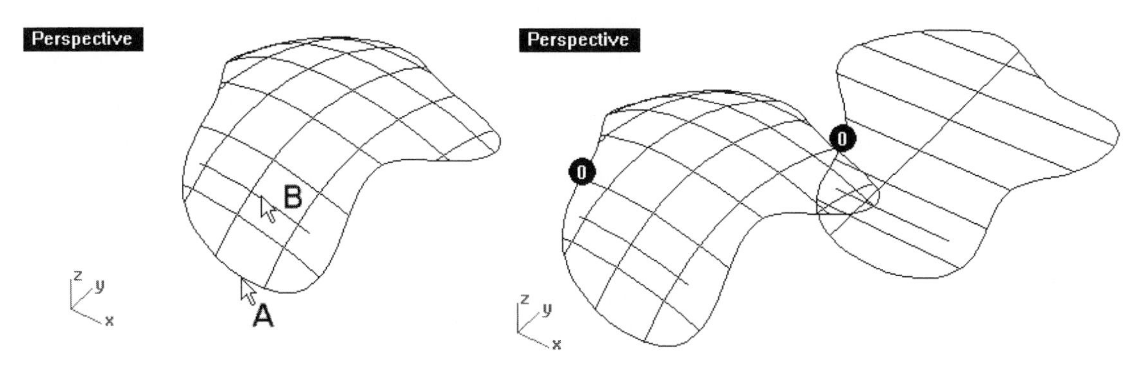

Figure 3–68. Free form surface (left) and smashed surface (right)

Flattening Curve Edges

Apart from smashing, it is possible to construct a 2D curve by projecting the edges of a developable surface through flattening. Perform the following steps:

1. Select File > Open and select the file Flatten.3dm from the Chapter 3 folder on the companion CD.
2. Right-click on the Unroll Developable Surface/Flatten Surface button on the Surface Tools toolbar, or type FlattenSrf at the command area.
3. Select edges A and B (Figure 3–69).
4. Press the ENTER key to accept the default sample spacing. A set of 2D curves is constructed.
5. Do not save your file.

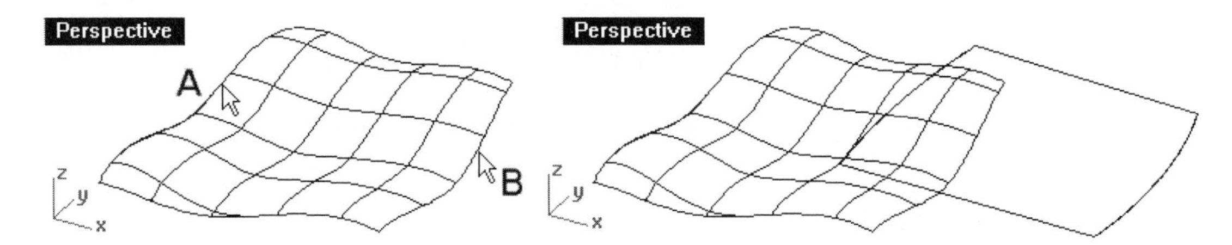

Figure 3–69. Developable surface (left) and edges flattened (right)

Rhino's Accuracy

One major advantage of using Rhino to construct free-form surfaces is that the framework of points/curves need not be accurately defined. For example, the sets of curves in a network curve surface need not intersect exactly, and the endpoints of the cross-section curves of swept surfaces need not lie exactly on the rail curves.

In other words, Rhino is an accurate NURBS surface tool on one hand and is a very flexible tool on the other hand.

Chapter Summary

For the sake of easy reference, we divide Rhino's NURBS surfaces logically into four categories.

The first surface category concerns the four major types of surfaces which are, in essence, common to most computer-aided design applications. They are extruded surfaces, revolved surfaces, swept surfaces, and lofted surfaces. Among them, extruded surface is further divided into six types (straight, tapered, to point, along curve, ribbon, and fin); revolved surface into two types (normal revolve and revolve rail); swept surface into two types (sweep along one rail and sweep along two rails); and lofted surface into three types (loft, edge curve, and curve network).

The second surface category includes Patch, Corner Points, Point Grid, Heightfield from Image, and Drape. The third surface category is planar surfaces. The fourth category is surfaces derived from existing surfaces. You can derive an offset surface, which offsets uniformly or un-uniformly. You can derive a flat pattern if the surface is a developable surface or have an approximated development by smashing.

In this chapter, you have examined various kinds of surfaces and learned how some of them are constructed from a set of curves and/or points. Because the curves and points are already given, constructing the surfaces is simple. However, making the curves and points can be a tedious job. Therefore, you will learn various techniques in point and curve construction in Chapters 4 through 6 and then continue with learning the fourth category of surfaces. However, if you are already familiar with Rhinoceros's curve construction and manipulation tools, you may proceed to Chapter 7 to continue with ways to manipulate NURBS surfaces.

Review Questions

1. List the four major kinds of Rhino surfaces that are common to most computer-aided design applications.
2. Apart from the free-form surfaces common to most computer aided design systems, what other kinds of free-form surfaces can you construct using Rhino?
3. How many ways can a planar surface be constructed? What are they?
4. List the two kinds of derived surfaces.

CHAPTER 4

Free-Form NURBS Curves and Point Objects

Introduction

This chapter illustrates various methods of constructing and manipulating free-form curves and point objects.

Objectives

After studying this chapter you should be able to:

- Use Rhino to construct and manipulate free-form curves and point objects

Overview

After learning how free-form surfaces can be constructed from a framework of points and/or curves in Chapter 3, this chapter and the next two chapters provide you with a solid understanding of curve and point construction. In this chapter, you will learn how to use Rhino as a tool for constructing and manipulating free-form curves and point objects. In Chapter 5, you will learn how to construct curves of regular pattern. Chapter 6 provides more information about Rhino curve tools. After equipping yourself with skills in 3D curve manipulation, in Chapters 7 and 8 you will continue to perform various methods of surface and solid construction and manipulation.

Rhino Curves

Common to most other contemporary surface modeling tools, Rhino uses NURBS mathematics to define curves and surfaces. In essence, a NURBS curve is a set of connected spline segments, and the joint between two contiguous spline segments is a knot. The degree of the polynomial equation, the control point location, and the weight of the control points determine the shape of each spline segment.

For ease of classification, we will divide Rhino's curves into two major types: free-form curves and curves with regular pattern such as line, arc, and circle. In this chapter, we will focus on Rhino's free-form curves and then proceed to curves of regular pattern in the next chapter.

Because all Rhino curves are NURBS curves, methods used to manipulate a free-form curve delineated in this chapter also apply to other types of curves.

To help you gain a deep understanding of NURBS curve manipulation and be confident in using it to design free-form surfaces, we will examine various ways to construct a free-form NURBS curve and methods to modify it.

Basic Curve Construction

When making surfaces, you select curves and points that are already constructed. But you need to specify a location to construct individual point objects and basic curves. To specify a location, you use your pointing device to select a point in one of the viewports, or input a set of coordinates at the command area. To help construct 3D points, you can use the elevator mode (holding down the Control key to select two locations from two viewports to compose a 3D point) and the planar mode (checking the Planar button on the Status bar).

As explained in Chapter 2, each viewport includes a construction plane. Selecting a point in one of the viewports specifies a location on the corresponding construction plane. To key in a set of coordinates, you use either the construction plane coordinate system or the World coordinate system.

The X and Y directions of the construction plane coordinates correspond to the X (red in color, by default) and Y (green in color, by default) axes of the construction plane. Therefore, typing the same construction plane coordinates (other than the origin) at the command area results in different locations in 3D space, depending on where the cursor is placed, for example, on the Top viewport as opposed to the Front viewport. In the following exercises, you will use the pointing device to select points in viewports.

Interpolated Curve and Control Point Curve

Using Rhino, there are many ways to construct a free-form curve. Two basic ways are specifying a set of points through which the curve has to interpolate and setting the control points of the curve.

Interpolated Curve

To construct an interpolated curve, you specify a number of points through which the curve has to interpolate. Practically, you may regard this type of curve as a best-fit curve through the selected points. As a result, this method is considered to be more user-friendly because you can better perceive the shape of the curve while constructing it. Perform the following steps:

1. Select File > Open and select the file FreeFormCurve01.3dm from the Chapter 4 folder on the companion CD.

In the file, you will find four point objects already constructed for you. Now you will construct a free-form curve passing through these points.

2. Check the Osnap button on the status bar.
3. In the Osnap dialog box, check the Point box.
4. Select Curve > Free-Form > Interpolate Points, or click on the Curve: Interpolate Points/Curve: By Handles button on the Curve toolbar.
5. Select points A, B, C, and D, shown in Figure 4–1, and press the ENTER key to terminate the command.

An interpolated curve is constructed passing through the selected points. Do not close the file.

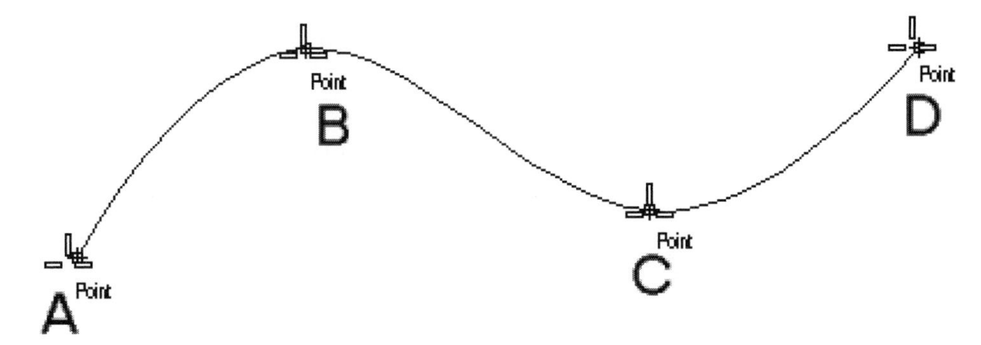

Figure 4–1. Interpolated curve being constructed

*(**NOTE:** You may go back to Chapter 2 to learn how to input 3D points by holding down the control key and clicking on two adjacent viewports.)*

Control Point Curve

The control point curve is more mathematically based. It is constructed by specifying the locations of the control points of the curve. You use this method to construct a curve when you know only the locations of the control points rather than the points through which the curve passes.

Control Point Location

Normally, control points lie outside the curve. In an open-loop curve, only the first and last control points coincide with the end points of the curve. These endpoints are also called anchor points. In a closed-loop curve, all control points lie outside the curve, an example of which is shown in Figure 4–2.

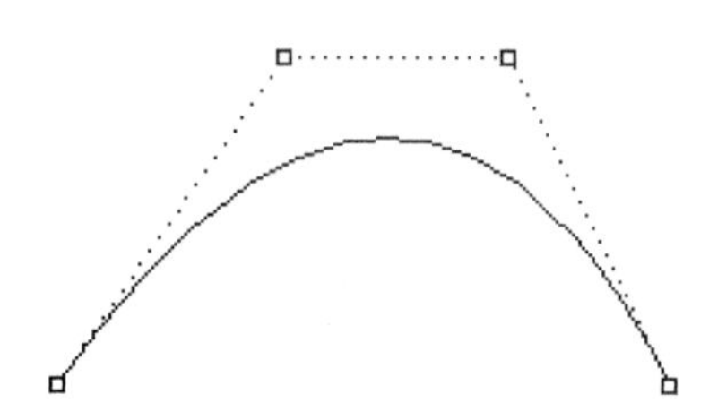

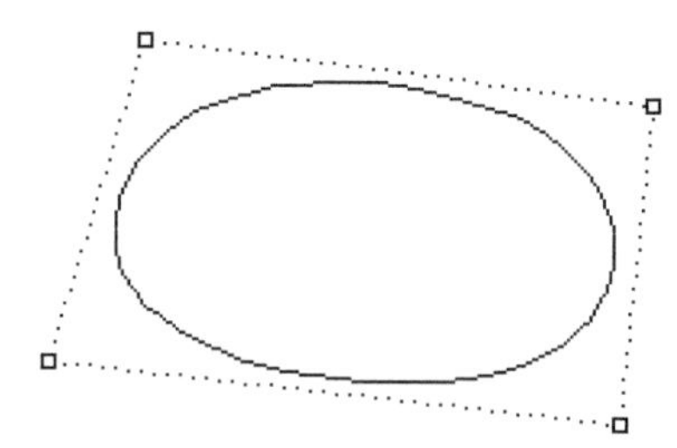

Figure 4–2. Control point locations of an open-loop curve and a closed-loop curve

Continue with the following steps to construct a control point curve using the point objects as control point locations.

6. Set current layer to Layer01.
7. Select Curve > Free-Form > Control Points, or click on the Control Point Curve/Curve Through Points button on the Curve toolbar.
8. Select points A, B, C, and D, shown in Figure 4–3, and then press the ENTER key to terminate the command.

A control point curve is constructed. Note the difference in shape between the interpolated curve and the control point curve.

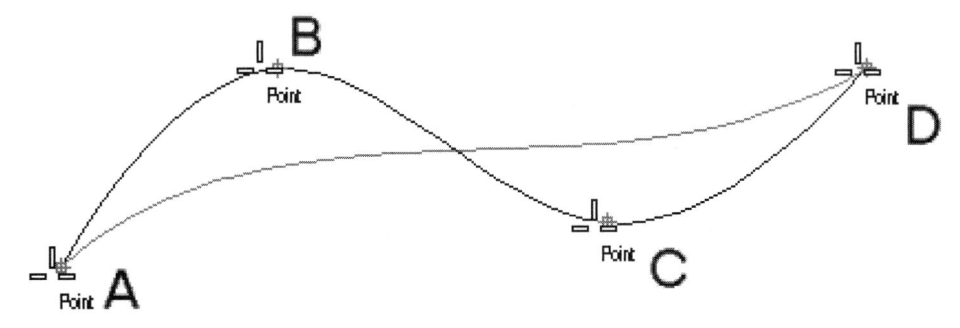

Figure 4–3. Control point curve constructed

Comparing Interpolated Curve and Control Point Curve

Regardless of the method you use to construct a curve, a set of control points governs the shape of the curve. To appreciate the difference between the interpolated curve and the control point curve in terms of the number of control points and their locations, let us turn on their control points, as follows:

9 Select Edit > Control Points > Control Points On, or click on the Control Points On/Points Off button on the Point Editing toolbar.

10 Select the free-form curves and press the ENTER key. The control points are turned on, as shown in Figure 4–4.

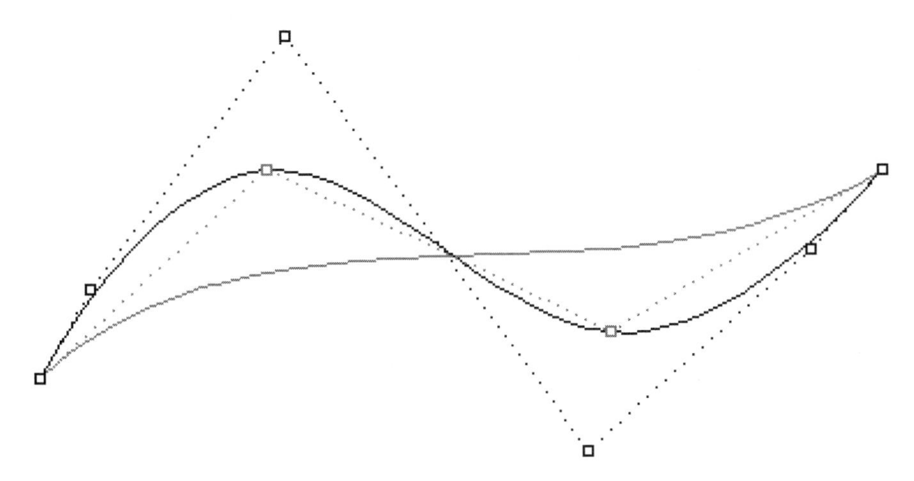

Figure 4–4. Control points turned on

The control point curve (curve constructed by specifying four control points) has the control point coincident with the point objects. The interpolated curve (curve constructed by specifying four interpolated points) has six control points, with the curve showing more segments. Excepting the end points, both curves do not pass through the control point locations. Turn off the control points, as follows.

11. Select Edit > Control Points > Control Points Off, or right-click on the Control Points On/Points Off button on the Point Editing toolbar.
12. Do not save your file.

Polynomial Degree Concepts

As we have said, a NURBS curve is a kind of polynomial spline, and the degree of a polynomial equation has a direct impact on the complexity of the shape of the curve. A higher polynomial degree curve has more control points than one with a lower polynomial degree. Basically, a line is a degree 1 NURBS curve, an arc is a degree 2 NURBS curve, and free-form curve is a NURBS curve of degree 3 or above.

Changing the Polynomial Degree of a Curve

One of the ways to modify a curve is to raise or reduce its polynomial degree. Simply raising the degree of a polynomial spline curve does not change the curve's shape, but it does increase the number of control points. In turn, more control points enable you to modify the curve to create a more complex shape. (You will learn how to modify the shape of a curve by manipulating its control points in the next paragraph.) On the other hand, reducing the degree decreases the number of control points and hence simplifies the shape of the curve. Perform the following steps:

1. Select File > Open and select the file FreeFormCurve02.3dm from the Chapter 4 folder on the companion CD.

Here, you have three identical curves. Their degree of polynomial is 3.

2. Select Edit > Control Points > Control Points On, or click on the Control Points On/Points Off button on the Point Editing toolbar.
3. Select the free-form curves A, B, and C, shown in Figure 4–5, and press the ENTER key. The control points are turned on.

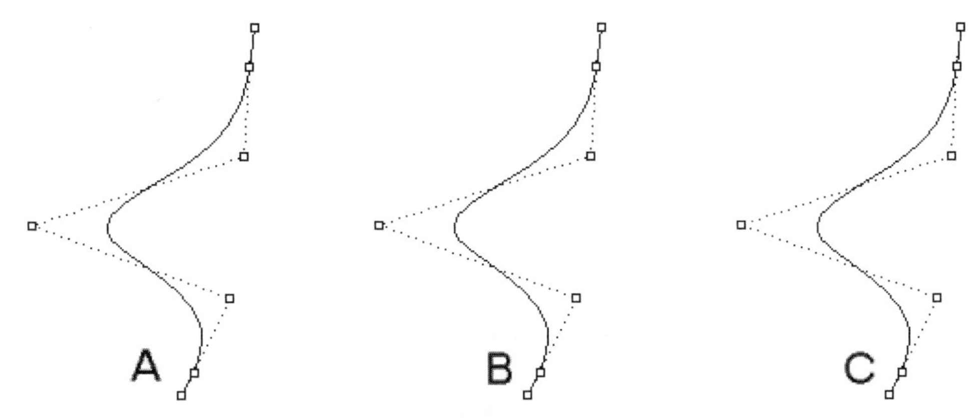

Figure 4–5. Control points turned on

4 Select Edit > Change Degree, or click on Change Degree on the Curve Tools toolbar.

5 Select curve A (Figure 4–6) and press the ENTER key.

6 Type 2 at the command prompt to change the degree of polynomial to 2.

7 Repeat the last command by pressing the ENTER key.

8 Select curve B (Figure 4–6) and press the ENTER key.

9 Type 5 at the command prompt to change the degree of polynomial to 5.

10 Do not save your file.

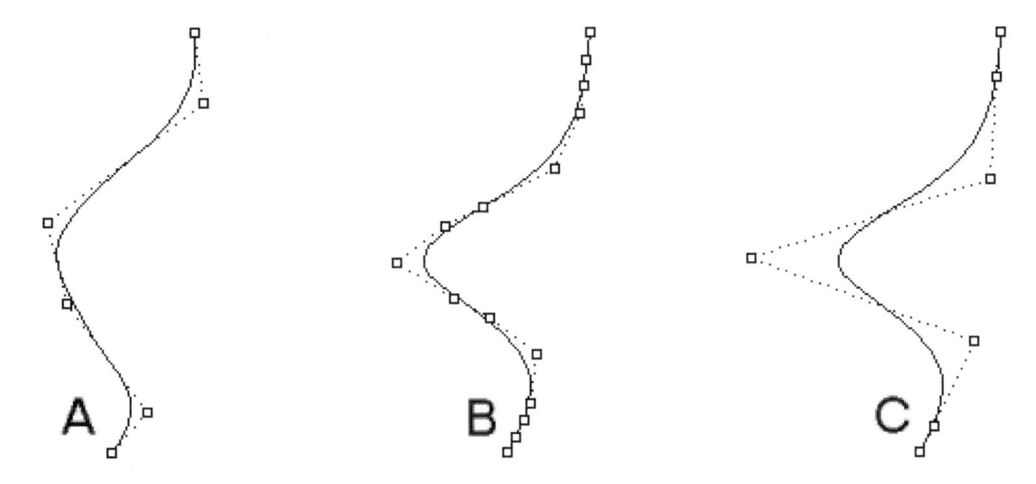

Figure 4–6. Degree of polynomial being changed

As can be seen, the number of control points in curve A is reduced and in curve B is increased. In regard to the shape of the curve, curve A is simplified but curve B remains unchanged.

Control Point Manipulation

Understanding that the degree of polynomial has a direct influence on the number of control points, you will now learn how to manipulate control points to modify a curve. You can change the location and the weight of the control points. You can also delete control points to simplify the shape of the curve.

Moving Control Points

A way to change the shape of a curve is to move the location of the curve's control points. You may select the control point, hold down the mouse button, and drag it to a new position. To move the control point in a more precise way, you can use the nudge keys, which is the combination of the ALT key and the arrow keys. To move the control point in three directions independent of the current construction plane, you can use the buttons on the Organic toolbar.

Perform the following steps:

1. Select File > Open and select the file FreeFormCurve03.3dm from the Chapter 4 folder on the companion CD.
2. Turn on the control points of the curve.
3. Click on control point A, indicated in Figure 4–7, and drag it to point B. Two other ways you can move the control points are holding down the ALT key and then pressing the arrow keys, and using the buttons on the Organic toolbar.

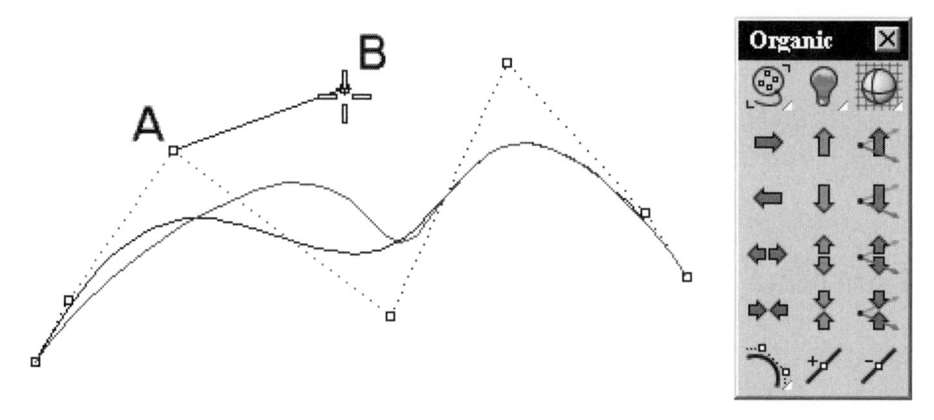

Figure 4–7. Dragging a control point to change the shape of the curve

Deleting Control Points

By deleting one or more control points of a curve, you simplify its shape. Continue with the following steps.

4 Select control point A, shown in Figure 4–8.

5 Press the DELETE key. The selected control point is deleted and the curve is simplified, in terms of the number of control points.

In addition to turning on a control point, selecting a control point, and pressing the DELETE key to remove a control point, you can click on the Remove a control point button on the Curve tools toolbar and then select a curve to remove unwanted control points.

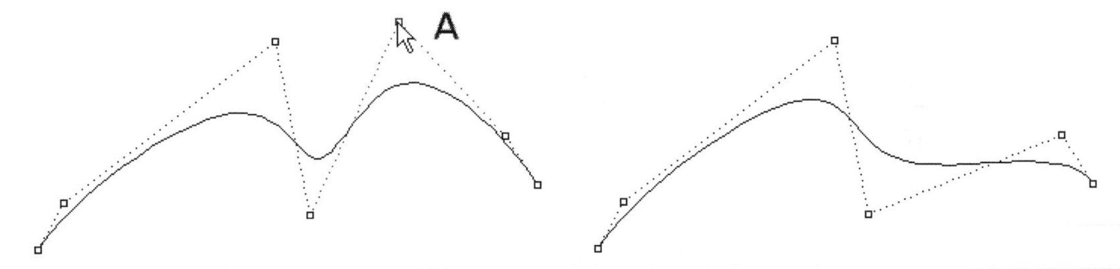

Figure 4–8. Control point to be deleted (left) and control point deleted (right)

Inserting Control Points

You can add control points to a curve as follows:

6 Select Edit Control Points > Insert Control Point, or click on the Insert a Control Point button on the Curve Tools toolbar.

7 Select curve A and then location B, shown in Figure 4–9. A control point is added to the curve.

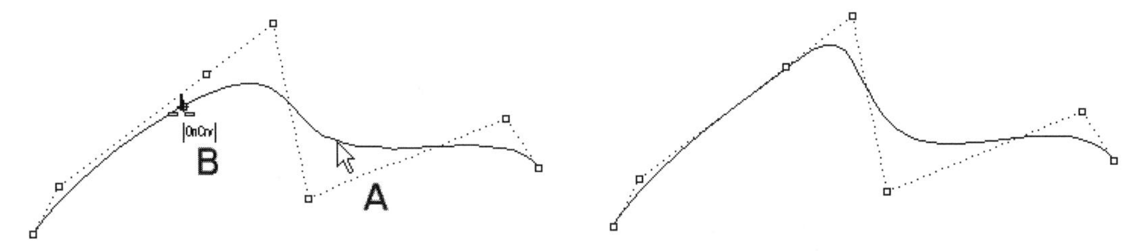

Figure 4–9. Control point being added (left) and control point added (right)

Adjusting Control Point Weight

The weight of control points has a significant effect on the shape of a spline segment. You can regard the weight of a control point as a pulling force that pulls the spline curve toward the control point. The higher the weight, the closer the curve will be pulled to the control point, an example of which is shown in Figure 4–10.

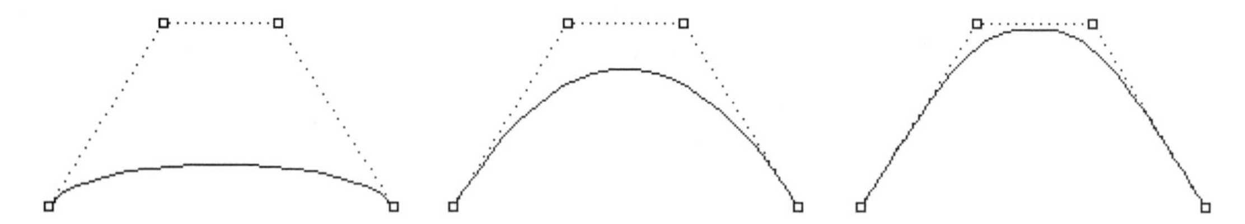

Figure 4–10. Control points with different weights (from left to right: weight 0.1, 5, and 7)

To edit the weight of a control point, continue with the following steps.

8 Select Edit > Control Points > Edit Weight, or click on the Edit Control Point Weight button on the Point Editing toolbar.

9 Select control point A (you can select multiple points) and press the ENTER key.

10 In the Set Control Point Weight dialog box, shown in Figure 4–11, set the weight of the selected point(s) to 0.2, and then click on the OK button.

11 Do not save your file.

Note that the curve is being pulled more closely to the selected control point.

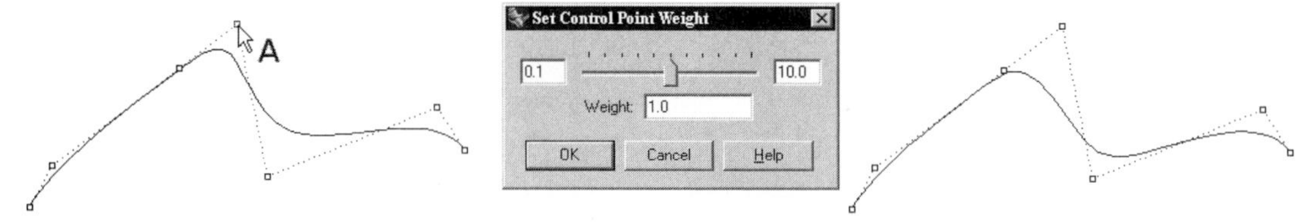

Figure 4–11. Point weight being modified in the Set Control Point Weight dialog box

Spline Segments

Because a NURBS curve is a polynomial spline with a number of segments connected end-to-end at the segments' endpoints, called knots, the number of segments has a direct impact on the number of control points. A curve with more segments has more control points than one with fewer segments. A way to modify the number of segments of a curve is to add or remove the knots of the curve.

Adding knots to a NURBS curve increases the number of spline segments without changing the shape of the curve. However, a curve with more segments has more control points. Subsequently, you can modify the curve to create a more complex shape.

Inserting Knot Points

A knot is the junction between two contiguous spline segments. One way to increase the complexity of the shape of a curve is to increase the number of spline segments by inserting knots along the curve. In the following, you will add a knot to the curve. Adding knots to a curve increases the number of spline segments. As a result, there will be more control points. However, the shape of the curve will not change until you manipulate the control points. Perform the following steps:

1. Select File > Open and select the file FreeFormCurve04.3dm from the Chapter 4 folder on the companion CD.
2. Turn on the control points of the curve.
3. Select Edit > Control Points > Insert Knot, or click on the Insert Knot/Insert Edit Point button on the Point Editing toolbar.
4. Select curve A, shown in Figure 4–12.
5. Select location B, shown in Figure 4–12, along the curve where you want to insert a knot. *(Note that you can insert multiple knot points.)*
6. Press the ENTER key to terminate the command. A knot is inserted.

Comparing Figure 4–12 (left) with Figure 4–12 (right), you will find that there is an increase in the number of control points. As for the shape of the curve, there is no change. Depending on the exact location where you insert the knot, Figure 4–12 (right) may not look the same as that displayed in your screen.

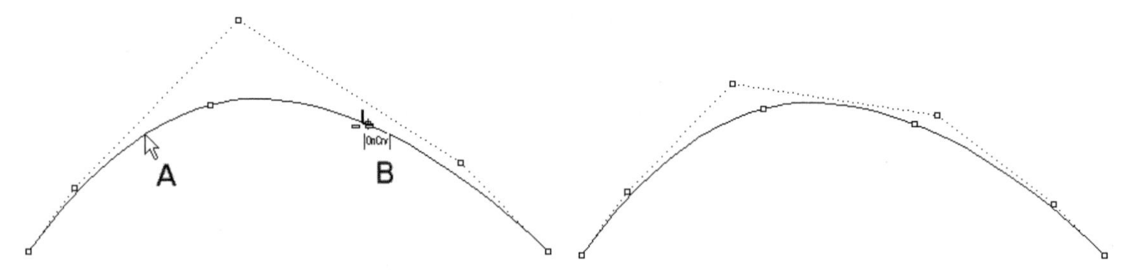

Figure 4–12. Knot being inserted in the curve (left) and number of control points increased (right)

Kink Point

A spline curve is analogous to bamboo: you can bend bamboo to a great degree and still retain a very smooth curve. If we continue to bend the bamboo further, it will fracture at a certain point. The fracture point is equivalent to a kink in a spline curve.

Mathematically, a kink point is a special type of knot on a curve in which the tangent direction of the contiguous spline segments is not the same. A kink occurs when you join two curves with different tangent directions, or when you explicitly add a kink to a curve.

A kink point is a special type of knot on a curve in which the tangent directions of contiguous spline segments are not the same. You add a kink point to a curve as follows.

7 Select Edit > Point Editing > Insert Kink, or click on the Insert Kink button on the Point Editing toolbar.

8 Select curve A, shown in Figure 4–13 (left).

9 Select location B, shown in Figure 4–13 (left), along the curve where you want to insert a kink. Note that you can insert multiple kink points.

10 Press the ENTER key to terminate the command. A kink is inserted.

11 Select kink point C, shown in Figure 4–13 (right), hold down the left mouse button, and drag the mouse to a new location to see the effect of a kink.

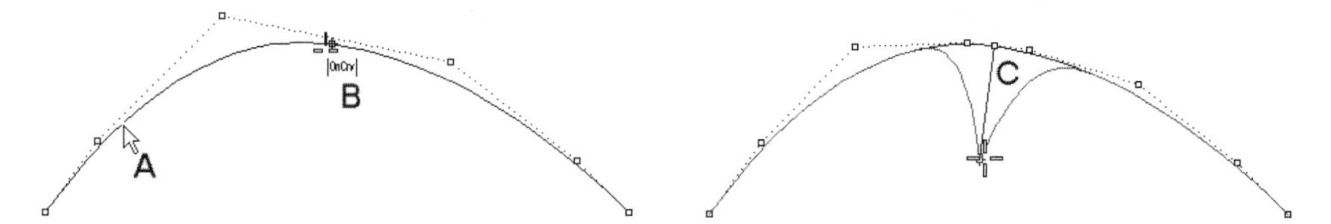

Figure 4–13. Kink point being inserted (left) and inserted kink point being moved (right)

Removal of Knot and Kink Points

To remove a kink point or to reduce the complexity of a curve, we remove the kink points or knots along the curve. Because removing knots on a curve reduces the complexity of the curve, the curve's shape will change. To remove a kink point or other type of knot from a curve, perform the following steps.

12 Select Edit > Control Points > Remove Knot, or click on the Remove Knot button on the Point Editing toolbar.

13 Select curve A, shown in Figure 4–14.

14 Select knots B and C, shown in Figure 4–14, and press the ENTER key.

15 Do not save your file.

A knot of the curve is removed, as shown in Figure 4–14 (right). After you remove a knot from a curve, the number of control points decreases. As a result, the shape of the curve is simplified and changed. Note that Figure 4–14 (right) may not be the same as that shown in your screen because the exact locations of the inserted kink and knot points are different.

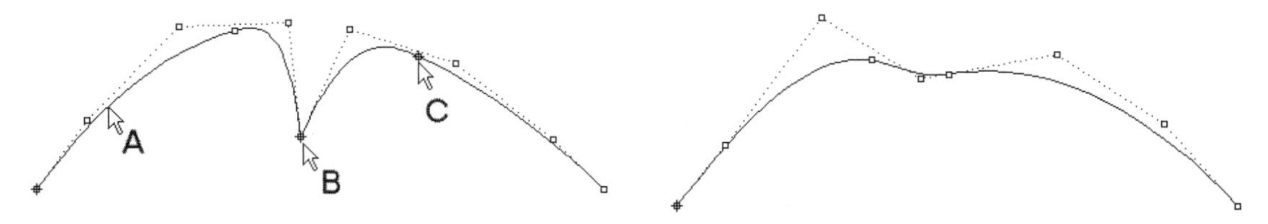

Figure 4–14. Knots being removed (left) and knot removed (right)

Spline Segment, Polynomial Degree, and Control Point

Because the number of control points of a NURBS curve depends on both the number of spline segments and the polynomial degree, you may increase the segment number or the polynomial degree to increase the number of control points. As a result, it is recommended that you keep the polynomial degree to 3 and increase or decrease the number of segments to modify the number of control points.

Periodic Curve

A periodic curve is a special kind of curve in that it is a closed curve with no kink point. Figure 4–15 shows a closed-loop curve with a kink and a periodic curve.

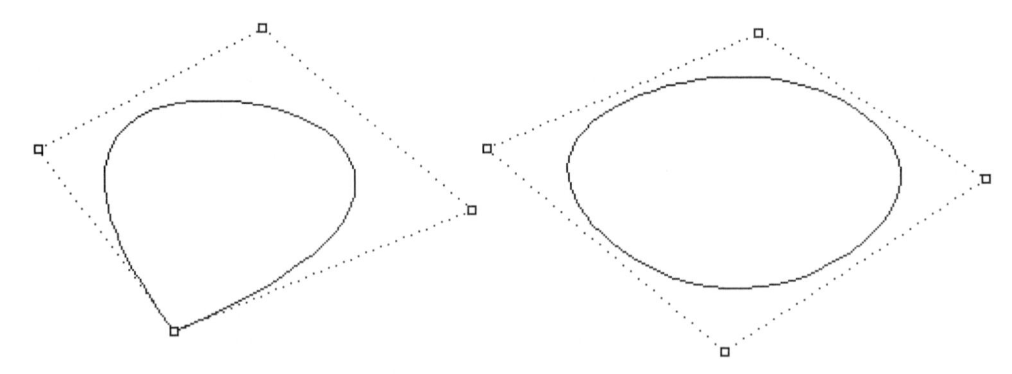

Figure 4–15. Closed-loop curve with a kink (not a periodic curve) and a periodic curve

To make a curve periodic, perform the following steps:

1. Select File > Open and select the file FreeFormCurve05.3dm from the Chapter 4 folder on the companion CD.
2. Turn on the control points so that you can see the effect of making the curve periodic.
3. Select Edit > Make Periodic, or click on the Make Periodic/Make Non-Periodic button on the Curve Tools toolbar.
4. Select curve A, shown in Figure 4–16, and press the ENTER key.
5. Select the Yes option on the command area to delete the original curve.

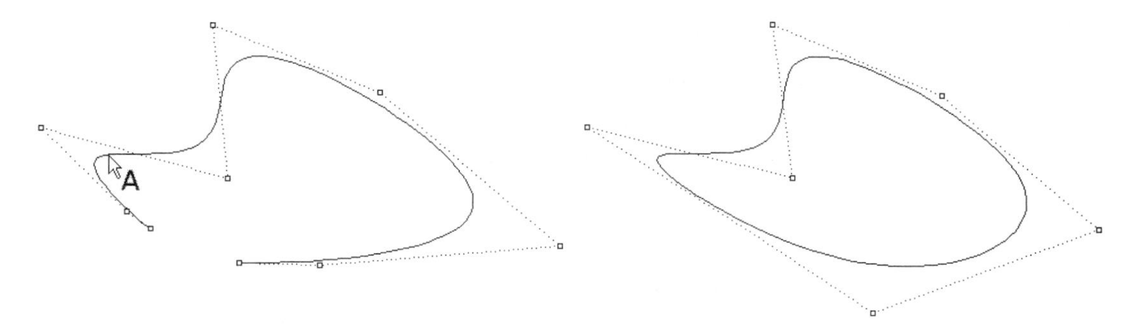

Figure 4–16. Original curve (left) and periodic curve (right)

*(**NOTE:** The same command can be applied to a surface to change it to a periodic surface. Details will be discussed in Chapter 7.)*

Adjusting the Seam of a Closed Curve

Although a periodic curve is a closed curve, it contains a seam point that is both the start and end points of the curve. If you construct a surface from one or more periodic curves, the location of the seam point may affect the final outcome of the surface. Therefore, you may want to adjust the seam point location, as follows:

6. Select Curve > Curve Edit Tools > Adjust Closed Curve Seam, or click on the Adjust Closed Curve Seam button on the Curve Tools toolbar.
7. Select curve A (Figure 4–17) and press the ENTER key.
8. Click on the seam point and drag it along the curve.
9. Press the ENTER key. The seam point location is changed.
10. Do not save your file.

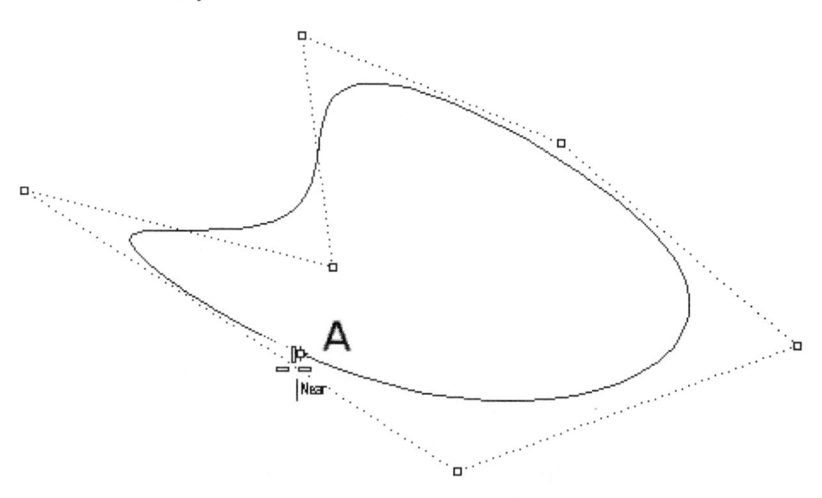

Figure 4–17. Seam point location being changed

Edit Point

Because control points are normally not lying along the curve, you may find it difficult to make a curve pass through designated locations by manipulating the control point. To add flexibility in modifying a NURBS curve, Rhino enables you to use edit points along the curve. Edit points are independent of degree, control points, and knots of a curve.

Edit Point Manipulation

Edit points lie along a curve. To change the shape of a curve, you display the edit points, select them, and drag them to new locations. To turn on the edit points of the curve and manipulate the points to modify the curve, perform the following steps:

1. Select File > Open and select the file FreeFormCurve06.3dm from the Chapter 4 folder on the companion CD.
2. Set object snap mode to point.
3. Select Edit > Control Points > Show Edit Points, or click on the Edit Points On/Points Off button on the Point Editing toolbar.
4. Select curve A, shown in Figure 4–18, and press the ENTER key.
5. Select edit point B (Figure 4–18), hold down the left mouse button, and drag the mouse to point C (Figure 4–18). The curve is modified.
6. Do not close the file.

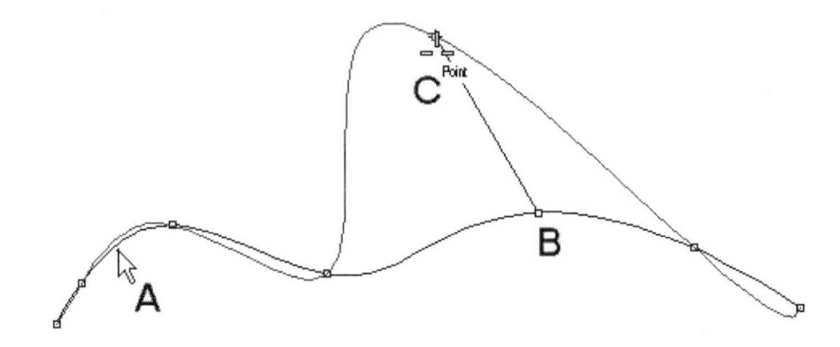

Figure 4–18. Selected edit point being dragged

Point Editing

The shape of a curve is controlled by the location and weight of control points. Except for the two anchor points at the end points of the curve, all the control points of a curve () lie outside the curve, so it may be more predictable to manipulate the edit points along the curve. Apart from manipulating the control points or the edit points, you can add or remove knots from a curve and add kink points to a curve.

Handlebar Editor

Edit points enable you to modify a curve only in a limited scope. To change the tangent direction and the location of any point along a curve, you use the handlebar. The handlebar is a special type of editing tool consisting of a point on the curve and two tangent lines. Moving the central point of the handlebar changes the location of a point along the curve. Selecting and dragging the end points of the handlebar changes the weight and tangent direction of the curve at a selected location. Continue with the following steps to modify the curve by using the handlebar editor.

7. Select Edit > Point Editing > Handlebar Editor, or click on the Handlebar Editor button on the Point Editing toolbar.
8. Select the curve, shown in Figure 4–19. The handlebar is displayed at the point where you selected the curve.
9. Click on an end point of the handlebar and drag, as shown in Figure 4–19.
10. Press the ENTER key.
11. The curve is modified. Using a handlebar to edit the shape of a curve, the curve profile changes but the location of the selected point at the handlebar does not change.
12. Do not save your file.

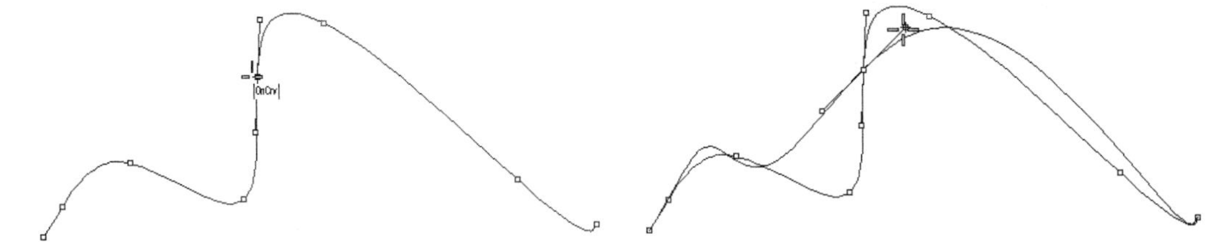

Figure 4–19. Handlebar being displayed (left) and handlebar being manipulated

Handle Curve

Having learnt how to use the handlebar to modify a curve, we will now introduce the use of handlebar to construct a curve, as follows:

1. Start a new file. Use the "Small Objects, Millimeters" template.
2. Maximize the Top viewport.
3. Select Curve > Free-Form > Handle Curve, or click on the Handle Curve button on the Curve toolbar

4 Click on location A, shown in Figure 4–20, to specify the first curve point location.

5 Click on location B (Figure 4–20) to specify the first handle location.

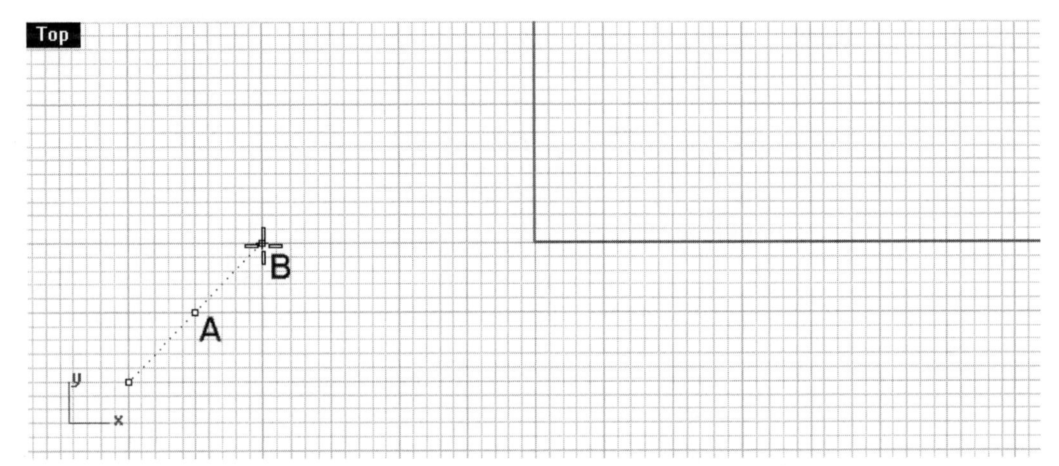

Figure 4–20. First curve point and handle location specified

6 Click on location C, shown in Figure 4–21, to specify the first curve point location.

7 Click on location D (Figure 4–21) to specify the first handle location.

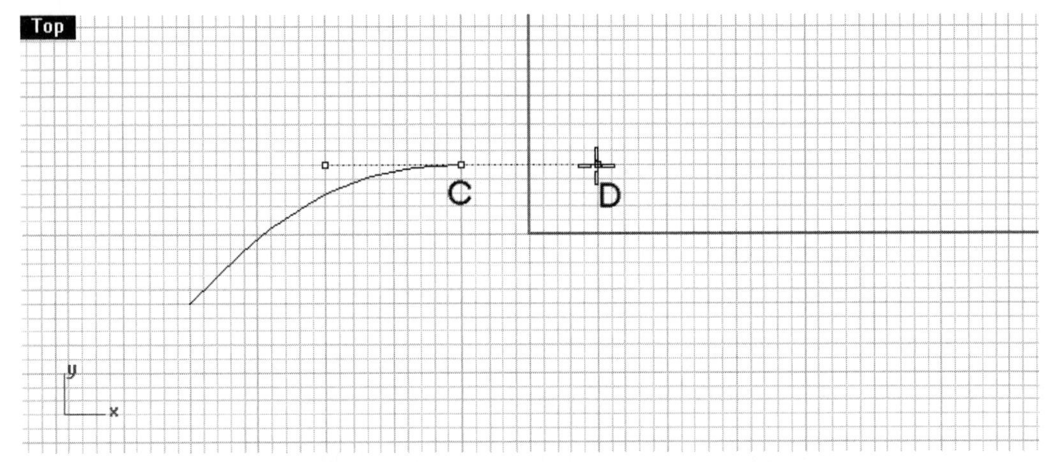

Figure 4–21. Second curve point and handle location specified

8 Click on location E, shown in Figure 4–22, to specify the first curve point location.

9 Click on location F (Figure 4–22) to specify the first handle location.

10 Press the ENTER key. A curve is constructed. (You may turn its control points to discover how complex the curve is.)
11 Do not save your file.

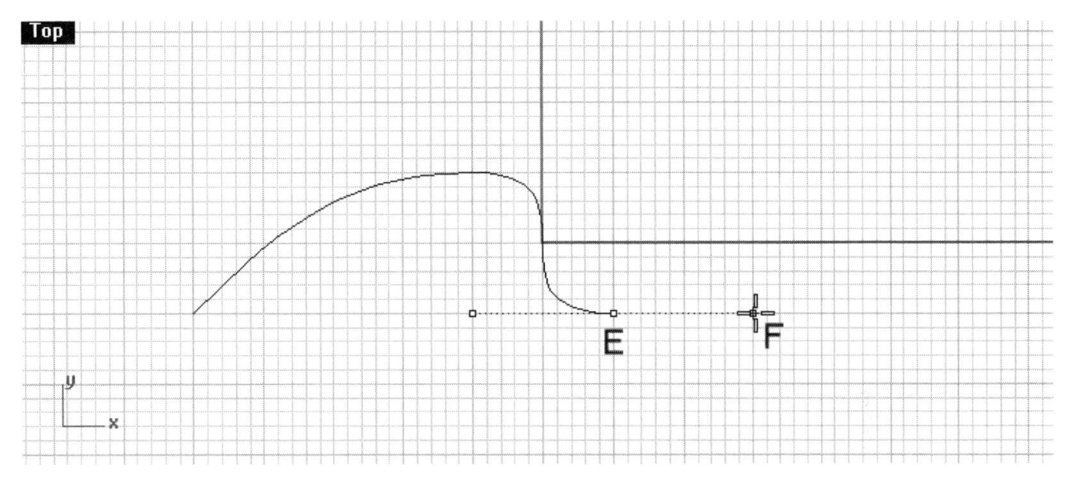

Figure 4–22. Third curve point and handle location specified

If the Control key is pressed while placing a handle point, the previous point will be moved to a new location.

Sketch Curve

Apart from specifying the interpolated points or the control points and using the handlebar, another way to construct a free-form sketch is to use sketching technique. With sketching technique, you use the mouse as a drafting pen to construct a curve. Perform the following steps to construct a sketch curve:

1 Start a new file. Use the "Small Objects, Millimeters" template.
2 Maximize the Top viewport.
3 Select Curve > Free-Form > Sketch, or click on the Sketch/Sketch On Surface button on the Curve toolbar.
4 Select location A, shown in Figure 4–23.
5 Hold down the left mouse button and drag the cursor to location B.
6 Release the mouse button.
7 Press the ENTER key. The curve is complete. Do not save your file.

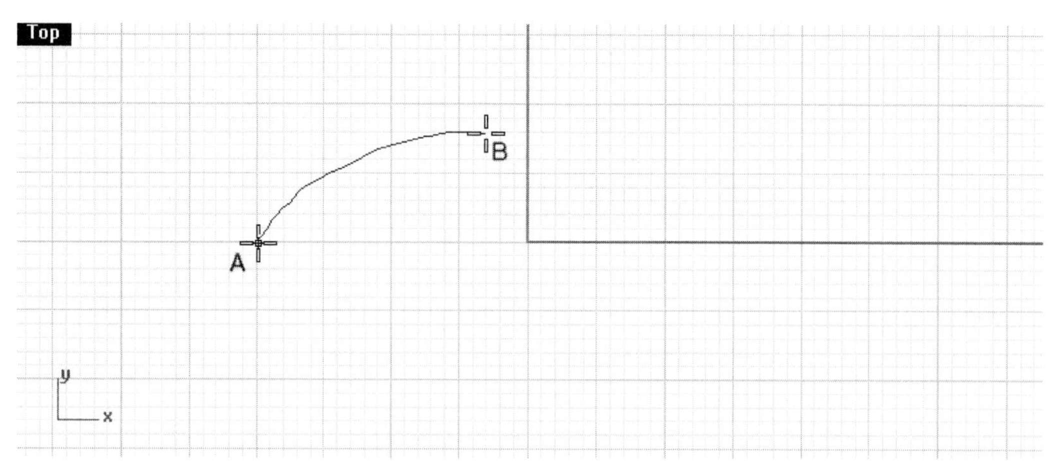

Figure 4–23. Sketch curve being constructed

A free-form sketch curve is constructed. Unlike the other three methods (specifying the control points, specifying the edit points, and using the handlebar), this method of constructing a curve lets you use your mouse as an electronic drawing pen. It is particularly useful if you embed a background bitmap in the viewport as a reference.

Zooming 1 to 1 Calibration

While you are constructing free-form sketches, you may better realize the actual size of your sketch by setting the zoom scale as follows:

1. Right click on the Zoom 1:1/Calibrate 1:1 Scale button on the View toolbar, or type Zoom1to1Calibrate at the command area.
2. Take out a ruler to measure the length of the horizontal blue bar of the Zoom 1 to 1 Calibration dialog box, shown in Figure 4–24.
3. Input the measured distance at the Length of bar box.
4. Click on the OK button. The zoom scale is set.
5. To zoom the display, click on the Zoom 1:1/Calibrate 1:1 scale button on the View toolbar.

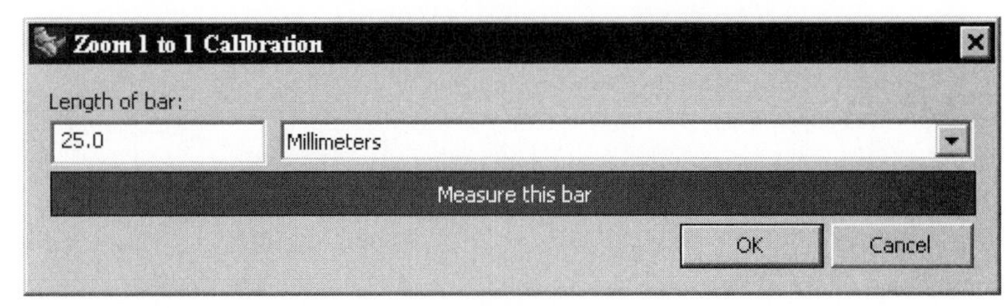

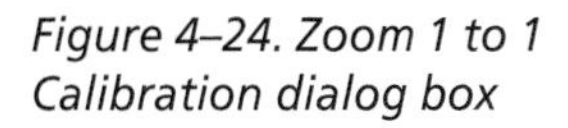
Figure 4–24. Zoom 1 to 1 Calibration dialog box

(***NOTE:*** *You may apply a scale factor while inputting the measure distance. This way, you can have a display that is zoomed to a known scale factor.)*

Tracing Background Images

Knowing that the fundamental issue of curve construction is to specify a location, you need to be able to use the pointing device to select a point or input the coordinates at the command area. To help determine the locations, you might sketch your design idea on graph paper. From the graph paper, you extract point locations. Another way of using the sketch is to scan it as a digital image and insert it in the viewport as a background image. If the image is properly scaled, you can directly sketch in the viewport.

When a background image is selected, point objects will be displayed at the four corners. You can click on these corner points to manipulate the image.

To try this way of sketching, perform the following steps:

1. Obtain a piece of graph paper.
2. Use a pencil to construct a free-hand sketch depicting the curve you want to construct in the computer.
3. Scan the sketch as a digital image.
4. After scanning, use an image processing application to trim away the unwanted portion of the image, and write down the dimensions of the grid lines in the image. For example, the bitmap image shown in Figure 4–25 is 30 mm wide and 35 mm tall.

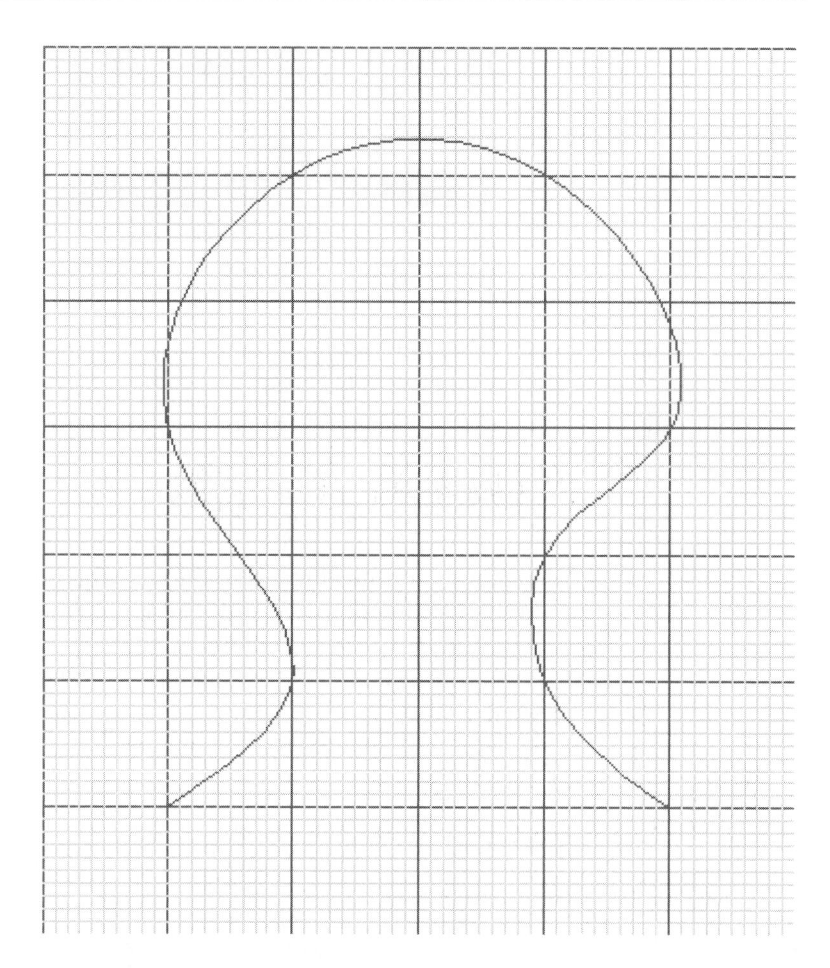

Figure 4–25. Bitmap image

Incorporate the digital image as a background image in the Front viewport by continuing with the following steps.

5. Start a new file. Use the "Small Objects, Millimeters" template.
6. Maximize the Top viewport.
7. Check the Snap button on the status bar to turn on snap mode.
8. Select View > Background Bitmap > Place, or click on the Place Background Bitmap button on the Background Bitmap toolbar.
9. In the Open Bitmap dialog box, select the bitmap file Background Image.tga from the Chapter 4 folder on the companion CD, and then click on the Open button.
10. Select a point on the Front viewport, as shown in Figure 4–26. Select a point 30 mm to the right of the first selected point, because the width of the bitmap used in this example is 30 mm wide.

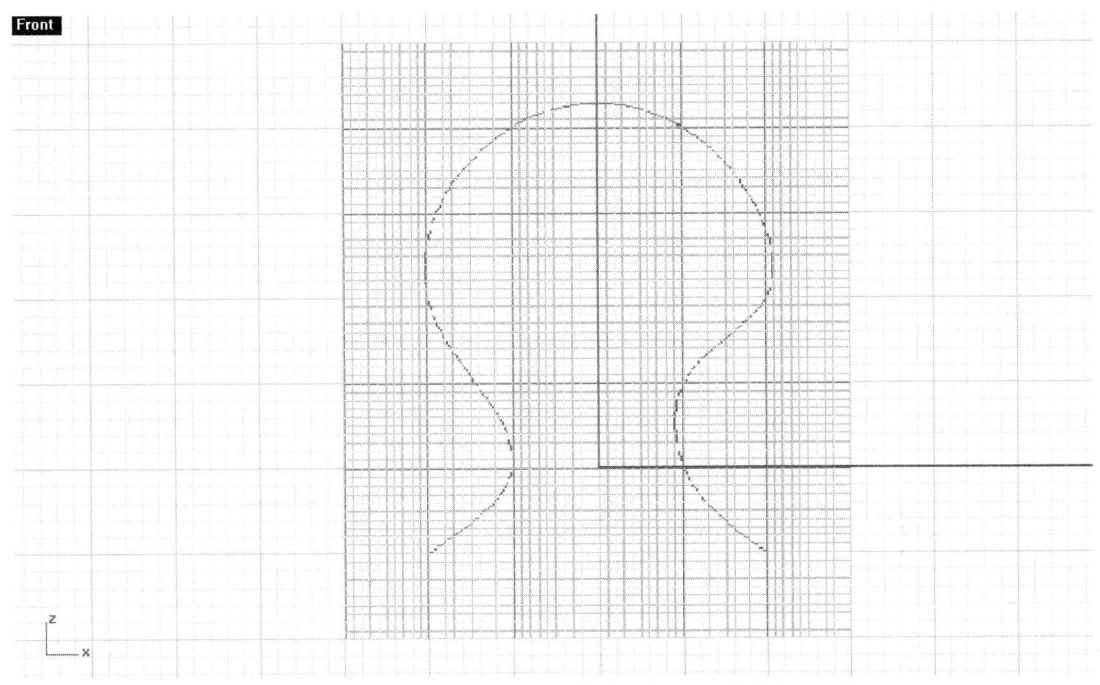

Figure 4–26. Image placed on the background of Front viewport

11 Select Curve > Free-Form > Sketch, or click on the Sketch button on the Curve toolbar. (Alternatively, you can use the Curve from Control Points option or the Interpolated Curve option to trace the images.)

12 Hold down the left mouse button to sketch a free-form curve.

13 Select View > Background Bitmap > Hide, or click on the Hide Background Bitmap button on the Background Bitmap toolbar.

14 Turn off grid display. A sketch is constructed and the bitmap is hidden You should note that the free-hand sketch constructed in this way is not smooth enough, and you may need to use the editing tools and transformation tools to modify it.

15 Do not save your file.

(**NOTE:** *If you save your file, the bitmap will also be saved in the Rhino file. Therefore, it is not necessary to save the bitmap file separately. Because the background image is saved in the Rhino file, you may save it back to your memory device as an individual bitmap file.*)

Although the bitmap file is saved in the Rhino file, there is still a link with the original bitmap file. Therefore, you may modify the source bitmap file and use the Refresh option of the BackgroundBitmap command to update the embedded bitmap.

Other Ways to Construct Free-Form Curves

There are five more ways to construct a free-form curve. You can construct an interpolated curve on a surface, sketch a curve on a surface, sketch a curve on a polygonal mesh object, fit a curve from a series of point objects, fit a curve from a polyline, and construct a shortest curve between two points on a surface.

Interpolated Curve on a Surface

You can construct an interpolated curve by selecting point locations on a selected surface, producing a curve on a surface. Perform the following steps:

1. Select File > Open and select the file FreeFormCurve07.3dm from the Chapter 4 folder on the companion CD.
2. Select Curve > Free-Form > Interpolate on Surface, or click on the Interpolate on Surface button on the Curve toolbar.
3. Select surface A, shown in Figure 4–27.
4. Select locations B, C, and D, shown in Figure 4–27, and then press the ENTER key. A free-form curve is constructed on the surface.

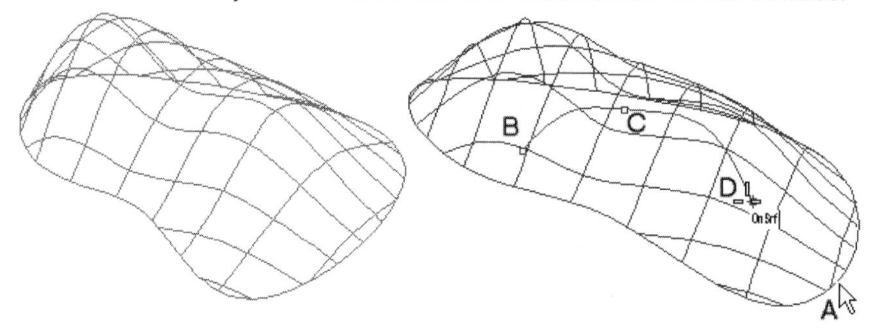

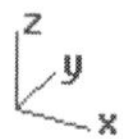

Figure 4–27. Free-Form curve being constructed on a surface

Free-Form Sketch Curve on a Surface

In addition to sketching on a construction plane, you can sketch on a surface. Continue with the following steps.

5. Select Curve > Free-Form > Sketch on Surface, or right-click on the Sketch/Sketch on Surface button on the Curve toolbar.
6. Select surface A, shown in Figure 4–28.
7. Select location B (Figure 4–28), hold down the mouse button, and drag to location C (Figure 4–28).
8. Release the mouse button when finished.

9 Press the ENTER key. A free-form sketch curve is constructed on the surface.

10 Do not save your file.

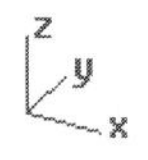

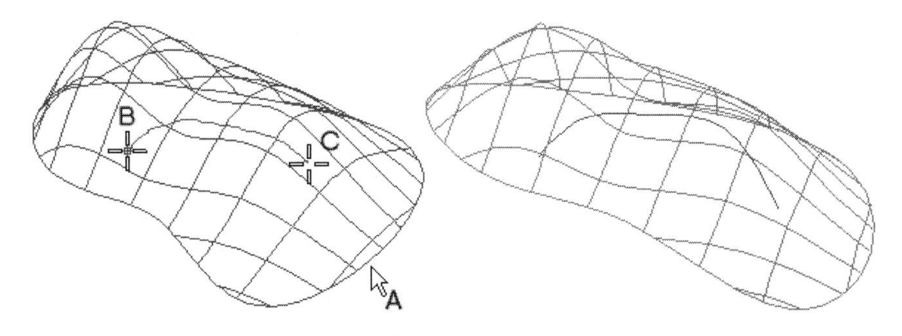

Figure 4–28. Free-form sketch curve being constructed on a surface

Free-Form Sketch Curve on a Polygon Mesh

In addition to sketching curves on NURBS surfaces, you can sketch a curve on a polygon mesh, as follows: (To reiterate, a polygon mesh is an approximation of a smooth surface via a set of planar polygons.)

1 Select File > Open and select the file FreeFormCurve08.3dm from the Chapter 4 folder on the companion CD.

2 Select Curve > Free-Form > Sketch on Polygon Mesh, or click on the Sketch on Polygon Mesh button on the Curve toolbar.

3 Select location A, shown in Figure 4–29, on the polygon mesh, hold down the mouse button, and drag it to location B(Figure 4–29).

4 Release the mouse button, and press the ENTER key. A free-form sketch curve is constructed on a polygon mesh.

5 Do not save your file.

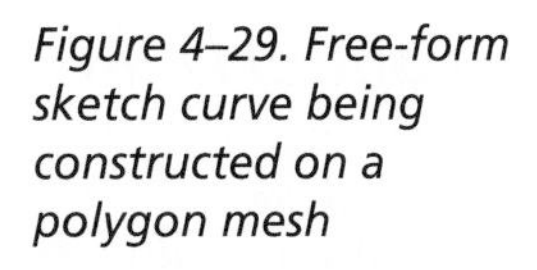

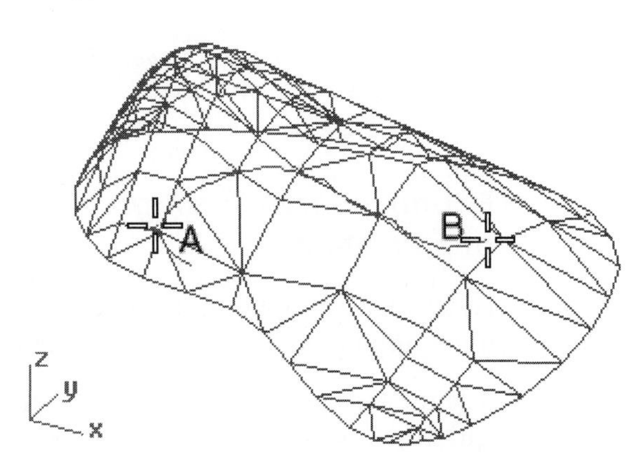

Figure 4–29. Free-form sketch curve being constructed on a polygon mesh

Fitting a Curve Through a Set of Points

If you already have constructed a set of point objects, you can select all of them collectively in one operation to construct a free-form curve, using the points as interpolated points or control points. Perform the following steps:

1. Select File > Open and select the file FreeFormCurve09.3dm from the Chapter 4 folder on the companion CD.
2. Select Curve > Free-Form > Fit to Points, or right-click on the Control Point Curve/Curve Through Points button on the Curve toolbar.
3. Select all the point objects, and press the ENTER key.

A preview of the curve is shown. *(Note that the setting of the curve depends on previous settings made to this command. The preview here shows a curve of degree 3 interpolating through the selected points.)*

4. If the Degree option is not 3 or above, select it from the command area and then type 3 to set the degree of polynomial to 3.
5. Select the CurveType option on the command area.
6. If CurveType = ControlPoint, select the Interpolated option on the command area to change to an interpolated curve. Otherwise, proceed to the next step.
7. Press the ENTER key. A curve is constructed. (See Figure 4–30.)
8. Do not save your file.

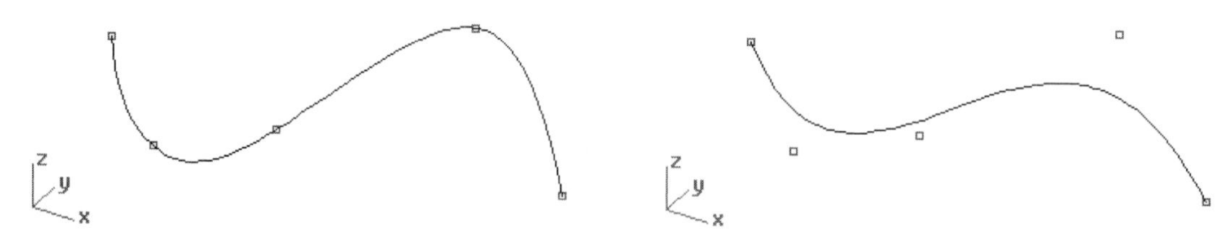

Figure 4–30. Interpolated curve constructed from a set of point objects (left) and control point curve constructed from a set of point objects

Fitting a Curve Through the Vertices or the Control Points of a Polyline

Other than using point objects, you can use the vertices of a polyline as the interpolated points or control points for making a curve.

If you have a polyline already constructed or imported from other applications, you can use the vertices of the polyline as interpolated points or control points to construct a free-form curve. (Ways to construct a polyline will be discussed in the next chapter.) Perform the following steps:

1. Select File > Open and select the file FreeFormCurve10.3dm from the Chapter 4 folder on the companion CD.
2. Select Curve > Free-Form > Fit to Polyline or click on Curve: Control Points from Polyline from the Curve toolbar.
3. Select polyline A, shown in Figure 4–31, and press the ENTER key.

If you use the toolbar icon to construct the curve, skip steps 4 and 5 and proceed to step 6.

4. If Degree = 3, proceed to the next step. Otherwise, select the Degree option on the command area and then type 3 at the command area.
5. If CurveType = ControlPoint, select the CurveType option on the command area and then select the Interpolated option on the command area. Otherwise, proceed to the next step.
6. Press the ENTER key. A curve is constructed.
7. Do not save your file.

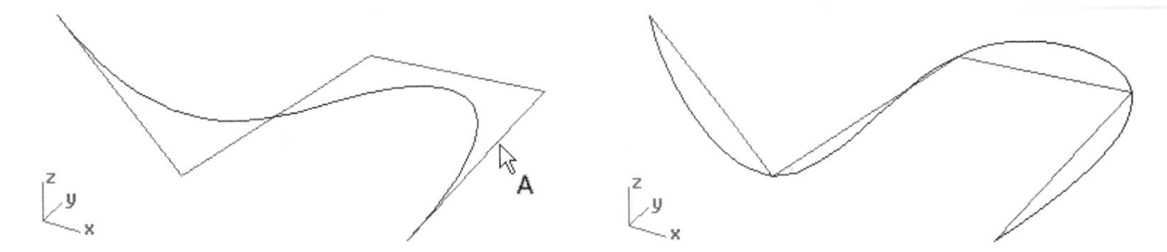

Figure 4–31. Control point curve constructed from the vertices of a polyline (left) and interpolated curve constructed from the vertices of a polyline (right)

Fitting a Curve Through the Control Points of a Surface

A surface has a matrix of control points in two orthogonal directions. You can use these control points as follows:

1. Select File > Open and select the file FreeFormCurve11.3dm from the Chapter 4 folder on the companion CD.
2. Type CurveThroughSrfControlPt at the command area.
3. Select surface A, shown in Figure 4–32. The control points of the surface are displayed.
4. Select the Direction option on the command area.
5. Select Both.

6. Select CurveType option on the command area.
7. Select Interpolated.
8. Select control point B and press the ENTER key. Two curves are constructed in two directions passing through the selected control point of the surface.
9. Do not save your file.

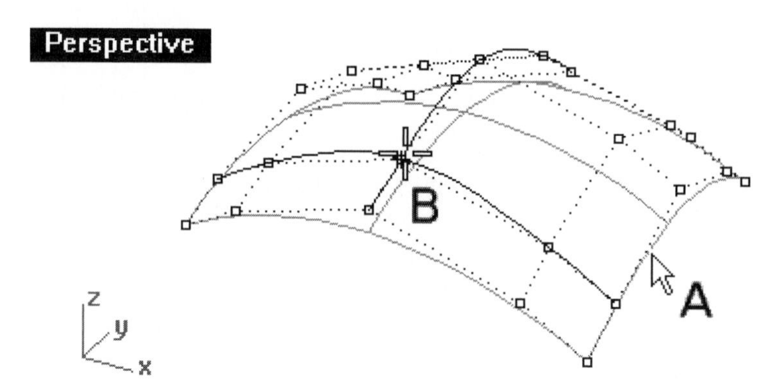

Figure 4–32. Curves constructed through the control points of a surface

Shortest Curve Between Two Points on a Surface

You can construct a free-form curve on a surface to depict the shortest path between two designated points. Perform the following steps:

1. Select File > Open and select the file ShortPath.3dm from the Chapter 4 folder on the companion CD.
2. Click on the Geodesic curve button on the Curve from Objects toolbar, or type ShortPath at the command area.
3. Select surface A, shown in Figure 4–33.
4. Set object snap mode to point.
5. Select points A and B. A free-form curve is constructed.

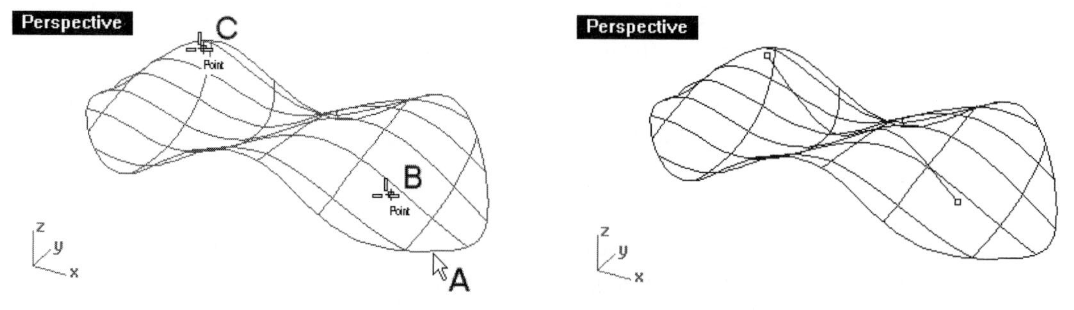

Figure 4–33. Surface and points being selected (left) and curve constructed (right)

Points and Point Clouds

A point is a node. A point cloud is a set of points. Using point objects, you specify locations in 3D space, define vertices and definition points of curves, and portray locations on a surface.

Multiple Points

You can construct point objects by specifying their locations. Typically, you can select a location on one of the construction planes, key in the coordinates of a point, or use a digitizer to directly input the coordinates of locations from a real physical object. Perform the following steps:

1. Start a new file. Use the "Small Objects, Millimeters" template.
2. Double-click on the Top viewport title to maximize the viewport.
3. Select Curve > Point Object > Multiple Points, or click on the Multiple Points button on the Point toolbar.
4. Select locations A, B, C, and D, as shown in Figure 4–34. (The exact location of points in this exercise is unimportant.)
5. Press the ENTER key to terminate the command. Four point objects are constructed.
6. Do not save your file.

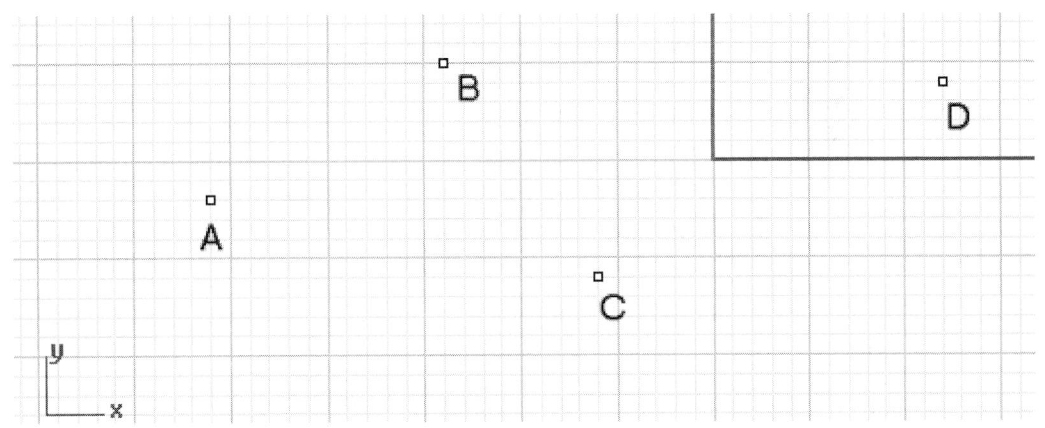

Figure 4–34. Four point objects

Points Near a Curve and Along a Curve

If you already have a curve, you can construct a point on the curve that is closest to a target object, points at the endpoints of the curve, a set of points along the curve at specified spacing, and a set of points along the curve at regular intervals. Perform the following steps:

1. Select File > Open and select the file Point01.3dm from the Chapter 4 folder on the companion CD.
2. Set object snap mode to Quad. (Quad means quadrant locations of circles and arcs.)
3. Select Curve > Point Object > Closest Point, or click on the Closest Point button on the Point toolbar.
4. Select curve A, shown in Figure 4–35, and press the ENTER key.
5. Select quadrant location B, shown in Figure 4–35. A point closest to the selected location is constructed along the curve.

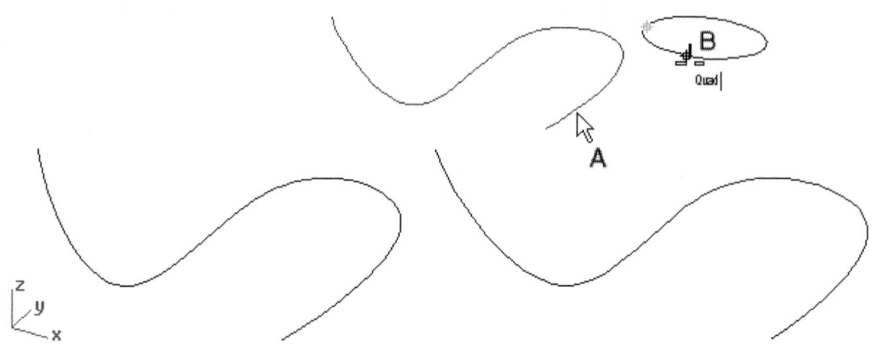

Figure 4–35. Point on a curve closest to the quadrant point of a circle being constructed

6. Select Curve > Point Object > Mark Curve Start, or click on the Mark Curve Start/Mark Curve End button on the Point toolbar.
7. Select curve A, shown in Figure 4–36, and press the ENTER key.
8. Select Curve > Point Object > Mark Curve End, or right-click on the Mark Curve Start/Mark Curve End button on the Point toolbar.
9. Select curve A, shown in Figure 4–36, and press the ENTER key. Points are constructed at the start point and end point of the curve.

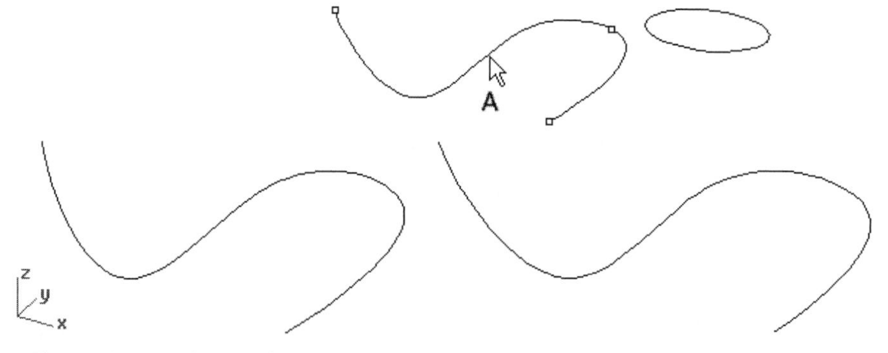

Figure 4–36. Points at the start point and end point of a curve constructed

To construct a set of point objects along a curve at regular intervals, perform the following steps.

10 Select Curve > Divide Curve by > Length of Segments, or left-click on the Divide Curve by Length/Divide Curve by Number of Segments button on the Point toolbar.
11 Select curve A, shown in Figure 4–37, and press the ENTER key.
12 Type 14 at the command area to specify a length. A set of points at 14-unit intervals is constructed along the curve. The first point object is constructed at the start point of the curve.
13 Select Curve > Divide Curve by > Number of Segments, or right-click on the Divide Curve by Length/Divide Curve by Number of Segments button on the Point toolbar.
14 If MarkEnds = No, select the MarkEnds option on the command area to change to MarkEnds = Yes.
15 Select curve B, shown in Figure 4–37, and press the ENTER key.
16 Type 7 at the command area to specify six segments. Eight points are constructed, as shown in Figure 4–37.
17 Do not save your file.

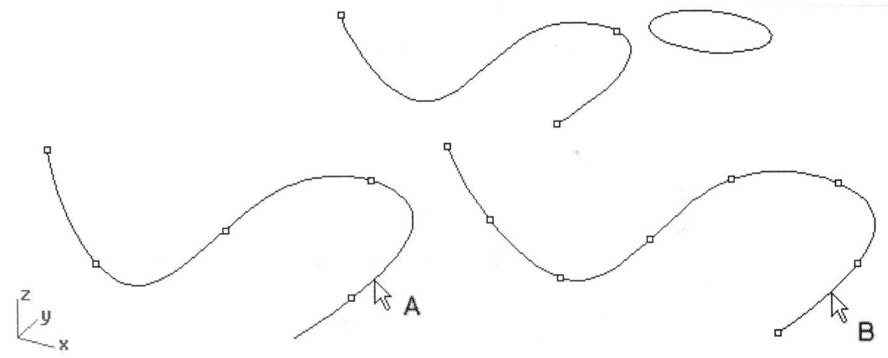

Figure 4–37. Points at 14-unit intervals and points equally spaced along the curve constructed

Drape Points

If you already have a surface or a polygon mesh, you can construct a matrix of point objects on them. Using these point objects, you can proceed to constructing new curves for making additional surfaces. Perform the following steps:

1 Select File > Open and select the file Point02.3dm from the Chapter 4 folder on the companion CD.

2. Select Curve > Point Object > Drape Points, or click on the Drape Point Grid Over Objects button on the Point toolbar.
3. Select points A and B, shown in Figure 4–38. A set of points is constructed.
4. Turn off layer Surface.
5. Do not save or close the file. You will continue to work on the point objects.

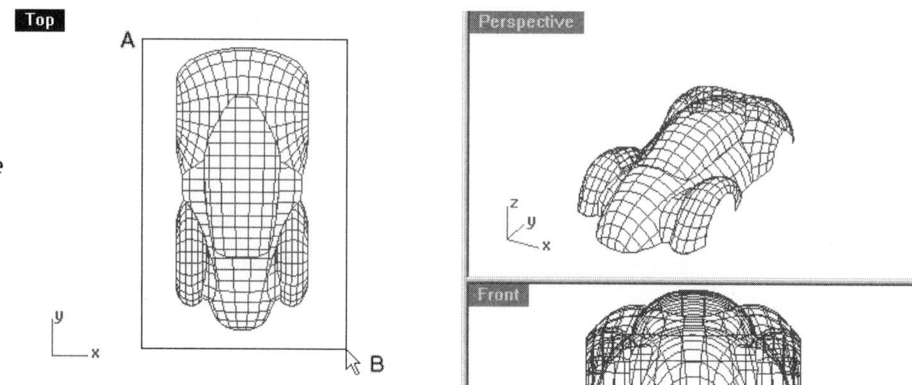

Figure 4–38. Drape points being constructed

Point Cloud Construction

A point cloud is a set of point objects grouped together. Point clouds are typical products of digitizing. Using point clouds, you can construct curves for making surfaces. To understand the use of point clouds, continue with the following steps.

6. Select Curve > Point Cloud > Create Point Cloud, or click on the Point Cloud button on the Point toolbar.
7. Select location A and B, shown in Figure 4–39, and press the ENTER key. A point cloud is constructed.

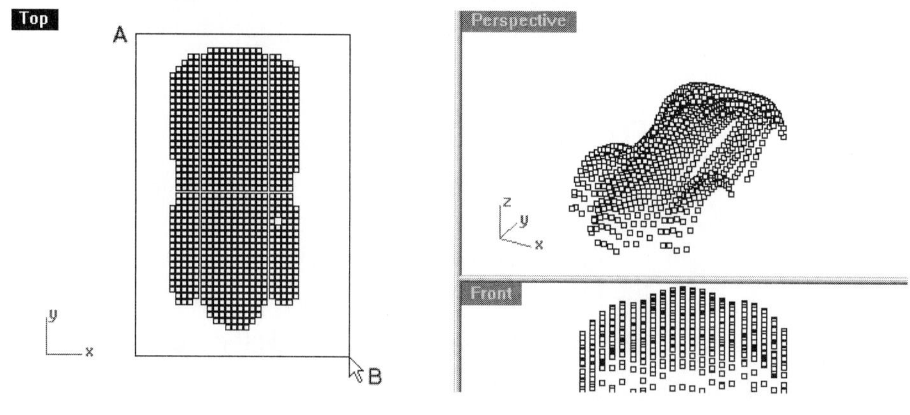

Figure 4–39. Point cloud being constructed from a set of selected points

8 To set the point cloud style, select Curve > Point Cloud > Point Cloud Style, and then select a style from the options available.

*(**NOTE:** You can add point objects to a point cloud by selecting Curve > Point Cloud > Add Points and remove point objects from a point cloud by selecting Curve > Point Cloud > Remove Points.)*

Curve Construction from Point Cloud

A way to reconstruct a surface from a point cloud is to construct sections across the point cloud. Continue with the following steps.

9 Set object snap mode to Point.

10 Select Curve > Point Cloud > Point Cloud Section, or click on the Point Cloud Section button on the Curve From Object toolbar.

11 Select the point cloud and press the ENTER key.

12 In the Point Cloud Section Options dialog (Figure 4–40), clear the Create polylines, if it is checked, and click on the OK button.

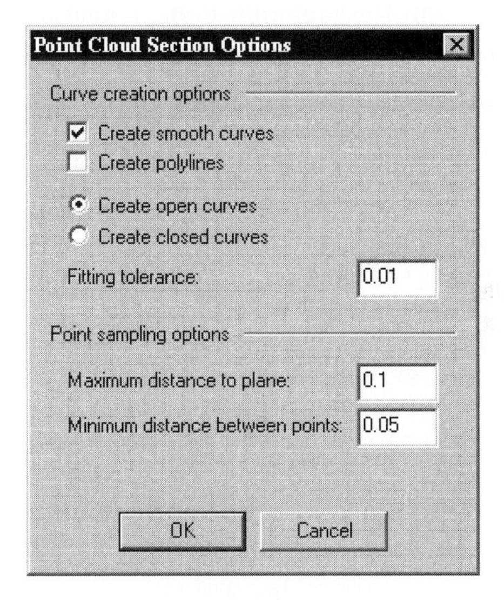

Figure 4–40. Point cloud section curves being constructed from a point cloud

13 Select location A and B, shown in Figure 4–41, to describe a section line across the point cloud. (If the section line is located too far from the points, a prompt "Too few points within minimum distance of requested section plane" will be displayed at the command area. If you see this message, try again.)

14 Continue to select pairs of start and end points.

15 Press the ENTER key when finished.

16 Do not save your file.

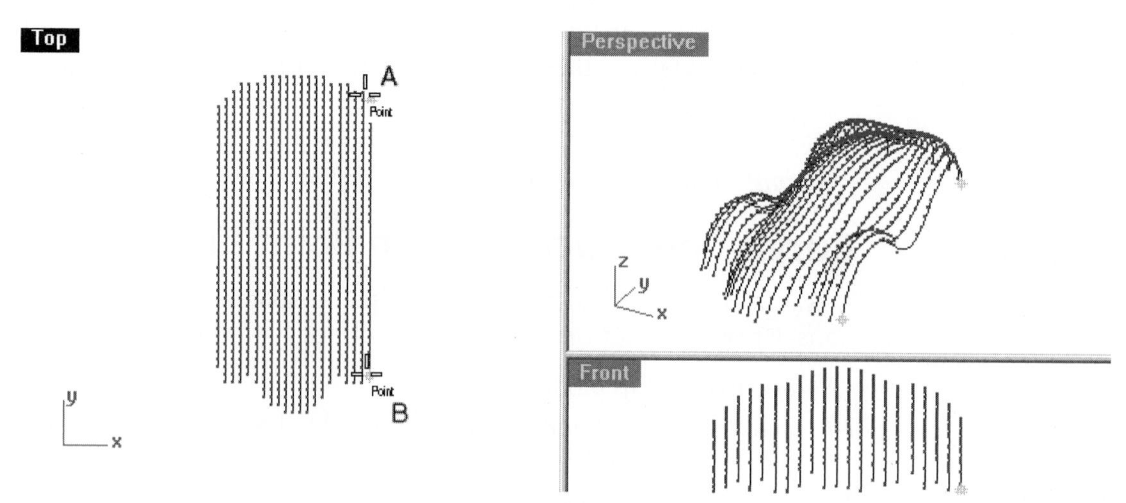

Figure 4–41. Point cloud curves

In the previous steps, constructing a point cloud from the surfaces and constructing point cloud curves serve to demonstrate how drape points, point clouds, and point cloud curves are constructed. In reality, point clouds are usually obtained by digitizing a real physical model for the purpose of reconstructing the surfaces of the model in the computer.

Chapter Summary

Rhino, like most modeling applications, uses NURBS (non-uniform rational basis spline) curves to construct free-form surfaces. Simply speaking, a NURBS curve is a set of polynomial spline segments. The shape of each spline segment in a curve is influenced by the degree of polynomial equation used to define the curve. For most designers, it is not necessary to know about the mathematics related to such equations. However, understanding the relationship between degree of polynomial and the shape of a curve is an advantage.

A line is degree 1, a circle is degree 2, and a free-form curve is degree 3 or above. The higher the degree, the more control points exist in a spline segment. For a curve of degree 3, each spline segment has four control points. Between two contiguous segments of a spline curve is a knot. Adding or removing knots increases or reduces the number of segments in a curve. This, in turn, has an influence on the total number of control points in a curve.

A special type of knot on a curve is a kink, which is a junction between two contiguous segments where the tangent directions and radii of curvature are not congruent. A closed curve without any kink is called a periodic curve. The overall shape of a NURBS curve is affected by the number of segments in the curve, the degree of polynomial equation, the location of control points, and the weight of the control points.

Points are as important as curves in making a framework for constructing surfaces. Point objects can be constructed from scratch or by digitizing.

Review Questions

1. State the characteristics and key features of a NURBS curve.
2. Illustrate the methods of editing a curve by manipulating its control points, edit points, knots, and degree of polynomial.
3. What is a point cloud and how is it used?

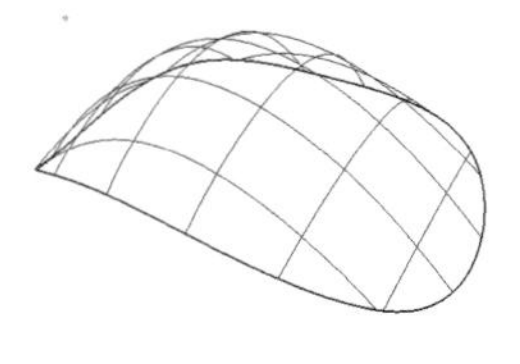

CHAPTER 5

Curves of Regular Pattern

Introduction

This chapter will delineate various kinds of curves with regular pattern that you may use in modeling.

Objectives

After studying this chapter, you should be able to

- ❒ Use various methods to construct regular pattern curves.

Overview

In addition to using free-form curves in your design, you may also need to deploy curves with regular pattern, such as lines, arcs, circles, and polygons. Using Rhino, you can construct 12 kinds of regular curves, categorized below in terms of their degree of polynomial.

- ❒ Degree 1 curves are line, polyline, rectangle, and polygon.
- ❒ Degree 2 curves are circle, arc, ellipse, parabola, hyperbola, and conic.
- ❒ Degree 3 curves are helix and spiral.

All Rhino's curves are NURBS curves, so you can modify them by increasing or reducing the degree of polynomial, manipulating the control points or edit points, adding or deleting control points, and adding or removing knots and kink points. You can also use the handlebar editor.

Line

A straight line is a degree 1 curve. You can construct it in 13 ways.

Single Line and Line Segments

Yon can construct a single line segment by specifying two end points or construct a series of individual line segments connected end to end by specifying their adjoining end points. Note that the line segments constructed are separate, individual line segments. Perform the following steps.

1. Select File > Open and select the file Line01.3dm from the Chapter 5 folder on the companion CD.
2. Set object snap mode to Point.
3. Select Curve > Line > Single Line, or click on the Line button on the Lines toolbar.
4. Select points A and B, as shown in Figure 5–1.
5. Select Curve > Line > Line Segments, or right-click on the Polyline/ Line Segments button on the Lines toolbar.
6. Select points C, D, E, and F, shown in Figure 5–1, and then press the ENTER key. A series of line segments is constructed.
7. Do not save your file.

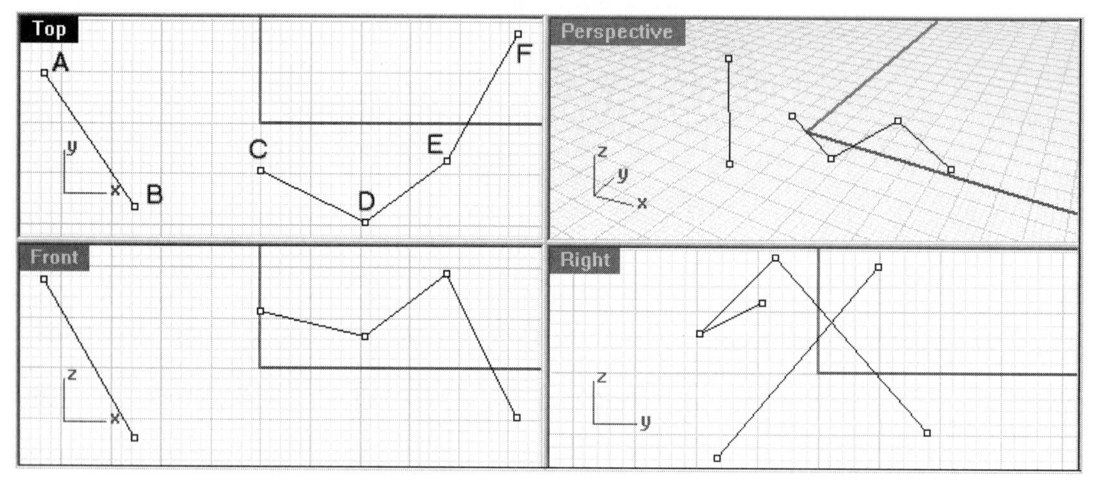

Figure 5–1. Single line and line segment constructed

Line Perpendicular/Tangent to a Curve and Line Perpendicular/Tangent to 2 Curves

You may construct a line perpendicular/tangent to a curve from a selected point and construct a line perpendicular/tangent to two curves. Depending on the location of the selected point and the shape of the curve (for a line perpendicular/tangent to a curve from a point) or the shape of the curves (for a line perpendicular/tangent to two curves), there may be no solutions at all or more than one solution. Perform the following steps:

1. Select File > Open and select the file Line02.3dm from the Chapter 5 folder on the companion CD.
2. Set object snap mode to Point, if it is not already set.
3. Select Curve > Line > Perpendicular from Curve, or right-click on the Line: Perpendicular from Curve/Line: Perpendicular to Curve button on the Lines toolbar.
4. Select point A and then curve B, shown in Figure 5–2. Exact location of curve B is unimportant.

A line perpendicular to a curve from a point is constructed.

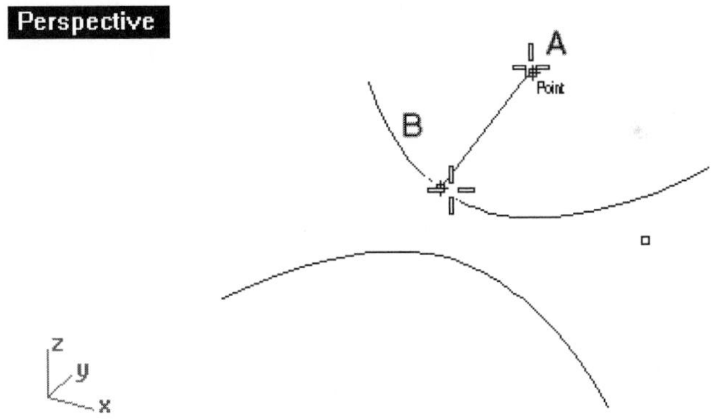

Figure 5–2. Line through a point and perpendicular to a curve being constructed

5. Select Curve > Line > Perpendicular to 2 Curves, or click on the Line: Perpendicular to Two Curves button on the Lines toolbar.
6. Select curve A and then curve B, shown in Figure 5–3.

A line perpendicular to two curves is constructed.

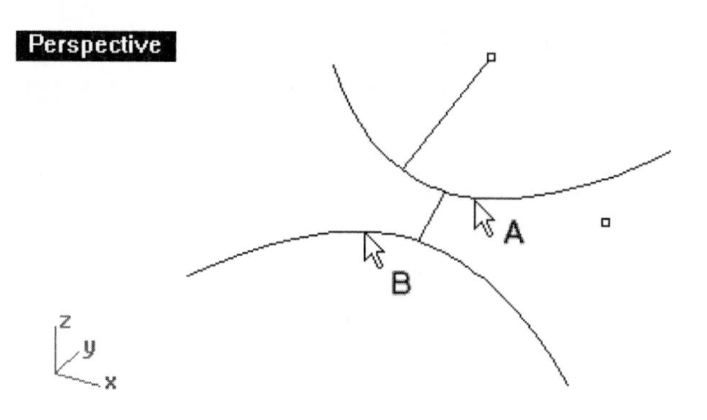

Figure 5–3. Line perpendicular to two curves being constructed

7 Select Curve > Line > Tangent from Curve, or click on the Line: Tangent from Curve button on the Lines toolbar.

8 Select curve B and then point A, shown in Figure 5–4.

A line tangent to a curve from a point is constructed.

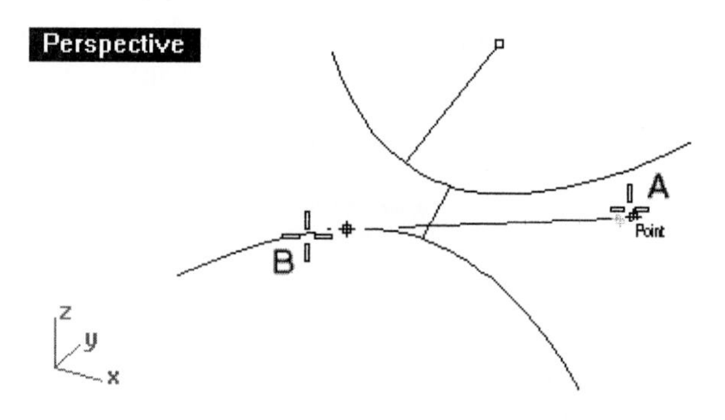

Figure 5–4. Line through a point and tangent to a curve being constructed

9 Select Curve > Line > Tangent to 2 Curves, or click on the Line: Tangent to Two Curves button on the Lines toolbar.

10 Select curves A and B, shown in Figure 5–5.

11 Do not save your file.

A line tangent to two curves is constructed.

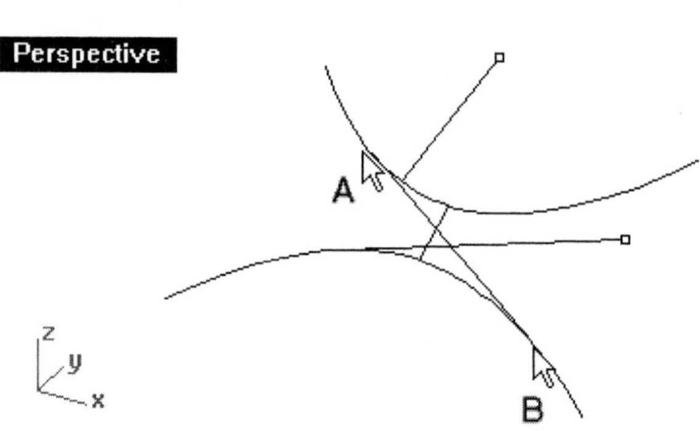

Figure 5–5. Line tangent to two curves being constructed

Line Tangent to a Curve and Perpendicular to another Curve

You may construct a line that is tangent to a curve and also perpendicular to another curve. Likewise, there may be no solutions or more than one solution to the problem.

Perform the following steps:

1. Select File > Open and select the file Line03.3dm from the Chapter 5 folder on the companion CD.
2. Select Curve > Line > Tangent, Perpendicular, or click on the Line: Tangent and Perpendicular to Curves button on the Lines toolbar.
3. Select curve A where the tangent point will be placed, as shown in Figure 5-6.
4. Select curve B where the perpendicular point will be placed, as shown in Figure 5-6.

A line tangent to a line and perpendicular to another line is constructed.

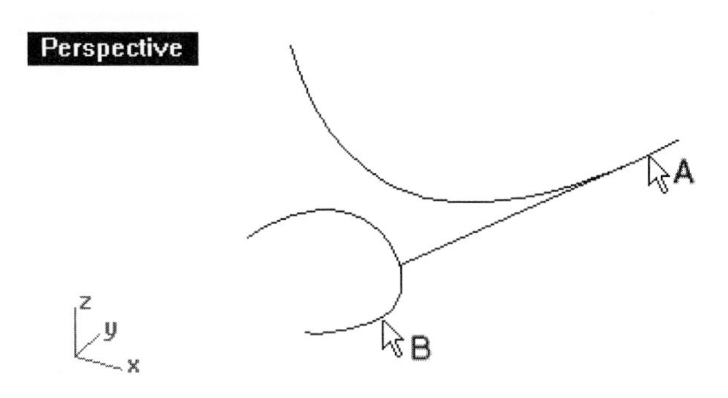

Figure 5-6. Line tangent to a curve and perpendicular to another curve being constructed

5. Do not save your file.

Line at an Angle, Line Bisecting 2 Lines, and Four-Point Line

You can construct a line at an angle to a selected reference line, construct a line bisecting two non-parallel lines, and construct a line in a direction defined by two points with the end points nearest to two selected points (i.e., a four-point line).

Perform the following steps:

1. Select File > Open and select the file Line04.3dm from the Chapter 5 folder on the companion CD.

2. Set object snap to End and Point.
3. Select Curve > Line > Angled, or left-click on the Line: Angled/ Line: Angled from Midpoint button on the Lines toolbar.
4. Select end points A and B, shown in Figure 5-7.
5. Type 45 to specify an angle.
6. Select point C, shown in Figure 5-7.

A line at 45 degrees to the selected line is constructed. It has an end point closest to the selected reference point. The angle is measured on a plane parallel to the active construction plane. Here, the perspective viewport's construction plane is the same as that of the top viewport.

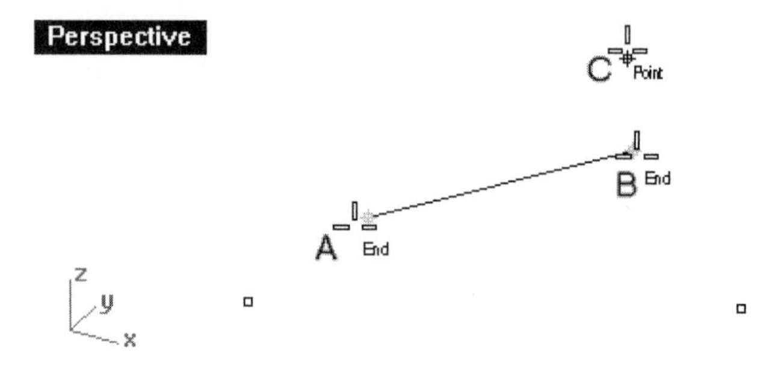

Figure 5-7. Line segment at an angle to a reference line being constructed

7. Select Curve > Line > Bisector, or left-click on the Line: Bisector/ Line: Bisector from Midpoint button on the Lines toolbar.
8. Select end points A and B, shown in Figure 5-8, to specify the first reference line (start of the bisector line and start of the angle to be bisected).
9. Select end points A and C, shown in Figure 5-8, to indicate the end of the angle to be bisected.
10. Select point D, shown in Figure 5-8.

A line bisector is constructed. Its end point is closest to the selected reference point D.

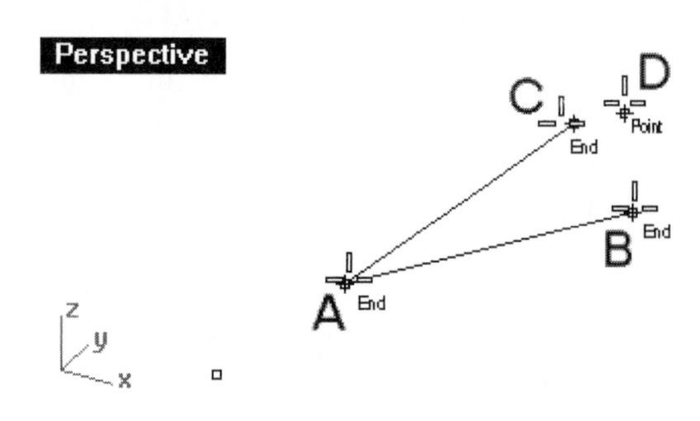

Figure 5-8. Bisector line being constructed

11 Select Curve > Line > From 4 Points, or left-click on the Line: by 4 Points/Line: by 4 Points from Midpoint button on the Lines toolbar.

12 Select points A and B, shown in Figure 5–9, to indicate a direction reference.

13 Select end points C and D, shown in Figure 5–9.

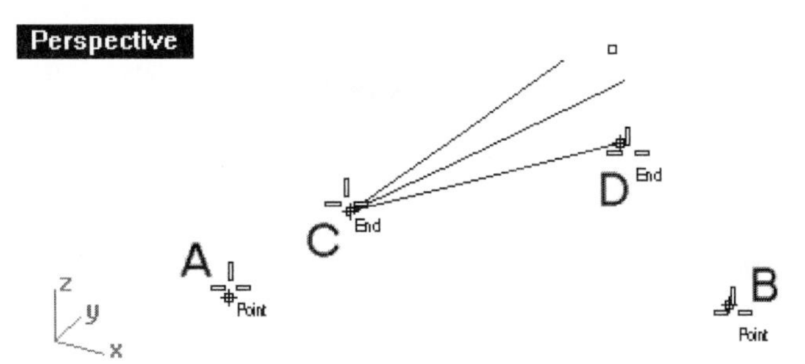

Figure 5–9. A four-point line being constructed

A line lying on the reference direction line and with its end points closest to two reference points is constructed. (See Figure 5–10.)

14 Do not save your file.

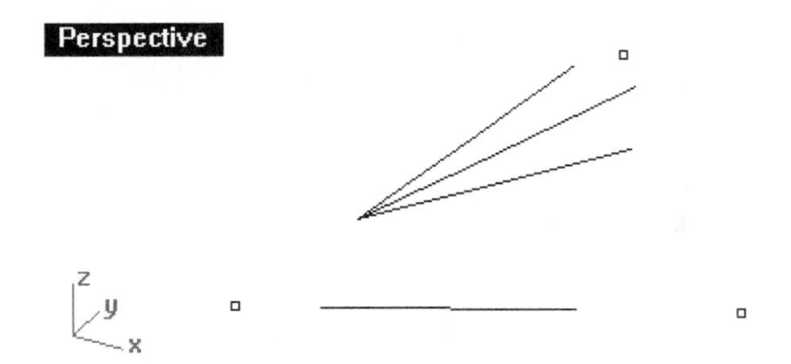

Figure 5–10. A four-point line constructed

Line Normal to a Surface and Line Vertical to a Construction Plane

You can construct a line normal to a surface, and you can construct a line perpendicular to a construction plane. Perform the following steps.

1 Select File > Open and select the file Line05.3dm from the Chapter 5 folder on the companion CD.

2 Set object snap mode to Int (intersection) and Point.

3 Select Curve > Line > Normal to Surface, or left-click on the Line: Surface Normal/Surface Normal Both Sides button on the Lines toolbar.

4 Select surface A, shown in Figure 5–11.
5 Select location B (isocurve's intersection), shown in Figure 5–11, to indicate the start point of the line on the surface.
6 Select reference point C, shown in Figure 5–11. A line normal to the surface is constructed. Its end point is closest to the selected reference point.

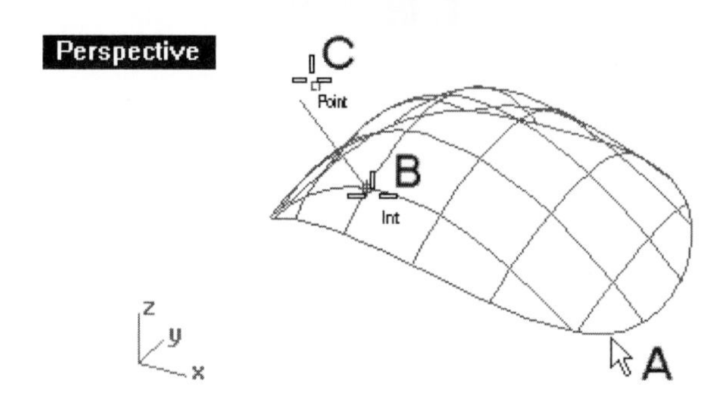

Figure 5–11. Normal line constructed

7 Select Curve > Line > Vertical to CPlane, or left-click on the Line: Vertical to CPlane/Line: Vertical to CPlane, from Midpoint button on the Lines toolbar.
8 Select end point A to specify the start point of the line, and select end point B to specify the reference point. This creates a line perpendicular to the construction plane, as shown in Figure 5–12. (Height is defined by the vertical distance between the first selected point and the second selected point.)
9 Do not save your file.

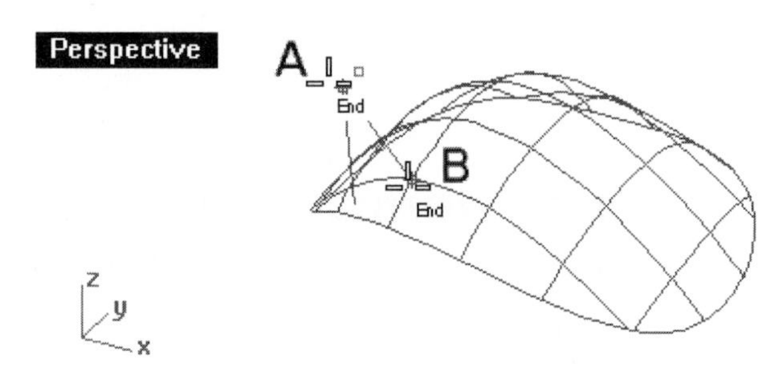

Figure 5–12. Line perpendicular to the construction plane being constructed

Line Passing Through a Set of Points

You can fit a line through a set of point objects and/or control points. Perform the following steps:

1 Select File > Open and select the file Line06.3dm from the Chapter 5 folder on the companion CD.

2. Click on the Line Through Points button on the Lines toolbar, or type LineThroughPt at the command area.
3. Referencing Figure 5–13, click on location A and drag to location B to select all the point objects, and press the ENTER key. A line is fitted.
4. Do not save your file.

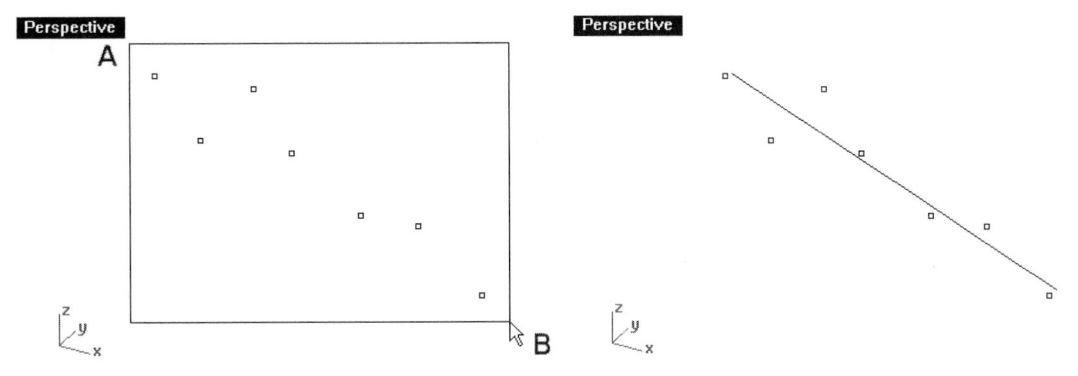

Figure 5–13. Point objects selected (left) and line fitted (right)

Polyline

A polyline is a series of contiguous line and/or arc segments joined together at their end points. In essence, you can explode a polyline to obtain a set of line and/or arc segments. Conversely, you can join a set of contiguous line and/or arc segments to become a polyline. You can construct a polyline in much the same way as constructing a series of line segments. You can also fit a polyline to a set of points, similar to fitting a curve to a set of points. Moreover, you can construct a polyline on a polygon mesh. To construct a polyline, perform the following steps.

1. Select File > Open and select the file Polyline01.3dm from the Chapter 5 folder on the companion CD.
2. Check the Osnap pane on the status bar to display the Osnap dialog box.
3. In the Osnap dialog box, check the Point box to establish object snap mode.
4. Select Curve > Polyline > Polyline, or left-click on the Polyline/Line Segments button on the Lines toolbar.
5. Select point A shown in Figure 5–14.
6. In the command area, if mode = arc, select the mode option to change to mode = line. Otherwise, proceed to the next step.

7 Select point B.

8 Select the mode option on the command area to change to mode = arc.

9 Select point C.

10 Select the mode option on the command area to change to mode = line.

11 Select point D.

12 Select the Close option on the command area. A closed polyline with line and arc segments is constructed.

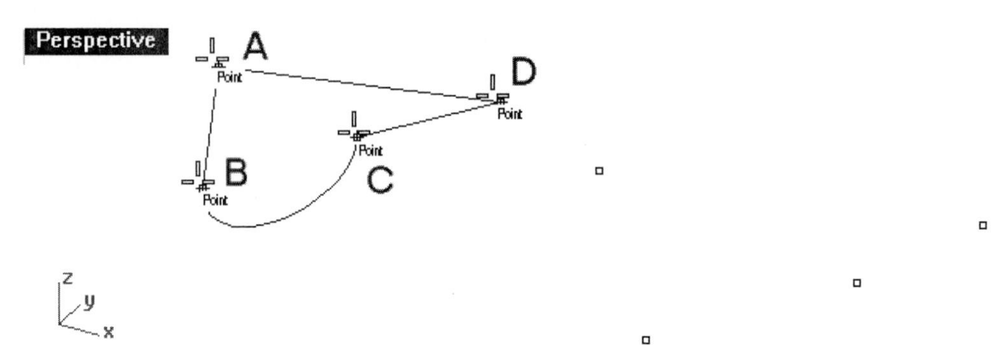

Figure 5–14. A closed polyline with line and arc segments being constructed

Polyline Through a Set of Points

You can construct a polyline to pass through a set of points. Perform the following steps.

13 Set the current layer to Layer02, and turn off Layer01.

14 Select Curve > Polyline > Polyline Through Points, or click on the Polyline: Through Points button on the Lines toolbar.

15 Click on A and drag to location B to select four point objects, and press the ENTER key.

Because this command is the same as Curve > Free-Form > Fit To Points, the degree of polynomial setting will determine the outcome.

16 If Degree option is not equal to 1, select it from the command area and then type 1 at the command area to set the degree of polynomial to 1. Otherwise, proceed to the next step. As will be explained in next chapter, the line segment's degree of polynomial is 1.

17 Press the ENTER key. A polyline is constructed.

18 Do not save your file.

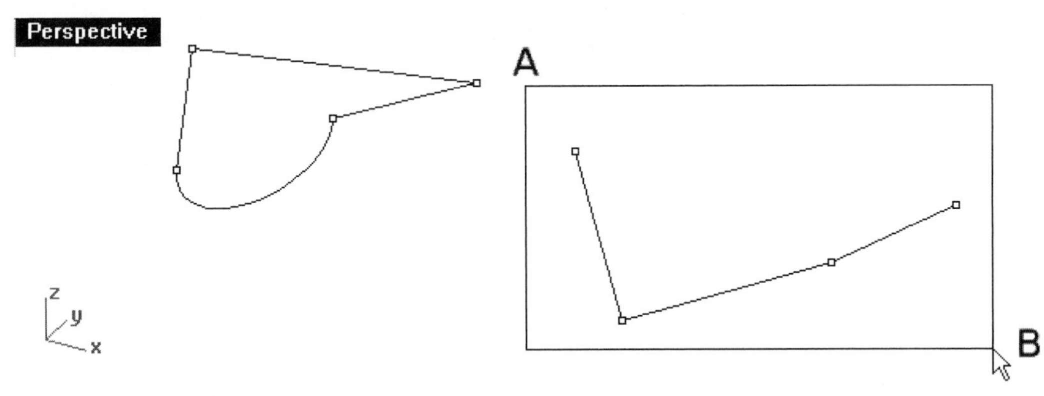

Figure 5–15.
A polyline constructed from a set of points

Polyline on a Polygon Mesh

To construct a polyline on a polygon mesh, perform the following steps:

1. Select File > Open and select the file Polyline02.3dm from the Chapter 5 folder on the companion CD.
2. Set object snap mode to Point.
3. Select Curve > Polyline > On Mesh, or click on the Polyline: On Mesh button on the Lines toolbar.
4. Select polygon mesh A, shown in Figure 5–16.
5. Select vertices B, C, and D, shown in Figure 5–16, and then press the ENTER key. A polyline is constructed on a polygon mesh.
6. Do not save your file.

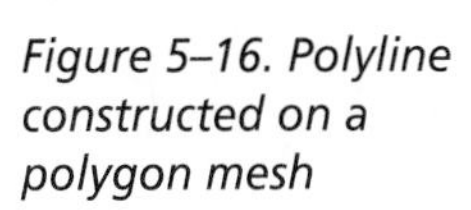
Figure 5–16. Polyline constructed on a polygon mesh

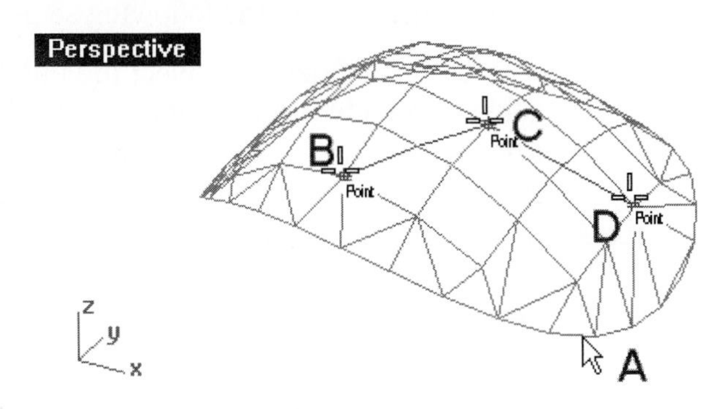

Connected Line Segments and Polyline

At a glance, the connected line segments and the polyline seem to be similar. In fact, the polyline is a single object and the connected line segments are separate line segments with their end points coincident with each other. To separate the polyline into individual line segments, you explode it by selecting Edit > Explode. To convert a set of connected line segments to a single polyline, you join the line segments by selecting Edit > Join.

Rectangle

Using Rhino, you can construct three types of rectangles. The first, consisting of four joined line segments, is a degree 1 curve. The second, consisting of four line segments and four arcs (filleted corners) joined, is a degree 2 curve. The third, consisting of four joined conic curves, is also a degree 2 curve. If you explode a rectangle, the first rectangle type will decompose into four separate line segments, the second will decompose into lines and arc segments, and the third will decompose into four conic curves.

Four Ways to Construct a Regular Rectangle

You can construct any type of rectangle in four ways: specifying its diagonal corners (Corner to Corner), specifying the center and a corner (CENTER, Corner), specifying an edge and a point on the opposite edge (3 Points), and specifying an edge and its height from the construction plane (Vertical). Perform the following steps:

1. Select File > Open and select the file Rectangle01.3dm from the Chapter 5 folder on the companion CD.
2. Check the Point box in the Osnap dialog box to establish object snap mode.
3. Select Curve > Rectangle > Corner to Corner, or click on the Rectangle: Corner to Corner button on the Rectangle toolbar.
4. Select points A and B, shown in Figure 5–17.
5. Select Curve > Rectangle > Center, Corner, or click on the Rectangle: CENTER, Corner button on the Rectangle toolbar.
6. Select points C and D, shown in Figure 5–17.
7. Select Curve > Rectangle > 3 Points, or click on the Rectangle: 3 Points button on the Rectangle toolbar.

8 Select points E, F, and G, shown in Figure 5–17.
9 Select Curve > Rectangle > Vertical, or click on the Rectangle: Vertical button on the Rectangle toolbar.
10 Select points H, J, and K, shown in Figure 5–17.
11 Do not save your file.

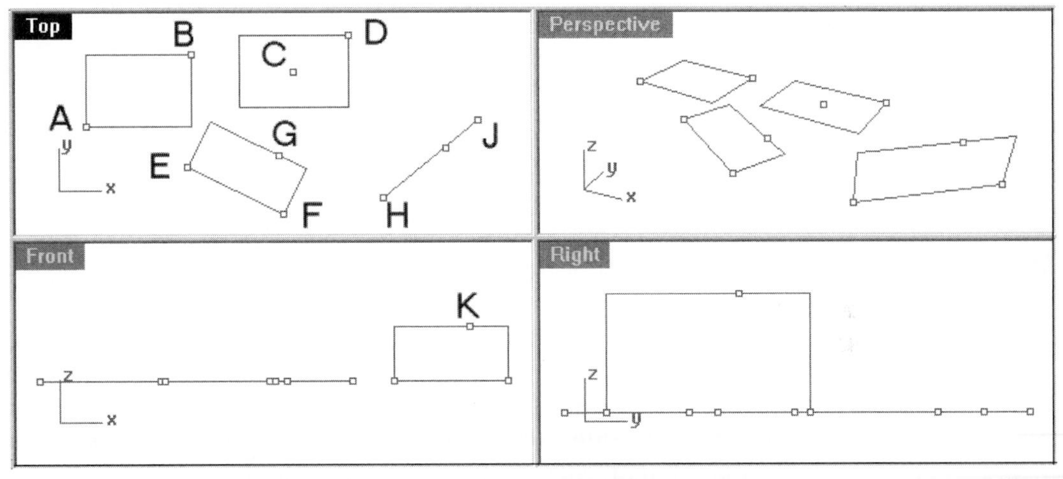

Figure 5–17. Four rectangles constructed

Rounded Rectangle and Conic Rectangle

However you construct a rectangle, you can have a filleted rectangle or conic rectangle instead of a rectangle with four line segments. Perform the following steps:

1 Select File > Open and select the file Rectangle02.3dm from the Chapter 5 folder on the companion CD.
2 Select Curve > Rectangle > Corner to Corner and select the Rounded option at the command area, or click on the Rounded Rectangle/Rounded Rectangle: Conic Corners button on the Rectangle toolbar.
3 Select points A and B, shown in Figure 5–18.
4 At the command area, if Corner = Conic, select the Corner option to change it to Arc.Otherwise, proceed to the next step.
5 Type 4 at the command area to specify the radius. A rectangle with circular arc corners is constructed.
6 Right-click on the Rounded Rectangle/Rounded Rectangle: Conic Corners button on the Rectangle toolbar.

7 Select locations C and D (Figure 5–18).
8 Select location E (Figure 5–18). A conic rectangle is constructed.
9 Do not save your file.

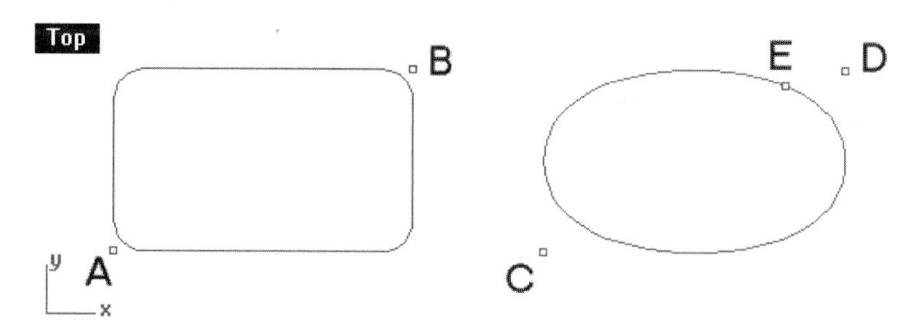

Figure 5–18. Filleted rectangle (left) and conic rectangle (right) constructed

Polygon

A polygon's polynomial degree is 1. It is a regular polygon in shape, consisting of a set of lines of equal length joined together. You can construct two types of polygons, regular and star-shaped, in several ways: specifying the number of sides, the center, and a corner; specifying the center and a midpoint of an edge; or specifying two end points of an edge.

To construct a star, you specify the number of sides, corner of the star, and the second radius of the star. In addition to constructing a polygon or star on the current construction plane, you can construct them around a curve.

Polygon and Polygon Around a Curve

To construct a regular polygon, star polygon, polygon around a curve, and star polygon around a curve, perform the following steps:

1 Select File > Open and select the file Polygon.3dm from the Chapter 5 folder on the companion CD.
2 Check the Point and End boxes in the Osnap dialog box to establish object snap mode.
3 Select Curve > Polygon > Center, Radius, or click on the Polygon: Center, Radius button on the Polygon toolbar.
4 Select the NumSides option on the command area.
5 Type 7 to specify the number of sides.
6 Select point A, shown Figure 5–19, to specify the center.
7 Select point B, shown in Figure 5–19, to specify the radius. A polygon specified by its center and the radius of inscribed circle is constructed. Radius of the inscribed circle is depicted by the distance between A and B.

8 Repeat the command and select Circumscribed option on the command area, or click on the Circumscribed Polygon: Center, Radius button on the Polygon toolbar.

9 Select points C and D, shown in Figure 5–19. A polygon specified by its center and the radius of circumscribed circle is constructed.

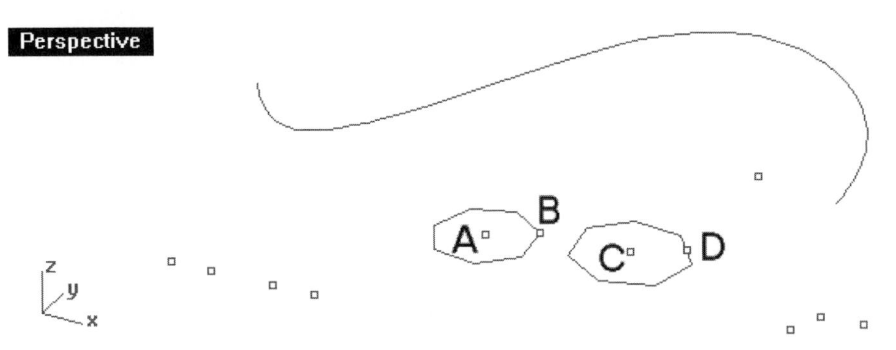

Figure 5–19. Inscribed polygon (left) and circumscribed polygon (right)

10 Select Curve > Polygon > By Edge or Polygon: Edge from the Polygon toolbar.

11 Select points A and B, shown in Figure 5–20.

12 Repeat the command.

13 Select Vertical option on the command area.

14 Select points C and D, shown in Figure 5–20.

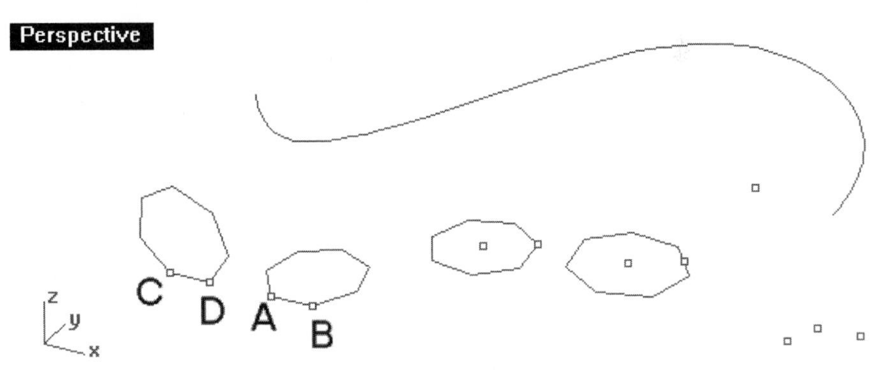

Figure 5–20. Edge polygon and vertical polyon constructed

15 Select Curve > Polygon > Star, or click on the Polygon: Star button on the Polygon toolbar.

16 Select point A, shown Figure 5–21, to indicate the center.

17 If you wish to change the number of sides, select the NumSides option on the command area and then type a value.

18 Select point B, shown in Figure 5–21, to indicate a corner point

19 Select point C, shown in Figure 5–21, to indicate the second radius.

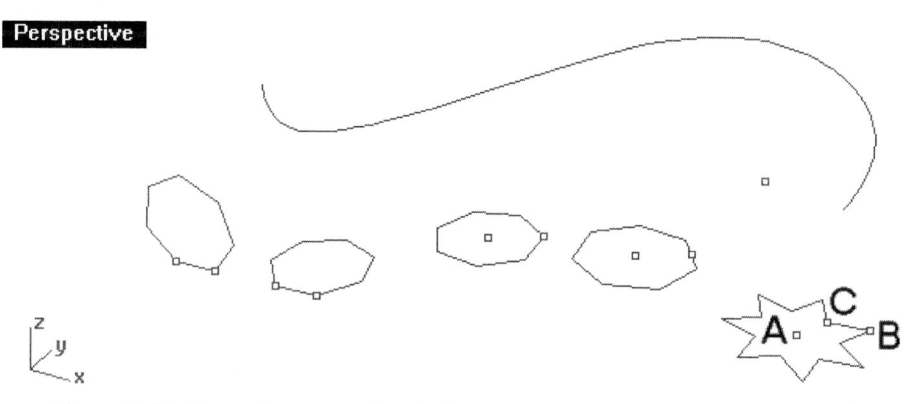

Figure 5–21. Star polygon constructed

20 Select Curve > Polygon > CENTER, Radius, or click on the Polygon: CENTER, Radius button on the Polygon toolbar.

21 Select the AroundCurve option on the command area.

22 Select curve A, shown in Figure 5–22.

23 Select end point B, shown in Figure 5–22, to indicate the corner location of the polygon.

24 Type 12 to specify the radius.

25 Select reference point C. A polygon is constructed around the selected curve. One of the corners is closest to the selected reference point.

26 Do not save your file.

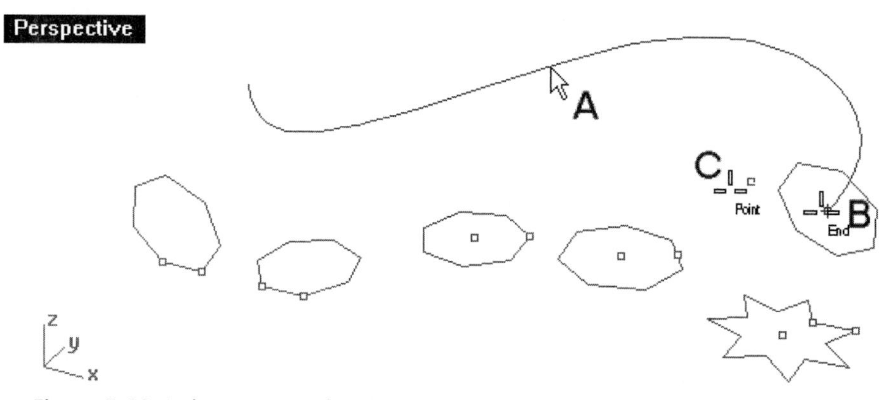

Figure 5–22. Polygon around a curve constructed

Circle

Using Rhino, you can construct two kinds of circles: true and deformable. A true circle is a degree 2 curve with four kink points at the quadrant positions. A deformable circle is a degree 3 periodic curve in the shape of a circle. You can construct a circle in many ways.

Center and Radius, Diameter, Three Points, and Multi-Points

Perform the following steps to construct a circle by specifying its center and radius, its diameter, three points on the circumference, and fit to a set of points.

1. Select File > Open and select the file Circle01.3dm from the Chapter 5 folder on the companion CD.
2. Select Curve > Circle > Center, Radius, or click on the Circle: Center, Radius button on the Circle toolbar.
3. Select points A and B, shown in Figure 5–23, to specify the center and a point on the circumference. A circle with its center at A and its circumference on point B is constructed.
4. Select Curve > Circle > Diameter, or click on the Circle: Diameter button on the Circle toolbar.
5. Select points C and D, shown in Figure 5–23. A circle with its diameter defined by points C and D is constructed.
6. Select Curve > Circle > 3 Points, or click on the Circle: 3 Points button on the Circle toolbar.
7. Select points E, F, and G, shown in Figure 5–23, to specify the center and a point on the circumference. A circle passing three points is constructed.
8. Select Curve > Circle > Fit Points, or click on the Circle: Fit Points button on the Circle toolbar.
9. Click on location H and drag J as shown in Figure 5–23.
10. Press the ENTER key. A circle fitting to the selected points is constructed.
11. Do not save your file.

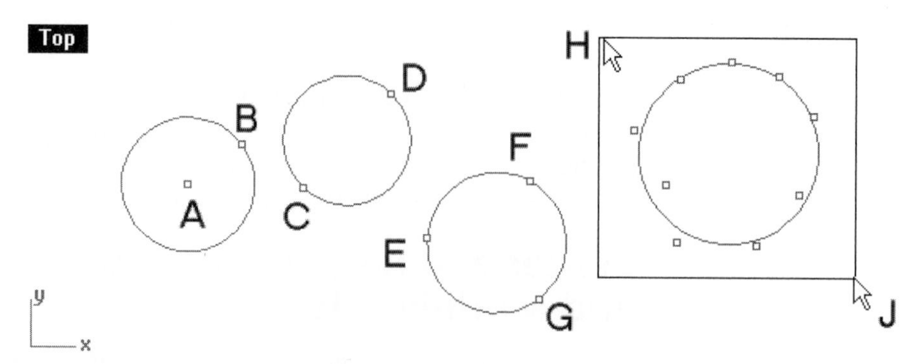

Figure 5–23. Circles constructed

Tangent Circle

You can construct a circle tangent to two curves or three curves. Perform the following steps:

1. Select File > Open and select the file Circle02.3dm from the Chapter 5 folder on the companion CD.
2. Select Curve > Circle > Tangent, Tangent, Radius, or click on the Circle: Tangent, Tangent, Radius button on the Circle toolbar.
3. Select curves A and B, shown in Figure 5–24.
4. Type 25 at the command area to specify the radius.
5. Select Curve > Circle > Tangent to 3 Curves, or click on the Circle: Tangent to 3 Curves button on the Circle toolbar.
6. Select curves A, B, and C, shown in Figure 5–24.
7. Do not save your file.

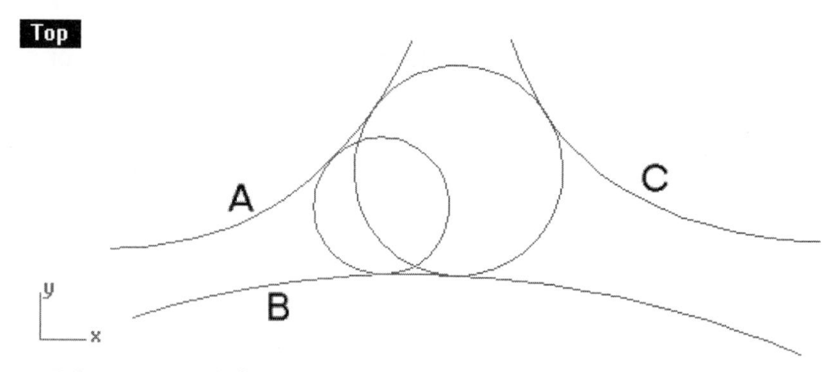

Figure 5–24. Tangent circles constructed

Vertical Circle and Circle Around a Curve

You can construct a circle vertical to the construction plane and a circle around a curve. Perform the following steps:

1. Select File > Open and select the file Circle03.3dm from the Chapter 5 folder on the companion CD.
2. Select Curve > Circle > Center, Radius and select Vertical option on the command area, or click on the Circle: Vertical to CPlane, Center, Radius button on the Circle toolbar.
3. Select points A and B, shown in Figure 5–25, to specify the center and the radius.
4. Select Curve > Circle > Center, Radius and select AroundCurve option on the command area, or click on the Circle: Around Curve button on the Circle toolbar.
5. Select curve C, shown in Figure 5–25.
6. Select point D, shown in Figure 5–25, to indicate the center location.
7. Type 40 at the command area to specify the radius.
8. Do not save your file.

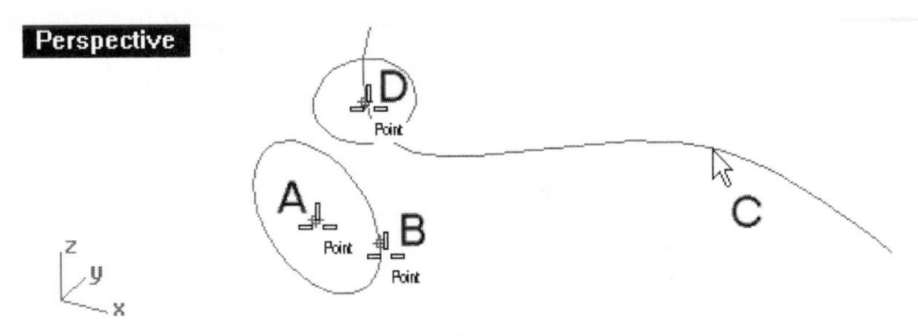

Figure 5–25. Vertical circle and circle around a curve constructed

Deformable Circle

As mentioned, a circle is a degree 2 curve. If you turn on its control points, you will find that it has four kink points. (Although it is a close loop curve, it is not a periodic curve because it has kink points.) However, you can use the deformable option to construct a deformable circle of degree 3 or above and without a kink point. In essence, a deformable circle is not a true circle. Instead, it is a free-form curve in the shape of a circle. Perform the following steps:

1. Select File > Open and select the file Circle04.3dm from the Chapter 5 folder on the companion CD.

2 Select Curve > Circle > Center, Radius, or click on the Circle: Center, Radius button on the Circle toolbar.
3 Select points A and B, shown in Figure 5–26, to specify the center and a point on the circumference.
4 Select Curve > Circle > Center, Radius, or click on the Circle: Center, Radius button on the Circle toolbar.
5 Select the Deformable option on the command area.
6 Select points C and D, shown in Figure 5–26, to specify the center and a point on the circumference.

Figure 5–26. True circle and deformable circle being constructed

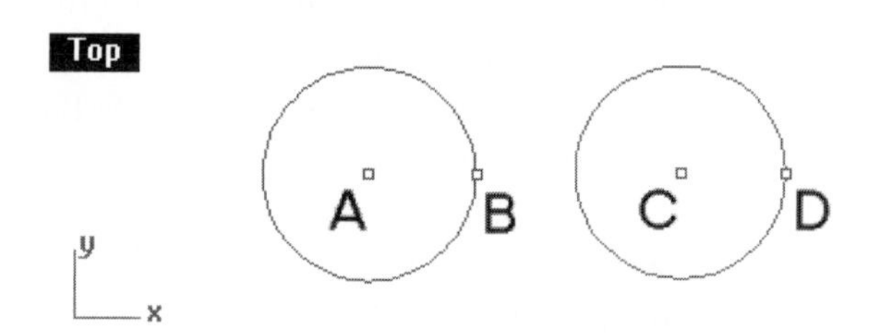

7 Select Edit > Control Point > Control Points On, or left-click on the Control Points On/Points Off button on the Point Editing toolbar.
8 Select circle A and B, shown in Figure 5–27, and press the ENTER key.

The control points are turned on. As shown, the true circle consists of four arc segments and four kink points, and the deformable circle is a free-form curve without any kink point.

9 Do not save your file.

Figure 5–27. True circle and deformable circle being constructed

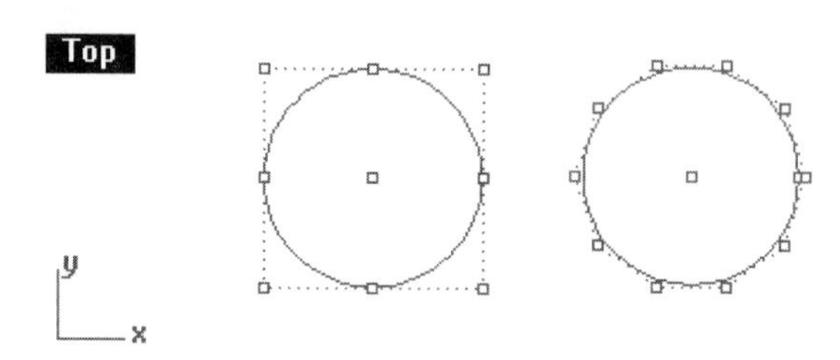

Arc

An arc is a degree 2 curve. There are seven ways to construct an arc, by specifying the following:

- the center point, the start point, and the included angle
- the start point, the end point, and the tangent direction at the start point

- the start point, the end point, and a point along the arc
- the start point, a point along the arc, and the end point
- the start point, the end point, and the radius of the arc
- two curves that the arc will be tangent to and the radius of the arc
- three curves that the arc will be tangent to

Perform the following steps to construct five arcs:

1. Select File > Open and select the file Arc01.3dm from the Chapter 5 folder on the companion CD.
2. Select Curve > Arc > Center, Start, Angle, or click on the Arc: Center, Start, Angle button on the Arc toolbar.
3. Select points A and B, shown in Figure 5–28, to specify the center and the start point of the arc.
4. Type 45 at the command area to specify the angle. An arc specified by center, start point, and arc angle is constructed.
5. Select Curve > Arc > Start, End, Point, or left-click on the Arc: Start, End, Point on Arc/Arc: Start, Point on Arc, End button on the Arc toolbar.
6. Select points, C, D, and E, shown in Figure 5–28. An arc specified by the start point, the end point, and a point on the arc is constructed.
7. Select Curve > Arc > Start, End, Direction, or left-click on the Arc: Start, End, Direction at Start/Arc: Start, Direction at Start, End button on the Arc toolbar.
8. Select points F, G, and H, shown in Figure 5–28, to specify the start point, the end point, and the direction of the arc at the start point. The third arc is constructed.
9. Select Curve > Arc > Start, End, Radius or click on the Arc: Start, End, Radius button on the Arc toolbar.
10. Select points J, K, and L, shown in Figure 5–28. J and K are the start and end points of the arc, and JL defines the radius of the arc.
11. Select Curve > Arc > Start, Point, End or right-click on the Arc: Start, End, Point on Arc/Arc: Start, Point on Arc, End button on the Arc toolbar.
12. Select points M, N, and P, shown in Figure 5–28. The fifth arc is constructed.
13. Do not save your file.

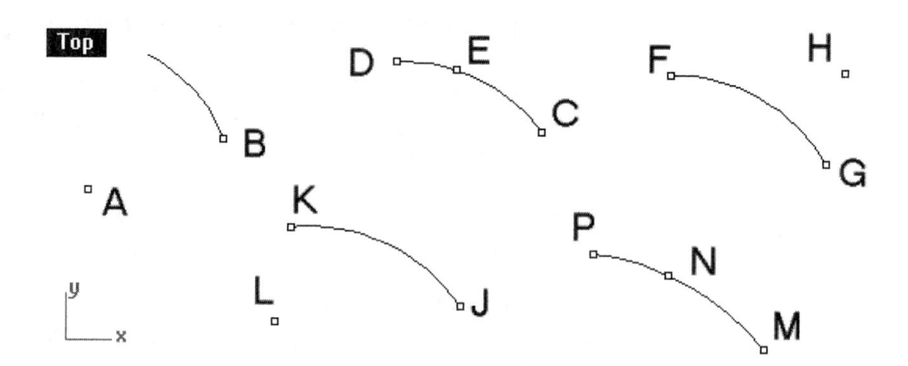

Figure 5–28. Varoius ways of arc construction

Perform the following steps to construct two more arcs:

1. Select File > Open and select the file Arc02.3dm from the Chapter 5 folder on the companion CD.
2. Select Curve > Arc > Tangent, Tangent, Radius or right-click on the Arc: Tangent to Curves/Arc: Tangent, Tangent, Radius button on the Arc toolbar.
3. Select curves A and B, shown in Figure 5–29.
4. Type 5 to specify the radius of the arc.
5. Because there are two possible solutions, click on location C. An arc tangent to two curves is constructed.
6. Select Curve > Arc > Tangent to 3 Curves or left-click on the Arc: Tangent to Curves/Arc: Tangent, Tangent, Radius button on the Arc toolbar.
7. Select curves D, E, and F, shown in Figure 5–29.
8. By moving the cursor around, you may find that there are six possible arcs.
9. Click on location G, Another arc is constructed.
10. Do not save your file.

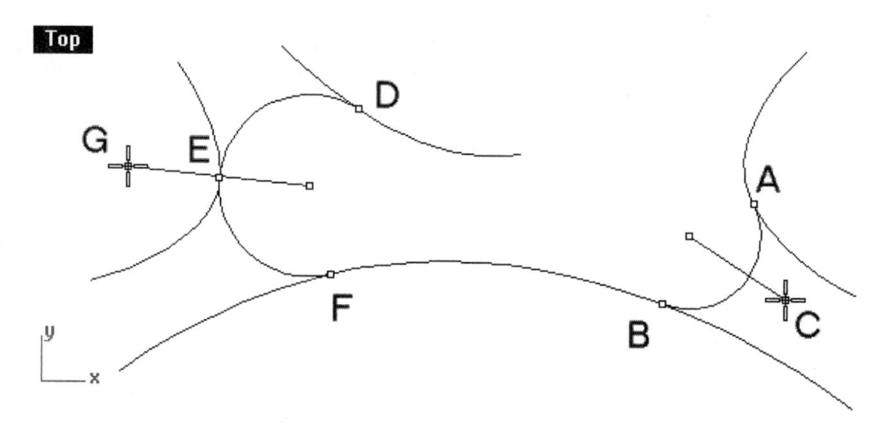

Figure 5–29. Arc tangent to three curves (left) and arc tangent to two curves (right)

Ellipse

Similar to a circle, a true ellipse is a degree 2 curve with four kink points. However, you can also construct a deformable ellipse which, in essence, is a degree 3 curve in the shape of an ellipse.

From Center, Diameter, From Foci, and Ellipse by Corners

You can construct an ellipse by specifying its center and its axes, by specifying one of its diameters and the second axis, and by specifying its foci and a point on the circumference of the ellipse. Perform the following steps by specifying two corners of a rectangle circumscribing the ellipse.

1. Select File > Open and select the file Ellipse01.3dm from the Chapter 5 folder on the companion CD.
2. Select Curve > Ellipse > From Center, or click on the Ellipse: From Center button on the Ellipse toolbar.
3. In the command area, select Deformable. (If you do not select this option, you will get a true ellipse.)
4. Select points A, B, and C, shown in Figure 5-30, to specify the center and two axes.
5. Select Curve > Ellipse > Diameter, or click on the Ellipse: Diameter button on the Ellipse toolbar.
6. Select points D and E to specify the diameter, and select point F to specify a point on the ellipse, as shown in Figure 5-30.
7. Select Curve > Ellipse > From Foci, or click on the Ellipse: From Foci button on the Ellipse toolbar.
8. Select points G and H to specify the foci, and select point J to specify a point on the ellipse, as shown in Figure 5-30.
9. Click on the Ellipse: by Corners button on the Ellipse toolbar.
10. Select points K and L to describe a rectangular region in which the ellipse is constructed, as shown in Figure 5-30.
11. Do not save your file.

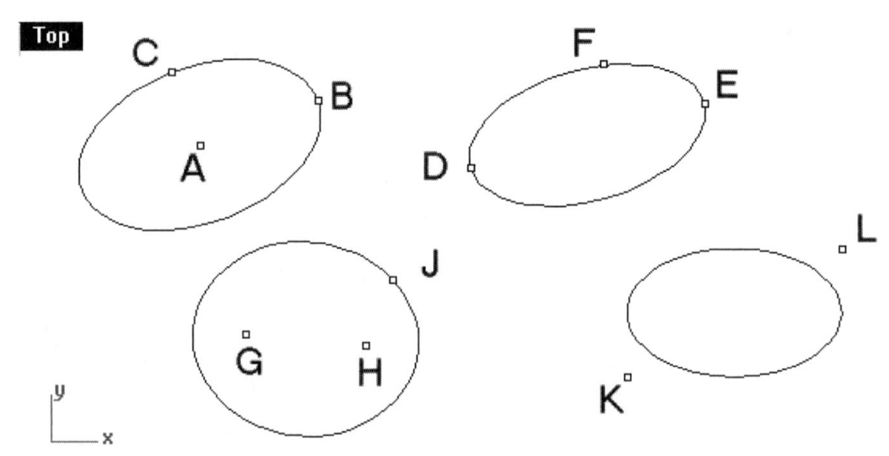

Figure 5–30. Ellipses constructed in various ways

Vertical Ellipse and Ellipse around a Curve

You can construct an ellipse vertical to the construction plane and an ellipse around a curve. Perform the following steps:

1. Select File > Open and select the file Ellipse02.3dm from the Chapter 5 folder on the companion CD.
2. Select Curve > Ellipse > Diameter, or click on the Ellipse: Diameter button on the Ellipse toolbar.
3. Select Vertical option on the command area.
4. Select points A and B in the Top viewport to specify the diameter, and select point C in the Front viewport to specify a point on the ellipse, as shown in Figure 5–31.
5. Select Curve > Ellipse > From Center and select AroundCurve option on the command area, or click on the Ellipse: Around Curve button on the Ellipse toolbar.
6. Select end point D (Figure 5–31) along the curve to indicate the center on the curve.
7. Select point E (Figure 5–31) to indicate a radius (distance between D and E).
8. Select point C, shown in Figure 5–31, to indicate the other radius (distance from the center to the selected point). An ellipse is constructed around a curve.
9. Do not save your file.

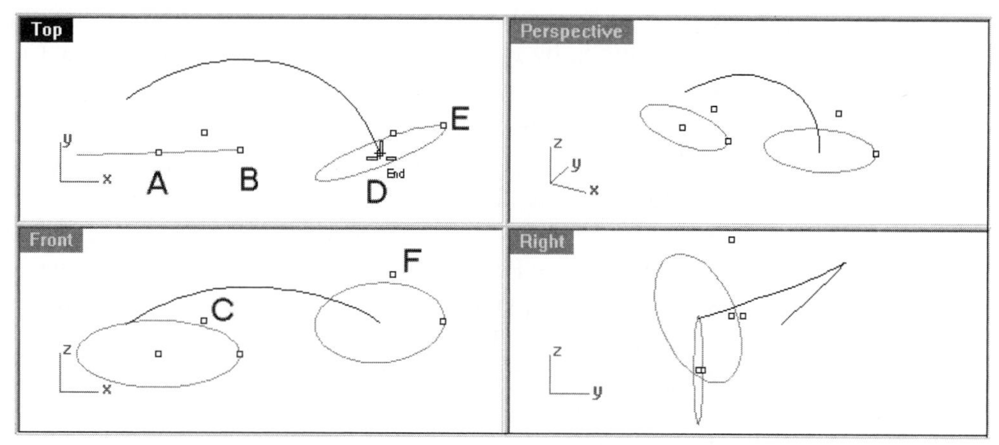

Figure 5–31. Ellipse vertical to the construction plane and ellipse around a curve being constructed

Deformable Ellipse

A true ellipse consists of four elliptical arcs joined at four kink points. A deformable ellipse is a free-form curve of degree 3 or above in the shape of an ellipse. Figure 5–32 shows the control points of a true ellipse and a deformable ellipse.

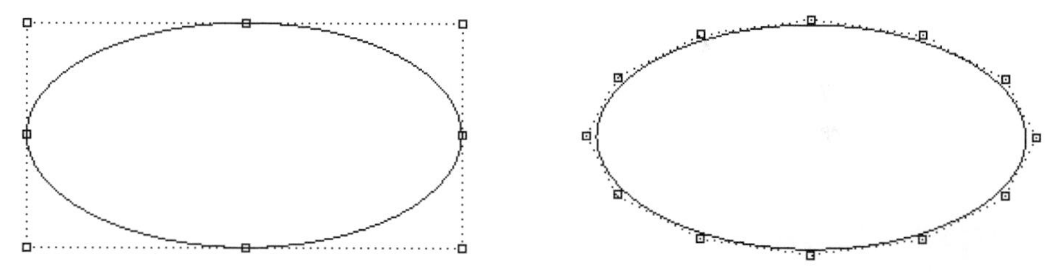

Figure 5–32. True ellipse (left) and deformable ellipse (right)

Parabola

A parabola is degree 2 curve. You can construct a parabola by specifying its focus and direction, and by specifying its vertex and focus. Perform the following steps:

1. Select File > Open and select the file Parabola.3dm from the Chapter 5 folder on the companion CD.

2. Select Curve > Parabola > Focus, Direction, or left-click on the Parabola: by Focus/Parabola by Vertex button on the Curve toolbar.
3. Select points A and B, shown in Figure 5–33, to specify the focus point and the direction.
4. Select point C, shown in Figure 5–33, to specify one of the end points of the parabola.
5. Select Curve > Parabola > Vertex, Focus, or right-click on the Parabola: by Focus/Parabola by Vertex button on the Curve toolbar.
6. Select points D and E, shown in Figure 5–33, to specify the focus point and the direction.
7. Select point F, shown in Figure 5–33, to specify a reference point. A parabola is constructed.
8. Do not save your file.

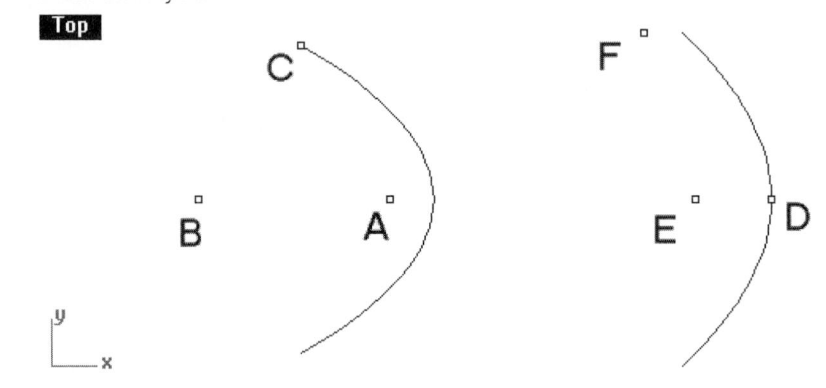

Figure 5–33. Parabola curves

Hyperbola

A hyperbola is also a degree 2 curve. Perform the following steps:

1. Select File > Open and select the file Hyperbola.3dm from the Chapter 5 folder on the companion CD.
2. Set object snap mode to Point.
3. Select Curve > Hyperbola > Center, Coefficient, or click on the Hyperbola button on the Curve toolbar and select the FromCoefficient option on the command area.
4. Select points A, B, and C, shown in Figure 5–34.
5. Select Curve > Hyperbola > From Foci, or click on the Hyperbola button on the Curve toolbar and select the FromFoci option on the command area.

6 Select points D, E, and F, shown in Figure 5–34.
7 Select Curve > Hyperbola > Vertex, Focus, or click on the Hyperbola button on the Curve toolbar and select the FromVertex option on the command area.
8 Select points G, H, and J, shown in Figure 5–34.
9 Three hyperbola curves are constructed. Do not save your file.

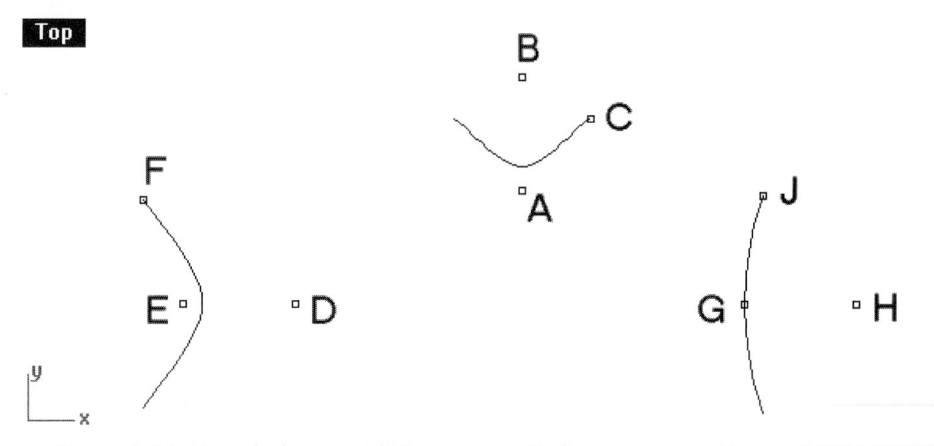

Figure 4–34. Hyperbola curves being constructed

Conic

A conic is also a degree 2 curve. You can construct it in several ways. You can construct a conic by specifying its end points, its apex, and a point on the conic. You can also construct a conic that is perpendicular to a selected curve and a conic that is tangent to two curves. Perform the following steps:

1 Select File > Open and select the file Conic.3dm from the Chapter 5 folder on the companion CD.
2 Set object snap mode to End and Point.
3 Select Curve > Conic, or left-click on the Conic/Conic: Perpendicular at Start button on the Curve toolbar.
4 Select points A and B (Figure 5–35) to indicate the end points of the conic.
5 Select point C (Figure 5–35) to indicate the apex location.
6 Select location D (Figure 5–35) to indicate a point on the conic curve.

7. Select Curve > Conic and select Perpendicular option on the command area, or right-click on the Conic/Conic: Perpendicular at Start button on the Curve toolbar.
8. Click on end point E (Figure 5–35).
9. Select the Perpendicular option on the command area.
10. Click on end point F (Figure 5–35).
11. Select point G. A conic perpendicular to two curves and passing through a reference point is constructed.
12. Select Curve > Conic and select Tangent on the command area, or left-click on the Conic: Tangent at Start/Conic: Tangent at Start, End button on the Curve toolbar.
13. Click on end point H (Figure 5–35) to specify the tangent end point.
14. Select the Tangent option on the command area. (You may skip this step if you right-click on the Conic: Tangent at Start/Conic: Tangent at Start, End button on the Curve toolbar to run the command.)
15. Select end point J (Figure 5–35) to specify the other end point.
16. Select point K (Figure 5–35). A conic tangent to two curves and passing through a reference point is constructed.
17. Do not save your file.

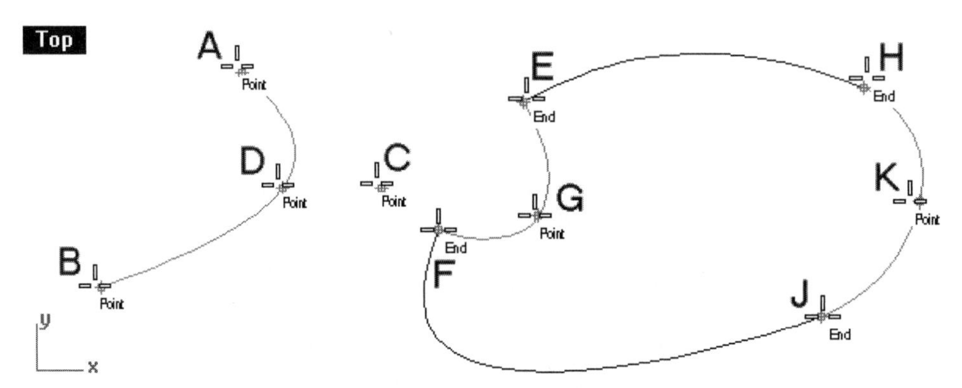

Figure 5–35. From left to right: Conic defined by end points, vertex, and a point on the curve, conic perpendicular to two curves, and conic tangent to two curves

Helix

A helix is a degree 3 curve. Using the Helix command, you can construct a helix with its axis lying on or perpendicular to the current construction plane. You can also construct a helix around a curve. The shape of a helix around a curve is the shape of the selected curve.

Helix with a Straight Axis

The simplest kind of helix curve is a helical coil having a straight axis. Perform the following steps:

1. Select File > Open and select the file Helix.3dm from the Chapter 5 folder on the companion CD.
2. Select Curve > Helix, or left-click on the Helix/Vertical Helix button on the Curve toolbar.
3. Select points A and B, shown in Figure 5–36, in the Top viewport to specify the axis end points.
4. Select the Turns option on the command area and type 3 at the command area to specify the number of coils.
5. Select point C (Figure 5–36) in the Top viewport to specify the radius of the helix.

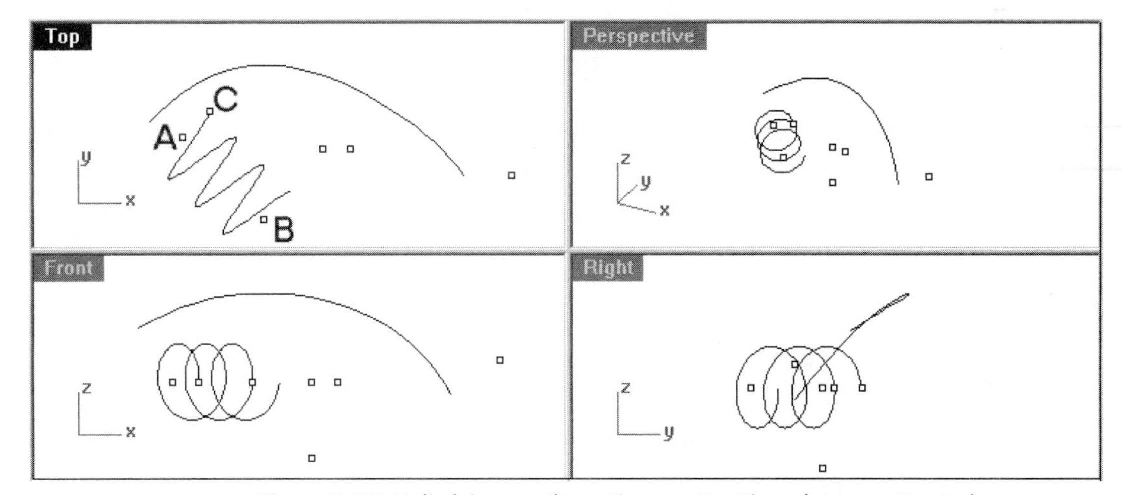

Figure 5–36. Helix lying on the active construction plane constructed

Vertical Helix

A vertical helix has its axis vertical to the construction plane. Continue with the following steps:

6. Select Curve > Helix and select Vertical option on the command area, or right-click on the Helix/Vertical Helix button on the Curve toolbar.
7. Select point D, shown in Figure 5–37, in the Top viewport to specify an axis end point.

8. Select point E (Figure 5–37) in the Front viewport to specify the other axis end point.
9. Select the Turns option on the command area, and type 4 at the command area to specify the number of coils.
10. Select point F (Figure 5–37) in the Top viewport to specify the radius.

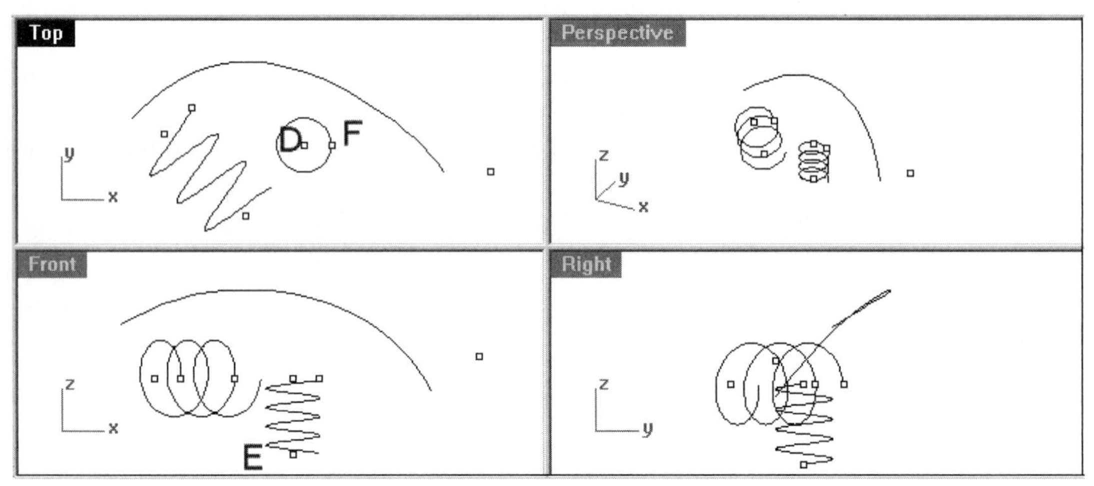

Figure 5–37. Helix vertical to the active construction plane constructed

Helix Around a Curve

A helix around a curve has its axis following the flow of a selected curve. Continue with the following steps:

11. Select Curve > Helix, or left-click on the Helix/Vertical Helix button on the Curve toolbar.
12. Select the AroundCurve option on the command area.
13. Select the free-form curve G, shown in Figure 5–38.
14. Type 5 at the command area to set the helix radius.
15. Select the Turns option on the command area and type 6 at the command area to specify the number of coils.
16. Select point H (Figure 5–38) to specify a reference.
17. Do not save your file.

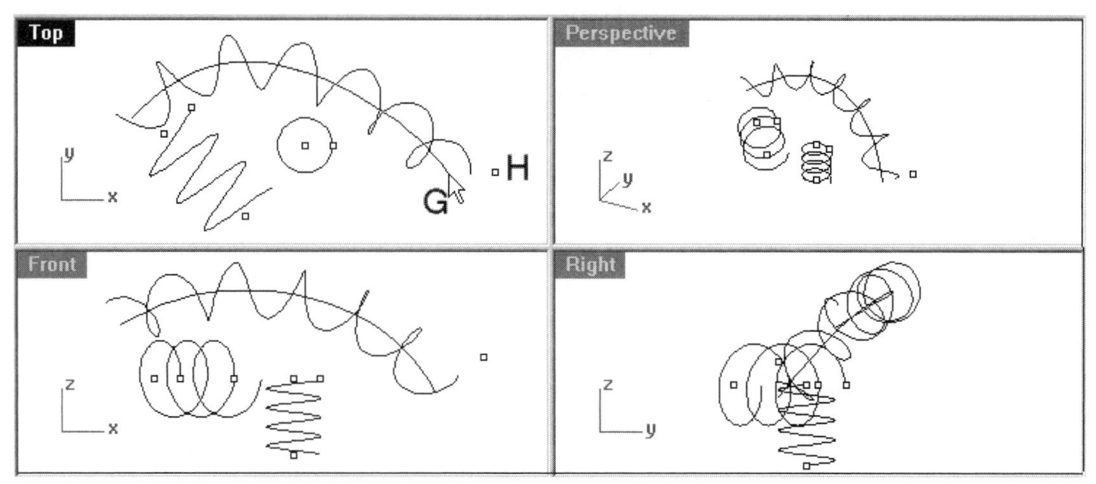

Figure 5–38. Helix around a curve constructed

Spiral

A spiral is a degree 3 curve. You can use the Spiral command to construct a 3D spiral or a flat spiral. You can also construct a 3D spiral around a curve.

Spiral with a Straight Axis and Flat Spiral

A spiral is similar to a helix in terms of being a coil around an axis. It differs from a helix in that the coil's radius is either increasing or decreasing. A flat spiral is a coil on a 2D plane. Perform the following steps:

1. Select File > Open and select the file Spiral.3dm from the Chapter 5 folder on the companion CD.
2. Select Curve > Spiral, or left-click on the Spiral/Flat Spiral button on the Curve toolbar.
3. Select points A and B, shown in Figure 5–39, to indicate the end points of the axis.
4. Select the Turns option on the command area and type 4 at the command area to specify the number of coils.
5. Select points C and D (Figure 5–39) to indicate the first and second radii.
6. Select Curve > Spiral and select Flat option on the command area, or right-click on the Spiral/Flat Spiral button on the Curve toolbar.

7 Select point E (Figure 5–39) to indicate the center of the spiral.
8 Select point F (Figure 5–39) to indicate the first radius.
9 Select the Turns option on the command area and type 5 at the command area to specify the number of coils.
10 Select point G (Figure 5–39) to indicate the second radius.

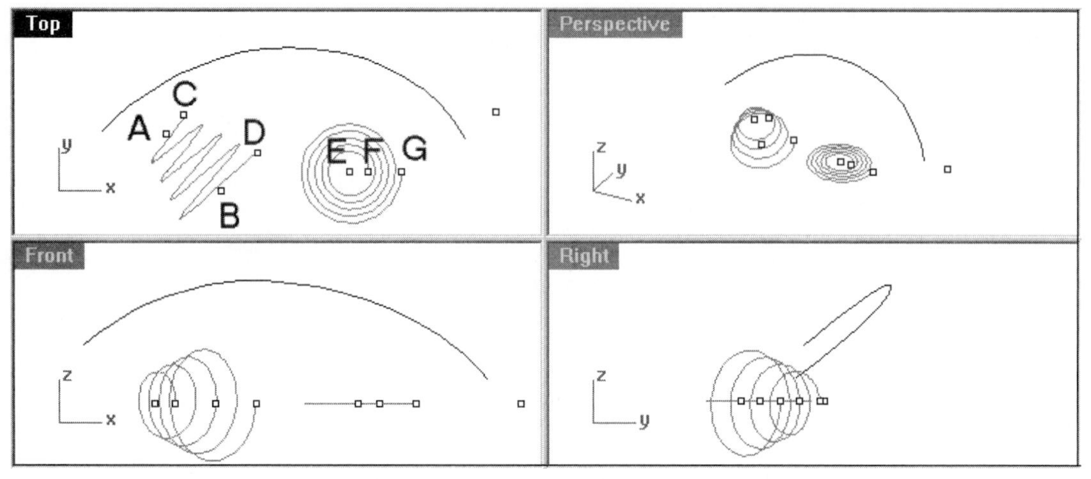

Figure 5–39. A straight spiral and flat spiral constructed

Spiral Around a Curve

You can construct a spiral around a curve, as follows:

11 Select Curve > Spiral, or click on the Spiral/Flat Spiral button on the Curve toolbar.
12 Select the AroundCurve option on the command area.
13 Select curve H, shown in Figure 5–40.
14 Select the Turns option on the command area, and type 6 to specify the number of coils.
15 Type 5 at the command area to specify the first radius.
16 Select point J (Figure 5–40) to specify a reference point.
17 Type 10 at the command area to specify the second radius. A spiral around the conic curve is constructed.
18 Do not save your file.

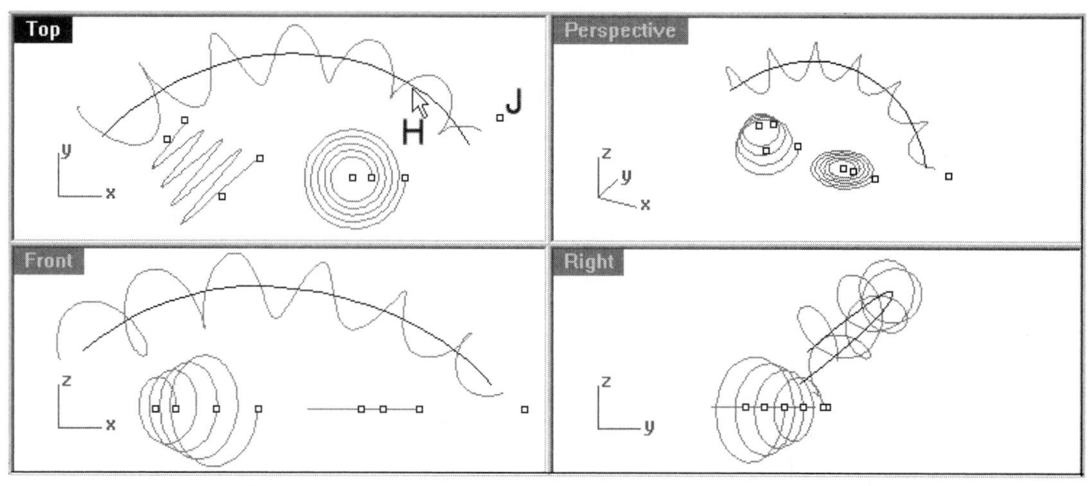

Figure 5–40. Spiral around a curve constructed

Degree of Polynomial and Point Editing

Because Rhino uses NURBS mathematics to define curves, the curve manipulating tools delineated in the previous chapter also apply to curves of regular pattern explained above. Naturally, you may change the degree of polynomial of the curves, manipulate the control points and edit points, and use the handlebar editor.

Because increasing the degree of polynomial will not change the shape of the curve, it is advisable to change the degree of polynomial of the curves to 3 after they are completed. To avoid having kink points in the curve, you should use deformable circle and deformable ellipse unless true circle and true ellipse are required in your design.

Chapter Summary

Apart from free-form curves of irregular pattern, you can use curves of regular pattern in surface construction. These curves are line, polyline, rectangle, polygon, circle, arc, ellipse, parabola, hyperbola, conic, helix, and spiral.

Regular curves may be classified by their degree of polynomial. Line, polyline, rectangle, and polygon are degree 1 curves. True circle, arc, true ellipse, parabola, hyperbola, and conic are degree 2 curves. In particular, true circle and true ellipse have kink points at their quadrant locations.

The deformable circle, deformable ellipse, helix, and spiral are degree 3 curves of regular pattern. To produce a free-form curve without a kink point in the shape of a circle or in the shape of an ellipse, you construct a deformable circle or deformable ellipse.

Because Rhino uses NURBS mathematics to define curves and surfaces, you can use all the editing tools delineated in Chapter 4 on curves of regular shapes. You can change the degree of polynomial, add or remove knots and kink points, adjust control point location and point weight, and add or remove control points. Naturally, you can use the handlebar editor as well.

Review Questions

1. Give a brief account of the types of curves you can construct using Rhino.
2. Classify the curves in accordance with their degree of polynomial.
3. Differentiate between deformable circle and deformable ellipse with true circle and true ellipse.

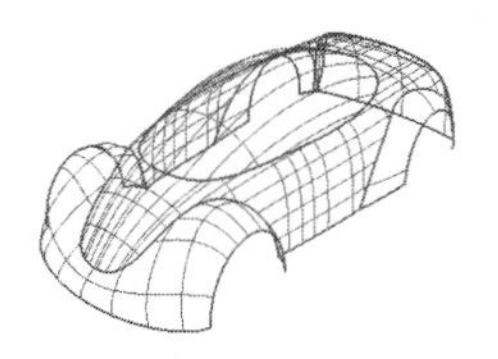

CHAPTER 6
Curve Manipulation

Introduction

This chapter explores various curve manipulation methods, including manipulation of a curve's length, treatment of two or more curves, refinement of curve profile, and deriving points and curves from existing objects.

Objectives

After studying this chapter, you should be able to do the following:

- ❒ Manipulate the length of a curve
- ❒ Treat two or more separate curves
- ❒ Refine curves
- ❒ Construct points and curves on existing objects

Overview

After learning how to construct free-form curves and curves of regular pattern in Chapters 4 and 5, this chapter addresses various curve manipulation tools, categorized into four groups. The first group deals with changing the length of a curve; the second group concerns the treatment of two or more separate curves; the third group explains ways to refine a curve; and the fourth group deals with deriving curves and point objects from existing objects.

Manipulating a Curve's Length

To meet design need, you may extend a curve to add extra length to it, trim a curve to remove unwanted portions, split a curve into two curves, and delete a portion of it, without changing the shape of the existing portion of the curve. If you have an open polyline, you can close it by adding a line segment.

Extending the Length of a Curve

You can extend a curve in several ways: by extending it to a boundary curve, dragging an end of the curve to a new position, adding a straight line segment to the curve, adding an arc segment to the curve, or extending it to meet the edge of a surface.

Extending to a Boundary

The simplest way of extending a curve is to extend it to a boundary. However, if the end point of a curve is too far from the boundary curve, a solution may not be available, in which case you will need to try other methods of extending the curve. Perform the following steps:

1. Select File > Open and select the file Extend01.3dm from the Chapter 6 folder on the companion CD.
2. Select Curve > Extend Curve > Extend Curve, or click on the Extend Curve button on the Extend toolbar.
3. Select curve A, shown in Figure 6–1, and press the ENTER key.
4. Select the Type option on the command area.

Four options are available: Natural, Line, Arc, and Smooth. Obviously, the Line and Arc options add a line or arc segment to the existing curve, and the Smooth option adds a free-form curve segment. The Natural option is quite special in that, if the original curve is a line, it adds a line segment; if the original curve is an arc, it adds an arc segment; and if the original curve is a free-form curve, it adds a free-form curve segment.

5. Select the Line option on the command area.
6. Select curve B to extend it to A.
7. Press the ENTER key to terminate the command. A curve is extended to a boundary curve.

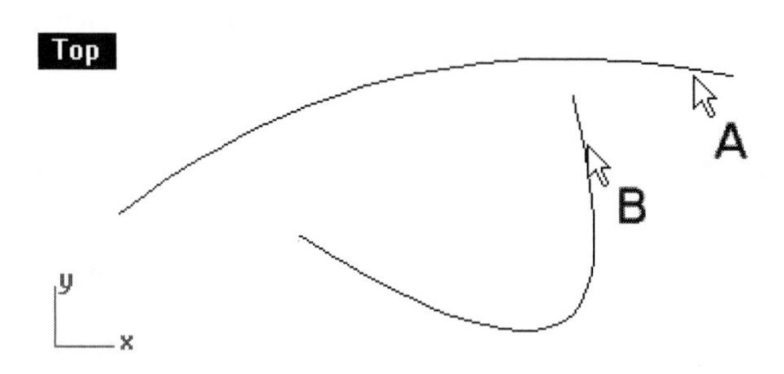

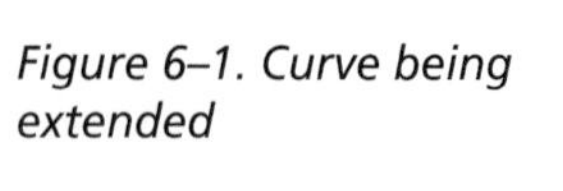
Figure 6–1. Curve being extended

Extend by Dragging

This method enables you to determine the end point location of the extended curve by dragging it to a new location. Continue with the following steps.

8 Click on the Extend Curve, Smooth button on the Extend toolbar. (The Smooth option is used.)

9 Select curve A (shown in Figure 6–2), and then select location B. (For the purpose of this tutorial, the exact location is unimportant.) The selected curve is extended by adding a spline segment at the end point.

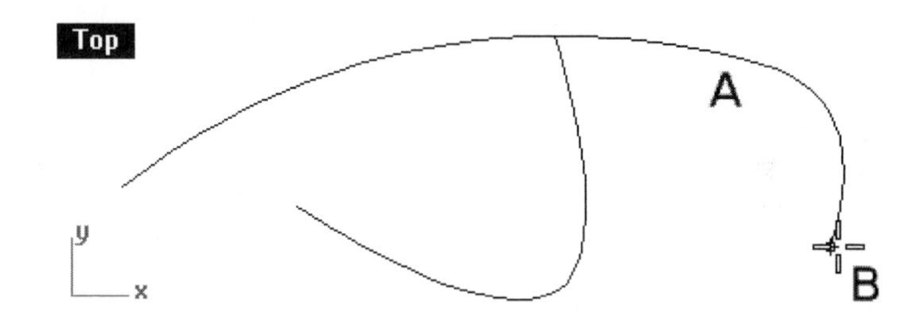

Figure 6–2. Curve being extended by adding a spline segment

Adding a Line Segment

This method enables you to add a line segment and determine the end point of the line segment. Continue with the following steps.

10 Select Curve > Extend Curve > By Line, or click on the Extend by Line button on the Extend toolbar.

11 Select curve A (shown in Figure 6–3), and then select location B. The selected curve is extended by attaching to it a tangent line segment.

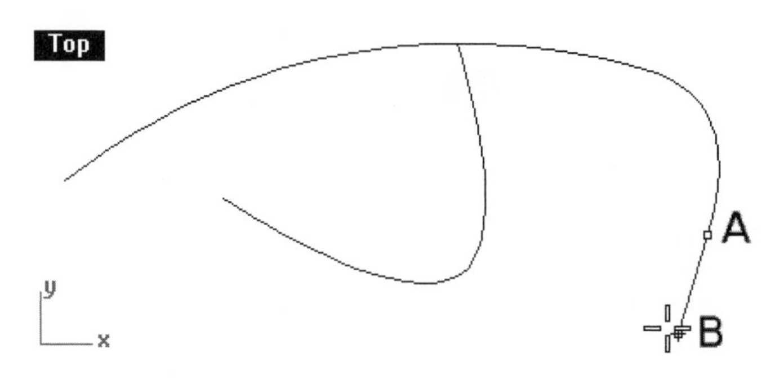

Figure 6–3. Curve being extended by adding a tangent line segment

Adding an Arc Segment with a Specified End Point

This method enables you to add an arc segment and determine the end point of the arc segment. Continue with the following steps.

12 Select Curve > Extend Curve > By Arc to Point, or click on the Extend by Arc to Point button on the Extend toolbar.

13 Select curve A (shown in Figure 6–4), and then select location B. The selected curve is extended by attaching to it a tangent arc segment defined by the arc's end point.

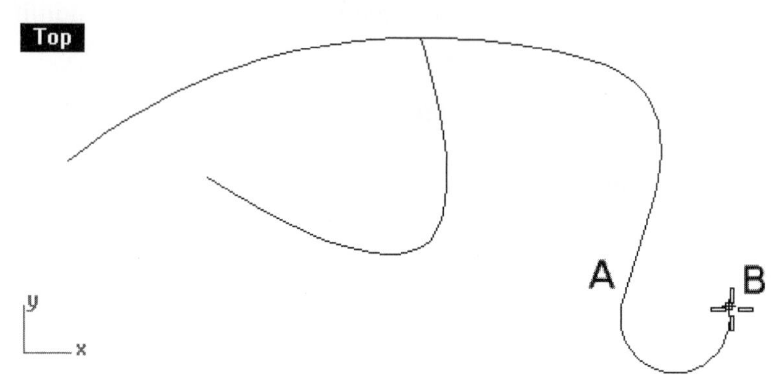

Figure 6–4. Curve being extended by adding a tangent arc segment

Adding an Arc Segment with Equal Radius of Curvature

This method enables you to add an arc segment having the same radius of curvature of the end point of the original curve.

14 Select Curve > Extend Curve > By Arc, or click on the Extend by Arc, keep radius button on the Extend toolbar.

15 Select curve A (shown in Figure 6–5) and then select location B. The selected curve is extended by attaching to it a tangent arc segment with a radius equal to the radius of curvature at the end point of the curve.

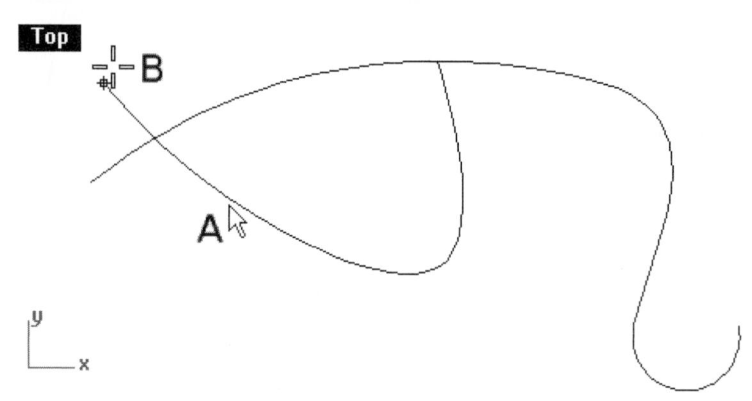

Figure 6–5. Curve being extended by adding a tangent arc segment

Adding an Arc Segment with Specified Radius

This method enables you to add an arc segment by specifying the radius and end point location of the arc. Continue with the following steps.

16 Select Curve > Extend Curve > By Arc with Center, or click on the Extend by Arc with Center button on the Extend toolbar.

17 Select curve A (shown in Figure 6–6).

18 Select location B (Figure 6–6) to specify the radius of the arc.

19 Select location C (Figure 6–6) to specify the end point of the arc. The selected curve is extended by attaching to it a tangent arc.

20 Do not save your file.

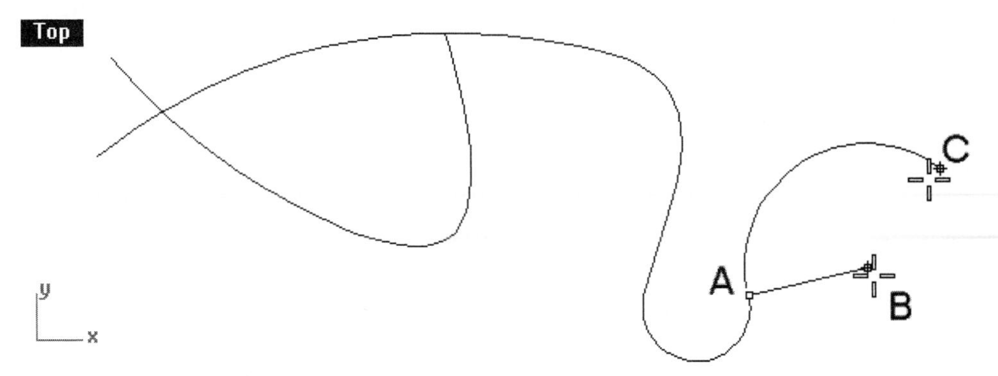

Figure 6–6. Curve being extended by adding a tangent arc segment of designated radius

Extending curve on a Surface

To extend a curve on a surface to the surface's boundary edge, perform the following steps:

1 Select File > Open and select the file Extend02.3dm from the Chapter 6 folder on the companion CD.

2 Select Curve > Extend Curve > Curve on Surface, or click on the Extend Curve on Surface button.

3 Select curve A and surface B (shown in Figure 6–7). The curve on the surface is extended to the boundary edges of the surface.

4 Do not save your file.

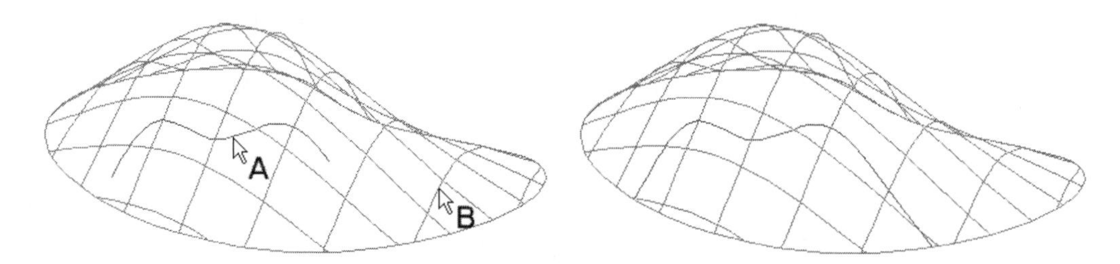

Figure 6–7. Curve being extended (left) and curve extended to the edge of the surface (right)

Trimming a Curve

Unwanted portions of a curve can be trimmed way. After trimming, the shape of the remaining portion of the curve remains unchanged. Perform the following steps:

1. Select File > Open and select the file TrimSplit.3dm from the Chapter 6 folder on the companion CD.
2. Select Edit > Trim, or click on the Trim/Untrim Surface button on the Main2 toolbar.
3. Select curve A, shown in Figure 6–8, and press the ENTER key. This is the cutting edge.
4. Select curve B (Figure 6–8). This is the portion to be trimmed away.
5. Press the ENTER key to terminate the command. The curve is trimmed.

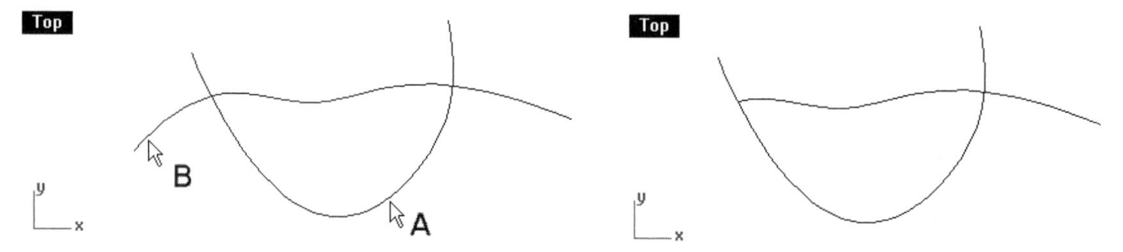

Figure 6–8. Curve being trimmed (left) and curve trimmed (right)

Splitting a Curve into Two Curves

You can split a curve into two curves and keep the shape of the split curves unchanged. Continue with the following steps.

6. Select Edit > Split, or click on the Split/Split Surface by Isocurve button on the Main1 toolbar.

7. Select curve A, shown in Figure 6–9, and press the ENTER key. This is the curve to be split.
8. Select curve B (Figure 6–9). This is the cutting object.
9. Press the ENTER key to terminate the command. Because curve B intersects curve A at two locations, curve A is split into three curves.
10. Do not save your file.

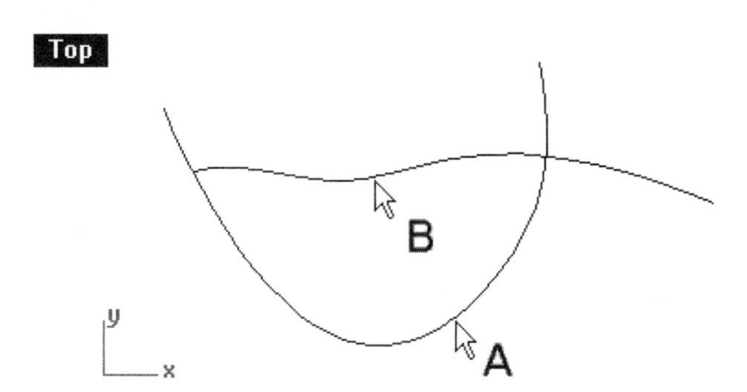

Figure 6–9. Curve being split

Deleting a Portion of a Curve

A portion of a curve can be deleted. As a result, two curves are derived from a single curve. Perform the following steps:

1. Select File > Open and select the file DelSubCurve.3dm from the Chapter 6 folder on the companion CD.
2. Set object snap mode to Point.
3. Select Curve > Curve Edit Tools > Delete Subcurve, or click on the Delete Sub Curve button on the Curve Tools toolbar.
4. Select curve A and then points B and C (shown in Figure 6–10). The portion of the curve between B and C is deleted.
5. Do not save your file.

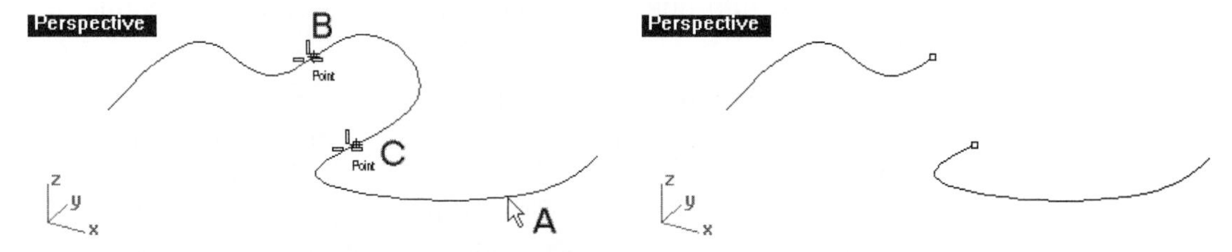

Figure 6–10. Original curve (left) and a portion deleted (right)

Closing an Open Polyline

An open polyline consists of line segments only and can be closed by adding a line segment to it, as follows:

1. Select File > Open and select the file ClosePolyline.3dm from the Chapter 6 folder on the companion CD.
2. Select Curve > Curve Edit Tools > Close Curve, or click on the Close Open Curves button on the Curve Tools toolbar.
3. Select polycurve A, shown in Figure 6-11, and press the ENTER key. A segment is added and the polycurve is closed.
4. Do not save your file.

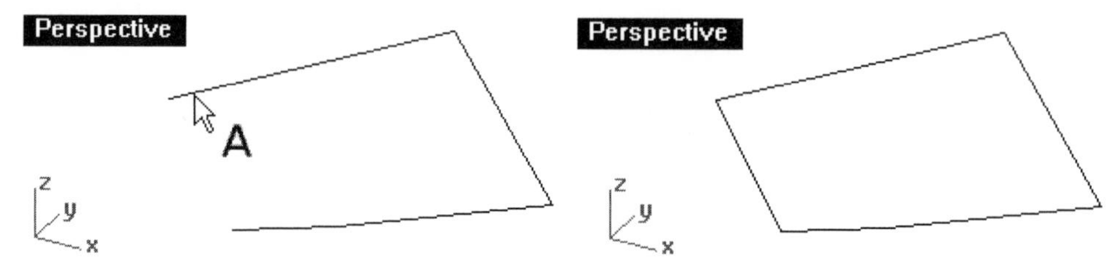

Figure 6-11. Open loop polycurve (left) and polycurve closed (right)

Treating Two or More Separate Curves

Chamfering, filleting, and blending are three basic ways to modify two separate curves to become a set of contiguous curves. Between the curves, chamfering adds a bevel edge which, in essence, is a straight line; filleting adds a tangent arc; and blending adds a smooth curve. As may be necessary, both chamfering and filleting extend or trim the existing curves to meet the bevelled edge or filleted arc.

In accordance with your design needs, you may have to join two or more curves into one, or explode a joined curve into individual curve components.

As explained in the previous chapter, with two contiguous curves meeting each other end point to end point, the continuity between them can be described as G0, G1, G2, G3, or G4. Typically, chamfered edge has G0 continuity, filleted edge has G1 continuity, and blended edge can have G2, G3, or G4 continuity. (See Appendix A for a more detailed explanation on degree of continuity.)

In addition to chamfering, filleting, and blending, you can also connect two coplanar curves to become a joined curve, modify a curve to match another curve, or modify two curves to match each other.

If curves intersect each other and there are closed regions between them, boundaries of such regions can be constructed using curve Boolean operations.

Chamfering

Chamfering joins two selected curves with a bevelled line. In essence, chamfering extends or trims the existing curves and inserts a line segment between them. While chamfering, you need to specify the distances of the end points of the inserted line segment measured from the imaginary intersection point of the two curves. Perform the following steps:

1. Select File > Open and select the file Joint.3dm from the Chapter 6 folder on the companion CD.
2. Select Curve > Chamfer Curves, or click on the Chamfer Curves button on the Curve Tools toolbar.
3. Select the Distance option on the command area.
4. Type 10 to set the first distance of the line segment from the imaginary intersection point of the two curves.
5. Type 5 to set the second distance. *(Note: By default, the second distance is equal to the first distance. Therefore, you may simply press the ENTER key to accept, if the second distance equals the first distance.)*
6. Select curves A and B, shown in Figure 6–12. Note that the first selected edge corresponds to the first chamfer distance. A chamfer curve is constructed.

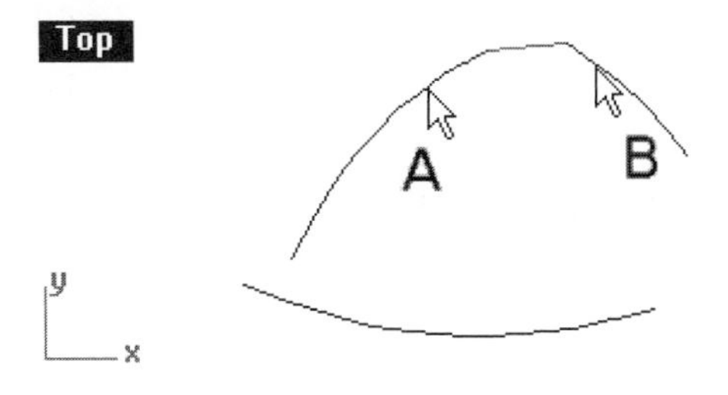

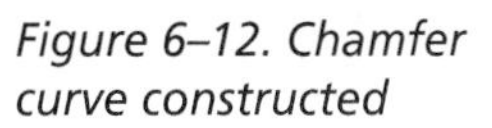

Figure 6–12. Chamfer curve constructed

Filleting

Filleting also joins two selected curves. It extends or trims the existing curves and adds a tangent arc segment between them. A filleted curve gives a G1 continuity. Continue with the following steps.

7 Select Curve > Fillet Curves, or click on the Fillet Curves button on the Curve Tools toolbar.

8 Select the Radius option on the command area.

9 Type 3 to specify a radius of 3 mm.

10 Select curves A and B, shown in Figure 6–13. The selected curves are trimmed/extended, and a fillet curve is constructed between them.

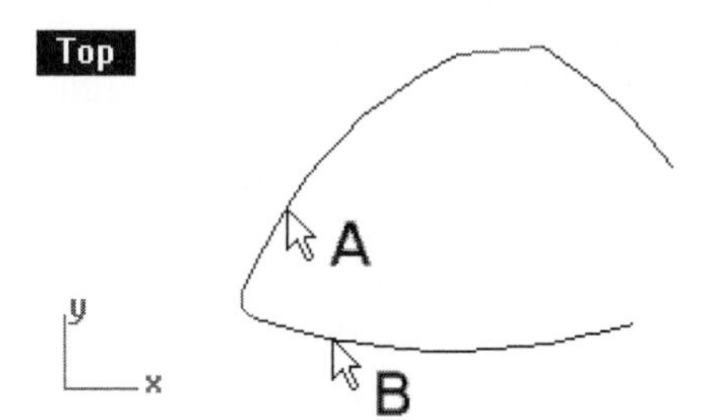

Figure 6–13. Fillet curve constructed

Filleting Vertices of a Polyline

If you have a polyline or a set of curves joined together end to end, you may fillet all its corners in a single operation, as follows:

11 Maximize the Front Viewport.

12 Select Curve > Fillet Corners, or click on the Fillet Corners button on the Curve Tools toolbar.

13 Select curve A, shown in Figure 6–14, and press the ENTER key.

14 Type 5 at the command area to specify the fillet radius.

In essence, all the corners of a polyline will be filleted. However, segments too short for the specified fillet radius will not be filleted. Here, a corner is not filleted for this reason.

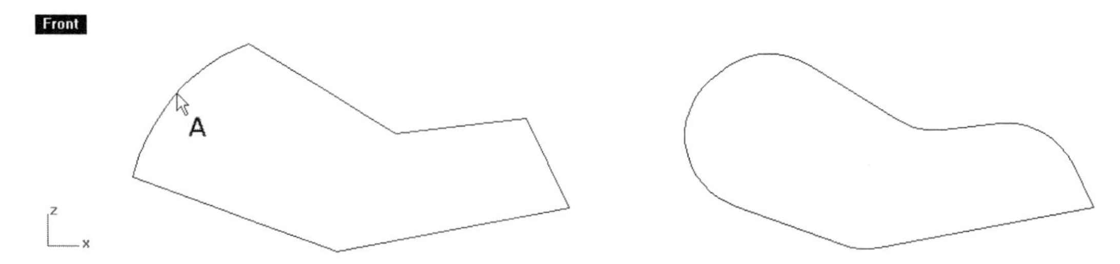

Figure 6–14. Polyline (left) and five corners of the polyline filleted (right)

Blending

To construct a curve to bridge two edges/curves with a continuity of G1, you construct a simple blend curve, as follows:

15 Maximize the Top viewport.

16 Select Curve > Blend Curve, or click on the Blend Curves/Blend Perpendicular to Two Curves button on the Curve Tools toolbar.

17 Select the Continuity option on the command area.

18 Select the Curvature continuity option, if it is not already selected.

19 Select curves A and B, shown in Figure 6–15. A continuous curvature curve is constructed.

Figure 6–15. Blend curve constructed

To construct a perpendicular and angular blended curve, continue with the following steps.

20 Maximize the Perspective viewport.

21 Set object snap mode to Point.

22 Select Curve > Blend Curve and select the Perpendicular option on the command area, or right-click on the Blend Curves/Blend Perpendicular to Two Curves button on the Curve Tools toolbar.

23 Select edge A, shown in Figure 6–16, and then click on point B.

24 Select edge C, shown in Figure 6–16, and then click on point D. A perpendicular blended curve is constructed.

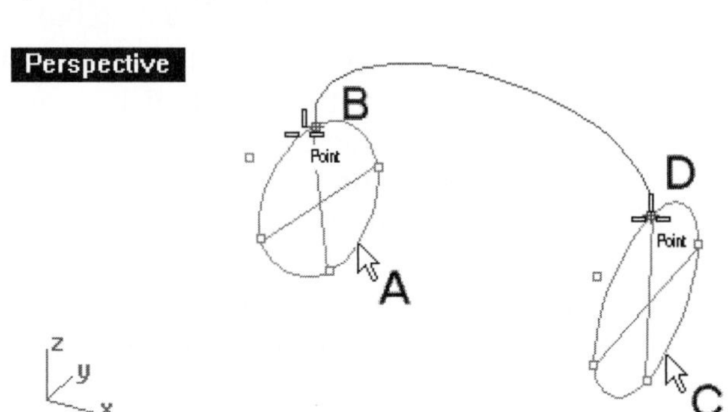

Figure 6–16. Perpendicular blend curve constructed

25 Select Curve > Blend Curve, or click on the Blend Curves/Blend Perpendicular to Two Curves button on the Curve Tools toolbar.
26 Select the AtAngle option on the command area.
27 Select edge A, shown in Figure 6–17, and then click on points B and C.
28 Select the AtAngle option on the command area.
29 Select edge D, shown in Figure 6–17, and then click on points E and F. An angular blended curve is constructed.
30 Do not save your file.

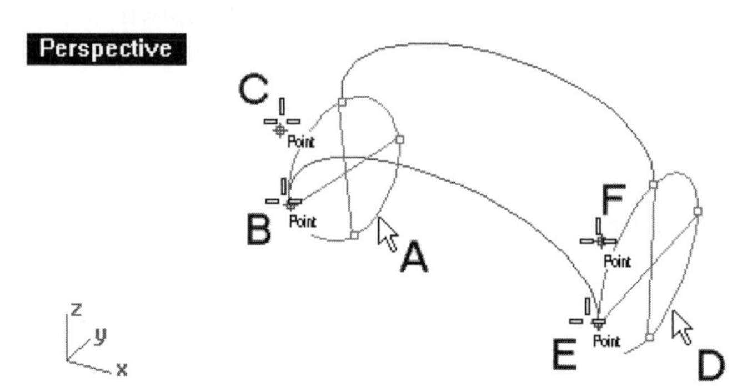

Figure 6–17. Angular blend curve constructed

Adjustable Curve Blend

To achieve a continuity beyond G2 and impose additional control on the shape while constructing a blend curve, you use the BlendCrv command. Perform the following steps:

1 Select File > Open and select the file AdjustableBlend.3dm from the Chapter 6 folder on the companion CD.
2 Click on the Adjustable Curve Blend button on the Curve Tools toolbar, or type BlendCrv on the command area.
3 Select curves A and B, shown in Figure 6–18.
4 If the Curvature Graph dialog box is not displayed, select the CurvatureDisplay option on the command area to change it to "yes."
5 In the Curvature Graph dialog box, click on the curve hair button and then select a color from the Select Color dialog box to choose a display for the curvature hairs that appear on the curves, then click on the Surface Hair button and select a color from the Select Color dialog box to choose a color for the surface hairs.
6 Click on the Display Scale and Density arrow buttons to adjust the curvature scale and density as may be necessary.
7 Click on the X mark of the dialog box to close it.

8 Select the control points and drag them to new locations to experience how the blend curve's shape is changed.
9 Press the ENTER key. A blend curve is constructed.

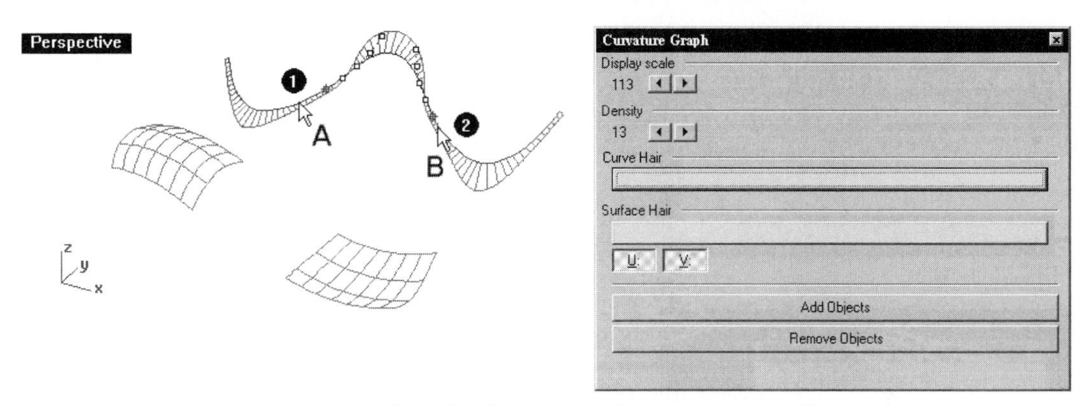

Figure 6–18. Adjustable blend curve being constructed between curves

In addition to blending between two selected curves, you can construct blend curve between two surfaces along selected locations on the surfaces' edges. Continue with the following steps.

10 Repeat the command.
11 Select the Edges option on the command area.
12 Select edges A and B, shown in Figure 6–19.
13 To adjust the location of the curve, click on the end point of the curve and drag to a new position.
14 To adjust the shape of the curve, click on the control points and drag to a new location.
15 Press the ENTER key. A curve is constructed.

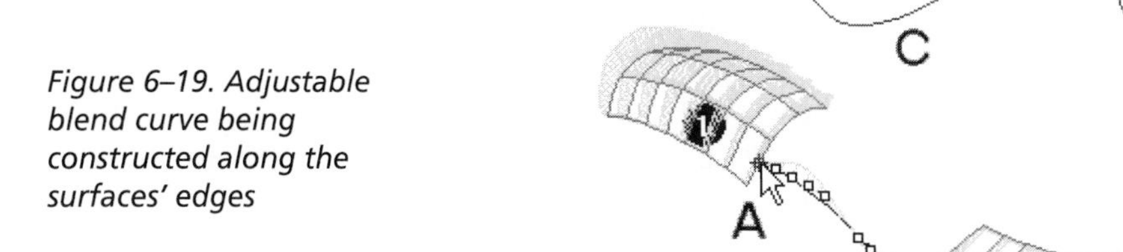

Figure 6–19. Adjustable blend curve being constructed along the surfaces' edges

Joining Contiguous Curves

To reflect our design intent, it is sometimes easier to construct a number of contiguous curves and then join them to form a single curve. Perform the following steps:

16 Select Edit > Join, or click on the Join button on the Main1 toolbar.

17 Select curves C, D, and E (Figure 6–19) in either clockwise or counter-clockwise direction, and press the ENTER key. The curves are joined.

Exploding a Joined Curve

To revert a joined curve into separate, individual curves, you can explode it. Continue with the following steps:

18 Select the joined curve A (shown in Figure 6–19) and press the ENTER key.

19 Do not save your file.

Connecting Two Coplanar Curves to Meet

A simple way to connect two coplanar curves to become a joined curve is to add line segments to their ends for them to meet. If the curves are arcs, you may optionally add arc segments instead of line segments. Perform the following steps:

1 Select File > Open and select the file Connect.3dm from the Chapter 6 folder on the companion CD.

2 Select Curve > Connect Curves, or click on the Connect button on the Extend toolbar.

3 If Join = No, select it on the command area to change it to Yes.

4 If ExtendArcsBy = Line, select it on the command area to change it to = Arc. Otherwise, proceed to next step.

5 Select curve A, shown in Figure 6–20.

6 Select ExtendArcsBy = Arc on the command area to change it to = Line.

7 Select curve B, shown in Figure 6–20. Curve A (extended by an arc segment) and curve B (extended by a line segment) are joined.

8 Do not save your file.

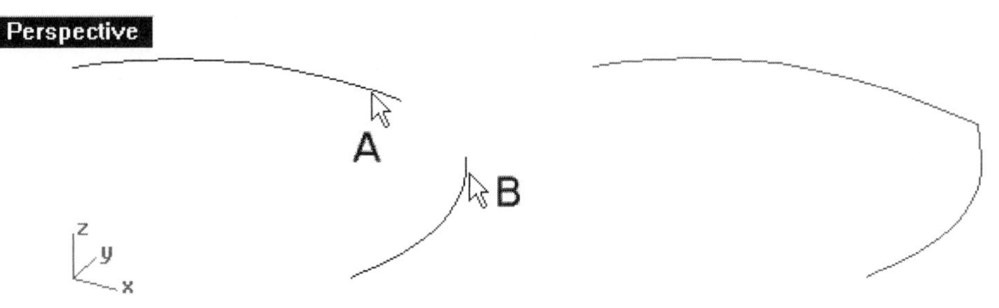

Figure 6–20. Curves selected (left) and curves connected (right)

Curve Matching

If we have two separate curves, we can modify the shape of one curve to make it match the other curve. Unlike blending, in which the original curves are not changed and a new curve is inserted, matching changes the shape of one of the curves. Between the curves, you have G2 continuity. Perform the following steps:

1. Select File > Open and select the file Match1.3dm from the Chapter 6 folder on the companion CD.
2. Select Curve > Curve Edit Tools > Match, or click on the Match Curve button on the Curve Tools toolbar.
3. Select curve A near its left end (shown in Figure 6–21).
4. Select curve B near its right end (shown in Figure 6–21).

This way, the left end of curve A will match to the right end of curve B.

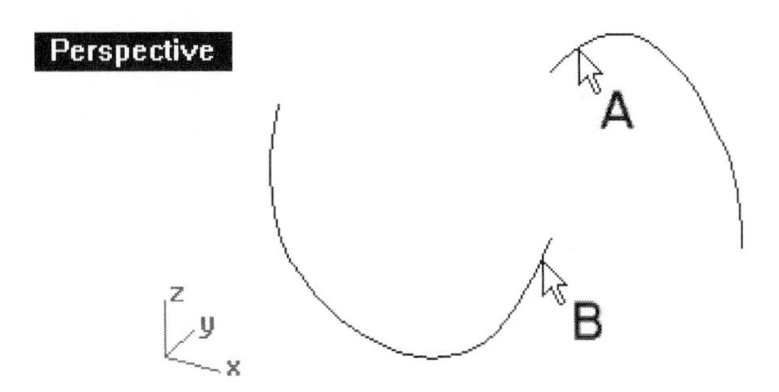

Figure 6–21. Freeform curves to be matched

The Match Curve dialog box includes several buttons. The Position, Tangency, and Curvature button concerns the continuity between the two curves after they are matched. If the Average curves button is not selected, only the first selected curve will be modified. If the Preserve other end button is selected, the shape of the other end of the modified curve(s) will be preserved as far as possible. The Join and Merge button enables the curves to be joined or merged. A join curve consists of two separate segments that can be exploded. A merge curve will become a single curve and cannot be exploded.

5 Try out various options in the Match Curve dialog box, shown in Figure 6–22.
6 Check the Curvature and Average curves, and then click on the OK button. Both curves are modified to match each other.
7 Do not save your file.

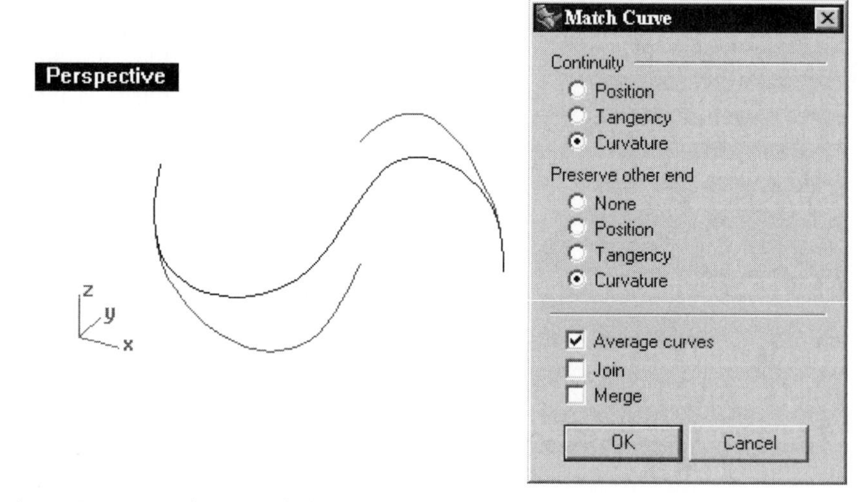

Figure 6–22. Match Curve dialog box and preview of matching

Matching 2 End Points of a Curve

Two end points of an open curve can be matched to form a periodic curve, as follows:

1 Select File > Open and select the file Match2.3dm from the Chapter 6 folder on the companion CD.
2 Select Curve > Curve Edit Tools > Match, or click on the Match Curve button on the Curve Tools toolbar.
3 Select curve A near one end (shown in Figure 6–23).
4 Select curve B near the other end (shown in Figure 6–23).
5 In the Match Curve dialog box, click on the OK button.

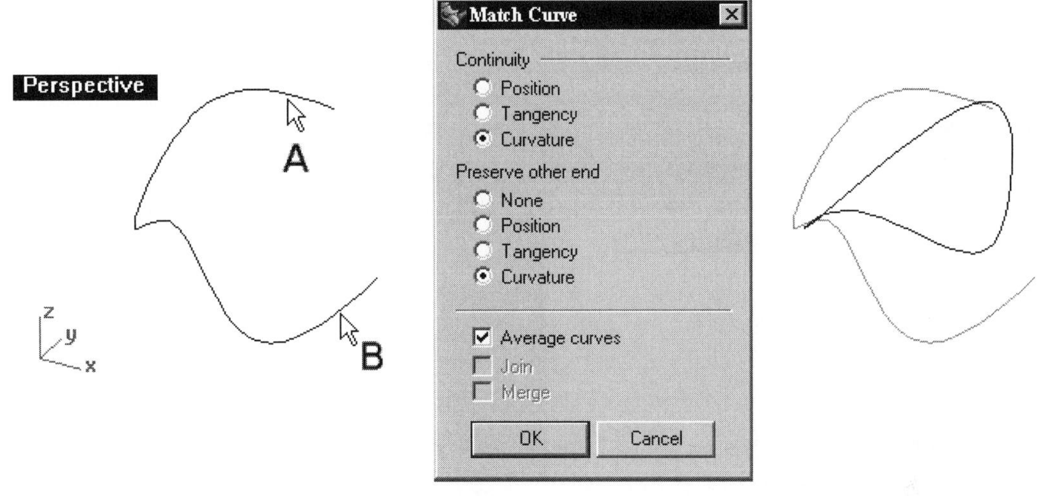

Figure 6–23. From left to right: Open curve, Match Curve dialog box, and preview of the matched curve

Curve Boolean

A quick way to construct closed-loop curves from two or more curves is to use Boolean operation. Perform the following steps:

1. Select File > Open and select the file CurveBoolean.3dm from the Chapter 6 folder on the companion CD.
2. Select Curve > Curve Edit Tools > Curve Boolean, or click on the Curve Boolean button from the Curve Tools toolbar.
3. Select curves A, B, C, and D, shown in Figure 6–24, and press the ENTER key.
4. Select the DeleteInput option on the command area to set it to Yes, if it is None. Otherwise, proceed to the next step.
5. Select the CombineRegion option on the command area to set it to Yes, if it is No. Otherwise, proceed to the next step.
6. Click on location E, shown inFigure 6–24. A region is highlighted.
7. Click on location F (Figure 6–24). Another region is highlighted.
8. Click on location G (Figure 6–24). A third region is highlighted and is combined with the second region.
9. Press the ENTER key. Two closed-loop curves are constructed from the input curves.
10. Turn off the Default layer to see the result.
11. Do not save your file.

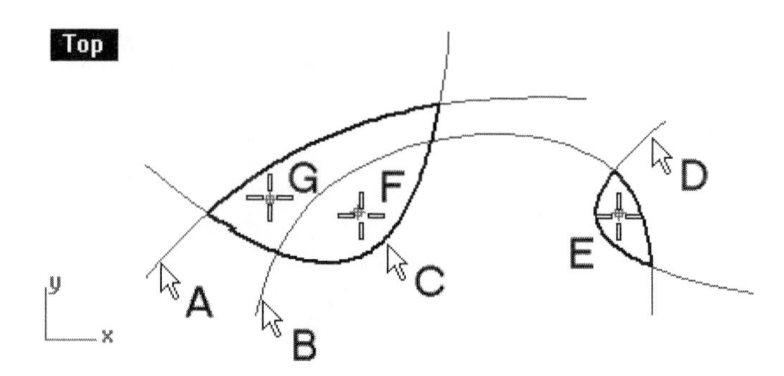

Figure 6–24. Closed loop curves being constructed

Curve Refinement Methods

A curve's shape can be refined in many ways: rebuilding by specifying its degree of polynomial and number of control points, rebuilding interactively, adjusting its end bulge without changing the tangent direction at its end points, soft editing, modifying while keeping the curve's length unchanged, moving individual segments if the curve is a polycurve, replacing a portion of a curve with a line segment, fairing, simplifying, refitting, and converting into line and arc segments.

Rebuilding a Curve

After you have finalized the shape of a curve (via various editing methods, such as those you performed previously), you might need to refine the curve by rebuilding it with a different number of control points and a different degree of polynomial. Rebuilding always removes kinks; it is a common technique used by designers to both simplify curves and remove potential problems with kinks. If you reduce the control point number, the curve simplifies and the shape changes. If you increase the control point number, the shape will not change until next time you manipulate the control points. Perform the following steps:

1. Select File > Open and select the file Rebuild.3dm from the Chapter 6 folder on the companion CD.
2. Select Curve > Control Points > Control Points On, or click on Control Points On button of the Point Editing toolbar.
3. Select curve A, shown in Figure 6–25, and press the ENTER key.
4. Select Edit > Rebuild, or click on the Rebuild button on the Curve Tools toolbar.
5. Select curve A, shown in Figure 6–25, and press the ENTER key.

6. In the Rebuild Curve dialog box, set the number of control points to 8. (The default value shown may not be the same as yours. This command always remembers the previous settings when last used.) Leave the degree unchanged.
7. Click on the Preview button. The maximum deviation of the rebuilt curve from the original curve is highlighted on the curve, and the deviation value is displayed in the dialog box.
8. If you are satisfied with the change, click on the OK button. The curve is rebuilt.

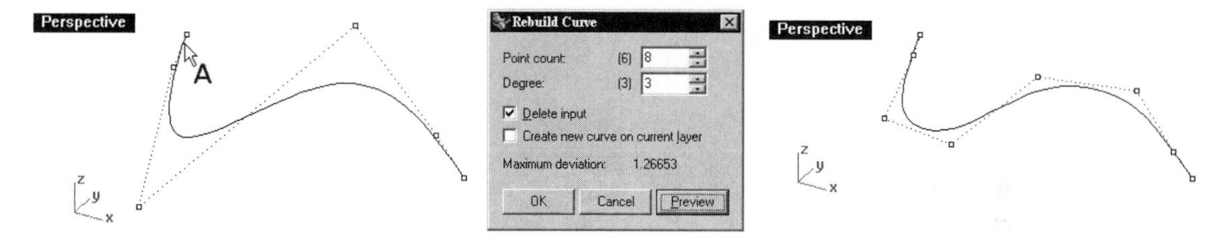

Figure 6–25. From left to right: Original curve, Rebuild Curve dialog box, and rebuilt curve

Rebuilding a Curve Interactively

To rebuild a curve interactively, continue as follows:

9. Click on the Rebuild Curves Non Uniform button on the Curve Tools toolbar, or type RebuildCrvNonUniform at the command area.
10. Select curve A, shown in Figure 6–26, and then press the ENTER key. *(Note: Which end you select on the curve has an effect on the outcome.)*
11. Select the MaxPointCount option on the command area.
12. Type 7 and press the ENTER key.
13. Press the ENTER key. The curve is rebuilt.
14. Do not save your file.

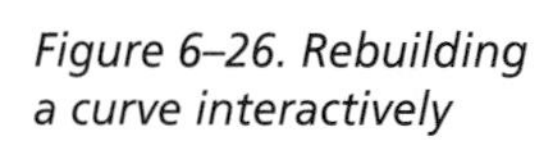
Figure 6–26. Rebuilding a curve interactively

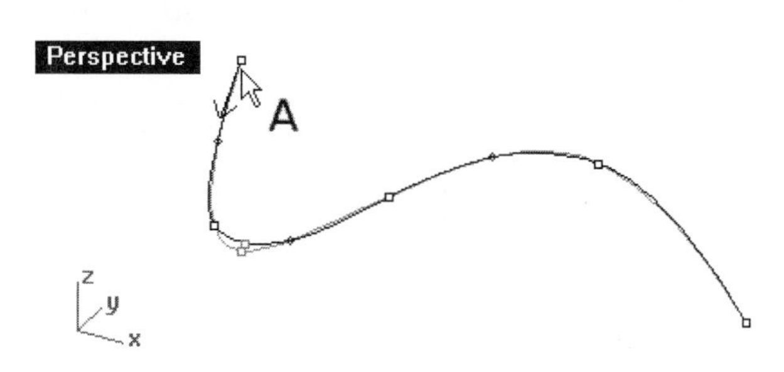

Adjusting a Curve's End Bulge

If you want to modify the shape of a curve and yet maintain the tangent directions at the end points, you adjust the curve's end bulge. Continue with the following steps:

1. Select File > Open and select the file EndBulge.3dm from the Chapter 6 folder on the companion CD.
2. Select Edit > Adjust End Bulge, or click on the Adjust End Bulge button on the Curve Tools toolbar.
3. Select curve A, shown in Figure 6–27.
4. Select and drag the control points B or C.
5. Press the ENTER key when finished. The end bulge is modified.
6. Do not save your file.

As shown, the two control points are restricted to translate along a straight line. As a result, the tangency at the end points of the curve is maintained even though the curve's shape is modified.

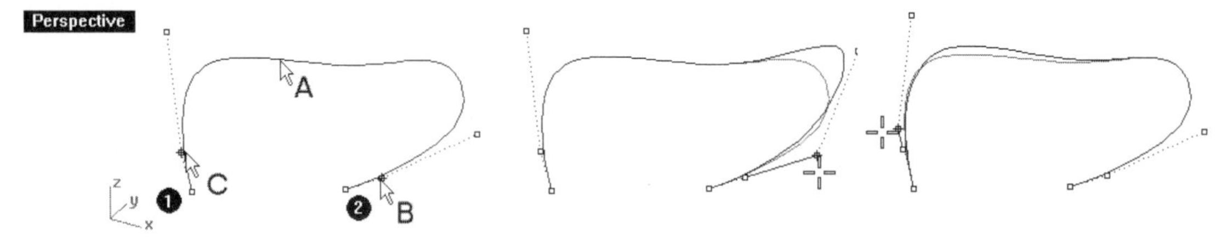

Figure 6–27. From left to right: original curve, control point B being dragged, and control point C being dragged

Soft Editing a Curve

A way to modify the shape of a curve is to soft edit it by picking a point on the curve and repositioning it to a new location. The term soft edit probably refers to the phenomenon that the curve, during soft editing, behaves like a soft object. To soft edit a curve, perform the following steps:

1. Select File > Open and select the file SoftEdit.Crv3dm from the Chapter 6 folder on the companion CD.
2. Select Curve > Curve Edit Tools > Soft Edit, or click on the Soft Edit Curve/Soft Edit Surface button on the Move toolbar.
3. If the FixEnds option is No, select it from the command area to change it to Yes.

4. If the Copy option is No, select it from the command area to change it to Yes.
5. Select the Distance option from the command area and set it to 20. This is the distance along the curve that behaves as "soft."
6. Select curve A, shown in Figure 6–28.
7. Type mid at the command area and pick midpoint B.
8. Select point C. The curve is copied and a copy is soft edited.
9. To appreciate the effect of the distance option, you may repeat soft editing the original curve with a distance of 50.
10. Do not save your file.

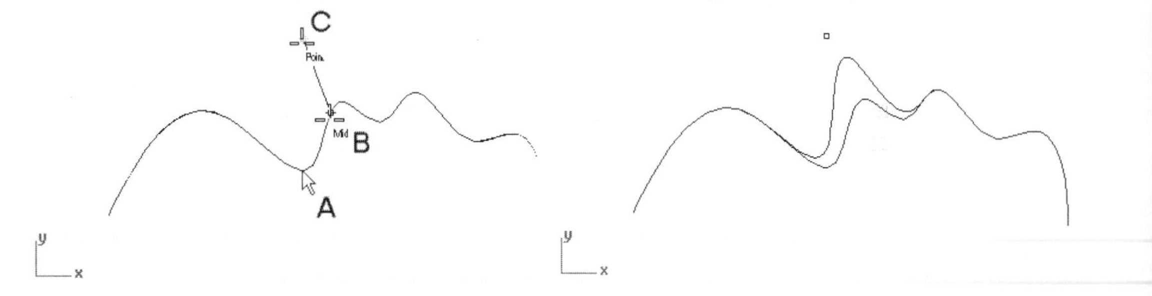

Figure 6–28. Original curve (left) and curve copied and soft edited (right)

Fixed Length Curve Edit

This process modifies the shape of a curve interactively without changing the length of the curve. Perform the following steps:

1. Select File > Open and select the file FixedLengthEdit.Crv3dm from the Chapter 6 folder on the companion CD.
2. Click on the Edit Curve with Fixed Length button on the Move toolbar, or type FixedLengthCrvEdit at the command area.
3. If the Copy option is No, select it from the command area to change it to Yes.
4. Select curve A, shown in Figure 6–29.
5. Check the Mid and Point boxes on the Osnap toolbar.
6. Select midpoint B and then point C. The curve is modified, but the length is kept unchanged.
7. Do not save your file.

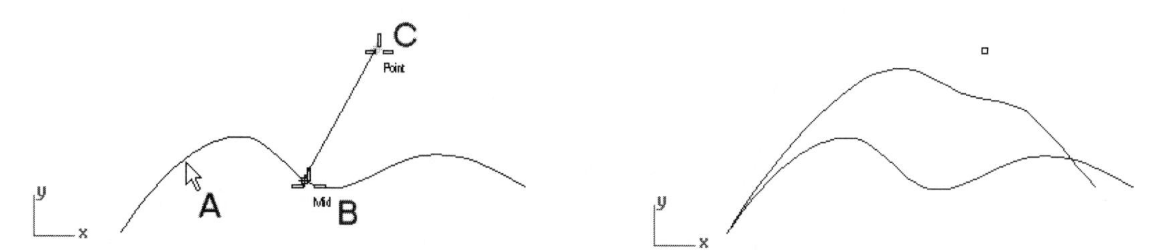

Figure 6–29. Original curve (left) and curve being modified (right)

Moving Segments of Polyline or Polycurve

When two or more line segments are joined together, you obtain a polyline. Naturally, you can construct a polyline by using the Polyline command. When two or more curve segments are joined together, you obtain a polycurve. Individual segment of a polyline or polycurve can be moved, as follows:

1. Select File > Open and select the file MoveSegment.3dm from the Chapter 6 folder on the companion CD.
2. Set object snap to point.
3. Click on the Move Sub Curve button on the Move toolbar, or type MoveCrv at the command area.
4. If the End option is Yes, select it on the command area to change it to No.
5. If Copy = Yes, select it on the command area to change it to No.
6. Click on segment A, shown in Figure 6–30.
7. Select point B and then point C. The selected polyline segment is moved.
8. Repeat the command.
9. Click on segment D, shown in Figure 6–30. (This is a polycurve constructed by joining three curves. Segment D is the central segment of the polycurve.)
10. Select point E and then point F. The selected curve segment is moved.

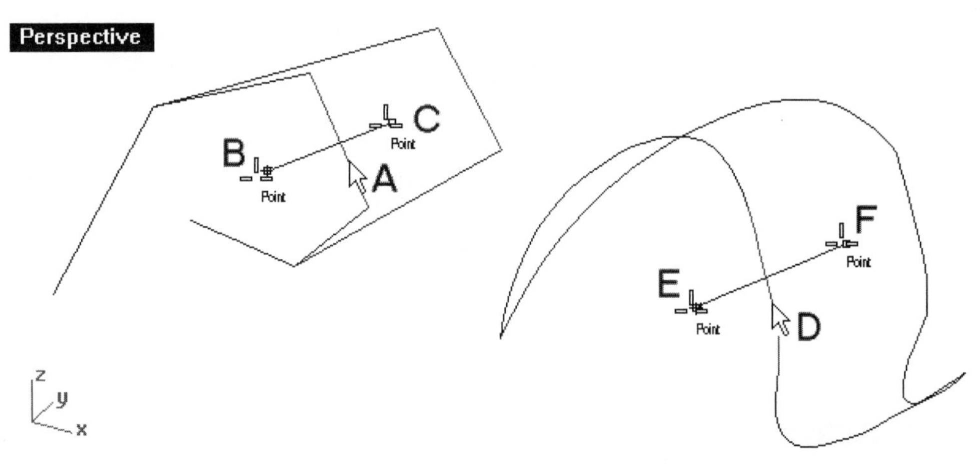

Figure 6–30. Segment being moved

11 Repeat the command.
12 Select the End option on the command area to change it to Yes.
13 Click on vertex A, shown in Figure 6–31.
14 Select point B. The polyline segment's end point is moved.
15 Repeat the command.
16 Click on joint C, shown Figure 6–31.
17 Select point D and then point E. The curve segment's end point is moved.
18 Do not save your file.

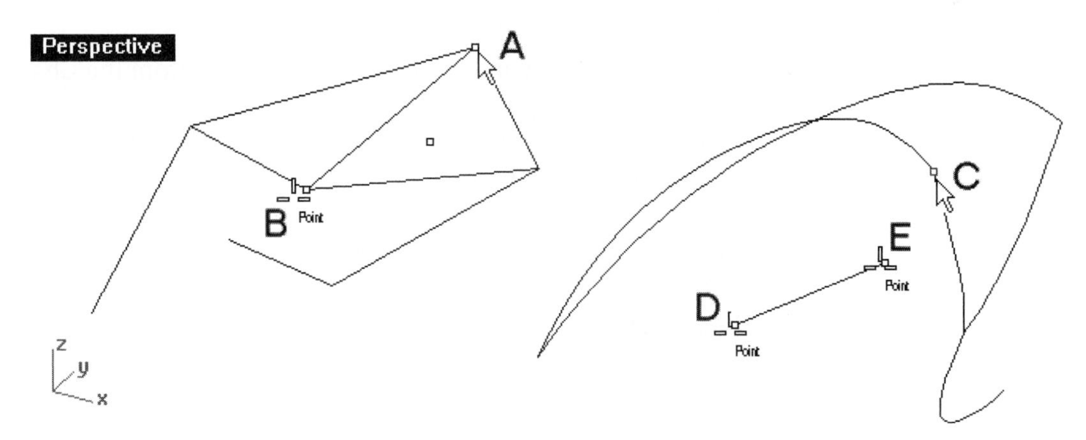

Figure 6–31. Segment end point being moved

Replacing a Portion of a Curve with a Line Segment

You can replace a portion of a curve with a line segment, as follows:

1. Select File > Open and select the file InsertLineSegment.3dm from the Chapter 6 folder on the companion CD.
2. Click on the Insert line into curve button on the Curve Tools toolbar.
3. Select curves A and then points B and C, shown in Figure 6-32. The portion BC is replaced by a line segment.
4. Do not save your file.

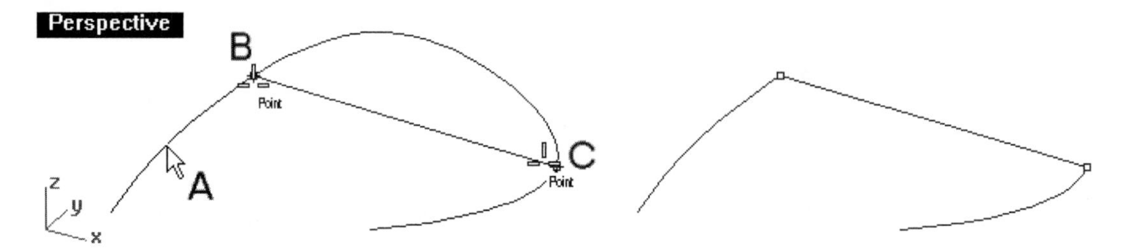

Figure 6-32. Curves selected (left) and curves connected (right)

Fairing a Curve

If you have a sketch curve or a digitized curve with large curvature deviation, you can smooth it. The process to remove large curvature variation is called fairing. Perform the following steps:

1. Select File > Open and select the file Fair.3dm from the Chapter 6 folder on the companion CD.
2. Select Curve > Curve Edit Tools > Fair, or click on the Fair button on the Curve Tools toolbar.
3. Select curve A (shown in Figure 6-33) and press the ENTER key.
4. Type 0.1 at the command line area to specify tolerance. The curve is faired.
5. Do not save your file.

Figure 6–33. Curve being faired (left) and faired curve (right)

Simplifying (Unifying the Polynomial of a Curve)

If you extend a curve by adding a line segment or an arc segment to it, the degree of polynomial of the added segment may not be congruent with the curve itself. The process to replace such line or arc segments in a curve with a true NURBS curve is called simplifying. In the following, you will add an arc segment to a curve and then simplify it:

1. Select File > Open and select the file Simplify.3dm from the Chapter 6 folder on the companion CD.
2. Select Curve > Extend Curve > By Line, or click on the Extend by Line button on the Extend toolbar.
3. Select curve A (shown in Figure 6–34).
4. Select location B (Figure 6–34) to indicate the end point of the added line segment. Note that for the purpose of this tutorial, the extended length is unimportant.

A line segment is added. Now you will simplify the curve, as follows:

5. Select Curve > Curve Edit Tools > Simply Lines and Arcs, or click on the Simplify Lines and Arcs button on the Curve Tools toolbar.
6. Select curve A (shown in Figure 6–34) and press the ENTER key. The extended line segment of the curve is now replaced with a true NURBS segment. You will not find much change on the shape of the curve because the curve takes on the shape of the line.
7. Do not save your file.

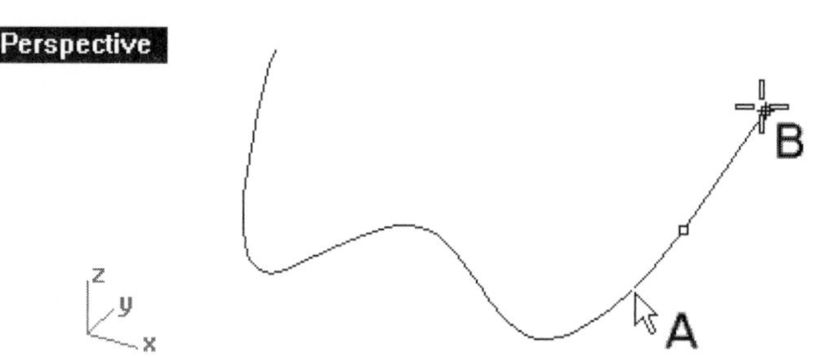

Figure 6–34. Curve being extended

Refitting a Curve from a Polyline

If we already have a polyline, we can "refit" a free-form curve to correspond to the polyline's tolerance setting. This is sometimes necessary when, for example, we input data from another system and, in the process, input polyline segments that do not correctly correspond with existing geometry. If we want to build a smooth surface, we need to refit the polyline to a free-form curve, as follows:

1. Select File > Open and select the file Refit.3dm from the Chapter 6 folder on the companion CD.
2. Select Curve > Curve Edit Tools > Refit to Tolerance, or click on the Refit Curve button on the Curve Tools toolbar.
3. Select polyline A, shown in Figure 6–35, and press the ENTER key. You may keep the original curve, or delete it using the Delete Input option. Here, the input curve is deleted.
4. Accept the default degree of polynomial by pressing the ENTER key. A free-form curve is refit from the selected polyline.
5. Do not save your file.

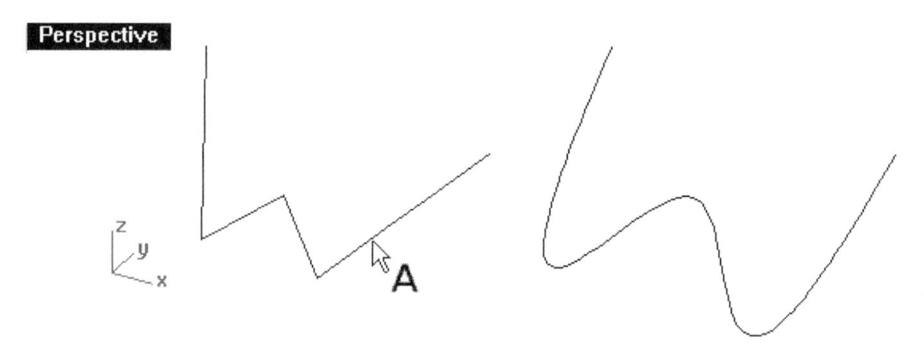

Figure 6–35. Polyline (left) and refitted curve (right)

Converting a Free-Form Curve to Line/Arc Segments

Contrary to refitting a polyline to a curve, you can convert a free-form curve to a polyline. Such inter-conversion is basically required when you import data from a system that outputs polylines and export to a system that reads polylines. If you are going to export curves to another computerized system that does not support splines, you may consider converting the splines to lines and arcs. If your curve to be simplified is composed of line and arc segments, you should use the SimplifyInput option to make sure that lines and arcs are converted accurately. Perform the following to convert a curve to a polyline and a curve to a set of joined arcs:

1. Select File > Open and select the file Convert.3dm from the Chapter 6 folder on the companion CD.
2. Select Curve > Convert > Curve to Lines, or click on the Convert Curve to Polyline button on the Lines toolbar.
3. Select curve A, shown in Figure 6–36, and press the ENTER key.
4. Select the Angle Tolerance option on the command area.
5. Type 25 to specify the angular tolerance.
6. Select the Tolerance option on the command area.
7. Type 5 to specify the tolerance.
8. Press the ENTER key. The curve is converted to a polyline. Smaller tolerance results in a higher number of line segments in the polyline.

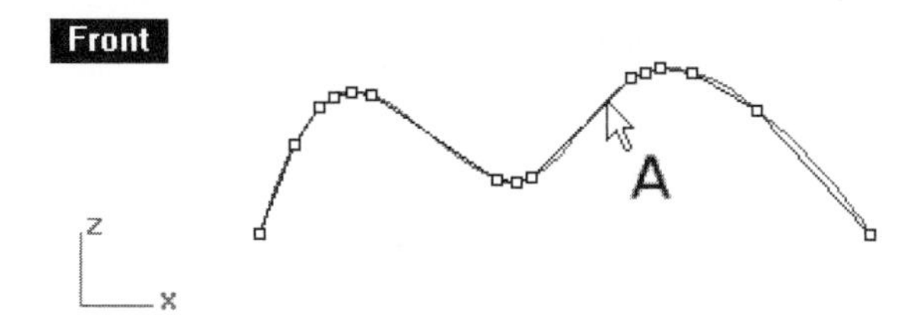

Figure 6–36. A free-form curve being converted to a polyline

9. Maximize the Right viewport.
10. Select Curve > Convert > Curve to Arcs, or click on the Convert Curve to Arcs button on the Arc toolbar.
11. Select curve A, shown in Figure 6–37, and then press the ENTER key.
12. Select the AngleTolerance option on the command area.
13. Type 15 to specify the angular tolerance.
14. Select the Tolerance option on the command area.

15. Type 5 to specify the tolerance.
16. Press the ENTER key. The curve is converted to a set of connected arcs. Smaller tolerance results in a higher number of line segments in the polyline.
17. Do not save your file.

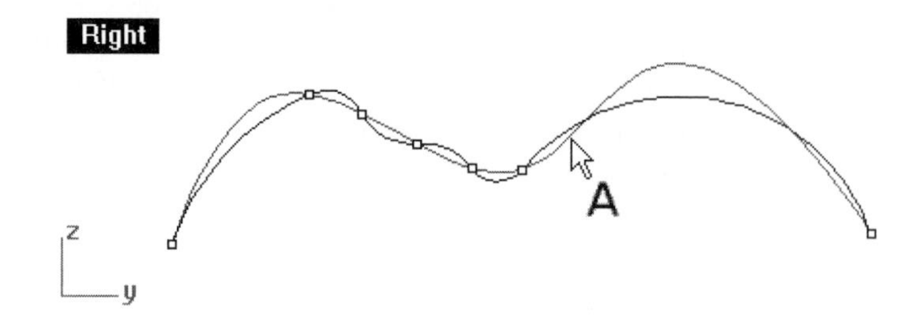

Figure 6–37. A free-form curve being converted to a set of connected arcs

Arc Curve

To construct a curve to interpolate among a set of points, and then convert the curve into a number of arc segments in one single operation, perform the following steps:

1. Select File > Open and select the file ArcCurve.3dm from the Chapter 6 folder on the companion CD.
2. Select the Arc Through Points button on the Arc toolbar.
3. Drag from A to B (Figure 6–38) to describe a rectangular area to select all the point objects.
4. Press the ENTER key. A curve is constructed and converted to a set of arc segments.
5. Do not save your file.

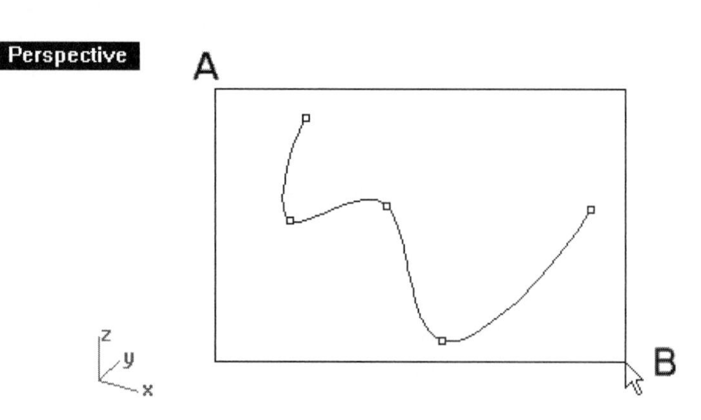

Figure 6–38. Curve constructed from point objects and converted to arc segments

Curves and Points from Existing Objects

Although it is logical to construct curves and/or points before creating surfaces and polysurfaces, detailing a 3D free-form model often requires intertwined construction of curves, points, and surfaces. Therefore, you may need to construct points and curves from existing surfaces, polysurfaces, and polygon meshes. Using these points and curves, you can develop and refine your design.

Offsetting a Curve

You can offset a curve to derive a new curve resembling the existing curve and at a specified distance from it. The offset distance is measured on a plane parallel to the active construction plane. Perform the following steps:

1. Select File > Open and select the file Offset01.3dm from the Chapter 6 folder on the companion CD.
2. Select Curve > Offset Curve, or click on the Offset Curve button on the Curve Tools toolbar.
3. Select the Distance option on the command area.
4. Type 4 to set offset distance.
5. Select curve A, shown in Figure 6–39.
6. Click on location B (Figure 6–39) to indicate the direction of offset. An offset curve is constructed.
7. Repeat the command.
8. Select curve C and click on location D (Figure 6–39). Another offset curve is constructed.

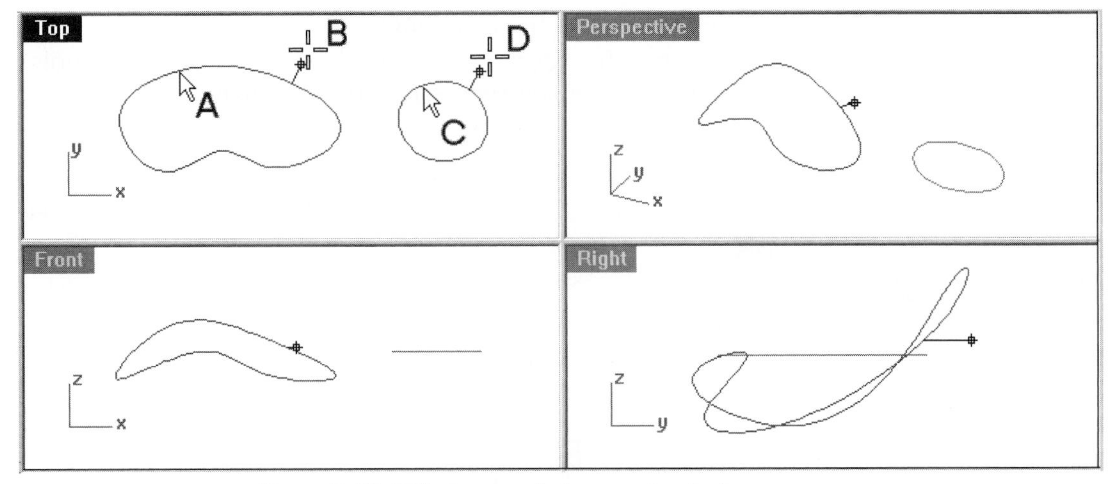

Figure 6–39. Offset curves being constructed

To appreciate how the outcome of offset is affected by the active construction plane, continue with the following steps.

9 Repeat the command.

10 Select curve A and click on location B, shown in Figure 6–40.

11 Repeat the command.

12 Select curve C and click on location D (Figure 6–40).

13 Do not save your file.

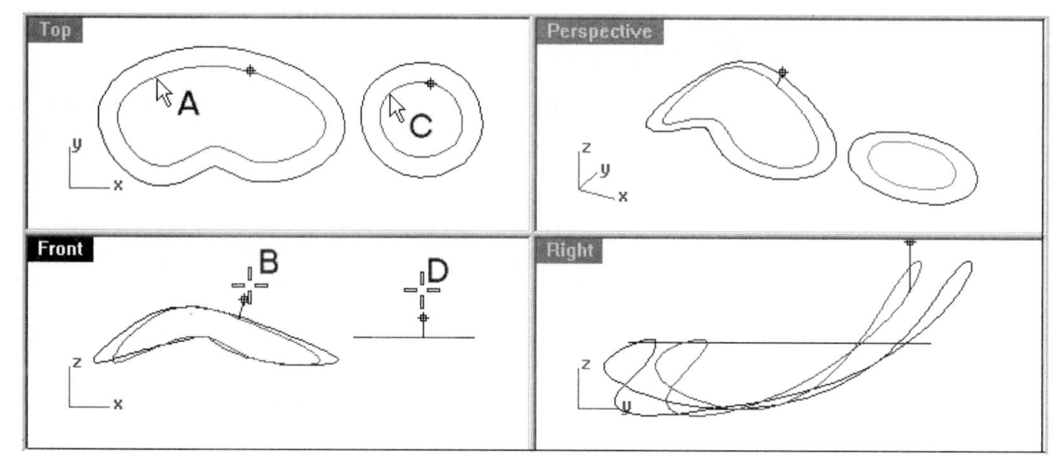

Figure 6–40. Second offset curve being constructed

Offsetting a Curve on a Surface

To offset a curve residing on a surface, perform the following steps:

1 Select File > Open and select the file Offset02.3dm from the Chapter 6 folder on the companion CD.

2 Click on the Offset curve on surface button on the Curve Tools toolbar, or type OffsetCrvOnSrf on the command area.

3 Select curve A, shown in Figure 6–41, and press the ENTER key.

4 Select surface B, shown in Figure 6–41.

5 Type 3 and press the ENTER key. An offset curve is constructed.

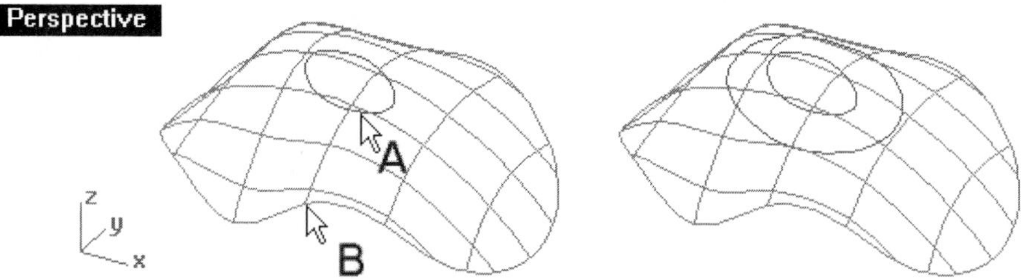

Figure 6–41. Curve on a surface (left) and curve offset on surface (right)

Offsetting a Curve in a Direction Normal to a Surface

Continue with the following steps to offset a curve normal to a surface.

6. Select Curve > Offset Normal to Surface or click on the Offset curve normal to surface button on the Curve Tools toolbar.
7. Select curve A, shown in Figure 6–42.
8. Select surface B.
9. Type 10 at the command area to specify the height. An offset curve is constructed.
10. Do not save your file.

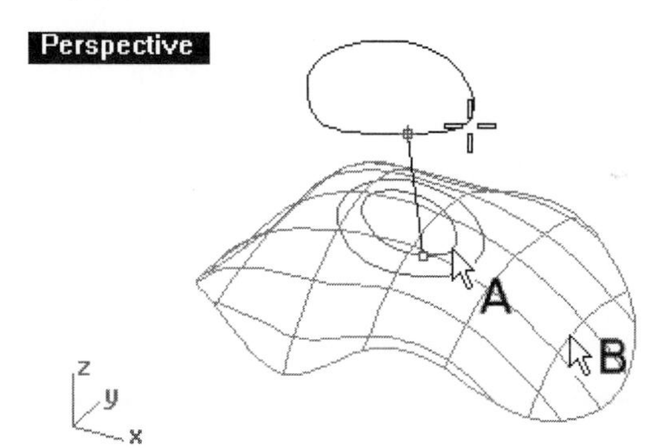

Figure 6–42. Curve offset normal to a surface being constructed

Mean Curve

You can construct a curve that is the mean of two selected curves, as follows:

1. Select File > Open and select the file MeanCurve.3dm from the Chapter 6 folder on the companion CD.
2. Select Curve > Mean Curve, or click on the Average 2 Curves button on the Curve toolbar.
3. Select curves A and B, shown in Figure 6–43. A mean curve is constructed.
4. Do not save your file.

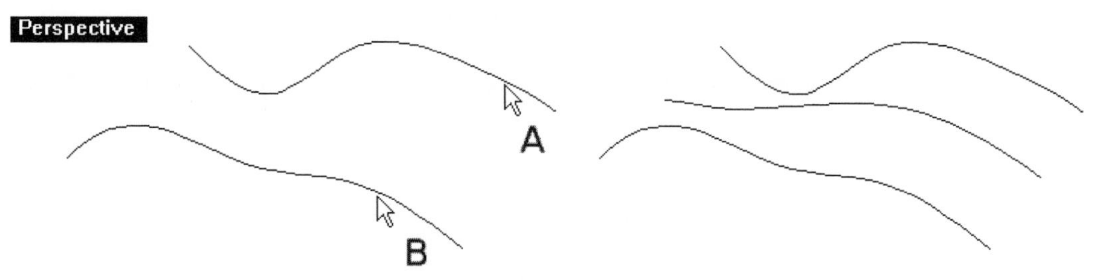

Figure 6–43. Two curves selected (left) and mean curve constructed (right)

Construction of a 3D Curve from Two Planar Curves

This method constructs a 3D curve. If you already have an idea of the shape of a 3D curve in two orthographic drawing views, you can first construct two planar curves residing on two adjacent viewports, and then derive the 3D curve from the planar curves.

Perform the following steps to construct a 3D curve from two planar curves.

1. Select File > Open and select the file CurveFrom2views.3dm from the Chapter 6 folder on the companion CD.
2. Select Curve > Curve from 2 Curves, or click on the Curve from 2 Curves button on the Curve Tools toolbar.
3. Select curves A and B. A 3D curve is derived, as shown in Figure 6–44.
4. Do not save your file.

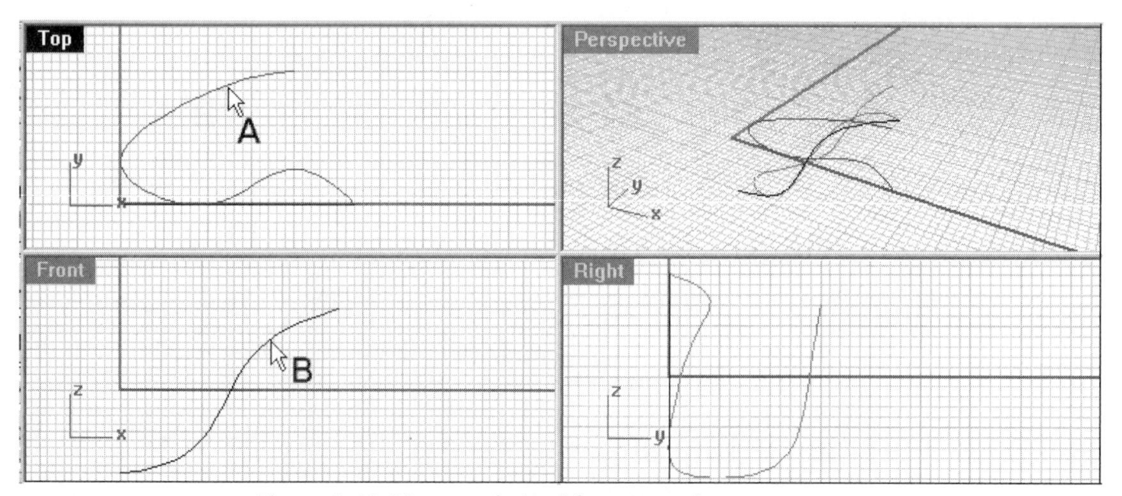

Figure 6–44. 3D curve derived from two planar curves

(Note: For better visualization of the 2D planar curves prior to constructing the 3D curve, you may synchronize the viewports.)

Construction of Cross-Section Profile Curves

A cross-section profile is a closed-loop, free-form curve interpolating the intersection points between a set of longitudinal curves and a specified section plane across the curves. The cross-section curves, together with the longitudinal curves, form the U and V curves for making a surface from a network of curves. Perform the following steps to construct cross-section profile curves:

1 Select File > Open and select the file CSecProfile.3dm from the Chapter 6 folder on the companion CD.

2 Select Curve > Cross Section Profiles, or click on the Curve from Cross Section Profiles button on the Curve Tools toolbar.

3 Select (in clockwise or counter-wise direction but not randomly) the profile curves A, B, C, and D, shown in Figure 6–45, and press the ENTER key.

4 Select locations E and F (Figure 6–45). A cross-section curve is constructed.

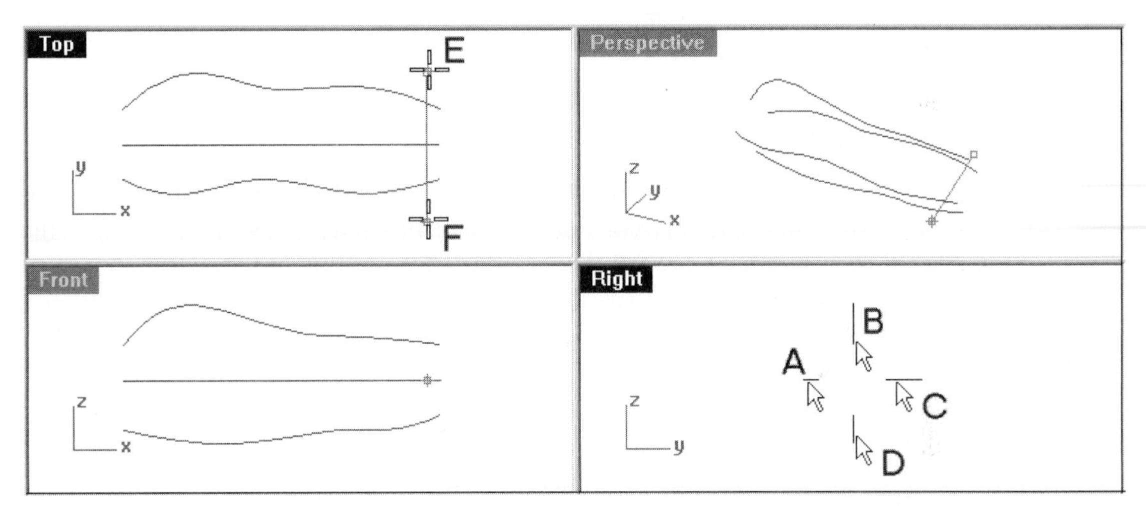

Figure 6–45. Profile curves selected and first cross-section curve being constructed

5 Select locations A and B (shown in Figure 6–46), locations C and D, locations E and F, locations G and H, and locations J and K. Five more cross-section curves are constructed.

6 Press the ENTER key to terminate the command.

7 Do not save your file.

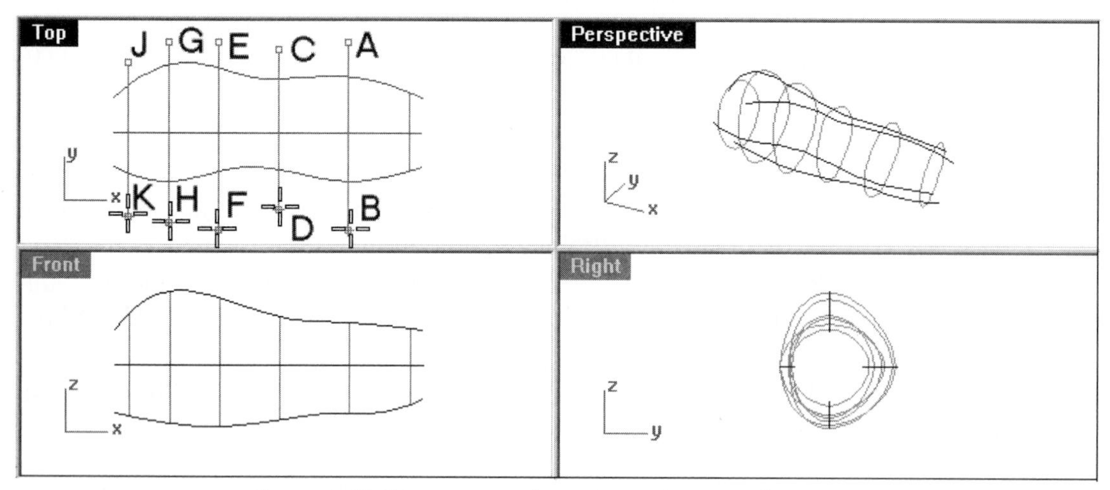

Figure 6–46. Five more section curves constructed

Projecting and Pulling a Curve to a Surface

Construction of 3D curves on a surface can be made easy by first constructing a 2D curve on one of the construction planes and then projecting it or pulling it toward the surface. The direction of projection is perpendicular to the construction plane, and the direction of pull is normal to the surface. To experiment with this functionality, perform the following steps:

1. Select File > Open and select the file ProjectPull.3dm from the Chapter 6 folder on the companion CD.
2. Click on the Top viewport to set it as the current viewport.
3. Select Curve > Curve From Objects > Project, or click on the Project to Surface button on the Curve From Object toolbar.
4. Select curve A (Figure 6–47) and press the ENTER key.
5. Select surface B (Figure 6–47) and press the ENTER key. A curve is projected.

If the History Manager is turned on prior to projecting, manipulation of the source curve will cause corresponding changes to the projected curve.

6. Click on the source curve A, and drag it to a new location. The projected curve will move also.

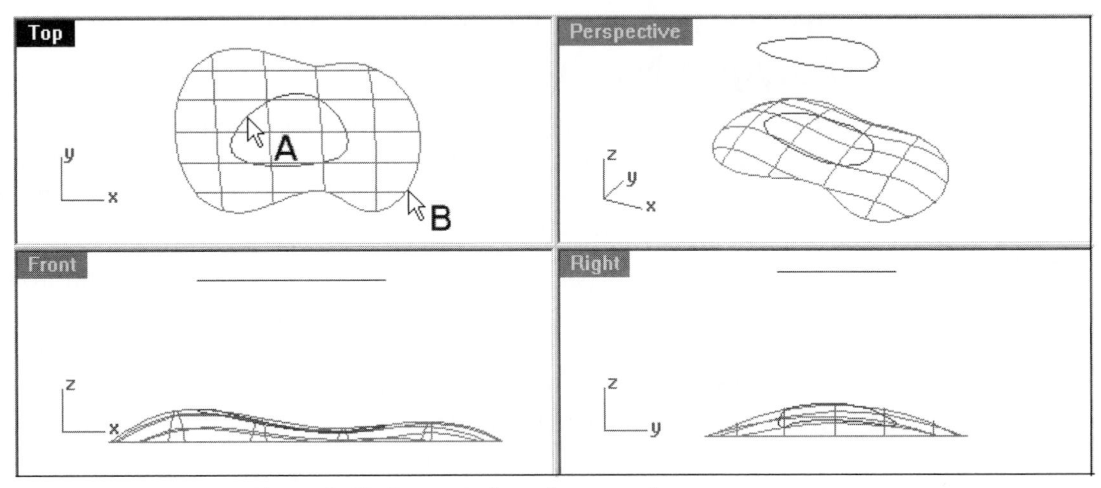

Figure 6–47. Curve projected onto surface

7. Select Curve > Curve From Objects > Pullback, or click on the Pull Curve to Surface button on the Curve From Objects toolbar.
8. Select curve A (Figure 6–48) and press the ENTER key.
9. Select surface B (Figure 6–48) and press the ENTER key.
10. Compare the result of pulling to result of projection.
11. Do not save your file.

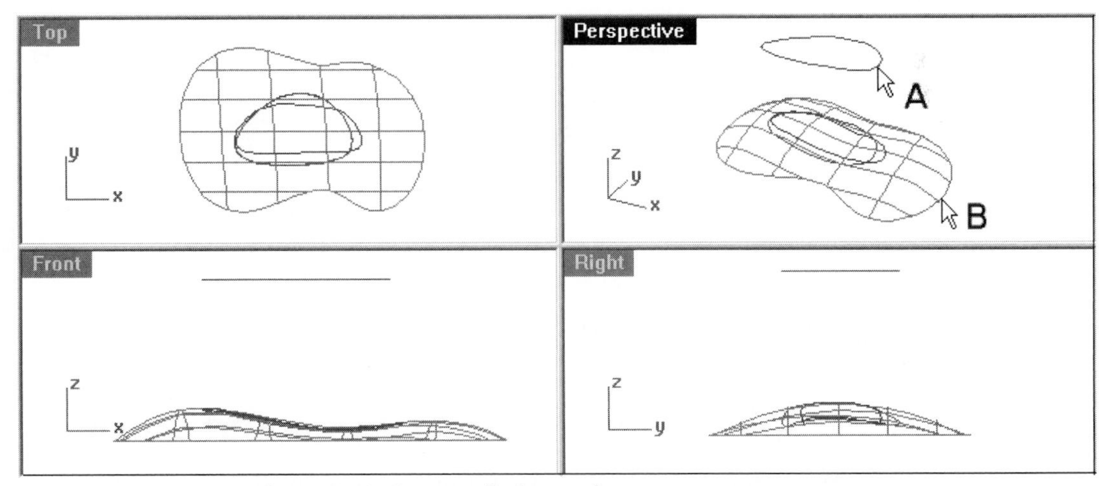

Figure 6–48. Curve pulled to surface

Projecting and Pulling a Curve to a Polygon Mesh

The same commands described in the previous section can be used to project/pull curves onto a polygon mesh. (Polygon mesh will be discussed in Chapter 9.) Because a polygon mesh is faceted, what you will get is a zigzag curve. Perform the following steps:

1. Select File > Open and select the file ProjectPullMesh.3dm from the Chapter 6 folder on the companion CD.
2. Select Curve > Curve From Objects > Project, or click on the Project to Surface button on the Curve From Object toolbar.
3. Select curve A, shown in Figure 6–49, and press the ENTER key.
4. Select polygon mesh B and press the ENTER key. The curve is projected onto the polygon mesh.
5. Select Curve > Curve From Objects > Pullback, or click on the Pull Curve to Surface button on the Curve From Objects toolbar.
6. Select curve C, shown in Figure 6–49, and press the ENTER key.
7. Select polygon mesh D and press the ENTER key. The curve is projected onto the polygon mesh.
8. Do not save your file.

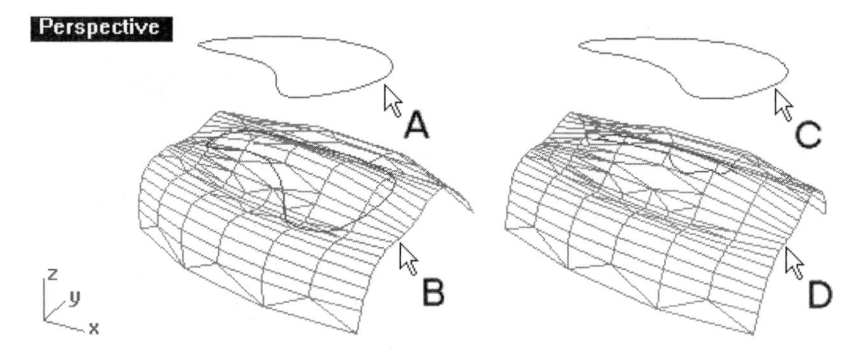

Figure 6–49. Curves projected and pulled to a polygon mesh

Duplication of Surface Edge and Border

You can construct a 3D curve from an edge or the border of a surface. Superficially, an edge seems to be analogous to a border. However, an open surface can have multiple edges but only a single border. For example, a rectangular surface has four edges but just one continuous border. Perform the following steps to duplicate a surface's edge and border:

1. Select File > Open and select the file EdgeBorder.3dm from the Chapter 6 folder on the companion CD.
2. Select Curve > Curve From Objects > Duplicate Edge, or click on the Duplicate Edge/Duplicate Mesh Edge button on the Curve From Objects toolbar.
3. Select edge A (shown in Figure 6–50) and press the ENTER key. An edge of the surface is duplicated.
4. Select Curve > Curve From Objects > Duplicate Border, or click on the Duplicate Border button on the Curve From Objects toolbar.
5. Select surface B (Figure 6–50) and press the ENTER key. The borders of the surface are duplicated.

To appreciate that an edge is duplicated, hide the surface, as follows:

6. Press the Esc key to ensure that nothing is selected.
7. Select Edit > Visibility > Hide, or click on the Hide Objects/ Show Objects button on the Visibility toolbar.
8. Select surfaces B and C (Figure 6–50) and press the ENTER key.

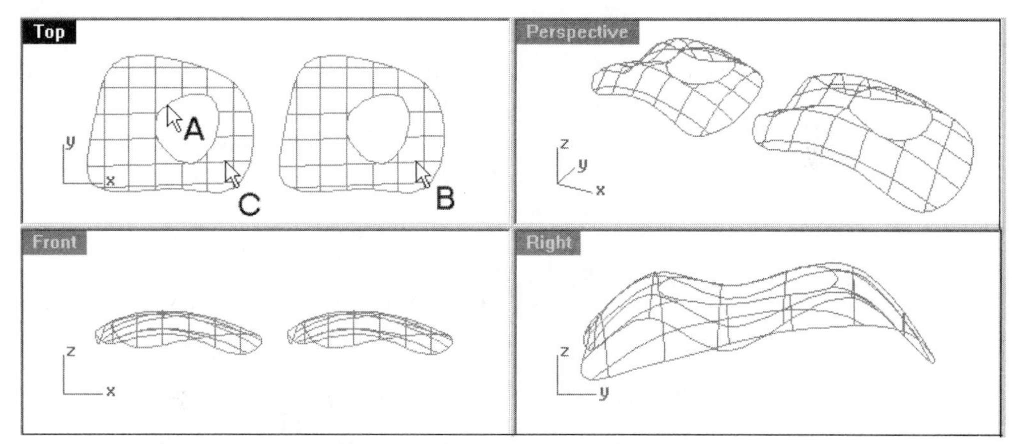

Figure 6–50. Surface edge being duplicated

9. The surface is hidden. (See Figure 6–51.) Do not save your file.

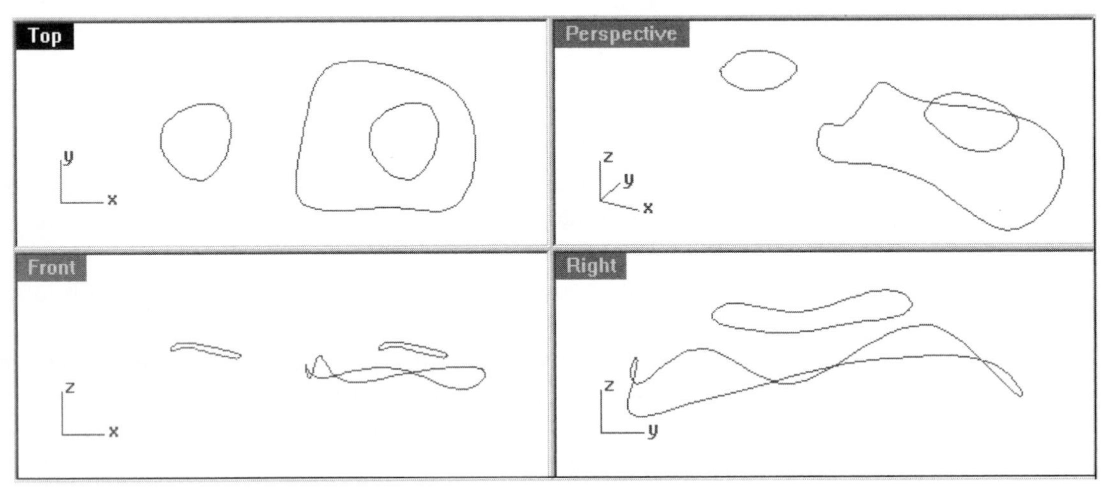

Figure 6–51. Surface border duplicated

Duplicating Border of Individual Surfaces of a Polysurface

To extract the border of a surface of a polysurface, perform the following steps:

1. Select File > Open and select the file Border.3dm from the Chapter 6 folder on the companion CD.
2. Select Curve > Curve From Objects > Duplicate Face Border, or click on the Duplicate Face Border button on the Curve From Objects toolbar.
3. Select face A of the polysurface, shown in Figure 6–52, and press the ENTER key. The border of a face of the polysurface is extracted.
4. Press the Esc key to ensure that nothing is selected.
5. Select Edit > Visibility > Hide, or click on the Hide Objects/ Show Objects button on the Visibility toolbar.
6. Select polysurface A (Figure 6–52) and press the ENTER key.
7. Do not save your file.

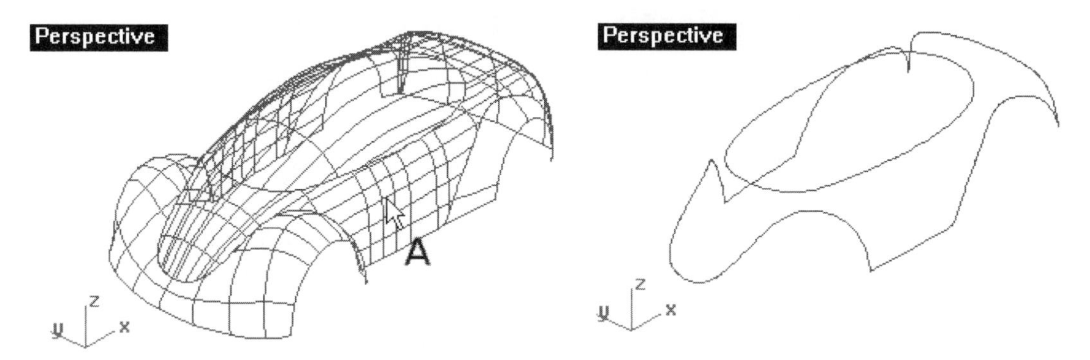

Figure 6–52. Surface border of a polysurface being extracted (left) and border extracted and polysurface hidden (right)

Duplicating Mesh Edge

Like the edges of a surface, the edges of a meshed object are not curves. If you want to make use of a meshed object's boundary edge for other drafting/construction purpose, you have to duplicate it as curves independent of the meshed object. Perform the following steps:

1. Select File > Open and select the file DuplicateBoundary.3dm from the Chapter 6 folder on the companion CD.
2. Click on the Duplicate mesh hole boundary button on the Mesh Tools toolbar.
3. Select edge A, shown in Figure 6–53. The edge boundary is duplicated.
4. Do not save your file.

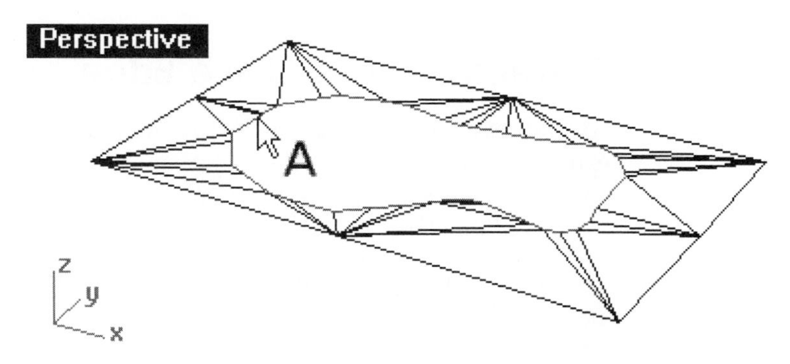

Figure 6–53. Edge boundary of a polygon mesh being duplicated

Duplicating a Mesh's Edge

You can duplicate a polygon mesh's edge in the form of a polyline, as follows:

1. Select File > Open and select the file DupMeshEdge.3dm from the Chapter 6 folder on the companion CD.
2. Select Curve > Curve From Objects > Duplicate Mesh Edge, or right-click on the Duplicate Edge/Duplicate Mesh Edge button on the Curve from Objects toolbar.
3. Select edge A and press the ENTER key. The selected edge is duplicated.
4. Repeat the command, select edge B, and press the ENTER key. Another edge is duplicated.
5. To appreciate the effect of edge duplication, turn off Layer Mesh. (See Figure 6–54.)
6. Do not save your file.

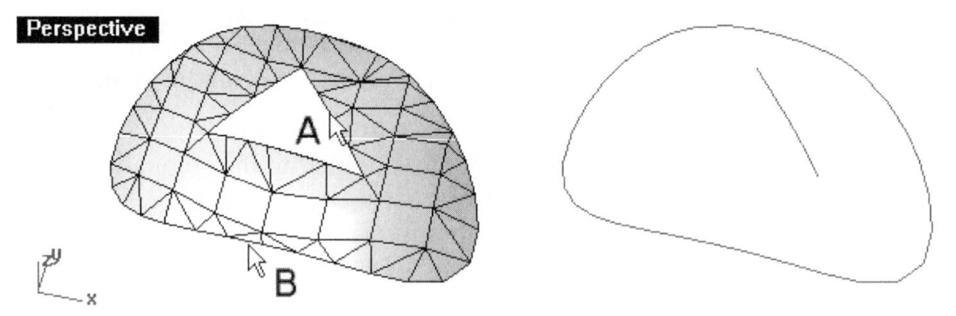

Figure 6–54. Edge being duplicated (left) and edge duplicated (right)

Extracting Mesh Face Edge

You can extract mesh face edges to obtain a sketch element. In essence, extraction of edges does not make any changes to the mesh; it simply constructs curve elements from selected edges of mesh faces. Perform the following steps:

1. Select File > Open and select the file ExtractMeshEdge.3dm from the Chapter 6 folder on the companion CD.
2. Click on the Extract Edges button on the Extract Mesh toolbar.
3. Select mesh A, shown in Figure 6–55.
4. In the Extract Edges dialog box, check the Un-welded box and click on the OK button.
5. Do not save your file.

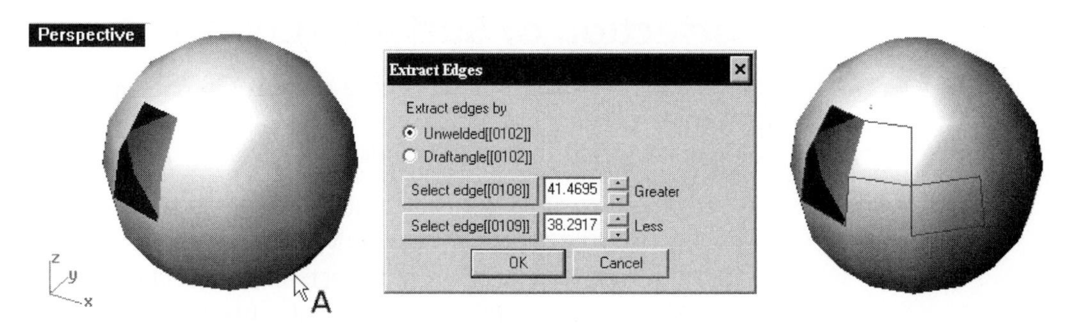

Figure 6–55. From left to right: Original mesh object, Extract Edges dialog box, and edges extracted

Constructing 2D Outlines of Surface and Polygon Mesh

An outline of selected surfaces and polygon meshes can be obtained by projecting the boundary at an angle perpendicular to the viewing direction of the active viewport. Perform the following steps:

1. Select File > Open and select the file MeshOutline.3dm from the Chapter 6 folder on the companion CD.
2. Select Curve > Curve From Objects > Mesh Outline, or click on the Mesh outline button on the Curve From Objects toolbar.
3. Select surface A and polygon mesh B in the Top viewport and press the ENTER key. (*Note*: This command is viewport dependent.) Boundaries are projected onto the construction plane of the Top viewport. (See Figure 6–56.)
4. Do not save your file.

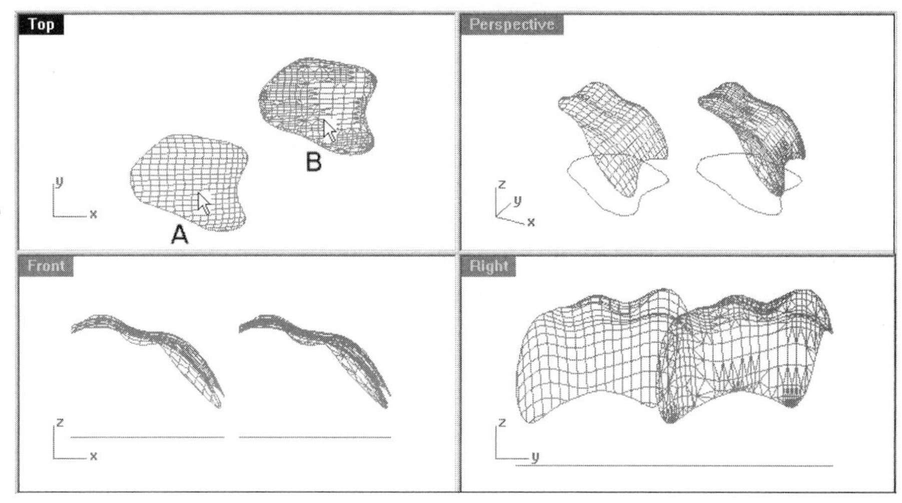

Figure 6–56. Outline projected

Intersection of Surfaces/Curves

Intersection of a surface and a curve produces a point object, and two intersected surfaces produce a curve. Turn on the History Manager to enable history, and perform the following steps:

1. Select File > Open and select the file Intersect.3dm from the Chapter 6 folder on the companion CD.
2. Select Curve > Curve From Objects > Intersection, or click on the Object Intersection button on the Curve From Objects toolbar.
3. Select curve A, and surfaces B and C (shown in Figure 6–57), and press the ENTER key. A curve is constructed at the intersection of surfaces B and C, and two points are constructed at the intersection of surface C and curve A.
4. Hide surfaces B and C and curve A to see the result, shown in Figure 6–57.
5. Do not save your file.

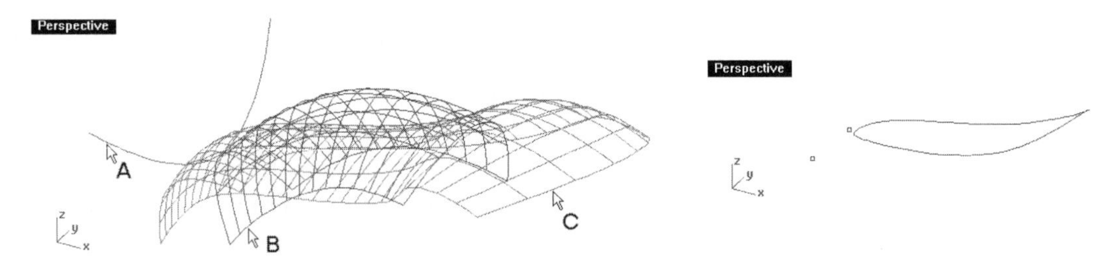

Figure 6–57. Intersection objects of surfaces and curves being constructed (left) and intersected points and curve (right)

Intersection of Polygon Meshes

You can construct a curve from the intersection of two polygon meshes by using the MeshIntersect command. Perform the following steps:

1. Select File > Open and select the file MeshIntersect.3dm from the Chapter 6 folder on the companion CD.
2. Click on the Mesh Intersect button on the Mesh toolbar, or type MeshIntersect at the command area.
3. Select polygon mesh A and B, shown in Figure 6–58. An intersection curve is constructed.

4. Hide meshes A and B.
5. Do not save your file.

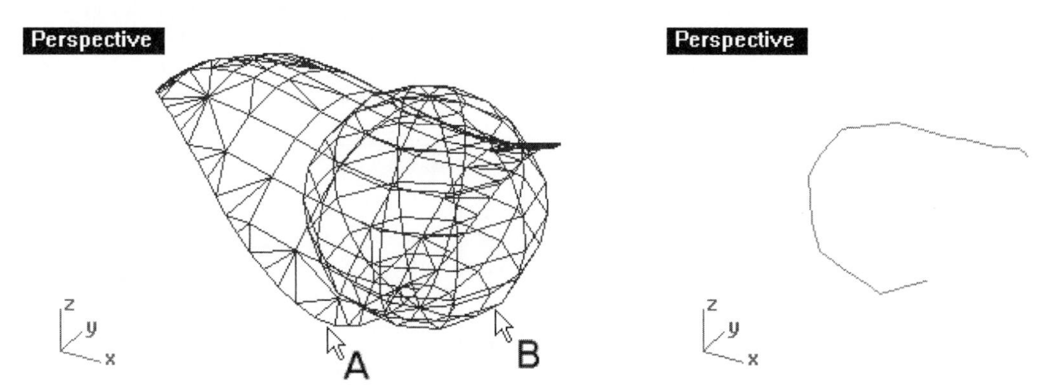

Figure 6–58. Polygon meshes (left) and intersection curve constructed (right)

Contour Lines and Section Lines of Surfaces

Contour lines and sections are both curves derived on a surface, and they are section lines cutting across a surface. Contour lines typically form a set of section lines spaced at a regular interval. To construct a set of contour lines across an ellipsoid surface, perform the following steps. Note that these commands also work on polygon meshes (you will learn polygon meshes in this Chapter 9). You can construct a NURBS surface from a set of contour lines, and you can construct sections through a polygon mesh. Perform the following steps:

1. Select File > Open and select the file ContourSection.3dm from the Chapter 6 folder on the companion CD.
2. Select Curve > Curve From Objects > Contour, or click on the Contour button on the Curve From Objects toolbar.
3. Select surface A, shown in Figure 6–59, and press the ENTER key.
4. Select points B and C (Figure 6–59) to indicate the contour plane base point and contour plane direction.
5. Type 3 at the command area to specify the between-contour lines.

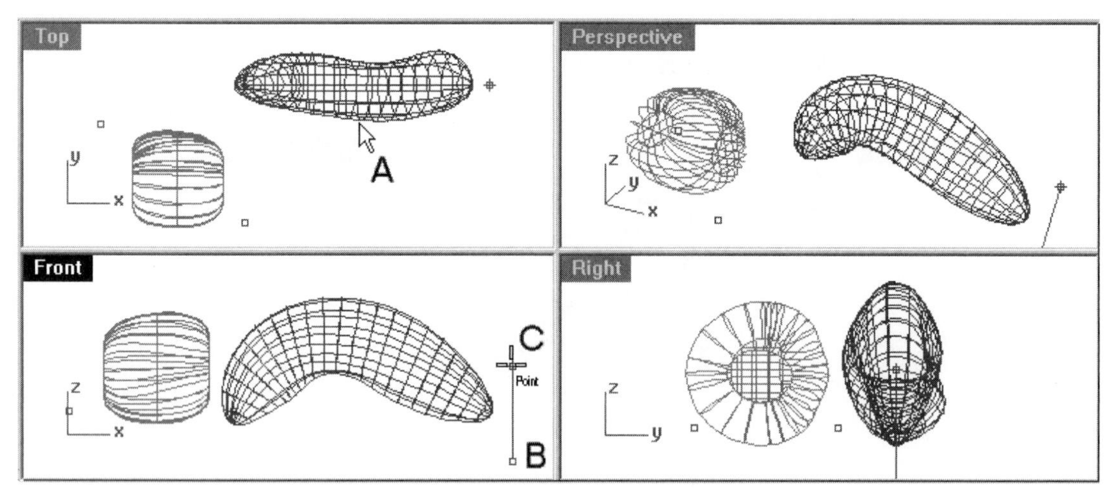

Figure 6–59. Contour lines being constructed

Contour lines are constructed. To construct sections, continue with the following steps.

6 Select Curve > Curve From Objects > Section, or click on the Section button on the Curve From Objects toolbar.
7 Select surface A, shown in Figure 6–60, and press the ENTER key.
8 Select points B and C (Figure 6–60) to define a section plane.
9 A section curve is constructed.
10 Press the ENTER key to terminate the command.

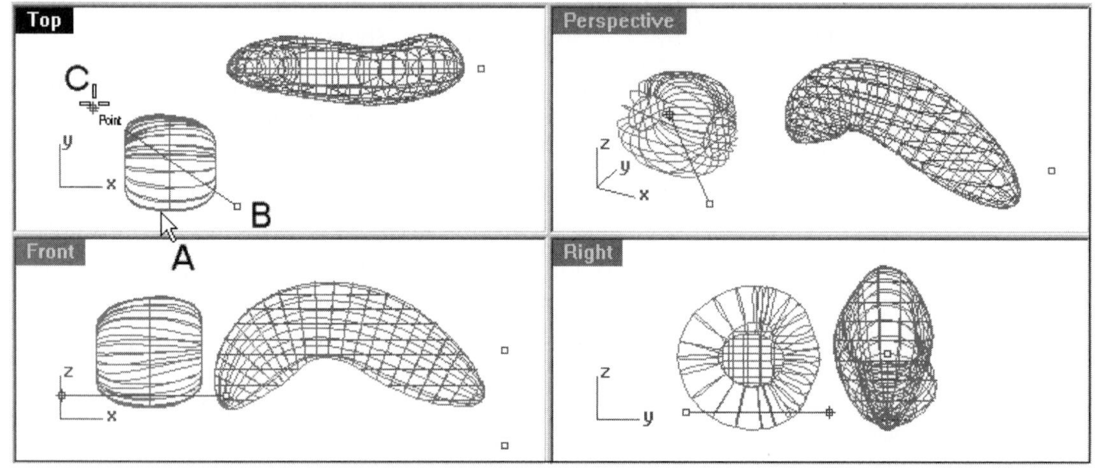

Figure 6–60. Section line being constructed

11 Turn off Layer01 to see the result. (See Figure 6–61.)
12 Do not save your file.

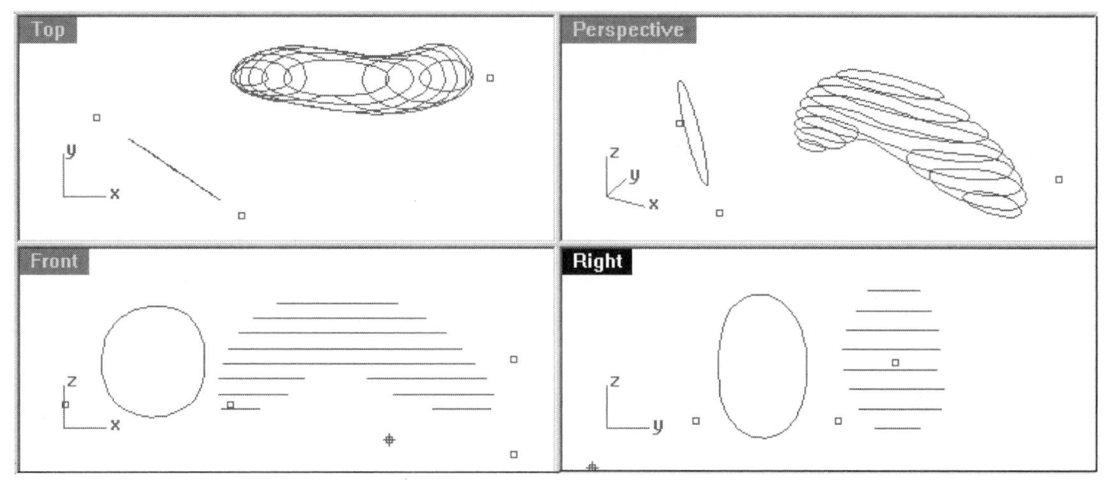

Figure 6–61. Contour lines and section line constructed

Silhouette of a Surface in a Viewport

The shape of the silhouette of a free-form surface resembles the edge of the shadow of the object projected in the viewing direction of the viewport. You see a different silhouette of the same object at different viewing angles. Perform the following steps:

1. Select File > Open and select the file Silhouette.3dm from the Chapter 6 folder on the companion CD.
2. Select Curve > Curve From Objects > Silhouette, or click on the Silhouette button on the Curve From Objects toolbar.
3. Select surface A, shown in Figure 6–62, and press the ENTER key. Note that this command is view dependent. The outcome will be quite different if you click on the Front viewport or the Right viewport.

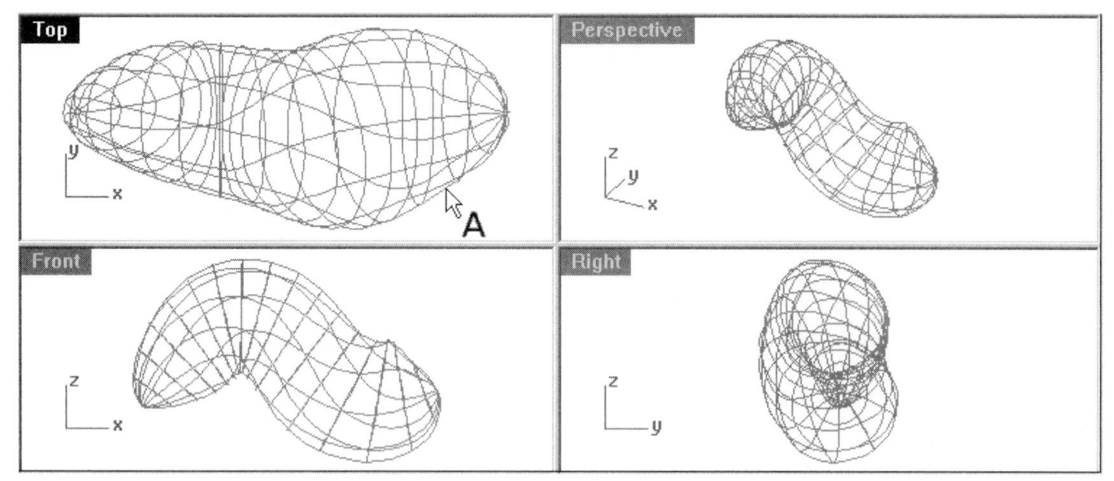

Figure 6–62. Silhouette being constructed

4. Press the Esc key to ensure that nothing is selected.
5. Hide the surface to see the result, shown in Figure 6–63.
6. Do not save your file.

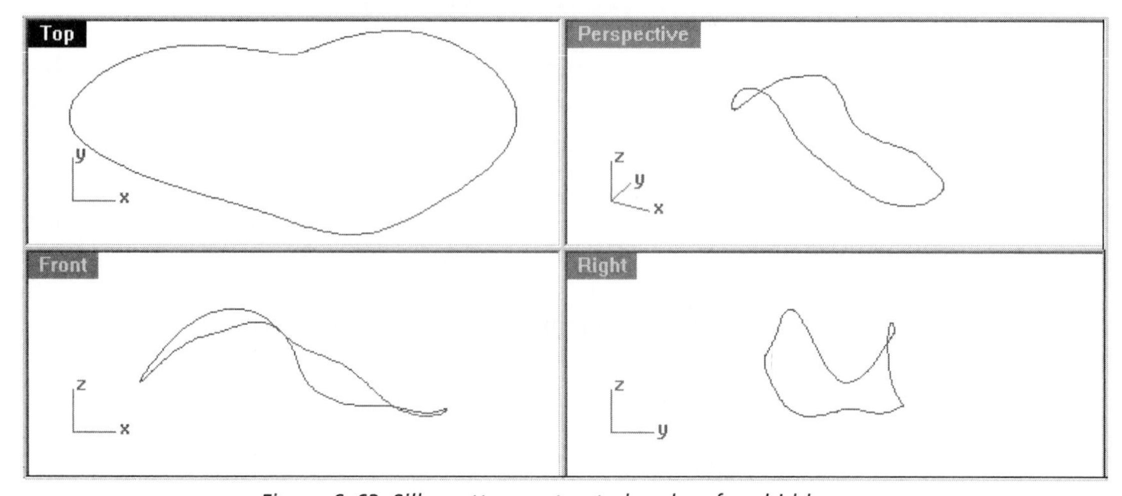

Figure 6–63. Silhouette constructed and surface hidden

Extracting Isocurves at Designated Locations of a Surface

Isocurve lines are directionally U and V lines displayed along a surface to help you visualize the profile and curvature of a surface. They are visual aids and not curves on the surface. To extract an isocurve at designated locations on the surface, perform the following steps:

1. Select File > Open and select the file Extract.3dm from the Chapter 6 folder on the companion CD.
2. Select Curve > Curve From Objects > Extract Isocurve, or click on the Extract Isocurve button on Curve From Objects toolbar.
3. Select surface A, shown in Figure 6–64.
4. Click on point B, shownFigure 6–64. (*Note*: You can use the INT osnap to snap to isocurve intersections.)
5. Select the Toggle option on the command area.
6. Click on point B again (Figure 6–64).
7. Press the ENTER key. Two isocurve lines are constructed.

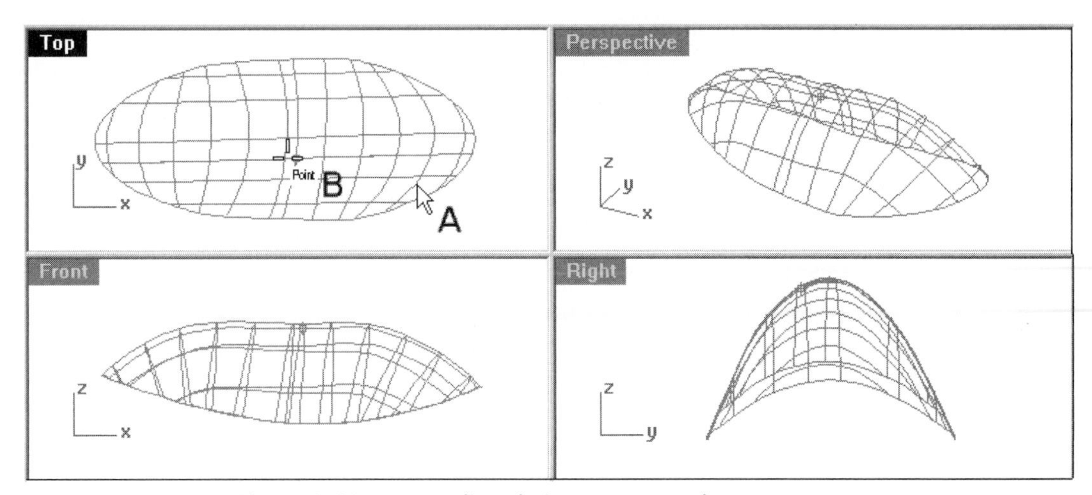

Figure 6–64. Isocurve lines being constructed

Extracting a Wireframe from Isocurves of a Surface

You can extract a wireframe from the isocurves of a surface for further development of your model. Continue with the following steps.

8. Select Curve > Curve From Objects > Extract Wireframe, or click on the Extract Wireframe button on the Curve From Objects toolbar.
9. Select surface A (shown in Figure 6–64) and press the ENTER key.

A wireframe is extracted from borders and isocurves of the surface. Because the extracted curves and the isocurve lines of the surface lie in the same location, you will not find any visual difference after you construct the curves. To see the difference, you can hide the surface and shade the viewport, as follows.

10. Press the Esc key to ensure that nothing is selected.

11 Hide the surface.

12 Right-click on the perspective viewport's label and select Shaded Display.

Extracting Control Points

Like curves, NURBS surfaces also have control points that govern their shape. To extract the control points of a surface and a curve, perform the following steps:

13 Unhide the surface and set the display to wireframe mode.

14 Select Curve > Curve From Objects > Extract Points, or click on the Extract Points button on the Curve From Objects toolbar.

15 Select surface A and curve B, shown in Figure 6–65, and press the ENTER key. Points are constructed at the control point locations of the surface and curve.

16 Do not save your file.

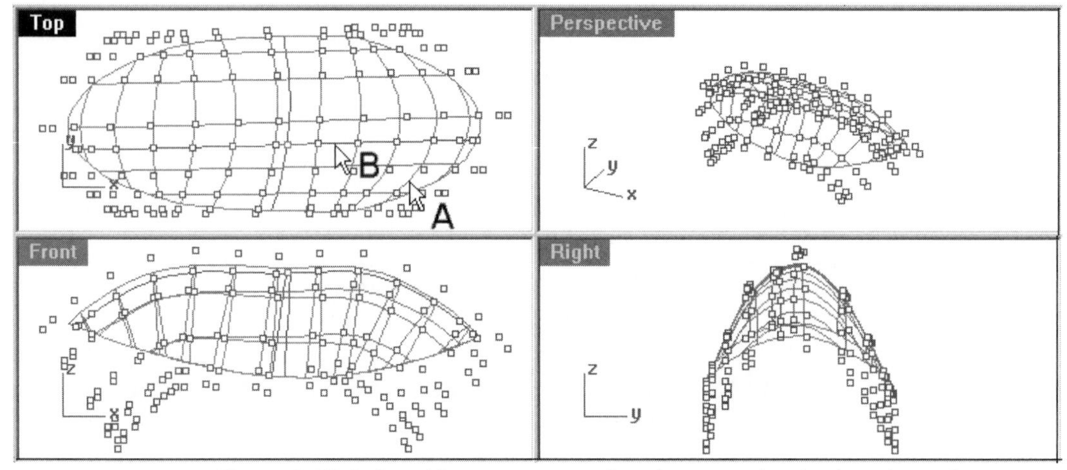

Figure 6–65. Point objects constructed at the control point locations of curve and surface

Extracting Sub-Curve of a Polycurve

When two or more curves are joined together, a polycurve is formed. To extract a curve element of a polycurve, perform the following steps:

1 Select File > Open and select the file SubCurve.3dm from the Chapter 6 folder on the companion CD.

2 Select Curve > Curve from Objects > Extract Curve, or click on the Extract Sub Curve button on the Curve Tools toolbar.
3 Select curve A, shown in Figure 6–66.
4 Click on location B.
5 Select the Copy option on the command area.
6 Press the ENTER key. A sub-curve is extracted.
7 Click on the extracted curve and drag it to location C to appreciate the effect.
8 Do not save your file.

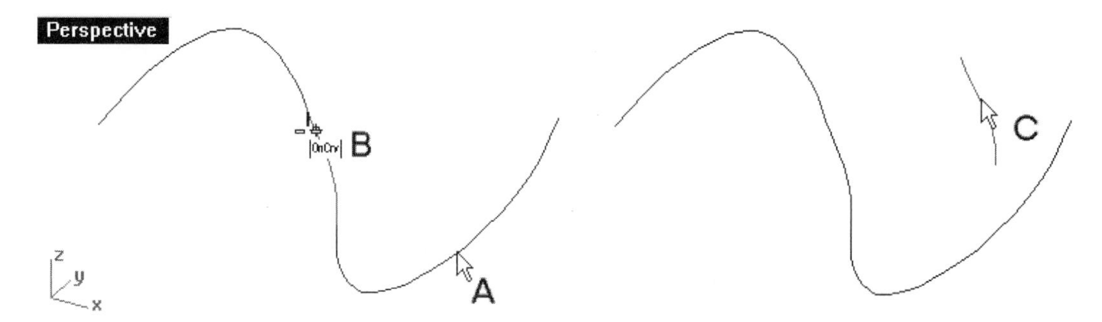

Figure 6–66. Polycurve (left) and subcurve extracted and dragged (right)

Creating Reference U and V Curves and Applying Planar Curves on Surfaces

To place a set of planar curves on a surface accurately with reference to the U and V orientation of the surface, you first construct a set of rectangular reference UV frame curves on the X-Y plane, construct a curve on the X-Y plane with reference to the reference UV curves, and apply the curves to the surface. Perform the following steps:

1 Select File > Open and select the file ApplyUV.3dm from the Chapter 6 folder on the companion CD.
2 Select Curve > Curve From Objects > Create UV Curves, or click on the Create UV Curves/Apply UV Curves button on the Curve From Objects toolbar.
3 Select surface A, shown in Figure 6–67.
4 A reference rectangle representing the U and V orientations of the selected surface is constructed on the XY plane.

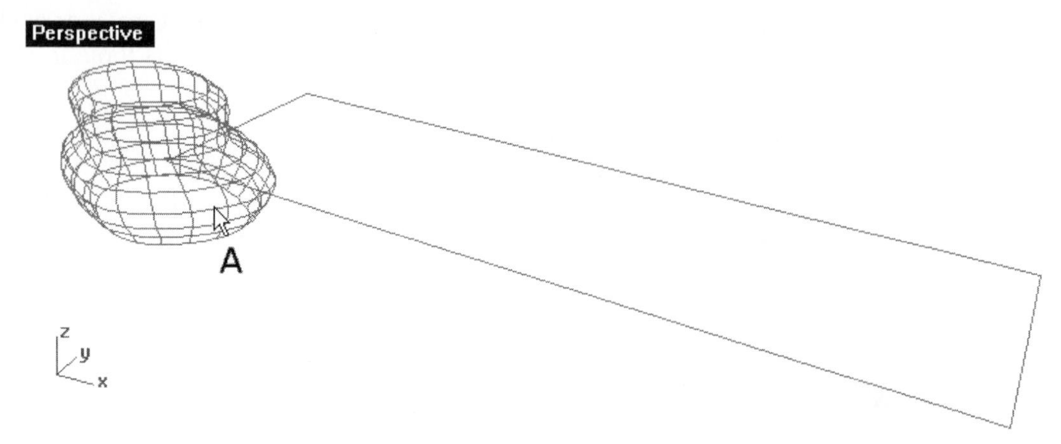

Figure 6–67. Reference UV curves

5 Referencing Figure 6–68, construct curve A relative to the reference UV curves. The exact shape of the curve is unimportant for this tutorial, as far as the curve is constructed within the reference rectangle.
6 Select Curve > From Objects > Apply UV Curves, or right-click on the Create UV Curves/Apply UV Curves button on the Curve From Objects toolbar.
7 Select curve A and B, shown in Figure 6–68, and press the ENTER key. It is important to select the reference curves as well as the curve to be applied.
8 Select surface C (Figure 6–68). Curves A and B are mapped onto the surface, as shown in Figure 6–62.
9 Do not save your file.

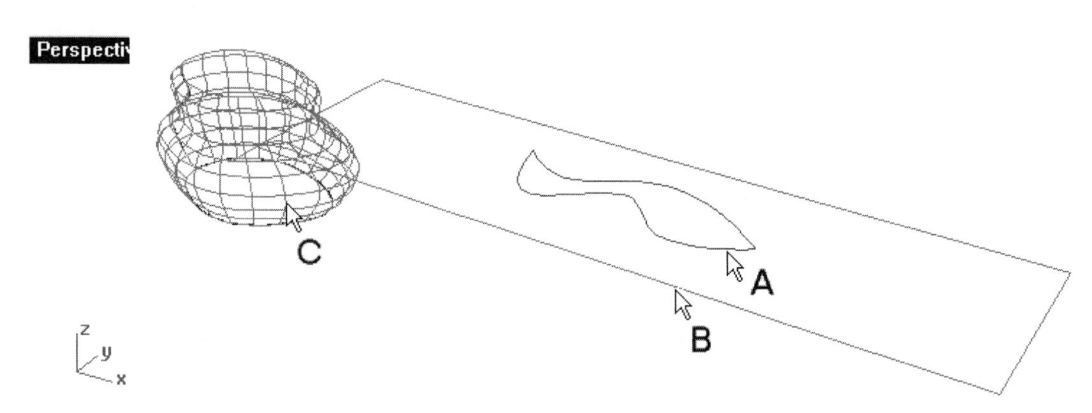

Figure 6–68. Curve applied to the surface

Chapter Summary

The basic way to manipulate a curve is to extend it (if it is not long enough), trim away the unwanted part (if necessary), and split it into two so that you get two contiguous curves with a smooth transition between them. You can also remove a portion of the curve to break it into two curves. You can also add a line segment to an open polyline to close it.

In between two curves, you can add a chamfer, a fillet, or a blended curve. You can also add line or arc segments to two coplanar curves for them to meet. Smoothness of the chamfer joint and connected joint are G0, fillet joint is G1, and blended joint can be G2, G3, or G4. Naturally, you can join two or more contiguous curves into one and explode a joined curve into its individual curve elements. If two or more curves intersect to form regions among them, you can extract boundaries of such regions.

There are many ways to refine a curve. You can rebuild it by specifying the degree of polynomial degree and the number of control points. As an alternative to blending two separate curves, which adds a curve between them, you can match a curve to another curve to obtain G2 continuity. To change the shape of a curve and yet maintain the tangent direction at the end points, you adjust end bulge. You can move individual segments of a polycurve, and you can replace a portion of a curve with a line segment. You can fair a curve to reduce large curvature deviation, simplify a curve consisting of segments of varying degree of polynomial, and refit an imported polyline to a free-form curve. If you are going to export free-form curves to other applications that do not support splines, you convert curves to line and arc segments.

From existing curves, you can derive curves and point objects. You can offset a curve to obtain another curve of similar shape at a specific distance from the source curve. Between two curves, you can construct a mean curve. To help produce a 3D curve, you can first construct two planar curves in two viewports to depict two orthographic views of the curve, and then derive a 3D curve from them. You can construct a set of cross-section curves from a set of longitudinal curves. As a result, you get a set of U and V curves, with the original longitudinal curves as the U curve and the cross-section curves as the V curves.

To ease the task of sketching or interpolating on a surface or polygon mesh, you can first construct a planar curve and then project or pull it onto a surface or a polygon mesh to obtain a curve on the surface or the polygon mesh. If edges and borders of a surface or polygon mesh are required for the development of some other geometric objects, you extract them. Curves and points produced from the intersection of two surfaces/two polygon meshes and intersection of a surface/polygon mesh and a curve are sometimes required for further design work. From an existing surface, you can construct contour curves, section curves, and silhouettes, and can extract isocurves, control points, and isocurve wireframe. You can also extract a portion of a curve.

To map a curve on a surface precisely, you can first create the UV curves of the surface on a plane, construct the curve to be mapped, and then apply the UV curve and the curve to be mapped to the surface.

Review Questions

1. In what ways can a curve's length be changed?
2. Outline the ways to treat two separate curves.
3. Illustrate how a curve's polynomial degree and fit tolerance can be modified.
4. Depict the methods of deriving curves from existing curves.

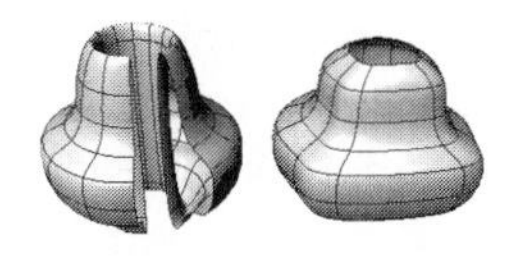

CHAPTER 7

NURBS Surface Manipulation

Introduction

After learning how various kinds of surfaces are constructed in Chapter 3 and how curves and points are constructed and manipulated in Chapters 4 through 6, this chapter examines various surface manipulation methods.

Objectives

After studying this chapter, you should be able to

- ❒ Manipulate the boundaries of surfaces
- ❒ Treat two or more adjacent surfaces
- ❒ Refine and modify a surface profile
- ❒ Manipulate surface edges

Overview

To help you master surface modeling techniques systematically, this chapter divides surface manipulation methods into four sections. The first section deals with surface boundary manipulation; the second section explains ways to treat two or more separate surfaces; the third section concerns surface refinement; and the final section details surface edge treatment.

Surface Boundary Manipulation

This section begins by explaining the concepts of surface trimming and delineating ways to trim a surface. It covers why a trimmed surface can be untrimmed, how the trimmed boundary can be detached, how a portion of the trimmed edge can be deleted, and how computer memory space can be saved by shrinking a trimmed surface. Following trimming and untrimming, this section describes how a surface can be split or, if there is kink along the surface, divided along creases. This section also depicts how an untrimmed surface can be extended to increase its size; extended perpendicularly to another surface, splitting it, and joining with it; and extended to become a periodic surface. Finally, this section explains one of the options of making a sweep surface from a kinked rail, in which the resulting surface can be two trimmed surfaces joined together.

Surface Trimming Concepts

To produce a smooth surface, it is necessary to use smooth defining curves and smooth boundary lines. However, most of the surfaces used to compose a design do not necessarily have smooth boundaries, although they have smooth profiles. If you use the irregular boundary edges to construct the surface directly, you get a surface with many sudden changes in curvature; the surface will not be smooth at all. To obtain a smooth surface with irregular edges, you build a larger surface from smooth wires, then trim the smooth surface with the irregular edges. The resulting surface is a trimmed surface.

Proper Way to Construct a Smooth Surface Profile

Figure 7–1 (left) shows the surface model of an automobile body panel. This is a smooth surface, but its boundary is irregular. Given this problem, you might intuitively use the boundary wires that you see as the defining wires to construct the surface model. If you did, you would probably get an irregular surface like the one shown in Figure 7–1 (right).

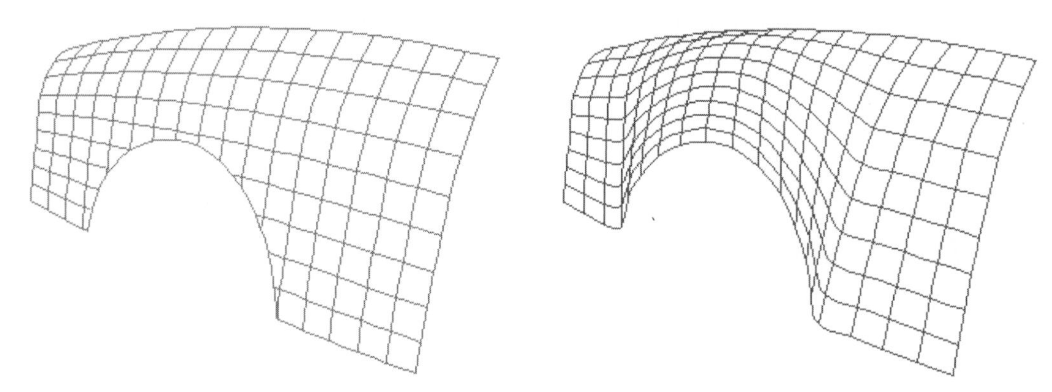

Figure 7–1. Smooth surface with irregular boundaries (left) and irregular surface defined by irregular boundaries (right)

Obviously, the surface shown in Figure 7–1 (right) is not the one we want. What has happened? The answer is that a set of irregular wires will generate an irregular surface. Unless the boundary lines are smooth wires, they cannot be used as defining wires for the surface.

To obtain a smooth surface with an irregular boundary, you have to perform two steps.

- You use a set of smooth wires to produce a smooth surface that is much larger than the required surface. This is called the base surface.
- You then use the irregular boundary wire to trim the smooth surface.

In the computer, a resulting trimmed surface consists of the original untrimmed surface (base surface) and the trim boundaries (trimmed edges). Although both of these are saved in the database, only the boundary and the remaining part of the trimmed surface are displayed. As a result, we obtain a smooth free-form surface with irregular boundaries. This is called a trimmed surface.

To produce a free-form surface that is large enough for subsequent trimming, define a set of wires that encompass the required surface. To make such wires, you need to be able to visualize the defining wires that are outside the required surface. In Figure 7–2, the construction of the smooth automobile body panel starts from a set of smooth wires. From the smooth wires, a smooth surface that is much larger than the required surface is made. To obtain the required surface, an irregular boundary is used to trim the large smooth surface.

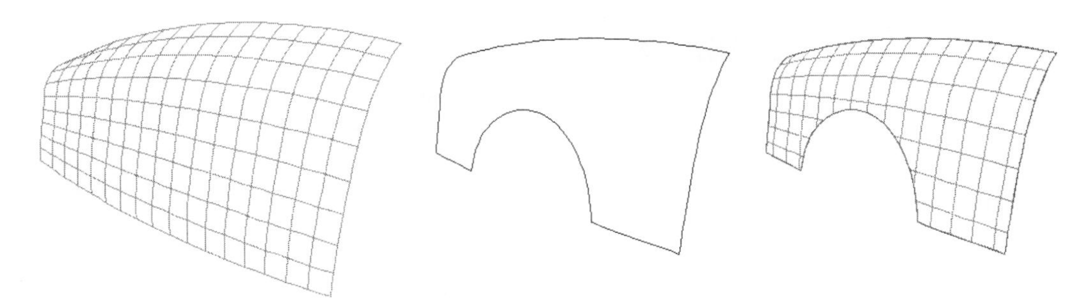

Figure 7–2. The untrimmed surface (left), the trimming boundary (center), and trimmed surface (right)

Methods to Trim a Surface

Basically, there are two methods to trim a surface or polysurface. (A polysurface is a set of contiguous surfaces joined together.) The first method is to construct a curve and use the curve to trim the surface. If the curve is not lying on the surface, the direction of trimming will be vertical to the active construction plane, as shown in Figure 7–3. Another way to trim a surface is to construct two intersecting surfaces and trim the unwanted portions of the surfaces away to form a sharp edge at the intersection, as shown in Figure 7–4. In both cases, the cutting object has to be long enough to cutt through the surface.

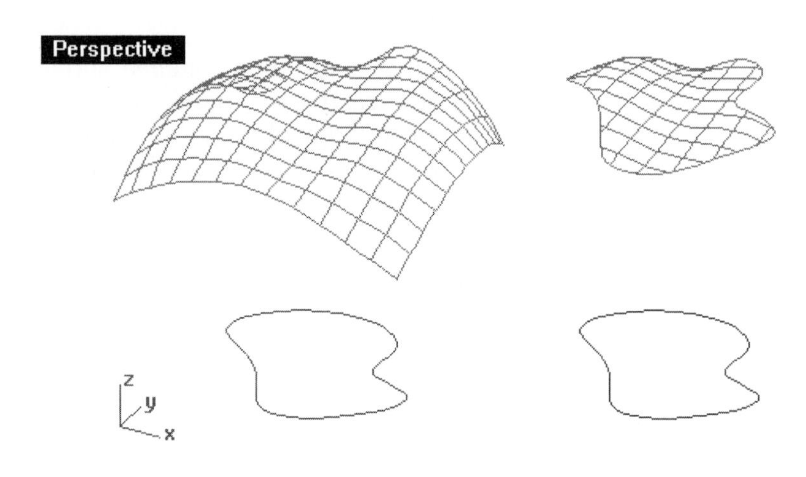

Figure 7–3. Surface and wire (left) and surface trimmed by projecting the wires in a direction perpendicular to the construction plane (right)

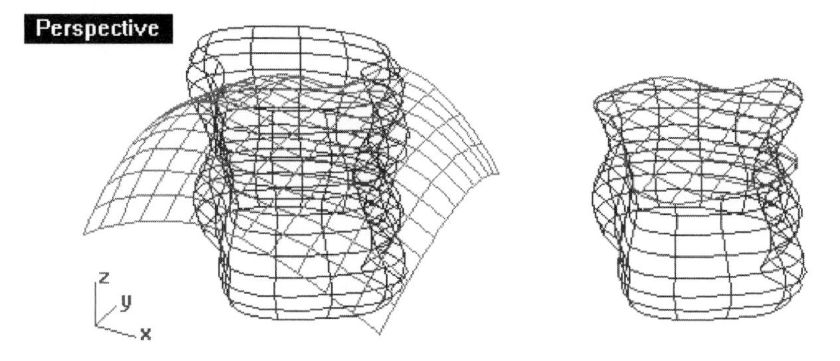

Figure 7–4. Two intersecting surfaces (left) and trimmed surfaces (right)

A surface can only be cut by a surface or curve that is long enough to pass through the surface. Perform the following steps:

1. Select File > Open and select the file TrimSurface01.3dm from the Chapter 7 folder on the companion CD.
2. Select Edit > Trim, or click on the Trim/Untrim Surface button on the main toolbar.
3. Select surface A and curve B, shown in Figure 7–5, as cutting objects and press the ENTER key.
4. Select locations C and D, shown in Figure 7–5, to indicate the portion to be trimmed away, and then press the ENTER key.

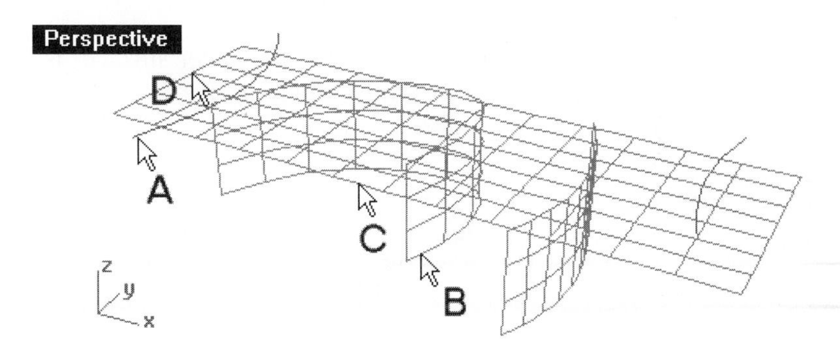

Figure 7–5. Surfaces being trimmed

5. Select Edit > Trim, or click on the Trim/Untrim Surface button on the main toolbar.
6. Select surface A and curve B (Figure 7–6) as cutting objects, and press the ENTER key.
7. Select location C, shown in Figure 7–6, to indicate the portion to be trimmed away, and then press the ENTER key. The surface is not trimmed, because the cutting objects are not long enough to cut through the target surface.
8. Do not save your file.

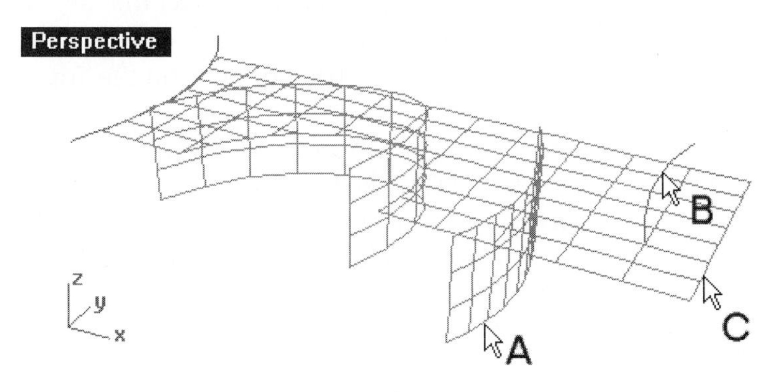

Figure 7–6. Cutting objects not long enough

If two surfaces of difference sizes intersect each other, the larger surface can trim the smaller one but not vice versa. Perform the following steps:

1. Select File > Open and select the file TrimSurface02.3dm from the Chapter 7 folder on the companion CD.
2. Select Edit > Trim, or click on the Trim/Untrim Surface button on the main toolbar.
3. Select surfaces A and B (Figure 7–7) as cutting objects, and press the ENTER key.
4. Select locations C and D, shown in Figure 7–7, to indicate the portion to be trimmed away, and then press the ENTER key.
5. Do not save your file.

Note that surface B (the narrower surface) is trimmed by surface A (the wider surface), but not vice versa.

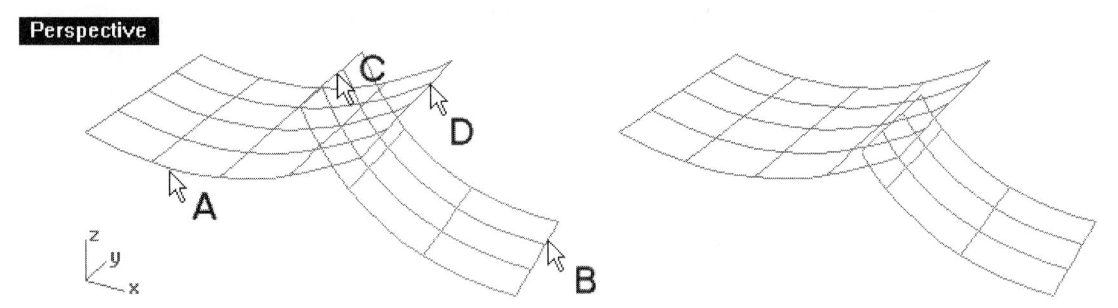

Figure 7–7. Surfaces being used as cut objects to trim each other (left) and outcome (right)

To have two surfaces trimming each other, you may need to first trim their edges. Perform the following steps:

1. Select File > Open and select the file TrimSurface03.3dm from the Chapter 7 folder on the companion CD.
2. Select Edit > Trim, or click on the Trim/Untrim Surface button on the main toolbar.
3. Select surfaces A and B (Figure 7–8) as cutting objects, and press the ENTER key.
4. Select locations C and D, shown in Figure 7–8, to indicate the portion to be trimmed away, and then press the ENTER key.

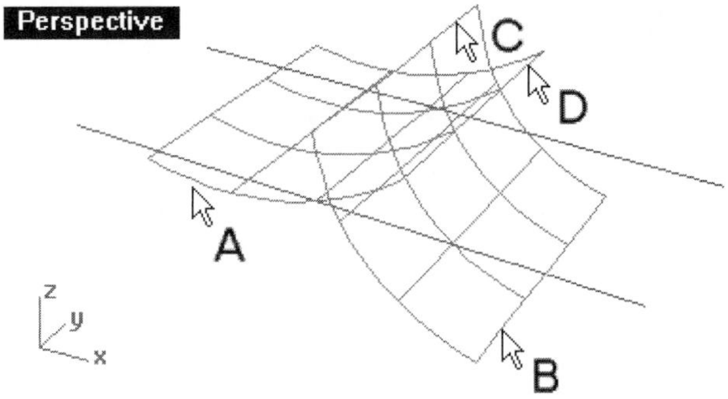

Figure 7–8. Surfaces being trimmed by curve and surface

You will find that both surfaces are not trimmed, because both A and B (the cutting objects) are not wide enough to pass through the objects to be trimmed. Continue with the following steps:

5 Maximize the Top viewport.

6 Select Edit > Trim, or click on the Trim/Untrim Surface button on the main toolbar.

7 Select curves A and B (Figure 7–9) as cutting objects, and press the ENTER key.

8 Select locations C, D, E, and F, shown in Figure 7–9, to indicate the portion to be trimmed away, and then press the ENTER key. The surfaces are trimmed by the curves. Now their widths are the same.

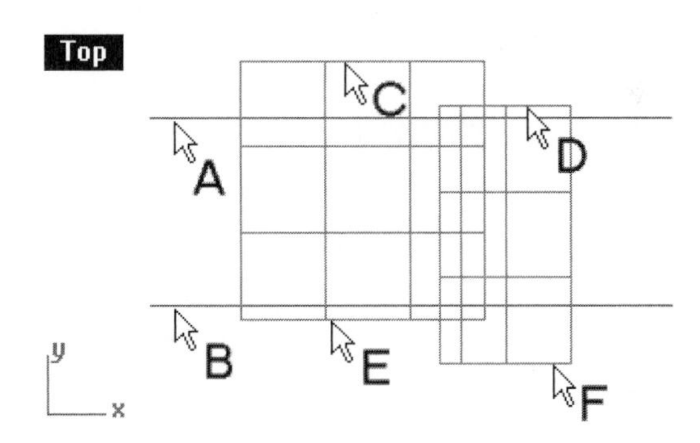

Figure 7–9. Surfaces being trimmed by curves

9 Maximize the Perspective viewport.

10 Select Edit > Trim, or click on the Trim/Untrim Surface button on the main toolbar.

11 Select surfaces A and B (shown in Figure 7–10) as cutting objects, and press the ENTER key.

12 Select locations C and D (shown in Figure 7–10) to indicate the portion to be trimmed away, and then press the ENTER key. The surfaces are trimmed.

13 Do not save your file.

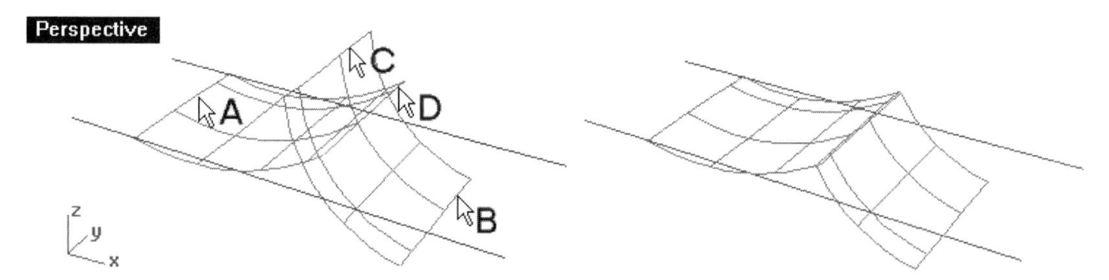

Figure 7–10. Surfaces trimming each other (left) and surfaces trimmed (right)

Untrimming a Trimmed Surface

In the course of refining your design, you may need to revert a trimmed surface back to its original untrimmed state. TBecause a trimmed surface in the computer consists of the original base surface and the trimmed edge, you can remove its trimmed boundary to change it back to its untrimmed state. This is called untrimming. Perform the following steps:

1. Select File > Open and select the file Trimboundary01.3dm from the Chapter 7 folder on the companion CD. In this file, there are four identical trimmed surfaces to be used for various operations and comparison.
2. Select Surface > Surface Edit Tools > Untrim, or click on the Untrim/Detach Trim button on the Surface Tools toolbar.
3. Select edge A, (shown in Figure 7–11. The surface is untrimmed.

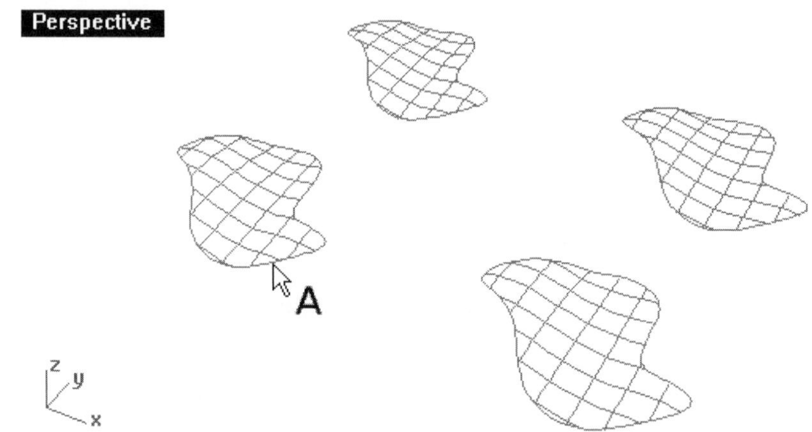

Figure 7–11. Four identical trimmed surfaces and one surface being untrimmed

If there is more than one trim boundary in a trimmed surface, you can use the All option to untrim all the edges in one operation. When the All option is selected and an edge is selected, all edge trims will be removed. If a hole is selected, all hole trims will be removed.

Detaching a Trimmed Boundary from a Trimmed Surface

While untrimming, you have the option of retaining the trimmed boundary as a set of curves. This is called detaching. To untrim a trimmed surface and retain the trimmed boundary curve, continue with the following steps.

4 Surface > Surface Edit Tools > Detach Trim, or right-click on the Untrim/Detach Trim button on the Surface Tools toolbar.

5 Select edge A, shown in Figure 7–12. The surface is untrimmed and the trimmed boundary is retained.

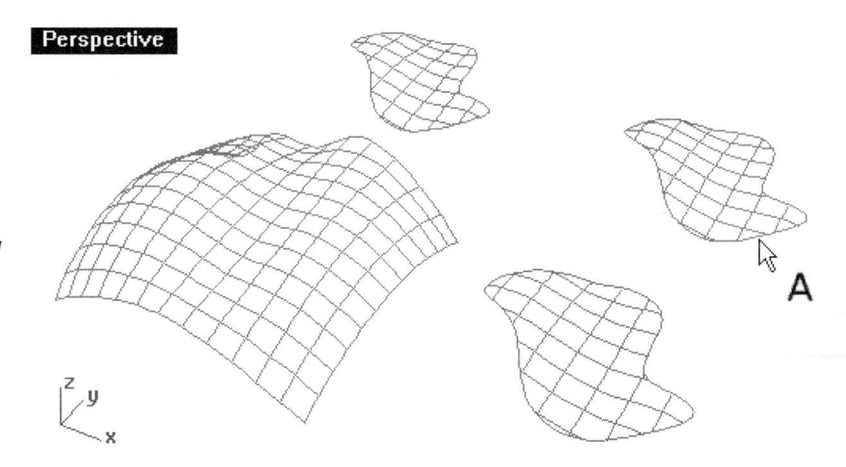

Figure 7–12. A surface untrimmed and a surface's boundary being detached

Shrinking a Trimmed Surface

To reduce the memory requirement to store a trimmed surface, you reduce the original untrimmed surface to its minimum size. This process is called shrinking. In essence, a trimmed surface retains its original smooth defining boundaries in the database while possessing a new trimmed boundary. The original surface is called the base surface, and the trimmed boundary is called the trim edge. Sometimes you might use a base surface that is much larger than required. If so, unnecessary memory space is wasted to store the unwanted part of the base surface. To reduce the memory used, you truncate the base surface of a trimmed surface. Continue with the following steps.

6 Select Surface > Surface Edit Tools > Shrink Trimmed Surface, or click on the Shrink Trimmed Surface/Shrink Trimmed Surface to Edge button on the Surface Tools toolbar.

7 Select edge A, shown in Figure 7–13, and press the ENTER key.

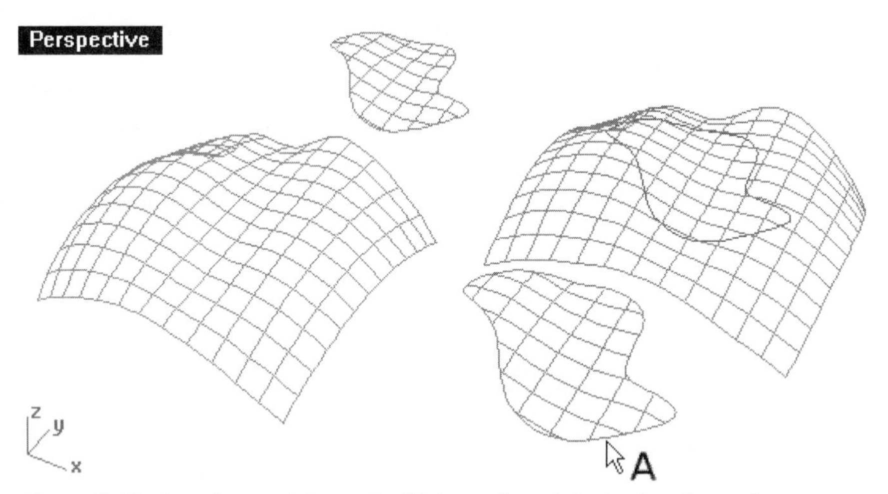

Figure 7–13. A surface untrimmed with boundary detached and a surface being shrunk

At a glance, both surfaces (the shrunk and the original) are the same. However, their base surfaces are different in size. To appreciate the effect of shrinking, continue with the following steps.

8 Surface > Surface Edit Tools > Detach Trim, or right-click on the Untrim/Detach Trim button on the Surface Tools toolbar.

9 Select edge A, (shown in Figure 7–13. The surface is untrimmed and the boundaries are retained. (See Figure 7–14.)

10 Do not save your file.

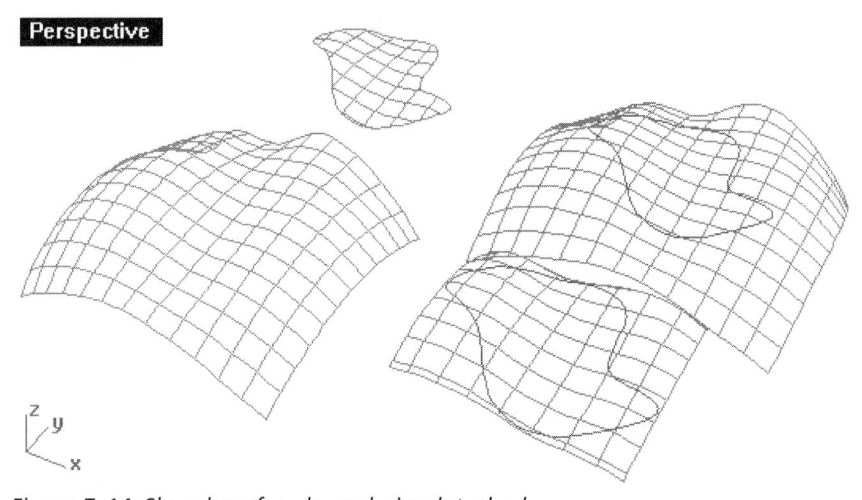

Figure 7–14. Shrunk surface boundaries detached

Note that shrinking a trimmed surface also makes subsequent fillet, blend, and sweep commands work better. Apart from a reduction in memory space, the topology of the control points of a shrunk, trimmed surface is also different from that of the same trimmed surface before it is shrunk.

Shrinking to Boundary Edge

To further reduce the size of the base surface of a trimmed surface, you may shrink the base surface to the trimming edge. However, this may cause problems later, when the surface boundaries are used for other inputs. Perform the following steps:

1. Select File > Open and select the file Trimboundary02.3dm from the Chapter 7 folder on the companion CD. In this file, there are four identical trimmed surfaces to be used for various operations and comparison.
2. Right-click on the Shrink Trimmed Surface/Shrink Trimmed Surface to Edge button on the Surface Tools toolbar, or type ShrinkTrimmedSrfToEdge at the command area.
3. Select surface A, shown in Figure 7–15, and press the ENTER key.
4. Select Surface > Surface Edit Tools > Shrink Trimmed Surface, or click on the Shrink Trimmed Surface button on the Surface Tools toolbar.
5. Select surface B, shown in Figure 7–15, and press the ENTER key.
6. Surface > Surface Edit Tools > Detach Trim, or right-click on the Untrim/Detach Trim button on the Surface Tools toolbar.
7. Select edges A and B, shown in Figure 7–15, and press the ENTER key. The surfaces are untrimmed and the trimmed boundaries are detached. As shown in Figure 7–16, the surface that is shrunk to the trimmed edge is smaller.
8. Do not save your file.

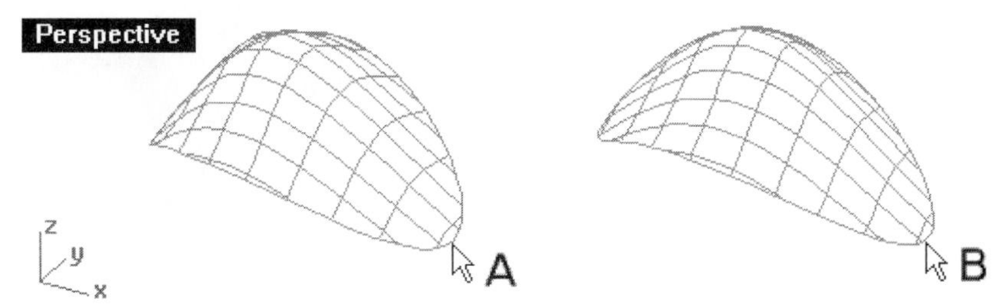

Figure 7–15. Two trimmed surfaces

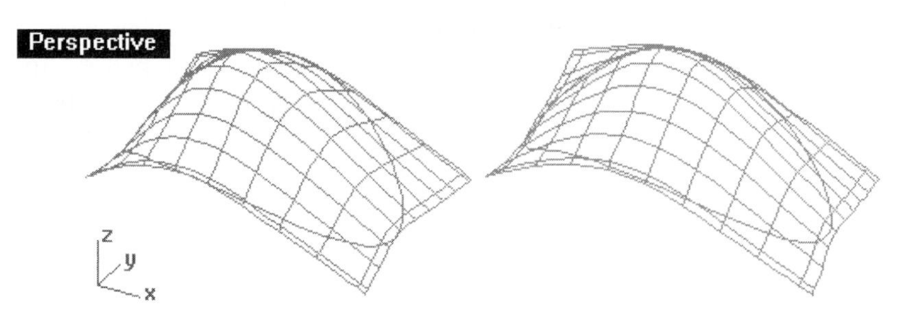

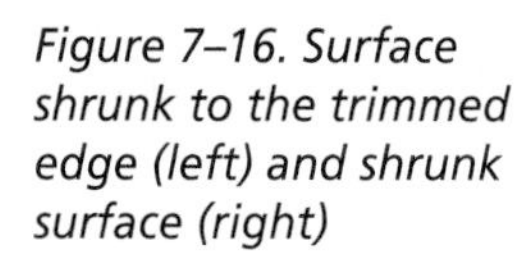

Figure 7–16. Surface shrunk to the trimmed edge (left) and shrunk surface (right)

Splitting One Surface into Two

Similar to the trimming process, you can split a surface into two contiguous surfaces with smooth transition. The splitting operation is very similar to trimming in that you need a cutting object. The difference is that the cutting object cuts the target object in two. Cutting objects can be curves or surfaces. A point to note when using curves not residing on the surfaces as cutting objects is that cutting is viewport dependent. Perform the following steps to split a surface using another surface and a curve:

1. Select File > Open and select the file SplitSurface.3dm from the Chapter 7 folder on the companion CD.
2. Select Edit > Split, or click on the Split button on the Main2 toolbar.
3. Select surface A, shown in Figure 7–17, as the object to be split, and then press the ENTER key.
4. Select surface B and curve C, shown in Figure 7–17, as splitting objects, and then press the ENTER key. The surface is split into three surfaces.
5. Do not save your file.

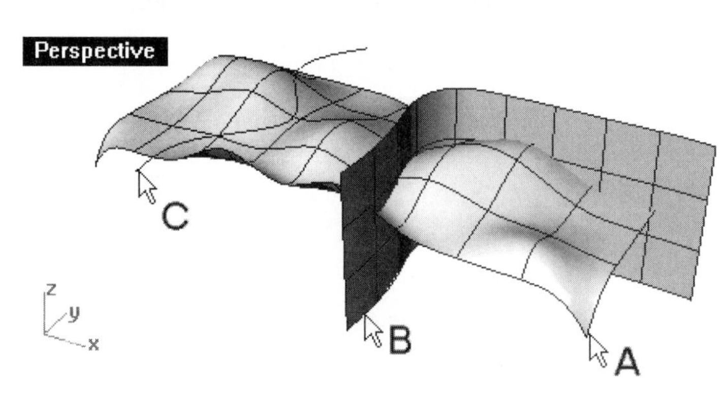

Figure 7–17. Surfaces being split

Splitting Along U and V Lines

A surface can be broken into two surfaces along a selected U or V line. After the break, the surfaces still maintain the original continuity, and the profiles and silhouettes of the broken surfaces will be the same as those of the original surface. Perform the following steps:

1. Select File > Open and select the file BreakSurface.3dm from the Chapter 7 folder on the companion CD.
2. Set object snap mode to Point objects. *(Note: You may use INT object snap to snap to isocurve intersection.)*
3. Surface > Surface Edit Tools > Split at Isocurve, or right-click on the Split/Split Surface by Isocurve button on the Main2 toolbar.
4. Select surface A (shown in Figure 7–18).
5. Select the Direction option from the command area.
6. Select the Both option from the command area.
7. Click on point B (shown in Figure 7–18) and press the ENTER key. The surface is split into four surfaces.
8. Do not save your file.

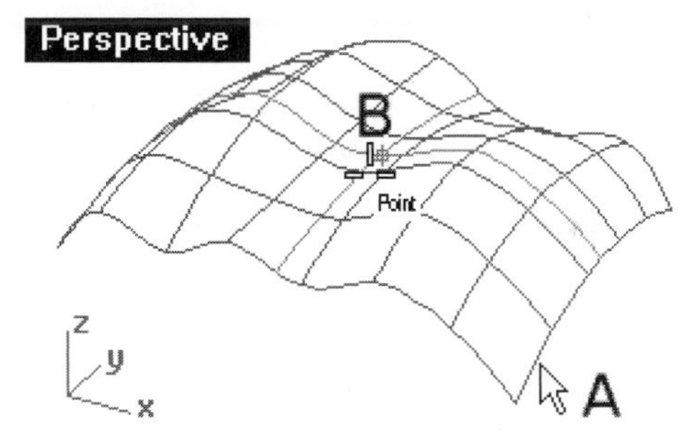

Figure 7–18. Surface being split along its isocurve

*(**NOTE:** You may use the INT object snap tool to snap to the isocurve intersection to split the surface along the isocurves.)*

Dividing a Surface Along its Creases

Creases are kinky edges on a single surface. If a surface possesses history and the defining curves are modified such that there are kink points along them, the resulting surfaces will have creases. If you want to remove the creases by dividing the surface along the creases, perform the following steps:

1. Select File > Open and select the file DivideAlongCreases.3dm from the Chapter 7 folder on the companion CD.
2. Select Surface > Surface Edit Tools > Divide Surfaces on Creases, or right-click on the Merge Surfaces/Divide Along Creases button on the Surface Tools toolbar.
3. Select surface A, shown in Figure 7–19, and press the ENTER key. The single surface is broken into a set of surfaces joined together, a polysurface. (We will discuss polysurface in the next section.)

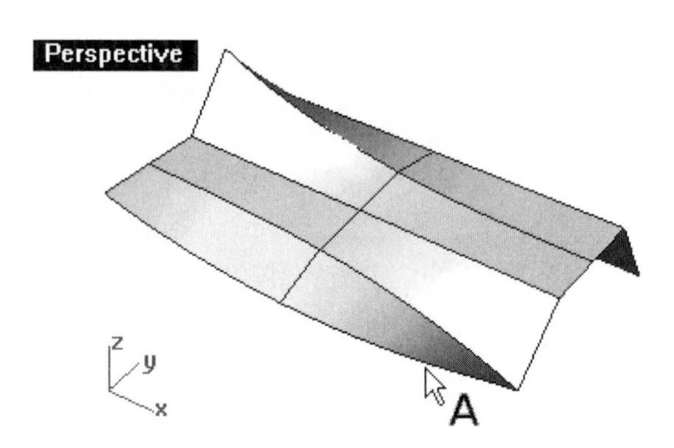

Figure 7–19. Surface being divided along its creases

To appreciate the difference, explode the surface into a set of surfaces and change their color.

4. Select Edit > Explode, or click on the Explode/Extract Surfaces button on the Main2 toolbar.
5. Select polysurface A, shown in Figure 7–19, or any part of the polysurface, and press the ENTER key. The polysurface is exploded.
6. Select Edit > Object Properties, or click on the Object Properties button on the Properties toolbar.
7. Select surface A (Figure 7–19) and press the ENTER key.
8. In the Properties dialog box, set the color to green and close the dialog box by clicking on the Checkmark in the upper right-hand corner.
9. Do not save your file.

Extending the Untrimmed Edge of a Surface

Contrary to trimming a surface to reduce its size, you can extend a surface to increase its size. However, it must be noted that a surface can only be extended if it is not trimmed, because the trimmed boundary edges are not the real edges of the original surface. Perform the following steps.

1. Select File > Open and select the file Extend.3dm from the Chapter 7 folder on the companion CD.
2. Select Surface > Extend Surface, or click on the Extend Untrimmed Surface/Extend Trimmed Surface button on the Surface Tools toolbar.
3. If Type = line, select Type option from the command area to set Type = Smooth. Otherwise, proceed to the next step.
4. Select edge A (shown in Figure 7–20).
5. Type 8 at the command area to set the extension factor. The surface is extended.
6. Do not save your file.

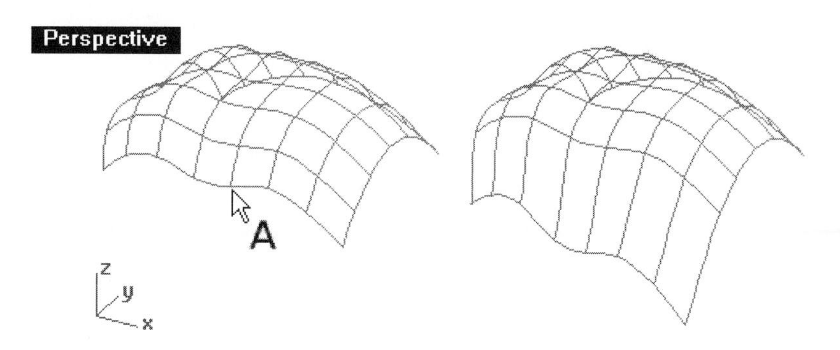

Figure 7–20. Surface being extended (left) and extended surface (right)

Extending the Trimmed Edge of a Surface

If a surface has three or more edges, its trimmed edge can be extended, as follows:

1. Select File > Open and select the file ExtendTrimmedEdge.3dm from the Chapter 7 folder on the companion CD.

There are two identical trimmed surfaces in this file: one for extending and one for reference.

2. Right-click on the Extend Untrimmed Surface/Extend Trimmed Surface button on the Surface Tools toolbar, or type ExtendT-rimmedSrf at the command area.
3. Select surface edge A, shown in Figure 7–21.
4. Type 5 at the command area to specify the extended length, and press the ENTER key. The trimmed edge of the surface is extended.

To discover the shape of the untrimmed surface, continue with the following steps:

5. Select Surface > Surface Edit Tools > Untrim, or click on the Untrim/Detach Trim button on the Surface Tools toolbar.
6. Select edge B, shown in Figure 7-21. The surface is untrimmed. See Figure 7-22 to compare the results.
7. Do not save your file.

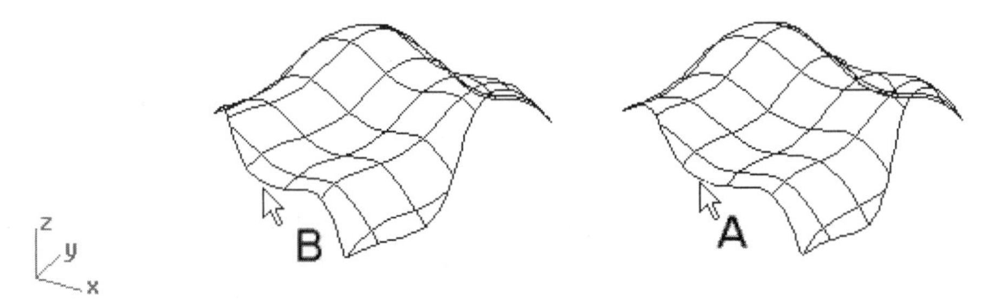

Figure 7-21. Two identical trimmed surfaces

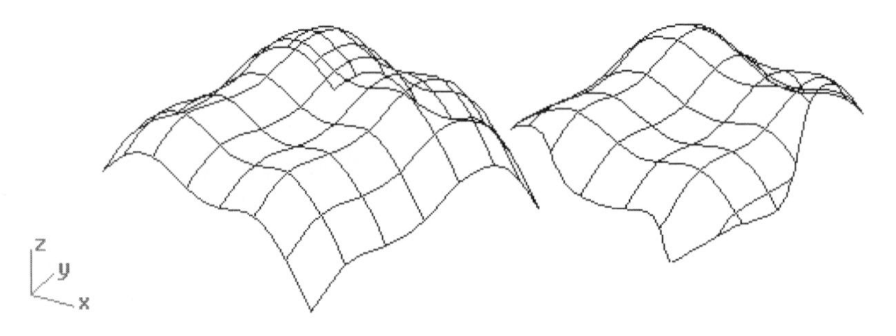

Figure 7-22. Surface untrimmed (left) and trimmed edge extended (right)

Extending a Surface to Make it Periodic

A periodic surface is a smooth, closed-loop surface. By making a surface periodic, the surface is extended to form a close loop. If the surface is a trimmed surface, it will be untrimmed automatically before extending to form a close loop. To appreciate how to make a surface periodic, perform the following steps.

1. Select File > Open and select the file Periodic.3dm from the Chapter 7 folder on the companion CD.

2 Select Edit > Make Periodic, or click on the Make Surface Periodic/ Make Surface Non-Periodic button on the Surface Tools toolbar.
3 Select edge A, shown in Figure 7–23, and press the ENTER key.
4 If the default is Yes for delete input, press the ENTER key. Otherwise, select Yes option from the command area. The surface is changed to a periodic surface.
5 Repeat the command.
6 Select edge B (Figure 7–23) and press the ENTER key.
7 Press the ENTER key again. The second surface is made periodic. (See Figure 7–24.)
8 Do not save your file.

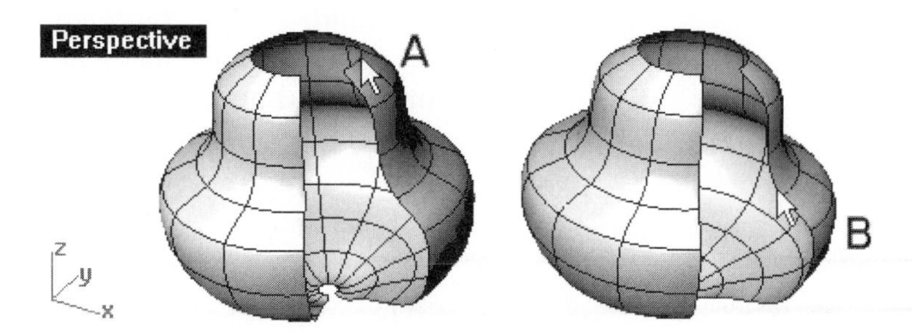

Figure 7–23. Surfaces being changed to periodic surfaces

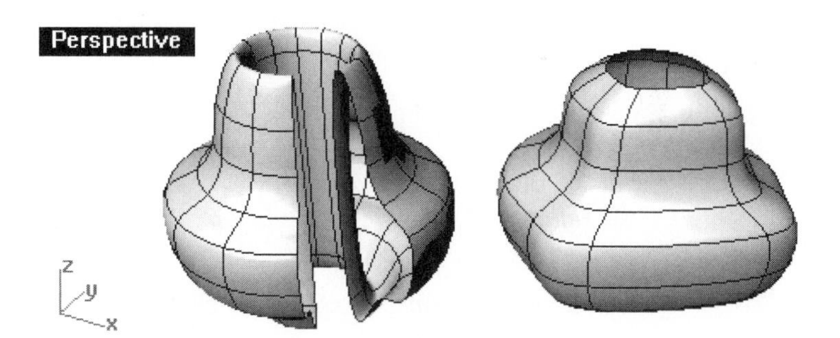

Figure 7–24. Surfaces made periodic

Surface Extension in Making a Sweep Surface from a Kinked Rail

Extending what you have learned about sweep surface construction in Chapter 3, if the rail for making a sweep surface has a kink in it, the resulting surface will be broken into two surfaces and joined together as a polysurface. Depending on the options selected while making the surface, you may have trimmed surfaces or mitered surfaces. Perform the following steps:

1. Select File > Open and select the file KinkRail.3dm from the Chapter 7 folder on the companion CD.
2. Select Surface > Sweep 1 Rail, or click on the Sweep 1 Rail button on the Surface toolbar.
3. Select curves A and B (Figure 7–25) and press the ENTER key.
4. Click on the OK button. A sweep surface consisting of two trimmed surfaces joined together is constructed
5. Repeat the command.
6. Select curves C and D (Figure 7–25) and press the ENTER key.
7. Check the Untrimmed Miters button and click on the OK button on the Sweep 1 Rail Options dialog box. A sweep surface consisting of two mitered surfaces joined together is constructed.

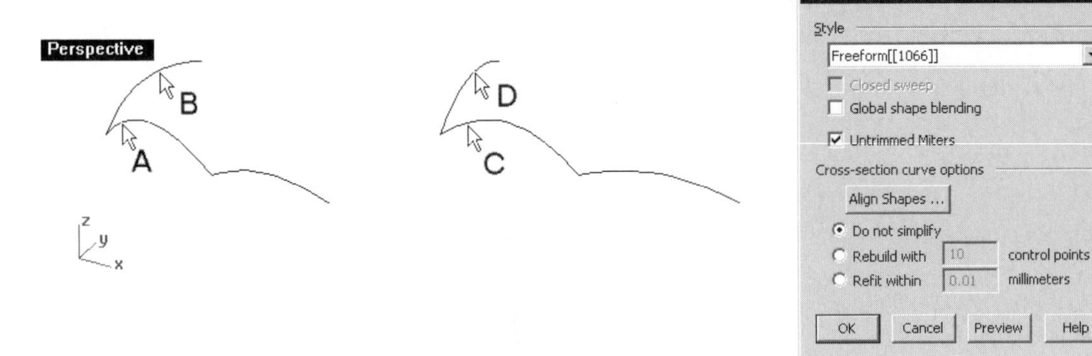

Figure 7–25. Sweep surfaces being constructed

8. Select Edit > Explode or click on the Explode button on the Main 2 toolbar.
9. Select A and B (Figure 7–26) and press the ENTER key. The surfaces are broken into four surfaces.
10. Select Surface > Surface Edit Tools > Untrim, or click on the Untrim button on the Surface Tools toolbar.
11. Select edge A (Figure 7–26). There are two surface edges. Select one of them.
12. Repeat the command.
13. Select edge A (Figure 7–26), the remaining edge.

14 Repeat the command two more times to untim the edges at B (Figure 7–26). The surfaces are untrimmed. Compare the result. (See Figure 7–27.)

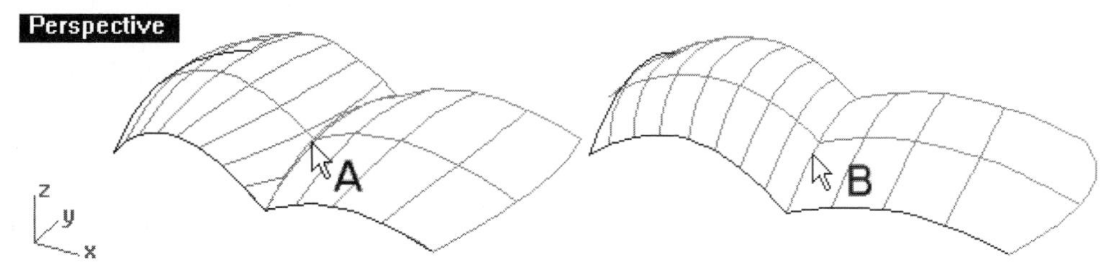

Figure 7–26. Surfaces being exploded and untrimmed

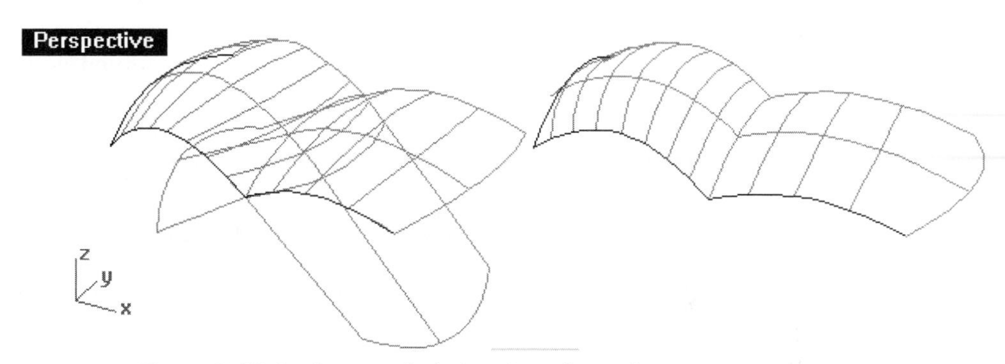

Figure 7–27. Surfaces exploded and untrimmed

Treating Two or More Separate Surfaces

In composing a surface model consisting of more than one surface, you can treat the joint between two consecutive surfaces by simply extending and trimming. Alternatively, you can insert a chamfer surface, a fillet surface, or a blended surface between the surfaces. The chamfer distance, fillet radius, or blend curvature along the joint can be constant or variable. The inserted chamfer, fillet, or blended surface, together with the original surfaces, are typical examples of five kinds of continuity between contiguous surfaces, with G0 for chamfer surface, G1 for fillet surface, and G2, G3, and G4 for blended surface.

Continuity Between Contiguous Surfaces

To reiterate, a G0 joint simply has edges of two contiguous surfaces meeting each other. Both the tangency direction and the radius of curvature at the joint are different. In a G1 joint, the tangency directions of the edges of two joined surface edges are the same but the radii of curvature are different. In a G2 joint, both the tangency directions and the radii of curvature of the edges are the same. For G3 continuity, the rate of change of curvature at the joint is constant. The smoothest joint is G4, in which the rate of change of the rate of change of curvature is constant.

Instead of inserting a surface, you can change the shape of one surface for it to match with the edge of the other one. To modify the shape of a surface, and yet maintain its degree of continuity with its contiguous surface, you modify its end bulge.

To handle two or more surfaces collectively, you join them into a single polysurface. The opposite of joining is exploding. If the surface boundaries between the contiguous surfaces are untrimmed edges, you may, instead of joining, merge them into a single surface. However, it must be noted that two surfaces merged together cannot be exploded into two surfaces.

Chamfering, Filleting, and Blending

Depending on styling requirement, you may insert a chamfer, fillet, or blended surface between two consecutive surfaces.

Constructing a Chamfer Surface

A chamfered surface is a flat surface that forms a bevelled edge between the two original surfaces. Because edges of the chamfered face and the contiguous surfaces simply join together without any tangent relationship nor congruent radius of curvature, the joint has G0 continuity.

Preferably, the original surfaces have to intersect each other. However, if they are not intersecting, the gap between them must be narrower than the bevelled edge. While chamfering, some of the original surfaces are trimmed away to have the chamfer surface fitted in between. As there are four possible chamfered surfaces on any given pair of intersecting surfaces, the location at which you select the original surface determines the location of the chamfered surface. Perform the following steps:

1 Select File > Open and select the file ChamferFillet.3dm from the Chapter 7 folder on the companion CD.

2. Select Surface > Chamfer Surface, or click on the Chamfer Surface button on the Surface Tools toolbar.
3. Select the Distances option from the command area.
4. Type 5 to set the first chamfer distance. The first chamfer distance applies to the first selected surface.
5. Type 4 to set the second chamfer distance.

By default, the second chamfer distance is set to be equal to the first chamfer distance. Therefore, you can simply press the ENTER key to accept if both chamfer distances are the same.

6. Select surfaces A and B (shown in Figure 7–28). A chamfer surface is constructed.

Remember, the location where you select the surfaces affects the outcome and, if there is a gap between the surfaces to be chamfered and the chamfer distance is too small in comparison to the gap's width, chamfer may not be successful.

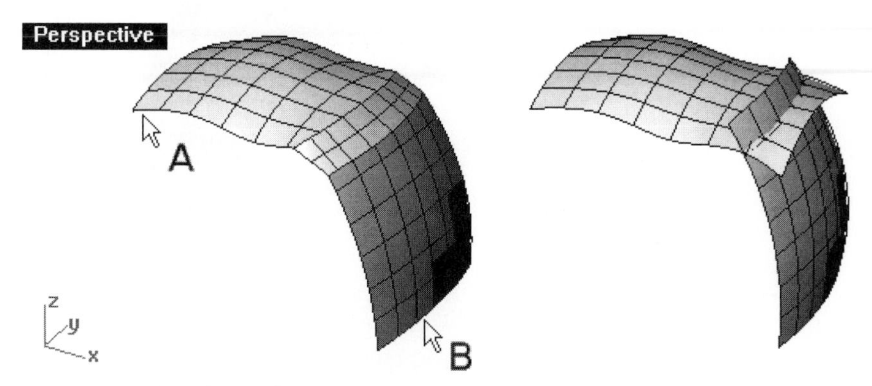

Figure 7–28. Chamfer surface constructed

Constructing a Fillet Surface

A filleted surface is has an arc-shaped cross-section. You may consider a filleted surface as being derived by rolling a ball between two surfaces. Because the joints between the added fillet surface and the original surfaces only have tangent relationship without having the same radius of curvature, filleted surface and contiguous surfaces have G1 continuity.

Like chamfering, the surfaces preferably have to intersect each other. If the surfaces in question are not intersecting, the widest gap between them must not be so wide that the rolling ball falls away from the surfaces. While filleting, you choose to trim the surfaces so that the filleted surface and the trimmed surfaces form a continuous surface profile or keep the original surfaces intact.

To construct a filleted surface, you specify the radius of the fillet and select two nonparallel surfaces. Like chamfering, any pair of intersecting surfaces has four possible filleted surfaces. Continue with the following steps:

7. Select Surface > Fillet Surface, or click on the Fillet Surface button on the Surface Tools toolbar.
8. Select the Radius option from the command area.
9. Type 7 at the command area.
10. Select surfaces A and B (shown in Figure 7–29). A filleted surface is constructed.
11. Do not save your file.

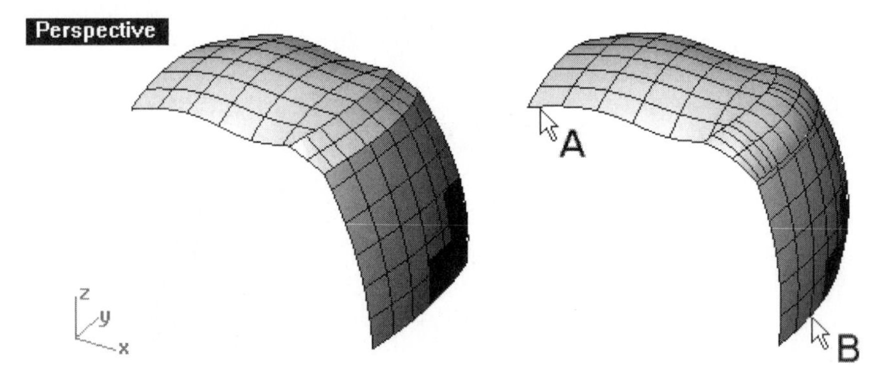

Figure 7–29. Fillet surface constructed

Constructing a Blended Surface

A blended surface produces a smooth transition between the two original surfaces, and the junction between the blended surface and contiguous surfaces has G2, G3, or G4 continuity. Perform the following steps.

1. Select File > Open and select the file Blend.3dm from the Chapter 7 folder on the companion CD.
2. Select Surface > Blend Surface, or click on the Blend Surface button on the Surface Tools toolbar.

While constructing a blended surface, you can select multiple contiguous surface edges, adjust end bulge, and set blend surface height.

3 Select edge A, shown in Figure 7–30, and press the ENTER key. This is the first set of edges.
4 Select edge B, shown in Figure 7–30, and press the ENTER key. This is the second set of edges.
5 Press the ENTER key again to accept the default location of the curve seam.
6 Select the Continuity_1 option on the command area.
7 Select the G4 option on the command area.
8 Select the Continuity_2 option on the command area.
9 Select the G4 option on the command area.

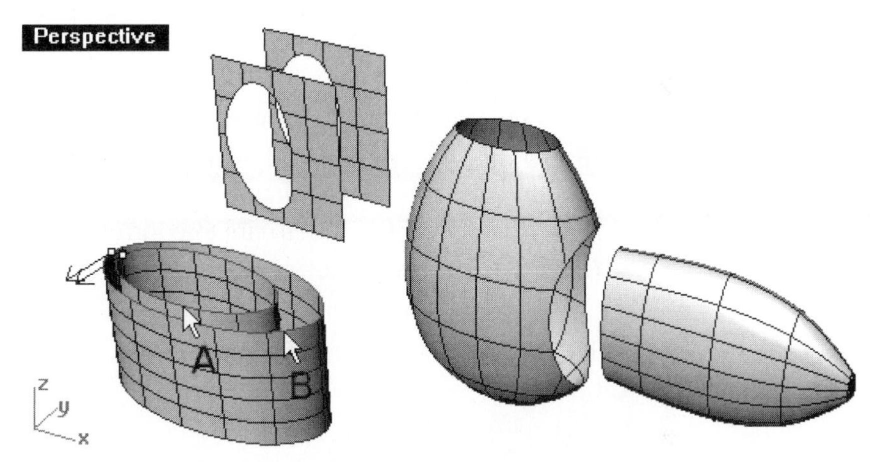

Figure 7–30. Surface edges selected and curve seam displayed

You can set continuity to G0, G1, G2, G3, or G4. If the continuity selected is set to G0, you produce a chamfered surface instead of a blended surface.

10 As necessary, click on the control points and drag them to change the shape of the blended surface.
11 In the Adjust Blend Bulge dialog box, click on the Preview button. A preview of the blend is displayed. (See Figure 7–31.)
12 Click on the Same height shapes button, and click on the OK button. A blend surface is constructed.

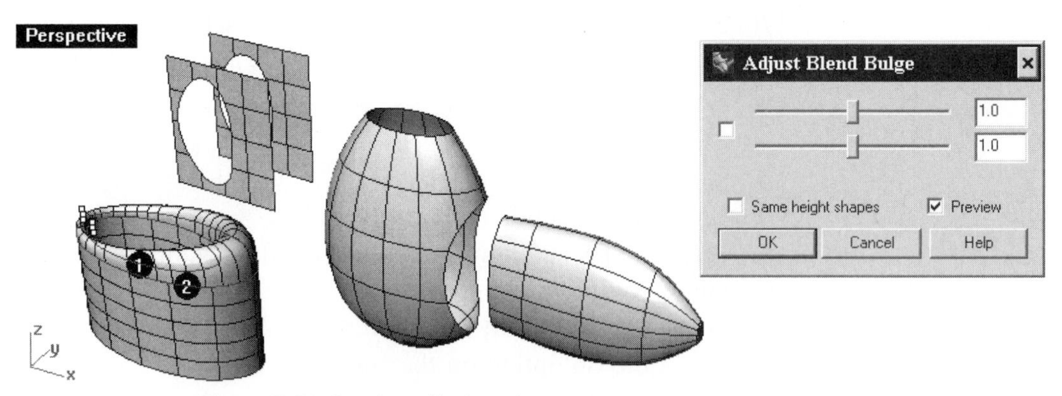

Figure 7–31. Preview displayed

13 Repeat the command.

14 Select surface edges A and B, shown in Figure 7–32, and press the ENTER key.

15 Press the ENTER key to accept the default seam location.

16 In the Adjust Blend Bulge dialog box, shown in Figure 7–32, click on the Same height shapes button and button A.

17 Drag one of the slider bars to control movements of both.

18 Click on the OK button. Another blend surface is constructed.

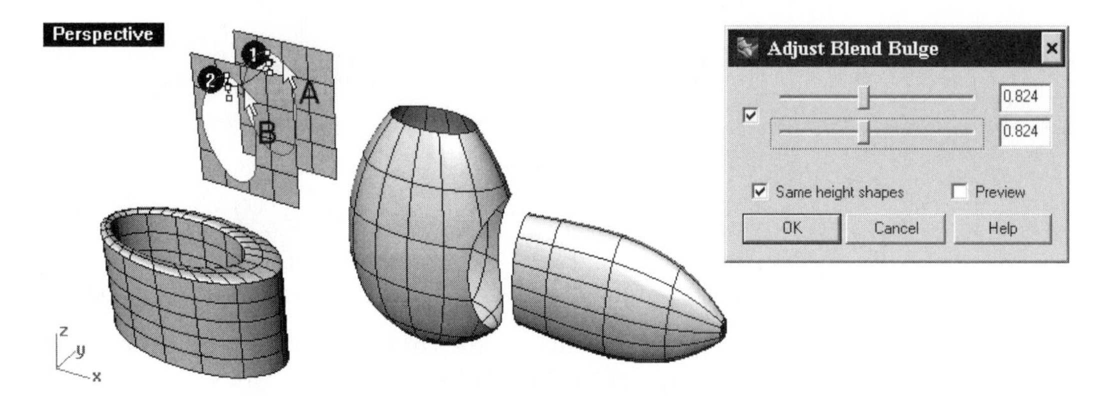

Figure 7–32. A blended surface with same height shape constructed and two surface edges selected

In the following, you are going to construct a blended surface to bridge two sets of edges.

19 Repeat the command again.

20 If AutoChain = No, select it on the command area to change it to Yes. Otherwise, proceed to the next step.

21 Select the ChainContinuity option on the command area.

22 Select the Tangency option on the command area.

23 Select edge A (Figure 7–33) and press the ENTER key. This is the first set of edges.

24 Select edge B (Figure 7–33) and press the ENTER key. This is the second set of edges.

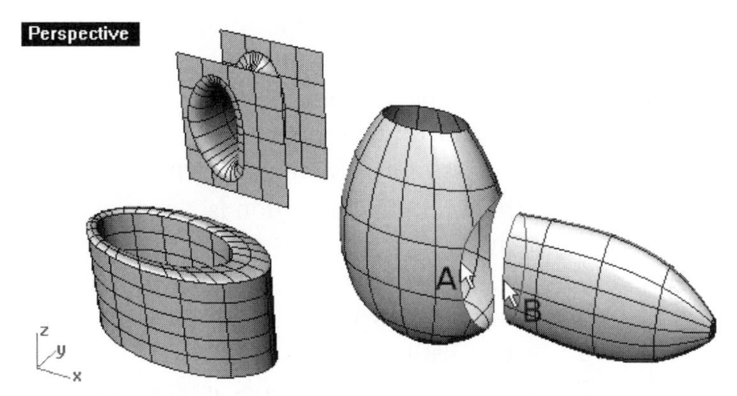

Figure 7–33. Second blend surface constructed and edges selected

Because there is a seam at the openings of the two original surfaces, each surface has two edges to blend.

25 In the Adjust Blend Bulge dialog box, click on the OK button. A blend surface is constructed. (See Figure 7–34.)

26 Do not save your file.

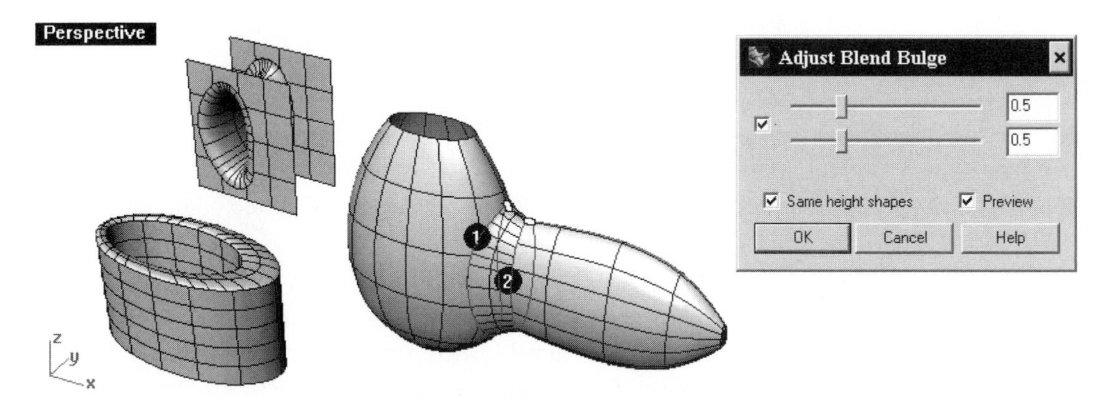

Figure 7–34. Blend surfaces constructed

Variable Chamfer, Fillet, and Blended Surfaces

If you have two intersecting surfaces, you can construct a variable distance chamfer, variable radius fillet, or variable curvature blended surface between them and appropriately trim the original surfaces. Perform the following steps:

1. Select File > Open and select the file Variable.3dm from the Chapter 7 folder on the companion CD.
2. Select Surface > Variable Fillet/Blend/Chamfer > Variable Chamfer Surfaces, or click on the Variable Radius Surface Chamfer button on the Surface Tools toolbar.
3. Select surfaces A and B, shown in Figure 7–35.
4. If the TrimAndJoin option is No, select it from the command area to change it to Yes. This way, the original surfaces will be trimmed and joined to the chamfer surface.
5. Select RailType option from the command area.
6. Select DistBetweenRails option.
7. Select handle C and type 6 at the command area to set the chamfer distance to 6 mm.
8. Select handle D and type 8 at the command area to set the chamfer distance to 8 mm.
9. Press the ENTER key. A variable chamfer is constructed.

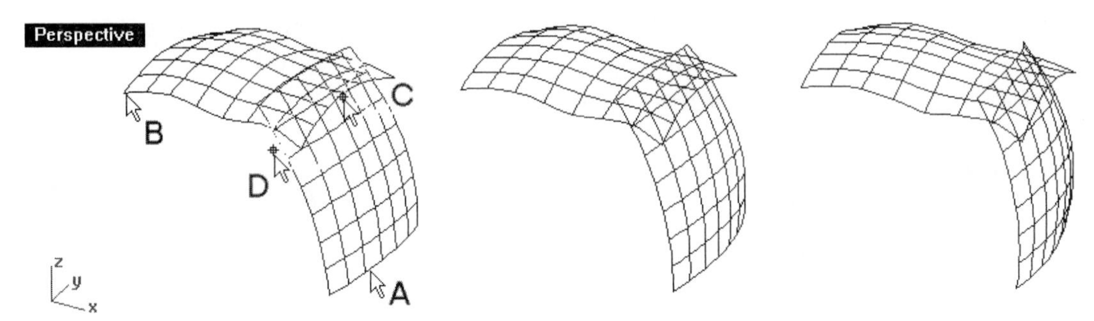

Figure 7–35. Variable chamfer surface being constructed

10. Select Surface > Variable Fillet/Blend/Chamfer > Variable Fillet Surfaces, or click on the Variable Radius Surface Fillet/Variable Radius Surface Blend button on the Surface Tools toolbar.
11. Select surfaces A and B, shown in Figure 7–36.
12. If the TrimAndJoin option is No, select it from the command area to change it to Yes.

13 Select RailType option from the command area.

14 Select RollingBall option.

15 Select handle C and type 5 at the command area to set the fillet distance to 5 mm.

16 Select handle D and type 7 at the command area to set the fillet distance to 7 mm.

17 Press the ENTER key. A variable fillet is constructed.

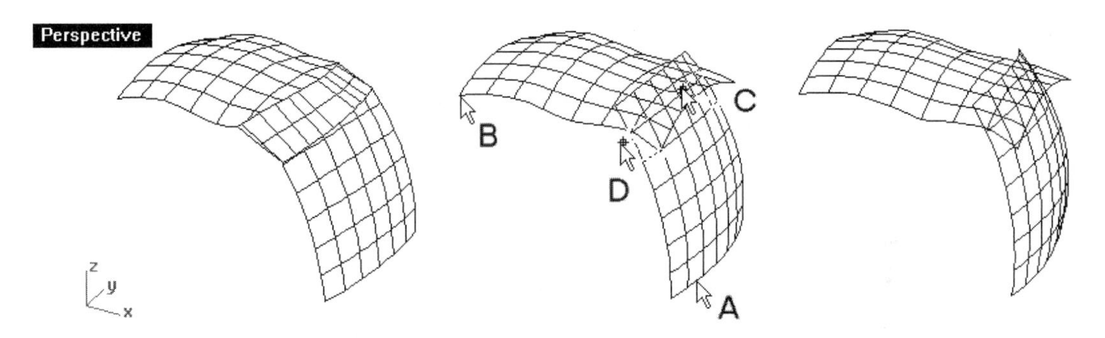

Figure 7–36. Variable chamfer surface constructed (left) and variable fillet surface being constructed (middle)

18 Select Surface > Variable Fillet/Blend/Chamfer > Variable Blend Surfaces, or right-click on the Variable Radius Surface Fillet/ Variable Radius Surface Blend button on the Surface Tools toolbar.

19 Select surfaces A and B, shown in Figure 7–37.

20 If the TrimAndJoin option is No, select it from the command area to change it to Yes.

21 Select DistFromEdge option from the command area.

22 Select RollingBall option.

23 Select handle C and type 7 at the command area to set the fillet distance to 7 mm.

24 Select handle D and type 9 at the command area to set the fillet distance to 9 mm.

25 Press the ENTER key. A variable blend is constructed. (See Figure 7–38.)

26 Do not save your file.

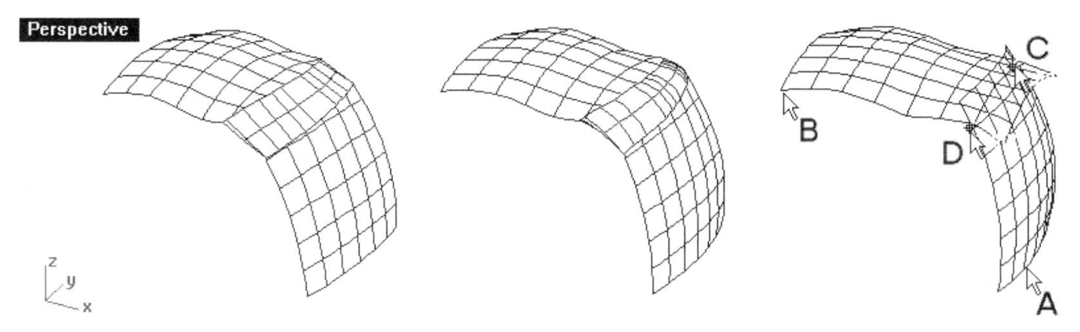

Figure 7–37. Variable radius fillet surface constructed (middle) and variable blend surface being constructed (right)

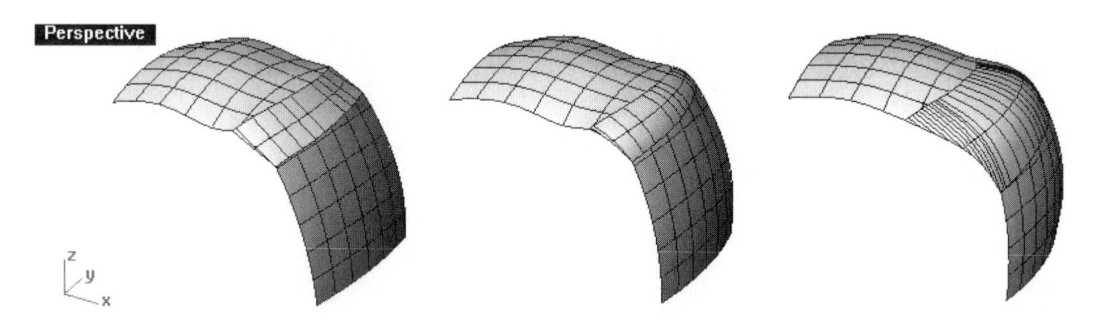

Figure 7–38. Rendered image of the variable chamfer, fillet, and blend surfaces

Connecting Two Surfaces

Connecting concerns extending or trimming two surfaces for them to meet at a sharp edge, as follows:

1. Select File > Open and select the file ConnectSrf.3dm from the Chapter 7 folder on the companion CD.
2. Select Surface > Connect Surfaces, or click on the Connect Surfaces on the Surface Tools toolbar.
3. Select surfaces A and B, shown in Figure 7–39.
4. Because the surfaces are separated, you have to select edges A and B. The surfaces are extended to meet.
5. Repeat the command.
6. Select surfaces C and D, shown in Figure 7–39. Because the surfaces are intersecting each other, they are trimmed and meet at a sharp edge.
7. Do not save your file.

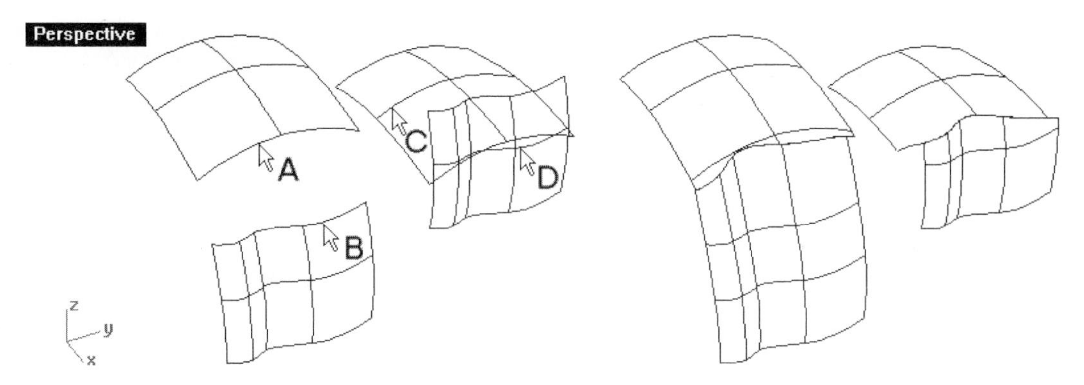

Figure 7–39. Surfaces to be connected (left) and connected surfaces (right)

Matching an Untrimmed Surface Edge with Another Surface Edge

Matching concerns extending an untrimmed edge of a surface for it to match the edge (either trimmed or untrimmed) of another surface to yield one of the three types of continuity (G0, G1, or G2) between the surfaces. Perform the following steps:

1. Select File > Open and select the file MatchSurface.3dm from the Chapter 7 folder on the companion CD.
2. Select Surface > Surface Edit Tools > Match, or click on the Match Surface button on the Surface Tools toolbar.
3. Select edge A, (shown in Figure 7–40, as the surface edge to be modified. Edges to be modified must be untrimmed edges.
4. Select edge B, (shown in Figure 7–40, as the target surface. Note that where you pick the surfaces is important in terms of eliminating twisted surfaces.
5. In the Match Surface dialog box, check the Curvature, Average surfaces, and Preserve opposite end check boxes and then click on the OK button. A surface's edge is changed to match the other, as shown in Figure 7–40.

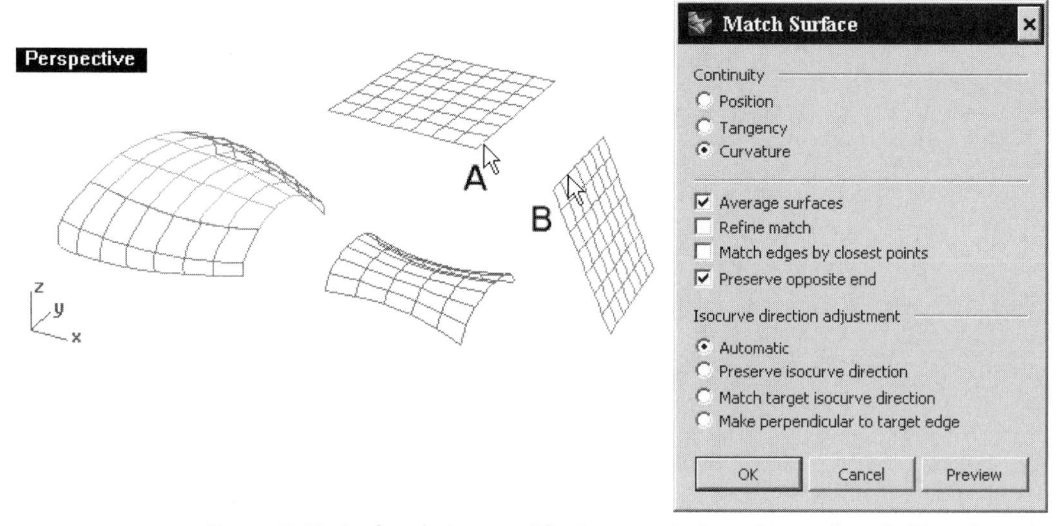

Figure 7–40. Surface being modified to match the other surface (left) and matched surface (right)

6 Repeat the command.

7 Select the MultipleMatches option on the command area.

8 Select edge A, shown in Figure 7–4.

9 Select the ChainEdges option on the command area.

10 Select edges B, C, and D, shown in Figure 7–41, and press the ENTER key.

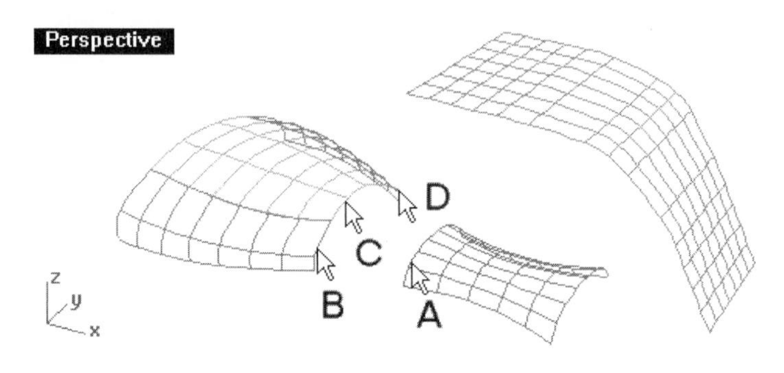

Figure 7–41. A planar surface matched to another planar surface and multiple edges selected for matching

11 In the Match Surface dialog box, the Average surfaces button is grayed out because there are multiple surfaces to be matched. Check the Curvature and Match edges by closest points buttons, and click on the OK button. A match surface is constructed. (See Figure 7–42.)

12 Do not save your file.

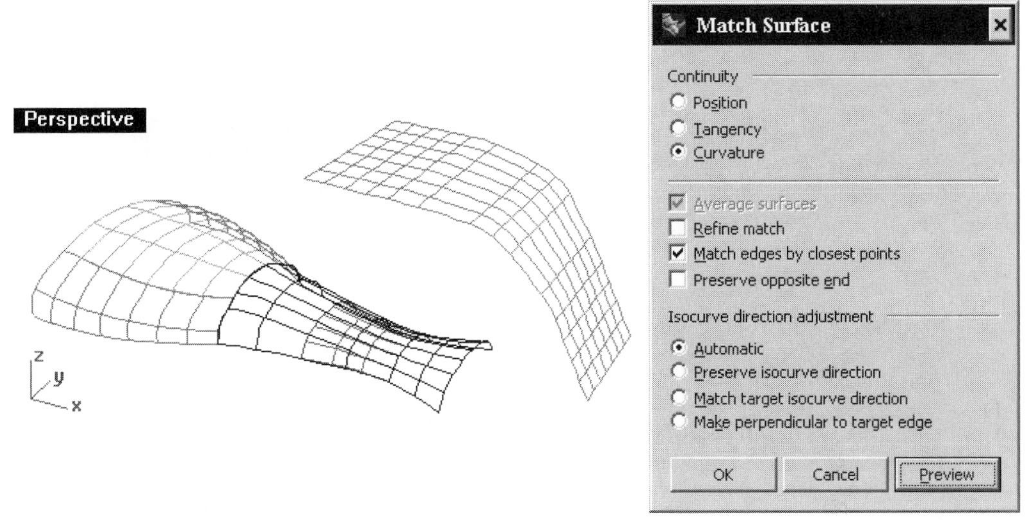

Figure 7–42. Match surface being constructed

Adjusting a Surface's End Bulge

If you have two contiguous surfaces and want to modify the shape of one of the surfaces without altering the continuity condition between the surfaces, you adjust the end bulge. Perform the following steps:

1. Select File > Open and select the file Endbulge.3dm from the Chapter 7 folder on the companion CD.
2. Select Surface > Surface Edit Tools > Adjust End Bulge, or click on the Adjust Surface End Bulge button on the Surface Tools toolbar.
3. Select surface edge A, shown in Figure 7–43. Because there are two contiguous surfaces sharing one common edge, select the surface edge in red.
4. Click on point B to specify the point to edit.
5. Click on points C and D to specify the start and end of region to edit.

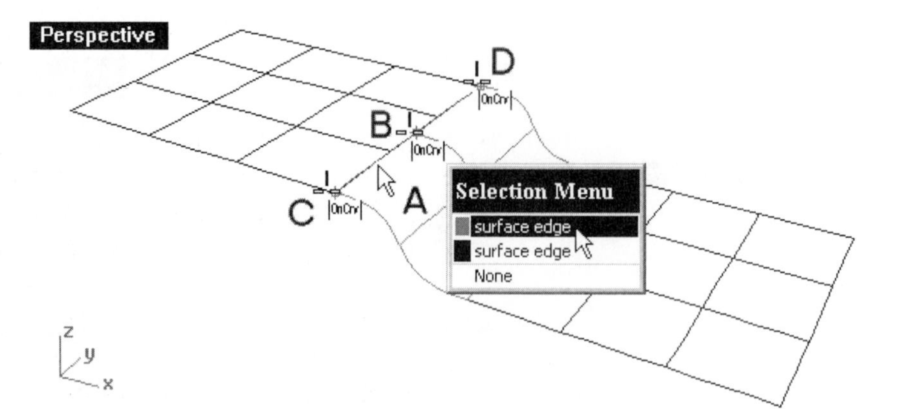

Figure 7–43. Surface edge, edit point, and edit region selected

6 Drag points A and B, shown in Figure 7–44, to modify the shape of the surface.

7 Press the ENTER key when done. The surface is modified.

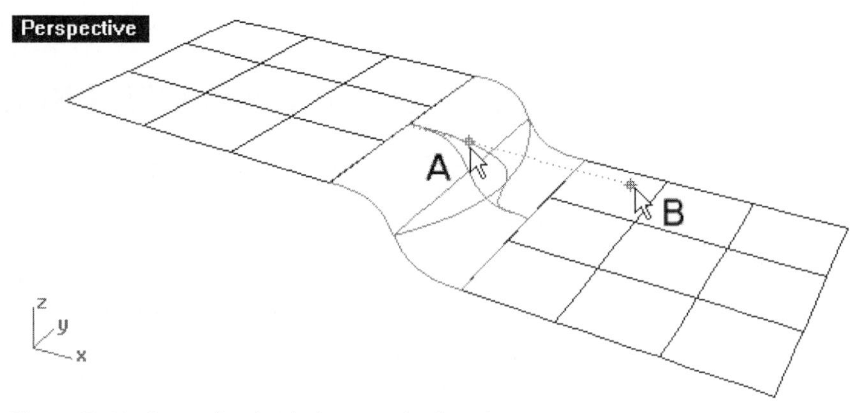

Figure 7–44. Control point being manipulated

Joining Contiguous Surfaces to Form a Polysurface

Joining two or more contiguous surfaces, you get a polysurface. It must be emphasized that the surfaces will not join unless they meet edge to edge within tolerance. Continue with the following steps.

8 Select Edit > Join, or click on the Join button on the Main1 toolbar.

9 Select the surface, shown in Figure 7–44, and press the ENTER key. The surfaces are joined to become a polysurface.

Note in the command area the report stating the result of joining.

Exploding Polysurfaces to Individual Surfaces

Contrary to joining, exploding a polysurface renders a set of individual surfaces. Continue with the following steps to explode the polysurface to revert it into a set of individual surfaces.

10 Select Edit > Explode, or click on the Explode/Extract Surfaces button on the Main2 toolbar.
11 Select the polysurface that you just constructed and press the ENTER key. The polysurface is exploded.
12 Do not save your file.

In Chapter 8, you will learn how to extract individual surfaces from a polysurface without exploding it.

Merging Contiguous Untrimmed Surfaces to Form a Single Surface

Contiguous surfaces sharing a common untrimmed edge can be merged to become a single surface. Note that unlike joining, merging results in a single surface, not a polysurface. Therefore, you cannot explode it into the surfaces you merged. Perform the following steps:

1 Select File > Open and select the file MergeSurface.3dm from the Chapter 7 folder on the companion CD.
2 Select Surface > Surface Edit Tools > Merge, or click on the Merge Surfaces button on the Surface Tools toolbar.
3 Select surfaces A and B, shown in Figure 7–45. The surfaces are merged into one (Figure 7–45). If the surfaces failed to merge, repeat the command, select Tolerance option from the command area, specify a larger joining tolerance value, and select the surfaces again.
4 Repeat the command.
5 Select B and C (Figure 7–45). The surfaces are merged.
6 Do not save your file.

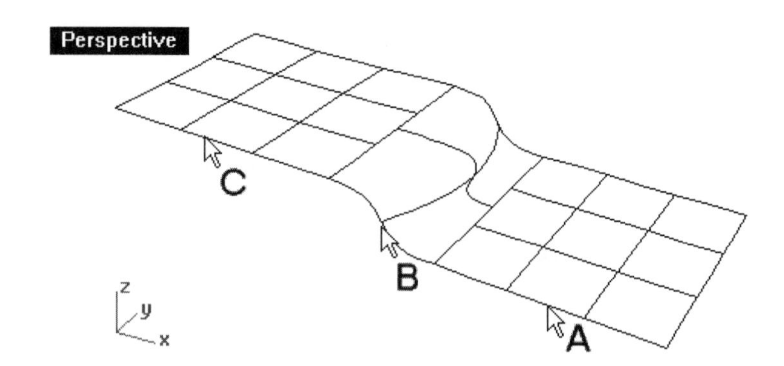

Figure 7–45. Two surfaces being merged at their untrimmed edges

Surface Profile Manipulation

This section deals with how the normal direction of a surface can be flipped, ways to modify the profile of a surface, rebuilding and reparameterizing a surface, and changing its degree of polynomial.

Flip Normal Direction

A surface has no thickness. To represent a 3D object in a computer, you need a number of surfaces. For the computerized downstream manufacturing operations to recognize which side of the surface represents a void and which side represents a volume, a normal vector is used. The direction of normal is determined by the direction of the curves, the curve patterns, and also the sequence of selection when you construct a surface from the curves. This sounds too complicated for us to memorize. Hence, you may simply disregard the normal direction when you first construct the surface and flip the normal in a later stage.

In Rhinoceros, the normal direction of a surface is depicted by a line normal to and at the corner of the surface. Perform the following steps:

1. Select File > Open and select the file Normal.3dm from the Chapter 7 folder on the companion CD.
2. Select Analyze > Direction, or click on Analyze Direction/Flip Direction button on the Main1 toolbar.
3. Select surface A, shown in Figure 7–46, and press the ENTER key.
4. Select the Flip option from the command area to flip the normal direction.
5. Press the ENTER key to terminate the command.
6. Do not save your file.

The U and V isocurves on a surface also have directions. Together with normal, a surface has three directions. Using this command, you can reverse the U and V directions, swap the U with V, and change the normal direction.

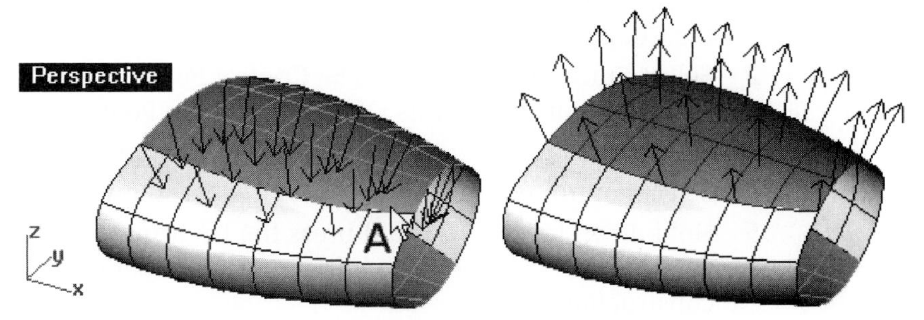

Figure 7–46. Normal directions displayed (left) and flipped (right)

(***NOTE:*** *The color of the normal directions can be changed by modifying the tracking color in the Colors tab of the Rhino Options dialog box.)*

Editing Surface Profile

To modify the shape of a surface, you can manipulate the locations of its control points by using the nudge keys, using the MoveUVN dialog box, or using the organic toolbar; adjusting a control point's weight; using the handlebar; adding or removing knots; adding or removing control points; and soft editing the surface.

Control Point Manipulation

One method of modifying the shape of a surface is to edit its control points. There are four ways to move a control point, as follows:

- Select the control point, hold down the mouse button, and drag it to a new position.
- Select the control point and use the nudge keys. (By default, the nudge keys are the Alt key plus the arrow keys.)
- Use the Move UVN command.
- Use any transform command.

Using the Nudge Keys

By holding down the ALT key, you can press the arrow key to move selected objects vertically or horizontally. To modify a control point of a surface using the nudge keys, perform the following steps:

1. Select File > Open and select the file SurfaceKnotPoint.3dm from the Chapter 7 folder on the companion CD.
2. Select Edit > Control Points > Control Points On, or click on the Control Points On/Points Off button on the Point Editing toolbar.
3. Select surface A, (shown in Figure 7–47, and press the ENTER key.
4. Select control point B (Figure 7–47) in the Top viewport.
5. Hold down the Alt key, and press the Up arrow key. The selected control point is moved in a horizontal plane parallel to the current construction plane.

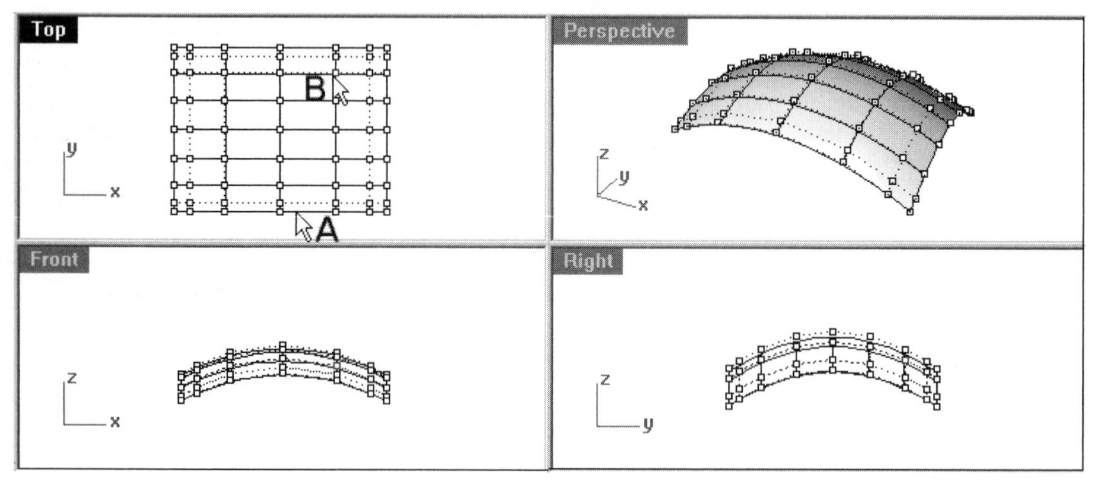

Figure 7–47. Control points of a surface being modified

Move Along U, V, and N Directions

As explained previously, U, V, and N are three directions on a surface. The U and V depict the two isocurve direction,s and the N depicts the normal direction. Continue with the following to use the MoveUVN command to move the control point's location in these three directions.

6. Maximize the Perspective viewport.
7. Select Transform > Move UVN, or click on the MoveUVN/Turn Move UVN Off button on the Point Editing toolbar.
8. Select control point A, (shown in Figure 7–48).

9 In the Move UVN dialog box, drag the N slider bar to move the control point in a direction normal to the surface, or drag the U and V bars to move the control point along the U and V directions of the surface.

10 Click on the X icon of the Move UVN dialog box to close it. *(Note: You may leave the dialog box on all the time.)*

Using the Organic Toolbar

To manipulate objects and, in particular, control points of a surface along the X, Y, and Z directions, you can use the buttons on the Organic Toolbar. Continue with the following steps.

11 Click on the Red, Green, or Black Arrow button on the Organic toolbar to move the control in X, Y, and Z direction. (See Figure 7–48.)

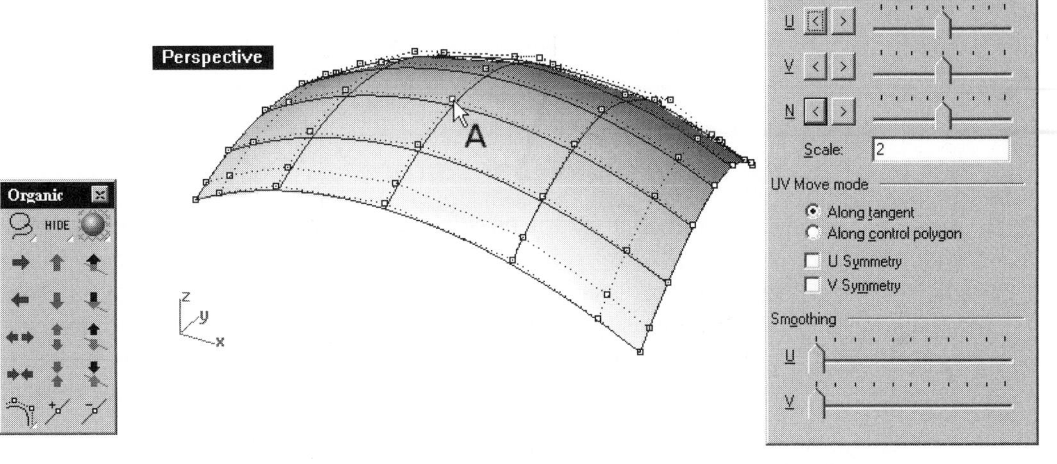

Figure 7–48. Control point being moved along U, V, and N directions

Adjusting Control Point Weight

Like the control points of a spline, the higher the weight, the closer the surface will be pulled to the control point. Continue with the following to change the control point's weight.

12 Select Edit > Control Points > Edit Weight, or click on the Edit Control Point Weight button on the Point Editing toolbar.

13 Select control point A (shown in Figure 7–49) and press the ENTER key.

14 In the Set Control Point Weight dialog box, drag and move the slider bar to increase or decrease the weight of the selected control point.
15 Click on the OK button. The surface is modified.
16 Do not save your file.

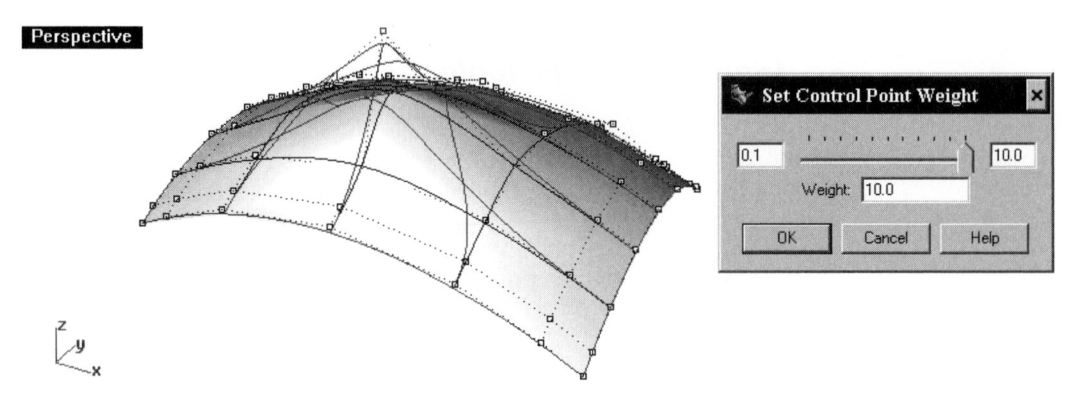

Figure 7–49. Control point's weight being modified

Using the Handlebar Editor

The Handlebar Editor enables you to modify the shape of a surface by adjusting the tangency of selected locations on the surface. To facilitate selection of specific locations on a surface, you may increase the isocurve density. To reiterate, isocurves are aids to help you visualize and select a surface. Changing the density does not affect the number of control points and the degree of polynomial. To edit a surface using the handlebar, perform the following steps:

1 Open the file SurfaceKnotPoint.3dm from the Chapter 7 folder on the companion CD again.
2 Select Edit > Control Points > Handlebar Editor, or click on the Handlebar Editor button on the Point Editing toolbar.
3 Select surface A and then location B (shown in Figure 7–50).
4 Select and drag one of the five handles of the handlebar to modify the shape of the surface at the selected location, as shown in Figure 7–50. Press the ENTER key when finished.
5 Do not save your file.

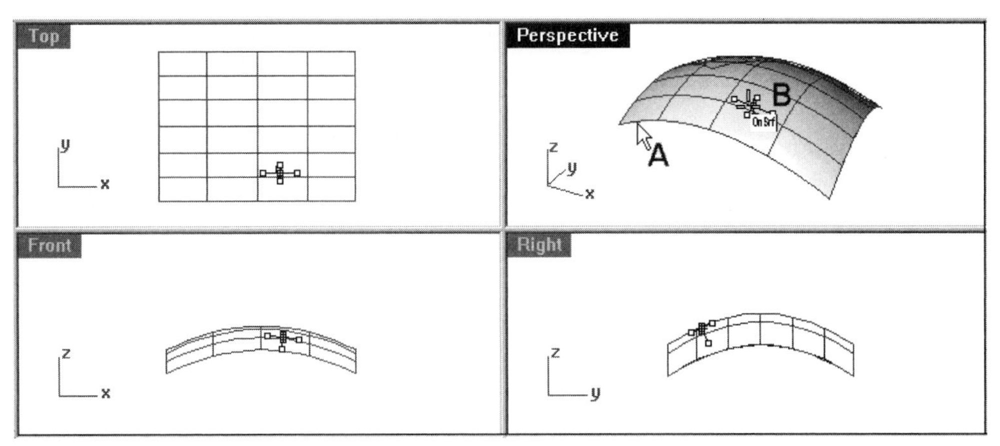

Figure 7–50. Handlebar editor being activated at selected location of a surface

Adding and Removing Knots

The number of knot points of a surface has a significant effect on how the surface will change in shape if a control point is moved. Adding knots increases the number of control points. The surface profile will not change until the next time you manipulate its control points. However, removal of a knot point simplifies the surface andcauses the surface profile to change. Perform the following steps to add knots to a surface:

1. Open the file SurfaceKnotPoint.3dm from the Chapter 7 folder on the companion CD again.
2. Select Edit > Control Points > Control Points On, or click on the Control Points On/Points Off button on the Point Editing toolbar.
3. Select surface A (shown in Figure 7–51) and press the ENTER key.
4. Select Edit > Control Points > Insert Knot, or click on the Insert Knot/Insert Edit Point button on the Point Editing toolbar.
5. Select surface A, shown in Figure 7–51.
6. Select the Direction option on the command area.
7. Select the Both option. A knot will be added in both U and V directions.
8. Select a location on the surface to add a knot point.
9. Continue to add as many knot points as may be required. Press the ENTER key when finished. The number of control points is increased accordingly.

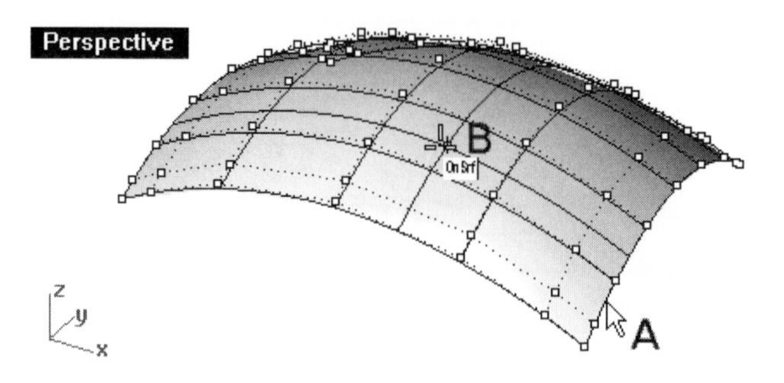

Figure 7–51. Knot point being added to a surface

To remove a knot from the surface, continue with the following steps.

10 Select Edit > Control Points > Remove Knot, or click on the Remove Knot button on the Point Editing toolbar.

11 Select surface A (shown in Figure 7–52).

12 If Direction = V, select it to change to Direction = U. Otherwise, proceed to the next step.

13 Select location B and press the ENTER key when finished. The selected knot in the U direction is removed.

14 Do not save your file.

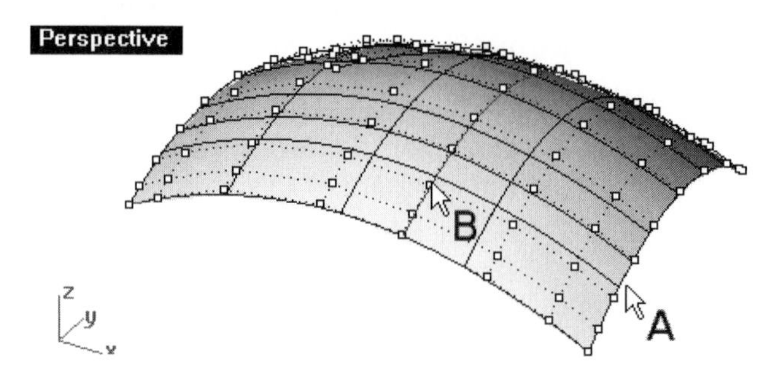

Figure 7–52. Knot point being removed

Control Point Insertion and Removal

Apart from inserting and removing knots, you can insert and remove control points of a surface. The surface can be a trimmed surface or an untrimmed surface. Perform the following steps:

1 Open the file SurfaceControlPoint.3dm from the Chapter 7 folder on the companion CD again.

2 Click on the Insert a control point button on the Surface Tools toolbar.

3. Select surface A, shown in Figure 7–53.
4. If Direction = U, select it on the command area to change to = V. Otherwise, proceed to the next step.
5. Click on location B. A control point is added in the V direction.

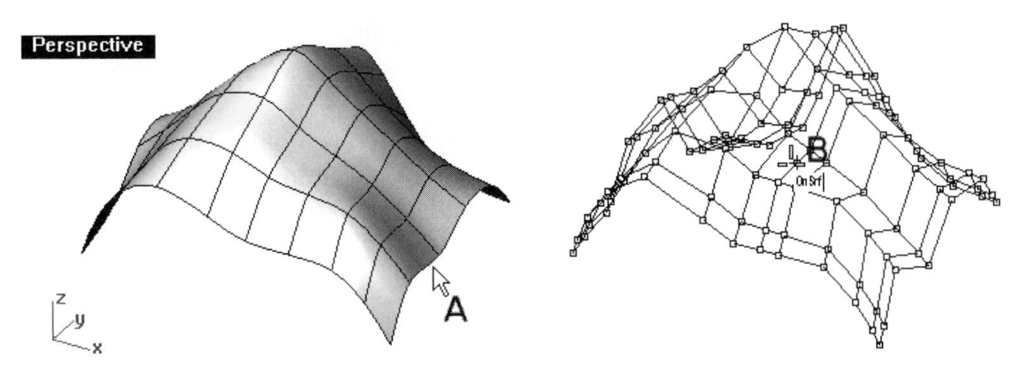

Figure 7–53. Surface being selected (left) and a control point being added (right)

6. Click on the Remove a control point button on the Surface Tools toolbar.
7. Select surface A, shown in Figure 7–54.
8. If Direction = U, select it on the command area to change to = V. Otherwise, proceed to the next step.
9. Click on location B. A control point in the V direction is removed.
10. Do not save your file.

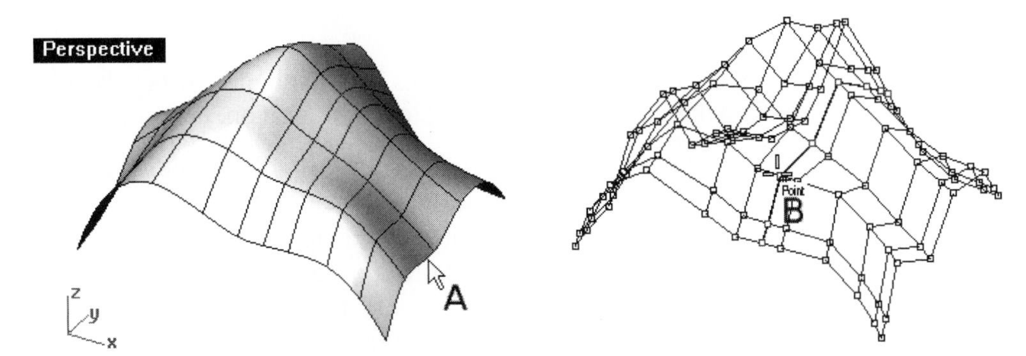

Figure 7–54. Surface being selected (left) and a control point being removed (right)

Hiding Control Points and Control Polygon

Control points and control polygon of surfaces that are behind other objects can be hidden or displayed, as follows:

1. Open the file CullControlPolygon.3dm from the Chapter 7 folder on the companion CD again.
2. Turn on the control points of the surfaces, as shown in Figure 7–55 (left).
3. Click on the Cull Control Polygon Backfaces button on the Point Editing toolbar, or type CullControlPolygon at the command area. Control points and control polygon behind other objects are hidden. See Figure 7–55 (right).
4. Do not save your file.

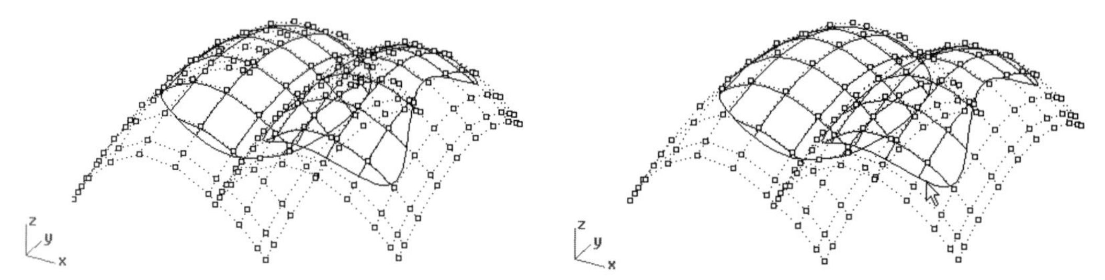

Figure 7–55. Control points and control polygon shown (left) and culled (right)

This command also works on polygon meshes that will be explained in Chapter 9.

Soft Editing a Surface

Soft editing a surface changes the shape of a surface by picking a point on the surface and moving it to a new location. Perform the following steps:

1. Select File > Open and select the file SoftEditSrf.3dm from the Chapter 7 folder on the companion CD.
2. Select Surface > Surface Edit Tools > Soft Edit, or right-click on the Soft Edit Curve/Soft Edit Surface button on the Move toolbar.
3. Select the U Distance and V Distance options one by one on the command area, and set their values to 5.
4. If FixEdges = No, select it on the command area to change it to Yes.
5. Select surface A, shown in Figure 7–56.
6. Select a point on the surface. Exact location is unimportant for this tutorial.

7 Hold down the Control key and click on location C.
8 With the Control key still held down, click on location D.
9 Press the ENTER key. The surface is soft edited. See Figure 7–57.
10 Do not save your file.

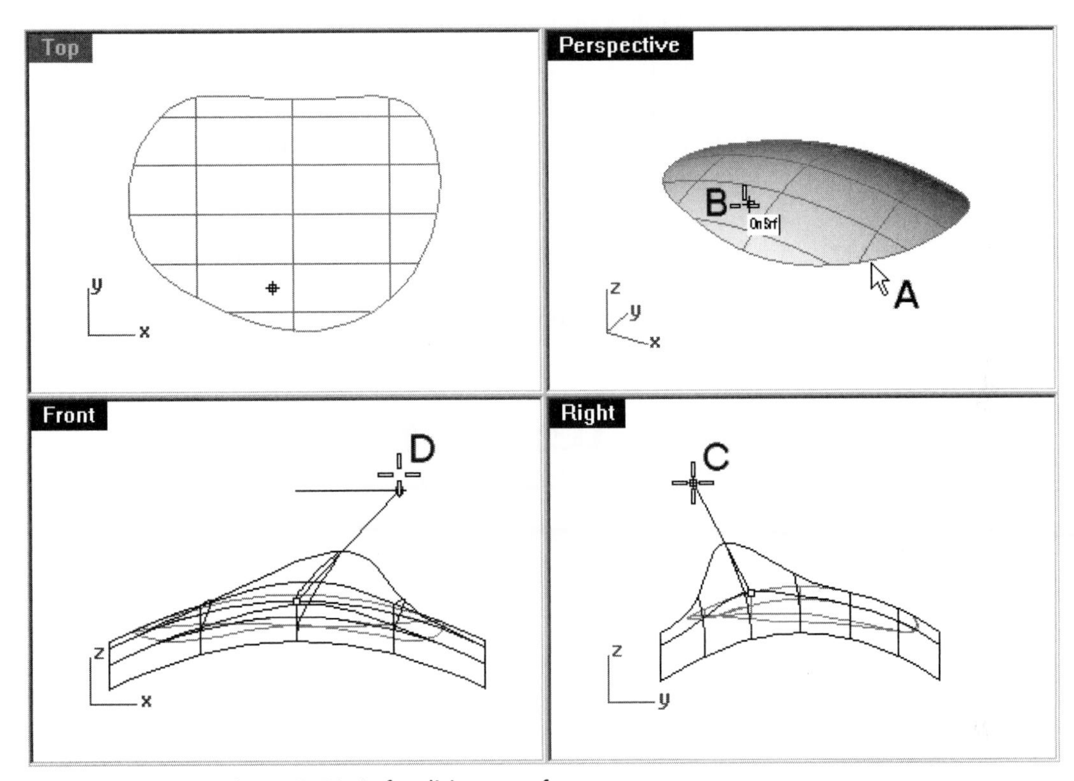

Figure 7–56. Soft editing a surface

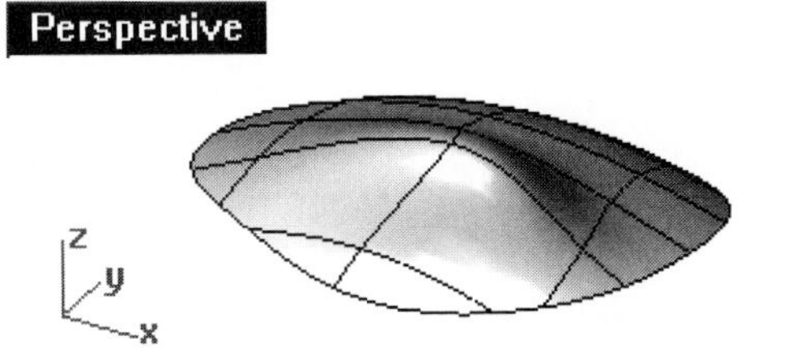

Figure 7–57. Surface's profile changed

Rebuilding

A way to improve the smoothness of a surface is to rebuild it. In rebuilding a surface, you can specify the number of control points and/or the degree of polynomials along the U and V directions of the surface. To rebuild a surface, perform the following steps:

1. Select File > Open and select the file SurfaceRebuild.3dm from the Chapter 7 folder on the companion CD.
2. Select Edit > Rebuild, or click on the Rebuild Surface button on the Surface Tools toolbar.
3. Select surface A (shown in Figure 7–58) and press the ENTER key.
4. In the Rebuild Surface dialog box, set the U and V point count to 6 and degree of polynomial to 4, and then click on the OK button. The surface is rebuilt, as shown in Figure 7–58.
5. Do not save your file.

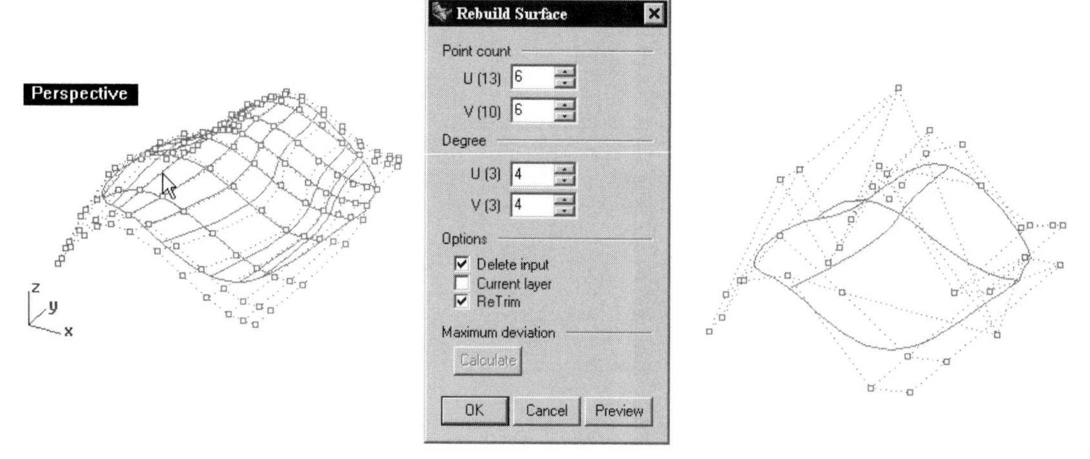

Figure 7–58. Surface being rebuilt (left) and rebuilt surface (right)

Reparameterizing Curves and Surfaces

Parameterization concerns recalculating the parameter space of a curve or a surface. This calculation process causes the parameter space of the objects to be approximately the same size as the object's original 3D geometry. Although there is not much change in the shape of the object, reparameterizing has two major advantages: texture mapping will become more accurate, and trimming can be done more properly. To reparameterize a surface, perform the following steps.

1. Open any file with a surface model to be reparameterized.
2. Type Reparameterize at the command area.
3. Select the surface and/or curve to be reparameterized, and press the ENTER key.
4. Select the Automatic option on the command area.

Changing Polynomial Degree

If you increase the degree of polynomial, the surface's shape will not change. However, if you reduce the degree of polynomial, the surface's shape is simplified. To change the polynomial degree of a surface along its U and V directions, perform the following steps:

1. Select File > Open and select the file SurfacePolynomial.3dm from the Chapter 7 folder on the companion CD.
2. Select Edit > Change Degree, or click on the Change Surface Degree button on the Surface Tools toolbar.
3. Select surface A (shown in Figure 7–59) and press the ENTER key.
4. Type 1 at the command area to set the degree of polynomial of the surface in the U direction.
5. Type 4 at the command area to set the degree of polynomial of the surface in the V direction. The surface's degree of polynomial is changed, as shown in Figure 7–59.
6. Do not save your file.

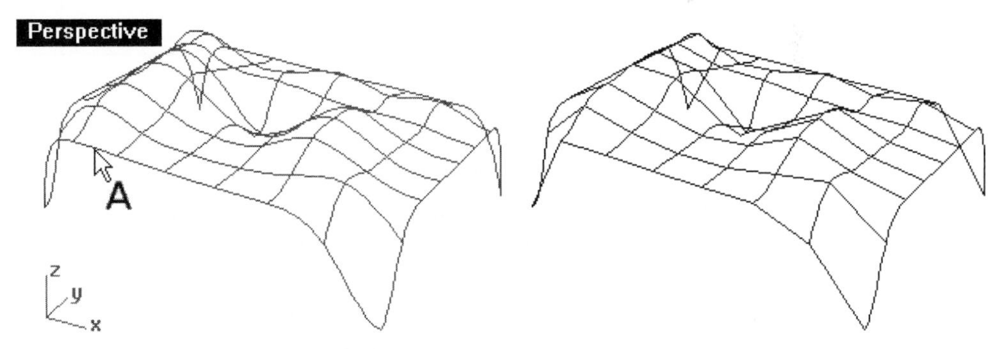

Figure 7–59. Degree of polynomial being changed (left) and degree of polynomial changed (right)

Surface Edge Manipulation

There are five ways to manipulate the edges of a surface. You can show selected edges to distinguish them from the isocurves. You can split an edge into two so that you can use a portion of the edge in some other operations. Contrary to splitting, you can merge split edges. You can join edges of two contiguous surfaces and you can rebuild an edge.

Show Edges

In the computer display, both surface edges and isocurves are displayed as curves. To see clearly the edges of a surface, you highlight them, as follows:

1. Select File > Open and select the file Edge.3dm from the Chapter 7 folder on the companion CD.
2. Select Analyze > Edge Tools > Show Edges, or click on the Show Edges/Edges Off button on the Edge Tools toolbar.
3. Select the surface A, shown in Figure 7–60, and press the ENTER key.

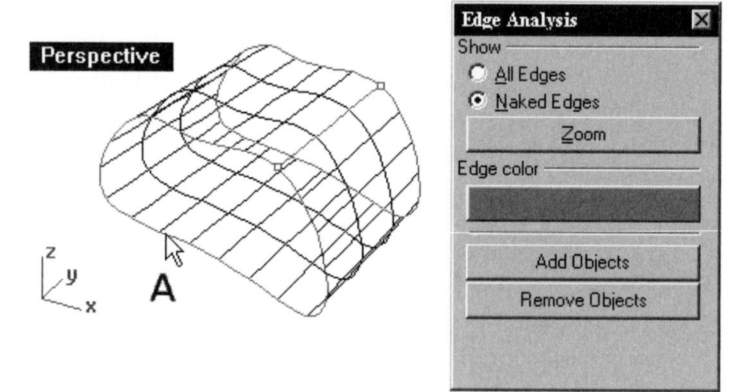

Figure 7–60. All edges highlight

4. In the Edge Analysis dialog box, click the All Edges box. This way, the seam edge is also highlighted.
5. If you wish to change the color of the highlighted edges, click on the Edge color button and select a color.
6. If you want to display edges of some other objects, click on the Add Objects button and select additional objects.
7. Click on the Remove object button and select objects to deselect them.
8. Click on Naked Edges in the Edge Analysis dialog box. The seam edge is not highlighted.

Zoom Naked Edge

To take a closer look at the naked edge, you can use the ZoomNaked command by clicking on the Zoom button of the Edge Analysis dialog box or typing ZoomNaked at the command area.

9 Close the Edge Analysis dialog box.

Splitting an Edge

By splitting an edge of a surface in two, you can make use of one of the edges to further construct surfaces. Continue with the following steps to split an edge.

10 Select Analyze > Edge Tools > Split Edge, or click on the Split Edge/Merge Edge button on the Edge Tools toolbar.

11 Select edge A (shown in Figure 7–61).

12 Select location B (shown in Figure 7–61). The selected edge is split.

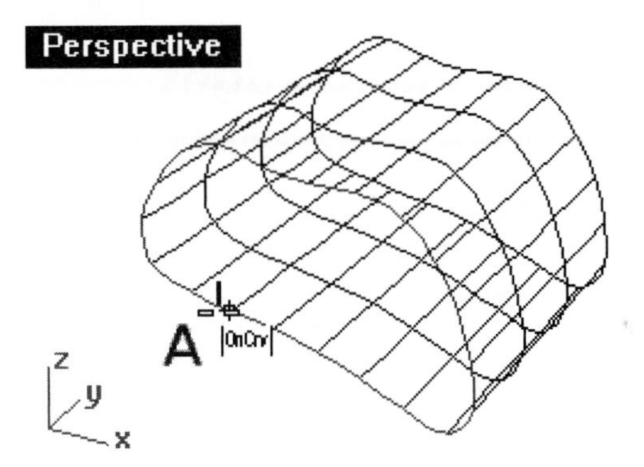

Figure 7–61. Edge split

To appreciate how a portion of an edge can be used, continue with the following steps.

13 Select Surface > Extrude > Straight, or click on the Extrude Straight button on the Extrude toolbar.

14 Select edge A (shown in Figure 7–62) and press the ENTER key.

15 Select location B (Figure 7–62) to indicate the extrusion height.

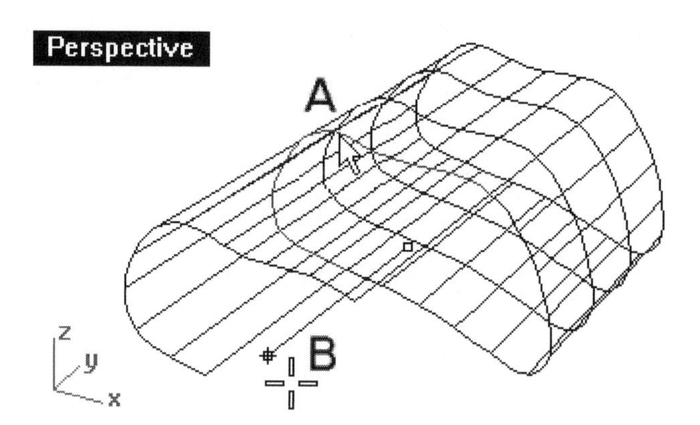

Figure 7–62. An edge being extruded

A split edge is extruded. To see how surfaces separated a small distance apart can be joined, you need to move the extruded surface, as follows.

16 Move the extruded surface A, as shown in Figure 7–63.

Merging Edges

The opposite of splitting an edge is merging split edges, as follows.

17 Select Analyze > Edge Tools > Merge Edge, or right-click on the Split Edge/Merge Edge button on the Edge Tools toolbar.

18 Select split edge B (shown in Figure 7–63).

19 Select all in the pop-up dialog box. The split edges are merged into a single edge.

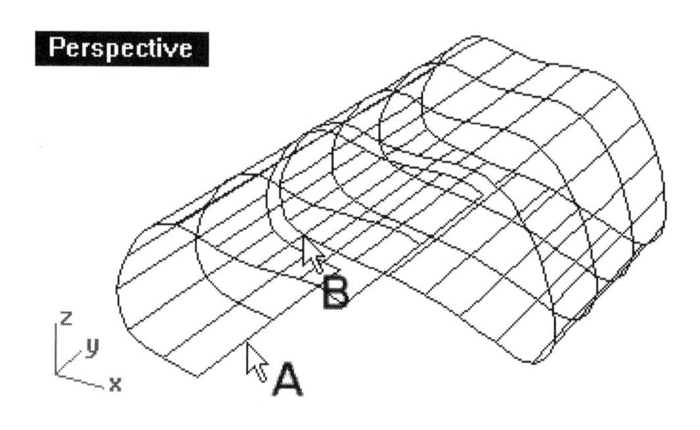

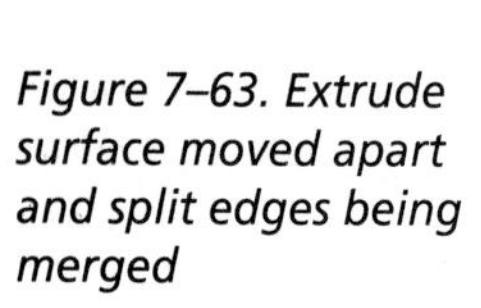

Figure 7–63. Extrude surface moved apart and split edges being merged

Joining Edges

Naked edges of two surfaces, although separated by a small distance, can be merged to become a single edge. Joining naked edges repairs any small gaps between contiguous surfaces. The display of the edges is forced closed. The geometry does not change. This command may work for models that are for rendering, but generally is not good enough for manufacturing. To join the naked edges of two surfaces, perform the following steps.

20 Select Analyze > Edge Tools > Join 2 Naked Edges, or click on the Join 2 Naked Edges button on the Edge Tools toolbar.

21 Select edges A and B (shown in Figure 7–64). A dialog box informing you of the tolerances between the edges is displayed.

22 In the Edge Joining dialog box, click on the OK button.

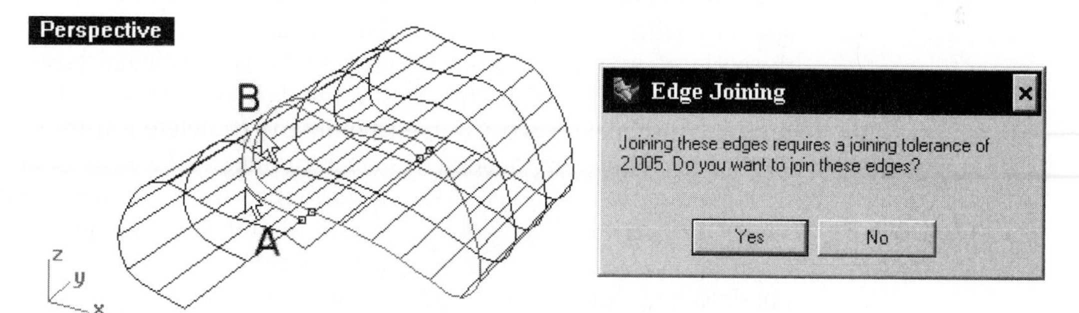

Figure 7–64. Edges being joined

Rebuilding Edges

To restore the original edge after you have joined edges, you explode the joined surface into individual surfaces and rebuild the edges, as follows.

23 Select Edit > Explode, or click on the Explode button on the Main2 toolbar.

24 Select the surface and press the ENTER key.

25 Select Analyze > Edge Tools > Rebuild Edges, or click on the Rebuild Edges button on the Edge Tools toolbar.

26 Select edge A (shown in Figure 7–65). Note that this command is also one of the first things to apply if your surface shows as a bad object.

27 Do not save your file.

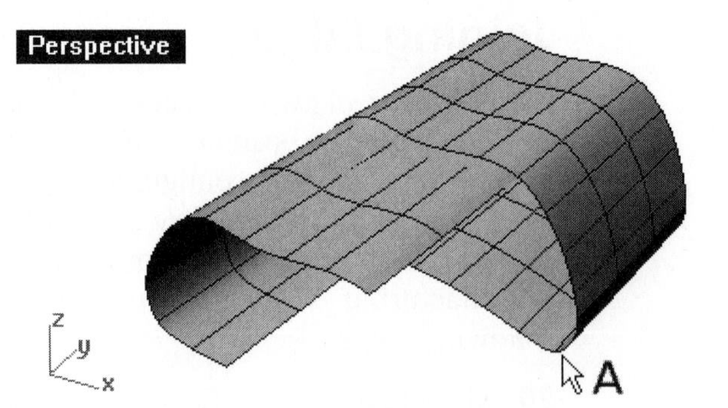

Figure 7–65. Polysurface exploded and edges being restored

Chapter Summary

Because smooth surfaces are constructed from smooth curves, smooth surfaces with an irregular boundary must be trimmed. To maintain the surface's profile, a trimmed surface retains its original base surface as well as the trimmed boundaries in the database. As a result, you can untrim a surface, detach trimmed boundary, or delete a trimmed edge. To reduce the memory needed to save a trimmed surface, you can shrink the base surface to reduce its size or further shrink to the trimmed edge. To obtain two surfaces with smooth continuity, you can split a surface by using a cutting object or by splitting it along its isocurves. Normally, possible kink lines in a surface automatically cause the surface to be broken into a number of surfaces. If a surface with kink lines is imported, you can break it into a set of surfaces along the creases. If a surface is too small in size, you can extend its untrimmed edges. Contrary to trimming, you can extend an untrimmed surface to increase its size, extend and form a periodic surface, and extend to a surface, splitting it and joining with it.

Apart from trimming and extending two consecutive surfaces to bridge them together, you can insert a constant or variable chamfer, fillet, or blended surface between them. These three kinds of surfaces, together with the source surfaces, are typical examples of five kinds of continuity: G0, G1, G2, G3, and G4. Another way of treating two separate surfaces is to modify one of them to match the other. To modify a surface without altering the tangency direction at its edges, you adjust its end bulge. Two or more contiguous surfaces connected at their edges can be joined to become a polysurface. If the connected edges are untrimmed, you can merge them to become a single surface. The reverse of joining surfaces to a polysurface is to explode it to become individual surfaces. Merged surfaces cannot be exploded.

Surface profiles can be edited in several ways. You can change the directions of a surface in terms of its U and V isocurves and its normal direction. To alter the shape of a surface, you can translate its control points in U, V, and N directions; adjust the weighting of the control points; use handlebar editor; add/remove knot points; and add/remove control points. Furthermore, you can rebuild a surface by specifying the point counts and degree of polynomial in the U and V directions. To refine a surface without changing its degree of polynomial and control point count, you reparameterize it.

To manipulate the edges of a surface, you display its edges, split an edge, merge a split edge, join edges, and rebuild edges.

Review Questions

1. Outline the ways to modify the boundaries of a surface.
2. How can you treat the junction between two separate surfaces?
3. Differentiate between joining and merging two surfaces.
4. List the ways you can refine or modify a surface's profile.
5. What are the ways you can manipulate the edge of a surface?

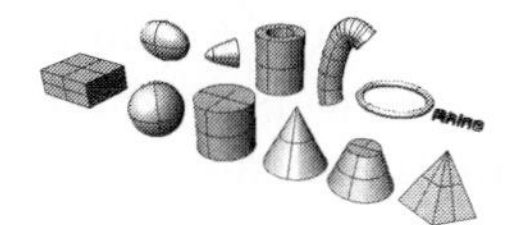

CHAPTER 8
Rhinoceros Polysurfaces and Solids

Introduction

This chapter details Rhino's polysurface and solid modeling methods.

Objectives

After studying this chapter you should be able to:

- ❒ Explain Rhino's solid modeling methods and relate them to its surface modeling tools
- ❒ Construct Rhinoceros solids of regular shapes and free-form shapes
- ❒ Combine two or more solid objects
- ❒ Detail and modify a solid object

Overview

As a NURBS surface modeling tool, Rhino represents a solid in the computer by using a closed-loop surface (such as a sphere or ellipsoid) or a polysurface (such as a box or a cylinder) that encloses a volume without an opening or gap. This chapter will describe various ways of using Rhinoceros to construct, manipulate, detail, and modify solid objects in the computer.

Rhino's Solid Modeling Method

There are many ways to represent a solid in the computer. Rhino uses a closed-loop surface or a closed polysurface without openings or gaps to depict a solid. This modeling method is particularly useful for constructing free-form objects because you can first decompose a complex free-form object into individual free-form surfaces and then construct individual surfaces one by one. Naturally, prior to composing a solid from the surfaces, you may need to trim the surfaces for them to enclose a volume without openings or intersections. To help you learn about using Rhino as a tool in solid modeling, Rhino's solid modeling tools are categorized in five groups.

- The first group covers ways to construct solids of regular geometric shapes.
- The second group explains how free-form solids can be composed.
- The third group covers how two or more Rhino solid objects can be combined.
- The fourth group covers ways to detail a solid object.
- The fifth group deals with various editing methods.

The fourth and fifth group applies to both polysurfaces and solids.

Sub-Object Selection

An individual surface in a polysurface or a solid is a sub-object. If you need to select these individual surfaces, you can hold down the Control and Shift keys simultaneously when clicking on the object.

Solids of Regular Geometric Shapes

Solid objects of regular geometric shapes range from simple 3D boxes (six planar surfaces joined together) to 3D text objects (combination of planar and curved surfaces). Among them, some are single closed-loop surfaces (sphere, ellipsoid, and torus) and some are polysurfaces joined together. Figure 8–1 shows the box, sphere, cylinder, cone, truncated cone, pyramid, ellipsoid, paraboloid, tube, pipe, torus, and text shapes.

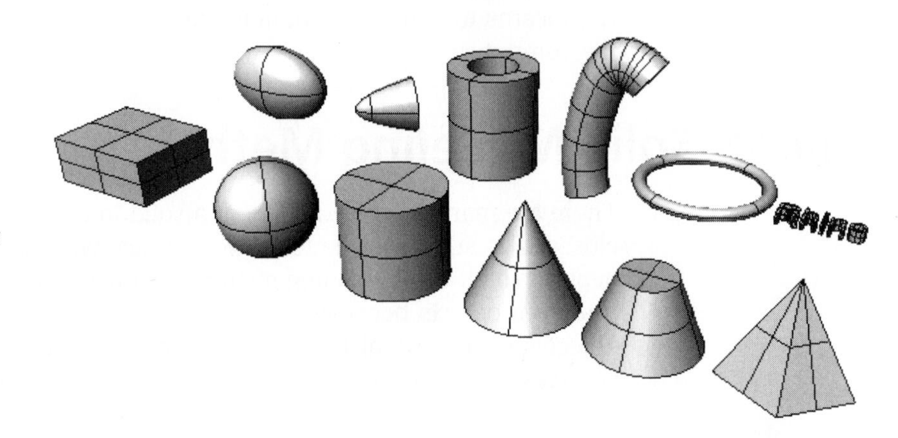

Figure 8–1. Rhino's primitive solid objects

Box

A box is a polysurface with six planar surfaces that are mutually perpendicular to one another. You can construct a box using one of five methods: specifying two corners of the box's base and the box's height; specifying the box's diagonal; using three points to specify the box's base and specifying the box's height; specifying two points in a viewport and one point in another viewport to define the box's base and specifying the box's height; and specifying the center point of the base, a corner point, and the height.

Specifying Corners and Height

Perform the following steps to construct a box by specifying corners of the base and the height:

1. Select File > Open and select the file Box.3dm from the Chapter 8 folder on the companion CD.
2. Set object snap mode to Point.
3. Select Solid > Box > Corner to Corner, Height, or click on the Box: Corner to Corner, Height button on the Box toolbar.
4. Select points A, B, and C (shown in Figure 8–2).

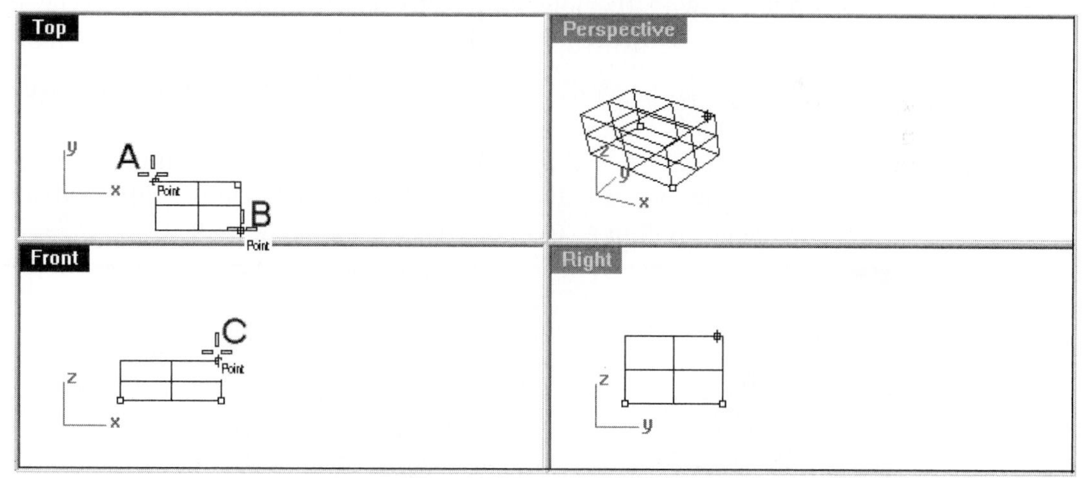

Figure 8–2. Solid box being constructed by specifying two corners of the base and height

Specifying Diagonal

Continue with the following steps to construct a box by specifying its diagonal points.

5 Turn off Layer 00 and turn on Layer 02. Do not change the current layer, which should be Layer 01.

6 Select Solid > Box > Diagonal, or click on Box: diagonal on the Box toolbar.

7 Select points A and B, shown in Figure 8–3.

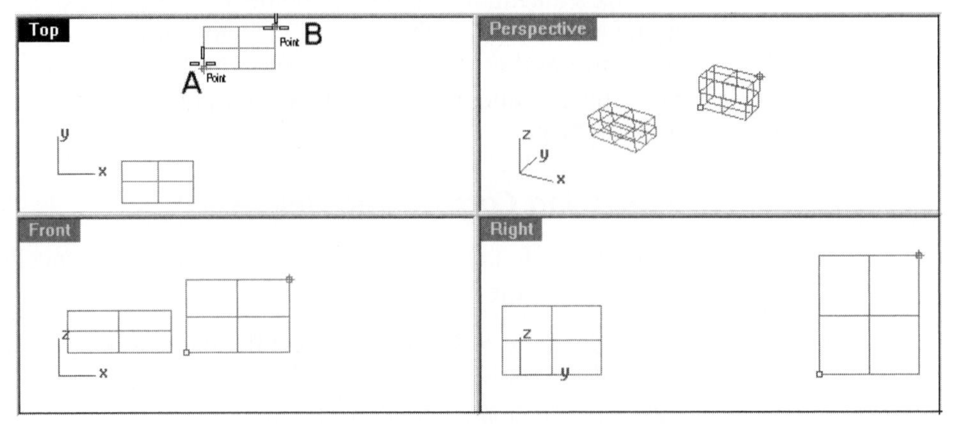

Figure 8–3. Solid box being constructed by specifying the diagonal

Specifying Three Points and Height

Continue with the following steps to construct a box by using three points to specify the base and specifying the height.

8 Turn off Layer 02 and turn on Layer 03. Again, keep the Layer 01 current.

9 Select Solid > Box > 3 points, height, or click on Box: 3 points, height/Box: Center, Corner, Height button on the Box toolbar.

10 Select points A, B, C, and D, shown in Figure 8–4.

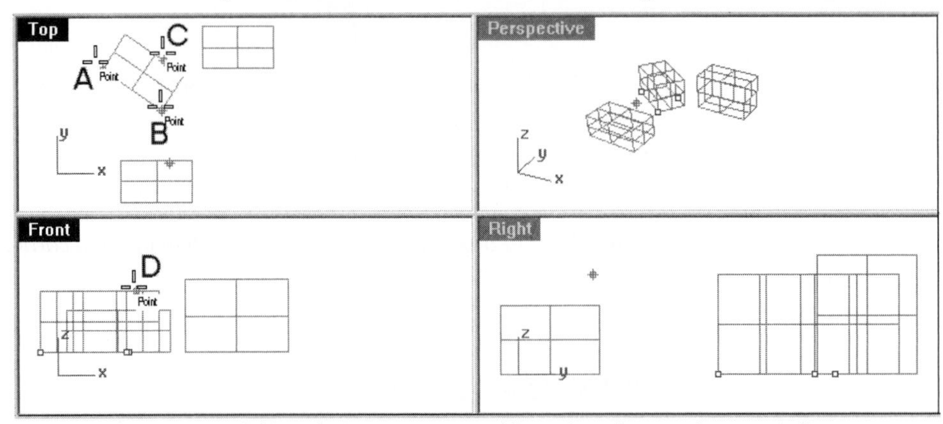

Figure 8–4. Solid box being constructed by using three points to define the base and specifying the height

Box with a Base Vertical to the Construction Plane

Continue with the following steps to construct a box vertical to the active construction plane.

11 Turn off Layer 03 and turn on Layer 04.

12 Select Solid > Box > Vertical Base.

13 Select points A, B, C, and D, shown in Figure 8-5.

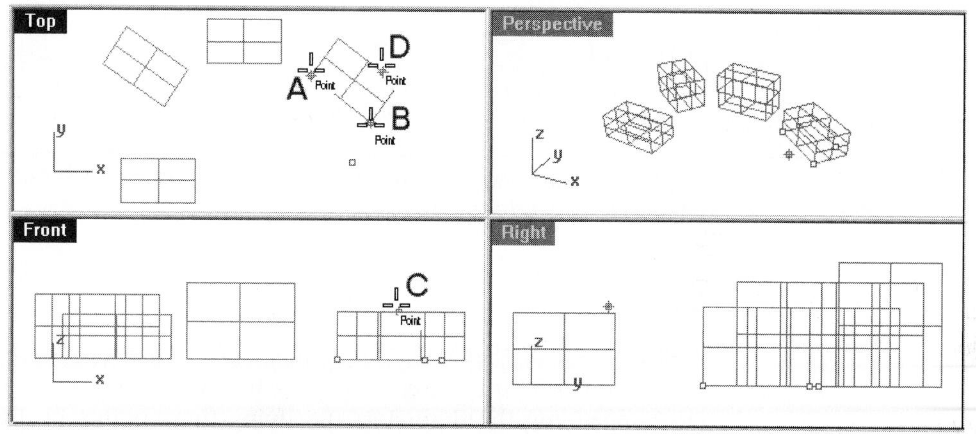

Figure 8–5. Solid box being constructed by constructing the box base vertical to a selected viewport and specifying the height

Specifying Center

Continue with the following steps to construct a box by specifying the center of the base, a corner, and the height.

14 Turn off Layer 04 and turn on Layer 05.

15 Select Solid > Box > Center of Base, Corner, Height, or right-click on the Box: 3 points, height/Box: Center, Corner, Height button on the Box toolbar.

16 Select points A, B, and C, shown in Figure 8-6.

17 Do not save your file.

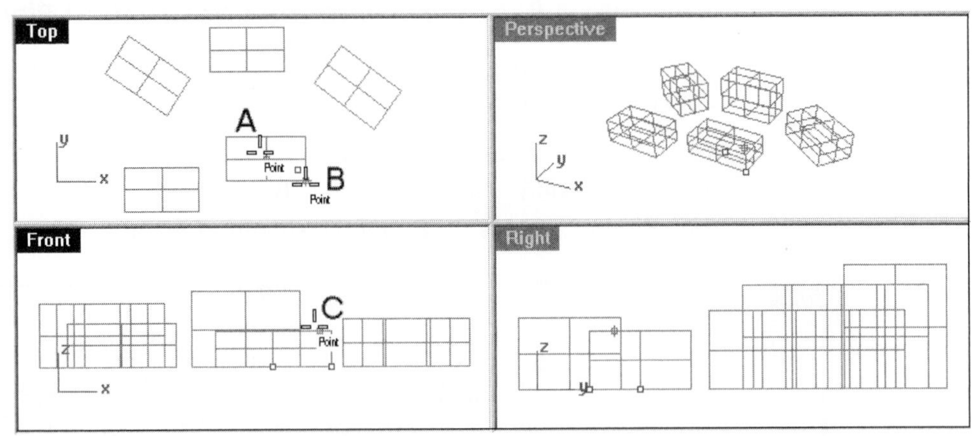

Figure 8–6. Solid box being constructed by specifying the center of the base, a corner, and the height

Sphere

A sphere is a single closed-loop surface. You can use one of seven methods to construct a sphere: specifying the center and a point on the surface; specifying the diameter; specifying three points on the surface of the sphere; specifying three tangent curves; selecting a curve and specifying the center along the curve and the radius; specifying four points on the surface of the sphere; and selecting a set of points.

Specifying Center and a Point on the Surface

Perform the following steps to construct a sphere by specifying the center and a point on the sphere's surface:

1. Select File > Open and select the file Sphere.3dm from the Chapter 8 folder on the companion CD.
2. Select Solid > Sphere > Center, Radius, or click on the Sphere: Center, Radius button on the Sphere toolbar.
3. Select points A and B, shown in Figure 8–7.

Specifying Diameter

Continue with the following steps to construct a sphere by specifying its diameter.

4. Select Solid > Sphere > 2 Points, or click on the Sphere: Diameter button on the Sphere toolbar.
5. Select points C and D, shown in Figure 8–7.

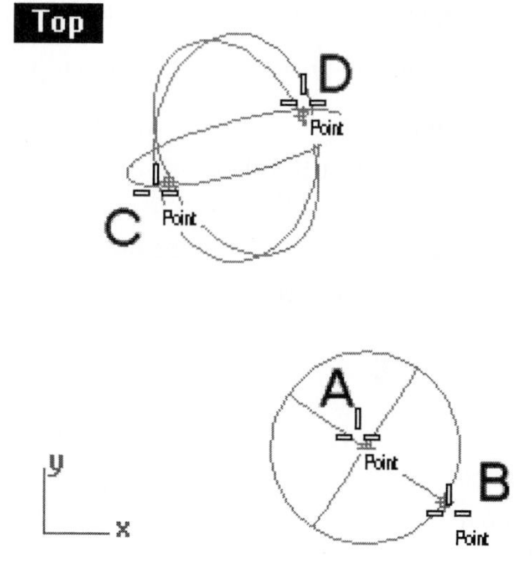

Figure 8–7. Sphere being constructed by specifying its center and a point on the surface and by specifying the diameter

Specifying Three Points

Continue with the following steps to construct a sphere by specifying three points on the surface.

6 Turn off Layer 00 and turn on Layer 02. Keep Layer 01 current.

7 Select Solid > Sphere > 3 Points, or click on the Sphere: 3 points button on the Sphere toolbar.

8 Select points A, B, and C, shown in Figure 8–8.

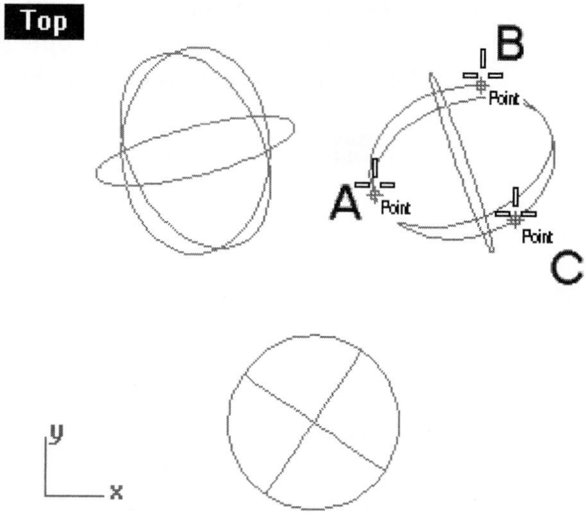

Figure 8–8. Sphere being constructed by specifying three points on the surface

Specifying Tangent Curves

Continue with the following steps by selecting three tangent curves.

9 Turn on Layer 03 and turn off Layer 02.

10 Select Solid > Sphere > Tangent to Curves, or click on the Sphere: From Circle Tangent to Curves button on the Sphere toolbar.

11 Select the Tangent option on the command area.

12 Select curves A, B, and C, shown in Figure 8–9.

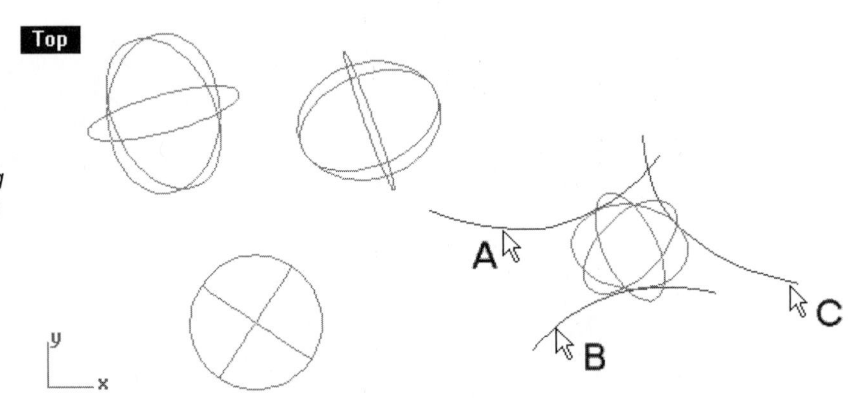

Figure 8–9. Sphere being constructed by selecting three tangent curves

Sphere Around a Curve

Continue with the following steps to construct a sphere around a curve.

13 Turn on Layer 04 and turn off Layer 03.

14 Select Solid > Sphere > Around Curve, or click on the Sphere: Around Curve button on the Sphere toolbar.

15 Select curve A and points B and C, shown in Figure 8–10.

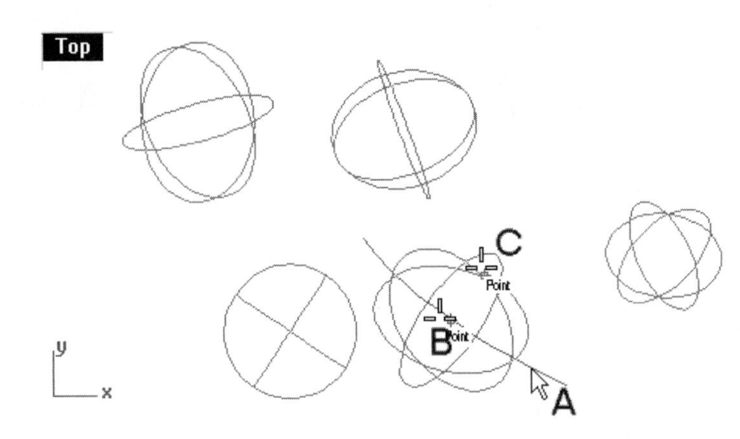

Figure 8–10. Sphere being constructed around a curve

Specifying Four Points

Continue with the following steps to construct a sphere by specifying four points. The first three specified points define a section of the sphere. The fourth point defines a point on the sphere.

16 Turn on Layer 05 and turn off Layer 04.

17 Select Solid > Sphere > 4 Points, or click on the Sphere: 4 Points button on the Sphere toolbar.

18 Select points A, B, C, and D, shown in Figure 8–11.

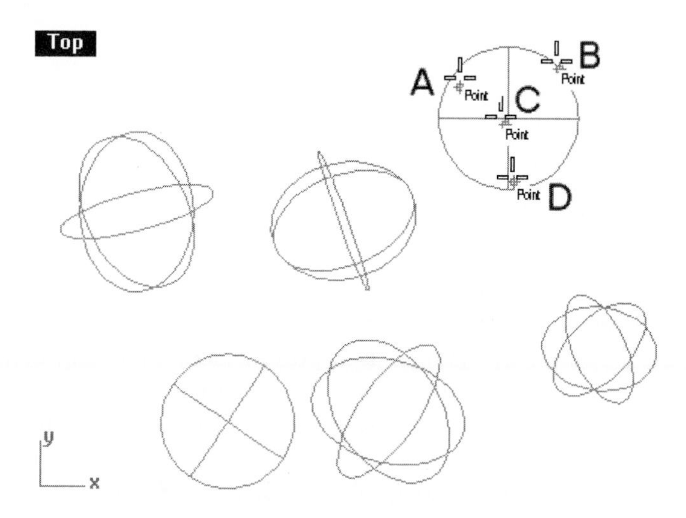

Figure 8–11. Sphere being constructed by specifying four points on the surface

Fitting to a Set of Points

Continue with the following steps to construct a sphere to best fit a set of selected points.

19 Turn on Layer 06 and turn off Layer 05.

20 Select Solid > Sphere > Fit Sphere to Points, or click on the Sphere: Fit Points button on the Sphere toolbar.

21 Click on location A (Figure 8–12) and drag to location B to select the point objects.

22 Do not save your file.

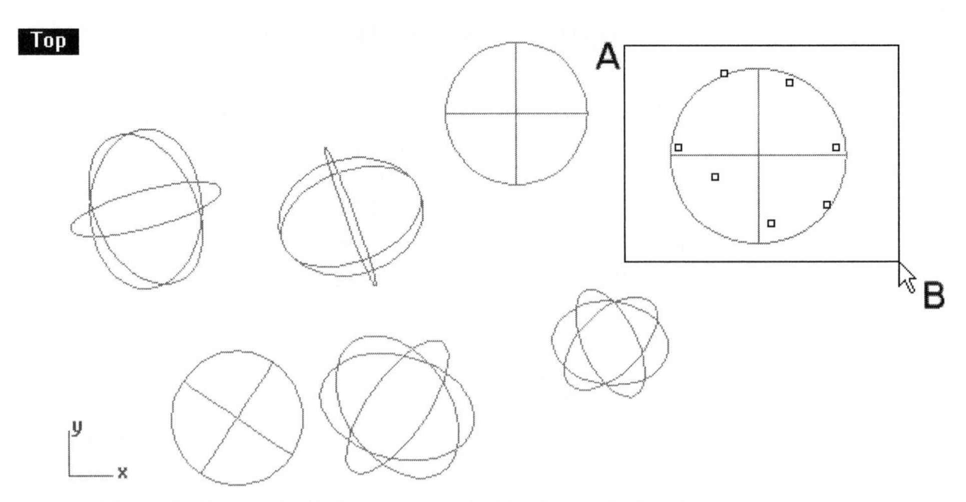

Figure 8–12. Sphere being constructed to fit a set of points

Cylinder

A cylinder is a polysurface consisting of three surfaces. The body is a cylindrical surface, and the top and bottom surfaces are circular planar surfaces. There are many ways to construct a cylinder.

Direction Constraints of Cylinder Axis

Direction constraint concerns the direction of the axis of the cylinder after its radius or diameter is specified. There are three direction constraints: None, Vertical, and Around Curve. If direction constraint is set to none, the other end point is freely determined. If direction constraint is set to vertical, the other axis end point is set to be perpendicular to the base of the cylinder. If you want to construct a cylinder around a curve, you set direction constraint to around curve. Perform the following steps:

1. Select File > Open and select the file Cylinder.3dm from the Chapter 8 folder on the companion CD.
2. Select Solid > Cylinder, or click on the Cylinder button on the Solid toolbar.
3. Select the DirectionalConstraint option on the command area.
4. Select the Vertical option on the command area.
5. Select points A, B, and C, shown in Figure 8–13. (A is the center, AB is the radius, and C's Z coordinate defines the height of the cylinder.
6. Repeat the command.

7 Select the DirectionalConstraint option on the command area.
8 Select the None option on the command area.
9 Select points D, E, and C, shown in Figure 8–13. (D is the center, DE is the radius, and C is the location of the other axis's end point.)

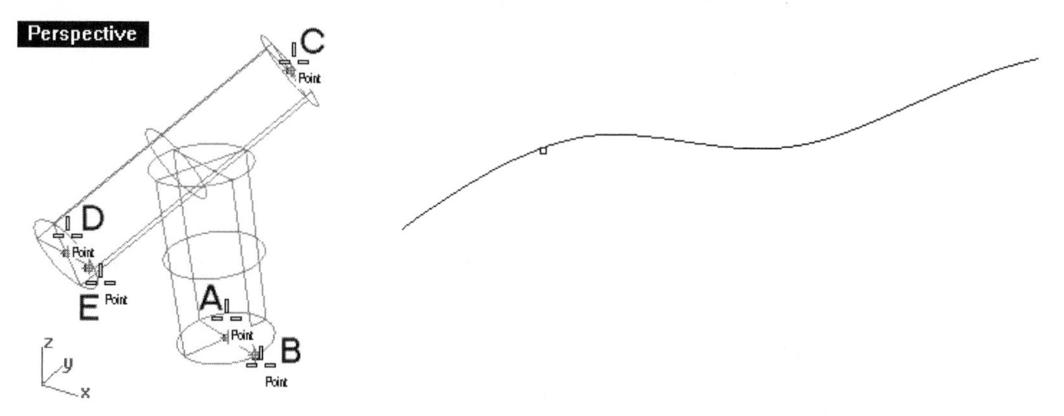

Figure 8–13. Vertical cylinder and cylinder with user defined axis end point being constructed

10 Repeat the command.
11 Select the DirectionalConstraint option on the command area.
12 Select the AroundCurve option on the command area.
13 Select curve A, shown in Figure 8–14.
14 Select point B to specify the center point.
15 Type 6 to specify the radius.
16 Type 30 to specify the height of the cylinder.

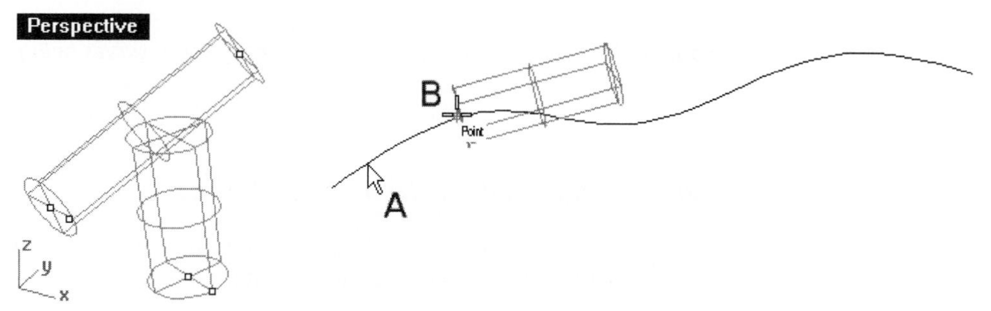

Figure 8–14. Cylinder with its base circle around a curve being constructed

Cylinder Base Tangent to Three Curves

Continue with the following steps to construct a cylinder with its base circle tangent to three curves.

17 Turn on Layer 02 and turn off Layer 00.

18 Select Solid > Cylinder, or click on the Cylinder button on the Solid toolbar.

19 Select the DirectionalConstraint option on the command area.

20 Select the None option on the command area.

21 Select the Tangent option on the command area.

22 Select curves A, B, and C shown in Figure 8–15.

23 Select point D. A cylinder with its axis end point closest to the selected point is constructed.

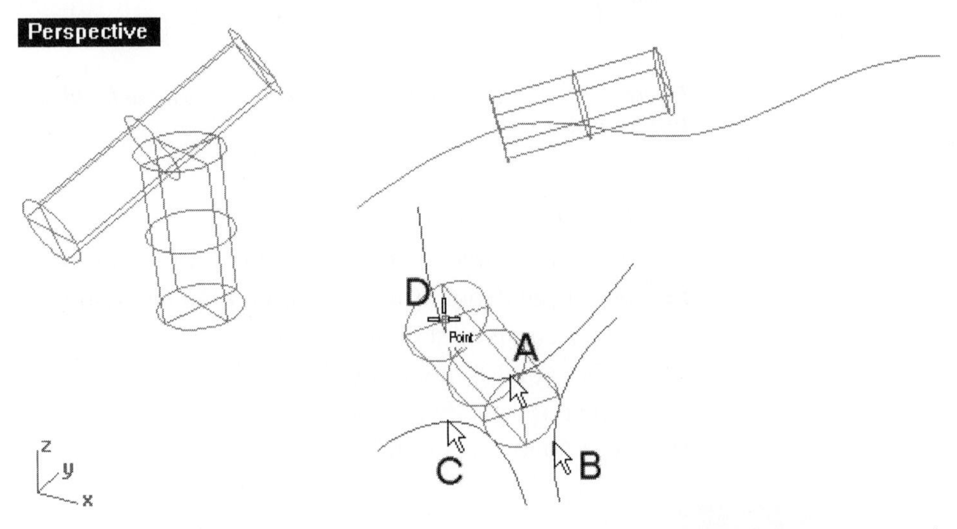

Figure 8–15. Cylinder with base circle tangent to three curves being constructed

Cylinder Base Fitted to a set of Points

Continue with the following steps to construct a cylinder with its base circle best fitted to a set of selected points.

24 Turn on Layer 03 and turn off Layer 02.

25 Select Solid > Cylinder, or click on the Cylinder button on the Solid toolbar.

26 Select the FitPoints option on the command area.

27 Click on A and drag to B (Figure 8–16) to select the point objects, and press the ENTER key.

28 Select point C. A cylinder with its axis end point closest to the selected point is constructed.

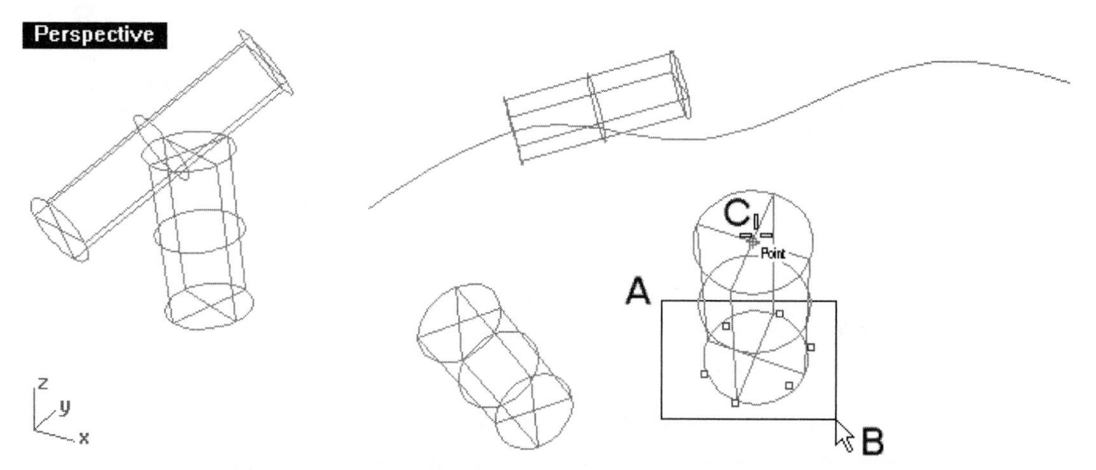

Figure 8–16. Cylinder with base circle best fitted to a set of points being constructed

Cylinder Base Passing Through Three Points

Continue with the following steps to construct a cylinder with its base circle defined by three selected points.

29 Turn on Layer 04 and turn off Layer 03.

30 Select Solid > Cylinder, or click on the Cylinder button on the Solid toolbar.

31 Select the 3Point option on the command area.

32 Select points A, B, and C, shown in Figure 8–17

33 Select point D. A cylinder with its axis end point closest to the selected point is constructed.

34 Do not save your file.

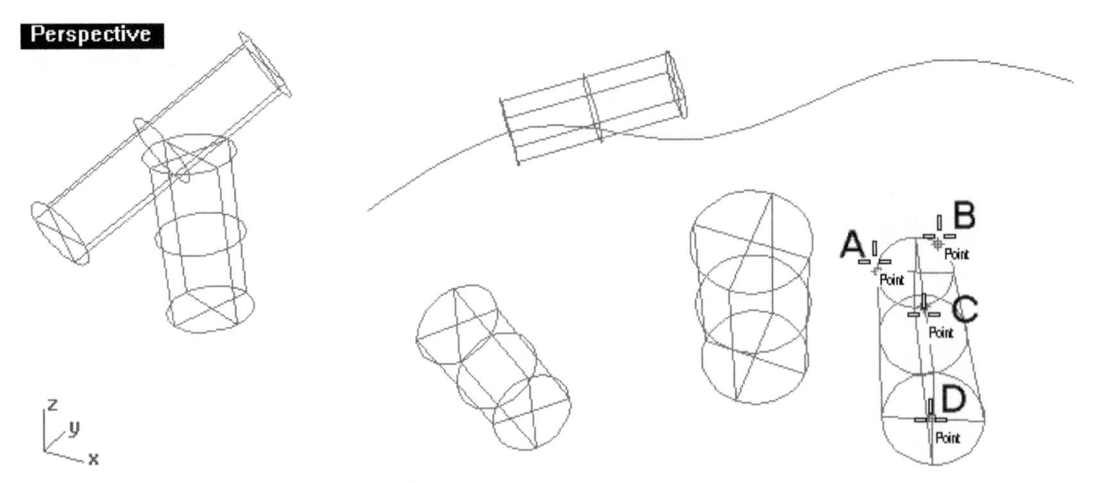

Figure 8–17. Cylinder with base circle defined by three selected points being constructed

Radius and Diameter Option

By default, the base circle is defined by a center point and a radius. If you want to specify the diameter, you should select the Diameter option from the command line.

Cone

A cone is a polysurface consisting of two joined surfaces. The slant surface is a closed surface, and the base is a circular planar surface.

Direction Constraints of Cone Axis

The methods to construct a cone are similar to those used to construct a cylinder. You can define direction constraint with the apex vertical to the base, with apex at a selected point, and with the axis of the base circle around a curve. You can also define a cone's base circle tangent to three curves, passing through three points, or best fitting to a set of points. Perform the following steps:

1. Select File > Open and select the file Cone.3dm from the Chapter 8 folder on the companion CD. (This template file is identical to the template file Cylinder.3dm.)
2. Select Solid > Cone, or click on the Cone button on the Solid toolbar.
3. Select the DirectionalConstraint option on the command area.
4. Select the Vertical option on the command area.

5. Select points A, B, and C, shown in Figure 8–18. (A is the center, AB is the radius, and C's Z coordinate defines the height of the cone.
6. Repeat the command.
7. Select the DirectionalConstraint option on the command area.
8. Select the None option on the command area.
9. Select points D, E, and C, shown in Figure 8–18. (D is the center, DE is the radius, and C is the location of the apex.)

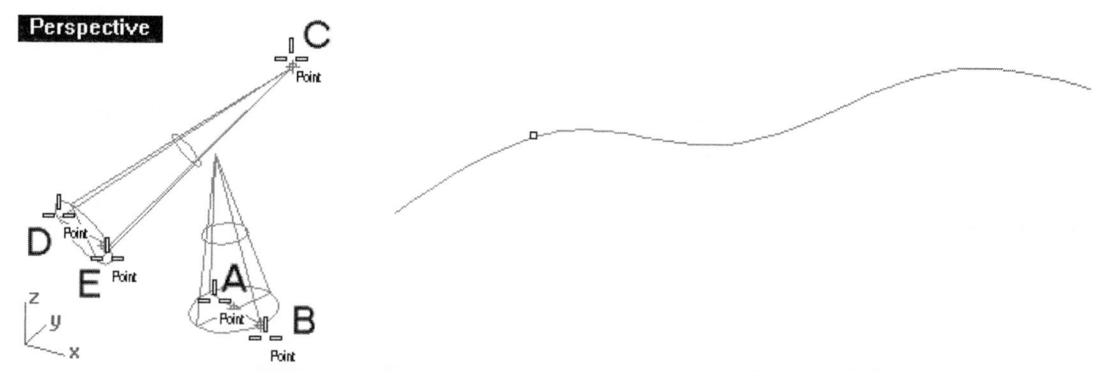

Figure 8–18. Vertical cone and cone with user defined apex being constructed

10. Repeat the command.
11. Select the DirectionalConstraint option on the command area.
12. Select the AroundCurve option on the command area.
13. Select curve A, shown in Figure 8–19.
14. Select point B to specify the center point.
15. Type 6 to specify the radius.
16. Type 30 to specify the height of the cone.

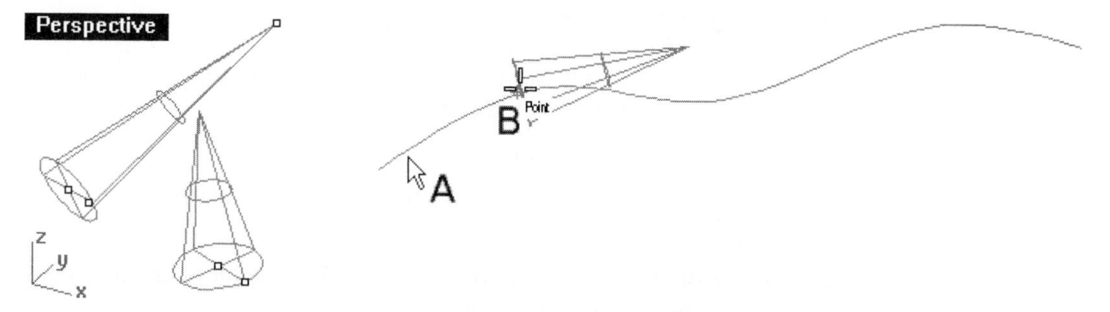

Figure 8–19. Cone with its base circle around a curve being constructed

Cone Base Tangent to Three Curves

Continue with the following steps to construct a cone with its base circle tangent to three curves.

17 Turn on Layer 02 and turn off Layer 00.

18 Select Solid > Cone, or click on the Cone button on the Solid toolbar.

19 Select the DirectionalConstraint option on the command area.

20 Select the None option on the command area.

21 Select the Tangent option on the command area.

22 Select curves A, B, and C, shown in Figure 8–20.

23 Select point D. A cone with its apex closest to the selected point is constructed.

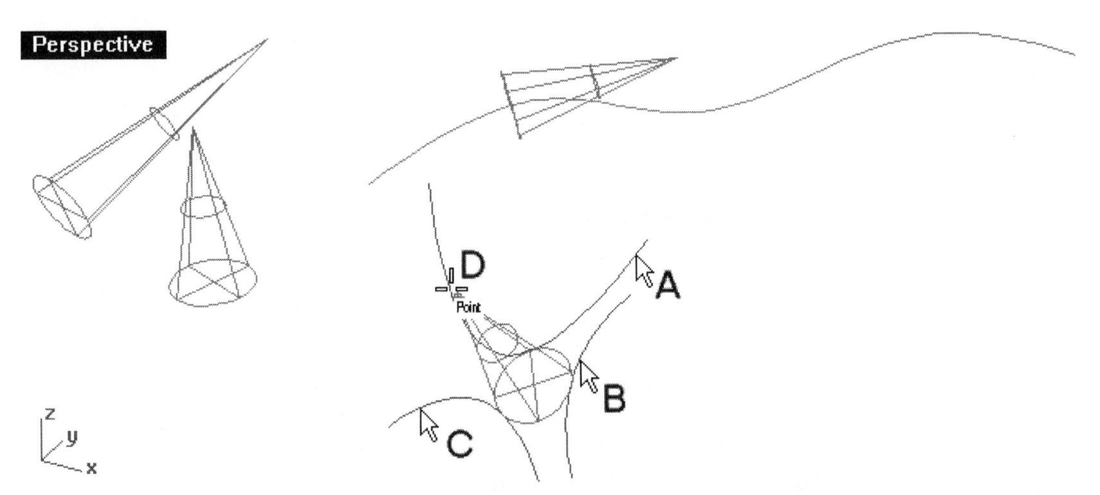

Figure 8–20. Cone with base circle tangent to three curves being constructed

Cone Base Fitted to a Set of Points

Continue with the following steps to construct a cone with its base circle best fitted to a set of selected points.

24 Turn on Layer 03 and turn off Layer 02.

25 Select Solid > Cone, or click on the Cone button on the Solid toolbar.

26 Select the FitPoints option on the command area.

27 Click on A and drag to B (Figure 8–21) to select the point objects, and press the ENTER key.

28 Select point C. A cone with its apex closest to the selected point is constructed.

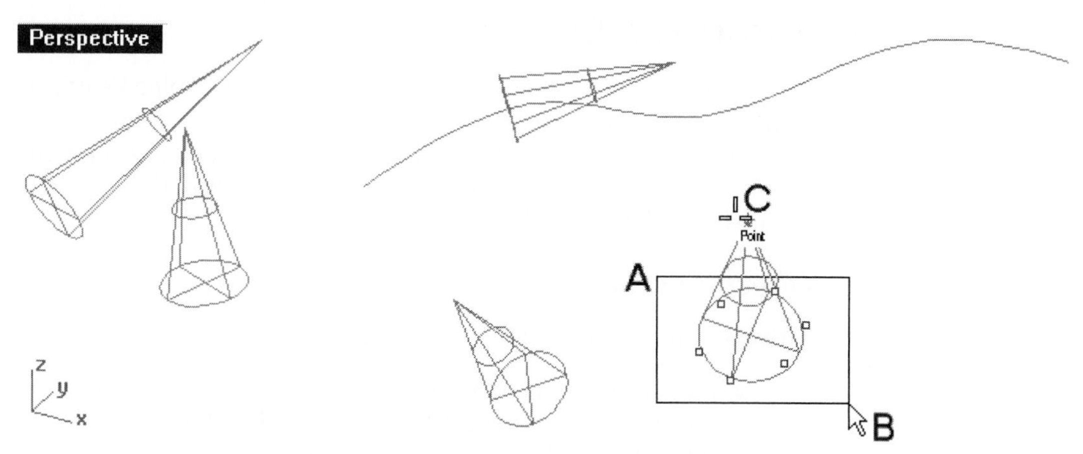

Figure 8–21. Cone with base circle best fitted to a set of points being constructed

Cone Base Passing Through Three Points

Continue with the following steps to construct a cone with its base circle defined by three selected points.

29 Turn on Layer 04 and turn off Layer 03.

30 Select Solid > Cone, or click on the Cone button on the Solid toolbar.

31 Select the 3Point option on the command area.

32 Select points A, B, and C, shown in Figure 8–22.

33 Select point D. A cone with its apex closest to the selected point is constructed.

34 Do not save your file.

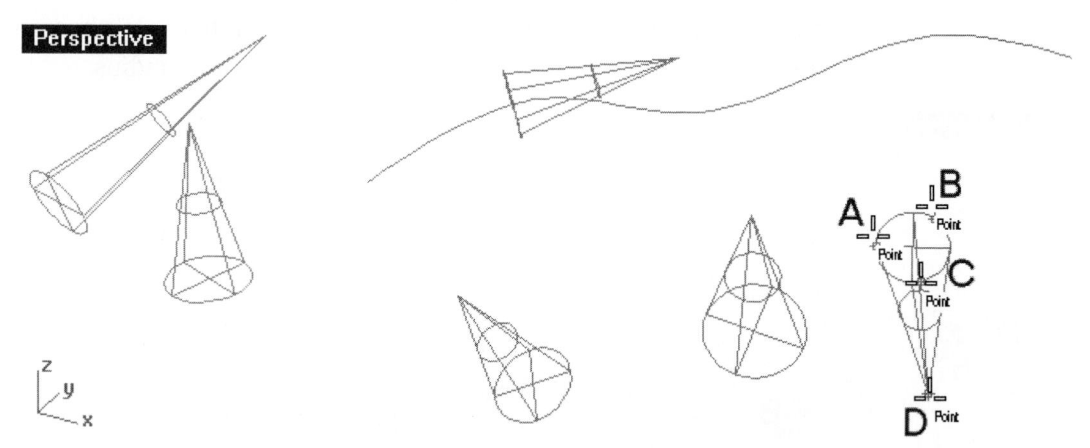

Figure 8–22. Cone with base circle defined by three selected points being constructed

Truncated Cone

A truncated cone is a polysurface consisting of three joined surfaces. The slant surface is a closed surface, and the top and bottom surfaces are circular planar surfaces. The methods to construct a truncated cone are very similar to constructing a cone, with the exception that a second radius has to be specified.

Direction Constraints of Truncated Cone's Axis

There are three directional constraints. Perform the following steps:

1. Select File > Open and select the file TCone.3dm from the Chapter 8 folder on the companion CD. (This template file is identical to the template files Cylinder.3dm and Cone.3dm.)
2. Select Solid > Truncated Cone, or click on the Truncated Cone button on the Solid toolbar.
3. Select the DirectionalConstraint option on the command area.
4. Select the Vertical option on the command area.
5. Select points A and B, shown in Figure 8–23, to define the base of the truncated cone.
6. Select point C to define the height of the cone.
7. Type 4 at the command area to specify the other radius. A truncated cone is constructed.
8. Repeat the command.
9. Select the DirectionalConstraint option on the command area.
10. Select the None option on the command area.
11. Select points D and E, shown in Figure 8–23, to specify the center and radius of the truncated cone.
12. Select point C to define the location of the other axis's end point.
13. Type 4 at the command area to specify the other radius.

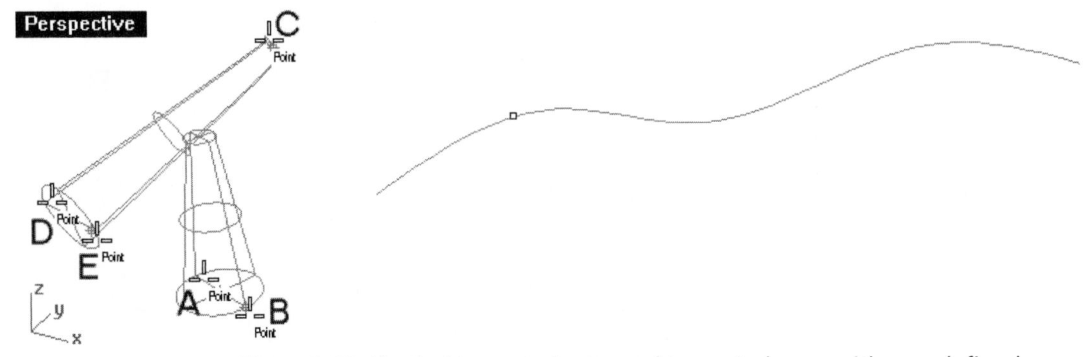

Figure 8–23. Vertical truncated cone and truncated cone with user defined second axis point being constructed

14. Repeat the command.
15. Select the DirectionalConstraint option on the command area.
16. Select the AroundCurve option on the command area.
17. Select curve A, shown in Figure 8–24.
18. Select point B to specify the center point.
19. Type 6 to specify the radius of the base circle.
20. Type 30 to specify the height of the truncated cone.
21. Type 4 to specify the radius of the second radius.

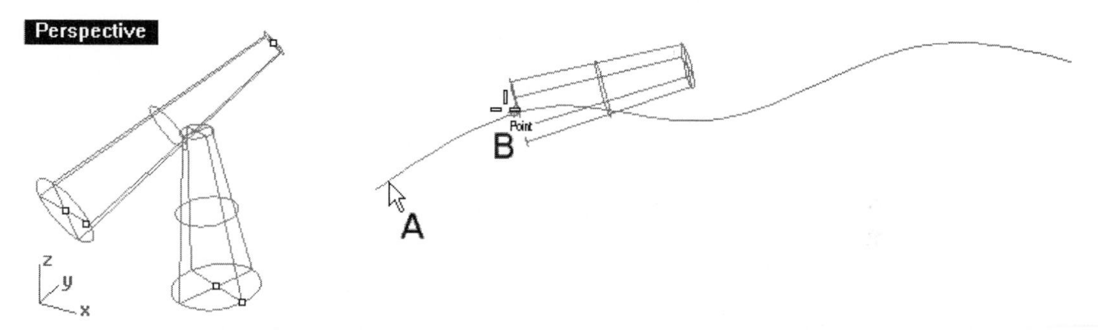

Figure 8–24. Truncated cone with its base circle around a curve being constructed

Truncated Cone Base Tangent to Three Curves

Continue with the following steps to construct a truncated cone with its base circle tangent to three curves.

22. Turn on Layer 02 and turn off Layer 00.
23. Select Solid > Truncated Cone, or click on the Truncated Cone button on the Solid toolbar.
24. Select the DirectionalConstraint option on the command area.
25. Select the None option on the command area.
26. Select the Tangent option on the command area.
27. Select curves A, B, and C shown in Figure 8–25.
28. Select point D to specify the height of the truncated cone.
29. Type 4 at the command area to specify the other radius. A truncated cone with its second axis point closest to the selected point is constructed.

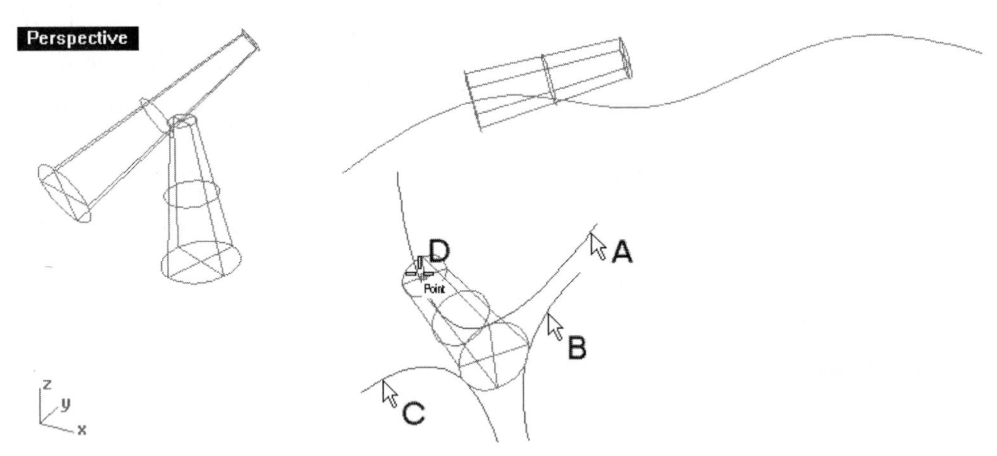

Figure 8–25. Cone with base circle tangent to three curves being constructed

Fitting to a Set of Points

Continue with the following steps to construct a truncated cone with its base circle best fitted to a set of selected points.

30 Turn on Layer 03 and turn off Layer 02.

31 Select Solid > Truncated Cone, or click on the Truncated Cone button on the Solid toolbar.

32 Select the FitPoints option on the command area.

33 Click on A and drag to B (Figure 8–26) to select the point objects, and press the ENTER key.

34 Select point C to specify the height of the truncated cone.

35 Type 4 at the command area to specify the radius at the other end. A truncated cone with second axis point closest to the selected point is constructed.

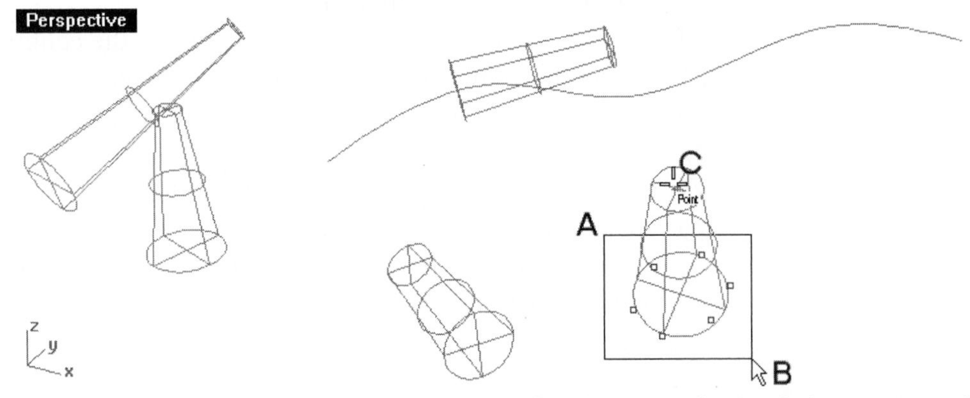

Figure 8–26. Cone with base circle best fitted to a set of points being constructed

Truncated Cone Base Passing Through Three Points

Continue with the following steps to construct a truncated cone with its base circle defined by three selected points.

36 Turn on Layer 04 and turn off Layer 03.

37 Select Solid > Truncated Cone, or click on the Truncated Cone button on the Solid toolbar.

38 Select the 3Point option on the command area.

39 Select points A, B, and C, shown in Figure 8–27.

40 Select point D to specify the height of the truncated cone.

41 Type 2 at the command area. A truncated cone with its second axis end point closest to the selected point is constructed.

42 Do not save your file.

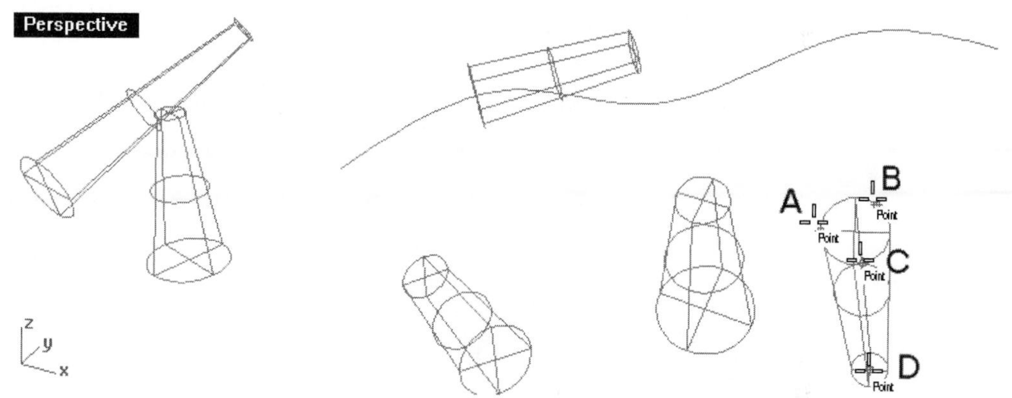

Figure 8–27. Cone with base circle defined by three selected points being constructed

Pyramid

A pyramid is a polysurface with a polygon base and a number of slant triangular surfaces that encloses a volume without opening or gap.

Direction Constraint of Pyramid Axis

Similar to constructing a cylinder, cone, and truncated cone, to construct a pyramid you need to select one of three ways to constrain the direction of the axis: none, vertical, and around curve. Perform the following steps.

1 Select File > Open and select the file Pyramid.3dm from the Chapter 8 folder on the companion CD.

2 Select Solid > Pyramid, or click on the Pyramid button on the Solid toolbar.
3 Select the DirectionConstraint option on the command area.
4 Select the Vertical option on the command area.

Regular Polygon and Star Polygon of Pyramid Base

The base of a pyramid is a polygon, which can be a regular polygon or a star polygon. Continue with the following steps.

5 Select the NumberSides option on the command area.
6 Type 5 to specify the number of sides.
7 Select points A, B, and C, shown in Figure 8–28. (Point A defines the center, point B is a corner of the base polygon, and point C defines the height.)
8 Repeat the command.
9 Select the Star option on the command area.
10 Select points D, E, F, and G, shown in Figure 8–28.

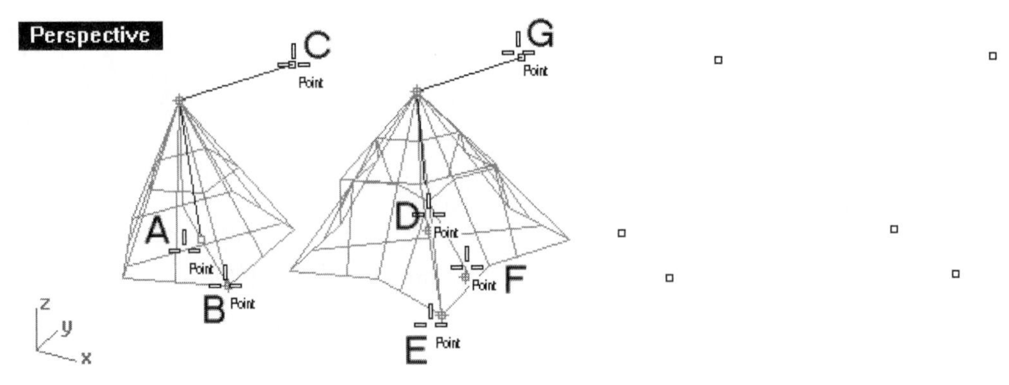

Figure 8–28. Two pyramids constructed

Edge, Inscribed, and Circumscribed Pyramid Base

There are three ways to define the base polygon of the pyramid: edge, inscribed circle, and circumscribed circle. The default way is to select a center point and the radius of the inscribed circle.

11 Repeat the command.
12 Select the Edge option on the command area.
13 Select points A, B, and C, shown in Figure 8–29.

14 Repeat the command.
15 Select the Circumscribed option on the command area.
16 Select points D, E, and F, shown in Figure 8–29.
17 Do not save your file.

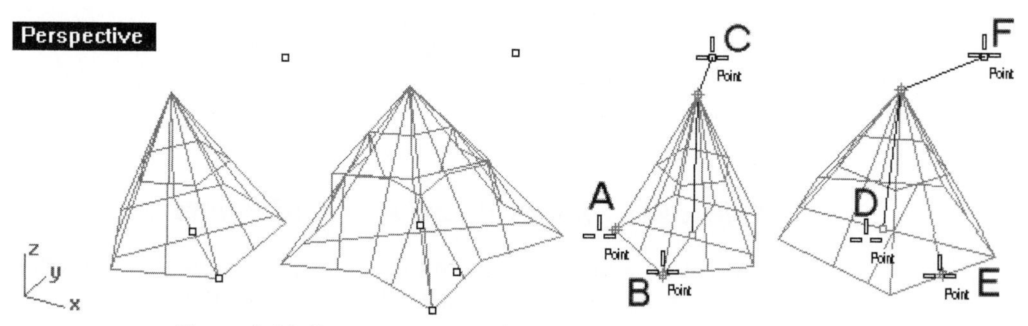

Figure 8–29. Two more pyramids constructed

Ellipsoid

An ellipsoid, like a sphere, is also a single closed-loop surface. You can construct an ellipsoid via five methods, as follows:

Ellipsoid's Center and Axis End Points

The first method creates an ellipsoid by specifying the ellipsoid's center and axis end points, as follows.

1 Select File > Open and select the file Ellipsoid.3dm from the Chapter 8 folder on the companion CD.
2 Select Solid > Ellipsoid > From Center, or click on the Ellipsoid: From Center button on the Ellipsoid toolbar.
3 Select points A and B, shown in Figure 8–30, to specify the center and an axis end point.
4 Type 5 at the command area, press the ENTER key, and click on location C to specify the second axis radius and direction.
5 Type 3 at the command area, press the ENTER key, and click on location D to specify the third axis radius and direction.

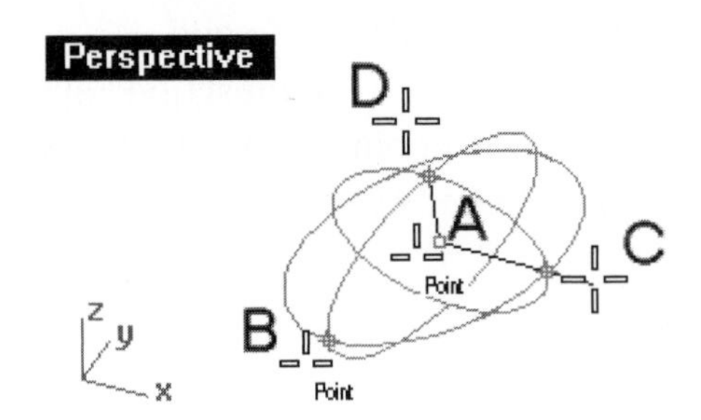

Figure 8–30. Ellipsoid specified by its center and axis end points

Ellipsoid's Diameter and Axis End Points

A second method creates an ellipsoid by specifying its diameter and two axis end points. Continue with the following steps.

6 Turn off Layer 00 and turn on Layer 02.

7 Click on the Ellipsoid: By Diameter button on the Ellipsoid toolbar.

8 Select points A and B, shown in Figure 8–31, to specify the center and an axis end point.

9 Type 10 at the command area, press the ENTER key, and click on location C to specify the second axis diameter and direction.

10 Type 6 at the command area, press the ENTER key, and click on location D to specify the third axis diameter and direction.

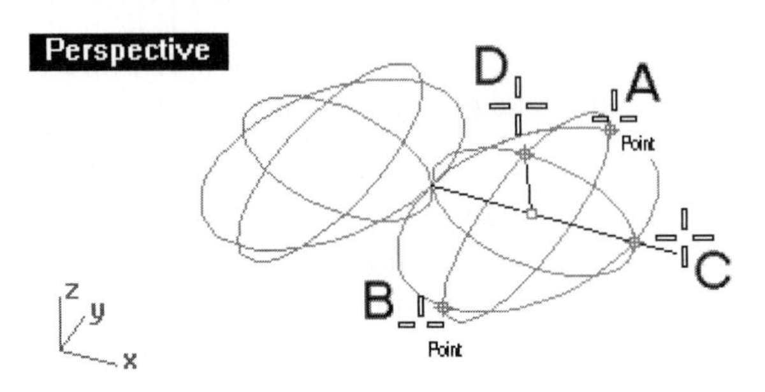

Figure 8–31. Ellipsoid specified by its diameter being constructed

Ellipsoid's Foci and a Point

A third method creates an ellipsoid by specifying its foci and a point on the ellipsoid. Continue with the following steps.

11 Turn on Layer 03 and turn off Layer 02

12 Select Solid > Ellipsoid > From Foci, or click on the Ellipsoid: From Foci button on the Ellipsoid toolbar.

13 Select points A and B (shown in Figure 8–32) to specify the foci, and then select point C to specify a point on the ellipsoid.

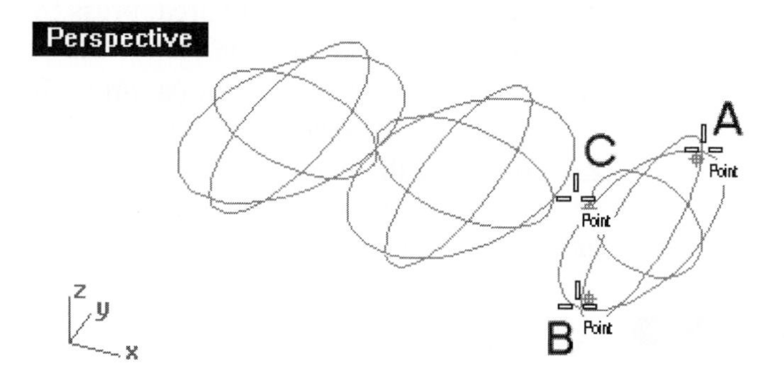

Figure 8–32. Ellipsoid specified by its foci being constructed

Ellipsoid's Bounding Box

A fourth method creates an ellipsoid by specifying a bounding box. Continue with the following steps.

14 Turn on Layer 04 and turn off Layer 03.

15 Click on the Ellipsoid: By Corners button on the Ellipsoid toolbar.

16 Select points A and B (shown in Figure 8–33).

17 Type 5 at the command area, and click on location C.

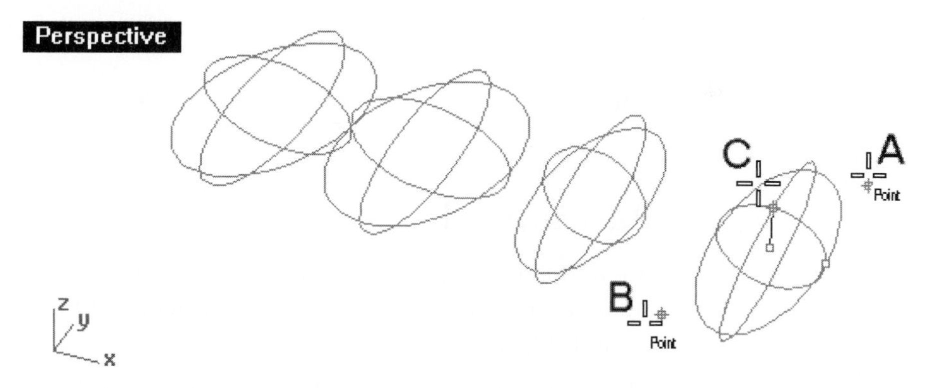

Figure 8–33. Ellipsoid specified by its bounding box being constructed

Ellipsoid on a Curve

A fifth method creates an ellipsoid on a curve. Continue with the following steps.

18 Turn on Layer 05 and turn off Layer 04.

19 Click on the Ellipsoid Around Curve button on the Ellipsoid toolbar.
20 Select point A (Figure 8–34) along the curve.
21 Type 5 at the command area, press the ENTER key, and click on location B to specify the first radius.
22 Type 4 at the command area, press the ENTER key, and click on location C to specify the second radius.
23 Type 3 at the command area, press the ENTER key, and click on location D to specify the third radius.
24 Do not save your file.

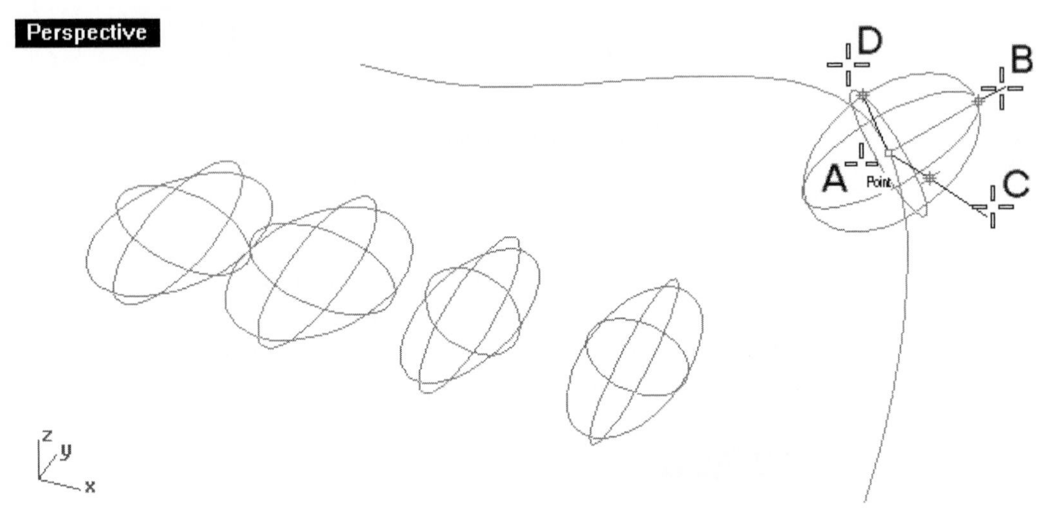

Figure 8–34. Ellipsoid on a curve being constructed

Paraboloid

A paraboloid is a polysurface consisting of a revolved parabola and a circular surface capping its end. To construct a paraboloid, perform the following steps:

1 Select File > Open and select the file Paraboloid.3dm from the Chapter 8 folder on the companion CD.
2 Select Solid > Paraboloid > Focus, Direction, or click on the Paraboloid/Paraboloid: Vertex, Focus button on the Solid toolbar.
3 If the prompt indicates Cap = No, select the Cap option on the command area to change it to Yes. *(Note: If Cap = No, an open surface will be constructed instead.)*
4 Select locations A, B, and C (shown in Figure 8–35) to indicate the focus, direction, and end point of the paraboloid.

5 Select Solid > Paraboloid > Vertex, Focus, or right-click on the Paraboloid/Paraboloid: Vertex, Focus button on the Solid toolbar.
6 Select locations D, E, and F (shown in Figure 8–35) to specify the vertex, focus, and an end point of the paraboloid. Two paraboloids are constructed.
7 Do not save your file.

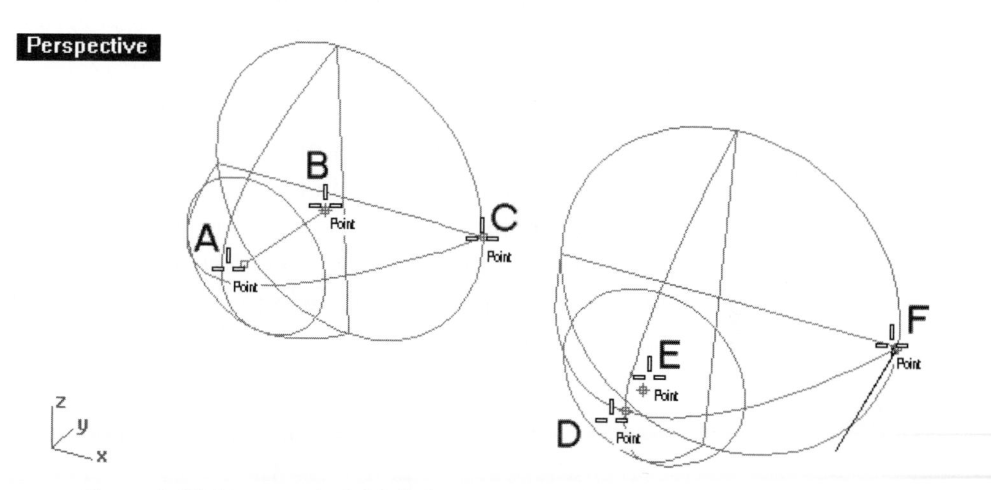

Figure 8–35. Two paraboloids being constructed

Tube

A tube is a polysurface consisting of two concentric cylindrical surfaces and two planar surfaces capping the ends. The options for constructing a tube are similar to constructing a cylinder, cone, and truncated cone.

Direction Constraints of Tube Axis

There are three directional constraints. Perform the following steps:

1 Select File > Open and select the file Tube.3dm from the Chapter 8 folder on the companion CD.
2 Select Solid > Tube, or click on the Tube button on the Solid toolbar.
3 Select the DirectionalConstraint option on the command area.
4 Select the Vertical option on the command area.
5 Select point A, shown in Figure 8–36.
6 Type 10 at the command area to specify the first radius.
7 Type 5 at the command area to specify the second radius.
8 Select point B to define the height of the cone.

9 Repeat the command.
10 Select the DirectionalConstraint option on the command area.
11 Select the None option on the command area.
12 Select point C, shown in Figure 8–36.
13 Type 10 at the command area to specify the first radius.
14 Type 5 at the command area to specify the second radius.
15 Select point B.
16 Repeat the command.
17 Select the DirectionalConstraint option on the command area.
18 Select the AroundCurve option on the command area.
19 Select curve D, shown in Figure 8–36.
20 Select point E to specify the center point.
21 Type 10 at the command area to specify the first radius.
22 Type 5 at the command area to specify the second radius.
23 Type 30 to specify the height of the tube.

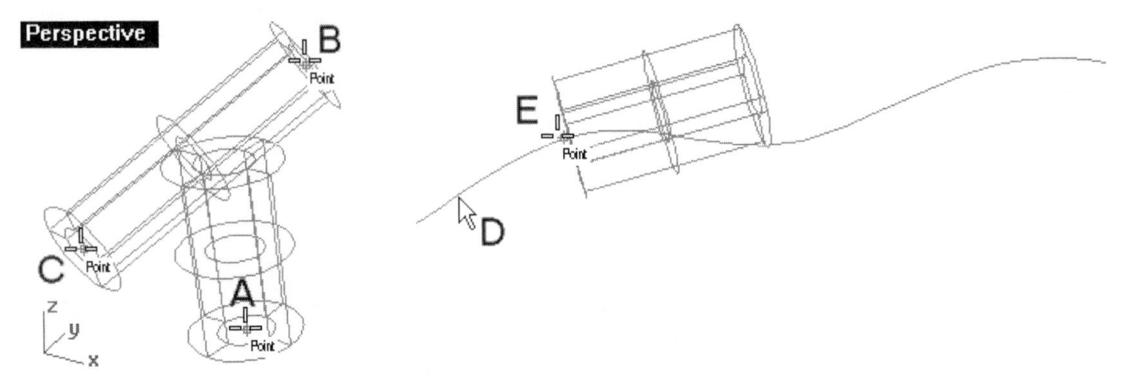

Figure 8–36. Tube in three directions being constructed

Axis of Tube's Base Tangent to Three Curves

Continue with the following steps to construct a tube with its base circle tangent to three curves.

24 Turn on Layer 02 and turn off Layer 00.
25 Select Solid > Tube, or click on the Tube button on the Solid toolbar.
26 Select the DirectionalConstraint option on the command area.
27 Select the None option on the command area.

28 Select the Tangent option on the command area.
29 Select curves A, B, and C, shown in Figure 8–37.
30 Type 3 at the command area to specify the second radius.
31 Select point D. A tube cone is constructed.

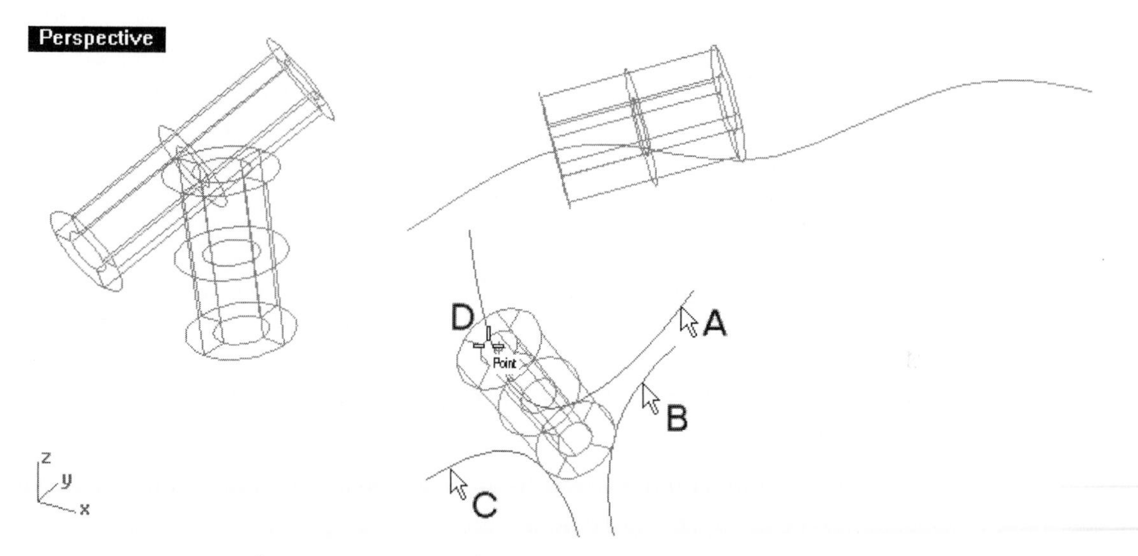

Figure 8–37. Cone with base circle tangent to three curves being constructed

Tube Base's Outside Diameter Fitted to a Set of Points

Continue with the following steps to construct a cone with its base circle best fitted to a set of selected points.

32 Turn on Layer 03 and turn off Layer 02.
33 Select Solid > Tube, or click on the Tube button on the Solid toolbar.
34 Select the FitPoints option on the command area.
35 Click on A and drag to B (Figure 8–38) to select the point objects, and press the ENTER key.
36 Type 3 at the command area to specify the radius at the other end.
37 Select point C.

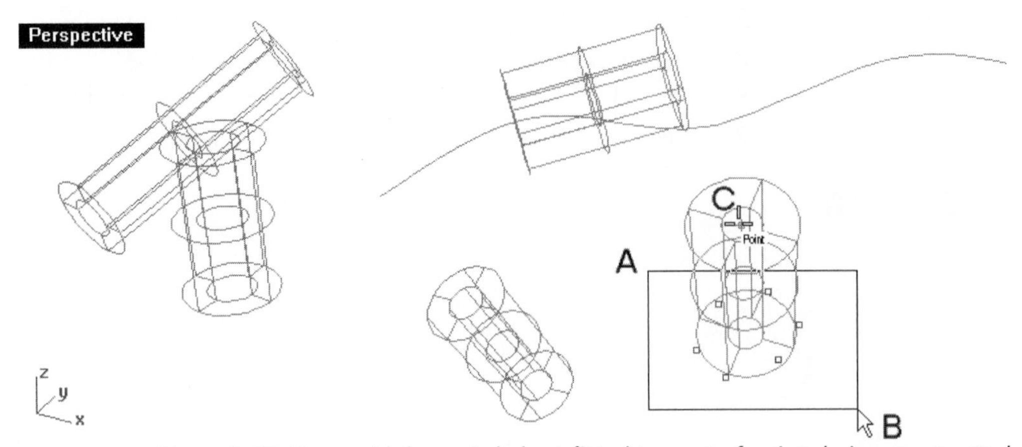

Figure 8–38. Cone with base circle best fitted to a set of points being constructed

Tube Base's Outside Diameter Passing Through Three Points

Continue with the following steps to construct a cone with its base circle defined by three selected points.

38 Turn on Layer 04 and turn off Layer 03.

39 Select Solid > Tube, or click on the Tube button on the Solid toolbar.

40 Select the 3Point option on the command area.

41 Select points A, B, and C, shown in Figure 8–39

42 Type 3 at the command area.

43 Select point D.

44 Do not save your file.

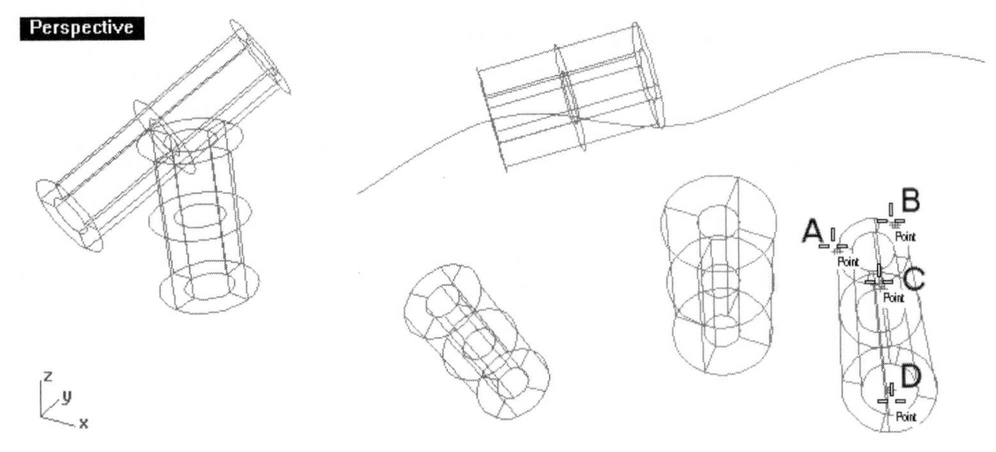

Figure 8–39. Cone with base circle defined by three selected points being constructed

Pipe

A pipe is a polysurface with a variable-radius, circular cross-section that runs along a curve or a chain of curve/edges. The end caps of the pipe can be flat or spherical. To speed curve chain selection, you can select the AutoChain option and use the ChainContinuity option to help control how smoothly the curve segments need to be connected to get selected. The end cap can be flat or rounded. Perform the following steps to construct a pipe with spherical round caps:

1. Select File > Open and select the file Pipe.3dm from the Chapter 8 folder on the companion CD.
2. Set object snap mode to End.
3. Select Solid > Pipe, or click on the Pipe, round cap/Pipe button on the Solid toolbar.
4. Select the ChainEdge option on the command area.
5. If AutoChain = No, select it to change it to Yes. Otherwise, proceed to next step.
6. If ChainContinuity is not Curvature, select it and change it to Curvature. Otherwise, proceed to the next step.
7. Select curve A, shown in Figure 8–40, and press the ENTER key.
8. If you use the pull down menu, select the Cap option and change it to Round. Otherwise, proceed to the next step.
9. Type 4 at the command area to specify the start radius. The end nearer to where you select the curve is the starting point.
10. Type 3 at the command area to specify the end radius.
11. Select end point B.
12. Type 2 at the command area to specify the radius at point B.
13. Select end point C.
14. Type 2 at the command area.
15. Press the ENTER key.
16. Do not save your file.

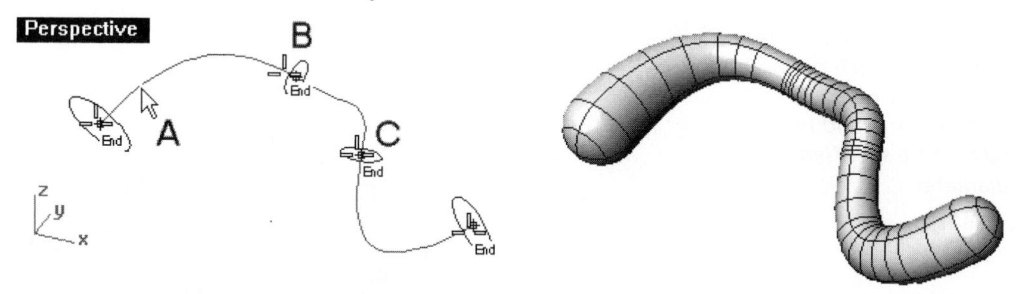

Figure 8–40. Pipe with variable radii and round caps being constructed (left) and pipe constructed (right)

Torus

A torus is single closed-loop surface in the shape of a donut, which resembles a cylinder bent in a ring. To construct a torus, you specify the pitch circle and the cylinder's radius or diameter. You can construct a torus in many ways.

Radius and Diameter of Torus Pitch Circle and Vertical Torus

The implicit circle on which the cylinder is bent is called the pitch circle. By default, the pitch circle is specified by its radius, and the pitch circle is placed horizontally on the active construction plane. However, you can also specify pitch diameter and have the pitch circle placed vertically to the construction plane. Perform the following steps:

1. Select File > Open and select the file Torus.3dm from the Chapter 8 folder on the companion CD.
2. Select Solid > Torus, or click on the Torus button on the Solid toolbar.
3. Select points A, B, and C, shown in Figure 8–41, where A is the center, AB is the radius of the pitch circle, and BC is the radius of the torus tube.
4. Repeat the command.
5. Select the Vertical option on the command area.
6. Select points D, E, and F, shown in Figure 8–41. A torus with its pitch circle vertical to the construction plane is constructed.
7. Repeat the command.
8. Select the Diameter option on the command area.
9. Select points G, H, and J, shown in Figure 8–41, where GH is the diameter of the pitch circle of the torus.

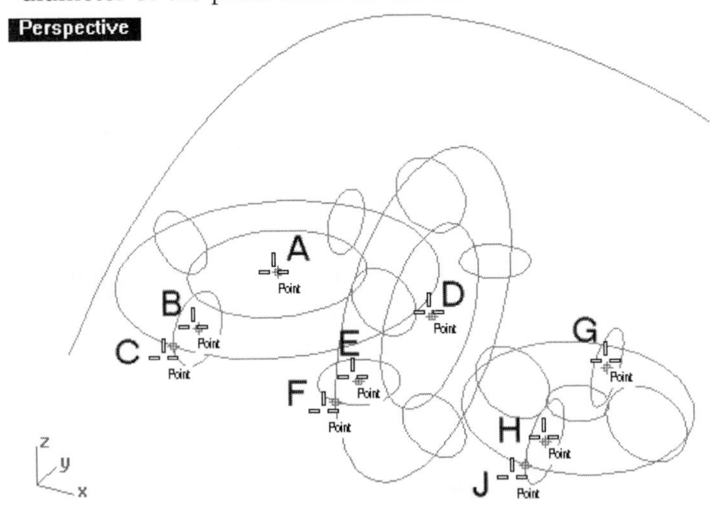

Figure 8–41. From left to right, torus pitch circle specified by radius, vertical torus, and torus pitch circle specified by diameter

Torus Around a Curve

Continue with the following steps to construct a torus with its axis around a curve.

10 Turn on Layer 02 and turn off Layer 00.

11 Select Solid > Torus, or click on the Torus button on the Solid toolbar.

12 Select the AroundCurve option on the command area.

13 Select curve A, shown in Figure 8–42.

14 Select point B on curve A.

15 Type 9 at the command area to specify the pitch radius.

16 Type 3 at the command area to specify the radius

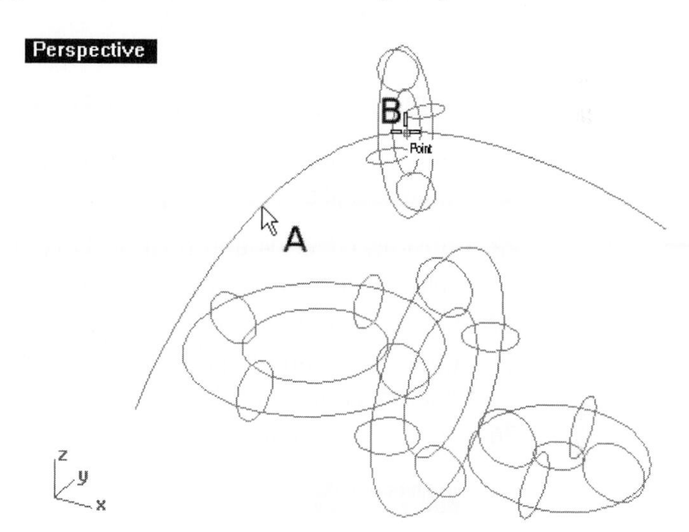

Figure 8–42. Torus with its pitch circle around a curve being constructed

Torus Pitch Circle Tangent to Three Curves

Continue with the following steps to construct a torus with its base circle tangent to three curves.

17 Turn on Layer 03 and turn off Layer 02.

18 Select Solid > Torus, or click on the Torus button on the Solid toolbar.

19 Select the Tangent option on the command area.

20 Select curves A, B, and C, shown in Figure 8–43.

21 Type 3 at the command area to specify the second radius.

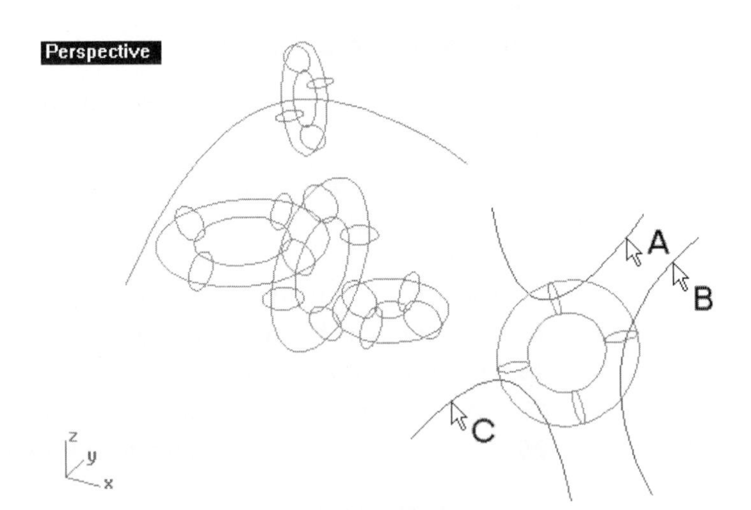

Figure 8–43. Torus with pitch circle tangent to three curves being constructed

Torus Pitch Circle Fitted to a Set of Points

Continue with the following steps to construct a torus with its base circle best fitted to a set of selected points.

22 Turn on Layer 04 and turn off Layer 03.

23 Select Solid > Torus, or click on the Torus button on the Solid toolbar.

24 Select the FitPoints option on the command area.

25 Click on A and drag to B (Figure 8–44) to select the point objects, and press the ENTER key.

26 Type 3 at the command area to specify the radius at the other end.

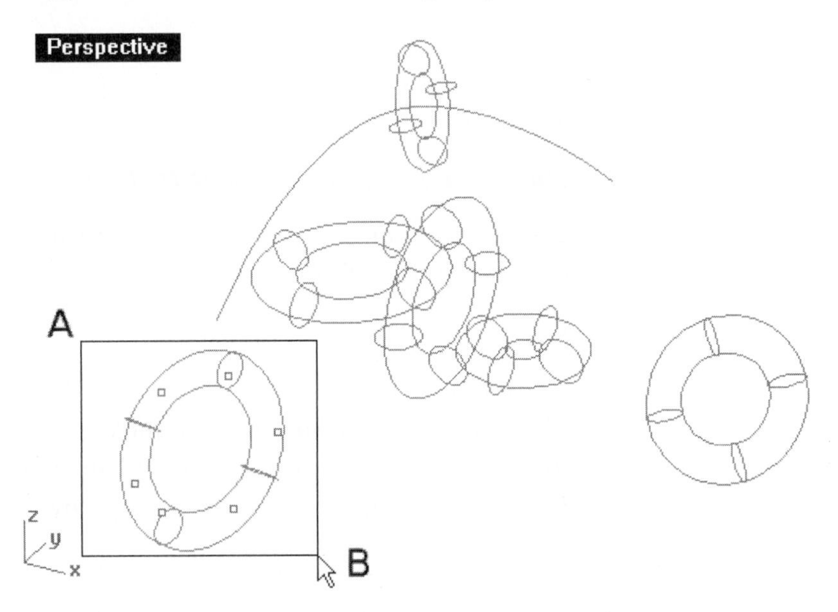

Figure 8–44. Torus with pitch circle best fitted to a set of points being constructed

Torus Pitch Circle Passing through Three Points

Continue with the following steps to construct a torus with its base circle defined by three selected points.

27 Turn on Layer 05 and turn off Layer 04.

28 Select Solid > Torus, or click on the Torus button on the Solid toolbar.

29 Select the 3Point option on the command area.

30 Select points A, B, and C, shown in Figure 8–45.

31 Type 3 at the command area.

32 Do not save your file.

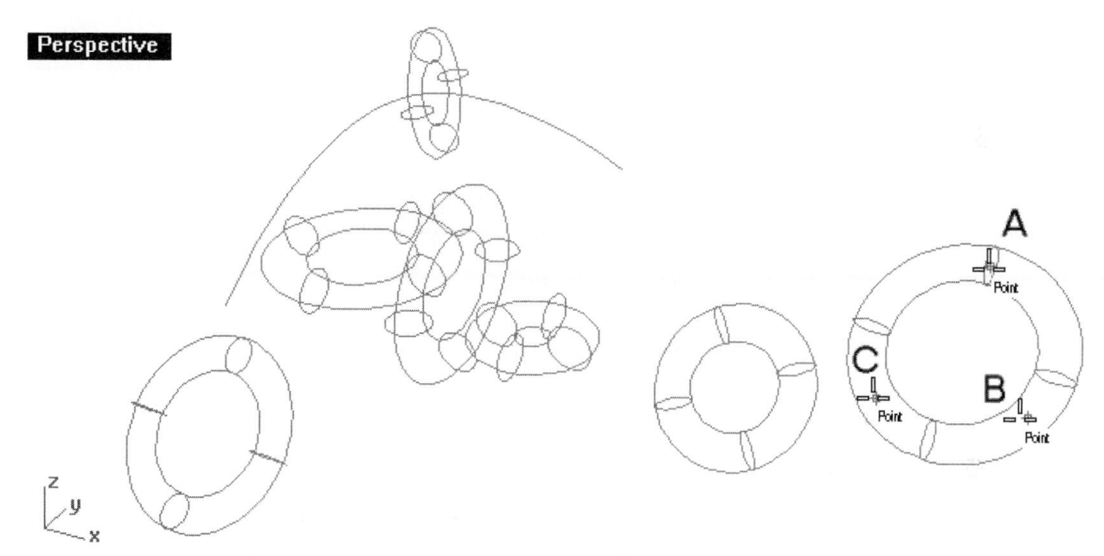

Figure 8–45. Torus with pitch circle defined by three selected points being constructed

Text

A text object is a polysurface consisting of three elements: top and bottom planar surfaces that resemble the text, and surfaces joining the top and bottom surfaces. To construct a text object, perform the following steps:

1 Start a new file. Use the "Small Objects, Millimeters" template.

2 Maximize the Perspective viewport.

3 Select Solid > Text, or click on the Text button on the Main1 toolbar.

4 In the Text Object dialog box, shown in Figure 8–46, type a text string, select a font and font style, specify text height and thickness, check the Solids box, and click on the OK button.

Figure 8–46. Text Object dialog box

5. Click on a location in the Perspective viewport. A text object is constructed. (See Figure 8–47.)
6. Do not save your file.

Figure 8–47. Text object constructed

Constructing Free-Form Solid Objects

In terms of aesthetic design, free-form objects are more commonly used than objects of regular shapes. Naturally, it is important to know how to construct free-form polysurfaces and solids.

The simplest way is to construct a free-form solid is to extrude a closed-planar curve. Another way is to extrude a surface, which in some other applications is referred to as thickening. The third way is to offset a polyline is to cap the ends and then extrude the curves. If you have a polysurface with planar openings, you can cap these planar holes. After capping, if the polysurface becomes a closed-loop polysurface, a solid is formed.

If the solid object consists of free-form surfaces, you can construct the individual surfaces using methods explained in previous chapters, and then compose a solid from the surfaces.

Extrude Planar Curve

As explained in Chapter 3, there are a number of ways to extrude a curve. If the curve to be extruded is a closed planar curve, the same command can be used to obtain a solid, which is a closed polysurface consisting of a surface extruded from the planar curve and two planar surfaces enclosing the extruded surface. Perform the following steps to construct an extruded solid object:

1. Select File > Open and select the file ExtrudePlanarCurve.3dm from the Chapter 8 folder on the companion CD.
2. Select Solid > Extrude Planar Curve > Straight, or click on the Extrude Closed Planar Curve button on the Extrude Solid toolbar.
3. Select curve A, shown in Figure 8–48, and press the ENTER key.
4. If Cap = No, select the Cap option on the command area to change to Yes.
5. Type 10 at the command area to set the extrude distance. An extruded solid is constructed.
6. Do not save your file.

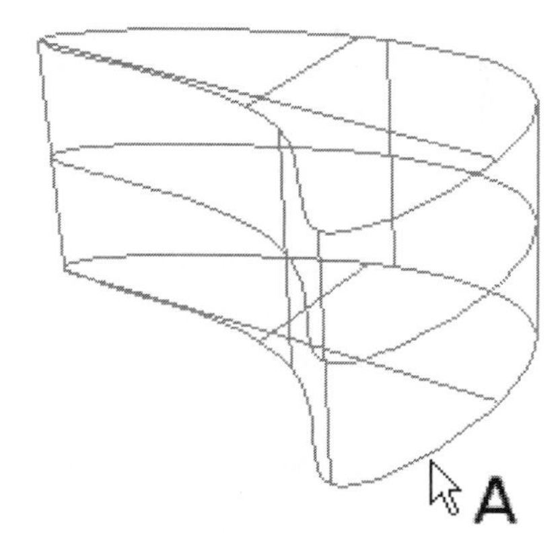

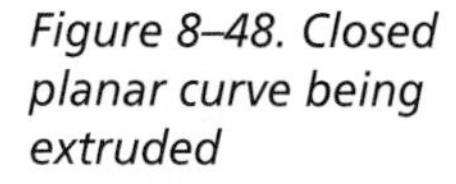
Figure 8–48. Closed planar curve being extruded

Note that this command is the same command used for constructing an extruded surface. Therefore, you can also indicate two points to specify an extrude direction, apply a taper angle in the extrusion process, extrude the curve in two directions, extrude the curve along a curve, extrude the curve along a sub-curve, and extrude the curve to a point. (See Figure 8–49.) Remember to set the “Cap” option to “Yes.”

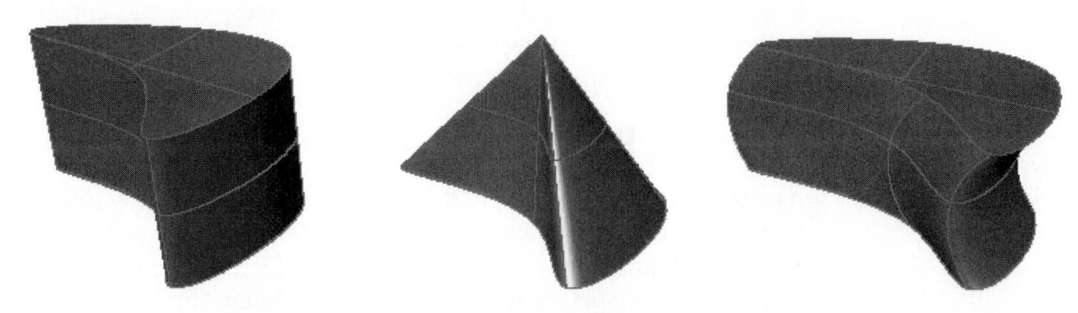

Figure 8–49. From left to right: Extrude in straight direction, extrude to a point, and extude along a curve

Extrude Surface

By extruding a surface, you construct a solid, which is a polysurface consisting of two copies of the original surface a distance apart, and a set of surfaces joining their edges. Perform the following steps:

1. Select File > Open and select the file ExtrudeSurface.3dm from the Chapter 8 folder on the companion CD.
2. Select Solid > Extrude Surface > Straight, or click on the Extrude Surface button on the Extrude Solid toolbar.
3. Select surface A, shown in Figure 8–50, and press the ENTER key.
4. Type 20 at the command area to specify the extrusion distance. The surface is extruded to a solid.
5. Select Solid > Extrude Surface > Along Curve, or click on the Extrude Surface Along Curve/Extrude Surface Along Sub Curve on the Extrude Solid toolbar.
6. Select surface B, shown in Figure 8–50, and press the ENTER key.
7. Select point C (Figure 8–50). The surface is extruded to a point.
8. Select Solid > Extrude Surface > To Point, or click on the Extrude Surface To Point button on the Extrude Solid toolbar.
9. Select surface D (Figure 8–50) and press the ENTER key.
10. Select curve E (Figure 8–50). The surface is extruded along a curve.
11. Select Solid > Extrude Surface to Boundary > Straight.
12. Select surface F (Figure 8–50) and press the ENTER key.
13. Select surface G (Figure 8–50) and press the ENTER key.
14. Click on any spot near surface F to indicate the direction.
15. Turn off the surface layer to appreciate the effect. (See Figure 8–51.)
16. Do not save your file.

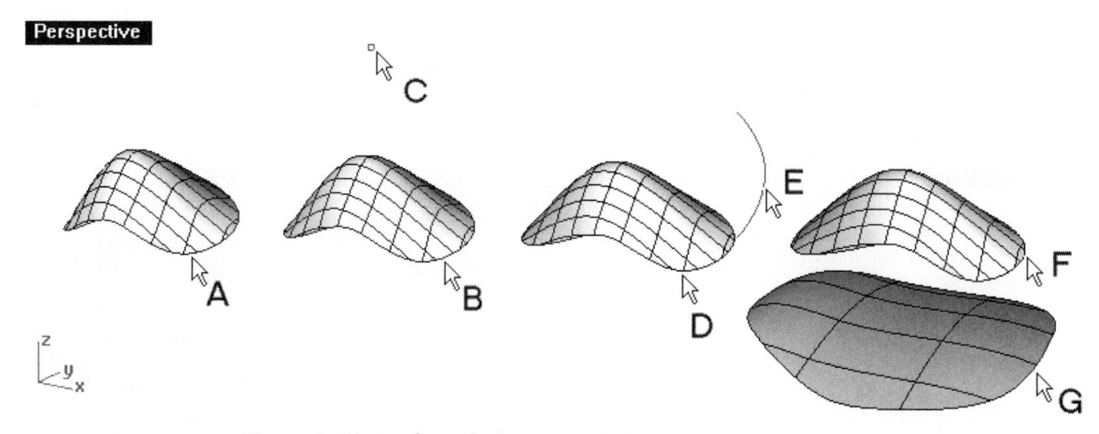

Figure 8–50. Surfaces being extruded

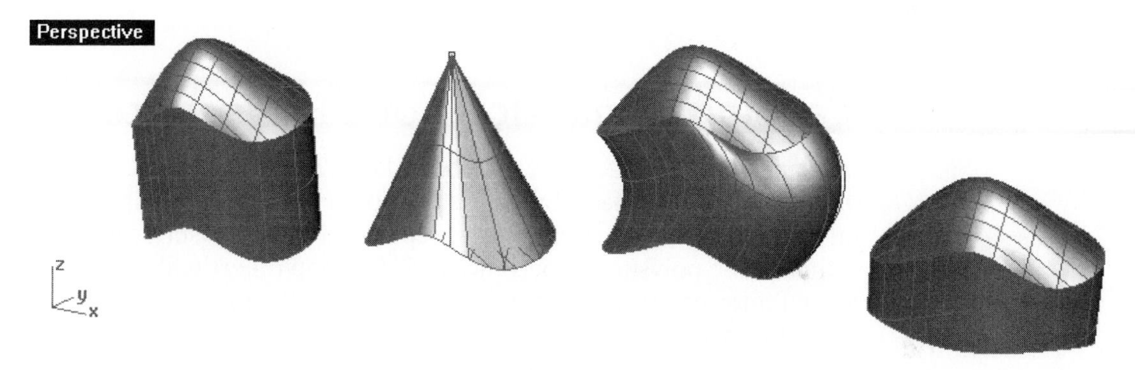

Figure 8–51. Surfaces extruded

Slab

To process of making a slab involves offsetting a polyline to produce a planar surface and then extruding the planar surface a specified distance. Perform the following steps:

1. Select File > Open and select the file Slab.3dm from the Chapter 8 folder on the companion CD.
2. Select Solid > Slab, or click on the Slab from Polyline button on the Extrude Solid toolbar.
3. Select polyline A, shown in Figure 8–52, and press the ENTER key.
4. Select the Distance option on the command area.
5. Type 5 to specify the offset distance.

6 Click on location B to indicate the offset direction.
7 Type 10 to specify the height of the slab.
8 Do not save your file.

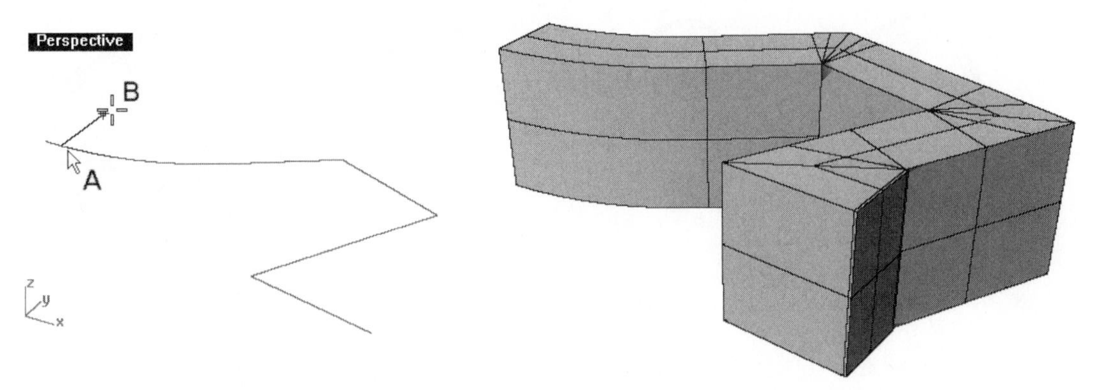

Figure 8–52. Polyline selected (left) and slab constructed (right)

Capping Planar Holes of a Polysurface

A planar hole is a hole with its boundary lying on a flat plane. Planar holes on a polysurface can be closed by capping, which is a process in which the planar openings are replaced by planar surfaces. If, after capping, the polysurface becomes a closed-loop object without openings or self-intersections, a solid is formed. Perform the following steps to cap a polysurface:

1 Select File > Open and select the file CapPlanarHole.3dm from the Chapter 8 folder on the companion CD.
2 Select Solid > Cap Planar Holes, or click on the Cap Planar Holes button on the Solid Tools toolbar.
3 Select surface A (shown in Figure 8–53) and press the ENTER key.
4 Do not save your file.

The surface is capped. Note that capping only works on planar holes. For holes that are not planar, you can use the Patch command to construct patch surfaces instead.

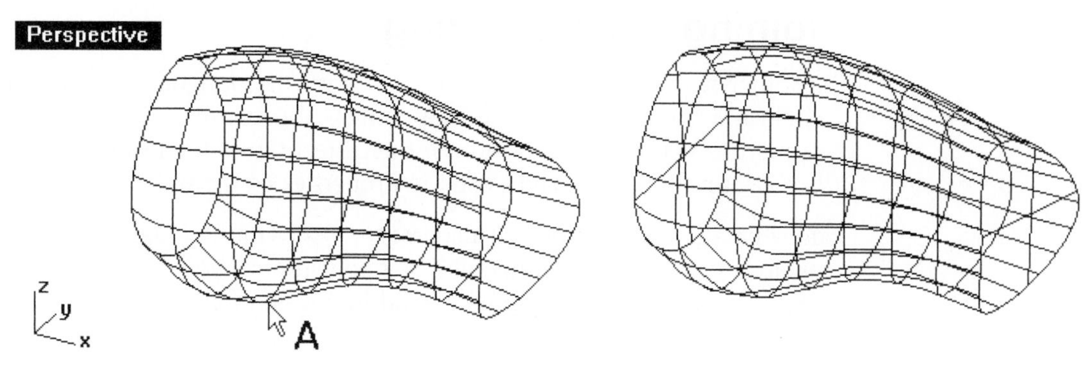

Figure 8–53. Surface being capped (left) and capped surface (right)

Solid from Closed-Loop Polysurface

You can construct a solid from a set of contiguous surfaces with common edges. Among the surfaces, there must not be any openings, gaps, or intersection except at their common edges and vertices. Perform the following steps:

1. Select File > Open and select the file CreateSolid.3dm from the Chapter 8 folder on the companion CD.
2. Select Solid > Create Solid, or click on the Create Solid button on the Solid Tools toolbar.
3. Click on A and drag to B, shown in Figure 8–54, and press the ENTER key. The surfaces area converted to a solid.
4. Do not save your file.

Figure 8–54. Solid being constructed from a set of surfaces

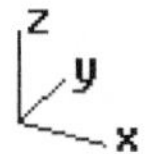

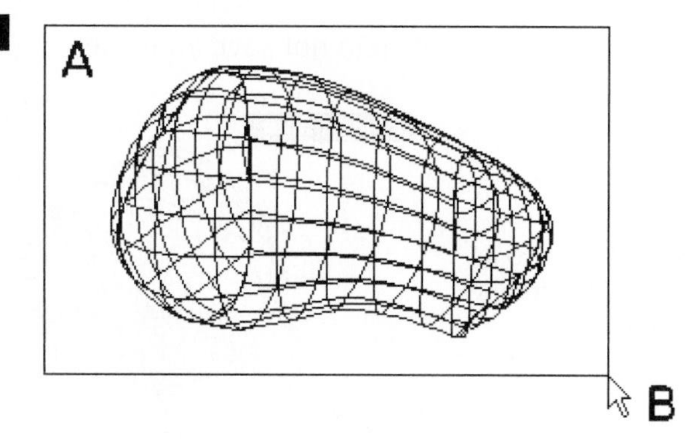

Joining and Exploding

Another way to form a solid is to join a set of contiguous surfaces that enclose a volume without any openings. You can examine an object's properties via the Detail button of the Object tab of the Properties dialog box. Naturally, you have to use the Properties command.

To separate a polysurface or a solid into a set of individual surfaces, you explode it. Because a Rhino solid is a polysurface, you can use the same Explode command to separate a solid into a set of surfaces.

Combining Rhino Solids

To compose a complex solid object, you can first construct two or more solid objects and then combine them using one of the three types of Boolean operations. Booleans generally work better if two objects overlap each other slightly.

Union

A union of a set of solids produces a solid that has the volume of all solids in the set. To construct a solid by uniting two solids, perform the following steps:

1. Select File > Open and select the file Union.3dm from the Chapter 8 folder on the companion CD.
2. Select Solid > Union, or click on the Boolean Union button on the Solid Tools toolbar.
3. Select solids A and B (shown in Figure 8–55) and press the ENTER key. The solids are united.
4. Do not save your file.

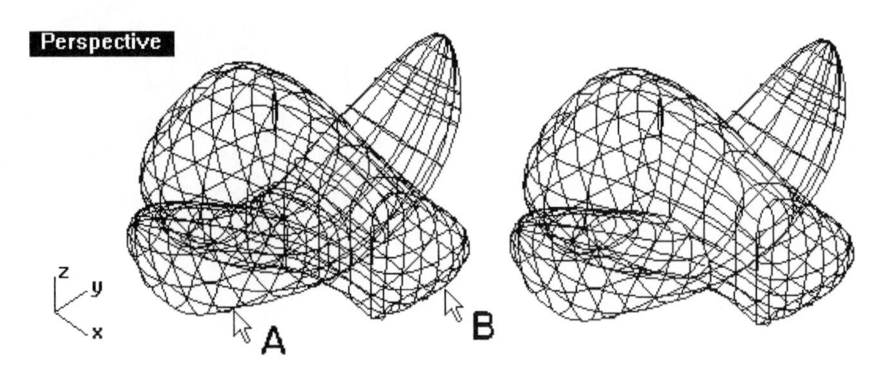

Figure 8–55. Two solids being united (left) and single solid (right)

Difference

A difference of two sets of solids produces a solid that has the volume contained in the first set of solids but not in the second set of solids. To construct a solid by subtracting one solid from another solid, continue with the following steps:

1. Select File > Open and select the file Subtract.3dm from the Chapter 8 folder on the companion CD.
2. Select Solid > Difference, or click on the Boolean Difference button on the Solid Tools toolbar.
3. Select solid A (shown in Figure 8–56) and press the ENTER key. (This is the first set of solids.)
4. Click on B and drag to C (shown in Figure 8–56) and press the ENTER key. (This is the second set of solids.) The second set of solids is subtracted from the first set of solids.
5. Do not save your file.

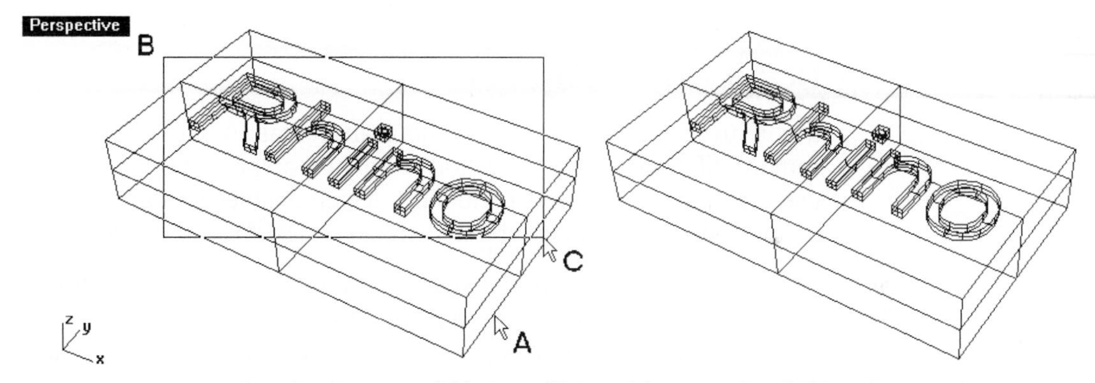

Figure 8–56. One solid being subtracted from another (left) and single solid (right)

Intersection

An intersection of two sets of solids produces a solid that has the volume contained in both the first and second sets of solids. To construct a solid that has the volume contained in two solids, continue with the following steps:

1. Select File > Open and select the file Intersect.3dm from the Chapter 8 folder on the companion CD.
2. Select Solid > Intersection, or click on the Boolean Intersection button on the Solid Tools toolbar.

3 Select solid A (shown in Figure 8–57) and press the ENTER key. (This is the first set of solids.)
4 Select solid B (shown in Figure 8–57) and press the ENTER key. (This is the second set of solids.) A solid of intersection is constructed.
5 Do not save your file.

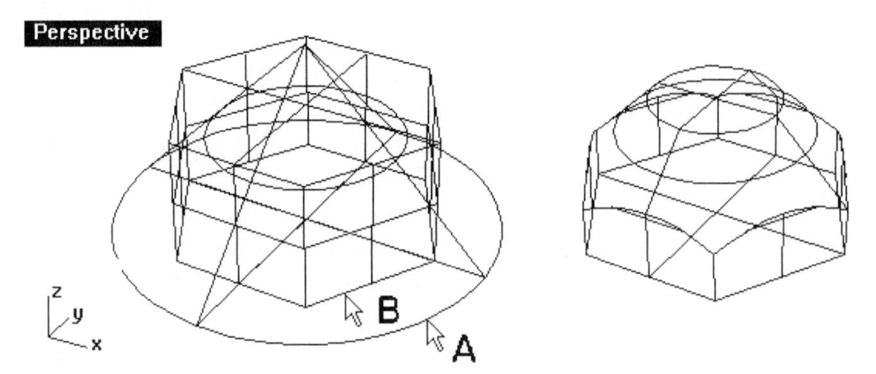

Figure 8–57. Two solids being intersected (left) and single solid (right)

Boolean 2 Objects

If you are not quite sure which Boolean operation should be performed (union, difference, or intersection) to obtain a specific outcome, you can use the Boolean 2 objects command, which enables you to cycle through various possibilities in combining two solids. Perform the following steps:

1 Select File > Open and select the file Boolean2Objects.3dm from the Chapter 8 folder on the companion CD.
2 Select Solid > Boolean Two Objects, or right-click on the Boolean Split/Boolean 2 Objects button on the Solid Tools toolbar.
3 Select objects A and B, shown in Figure 8–58.
4 Click to cycle through various possibilities.
5 Press the ENTER key to accept.
6 The objects are combined, using one of the Boolean operations chosen. Do not save your file.

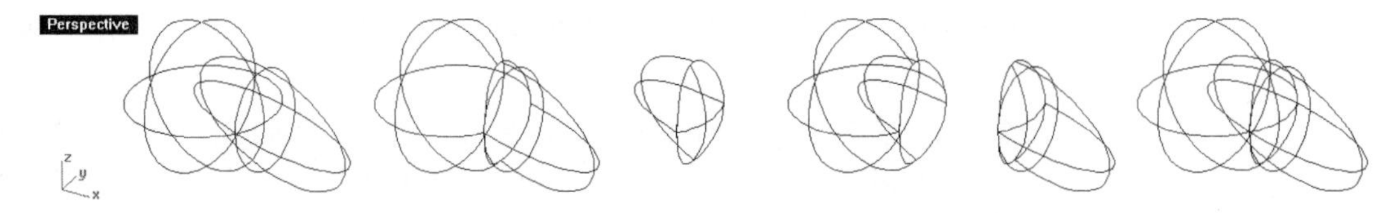

Figure 8–58. From left to right: two objects being booleaned and five possibilities (join, intersect, subtract, subtract, and split)

Detailing a Solid

To detail a Rhino solid, you can treat the edges by constructing constant/variable chamfers, fillets, and blends; you can construct bosses or a ribs on selected faces; and you can make holes in various ways.

Edge Treatment

There are three ways to treat the edges of a polysurface or a solid. You can construct a variable distance chamfer, a constant/variable radius fillet, or a constant/variable radius blend.

Variable Chamfer Edge

You can construct a variable distance chamfer surface on edges of a polysurface or solid, as follows:

1. Select File > Open and select the file VariableChamfer.3dm from the Chapter 8 folder on the companion CD.
2. Select Solid > Fillet Edge > Chamfer Edge, or click on the Variable Radius Chamfer button on the Solid Tools toolbar.
3. Select the CurrentChamferDistance option on the command area.
4. Type 6 to set the default chamfer distance for all selected edges. *(Note: The value affects edges selected subsequently.)*
5. Select edges A, B, C, D, and E, shown in Figure 8–59, and press the ENTER key.

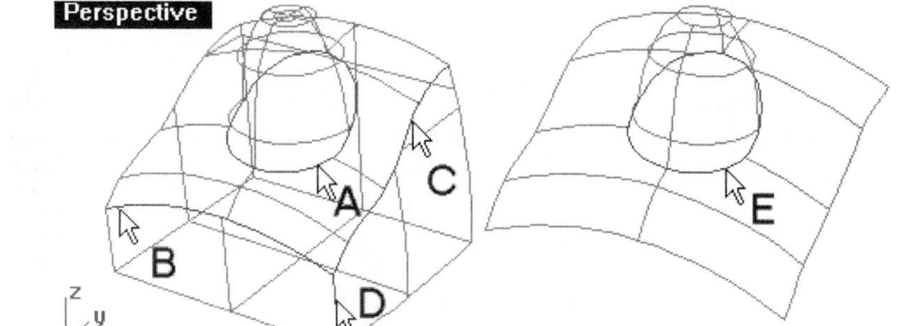

Figure 8–59. Edges of a solid (left) and edge of a polysurface (right) selected

6. Select the AddHandle option on the command area.
7. Click on location A (Figure 8–60) and press the ENTER key. A new handle is added. (Exact location is unimportant.)
8. Click on the added handle B (Figure 8–60).

9 Type 10 at the command area to set the chamfer distance at the added handle.

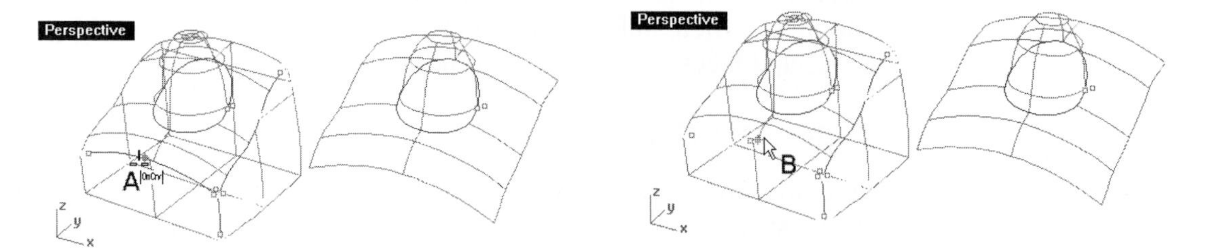

Figure 8–60. Handle added and being modified

10 Select the RailType on the command area.
11 From the three options, select DistFromEdge.
12 Select the Preview option to display a preview.
13 Press the ENTER key. Variable chamfer edges are constructed. (See Figure 8–61.)
14 Do not save your file.

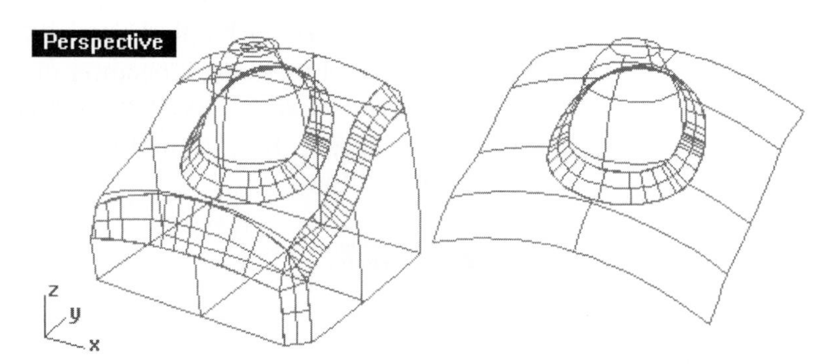

Figure 8–61. Preview (left) and edges chamfered (right)

Variable Fillet Edge

You can construct variable radius fillet surface on edges of a polysurface or solid, as follows:

1 Select File > Open and select the file VariableFillet.3dm from the Chapter 8 folder on the companion CD.
2 Select Solid > Fillet Edge > Fillet Edge, or click on the Variable Radius Fillet/Variable Radius Blend button on the Solid Tools toolbar.
3 Select the Current Radius option on the command area.

4 Type 10 at the command area to set default value.

5 Select edges A, B, C, D, and E, shown in Figure 8–62, and press the ENTER key.

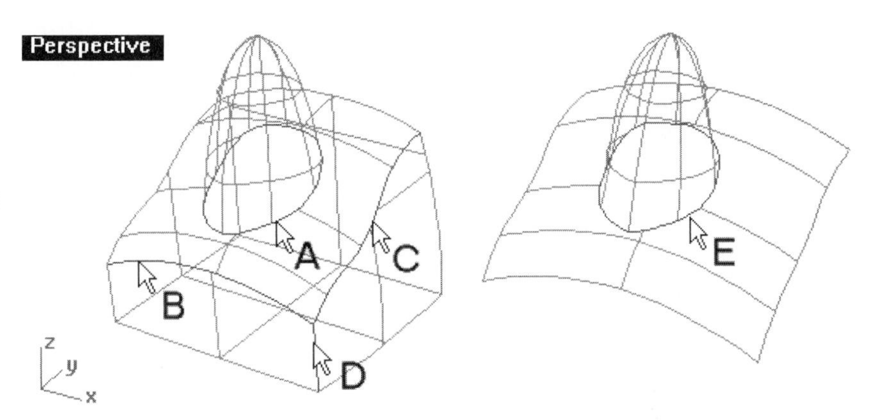

Figure 8–62. Edges of a solid (left) and edge of a polysurface (right) selected

6 Select the AddHandle option on the command area.

7 Click on edge A, shown in Figure 8–63, and press the ENTER key to add a handle.

8 Click on the added handle B, shown in Figure 8–63.

9 Type 6 at the command area to set the handle's value.

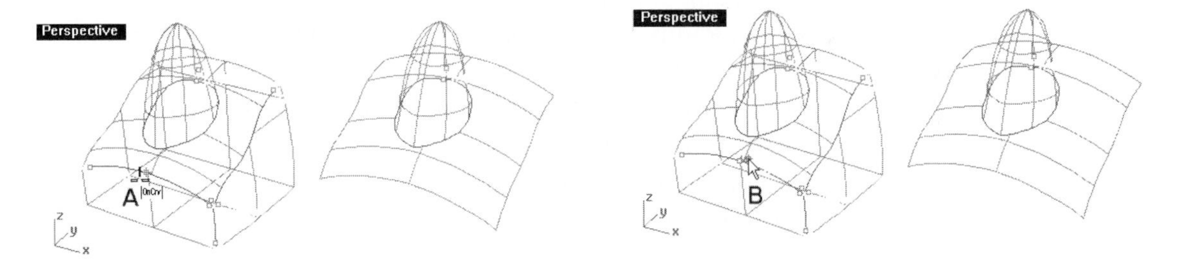

Figure 8–63. A handle added (left) and handle selected (right)

10 Select the RailType option on the command area.

11 Select the DistBetweenRails option.

12 Select the Preview option.

13 If you are satisfied with the preview's edges, press the ENTER key. Edges are filleted. (See Figure 8–64.)

14 Do not save your file.

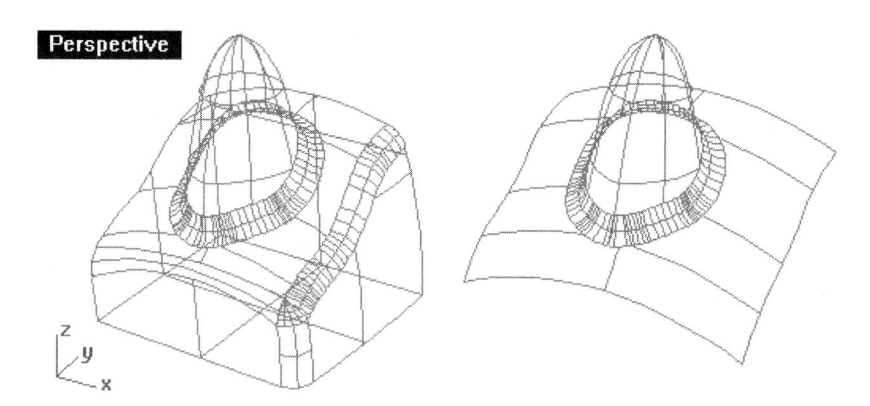

Figure 8–64. Preview (left) and final filleted edges (right)

Variable Blend Edge

A blended edge has G2 continuity. Not only are the tangency directions of the contiguous surfaces the same, the radii of curvature are also identical. Perform the following steps:

1. Select File > Open and select the file VariableBlend.3dm from the Chapter 8 folder on the companion CD.
2. Select Solid > Fillet Edge > Blend Edge, or right-click on the Variable Radius Fillet/Variable Radius Blend button on the Solid Tools toolbar.
3. Select the Current Radius option on the command area.
4. Type 6 at the command area.
5. Select edges A, B, C, D, and E, shown in Figure 8–65, and press the ENTER key.

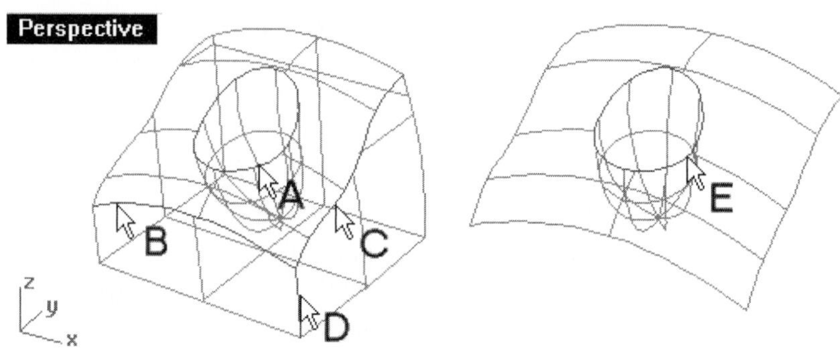

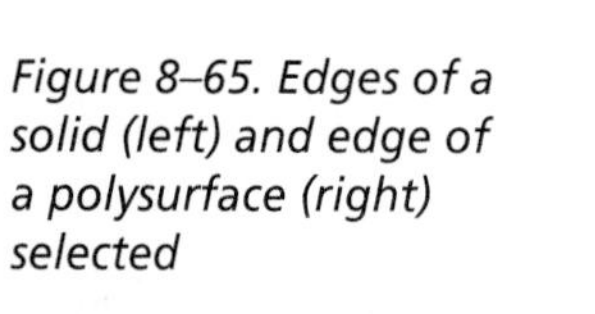

Figure 8–65. Edges of a solid (left) and edge of a polysurface (right) selected

6 Select the AddHandle option on the command area.
7 Click on edge A, shown in Figure 8–66, and press the ENTER key to add a handle.
8 Click on the added handle B, shown in Figure 8–66.
9 Type 10 at the command area to set the handle's value.

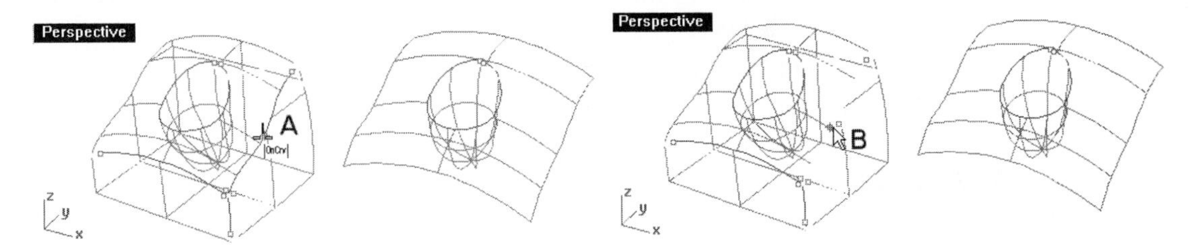

Figure 8–66. Handle being added (left) and handle selected (right)

10 Select the RailType option on the command area.
11 Select the Rolling Ball option.
12 Select the Preview option.
13 If you are satisfied with the preview's edges, press the ENTER key. Edges are filleted. (See Figure 8–67.)
14 Do not save your file.

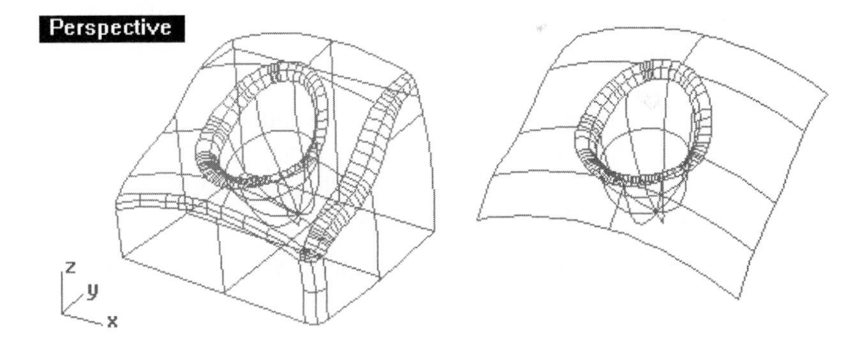

Figure 8–67. Preview (left) and edges blended (right)

Boss

A boss is constructed by extruding a planar curve in 3D space to a boundary solid object. Optionally, you may include a draft angle. The boss constructed will be united with the boundary solid. Perform the following steps:

1 Select File > Open and select the file Boss.3dm from the Chapter 8 folder on the companion CD.

2 Select Solid > Boss, or click on the Boss button on the Extrude Solid toolbar.
3 Select the Mode option on the command area.
4 Select the Draft Angle option on the command area.
5 Type 3 at the command area to specify the draft angle.
6 Select closed planar curve A, shown in Figure 8–68, and press the ENTER key.
7 Select solid B, shown in Figure 8–68. A boss is constructed.
8 Do not save your file.

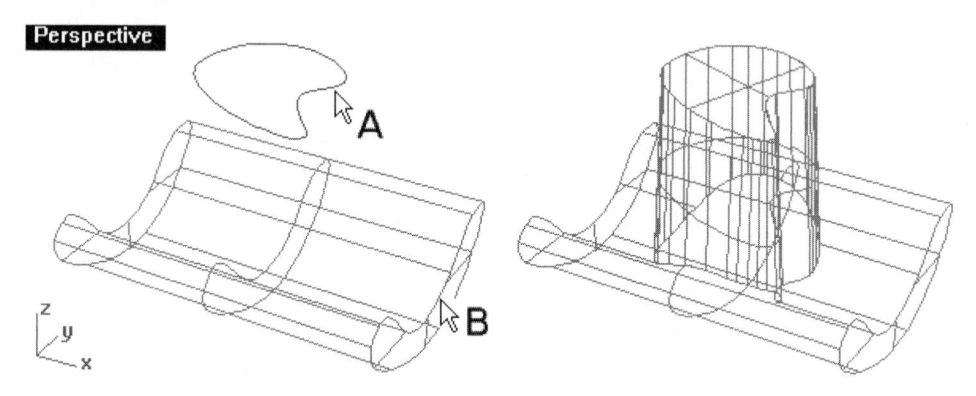

Figure 8–68. Boss being constructed (left) and boss constructed (right)

Rib

A rib is constructed by offsetting a planar curve and then extruding it towards a boundary solid. The shape of a rib is quite similar to a slab solid. While constructing the rib, you may include a draft angle. Perform the following steps:

1 Select File > Open and select the file Rib.3dm from the Chapter 8 folder on the companion CD.
2 Select Solid > Rib, or click on the Rib button on the Extrude Solid toolbar.
3 Select the Distance option on the command area.
4 Type 3 at the command area to specify the offset distance.
5 Select the Mode option on the command area.
6 Select the DraftAngle option on the command area.
7 Type 1 at the command area to specify the draft angle.

8. Select curve A, shown in Figure 8–69, and press the ENTER key.
9. Select solid B, shown in Figure 8–69. A rib with a draft angle is constructed.
10. Do not save your file.

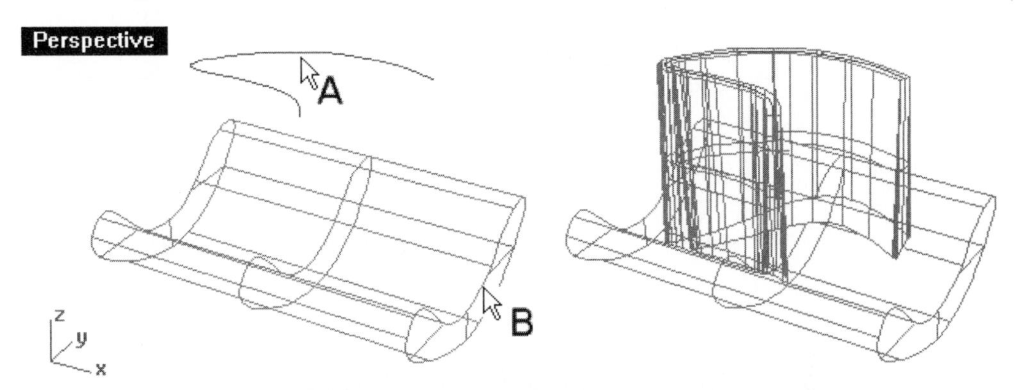

Figure 8–69. Rib being constructed (left) and rib constructed (right)

Hole

There are two general kinds of holes: round holes and holes of irregular shape. To construct a hole of irregular shape, you construct a planar closed curve to define the cross-section and then extrude it towards a solid or polysurface. Perform the following steps to construct a hole of irregular shape:

1. Select File > Open and select the file Hole.3dm from the Chapter 8 folder on the companion CD.
2. Select Solid > Solid Edit Tools > Holes > Make Hole, or click on the Make Hole/Place Hole button on the Holes toolbar.
3. Select curve A, shown in Figure 8–70.
4. Select closed polysurface B, shown in Figure 8–70.
5. Type 12 at the command area to indicate the depth of cut.
6. Click on location C, shown in Figure 8–70. A hole is cut.
7. Do not save your file.

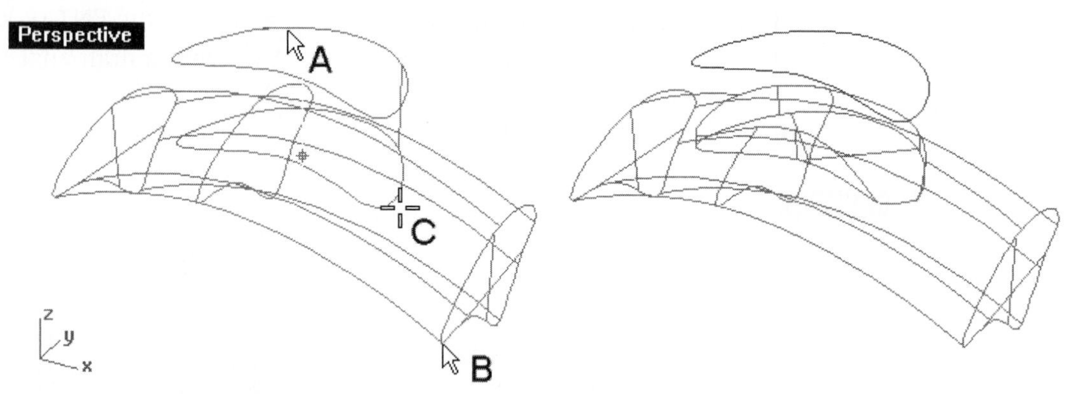

Figure 8–70. Closed planar curve being used to cut a solid (left) and hole cut (right)

Placing Multiple Holes

If a number of holes is required to be made on a solid, you can first construct a planar curve to depict the shape of the hole, and then construct multiple copies on a selected surface of a solid or polysurface. Perform the following steps:

1. Select File > Open and select the file PlaceHole.3dm from the Chapter 8 folder on the companion CD.
2. Select Solid > Solid Edit Tools > Holes > Place Hole, or right-click on the Make Hole/Place Hole button on the Holes toolbar.
3. Select planar curve A, shown in Figure 8–71.
4. Click on a point along the curve to specify the base point. Exact location is unimportant for this tutorial.
5. Select target surface B.
6. Select target point C.
7. Type 4 at the command area to specify the depth.
8. Press the ENTER key to accept the default rotation angle.
9. Select points D, E, and F, and press the ENTER key. Four holes are placed.
10. Do not save your file.

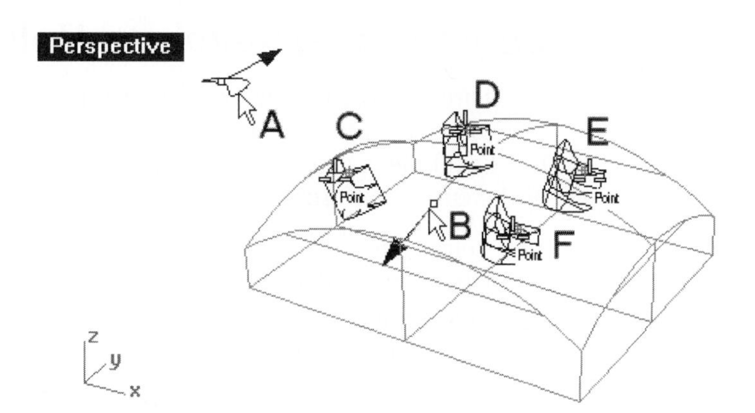

Figure 8–71. Mulitple copies of hole being placed on a surface of a solid

Constructing Round Holes

If the hole to be constructed is a round hole, you do not have to construct a circle to depict the hole's cross-section. You simply specify a radius and location on a surface of a solid or polysurface. Perform the following steps:

1. Select File > Open and select the file RoundHole.3dm from the Chapter 8 folder on the companion CD.
2. Select Solid > Solid Edit Tools > Holes > Round Hole, or click on the Round Hole/Place Hole button on the Holes toolbar.
3. Select surface A, shown in Figure 8–72.
4. Select the Depth option on the command area.
5. Type 5 to specify the depth of the hole.
6. Select the Radius option on the command area.
7. Type 3 to specify the radius.
8. Select point B, C, and D, and press the ENTER key. Three holes are placed.
9. Do not save your file.

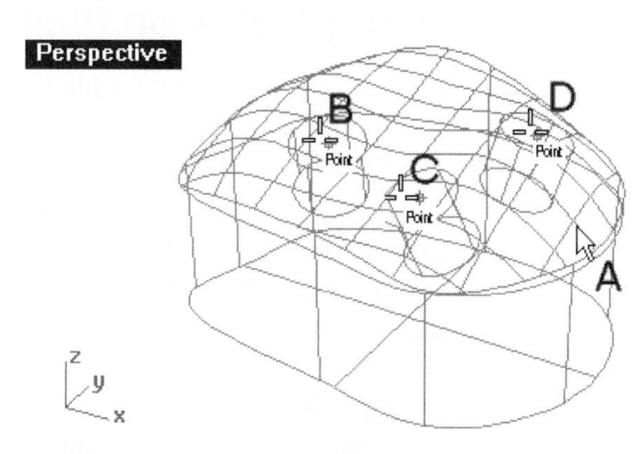

Figure 8–72. Hole being placed on a surface of a solid

Constructing Revolved Hole

Apart from constructing round holes and holes of irregular shape, you can construct a hole by revolving a curve. The axis of revolution of the revolved hole is determined by an imaginary line joining two end points of the curve. Perform the following steps:

1. Select File > Open and select the file RevolveHole.3dm from the Chapter 8 folder on the companion CD.
2. Select Solid > Solid Edit Tools > Holes > Revolved Hole, or click on the Revolved Hole button on the Holes toolbar.
3. Select curve A, shown in Figure 8–73.
4. Select endpoint B as the base point.
5. Select target face C
6. Select point D and press the ENTER key. A revolved hole is constructed. *(Note: You can place multiple copies.)*
7. Repeat the command, using curve E as the profile curve, end point F as the base point, and point G of surface C as the target.
8. Do not save your file.

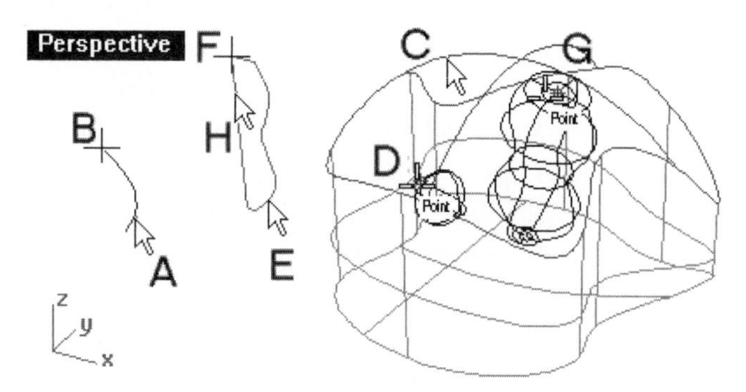

Figure 8–73. Revolved holes being placed

Arraying Holes on Planar Face

Holes already constructed on a planar face of a solid or polysurface can be repeated by arraying in a rectangular pattern or a polar pattern. Perform the following steps:

1. Select File > Open and select the file ArrayHole.3dm from the Chapter 8 folder on the companion CD.
2. Select Solid > Solid Edit Tools > Holes > Array Hole, or click on the Array Hole button on the Holes toolbar.
3. Select edge A, shown in Figure 8–74.
4. Type 4 at the command area to specify the number of holes in direction A. "A" is the first direction.

5 Type 3 at the command area to specify the number of holes in B direction. "B" is the second direction.
6 If the Rectangular option shown on the command area is Yes, click on it to change to No.
7 Click on end points B and C to specify direction A.
8 Click on end points B and D to specify direction B.
9 Select the ASpacing option on the command area.
10 Type 5 to specify the spacing along direction A.
11 Select the BSpacing option on the command area.
12 Type 4 to specify the spacing along direction B.
13 Press the ENTER key. A rectangular pattern of holes is constructed.

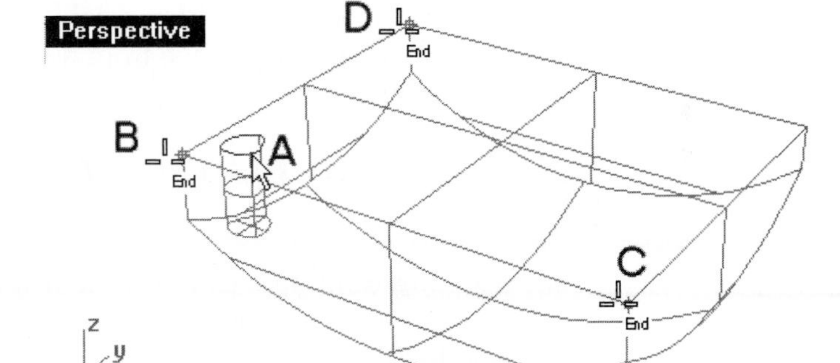

Figure 8–74. Rectangular pattern of holes being constructed

14 Select Solid > Solid Edit Tools > Holes > Array Hole Polar, or click on the Array Hole Polar button on the Holes toolbar.
15 Select edge A, shown in Figure 8–75.
16 Select endpoint B as the center of array.
17 Type 4 to specify the number of holes.
18 Type -35 to specify the included angle.
19 Press the ENTER key. A polar pattern of holes is constructed.
20 Do not save your file.

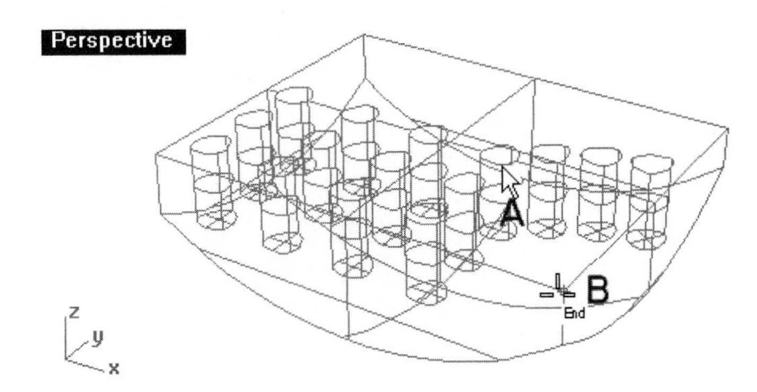

Figure 8–75. Polar pattern of holes constructed

When arraying, the following points have to be noted: Because this command simply repeats a hole that is already constructed, a rectangular or polar pattern of holes in a solid with curved surface on the opposite side of the hole may not result with a set of through holes. To avoid an unexpected outcome, do not have the repeated copies of holes overlap each other.

Editing Solids

There are several ways to edit solids. You can merge coplanar faces resulting from Boolean operations. You can split a solid by using a surface or a curve. To change the shape, you can manipulate its faces or its edges. If a surface is needed from a solid or polysurface, you can extract it. However, extracting a surface from a solid separates the surface from the solid or polysurface, leaving an opening.

Solid Cleaning Up: Merging Co-Planar Faces

Solids resulting from application of Boolean operations may have overlapping co-planar faces. You can clean up the solid by merging co-planar faces, as follows:

1. Select File > Open and select the file MergeFaces.3dm from the Chapter 8 folder on the companion CD.
2. Select Solid > Union, or click on the Boolean Union button on the Solid Tools toolbar.
3. Select solids A and B (shown in Figure 8–75) and press the ENTER key. The solids are united. Note that overlapped faces resulted from the Boolean operation.
4. Select Solid > Solid Edit Tools > Faces > Merge face, or click on the Merge two coplanar faces/Merge all coplanar faces on the Solid Editing toolbar.
5. Select faces C and D, shown in Figure 8–76.
6. Repeat the command.
7. Select faces C and E, shown in Figure 8–76.
8. Right-click on the Merge two coplanar faces/Merge all coplanar faces button on the Solid Tools toolbar
9. Select solid F, shown in Figure 8–76. All coplanar faces are merged. (See Figure 8–77.)
10. Do not save your file.

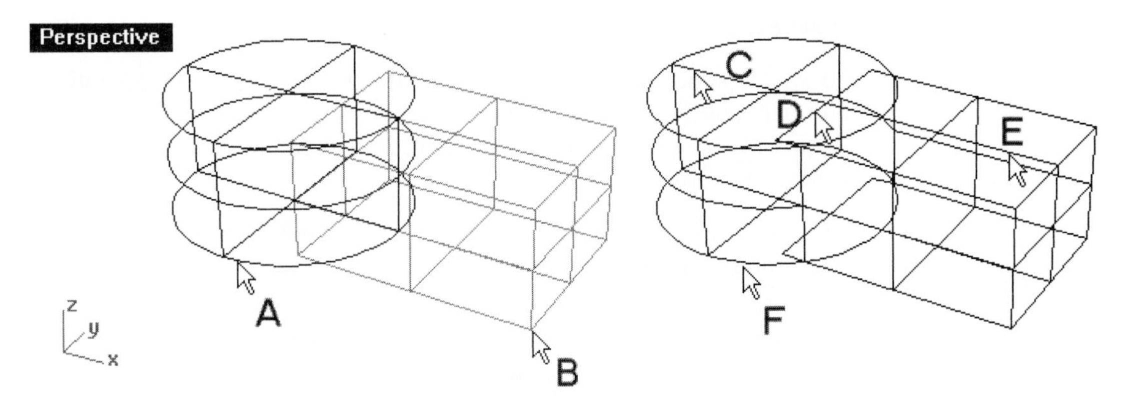

Figure 8–76. Two solids (left) and united solid (right)

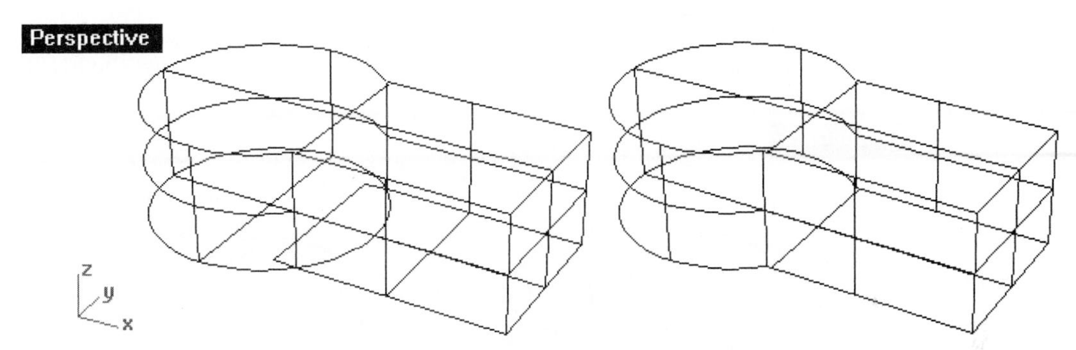

Figure 8–77. Top faces of the solid merged (left) and all faces of the solid merged (right)

Splitting and Trimming a Polysurface or a Solid

To split or trim a polysurface or a solid, you can use a curve, a surface, or a polysurface as cutting objects. If the cutting object is a curve and is residing on the surface or polysurface, the curve will be pulled back to cut. If the cutting curve is not on the surface or polysurface, the curve will be extruded through the surface or polysurface in a direction perpendicular to the active construction plane. Like splitting or trimming a surface, the cutting object, if it is an open-looped surface or polysurface, has to pass through the entire polysurface for trimming action to carry out. However, splitting or trimming a closed-loop surface or polysurface will produce an open-loop surface or polysurface, which is no longer a solid.

Boolean Splitting and Splitting

To split a solid into two solids, you use the Boolean split operation. To compare splitting a solid by using the BooleanSplit command on the Solid Tools toolbar with the Split command on the Main2 toolbar, perform the following steps:

1. Select File > Open and select the file Split.3dm from the Chapter 8 folder on the companion CD.
2. Select Solid > Boolean Split, or click on the Boolean Split/ Boolean 2 Objects button on the Solid Tools toolbar.
3. Select solid A, shown in Figure 8–78, and press the ENTER key.
4. Select surface B, shown in Figure 8–78, and press the ENTER key.
5. Select Edit > Split, or click on the Split button on the Main2 toolbar.
6. Select solid C, shown in Figure 8–78, and press the ENTER key.
7. Select surface D, shown in Figure 8–78, and press the ENTER key.
8. Select Edit > Visibility > Hide.
9. Select A, B, C, and D, shown in Figure 8–78, and press the ENTER key.

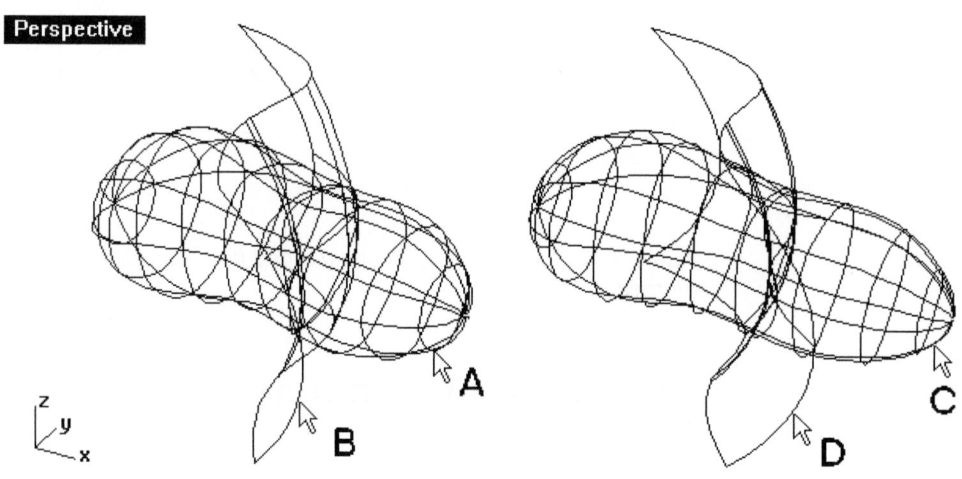

Figure 8–78. Solids being split in two ways

10. Shade the display. (See Figure 8–79.)
11. Do not save your file.

The split command leaves the split polysurface open, but the Boolean Split command uses the splitting object to patch the opening.

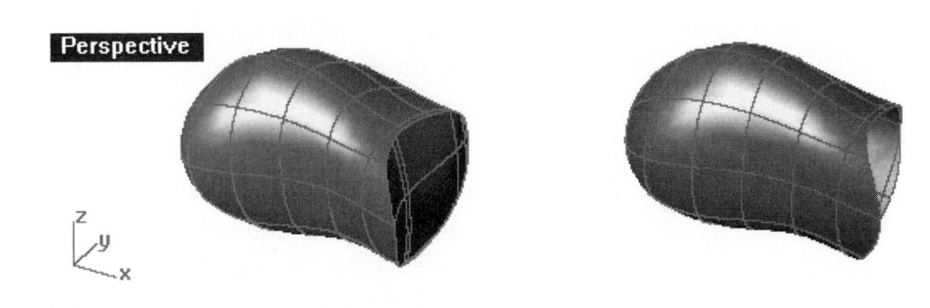

Figure 8–79. Objects removed and viewport shaded

Trimming and Wire-Cutting a Solid

A curve can be used to cut or trim a solid in two ways. Using the trim command on the Edit menu, the solid will become an open-loop polysurface after trimming. By wire-cutting, the resulting solid remains a closed-loop polysurface. Perform the following steps:

1. Select File > Open and select the file WireCut.3dm from the Chapter 8 folder on the companion CD.
2. Select Edit > Trim or click on Trim/Untrim Surface button on the Main1 toolbar.
3. Select curves A and B, shown in Figure 8–80, and press the ENTER key.
4. Select C and D, shown in Figure 8–80, and press the ENTER key. The solids are trimmed and they become open-loop surfaces/polysurface.

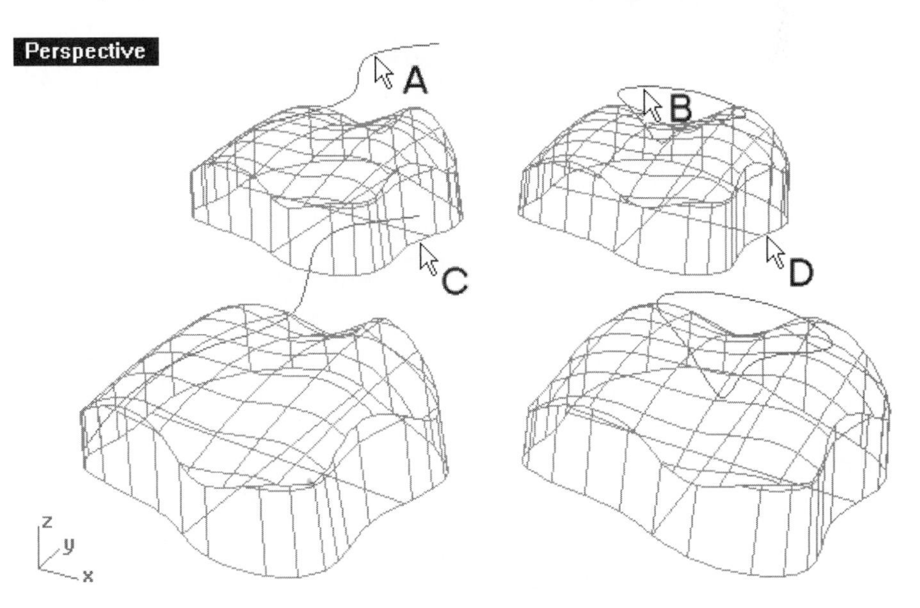

Figure 8–80. Solids being trimmed

5. Select Solid > Solid Edit Tools > Wire Cut, or click on the Wire Cut button on the Solid Editing toolbar.
6. Select curve A, shown in Figure 8–81.
7. Select solid B, shown in Figure 8–81.
8. Press the ENTER key to cut through the solid.
9. If the outer portion of the solid is highlighted, press the ENTER key to accept. If the inner portion of the solid is highlighted, select the Flip option on the command area and then press the ENTER key. The outer portion is wire-cut away.
10. Repeat the command.
11. Select curve C, shown in Figure 8–81.
12. Select solid D, shown in Figure 8–81.
13. Press the ENTER key to cut through the solid.
14. Select the Direction option the command area.
15. Select the X option on the command area.
16. Type 30 at the command area to specify the cut width in X direction.
17. Press the ENTER key to accept.
18. If the right-side portion of the solid is highlighted, press the ENTER key to accept. If the left-side portion of the solid is highlighted, select the Flip option on the command area and then press the ENTER key. The solid is wire-cut. (See Figure 8–81.)
19. Do not save your file.

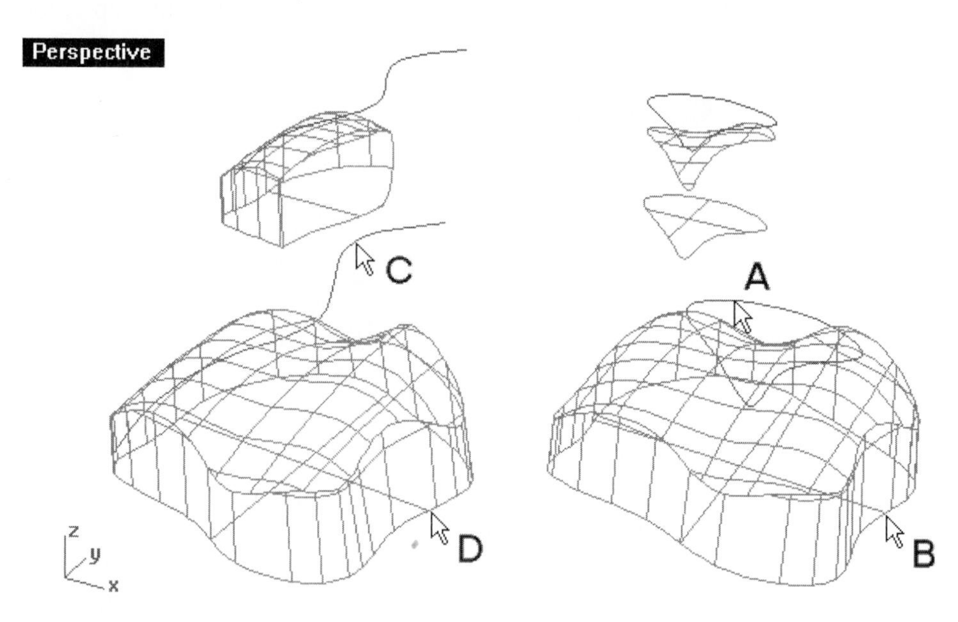

Figure 8–81. Solids being wire-cut

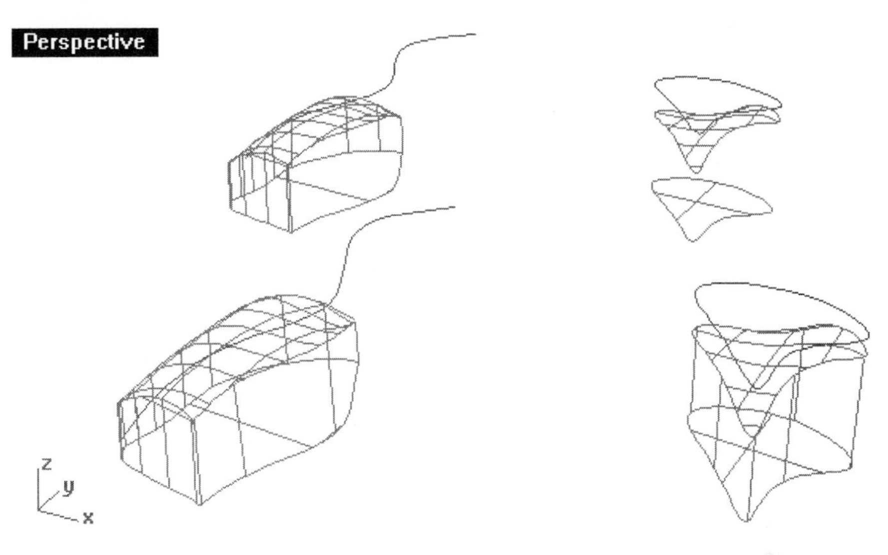

Figure 8–82. Solids wire-cut

Manipulating Faces of a Solid

A solid's shape can be modified by moving, rotating, shearing, or extruding its faces. Because faces adjacent to the selected faces have to change in shape to respond, these operations may not always be successful if one of the adjacent faces cannot be changed

Face Moving

A way to modify the shape of a solid is to move its face(s) to a new location. When a face or a number of faces are moved, its adjacent faces have to be stretched. If one of the adjacent faces cannot be stretched, the moving operation will be unsuccessful. Perform the following steps:

1. Select File > Open and select the file MoveFace.3dm from the Chapter 8 folder on the companion CD.
2. Select Solid > Solid Edit Tools > Faces > Move Face, or click on the Move Face/Move Untrimmed Face button on the Solid Editing toolbar.
3. Select face A, shown in Figure 8–83, and press the ENTER key.
4. Click on the Ortho button on the status bar to turn on ortho mode.
5. Click on locations B and C. The selected face is moved.
6. Do not save your file.

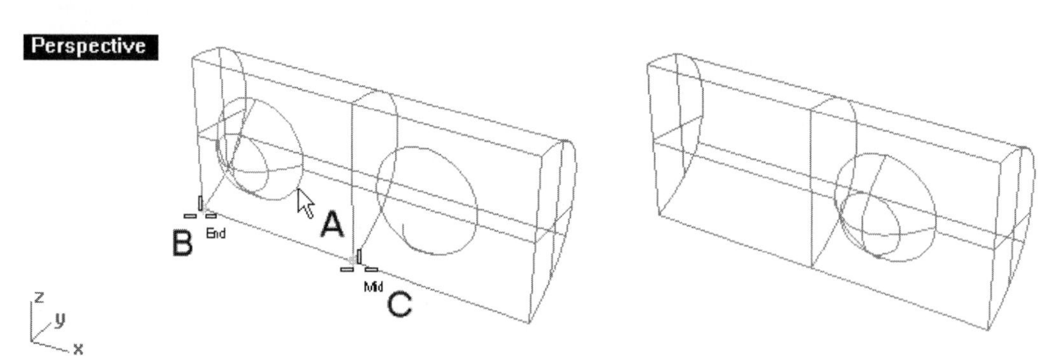

Figure 8–83. Face selected (left) and face moved (right)

Face Moving to a Boundary

A face can be moved to a boundary, thus changing the shape of the solid. Perform the following steps to understand this process:

1. Select File > Open and select the file MovetoBoundary.3dm from the Chapter 8 folder on the companion CD.
2. Select Solid > Solid Edit Tools > Faces > Move Face to Boundary, or click on the Move Face to a Boundary button on the Solid Editing toolbar
3. Select face A (Figure 8–84) and press the ENTER key.
4. Select the ToBoundary option on the command area.
5. Select surface B.
6. Turn off the Boundary layer to see the effect.
7. Do not save your file.

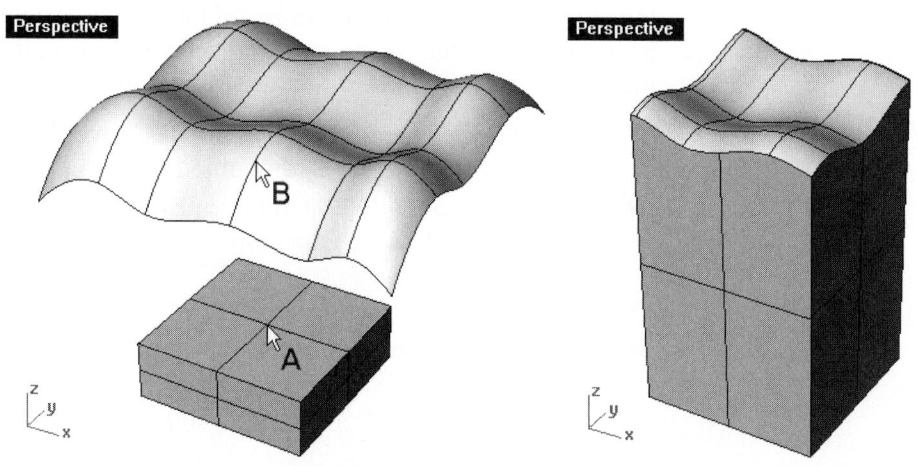

Figure 8–84. Face being moved (left) and face moved to boundary surface (right)

Face Rotation

Rotating selected faces of a solid around an axis is a way to modify a solid, as follows:

1. Select File > Open and select the file RotateFace.3dm from the Chapter 8 folder on the companion CD.
2. Select Solid > Solid Edit Tools > Faces > Rotate Face, or click on the Rotate Face/Shear Face button on the Solid Editing toolbar.
3. Select face A, shown in Figure 8–85, and press the ENTER key.
4. Select the 3dAxis option on the command area.
5. Select end points B and C.
6. Type 15 at the command area.
7. The selected face is rotated. Do not save your file.

Perspective

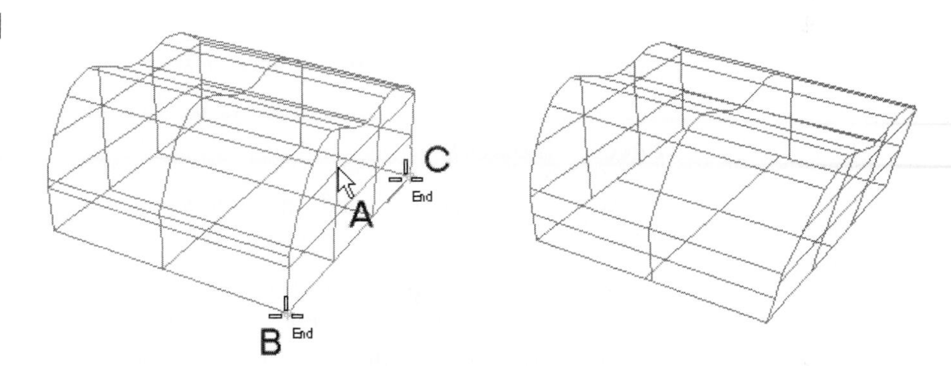

Figure 8–85. Face being selected (left) and face rotated (right)

Face Shearing

Some surfaces can be deformed by shifting at a specific angle through shearing. Because shearing applies only locally to selected surfaces, the process may not always be successful if any edge is moved out of face tolerance. Perform the following steps:

1. Select File > Open and select the file ShearFace.3dm from the Chapter 8 folder on the companion CD.
2. Select Solid > Solid Edit Tools > Faces > Shear Face, or right-click on the Rotate Face/Shear Face button on the Solid Editing toolbar.
3. Select face A, shown in Figure 8–86, and press the ENTER key.
4. Select the 3DAxis option on the command area.

5 Select end points B and C.
6 Type 15 at the command area. The selected face is sheared for an angle of 15 degrees.
7 Do not save your file.

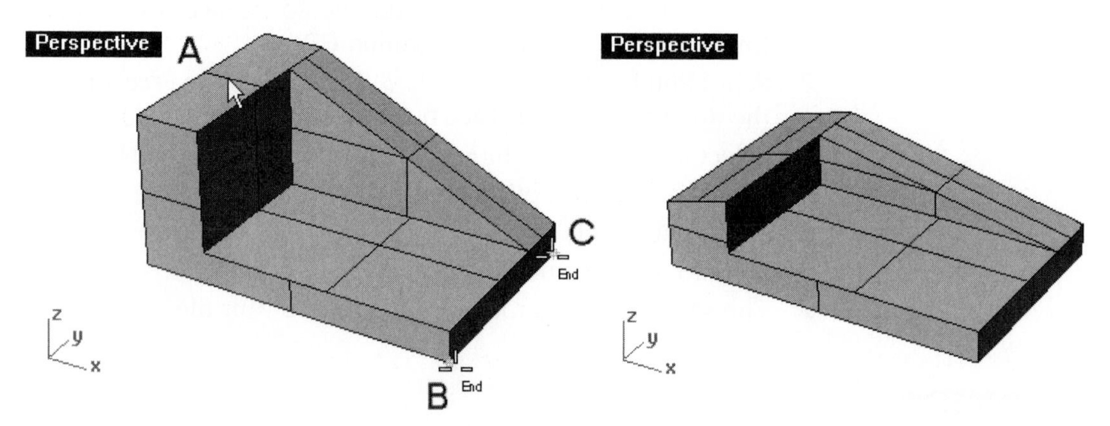

Figure 8–86. Face selected (left) and face sheared (right)

Face Extrusion

By extruding a face of a solid, the solid's shape is modified. Perform the following steps:

1 Select File > Open and select the file ExtrudeSide.3dm from the Chapter 8 folder on the companion CD.
2 Click on the Extrude Face/Extrude Face along path button on the Solid Editing toolbar.
3 Select face A, shown in Figure 8–87, and press the ENTER key.
4 Uncheck the Ortho button on the status bar.
5 Click on points A and B. The selected face is extruded.
6 Do not save your file.

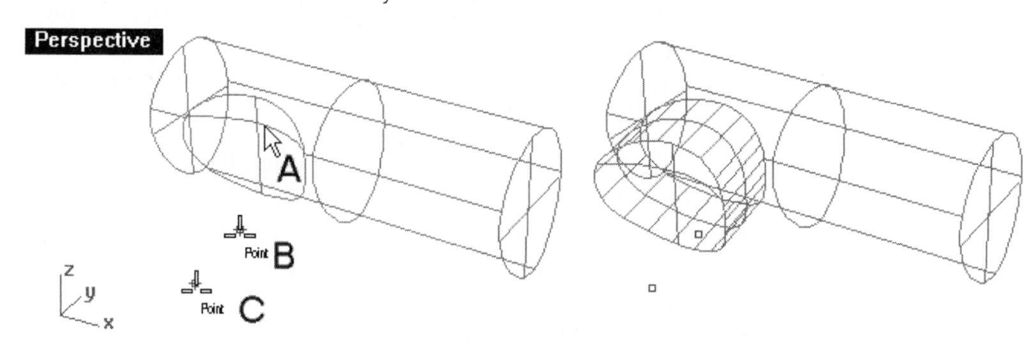

Figure 8–87. Face selected (left) and face extruded (right)

Face Extrusion Along a Curve

This process is similar to extruding a curve or a surface along a curve. Here, selected surfaces are extruded along a curve. Perform the following steps:

1. Select File > Open and select the file ExtrudeFaceAlongCurve.3dm from the Chapter 8 folder on the companion CD.
2. Right-click on the Extrude Face/Extrude Face along path button on the Solid Editing toolbar.
3. Select face A, shown in Figure 8–88, and press the ENTER key.
4. Select curve B. The selected face is extruded along a path, and adjacent surfaces are modified.
5. Do not save your file.

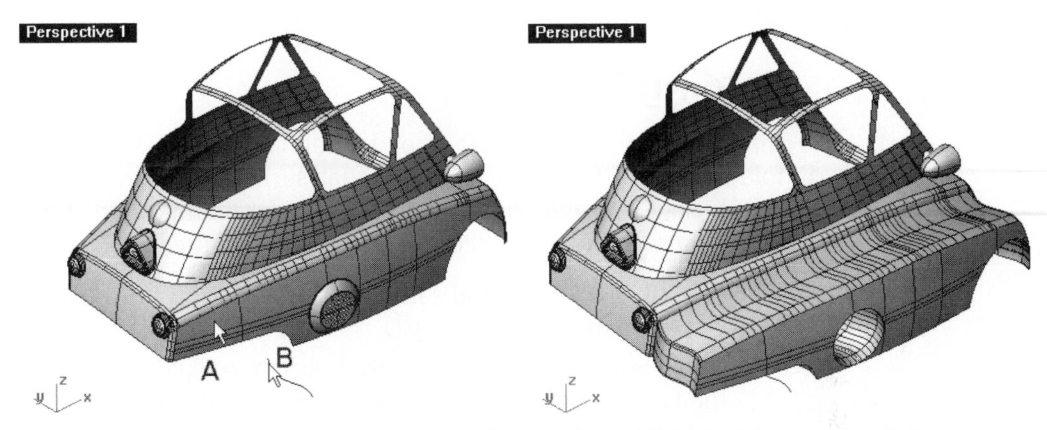

Figure 8–88. Face and curve being selected (left) and face extruded along a curve

Splitting a Face for Subsequent Operations

If you want to manipulate (move, rotate, or extrude) a portion of a surface instead of an entire face, you first have to split the face in two, as follows:

1. Select File > Open and select the file SplitFace.3dm from the Chapter 8 folder on the companion CD.
2. Select Solid > Solid Edit Tools > Faces > Split Face, or click on the Split planar face button on the Solid Editing toolbar.
3. Select face A, shown in Figure 8–89, and press the ENTER key.
4. Select the Curve option on the command area.
5. Select curve B. The curve is pulled to the face to split the face into two.
6. Click on the Extrude Side/Extrude Side along path button on the Solid Editing toolbar.

7 Select face C, shown in Figure 8–89, and press the ENTER key.
8 Select points D and E to specify the extrusion distance.
9 Do not save your file.

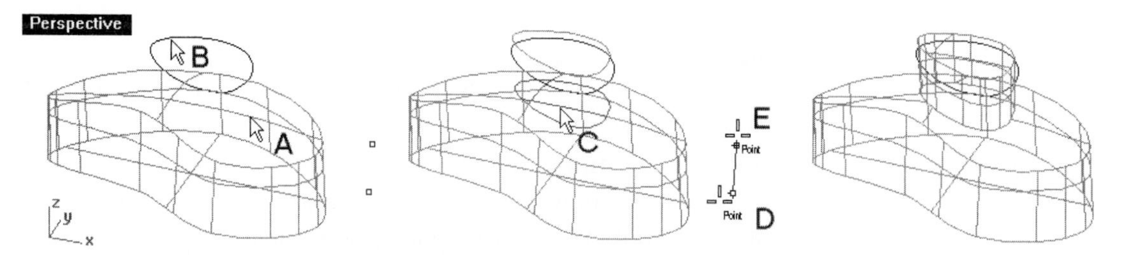

Figure 8–89. From left to right, Face being split, face split and being extruded, and split face extruded

Folding Faces

An individual planar surface or planar face of a polysurface can be folded along a folding axis. If a planar surface is folded, the surface is simply split into two and rotated around the folding axis. If a face of a polysurface is folded, adjacent faces will be modified.

1 Select File > Open and select the file Fold.3dm from the Chapter 8 folder on the companion CD.
2 Select Solid > Solid Edit Tools > Faces > Fold Face, or click on the Fold planar face button on the Solid Editing toolbar.
3 Select face A, shown in Figure 8–90, and press the ENTER key.
4 Select points B and C.
5 Select the Symmetrical option on the command area.
6 Type 25 at the command area. The surface is folded.
7 Repeat the command.
8 Select face D, shown in Figure 8–90, and press the ENTER key.
9 Select points E and F.
10 Select the Symmetrical option on the command area.
11 Type 25 at the command area. The surface is folded.
12 Do not save your file.

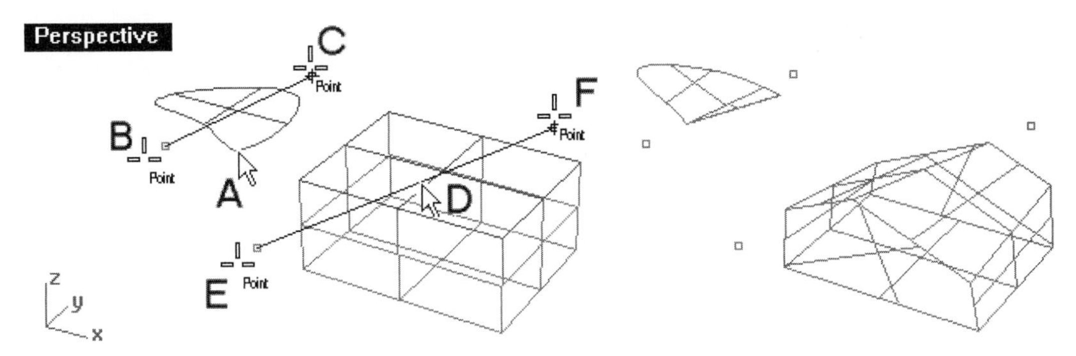

Figure 8–90. Faces being folded (left) and faces folded (right)

Rotating, Moving, and Scaling Edges of a Solid

A solid's shape can be modified by manipulating its edges. Naturally, faces adjacent to the manipulated edges have to change in shape. Therefore, edge manipulation may not always be successful if one of the affected faces cannot be changed.

Edge Rotation

Selected edges of a solid can be rotated around an axis to change the shape of a solid. Perform the following steps:

1. Select File > Open and select the file Edge.3dm from the Chapter 8 folder on the companion CD.
2. Select Solid > Solid Edit Tools > Edges > Rotate Edges, or Click on the Rotate Linear Edge button on the Solid Editing toolbar.
3. Select edge A, shown in Figure 8–91, and press the ENTER key.
4. Select point A and then point B to specify the axis of rotation. The sequence of selection affects the direction of rotation.
5. Type 15 at the command area. The edge is rotated and adjacent faces are modified.

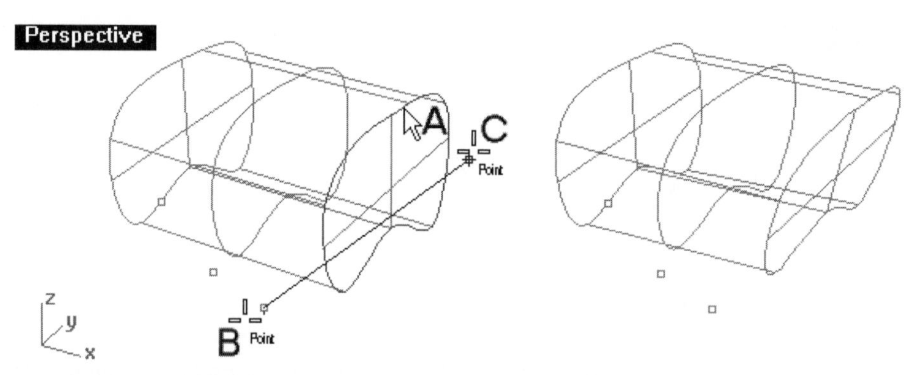

Figure 8–91. Edge being rotated (left) and edge rotated (right)

Edge Moving

If one or more edges are moved, the shape of the solid will be changed. Continue with the following steps.

6 Select Solid > Solid Edit Tools > Edges > Move Edges, or Click on the Move Edge/Move Untrimmed Edge button on the Solid Editing toolbar.
7 Select edge A, shown in Figure 8–92, and press the ENTER key.
8 Select point B and then point C. The edge is moved.

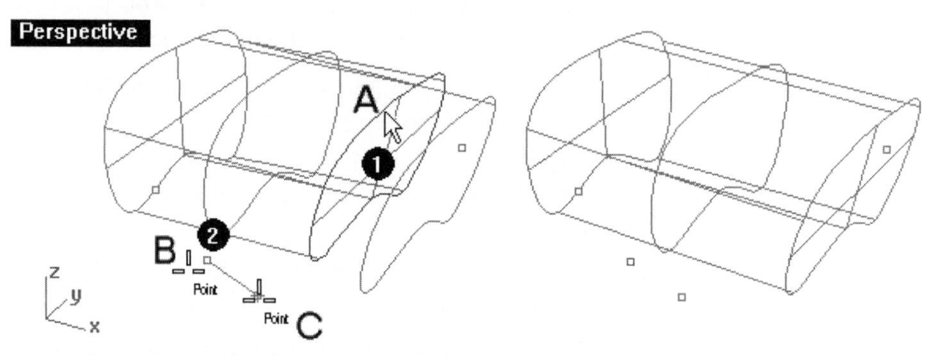

Figure 8–92. Edge being moved (left) and edge moved (right)

Edge Scaling

Scaling an edge changes the edge's length. As a result, the shape of the solid is changed. Continue with the following steps.

9. Select Solid > Solid Edit Tools > Edges > Scale Edge, or click on the Scale edge button on the Solid Editing toolbar.
10. Select edge A, shown in Figure 8–93, and press the ENTER key.
11. Select point B as the origin point.
12. Type 0.5 at the command area to specify the scale factor. The edge is scaled.
13. Do not save your file.

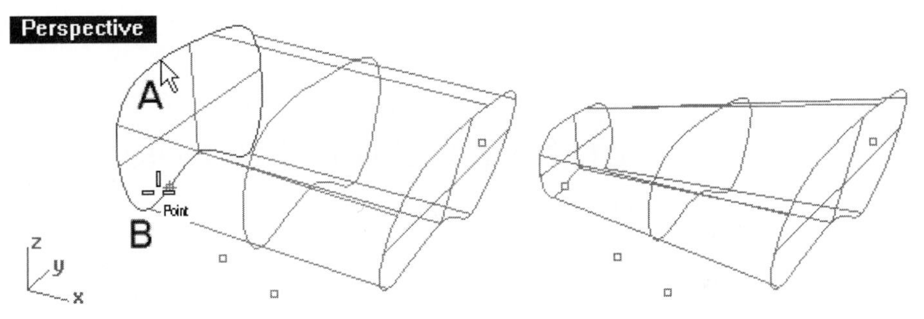

Figure 8–93. Edge being scaled (left) and edge scaled (right)

Moving Untrimmed Edges

An untrimmed edge of a surface or a polysurface can be moved, as follows:

1. Select File > Open and select the file MoveUntrimmedEdge.3dm from the Chapter 8 folder on the companion CD.
2. Right-click on the Move Edge/Move Untrimmed Edge button on the Solid Editing toolbar.
3. Select edge A, shown in Figure 8–94. This is an untrimmed edge of a surface.
4. Select points B and C to define the distance and direction of move. The edge is moved.
5. Repeat the command.
6. Select edge D. This is the untrimmed edge of a polysurface.
7. Select points B and C. The edge is moved.
8. Do not save your file.

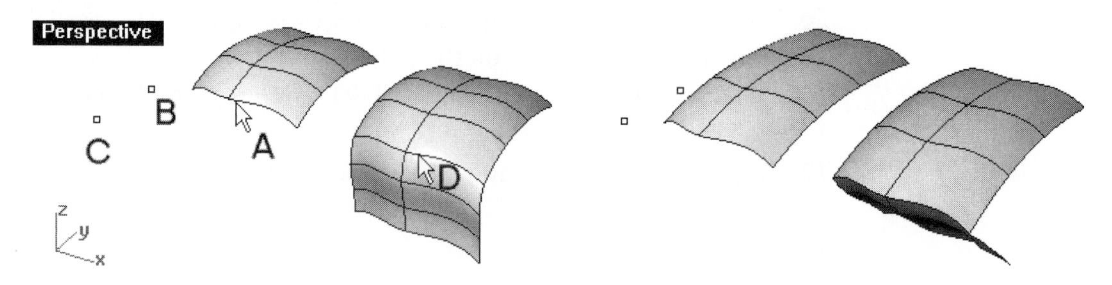

Figure 8–94. Untrimmed edges being selected one by one (left) and edges moved (right)

Rotating and Moving Holes

Holes residing on planar faces can be relocated by moving or rotation.

1. Select File > Open and select the file TranslateHole.3dm from the Chapter 8 folder on the companion CD.
2. Select Solid > Solid Edit Tools > Holes > Rotate Hole, or click on the Rotate Hole button on the Holes toolbar.
3. Select edge A, shown in Figure 8–95.
4. Select point B.
5. Type -90 at the command area. The hole is rotated 90 degrees counter-clockwise.

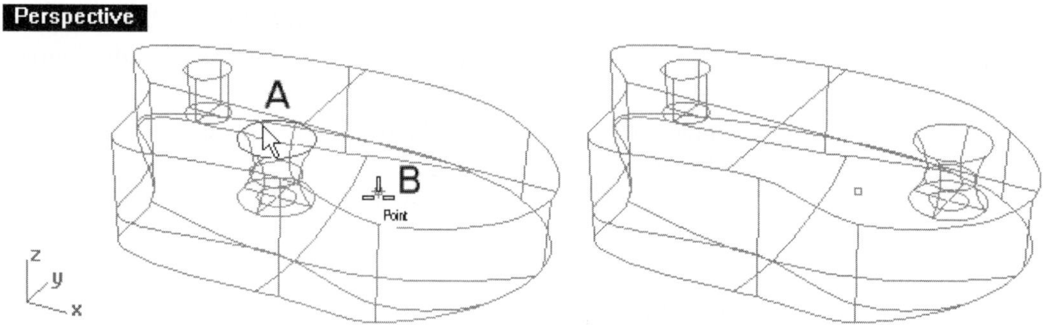

Figure 8–95. Hole being rotated (left) and hole rotated (right)

Continue with the following steps to move a hole.

6. Select Solid > Solid Edit Tools > Holes > Move Hole, or Click on the Move Hole/Copy Hole button on the Holes toolbar.
7. Select edge A, shown in Figure 8–96.
8. Select point B.

9 Type r10 < 0 at the command area. The hole is moved a distance of 10 units in 0 degree direction.

10 Do not save your file.

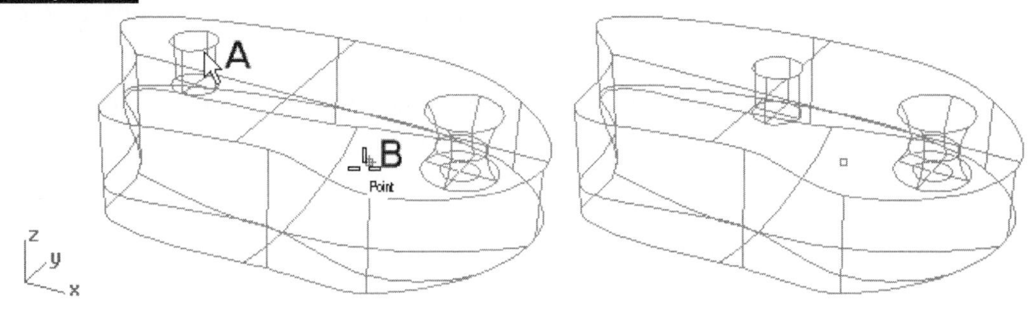

Figure 8–96. Hole being moved (left) and hole moved (right)

Copying Holes

Right-clicking the Move Hole/Copy Hole button on the Holes toolbar will copy a hole instead of moving it.

Deleting Holes

Unwanted holes in a solid can be deleted, as follows:

1 Select File > Open and select the file DeleteHole.3dm from the Chapter 8 folder on the companion CD.

2 Select Solid > Solid Edit Tools > Holes > Delete Hole, or click on the Delete Hole button on the Holes toolbar.

3 Select hole edges A, B, and C, shown in Figure 8–97. The holes are deleted. Press the ENTER key to exit the command.

4 Do not save your file.

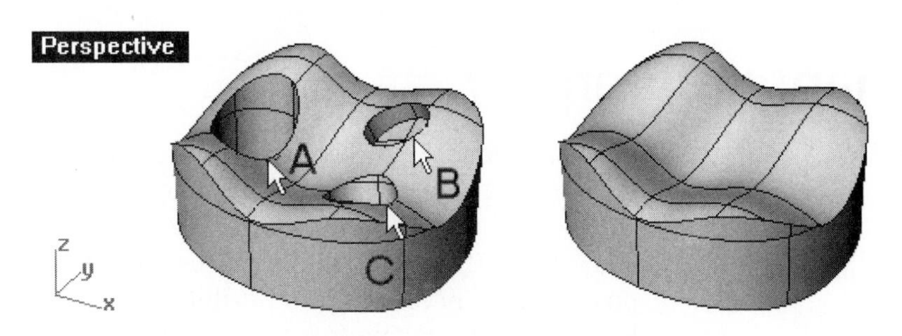

Figure 8–97. Original solid (left) and holes deleted (right)

Extracting a Surface from a Solid or Polysurface

Unlike exploding, which breaks down a joined polysurface into individual surfaces, extracting separates the selected surfaces from the polysurface, leaving the remaining surfaces joined. To extract a surface from a solid, perform the following steps:

1. Select File > Open and select the file ExtractSurface.3dm from the Chapter 8 folder on the companion CD.
2. Select Solid > Extract Surface, or click on the Extract Surface button on the Solid Tools toolbar.
3. Select surface A (shown in Figure 8–98) and press the ENTER key. A surface is extracted.
4. Select Edit > Visibility > Hide, or click on the Hide/Show button on the Visibility toolbar.
5. Select the extracted surface, if it is not already selected. The extracted surface is hidden.
6. Do not save your file.

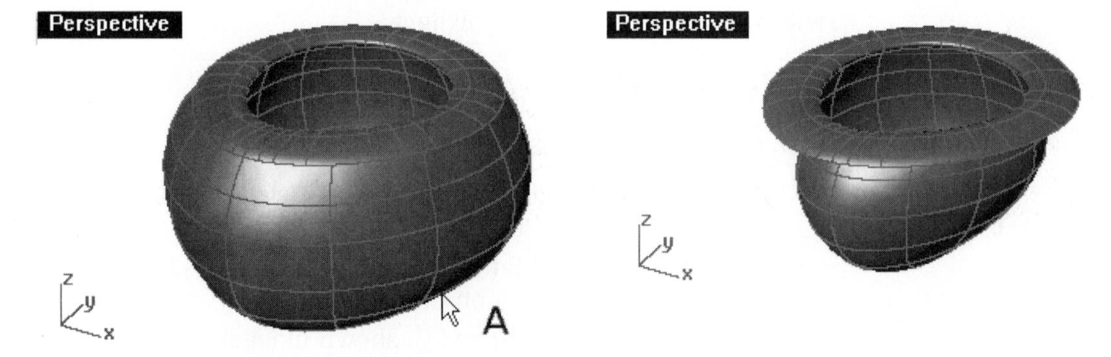

Figure 8–98. Surface being extracted from a polysurface (left) and extracted surface hidden (right)

Chapter Summary

A Rhino solid is a closed-loop NURBS surface or polysurface that encloses a volume with no gap or opening. In essence, you can use all the NURBS curves and surface tools delineated in previous chapters to construct free-form NURBS surfaces and then use these surfaces to compose a solid. Because of its flexibility and ease of use, Rhino is particularly useful in making free-form solids.

The simplest way to construct a solid is to construct solids of regular pattern, including box, sphere, cylinder, cone, truncated cone, pyramid, ellipsoid, paraboloid, tube, pipe, torus, and text.

In addition to constructing solids of regular shapes, you can extrude a planar curve or a free-form surface to obtain an extruded solid, construct a slab solid from a curve, cap planar holes of a polysurface to convert it to a solid, and construct a solid from a set of contiguous surfaces that encloses a volume without any gap or opening.

To combine two or more solids to form a more complex solid, you use Boolean operations. To detail a solid, you can add variable radius fillet, variable distance chamfer, variable blend, boss, rib, and hole.

To clean up overlapped coplanar surfaces in a solid resulting from any Boolean operations, you merge them into a single face. If required, you can split a solid into two by using a surface or a curve. To modify a solid, you can manipulate its faces and edges, and move the location of the holes. Because a Rhino solid is a polysurface, you can extract a surface from a solid or a polysurface. However, extracting a surface reduces a solid to an open loop polysurface or surface.

Review Questions

1. Explain, in terms of Rhino definition, the difference between a polysurface and a solid.
2. List the kinds of regular shape solids available in Rhino.
3. How can solids of free-form shape be constructed in Rhino?
4. What are the ways to combine two or more solid objects into a single solid object?
5. In what ways can a solid be detailed?
6. Outline the ways to edit a Rhino solid.

CHAPTER 9

Polygon Meshes

Introduction

This chapter explains the ways to construct and manipulate polygon meshes that are used to approximate free-form objects.

Objectives

After studying this chapter, you should be able to

- ❒ Construct primitive meshed objects from scratch
- ❒ Derive polygon meshes from existing objects
- ❒ Combine and separate polygon meshes
- ❒ Manipulate mesh faces, edges, and vertices

Overview

As previously explained, there are two ways to represent a surface in the computer: using NURBS surfaces to exactly represent surfaces or using polygon meshes to approximate surfaces.

Naturally, deployment of NURBS surfaces is the preferred method for designing and constructing 3D objects in the computer because this is the most accurate way of representing free-form objects. However, being an approximation of a smooth surface, polygon mesh is still used in many computer applications. In particular, rapid prototyping models in STL file format are a typical application of polygon meshes as a means of surface representation. In addition, when you import surfaces from other computerized applications, you may often import a set of polygon meshes. Therefore, you need to know how to deal with such meshes.

In this chapter, you will first learn how to construct mesh geometry of regular shapes. Then you will learn how to derive meshed objects from existing objects and how to combine meshed objects. Finally, and most importantly, you will learn how to edit meshed objects. Knowing how to edit meshed objects is particularly useful in repairing polygon meshes for rapid prototyping.

Constructing Polygon Mesh Primitives

Mesh primitives are meshed objects of basic geometric shapes. Although it is not advisable to model in polygon mesh, learning how to construct these primitives lets you gain a basic understanding of meshed object construction from scratch. In regard to basic geometric shapes, the simplest meshed objects are rectangular mesh planes and 3D mesh faces. In addition, there are seven types of mesh primitives: box, sphere, cylinder, cone, truncated cone, ellipsoid, and torus. The following sections take you through the construction process for each of these primitives.

Mesh Density Setting

Because a polygon mesh is a set of planar polygons that approximates a surface, the number of polygons used in the mesh has a direct impact on the accuracy of representation and the file size. A higher number of constituent polygons provides a more accurate model. However, file size increases quickly with a decrease in polygon size (smaller size meaning a greater number of polygons in the mesh). Hence, you need to consider using the most appropriate mesh density to provide an optimal balance between the representational accuracy of the model and the size of the file.

Before constructing polygon meshes from scratch, you have to decide how dense you want the polygon mesh to be. After a polygon mesh is constructed, increasing the mesh density, if possible, does not increase the accuracy of the model. Conversely, you can reduce the mesh density after the model is constructed. However, reduction of mesh density simplifies the shape of the mesh.

Mesh 3D Face

A mesh 3D face has four corner points, which can be a planar quadrilateral or a 3D mesh face consisting of two triangular planes. The use of mesh 3D face is best applied to filling openings of existing polygon mesh, converting them to a closed-loop object for rapid prototyping. Because a 3D face is a polygon mesh of simplest structure, there is no mesh density to manipulate. Perform the following steps:

1. Select File > Open and select the file 3DFace.3dm from the Chapter 9 folder on the companion CD.
2. Select Mesh > Polygon Mesh Primitives > 3-D Face, or click on the Single Mesh Face button on the Mesh toolbar.
3. Select points A, B, C, and D, shown in Figure 9–1. A planar 3D face mesh is constructed.

4 Repeat the command.
5 Select points E, F, G, and H, shown in Figure 9–1. A three-dimensional 3D face mesh is constructed.
6 Do not save your file.

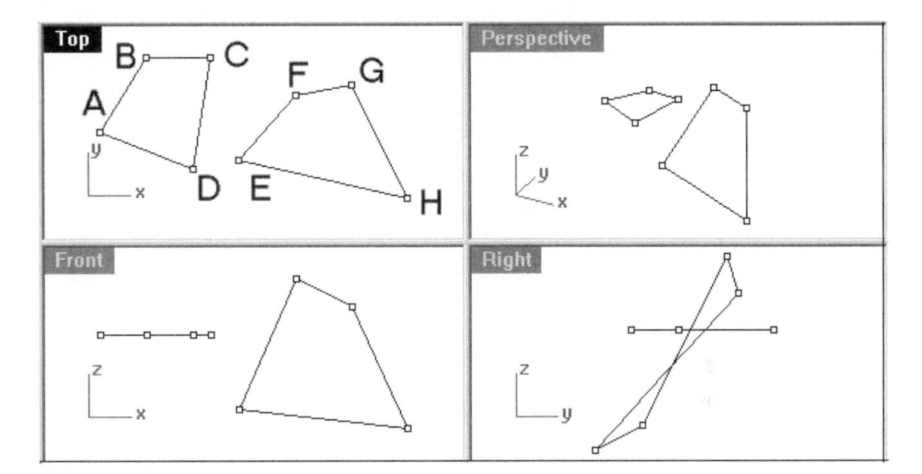

Figure 9–1. 3D faces being constructed

Mesh Rectangular Plane

A mesh rectangular plane is a planar mesh in the shape of a rectangle. There are four ways to construct a mesh rectangular plane:

- By specifying the diagonal points
- By specifying two points to depict the base of the rectangle and a third point to depict the height of the rectangle
- By specifying two points to depict the base of the rectangle and a third point in the adjacent construction plane to construct a plane vertical to the active viewport
- By specifying the center and a corner of the rectangle

Mesh Density

Mesh density of a rectangular planar mesh is defined by two options: XFaces and YFaces. They determine the number of polygon mesh in the X and Y directions. To manipulate these options, select them from the command area while running the command.

Perform the following steps to construct rectangular mesh planes in various ways:

1 Select File > Open and select the file PlaneMesh.3dm from the Chapter 9 folder on the companion CD.

2. Select Mesh > Polygon Mesh Primitives > Plane, or click on the Mesh Plane button on the Mesh toolbar.
3. Select points A and B, shown in Figure 9–2.
4. Repeat the command and select the 3Point option.
5. Select points C, D, and E, shown in Figure 9–2.
6. Repeat the command and use the Vertical option.
7. Select points F, G, and H, shown in Figure 9–2.
8. Repeat the command and use the Center option.
9. Select points J and K, shown in Figure 9–2.
10. Do not save your file.

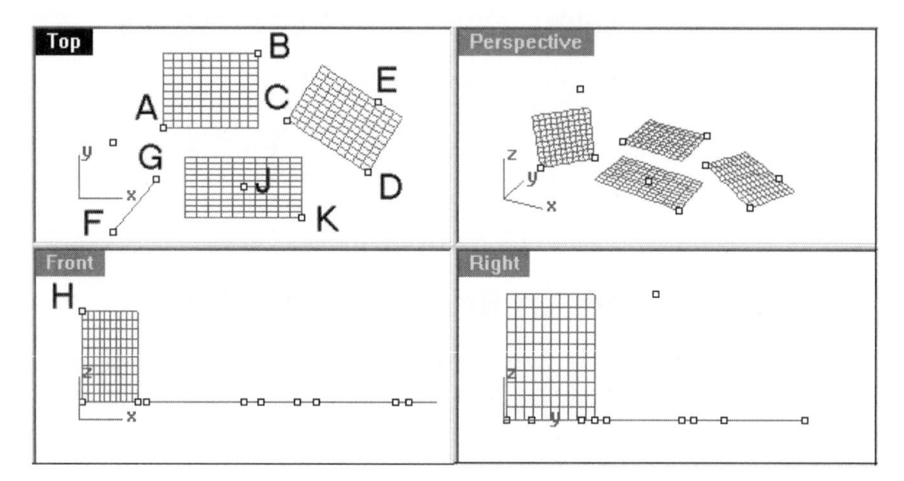

Figure 9–2. Mesh rectangular planes being constructed

Mesh Box

A mesh box consists of six rectangular planar meshes joined together. The methods to construct a mesh box are similar to the methods to construct a NURBS surface box, explained in Chapter 9:

- By specifying the corners of the base rectangle and the height
- By specifying the diagonal of the box
- By using three points to specify the base rectangle and specifying the height
- By using two points on a viewport and a point in an adjacent viewport to specify the base rectangle and the height
- By specifying the center of the base rectangle, a corner of the base rectangle, and the height

Mesh Density

Mesh density of a rectangular planar mesh is defined by three options: XFaces, YFaces, and ZFaces. Perform the following steps to construct mesh boxes in various ways:

1. Select File > Open and select the file BoxMesh.3dm from the Chapter 9 folder on the companion CD.
2. Select Mesh > Polygon Mesh Primitives > Box, or click on the Mesh Box button on the Mesh toolbar.
3. Select points A and B, shown in Figure 9–3, and type 15 at the command area to specify the height. A mesh box is constructed.
4. Repeat the command.
5. Select the 3Point option on the command area.
6. Select points C, D, and E, shown in Figure 9–3, and type 10 at the command area to specify the height. The second mesh box is constructed.
7. Repeat the command.
8. Select the Diagonal option.
9. Select points F and G, shown in Figure 9–3. The third mesh box is constructed.
10. Repeat the command.
11. Select the Vertical option.
12. Select points H and J shown in Figure 9–3.
13. Type 15 to specify the vertical height.
14. Click on location K to indicate the direction.
15. Type 10. A vertical mesh box is constructed.
16. Repeat the command.
17. Select the Center option on the command area.
18. Select points L and M.
19. Type 7 at the command area.
20. Do not save your file.

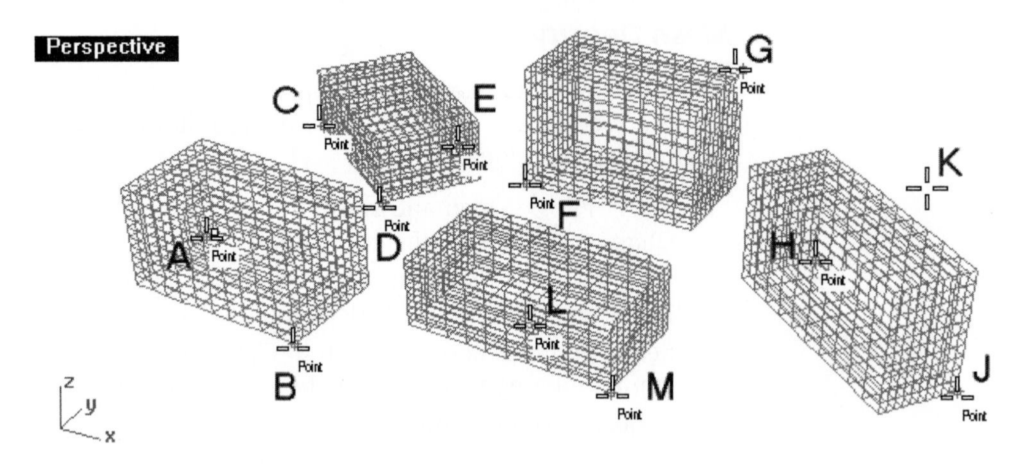

Figure 9–3. Mesh boxes constructed

Mesh Sphere

A mesh sphere is a single closed-loop meshed object. The ways to construct a mesh sphere are the same as those described in Chapter 9 to construct a NURBS sphere:

- By specifying the center and a point on the sphere
- By specifying the diameter of the sphere
- By specifying three points to define a section passing through the center of the sphere
- By specifying three tangent curves
- By specifying a point on a curve and the radius/diameter of the sphere
- By specifying four points on the surface of the sphere
- By fitting the surface of the sphere to a set of points

Mesh Density

A sphere's mesh density is defined by two options: VerticalFaces and AroundFaces

Perform the following steps to construct mesh spheres in various ways:

1. Select File > Open and select the file SphereMesh.3dm from the Chapter 9 folder on the companion CD.
2. Select Mesh > Polygon Mesh Primitives > Sphere, or click on the Mesh Sphere button on the Mesh toolbar.

3 Referencing Figure 9–4 and using the default option, construct a mesh sphere with its center at A and radius AB; using the 2Point option, construct a mesh sphere with its diameter CD; and using the 3Point option, construct a mesh sphere with a section across the center passing through E, F, and G.

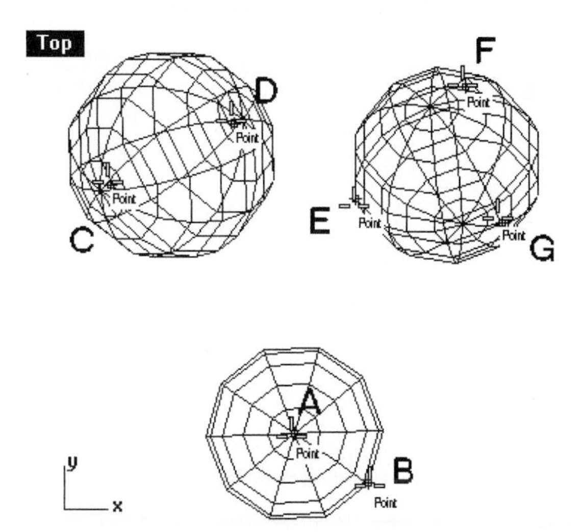

Figure 9–4. From left to right: Mesh sphere by specifying the diameter, mesh sphere by specifying the center and radius, and mesh sphere by specifying three points on the surface

4 Turn on Layer 02 and turn off Layer 00.

5 Referencing Figure 9–5 and using the Tangent option, construct a mesh sphere tangent to curves A, B, and C and, using the Around-Curve option, construct a mesh sphere around curve D, center at E, and radius EF.

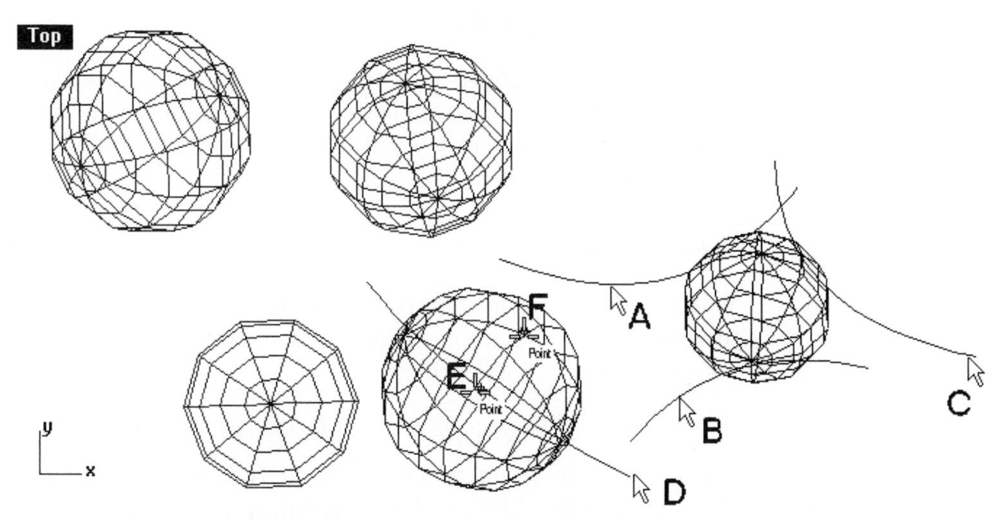

Figure 9–5. Mesh sphere around a curve and mesh sphere tangent to three curves constructed

6 Turn on Layer 03 and turn off Layer 00.

7 Referencing Figure 9–6 and using the 4Point option, construct a mesh sphere passing through points A, B, C, and D and, using the FitPoints option, construct a mesh sphere fitted to a set of points enclosed by the rectangle DE.

8 Do not save your file.

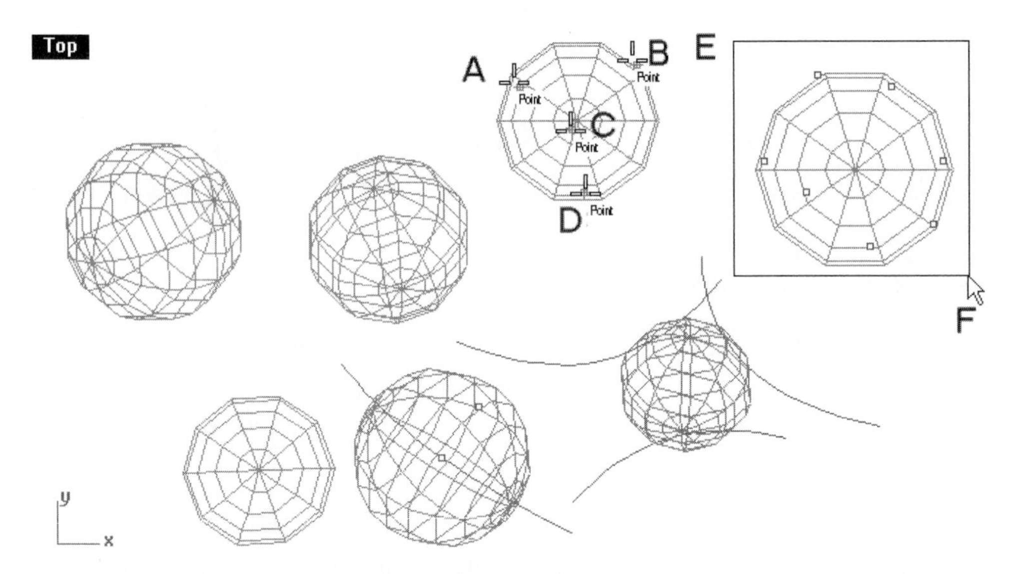

Figure 9–6. Mesh sphere passing through four points and mesh sphere fitted to a set of points constructed

Mesh Cylinder

A mesh cylinder consists of three polygon meshes, a cylindrical mesh and two circular planar meshes, joined together. The methods to construct a mesh cylinder are the same as those used to construct a NURBS surface cylinder.

Mesh Density

Mesh density can be set while constructing a meshed object. The mesh density in two directions of the polygon mesh is defined by two options: VerticalFaces and AroundFaces.

Perform the following steps to construct mesh cylinders in various ways:

1 Select File > Open and select the file CylinderMesh.3dm from the Chapter 9 folder on the companion CD.

2 Select Mesh > Polygon Mesh Primitives > Cylinder, or click on the Mesh Cylinder button on the Mesh toolbar.
3 Select VerticalFaces option on the command area.
4 Type 8 at the command area to specify the number of meshes in the vertical direction.
5 Select AroundFaces option on the command area.
6 Type 8 at the command area to specify the number of meshes in the second direction.

Direction Constraints

Axis direction of a mesh cylinder can be defined in three ways. Continue with the following steps.

7 Select the DirectionConstraint option on the command area.
8 Select the Vertical option on the command area.
9 Set object snap mode to Point.
10 Select point A, shown in Figure 9–7, to specify the cylinder's center.
11 Type 10 at the command area to specify the radius. (You can specify diameter instead of radius by selecting the Diameter option on the command area.)
12 Select point B to use the point's z-coordinate as the height of the cylinder.
13 Repeat the command.
14 Select the DirectionConstraint option on the command area.
15 Select the None option.
16 Select point C, shown in Figure 9–7, to specify the cylinder's center.
17 Type 10 at the command area to specify the radius.
18 Select point C to define the other end point of the cylinder.

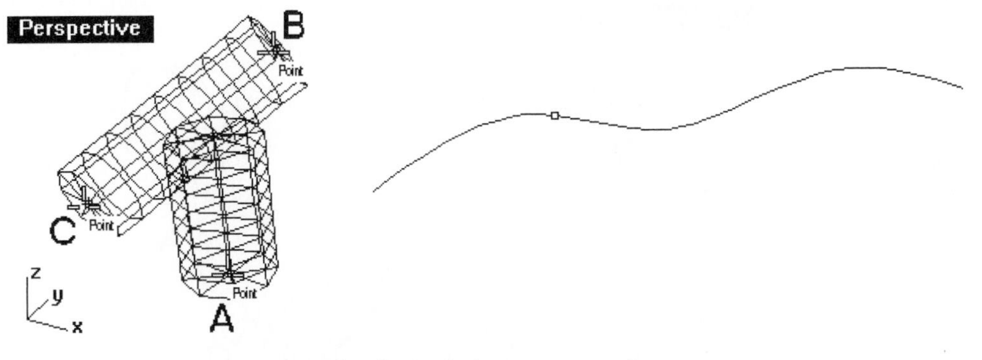

Figure 9–7. Two mesh cylinder being constructed

19 Repeat the command.
20 Select the DirectionConstraint option on the command area.
21 Select the AroundCurve option on the command area.
22 Select curve A and then point B, shown in Figure 9–8.
23 Type 10 at the command area to specify the radius.
24 Type 25 at the command area to specify the height.

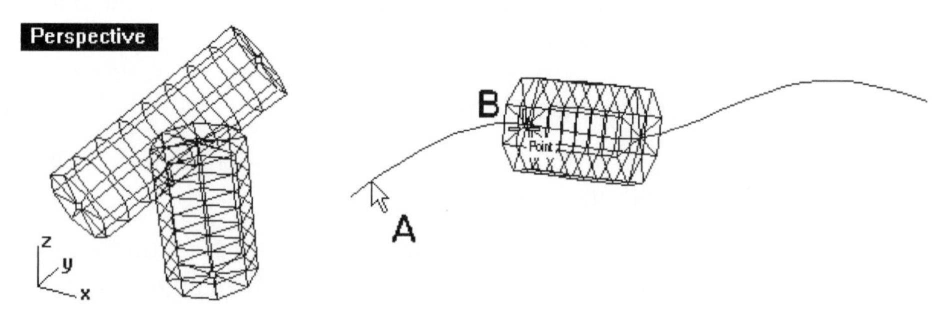

Figure 9–8. Mesh cylinder around a curve being constructed

Tangent Cylinder

The base circle of a mesh cylinder can be made tangent to three curves. Continue with the following steps.

25 Turn on Layer 02 and turn off Layer 00.
26 Select Mesh > Polygon Mesh Primitives > Cylinder, or click on the Mesh Cylinder button on the Mesh toolbar.
27 Select the Tangent option on the command area.
28 Select curves A, B, and C, shown in Figure 9–9.
29 Type 25 at the command area to specify the height. *(Note that the sequence of selection of the tangent curves has an effect on the direction of the cylinder.)*

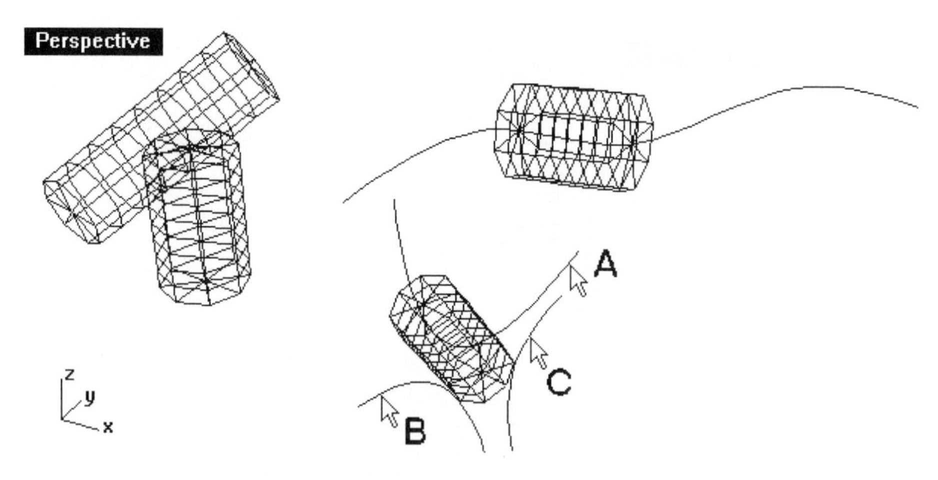

Figure 9–9. Tangent mesh cylinder being constructed

Mesh Cylinder with Base Circle Fitted to a Set of Points

The base circle of a mesh cylinder can be made to fit closest to a set of points. Continue with the following steps.

30 Turn on Layer 03 and turn off Layer 02.

31 Select Mesh > Polygon Mesh Primitives > Cylinder, or click on the Mesh Cylinder button on the Mesh toolbar.

32 Select the FitPoints option on the command area.

33 Click on A and drag to B (Figure 9–10) to select the point objects, and press the ENTER key.

34 Select point C. A cylinder with its axis end point closest to the selected point is constructed.

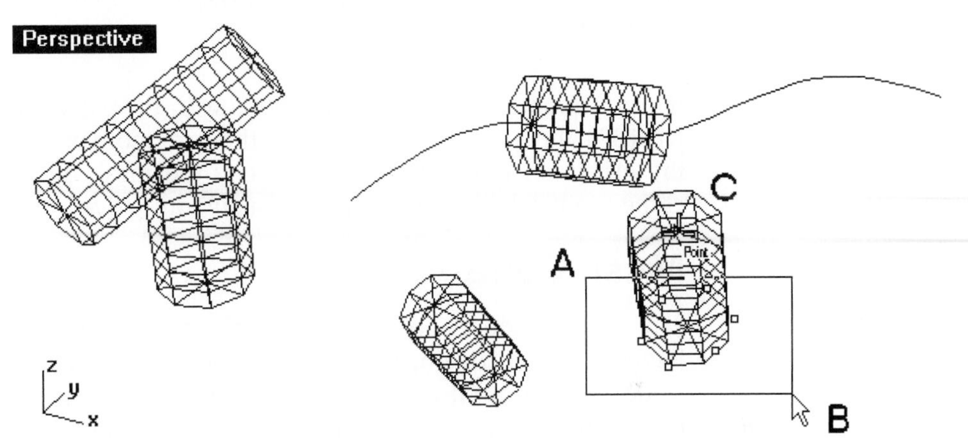

Figure 9–10. Mesh cylinder with its base circle fitted to a set of points being constructed

Mesh Cylinder with Base Circle Passing Through Three Points

The base circle of a mesh cylinder can be made to pass through three points. Continue with the following steps.

35 Turn on Layer 04 and turn off Layer 03.

36 Select Mesh > Polygon Mesh Primitives > Cylinder, or click on the Mesh Cylinder button on the Mesh toolbar.

37 Select the 3Point option on the command area.

38 Select points A, B, and C, shown in Figure 9–11.

39 Select point D. A cylinder with its axis end point closest to the selected point is constructed.

40 Do not save your file.

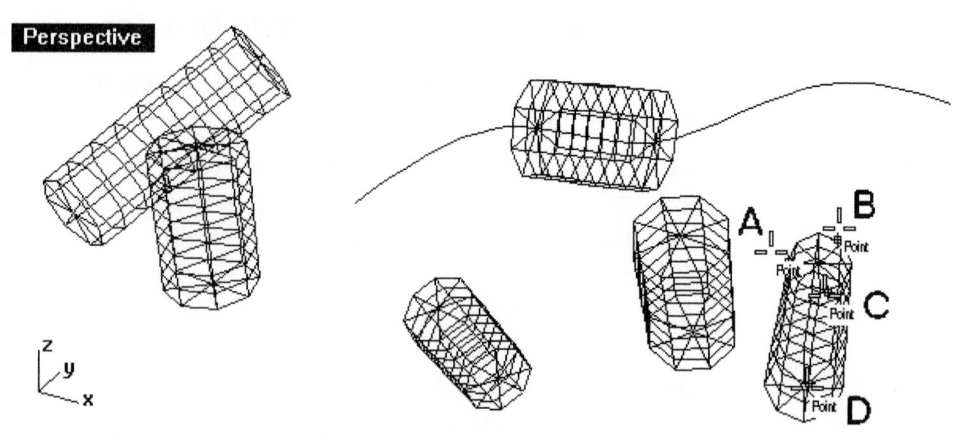

Figure 9–11. Mesh cylinder with its base circle passing through three points being constructed

If you compare the above delineations with the explanations for making NURBS cylinders, you will find that they are more or less the same, with the exception that you have to specify the mesh density and the outcome is a meshed object instead of a smooth polysurface.

Mesh Cone

A mesh cone consists of a slant conical mesh and a circular planar mesh joined together. Again, the methods to construct a mesh cone are the same as those used to construct a NURBS cone. Apart from specifying the mesh density, you can construct a cone without direction constraint, with vertical constraint, with its base circle around a curve, with its base circle tangent to three curves, with its base circle fitted to a set of points, and with its base circle passing through three points.

Mesh Density

A mesh cone's mesh density is defined by two options: VerticalFaces and AroundFaces.

Perform the following steps to construct mesh cones in various ways:

1. Select File > Open and select the file CylinderMesh.3dm from the Chapter 9 folder on the companion CD.
2. Select Mesh > Polygon Mesh Primitives > Cone, or click on the Mesh Cone button on the Mesh toolbar.
3. Use the Vertical direction constraint option to construct a mesh cone (Cone A in Figure 9–12) with a radius of 10 mm and height of 25 mm.

4. Repeat the command and use the Around curve direction constraint option to construct a cone (Cone B in Figure 9–12) with a base radius of 10 mm and a height of 25 mm.
5. Repeat the command and use the None direction constraint option to construct another mesh cone (Cone C in Figure 9–12) with a radius of 10 mm.
6. Turn on Layer 02 and turn off Layer 00.
7. Construct a mesh cone (Cone D in Figure 9–12) tangent to three curves. The height of the cone is 25 mm.
8. Turn on Layer 03 and turn off Layer 02.
9. Construct a mesh cone (Cone E in Figure 9–12) with its base circle fitted to a set of points. The height of the cone is also 25 mm.
10. Turn on Layer 04 and turn off Layer 03.
11. Construct a mesh cone (Cone F in Figure 9–12) with its base passing through three points.
12. Do not save your file.

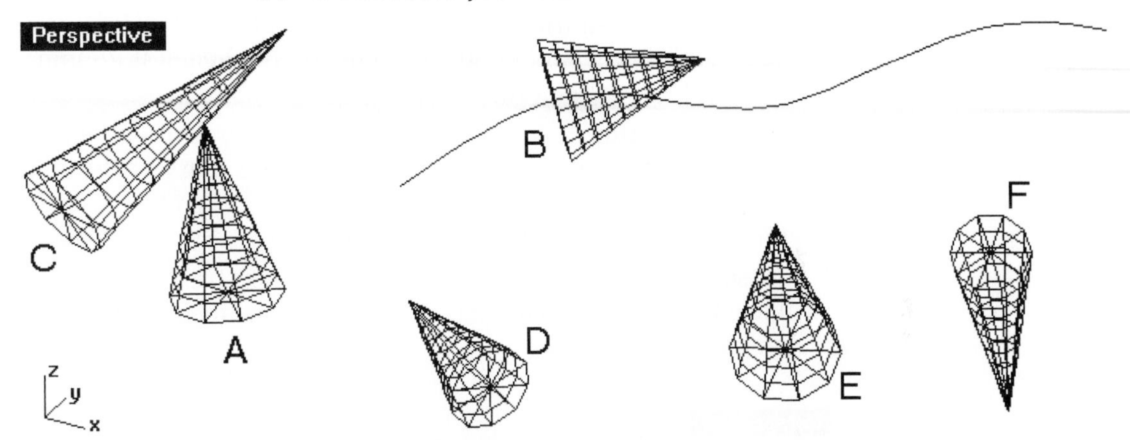

Figure 9–12. From left to right: cone without direction constraint, cone with vertical constraint, cone with its base circle tangent to three curves, cone with its base circle around a curve, cone with its base circle fitted to a set of points, and cone with its base circle passing through three points

Mesh Truncated Cone

A mesh cone consists of a slant conical mesh and two circular planar meshes joined together. The methods to construct a mesh truncated cone are the same as those used for constructing truncated NURBS surface cone.

Mesh Density

A truncated cone's mesh density is defined by two options: VerticalFaces and AroundFaces.

Perform the following steps to construct truncated mesh cones in various ways:

1. Select File > Open and select the file TConeMesh.3dm from the Chapter 9 folder on the companion CD.
2. Select Mesh > Polygon Mesh Primitives > Truncated Cone, or click on the Mesh TCone button on the Mesh toolbar.
3. Construct a truncated cone (A in Figure 9–13) with vertical direction constraint, having a base radius of 10 mm, a height of 32 mm, and a radius of 6 mm at the end.
4. Repeat the command to construct a truncated cone (B in Figure 9–13) without direction constraint, having a base radius of 10 mm and radius of 6 mm at the end.
5. Repeat the command to construct a truncated cone (C in Figure 9–13) around a curve. The base radius is 10 mm, the height is 20, and the end radius is 6 mm.
6. Turn on Layer 02 and turn off Layer 00.
7. Construct a mesh truncated cone (D in Figure 9–13) tangent to three curves. The end radius is 4 mm and the height is 25 mm.
8. Turn on Layer 03 and turn off Layer 02.
9. Construct a truncated mesh cone (E in Figure 9–13) with its base circle fitted to a set of points. The height is 25 mm and the end radius is 4 mm.
10. Turn on Layer 04 and turn off Layer 03.
11. Construct a mesh cone (F in Figure 9–13) with its base passing through three points. The end radius is 2 mm and the height is 25 mm.
12. Do not save your file.

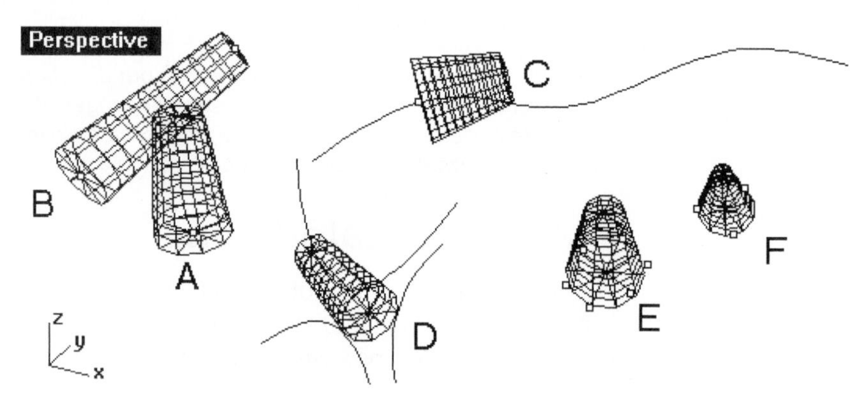

Figure 9–13. From left to right: truncated cone without direction constraint, with vertical constraint, with base circle tangent to three curves, with base circle around a curve, with base circle fitted to a set of points, and with base circle passing through three points

Mesh Ellipsoid

A mesh ellipsoid is a single meshed object. The methods to construct a mesh ellipsoid are the same as those for constructing a NURBS surface ellipsoid.

Mesh Density

A mesh ellipsoid's density is defined by two options: 1stDirFaces and 2ndDirFaces.

Perform the following steps to construct mesh ellipsoids in various ways:

1. Select File > Open and select the file EllipsoidMesh.3dm from the Chapter 9 folder on the companion CD.
2. Select Mesh > Polygon Mesh Primitives > Ellipsoid, or click on the Ellipsoid from Center button on the Mesh toolbar.
3. Select points A and B, shown in Figure 9–14, to specify the center and an axis end point.
4. Type 5 at the command area, press the ENTER key, and click on any location on the current construction plane to specify the second axis radius and direction.
5. Type 3 at the command area and press the ENTER key.
6. Repeat the command.
7. Select the Diameter option on the command area.
8. Select points C and D, shown in Figure 9–14, to specify the center and an axis end point.
9. Type 5 at the command area, press the ENTER key, and click anywhere on the active construction plane to specify the second axis diameter and direction.
10. Type 3 at the command area and press the ENTER key. A ellipsoid polygon mesh is constructed.

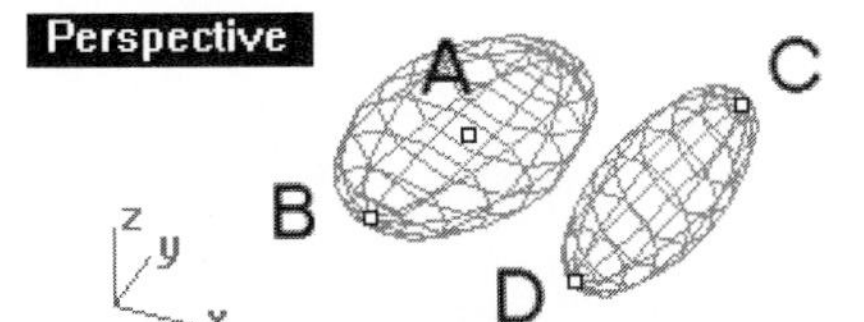

Figure 9–14. Mesh ellipsoid specified by radii (left) and mesh ellipsoid specified by diameters

11. Turn on Layer 02 and turn off Layer 00
12. Select Mesh > Polygon Mesh Primitives, or click on the Ellipsoid from Center button on the Mesh toolbar.
13. Select the From Foci option on the command area.
14. Select points A and B (shown in Figure 9–15) to specify the foci, and then select point C to specify a point on the ellipsoid.

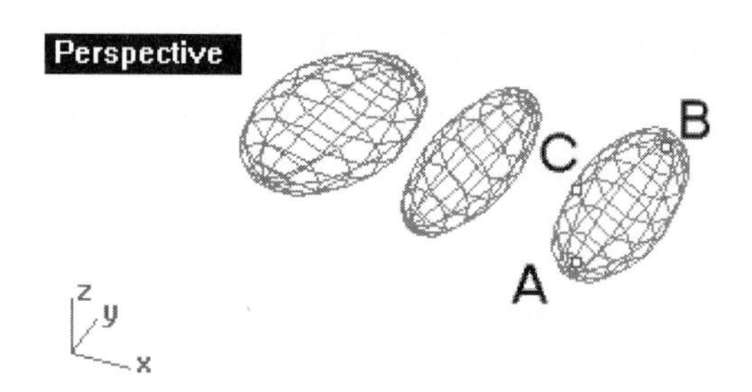

Figure 9–15. Mesh ellipsoid specified by foci and mesh ellilpsoid specified by corners constructed

15 Turn on Layer 03 and turn off Layer 02.
16 Select Mesh > Polygon Mesh Primitives > Ellipsoid, or click on the Ellipsoid from Center button on the Mesh toolbar.
17 Select the Around Curve option on the command area.
18 Select point A (Figure 9–16) along the curve to specify the center of the ellipsoid.
19 Type 5 at the command area, press the ENTER key, and click on location B to specify the first radius.
20 Type 4 at the command area, press the ENTER key, and click on location C to specify the second radius.
21 Type 3 at the command area and press the ENTER key.
22 Do not save your file.

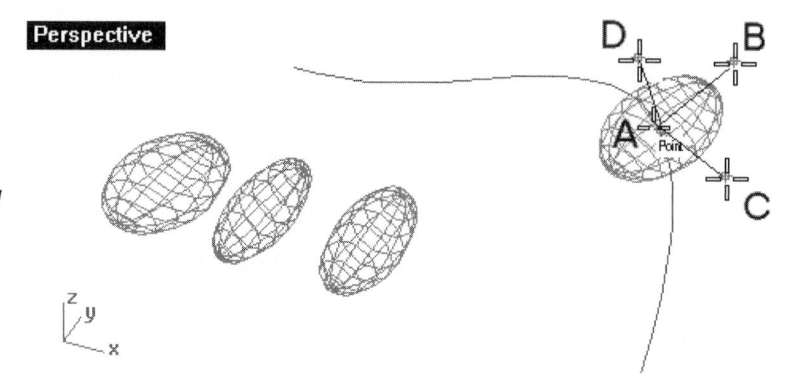

Figure 9–16. Mesh ellipsoid around a curve constructed

Mesh Torus

A mesh torus is a single closed-loop meshed object and the ways to construct a mesh torus are the same as those for making a NURBS torus.

Mesh Density

The density of a mesh torus is defined by two options: VerticalFaces and AroundFaces.

Perform the following steps to construct mesh tori in various ways:

1. Select File > Open and select the file TorusMesh.3dm from the Chapter 9 folder on the companion CD.
2. Select Mesh > Polygon Mesh Primitives > Torus, or click on the Mesh Torus button on the Mesh toolbar.
3. Select points A and B, shown in Figure 9–17. Here A is the center and AB is the radius of the pitch circle.
4. Type 4 at the command area to specify the radius of the torus tube.
5. Repeat the command.
6. Select the Vertical option on the command area.
7. Select points C and D, shown in Figure 9–17.
8. Type 4 at the command area.
9. Repeat the command.
10. Select the 2Point option on the command area.
11. Select points E and F, shown in Figure 9–17.
12. Type 4 at the command area.

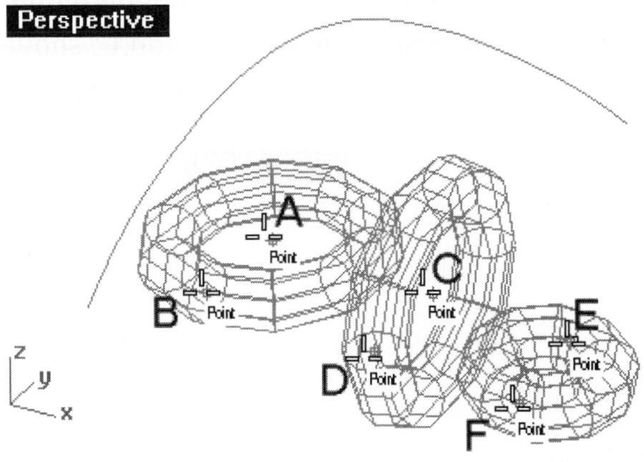

Figure 9–17. From left to right, torus pitch circle specified by radius, vertical torus, and torus pitch circle specified by diameter

13. Turn on Layer 02 and turn off Layer 00.
14. Select Mesh > Polygon Mesh Primitives > Torus, or click on the Mesh Torus button on the Mesh toolbar.
15. Select the AroundCurve option on the command area.
16. Select curve A, shown in Figure 9–18.
17. Select point B on curve A.
18. Type 9 at the command area to specify the pitch radius.
19. Type 3 at the command area to specify the radius

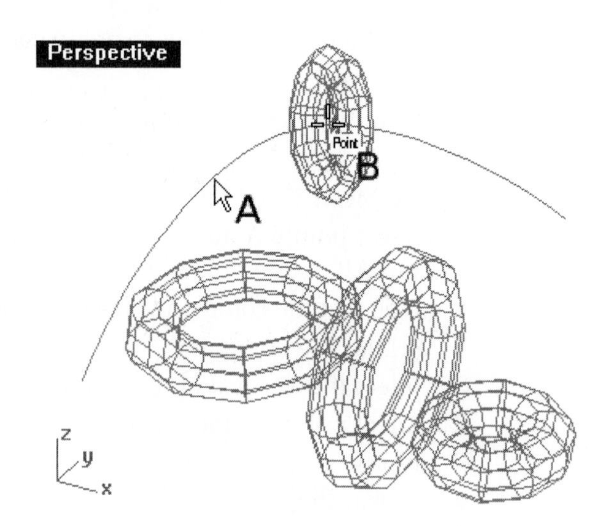

Figure 9–18. Mesh torus with its pitch circle around a curve being constructed

20 Turn on Layer 03 and turn off Layer 02.

21 Select Mesh > Polygon Mesh Primitives > Torus, or click on the Mesh Torus button on the Mesh toolbar.

22 Select the Tangent option on the command area.

23 Select curves A, B, and C, shown in Figure 9–19.

24 Type 3 at the command area to specify the second radius.

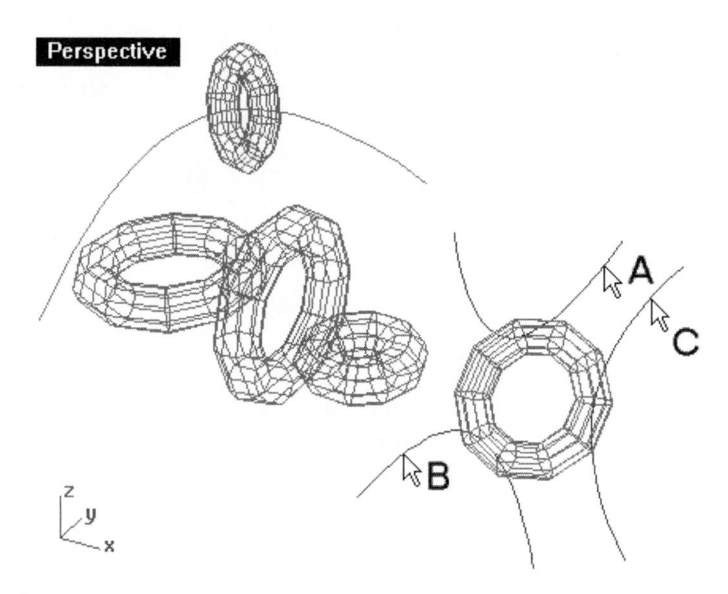

Figure 9–19. Mesh torus with pitch circle tangent to three curves being constructed

25 Turn on Layer 04 and turn off Layer 03.

26 Select Mesh > Polygon Mesh Primitives > Torus, or click on the Mesh Torus button on the Mesh toolbar.

27 Select the FitPoints option on the command area.

28 Click on A and drag to B (Figure 9–20) to select the point objects, and press the ENTER key.

29 Type 3 at the command area to specify the radius at the other end.

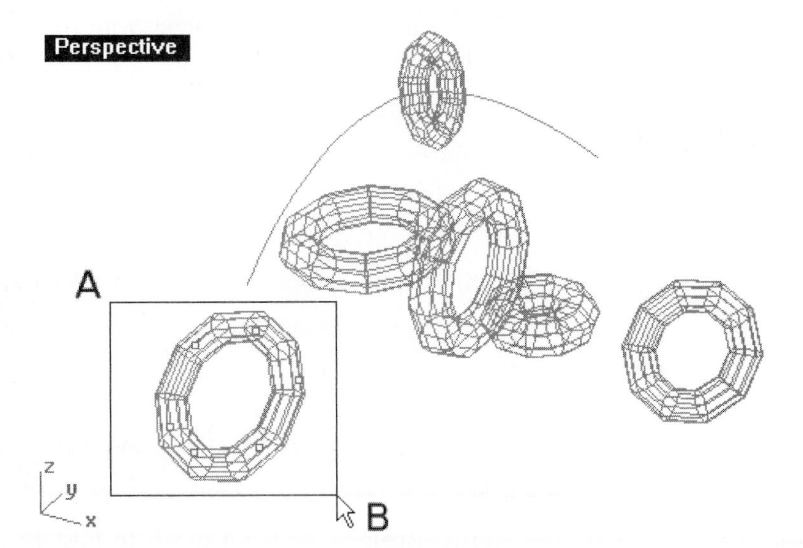

Figure 9–20. Mesh torus with pitch circle best fitted to a set of points being constructed

30 Turn on Layer 05 and turn off Layer 04.

31 Select Mesh > Polygon Mesh Primitives > Torus, or click on the Mesh Torus button on the Mesh toolbar.

32 Select the 3Point option on the command area.

33 Select points A, B, and C, shown in Figure 9–21.

34 Type 3 at the command area.

35 Do not save your file.

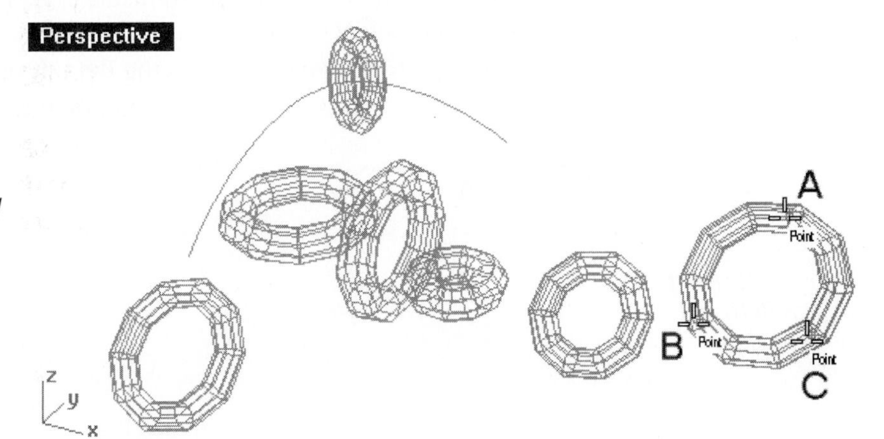

Figure 9–21. Mesh torus with pitch circle defined by three selected points being constructed

Constructing Polygon Meshes from Existing Objects

After learning ways to construct polygon mesh from scratch, you will now learn how to construct polygon mesh from existing objects, which can be done in six ways:

- By using a closed planar curve as boundary
- By using a closed curve as boundary and, optionally, curves and points to define mesh details
- By using the profile of a NURBS surface
- By using the control points of a NURBS surface
- By using the heightfield of a bitmap as vertices
- By offsetting an existing polygon mesh

Constructing a Polygon Mesh from a Set of Points

You can construct a polygon mesh to best fit to a set of point objects. This method is useful when you already have a cloud of points obtained by digitizing an object. Perform the following steps to construct a polygon mesh from a set of points:

1. Select File > Open and select the file PointMesh.3dm from the Chapter 9 folder on the companion CD.
2. Select Mesh > Mesh from Points, or click on the Mesh from points button on the Mesh Tools toolbar.
3. Click on location A and drag to location B, shown in Figure 9–22, and press the ENTER key.
4. Press the ENTER key to accept the default options.
5. A polygon mesh is constructed. Do not save your file.

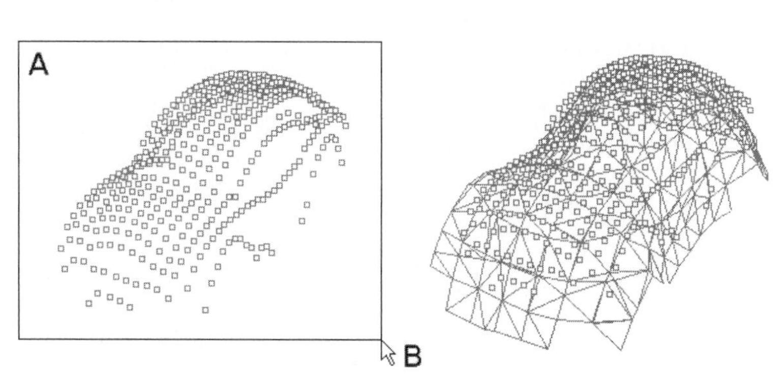

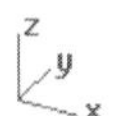

Figure 9–22. Point objects (left) and polygon mesh (right)

Constructing a Polygon Mesh from a Closed Planar Curve

You can construct a planar polygon mesh from a closed curve, which can be polyline or a NURBS curve. Perform the following steps:

1. Select File > Open and select the file PlanarMesh.3dm from the Chapter 9 folder on the companion CD.
2. Right-click on the Mesh from Closed Curve/Mesh from Planar Curve button on the Mesh toolbar, or type PlanarMesh at the command area.
3. Select curves A and B, shown in Figure 9–23, and press the ENTER key. Two planar mesh objects are constructed.
4. Do not save your file.

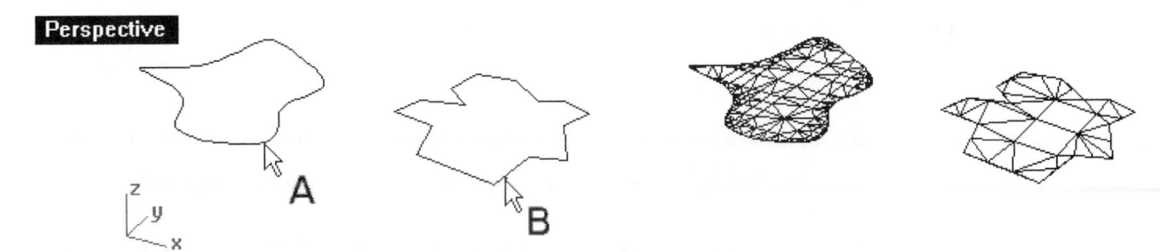

Figure 9–23. NURBS curve and polyline (left) and planar meshes constructed (right)

Constructing a Polygon Mesh from a Closed Polyline

You can construct a polygon mesh from a closed polyline or a curve, which can be planar or 3D. If a curve is used, a polyline will first be approximated to a polyline prior to outputting a mesh. The resulting polygon mesh will use the end points of the segments of the polyline as vertices of the meshes. Perform the following steps from a closed polyline:

1. Select File > Open and select the file PolylineMesh.3dm from the Chapter 9 folder on the companion CD.
2. Select Mesh > Polygon Mesh > From Closed Polyline, or click on the Mesh from Closed Curve/Mesh from Planar Curve button on the Mesh toolbar.
3. Select polyline A (shown in Figure 9–24). A polygon mesh is constructed.

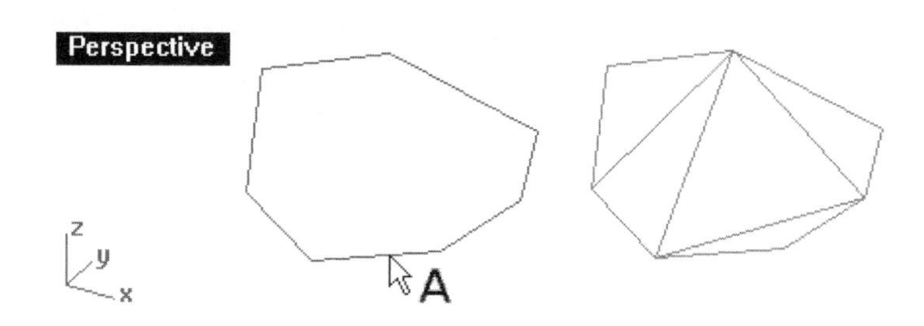

Figure 9–24. Closed polyline (left) and polygon mesh from the polyline (right)

Constructing a Mesh Patch

Constructing a mesh patch is similar to making a NURBS surface patch. You select curves and points as defining objects. Perform the following steps:

1. Select File > Open and select the file MeshPatch1.3dm from the Chapter 9 folder on the companion CD.
2. Click on the Mesh Patch button on the Mesh toolbar, or type MeshPatch at the command area.
3. Select curves A, B, C, and D and point E (Figure 9–25) and press the ENTER key.
4. Select curve B and press the ENTER key. This is the inner curve.
5. Select curve A. A mesh patch is constructed. Do not save your file.

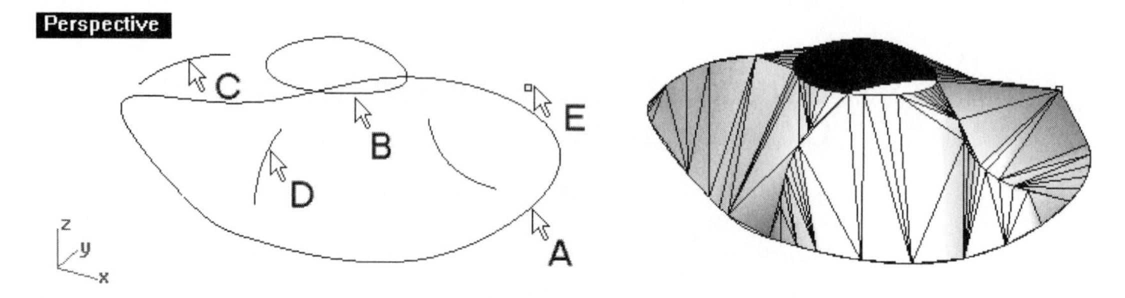

Figure 9–25. Curves and points (left) and mesh patch constructed (right)

To appreciate how point objects can be used in making a mesh object, perform the following steps:

1. Select File > Open and select the file MeshPatch2.3dm from the Chapter 9 folder on the companion CD.
2. Click on the Mesh Patch button on the Mesh toolbar, or type MeshPatch at the command area.

3. Click on location B and drag to location C (Figure 9–26) to select all the point objects, and press the ENTER key.
4. Because there is no inner curve, press the ENTER key.
5. Select curve A. A mesh patch is constructed. Do not save your file.

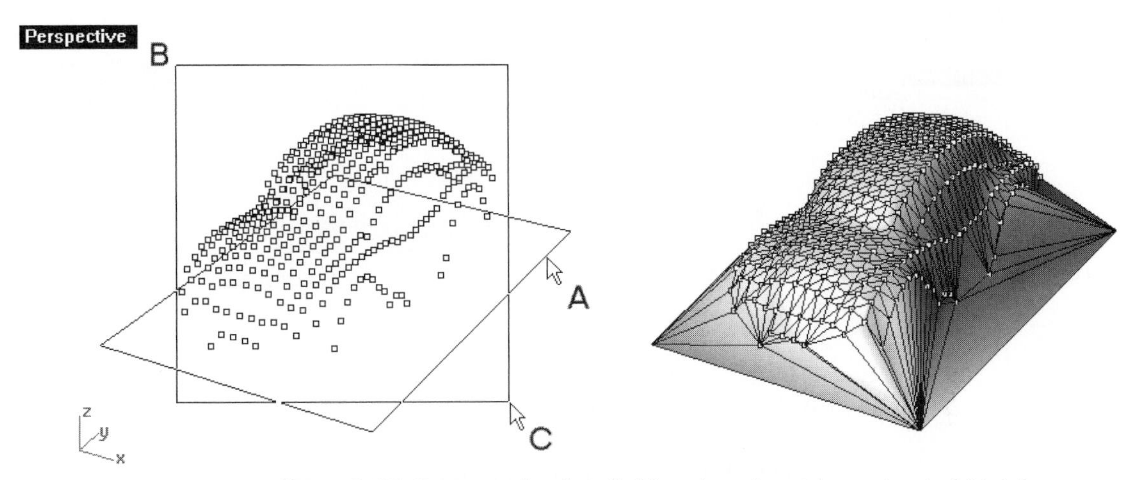

Figure 9–26. Curves and points (left) and mesh patch constructed (right)

Constructing a Polygon Mesh from a NURBS Surface

NURBS surfaces and polygon meshes are two different types of objects. A NURBS surface is an exact representation, but the polygon mesh is an approximated representation of the 3D object.

In previous chapters, you already learned various ways to construct a NURBS surface model of complex shapes. Now, if you want to construct a polygon mesh of very complex shape, you may think about first building the model in NURBS surface or polysurface and then construct a polygon mesh from it. Perform the following steps to construct a mesh polygon from a NURBS surface:

1. Select File > Open and select the file MeshfromSurface.3dm from the Chapter 9 folder on the companion CD.
2. Select Mesh > From NURBS Objects, or click on the Mesh from Surface/Polysurface button on the Mesh toolbar.
3. Select surface A (shown in Figure 9–27) and press the ENTER key.
4. In the Polygon Mesh Options dialog box, set the polygon density by moving the slider bar, and then click on the OK button. Alternatively, you can click on the Detailed Controls button and manipulate various parameters.

5. A polygon mesh is constructed. Do not save your file.

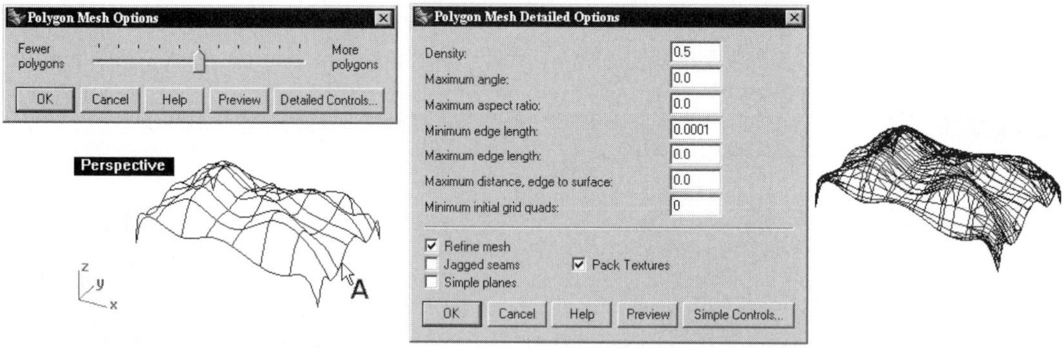

Figure 9–27. NURBS surface (left) and polygon mesh constructed from the surface (right)

If you now set the display to shaded, you will not find any difference between the original NURBS surface and the derived polygon mesh, because shading and rendering are both making use of meshes.

Constructing a Polygon Mesh Through the Control Points of a NURBS Surface

Apart from approximating the profile of a NURBS surface, you can use the NURBS surface control points as vertices to construct a polygon mesh. Perform the following steps to construct a polygon mesh from the control points of a NURBS surface.

1. Select File > Open and select the file MeshfromControlPolygon.3dm from the Chapter 9 folder on the companion CD.
2. Select Mesh > From NURBS Control Polygon, or click on the Extract Mesh from NURBS Control Polygon button on the Mesh toolbar.
3. Select surface A (shown in Figure 9–28) and press the ENTER key.
4. A polygon mesh is constructed. Do not save your file.

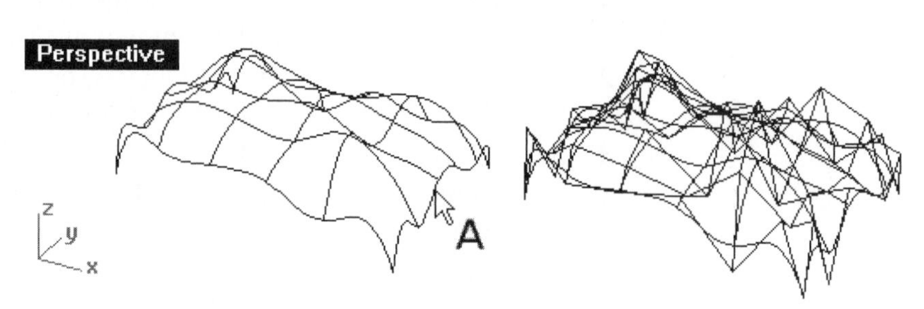

Figure 9–28. Surface (left) and polygon mesh constructed from the control vertices (right)

Constructing a Polygon Mesh from an Image's Heightfield

Similar to constructing a surface from a raster image, you can use the heightfield of an image to construct a polygon mesh. If the purpose of using the heightfield of an image is to construct a rapid prototype, you can directly construct a polygon mesh from the image, rather than first constructing a NURBS surface and then exporting/converting to polygon mesh. Perform the following steps to construct a polygon mesh from the heightfield of an image:

1. Select File > Open and select the file HeightfieldMesh.3dm from the Chapter 9 folder on the companion CD.
2. Select Mesh > Mesh Heightfield, or click on the Mesh heightfield from Image button on the Mesh Tools toolbar.
3. Select the file car.tga from the Chapter 9 folder on the companion CD.
4. Select points A and B, shown in Figure 9–29.
5. Select the GridSize option on the command area.
6. Select the Width option on the command area.
7. Type 200 at the command area.
8. Select the Height option on the command area.
9. Type 200 at the command area.
10. Press the ENTER key.
11. Select the Elevation option on the command area.
12. Select the Value factor on the command area. (This option determines the height values of the polygon mesh.)
13. Type 0.01 at the command area.
14. Press the ENTER key twice.
15. A polygon mesh is constructed from the heightfield of an image. Do not save your file.

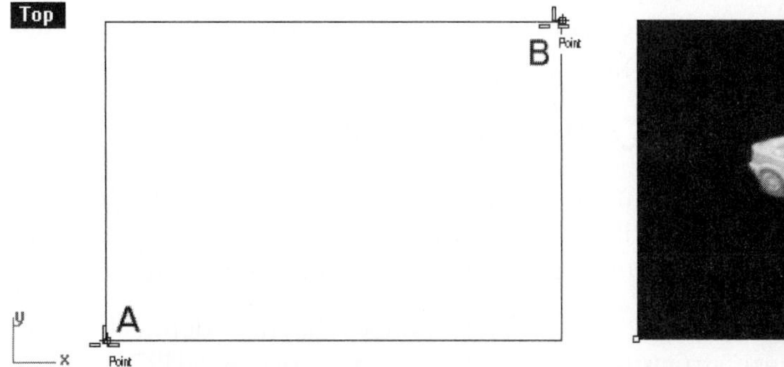

Figure 9–29. Two points selected (left) and polygon mesh constructed (right)

Constructing a Mesh by Offsetting a Mesh

By offsetting an existing polygon mesh, you obtain another polygon mesh. Using the cap option, you can construct a polygon mesh that encloses a volume from a single polygon mesh. Perform the following steps to construct an offset polygon mesh from an existing polygon mesh:

1. Select File > Open and select the file OffsetMesh.3dm from the Chapter 9 folder on the companion CD.
2. Select Mesh > Offset Mesh, or click on the Offset mesh button on the Mesh Tools toolbar.
3. Select mesh A, shown in Figure 9–30, and press the ENTER key.
4. Select the PickOffsetDistance option on the command area.
5. In the Offset Mesh dialog box, set offset distance to 2.
6. Click on the FlipAll button as necessary to change the offset direction.
7. Click on the OK button.
8. An offset mesh is constructed. Do not save your file.

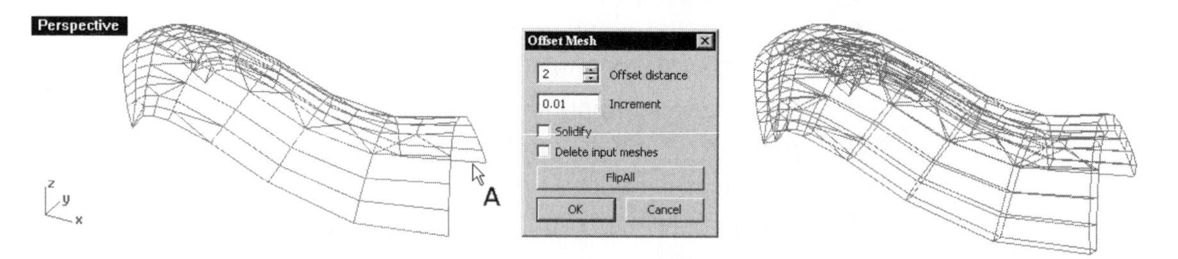

Figure 9–30. Original polygon mesh (left) and offset with end caps polygon mesh (right)

Mapping Polygon Meshes

You can map a polygon mesh object to surface in two ways: in accordance with vertex count and mesh structure and in accordance with the surface's UVN.

Vertex Count and Structure

In many animation programs that use polygon meshes to represent 3D objects, morphing of two different shapes requires the polygon meshes to have an identical vertex count and mesh structure.

To construct two different shapes having identical vertex count and mesh structure, you construct two shapes as NURBS surfaces, construct a polygon mesh from a surface, and map the polygon mesh to the other surface. Perform the following steps to map a polygon mesh to a surface:

1. Select File > Open and select the file ApplyMesh.3dm from the Chapter 9 folder on the companion CD.

This file contains a polygon mesh constructed from a NURBS surface. To obtain another polygon mesh of identical vertex count and mesh structure, you will map the polygon mesh to another NURBS surface of different shape.

2. Select Mesh > Apply to Surface, or click on the Apply Mesh to NURBS Surface button on the Mesh toolbar.
3. Select mesh A (shown in Figure 9–31).
4. Select surface B (shown in Figure 9–31).
5. The polygon mesh constructed from a NURBS surface is applied to another surface, as shown in Figure 9–32. Do not save your file.

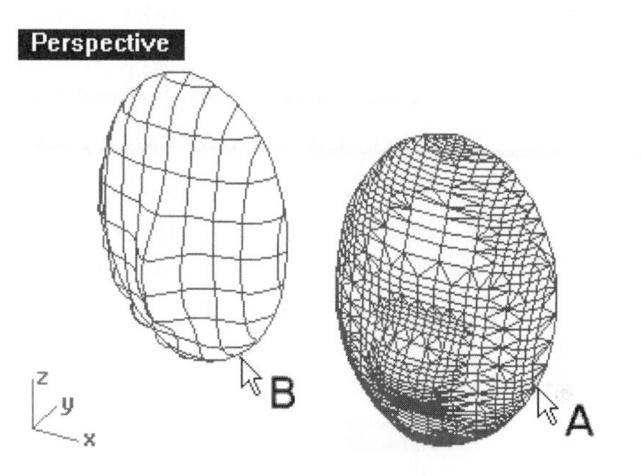

Figure 9–31. Polygon mesh being applied to a surface

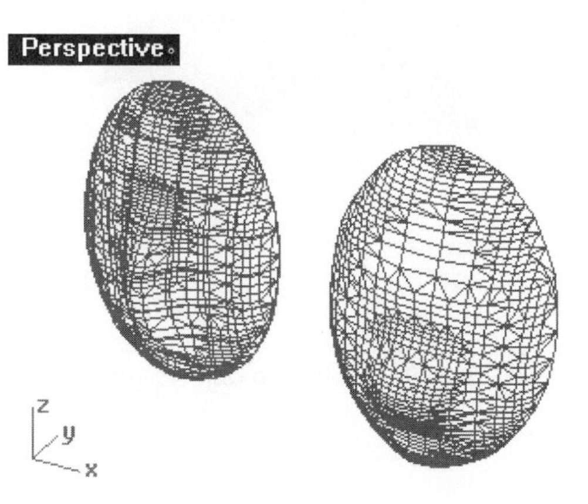

Figure 9–32. Polygon mesh applied to a surface

Surface's UVN

To map meshed objects and point objects accurately onto a surface in accordance with the surface's u and v coordinates, you use the ApplyMeshUVN command, as follows:

1. Select File > Open and select the file ApplyMeshUVN.3dm from the Chapter 9 folder on the companion CD.
2. Select Mesh > Apply Mesh UVN, or click on the Apply Mesh UVN button on the Mesh Tools toolbar.
3. Click on A and drag to B, shown in Figure 9–33, and press the ENTER key.
4. Select the VerticalScale option on the command area.
5. Type 0.5 at the command area to specify a vertical scale factor.
6. Select surface C (Figure 9–33). The selected meshed objects and point objects are mapped onto the surface. (See Figure 9–34.)
7. Repeat the command.
8. Click on A and drag to B, shown in Figure 9–33, and press the ENTER key.
9. Select the VerticalScale option on the command area.
10. Type 2 at the command area to specify a vertical scale factor.
11. Select surface D (Figure 9–33). The selected meshed objects and point objects are mapped onto the surface. (See Figure 9–34.)
12. Do not save your file.

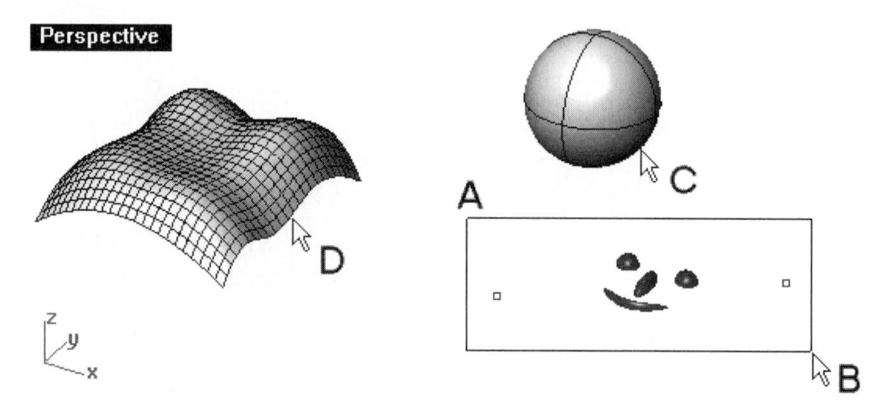

Figure 9–33. Meshed objects and point objects being mapped

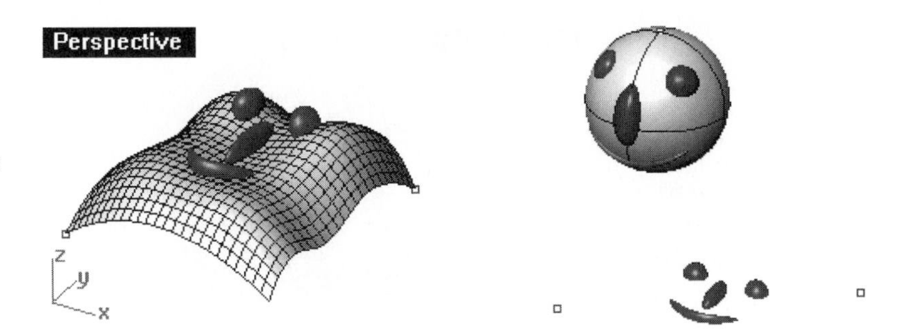

Figure 9–34. Meshed objects and point objects mapped and scaled vertically

Combining and Separating

More complex meshed objects can be composed by combining two or more individual meshed objects using Boolean operations (union, difference, and intersect). Apart from using Boolean operations, you can join contiguous meshed objects. Conversely, you can explode joined objects to revert into individual meshed objects. To split or trim meshed objects, you use a curve, surface, polysurface, or polygon mesh.

Boolean Operations

There are three basic Boolean operations: union, difference, and intersection. To carry out these operations, the polygon meshes have to overlap each other in 3D space.

Mesh Union

A union of a set of polygon meshes produces a polygon mesh that has the volume of all polygon meshes in the set. Any overlapped portions of the polygon meshes will be trimmed away, and the final model will be a single polygon mesh. To construct a polygon mesh by uniting two polygon meshes, perform the following steps:

1. Select File > Open and select the file UnionMesh.3dm from the Chapter 9 folder on the companion CD.
2. Select Mesh > Mesh Boolean > Union, or click on the Mesh Boolean Union button on the Mesh Booleans toolbar.
3. Select polygon meshes A and B, shown in Figure 9–35, and press the ENTER key.
4. Two polygon meshes are united into one. Do not save your file.

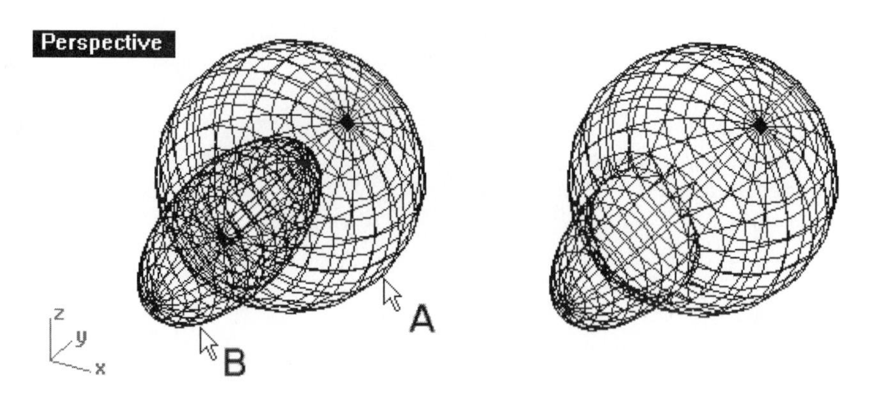

Figure 9–35. Two polygon meshed (left) and polygon meshes united into one (right)

Mesh Difference

A difference of two sets of polygon meshes produces a polygon mesh that has the volume contained in the first set of polygon meshes but not the second set of polygon meshes. In the operation, the second set of polygon meshes is subtracted from the first set of polygon meshes, and the result is a single polygon mesh. Perform the following steps to subtract a polygon mesh from another polygon mesh:

1. Select File > Open and select the file DifferenceMesh.3dm from the Chapter 9 folder on the companion CD.
2. Select Mesh > Mesh Boolean > Difference, and click on the Mesh Boolean Difference button on the Mesh Booleans toolbar.
3. Select mesh A, shown in Figure 9–36, and press the ENTER key. This is the first set of polygon mesh.
4. Select mesh B, shown in Figure 9–36, and press the ENTER key. This is the second set of polygon mesh.
5. The second set of polygon mesh is subtracted from the first set of polygon mesh. Do not save your file.

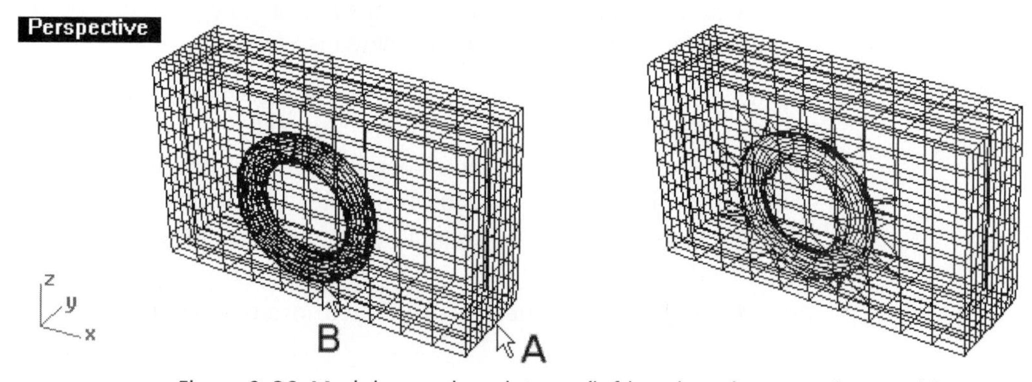

Figure 9–36. Mesh box and mesh torus (left) and mesh torus subtracted from mesh box (right)

Mesh Intersection

An intersection of two sets of polygon meshes produces a polygon mesh that has the volume contained in the first and second sets of polygon mesh. In other words, the volume common to two sets of polygon meshes will be retained, and the result is a single polygon mesh. Perform the following steps to construct an intersection polygon mesh from a pair of polygon meshes:

1. Select File > Open and select the file IntersectMesh.3dm from the Chapter 9 folder on the companion CD.
2. Select Mesh > Mesh Boolean > Intersection, or click on the Mesh Boolean Intersection button on the Mesh Booleans toolbar.
3. Select mesh A, shown in Figure 9–37, and press the ENTER key. This is the first set of polygon mesh.
4. Select mesh B, shown in Figure 9–37, and press the ENTER key. This is the second set of polygon mesh.
5. An intersection of two polygon meshes is constructed. Do not save your file.

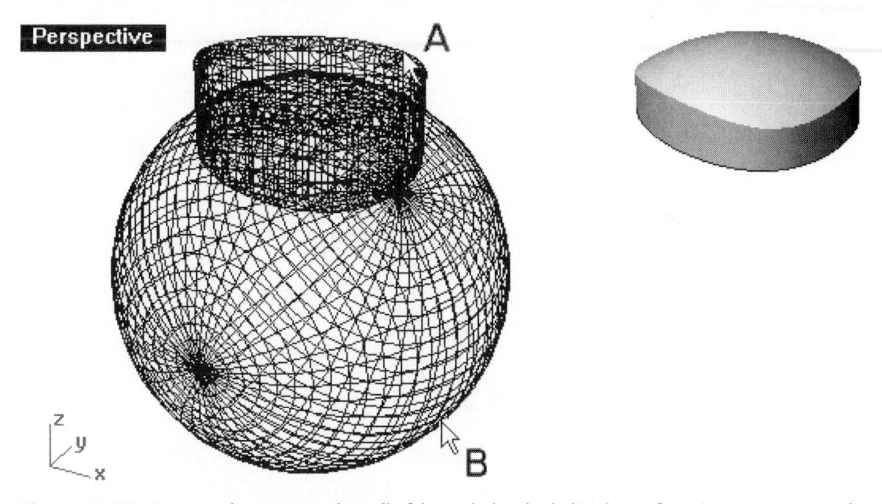

Figure 9–37. Two polygon meshes (left) and shaded display of an intersection of the polygon meshes (right)

Joining and Exploding Meshes

Quite often, you may have to join several meshed objects sharing common boundaries into a single meshed object. The opposite of joining is exploding joined meshes to obtain individual polygon meshed objects.

Perform the following steps to join two contiguous meshed objects:

1. Select File > Open and select the file JoinExplode.3dm from the Chapter 9 folder on the companion CD.
2. Select Edit > Join, or click on the Join button on the Main1 toolbar.
3. Select meshed objects A, B, and C (Figure 9–38) and press the ENTER key. The selected meshes are joined.

Continue with the following steps to explode joined meshed objects.

4. Select Edit > Explode, or click on the Explode button on the Main2 toolbar.
5. Select the joined polygon meshes and press the ENTER key. The meshes are exploded into individual polygon meshes.
6. Do not save your file.

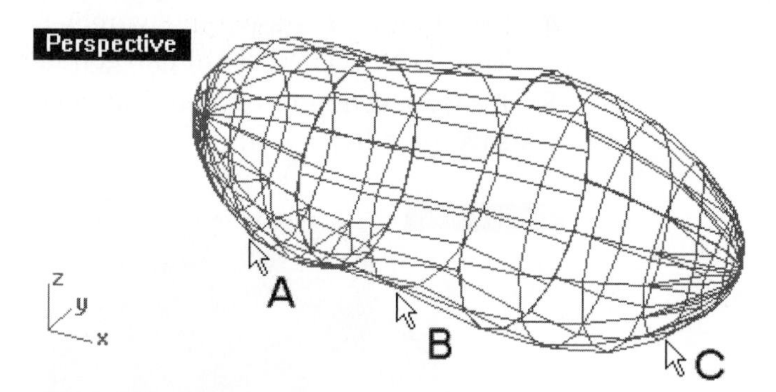

Figure 9–38. Meshed objects being joined

Mesh Splitting

A polygon mesh can be split into two in two basic ways: Boolean split and splitting by a curve, surface, or polygon mesh.

Boolean Split

With the Boolean split method, you use a polygon mesh, surface, or polysurface to split a polygon mesh into two; the cutting object has to be large enough to cut through the polygon mesh. If the polygon mesh to be split is a closed-loop polygon mesh, the openings after splitting will be covered by a polygon mesh face resembling the shape of the cutting object. As a result, you obtain two closed-loop polygon meshes. Perform the following steps:

1. Select File > Open and select the file BooleanSplit.3dm from the Chapter 9 folder on the companion CD.
2. Select Mesh > Mesh Boolean > Boolean Split, or click on the Mesh Boolean Split button on the Mesh Booleans toolbar.

3 Select mesh A, shown in Figure 9–39, and press the ENTER key. This is the mesh to be split.
4 Select mesh B, shown in Figure 9–39, and press the ENTER key. This is the cutting mesh.
5 Mesh A is split into two meshes. Hide A and B.
6 Shade the viewport to appreciate the result. Do not save your file.

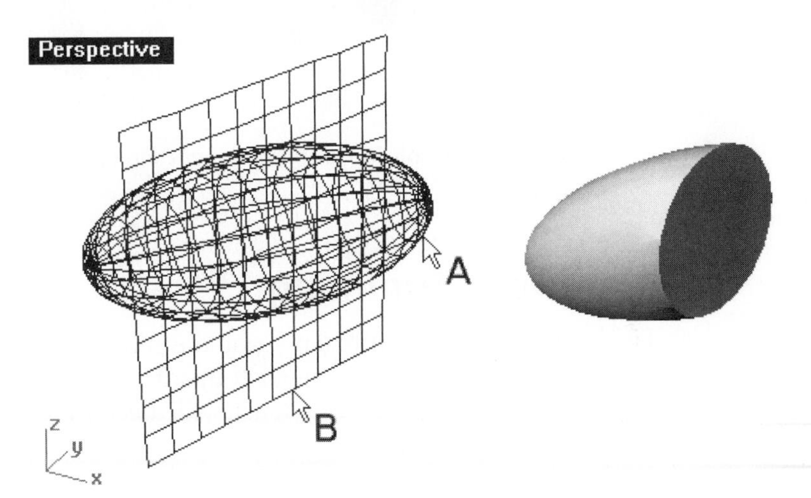

Figure 9–39. Two polygon meshes (left) and a polygon mesh split and two polygon mesh hidden (right)

Splitting by a Curve, Surface, or Polygon Mesh

The second method is to use a curve, surface, or polygon mesh as cutting object. If a curve is used as the splitting tool, the curve is projected vertically onto the active construction plane, splitting the polygon mesh into two. Unlike Boolean split, the openings caused by splitting will not be filled up. Therefore, a closed-loop polygon mesh will become an open-loop polygon mesh after splitting this way. Perform the following steps:

1 Select File > Open and select the file MeshSplitTrim.3dm from the Chapter 9 folder on the companion CD.
2 Select Mesh > Mesh Edit Tools > Mesh Split, or click on the Mesh Split button on the Mesh toolbar.
3 Select polygon meshes A, B, and C, shown in Figure 9–40, and press the ENTER key. These are the polygon meshes to be split.
4 Select curve D, surface E, and polygon mesh F, shown in Figure 9–40, and press the ENTER key. These are the splitting tools. The polygon meshes are split into two.
5 To appreciate the effect of splitting, move polygon meshes A, B, and C a small distance apart from their original position.
6 Do not save your file.

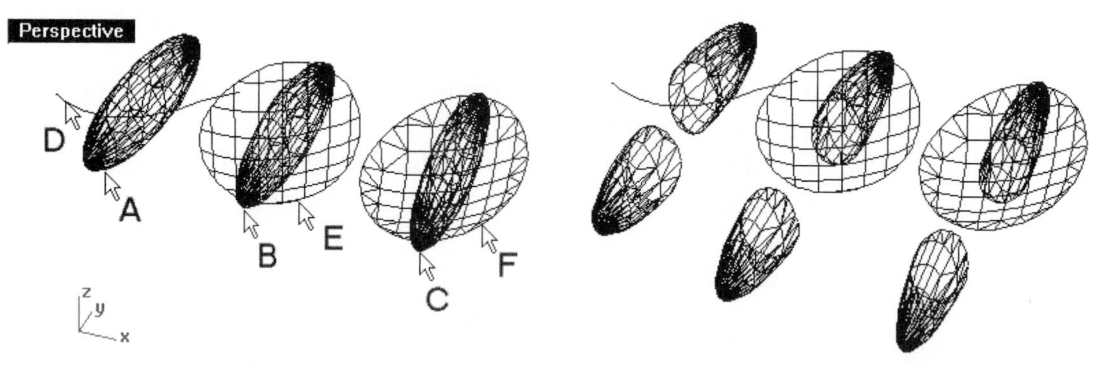

Figure 9–40. Polygon meshes and splitting tools (left) and split polygon meshes moved apart (right)

Mesh Trimming

You can split or trim a polygon mesh by using a curve, surface, or polygon mesh. Again, the opening left by trimming will not be filled up. Therefore, a closed-loop polygon mesh will become an open-oop polygon mesh after trimming.

1. Select File > Open and select the file MeshSplitTrim.3dm from the Chapter 9 folder on the companion CD.
2. Select Mesh > Mesh Edit Tools > Mesh Trim, or click on the Mesh Trim button on the Mesh toolbar.
3. Select curve A, surface B, and polygon mesh C, shown in Figure 9–41, and press the ENTER key. These are the trimming tools.
4. Select polygon meshes D, E, and F, and press the ENTER key. The selected portions of the polygon meshes are trimmed.
5. Do not save your file.

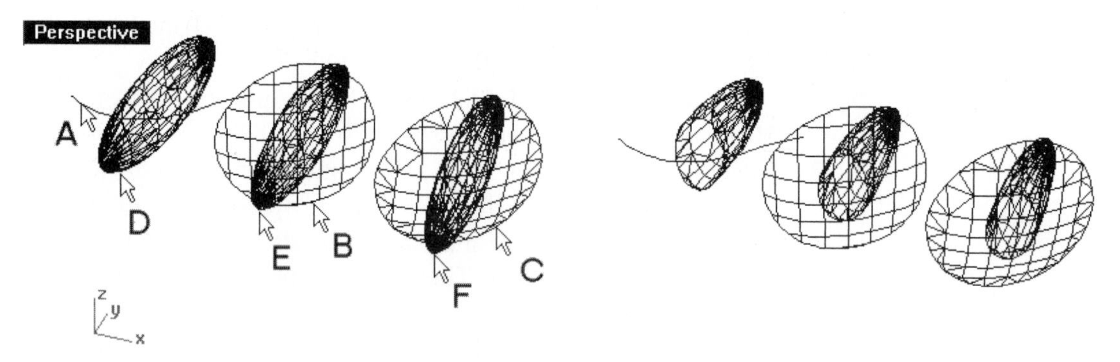

Figure 9–41. Polygon meshes and trimming tools (left) and polygon meshes trimmed (right)

Manipulating Mesh Faces

Having learned to make primitive polygon meshes, construct polygon meshes from existing objects, and combine and separate polygon meshes, you will now work on the most important part of this chapter—editing polygon meshes. This information will help you when manipulating existing polygon mesh, in particular, and repairing polygon meshes for rapid prototyping.

Reducing Mesh Density

As explained earlier, a high polygon mesh density results in a large file size. If the polygon mesh is overly dense, you can reduce its density. Decreasing the polygon count means simplifying the mesh. To reduce a polygon density, perform the following steps. This process is irreversible, meaning that if you reduce the polygon count and later want to increase the count, you have to rebuild the mesh.

1. Select File > Open and select the file ReduceMesh.3dm from the Chapter 9 folder on the companion CD.
2. Select Mesh > Mesh Edit Tools > Collapse > Reduce Vertex Count, or click on the Reduce Mesh Polygon Count/Convert Quad Meshes to Triangles button on the Mesh toolbar.
3. Select mesh A, shown in Figure 9–42, and press the ENTER key.
4. In the Reduce Mesh Options dialog box, set reduction percentage to 90, and click on the OK button. Note that you can first click on the Preview button to discover the outcome before confirming by clicking the OK button.
5. Do not save your file.

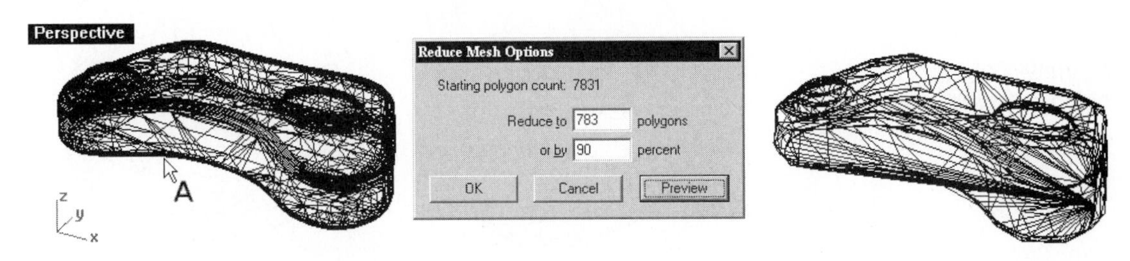

Figure 9–42. From left to right: Original meshed object, Reduce Mesh Options dialog box, and mesh count reduced

Collapsing Mesh Elements

Reduction of a meshed object's mesh count decreases the number of meshes evenly throughout the entire object globally. If you want to reduce mesh count in a more controlled way locally, you should use the mesh collapsing tools, which remove the selected quadrilateral mesh element's edge to turn it into a triangular mesh or remove selected mesh elements. When a mesh element is removed from the meshed object, unless the element lies on the boundary edge of the meshed object, those mesh elements adjacent to the removed element will be adjusted automatically to fill the gap left by the removed element. Therefore, do not confuse collapsing mesh elements with deleting mesh elements, which will be discussed later in this chapter.

Collapsing Mesh Vertex

You can select a vertex from a meshed object and collapse it to another selected vertex. By doing so, meshes sharing the selected vertex will be affected.

1. Select File > Open and select the file Collapse.3dm from the Chapter 9 folder on the companion CD.
2. Select Mesh > Mesh Edit Tools > Collapse > Vertex, or click on the Collapse Vertex button on the Collapse Mesh toolbar.
3. Select vertices A and B, shown in Figure 9–43, and press the ENTER key. The selected vertices are collapsed.
4. Do not save your file.

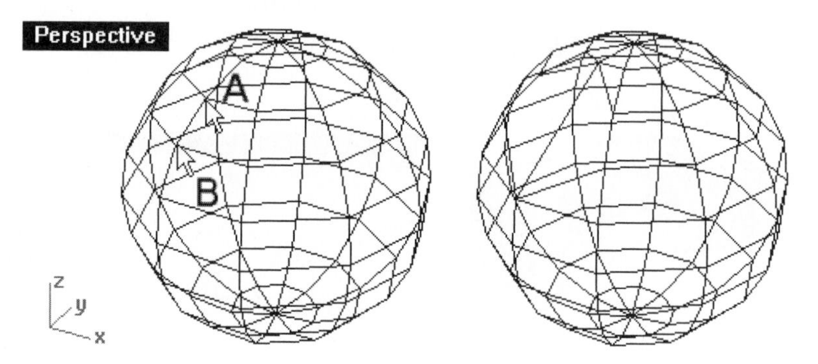

Figure 9–43. Vertices being collapsed

Collapsing Mesh Edge

Collapsing a mesh edge removes an edge of the meshed object. Any mesh element sharing that removed edge will be affected. If the affected edge is a quadrilateral edge, it will become a triangular mesh. If it is a triangular mesh, removing an edge means removing the mesh element.

1. Select File > Open and select the file Collapse.3dm from the Chapter 9 folder on the companion CD.
2. Select Mesh > Mesh Edit Tools > Collapse > Edge, or click on the Collapse Edge button on the Collapse Mesh toolbar.
3. Select edge A and then edge B, shown in Figure 9–44, and press the ENTER key.
4. Do not save your file.

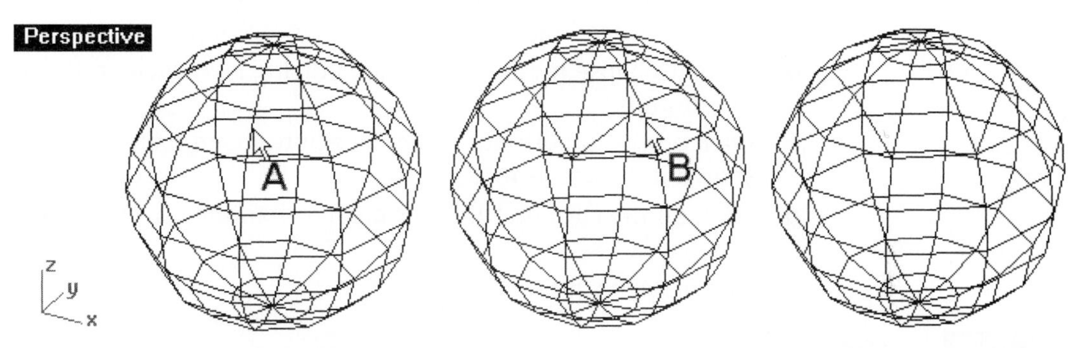

Figure 9–44. From left to right: original meshed object, an edge collapsed, and two edges collapsed

Collapsing Mesh Face

As the name implies, collapsing mesh face concerns the removal of the selected mesh element.

1. Select File > Open and select the file Collapse.3dm from the Chapter 9 folder on the companion CD.
2. Select Mesh > Mesh Edit Tools > Collapse > Face, or click on the Collapse Face button on the Collapse Mesh toolbar.
3. Select face A, shown in Figure 9–45, and press the ENTER key.
4. Do not save your file.

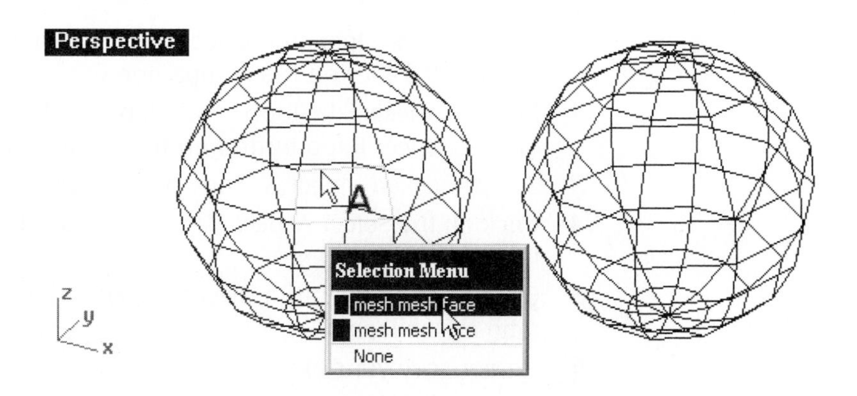

Figure 9–45. Mesh face selected (left) and face collapsed (right)

Collapsing Mesh Element by Mesh Area

To have meshes collapsed by specifying a mesh element's area, perform the following steps:

1. Select File > Open and select the file Collapse.3dm from the Chapter 9 folder on the companion CD.
2. Select Mesh > Mesh Edit Tools > Collapse > By Area, or click on the Face Area button on the Collapse Mesh toolbar.
3. Select meshed object A, shown in Figure 9-46.
4. In the Collapse mesh faces by area dialog box, click on the Select face button.
5. Click on the upper-right Select face button and then select face B, shown in Figure 9-46.
6. Click on the OK button. Faces greater than the selected face are collapsed.
7. Do not save your file.

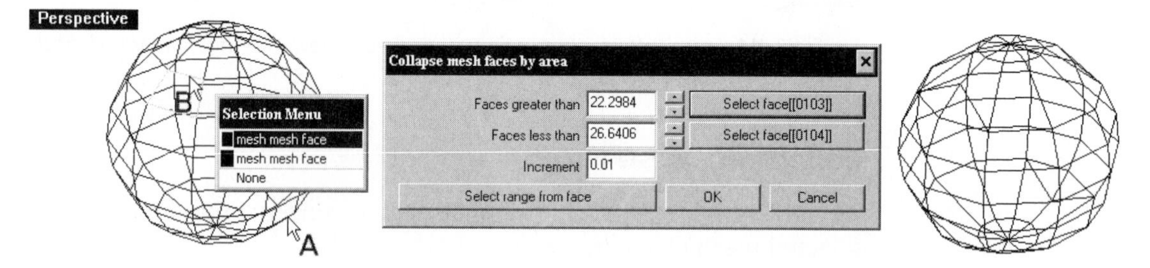

Figure 9-46. From left to right: meshed object and a face selected, Collapse mesh faces by area dialog box, and collapsed meshed object

Collapsing Mesh Element by Mesh's Aspect Ratio

A method to remove mesh element is to state the aspect ratio of mesh elements to be collapsed.

1. Select File > Open and select the file Collapse.3dm from the Chapter 9 folder on the companion CD.
2. Mesh > Mesh Edit Tools > Collapse > By Aspect Ratio, or click on the Aspect Ratio button on the Collapse Mesh toolbar.
3. Select meshed object A, shown in Figure 9-47.
4. Click on the Select Aspect Ratio From Face button, and then select face B.
5. Click on the OK button. Faces equal to the selected face's aspect ratio are collapsed.
6. Do not save your file.

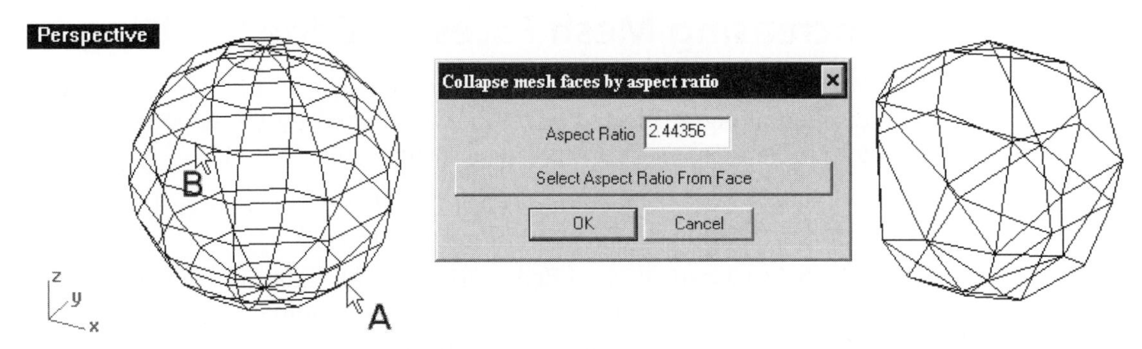

Figure 9–47. From left to right: meshed object and mesh face selected, Collapse mesh face by aspect ratio dialog box, and collapsed meshed object

Collapsing Mesh Element by Edge Length

You can specify mesh elements' edge length to collapse, as follows:

1. Select File > Open and select the file Collapse.3dm from the Chapter 9 folder on the companion CD.
2. Select Mesh > Mesh Edit Tools > Collapse > By Edge Length, or click on the Edge Length button on the Collapse Mesh toolbar.
3. Select meshed object A, shown in Figure 9–48.
4. Click on the Select Edge button of the Collapse mesh faces by edge length dialog box.
5. Select edge B.
6. Click on the OK button. Edges longer than the selected edge are collapsed.
7. Do not save your file.

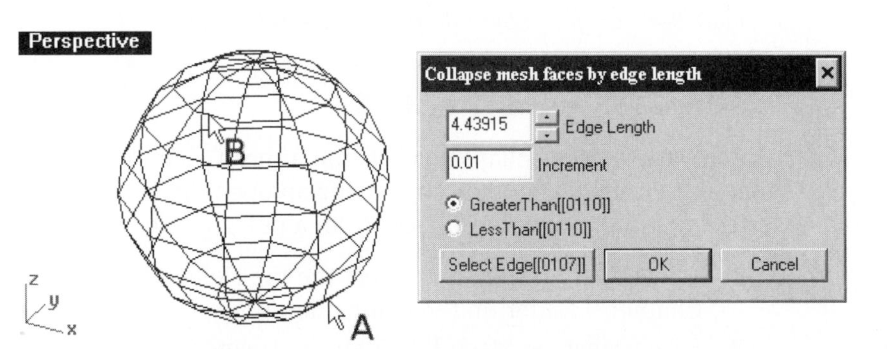

Figure 9–48. From left to right: Meshed object and edge selected, Collapse mesh faces by edge length dialog box, and collapsed meshed object

Increasing Mesh Faces by Edge Splitting

You can split the edge of a mesh to increase the number of mesh. By clicking along a selected edge, meshes shared by the edge will be split into more meshes. Perform the following steps:

1. Select File > Open and select the file SplitMeshEdge.3dm from the Chapter 9 folder on the companion CD.
2. Select Mesh > Mesh Repair Tools > Split Edge, or click on the Split a mesh edge button on the Mesh Tools toolbar.
3. Select edge A, shown Figure 9–49.
4. Select a location near the midpoint of edge A, and press the ENTER key.
5. Select edge B, shown Figure 9–49.
6. Select a location near the midpoint of edge A, and press the ENTER key.
7. Press the ENTER key.
8. Do not save your file.

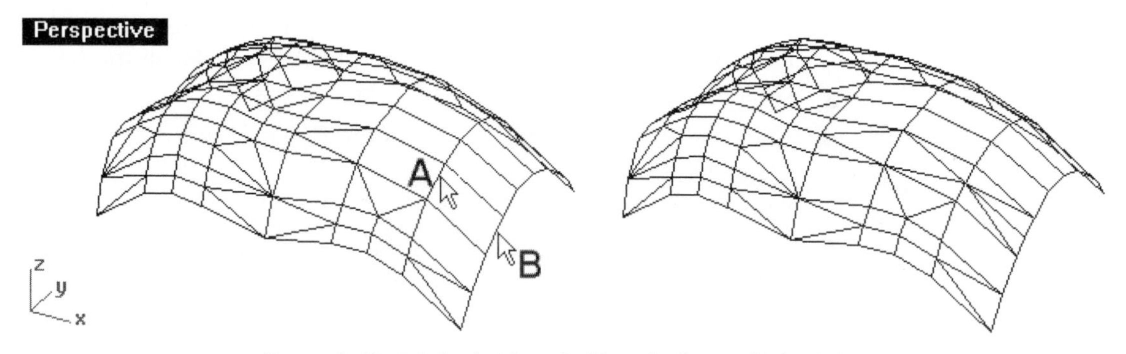

Figure 9–49. Original object (left) and edges split (right)

Deleting Mesh Faces

Unlike mesh face collapsing, deleting a mesh face removes a mesh element from the polygon mesh, leaving an opening. Perform the following steps to remove four mesh faces from a polygon mesh sphere:

1. Select File > Open and select the file DeleteMesh.3dm from the Chapter 9 folder on the companion CD.
2. Select Mesh > Mesh Edit Tools > Delete Mesh Faces, or click on the Delete mesh faces button on the Mesh Tools toolbar.
3. Select faces A, B, C, and D, shown in Figure 9–50, and press the ENTER key. The faces are deleted.
4. Do not save your file.

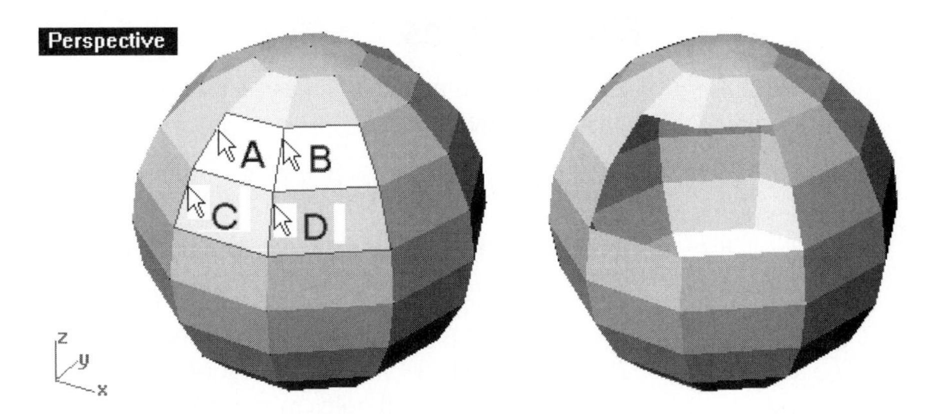

Figure 9–50. Mesh faces selected (left) and faces deleted (right)

Patching Single Mesh Face

Patching refers to filling up openings by adding individual triangular or quadrilateral mesh faces. You can construct triangular mesh face by selecting an edge and a vertex. If you select two edges, you construct a quadrilateral face. Perform the following steps:

1. Select File > Open and select the file PatchFace.3dm from the Chapter 9 folder on the companion CD.
2. Select Mesh > Mesh Repair Tools > Patch Single Face, or click on the Add a mesh face button on the Mesh Tools toolbar.
3. If JoinMesh = No, select JoinMesh option on the command area to change it to Yes. Otherwise, proceed to the next step.
4. Select edge A and vertex B, shown in Figure 9–51. A triangular mesh face is patched.
5. Repeat the command.
6. Select edges C and D. A quadrilateral mesh face is patched.
7. Do not save your file.

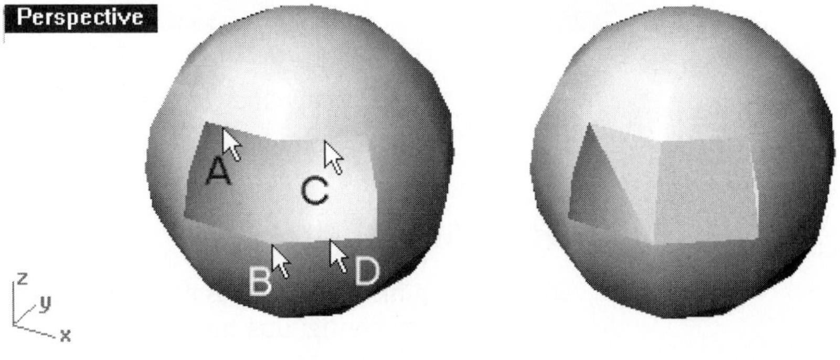

Figure 9–51. Meshed object with a hole (left) and two mesh faces patched

Filling Holes

Instead of patching single faces on a meshed object, you can fill selected holes or all the holes.

Filling Individual Holes

To fill individual holes of a meshed object, perform the following step:

1. Select File > Open and select the file FillHole.3dm from the Chapter 9 folder on the companion CD.
2. Select Mesh > Mesh Repair Tools > Fill Hole, or click on the Fill Mesh Hole/Fill all holes in mesh button on the Mesh Tools toolbar.
3. Select edge A, shown in Figure 9–52. A hole is filled with triangular meshes.

Filling All Holes

To fill all the holes in a meshed object, continue with the following steps.

4. Select Mesh > Mesh Repair Tools > Fill Holes, or right-click on the Fill Mesh Hole/Fill all holes in mesh button on the Mesh Tools toolbar.
5. Select meshed object B, shown in Figure 9–52. All the holes are filled.
6. Do not save your file.

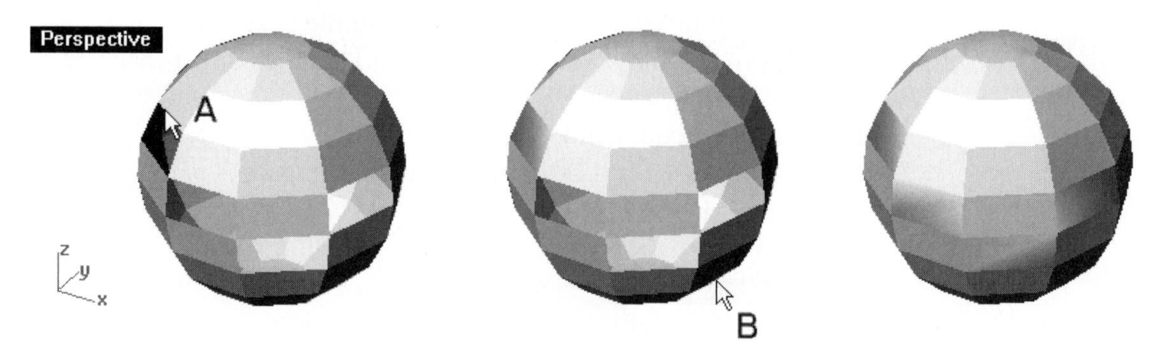

Figure 9–52. From left to right: Original meshed object, a grahole filled, and all holes filled

Matching Edges

Sometimes, a meshed object is not watertight because there are misaligned edges between contiguous mesh elements in the meshed object, leaving some openings along the edges. To repair such meshed objects, you can use the MatchMeshEdge command to match vertices within a specified tolerance and split the edges to make the edges match.

1. Select File > Open and select the file MatchMeshEdge.3dm from the Chapter 9 folder on the companion CD. This meshed object consists of four polygon meshes joined together. Along the joints, there are gaps.
2. Select Mesh > Mesh Repair Tools > Match Mesh Edge, or click on the Match mesh edges button on the Mesh Tools toolbar.
3. Select the DistanceToAdjust option on the command area.
4. Type 4 at the command area.
5. Select the PickEdges option on the command area.
6. Select edge A, shown in Figure 9–53, and press the ENTER key. The edges of the mesh elements along the selected edge are adjusted and matched.
7. Repeat the command.
8. Select the meshed object and press the ENTER key. All edges are adjusted to match.
9. Do not save your file.

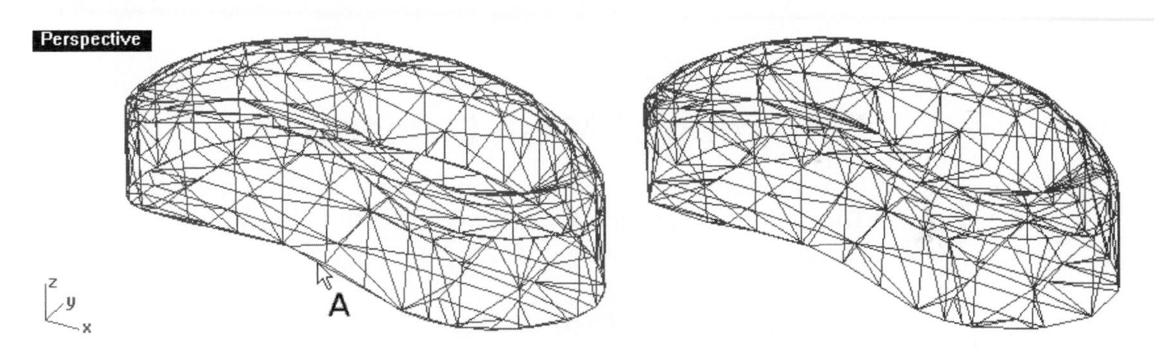

Figure 9–53. Original meshed object (left) and matched object (right)

Aligning Vertices

A way to close small gaps in a meshed object is to specify a tolerance value for vertices within such value to align into a single vertex. Care must be taken not to specify excessively large tolerance. Otherwise, too many mesh vertices will be collapsed and the polygon mesh will be over simplified. Perform the following steps:

1. Select File > Open and select the file AlignMeshVertices.3dm from the Chapter 9 folder on the companion CD. This meshed object consists of four polygon meshes joined together. Along the joints, there are gaps.

2. Select Mesh > Mesh Repair Tools > Align Mesh Vertices, or click on the Align mesh vertices to tolerance button on the Mesh Tool toolbar.
3. Select the DistanceToAdjust option on the command area.
4. Type 1 at the command area.
5. Select meshed object A, shown in Figure 9–54. Vertices within the tolerance region are adjusted.
6. Repeat the command with a tolerance of 2 units. As can be seen, the meshed object is now over simplified.
7. Do not save the drawing.

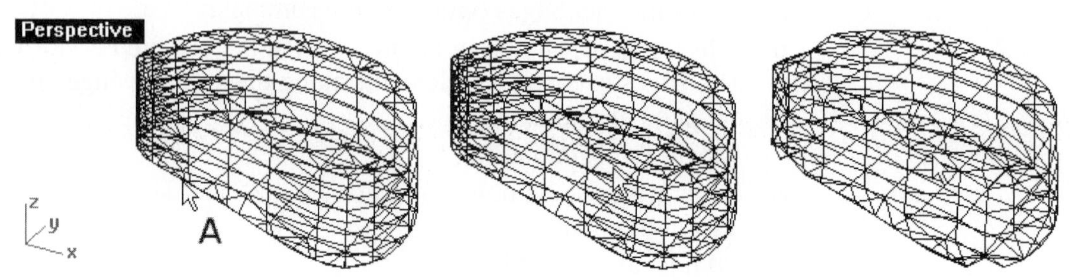

Figure 9–54. From left to right: Original meshed object, vertices within 1 unit tolerance collapsed, and vertices within 2 units tolerance collapsed

Extraction of Mesh Elements

Extraction involves isolating selected mesh faces from a polygon mesh. The extracted elements are removed from the original mesh as separate polygon mesh. They are not deleted.

Extracting Mesh Face

You can extract selected faces of a meshed object. Extraction of mesh face isolates the mesh faces from the main body of the mesh object, leaving holes in the original object. Perform the following steps:

1. Select File > Open and select the file Extract01.3dm from the Chapter 9 folder on the companion CD.
2. Select Mesh > Mesh Edit Tools > Extract > Faces, or click on the Faces button on the Extract Mesh toolbar.
3. Select mesh face A and B, shown in Figure 9–55, and press the ENTER key. The selected mesh faces are extracted (detached and isolated) from the original mesh object.

4 To realize that the faces are extracted, press the DELETE key to delete the extracted faces and shade the viewport.
5 Do not save your file.

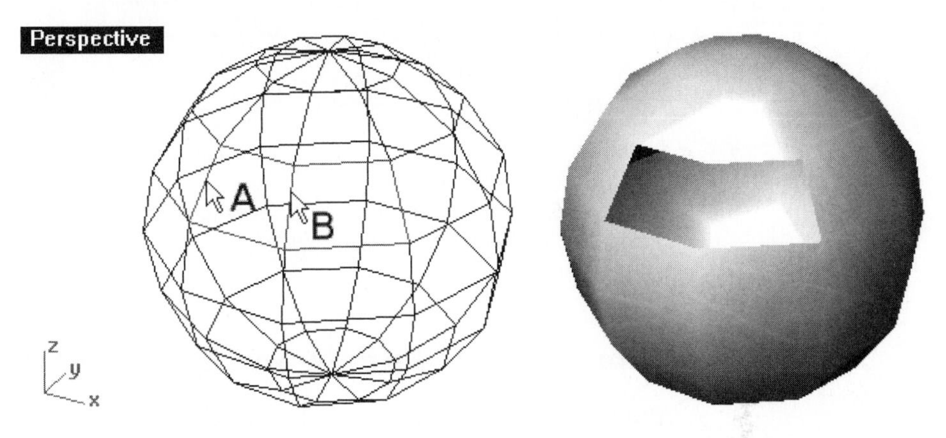

Figure 9–55. Faces being extracted (left) and extracted faces deleted and viewport shaded.

Extracting Mesh Faces by Specifying Mesh Face Area

Apart from extracting individual mesh faces, you can extract a set of mesh faces collectively by specifying mesh area.

1 Select File > Open and select the file Extract01.3dm from the Chapter 9 folder on the companion CD. This is the same file as the previous tutorial.
2 Select Mesh > Mesh Edit Tools > Extract > By Area, or click on the By Area button on the Extract Mesh toolbar.
3 Select meshed object A, shown in Figure 9–56.
4 In the Extract mesh faces by area dialog box, click on the Select range from face button.
5 Select face B, shown in Figure 9–56.
6 Click on the OK button. Mesh faces within area range are extracted.
7 Delete the extracted mesh faces, and shade the viewport to appreciate the result.
8 Do not save your file.

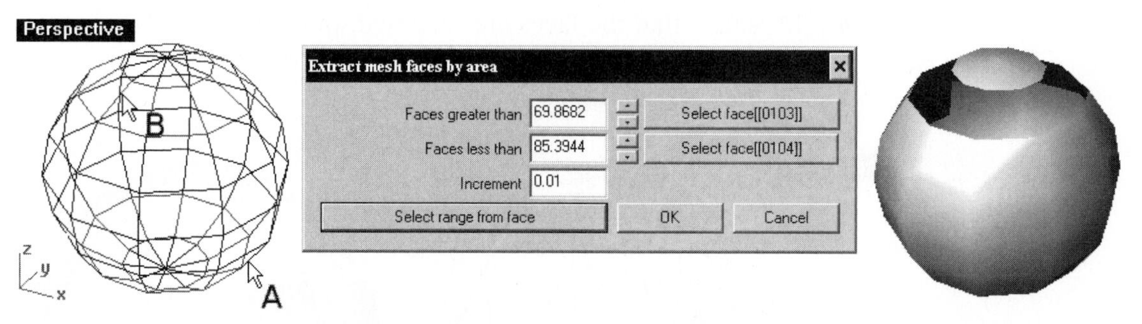

Figure 9–56. From left to right: mesh faces selected, Extract mesh faces by area dialog box, and extracted faces deleted

Extracting Mesh Faces by Specifying Mesh Aspect Ratio

Another way of extracting a set of mesh faces is to specify mesh aspect ratio. Perform the following steps:

1. Select File > Open and select the file Extract01.3dm from the Chapter 9 folder on the companion CD.
2. Select Mesh > Mesh Edit Tools > Extract > By Aspect Ratio, or click on the By Aspect Ratio button on the Extract Mesh toolbar.
3. Select meshed object A, shown in Figure 9–57.
4. Click on the Select Aspect Ratio From Face button of the Extract mesh faces by aspect ratio dialog box.
5. Select mesh face B.
6. Click on the OK button.
7. Delete the extracted faces and shade the viewport.
8. Do not save your file.

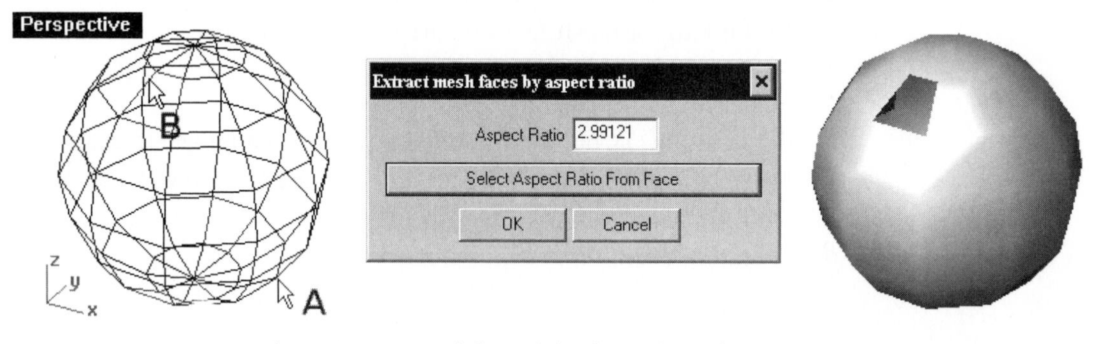

Figure 9–57. From left to right: face selected, Extract mesh faces by aspect ratio dialog box, and extracted faces deleted

Extracting Mesh Faces by Specifying Mesh Draft Angle

In accordance with mesh draft angle, you can extract a set of mesh faces. Perform the following steps:

1. Select File > Open and select the file Extract01.3dm from the Chapter 9 folder on the companion CD.
2. Select Mesh > Mesh Edit Tools > Extract > By Draft Angle, or click on the By draft angle button on the Extract Mesh toolbar.
3. Select meshed object A, shown in Figure 9–58.
4. Set End angle from camera direction to 30 degree in the Extract Mesh Faces by Draft Angle dialog box.
5. Click on the OK button.
6. Delete the extracted faces and shade the viewport.
7. Do not save your file.

Figure 9–58. From left to right: Meshed object, Extract Mesh Faces By Draft Angle dialog box, and extracted faces deleted

Extracting Mesh Faces by Specifying Mesh Face Edge Length

Mesh faces in a mesh object can be extracted by specifying mesh face aspect ratio. Perform the following steps:

1. Select File > Open and select the file Extract01.3dm from the Chapter 9 folder on the companion CD.
2. Select Mesh > Mesh Edit Tools > Extract > By Edge Length, or click on the By Edge Length button on the Extract Mesh toolbar.
3. Select mesh object A, shown in Figure 9–59.
4. In the Extract mesh faces by edge length dialog box, check the Greater Than button and click on the Select Edge button.
5. Select edge B, shown in Figure 9–59.
6. Click on the OK button.
7. Delete the extracted faces to see the result.
8. Do not save your file.

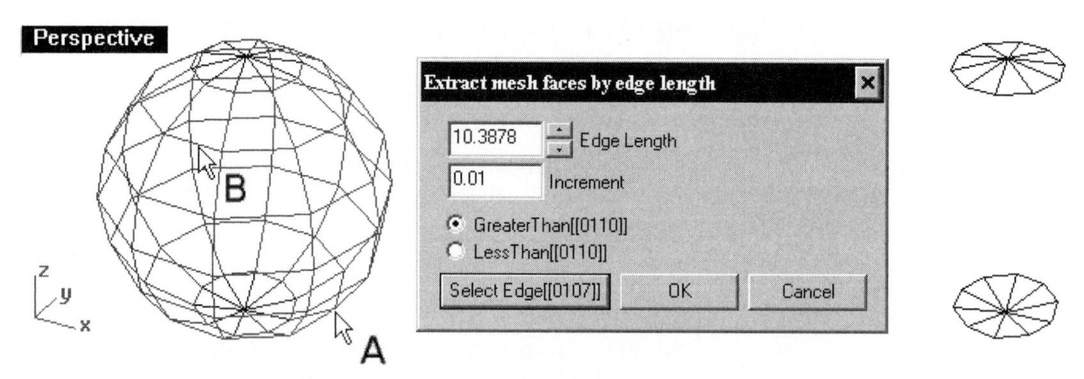

Figure 9–59. From left to right: mesh face selected, Extract mesh faces by edge length dialog box, and extracted faces deleted

Extracting Connected Mesh Faces

You can extract mesh faces that are connected to a selected face within a specified break angle. Perform the following steps:

1. Select File > Open and select the file Extract02.3dm from the Chapter 9 folder on the companion CD.
2. Select Mesh > Mesh Edit Tools > Extract > Connected, or click on the Connected Faces button on the Extract Mesh toolbar.
3. Select face A, shown in Figure 9–60.
4. In the Extract Connected Mesh Faces dialog box, set angle value to 35, check the Less than box, and click on the OK button.
5. Delete the extracted mesh faces to appreciate the result.
6. Do not save your file.

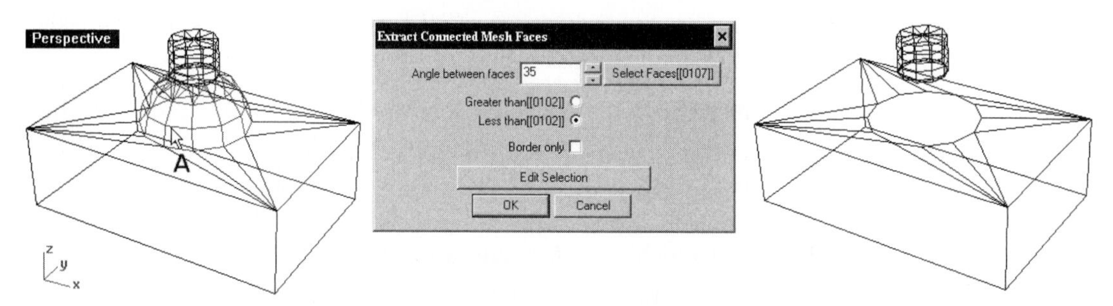

Figure 9–60. From left to right: face selected, Extract Connected Mesh Faces dialog box, and extracted faces deleted

Extracting Mesh Faces Bounded by Unwelded Edges

You can extract faces of a meshed object that are not welded. Perform the following steps:

1. Select File > Open and select the file Extract03.3dm from the Chapter 9 folder on the companion CD.
2. Select Mesh > Mesh Edit Tools > Extract > Part, or click on the Extract Part button on the Extract Mesh toolbar.
3. Select face A, shown in Figure 9–61
4. Delete the extracted faces to see the result.
5. Do not save your file.

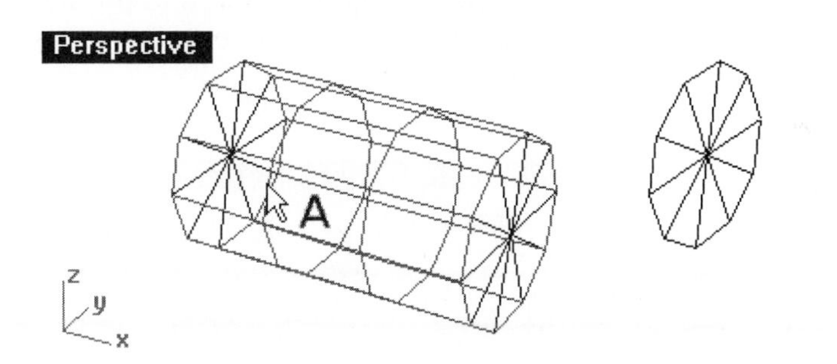

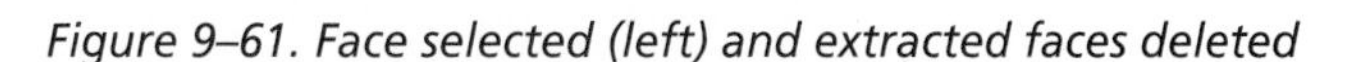

Figure 9–61. Face selected (left) and extracted faces deleted

Converting Mesh Faces

Polygon mesh can be manifested as quadrilateral mesh elements or triangular mesh elements. You can convert all the meshes of a meshed object to triangular meshes or only convert non-planar quadrilateral meshes to triangular meshes. Conversely, you can convert triangular mesh faces to quadrilateral mesh faces.

Converting Non-Planar Quadrilateral Mesh Faces to Triangular Mesh Faces

To convert only non-planar quadrilateral mesh elements of a meshed object to triangular mesh elements, perform the following steps:

1. Select File > Open and select the file TriangulateMesh.3dm from the Chapter 9 folder on the companion CD.
2. Select Mesh > Mesh Edit Tools > Triangulate Non-Planar Quads, or right-click on the Triangular Mesh/Triangulate Non-Planar Quads on the Mesh toolbar.

3. Select mesh A, shown in Figure 9–62.
4. Click on the OK button of the Triangulate Non-Planar Quads dialog box. Non-planar quadrilateral meshes are converted to triangular meshes.

Converting All Mesh Faces to Triangular Mesh Faces

To convert all quadrilateral mesh elements of a meshed object to triangular mesh elements, continue with the following steps:

5. Click on the Triangular Mesh/Triangulate Non-Planar Quads on the Mesh toolbar, or type TriangulateMesh at the command area.
6. Select B, shown in Figure 9–62, and press the ENTER key. All quadrilateral meshes are converted to triangular meshes. (See Figure 9–63.)

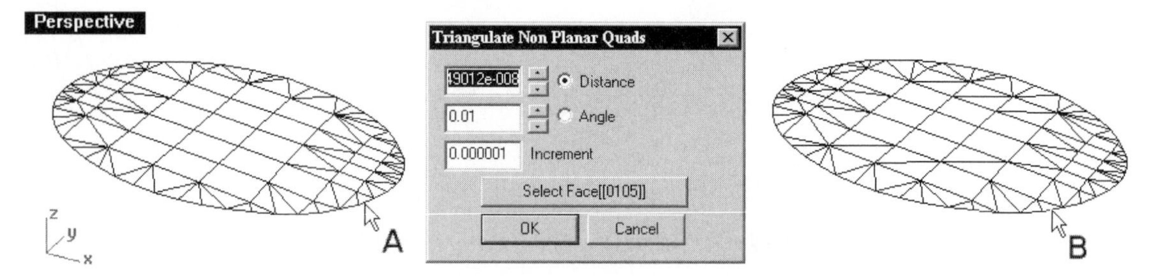

Figure 9–62. From left to right: Original meshed object, Triangulate Non-Planar Quads dialog box, and non-planar quadrilateral meshes changed to triangular meshes

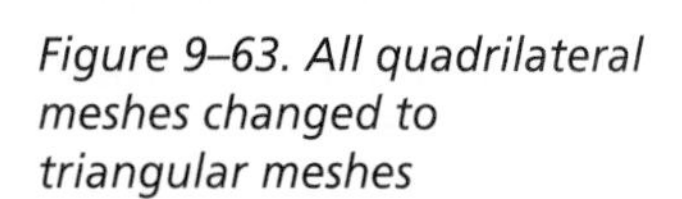

Figure 9–63. All quadrilateral meshes changed to triangular meshes

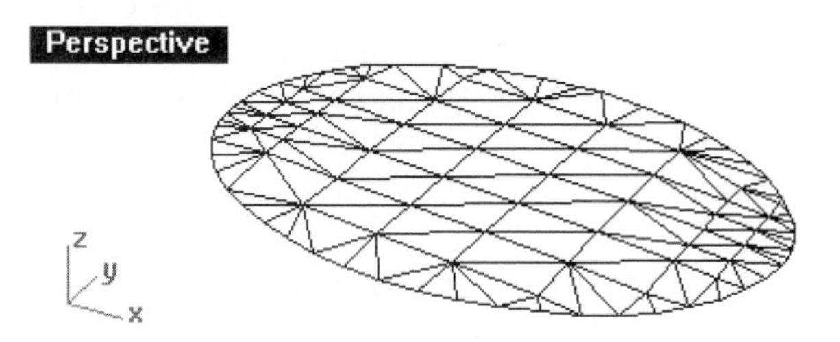

Converting Triangular Mesh Faces to Quadrilateral Mesh Faces

Contrary to converting quadrilateral mesh faces to triangular mesh faces, you can convert triangular mesh faces to quadrilateral mesh faces. However, not all triangular mesh faces can be converted. Continue with the following steps.

7 Click on the Quadrangulate Mesh button on the Mesh toolbar, or type QuadrangulateMesh at the command area.
8 Select the polygon mesh and press the ENTER key. Possible mesh faces are converted to quadrilateral faces, as shown in Figure 9–64.
9 Do not save your file.

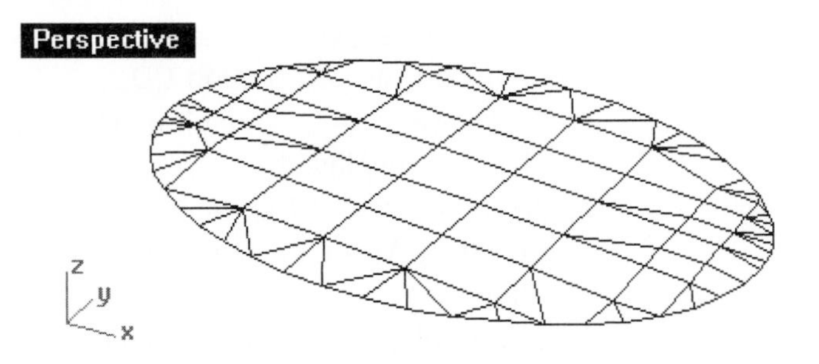

Figure 9–64. Some faces converted to quadrilateral faces

Rearranging Mesh Faces by Swapping Edges

You can improve mesh organization by swapping individual mesh edges. Corners of the selected triangle will swap.

1 Select File > Open and select the file SwapMeshEdge.3dm from the Chapter 9 folder on the companion CD.
2 Select Mesh > Mesh Repair Tools > Swap Edge, or click on the Swap mesh edge button on the Mesh Tools toolbar.
3 Select mesh edges A and B, shown in Figure 9–65, and press the ENTER key. The related meshes are rearranged. You can continue to swap other edges as appropriate.
4 Do not save your file.

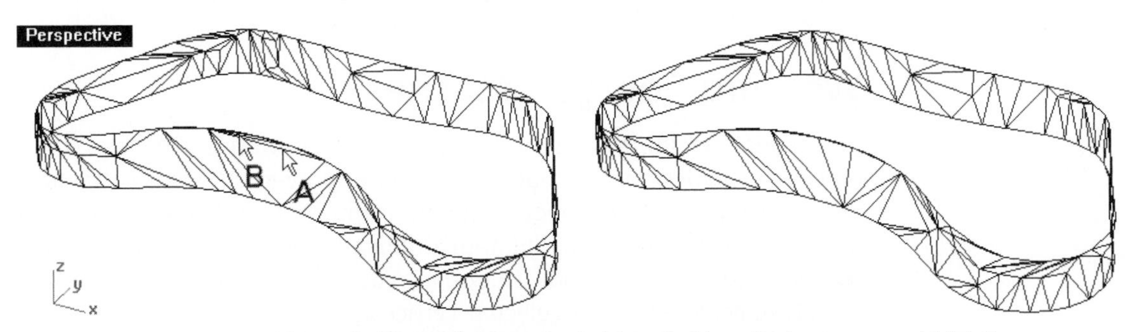

Figure 9–65. Original meshed object (left) and edges swapped (right)

Unifying Normal Direction

Quite often, a polygon mesh model is constructed from a set of polygon meshes joined together. Naturally, the normal directions of each individual polygon mesh may not be congruent. To unify the normal directions of the joined meshes, perform the following steps:

1. Select File > Open and select the file Normal.3dm from the Chapter 9 folder on the companion CD.
2. Select Analyze > Direction. The normal direction of the meshed object is displayed.
3. Select Mesh > Mesh Repair Tools > Unify Normals, or click on the Unify Mesh Normals/Flip Mesh Normals button on the Mesh toolbar.
4. Select polygon mesh A, shown in Figure 9–66. The normal directions of the meshes are unified.
5. Select Analyze > Direction.
6. Click on the Flip option on the command area as necessary.
7. Do not save your file.

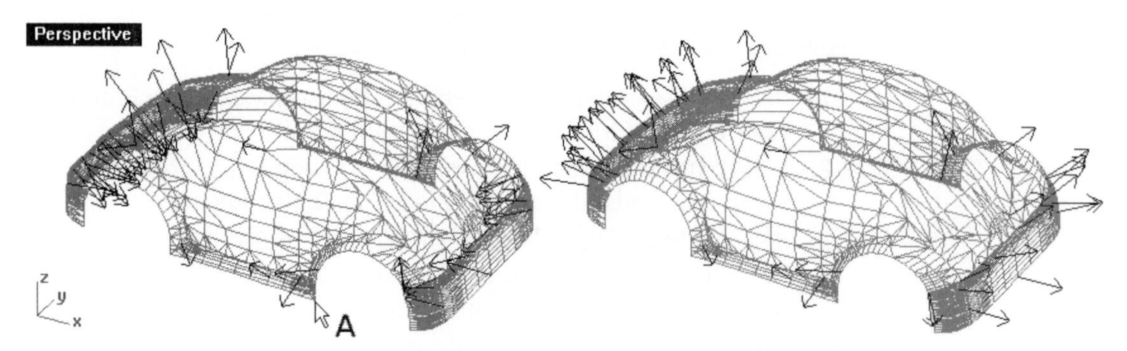

Figure 9–66. Normal directions (left) and normal directions unified (right)

Welding Polygon Meshes

At a glance, welding seems to be analogous to joining. However, these processes are different in that a joined set of polygon meshes still retains the common edges between contiguous meshes and a welded set of polygon meshes has its contiguous duplicated edges removed. To make rendering and shading look smoother, you can weld the entire meshed object, selected edges, or selected vertices.

Welding

To weld a meshed object, you have to specify angle tolerance that governs the angle between contiguous mesh elements. Perform the following steps:

1. Select File > Open and select the file Weld.3dm from the Chapter 9 folder on the companion CD.
2. Select Mesh > Mesh Edit Tools > Weld, or click on the Weld Mesh/Unweld Mesh button on the Welding toolbar.
3. Select meshed object A, shown in Figure 9–67, and press the ENTER key.
4. Type 35 at the command area to specify the angle tolerance.

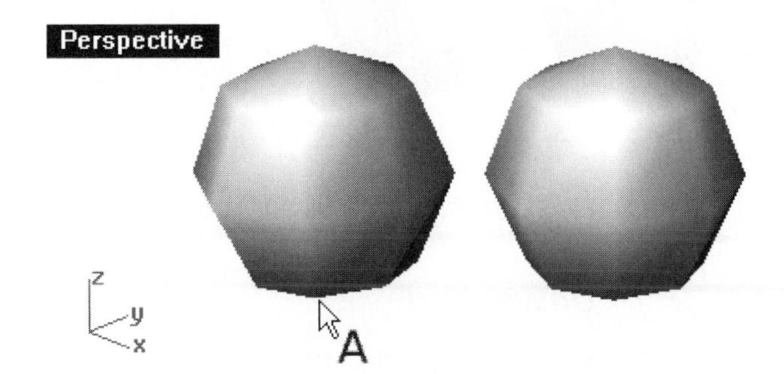

Figure 9–67. Polygon meshed object being selected

Welding Edges

To weld selected edges of a meshed object, continue with the following steps.

5. Click on the Weld Mesh Edge/Unweld Mesh Edge button on the Welding toolbar, or type WeldEdge at the command area.
6. Select edges A, B, C, D, E, F, G, H, J, K, L, and M, shown in Figure 9–68. The selected edges are welded.

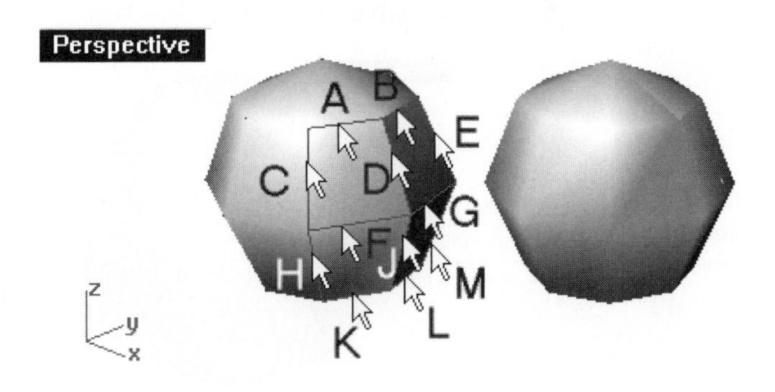

Figure 9–68. Edges being welded

Welding Vertices

To weld selected vertices, continue with the following steps.

7 Select Mesh > Mesh Edit Tools > Weld Selected Vertices, or click on the Weld Mesh Vertices button on the Welding toolbar.

8 Select vertices A, B, C, D, E, F, G, H, J, and K shown in Figure 9–69 and press the ENTER key. The selected vertices are welded.

9 Do not save your file.

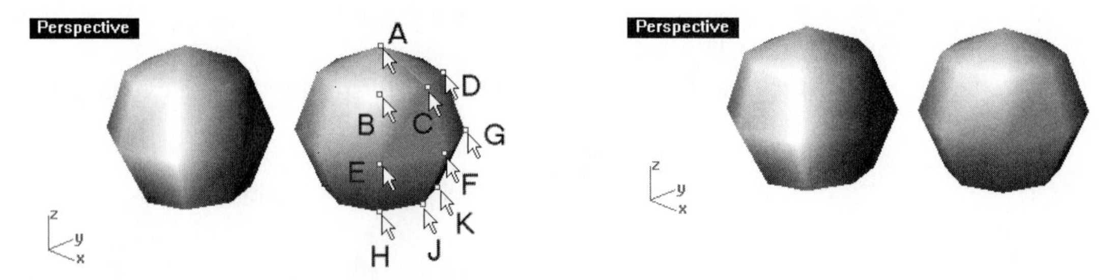

Figure 9–69. Vertices being welded (left) and vertices welded (right)

Unwelding

You can unweld vertices in accordance to a specified break angle At the unwelded joints, no face will share a vertex with another. Perform the following steps:

1 Select File > Open and select the file Unweld.3dm from the Chapter 9 folder on the companion CD.

2 Select Mesh > Mesh Edit Tools > Unweld, or right-click on the Weld Mesh/Unweld Mesh button on the Welding toolbar.

3 Select meshed object A, shown in Figure 9–70, and press the ENTER key.

4 Type 10 to set the break angle. The meshed object is unwelded.

5 Do not save your file.

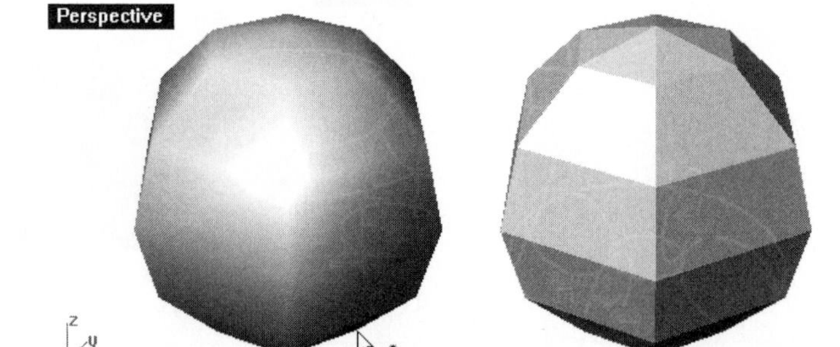

Figure 9–70. Original meshed object (left) and unwelded object (right)

General Repair Tools

Apart from the manipulation tools delineated above, you can improve a mesh object by rebuilding the entire mesh object, rebuilding the mesh normals, extracting bad mesh faces or duplicated faces, culling bad faces, and splitting disjoint mesh.

Rebuilding Mesh

If mesh is not acting properly, you may rebuild it. Rebuilding polygon mesh concerns stripping all information from a mesh and returning only the mesh geometry. You can rebuild a mesh object by:

- Selecting Mesh > Mesh Repair Tools > Rebuild Mesh, or
- Clicking on the Rebuild Mesh button on the Mesh Tools toolbar.

Rebuilding Mesh Normal

This is a way to improve the quality of a meshed object. The mesh normals are first removed, then the face and vertex normals are reconstructed based on the orientation of the faces. Note that, after rebuilding, the direction of normal will not change. Normal can be rebuilt by:

- Selecting Mesh > Mesh Repair Tools > Rebuild Mesh Normals, or
- Clicking on the Rebuild Mesh Normals button on the Mesh Tools toolbar.

Extracting Duplicated Mesh Face

To remove duplicated mesh faces from a mesh object, click on the Duplicate Faces button on the Extract Mesh toolbar and select the mesh object.

Culling Mesh of Zero Area

A meshed object may be invalid because one or more of its mesh elements have a zero area. To remove all such elements from the meshed object, you cull degenerated faces by:

- Selecting Mesh > Mesh Edit Tools > Cull Degenerate Mesh Faces or
- Clicking on the Cull degenerate mesh faces on the Mesh Tools toolbar.

Splitting Disjoint Mesh

If a polygon mesh containsare any disjoint mesh elements, one method to repair the polygon mesh is to first split the disjoint mesh element and then do subsequent repair work. To split disjoint mesh element, click on the Split disjoint mesh button on the Mesh Tools toolbar and select the mesh object.

Chapter Summary

Although NURBS surfaces are the mainstream surface type in design and manufacturing, some systems still use polygon meshes to represent a free-form body, in particular, rapid prototyping systems. To cope with these systems, you may need to know how to construct and manipulate polygon meshes.

A polygon mesh approximates a smooth surface by employing a set of small planar polygonal faces. Accuracy of representation is inversely proportional to the size of the polygon—the smaller the polygon size, the more accurate the surface. One major disadvantage of the polygon mesh is that file size increases tremendously with a decrease in polygon size. Another disadvantage is that a surface can never be accurately represented with a mesh, no matter how small the polygon size.

You learned how to construct various kinds of polygon meshes from scratch. These polygon meshes are single mesh face, rectangular mesh plane, mesh box, mesh sphere, mesh cylinder, mesh cone, mesh truncated cone, mesh ellipsoid, and mesh torus.

You learned ways to derive polygon meshes from a set of points, from a closed polyline, from the profile of a surface/polysurface, from the control points of a surface, and from the heightfield of an image. You also learned how to offset a polygon mesh, map a polygon mesh to a surface, and apply polygon meshes and point objects to a surface's U, V, and N directions.

To combine two or more polygon meshes, you use Boolean union, difference, and intersection operations. Without changing the shape of the polygon meshes, you join contiguous polygon meshes and explode joined polygon meshes. In addition, you can split a polygon mesh into two and trim away unwanted portion of a polygon mesh.

Finally, you learned various ways to manipulate individual mesh elements of a polygon mesh, which is particularly useful in repairing polygon meshes for rapid prototyping.

Review Questions

1 Outline the methods of constructing various kinds of polygon meshes.

2 List the ways to construct polygon meshes from existing objects.

3 How can two or more polygon meshes be combined?

4 In what ways can a polygon be split and trimmed?

5 List the ways to manipulate elements of a polygon mesh.

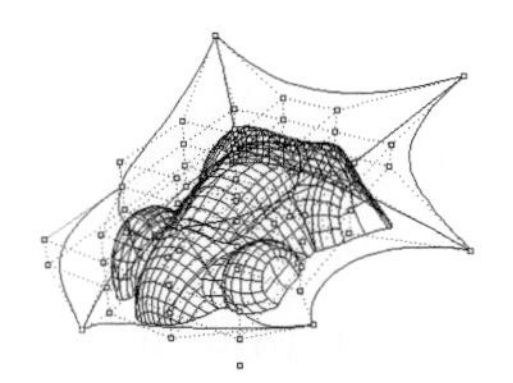

CHAPTER 10

Object Transformation

Introduction

This chapter explores how curves, surfaces, and polygon meshes are transformed in terms of translation and deformation in order to achieve special modeling effects.

Objectives

After studying this chapter you should be able to:

- Translate curves, surfaces, and polygon meshes
- Deform curves, surfaces, and polygon meshes

Overview

An advanced way of modeling is to translate and deform curves, surfaces, and polygon meshes to achieve specific design outcomes. In essence, translation of objects concerns relocating selected objects to a new location or scaling selected objects without changing the basic shape of the objects. On the other hand, deformation changes an existing object's basic shapes.

Translation

Building on the basic translation methods (moving, copying, rotating, scaling, and mirroring) depicted in Chapter 2, this section will explain the following translation processes: soft moving, aligning, orienting, remapping, arraying, and making symmetric objects. In addition, history update will be reviewed.

Soft Moving

You can move a set of objects using a falloff value, so that objects farther away from the center will move less. Perform the following steps:

1. Select File > Open and select the file SoftMove.3dm from the Chapter 10 folder on the companion CD.
2. Select Transform > Soft Move, or click on the Soft Move button on the Move toolbar.
3. Select all the sphere objects, and press the ENTER key.
4. Click on location A (shown in Figure 10–1) to specify the center of move.
5. Click on location B (Figure 10–1) to specify the radius affected by the move.
6. Click on location C (Figure 10–2) to specify the offset point, and press the ENTER key. The objects area moved.
7. Do not save your file.

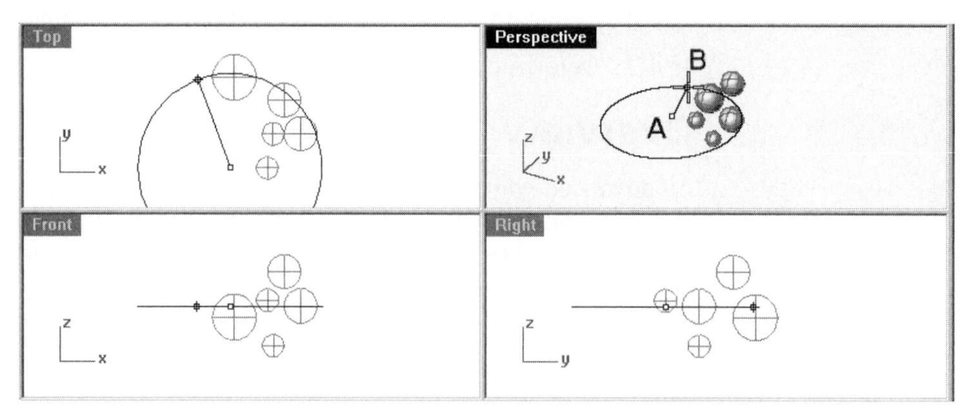

Figure 10–1. Point to move from and radius point selected

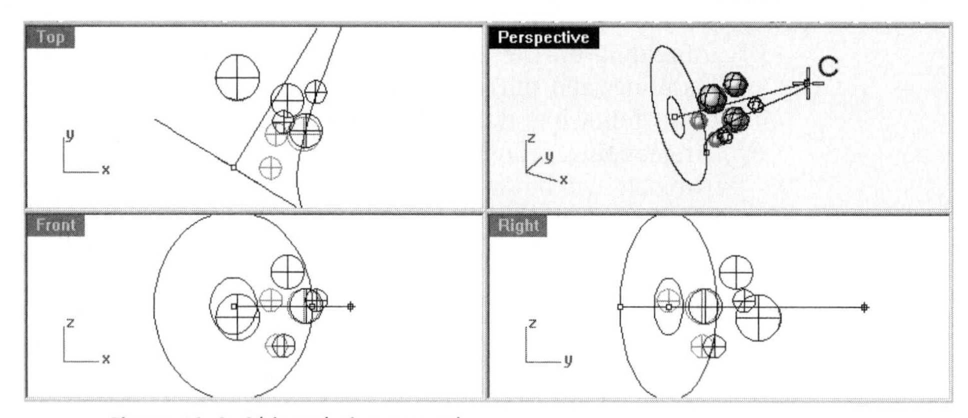

Figure 10–2. Objects being moved

Aligning Curves

Planar curves residing on one of the world construction planes (top, front, and right) can be aligned, as follows:

1. Select File > Open and select the file AlignProfile.3dm from the Chapter 10 folder on the companion CD.
2. Type AlignProfiles at the command area.
3. Select curve A, shown in Figure 10–3. (This is the curve you will align to.)
4. Select curve B. (This is the curve to change.) The second curve is modified to align to the first curve.
5. Do not save your file.

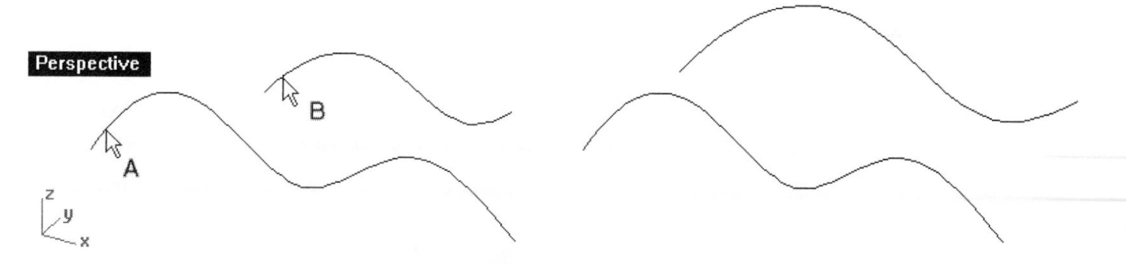

Figure 10–3. Original curves (left) and aligned curve (right)

Aligning Objects

Objects can be aligned by their bounding boxes, as follows:

1. Select File > Open and select the file Align.3dm from the Chapter 10 folder on the companion CD.
2. Select Transform > Align, or click on the Align Objects button on the Transform toolbar.
3. Click on the Top viewport to make it the active viewport.
4. Select objects A, B, and C (Figure 10–4) and press the ENTER key.
5. Select the VertCenter option on the command area.
6. Repeat the command.
7. Click on the Right viewport, select all three objects, and press the ENTER key.
8. Select the VertCenter option on the command area. Objects are aligned, as shown in Figure 10–5.
9. Do not save your file.

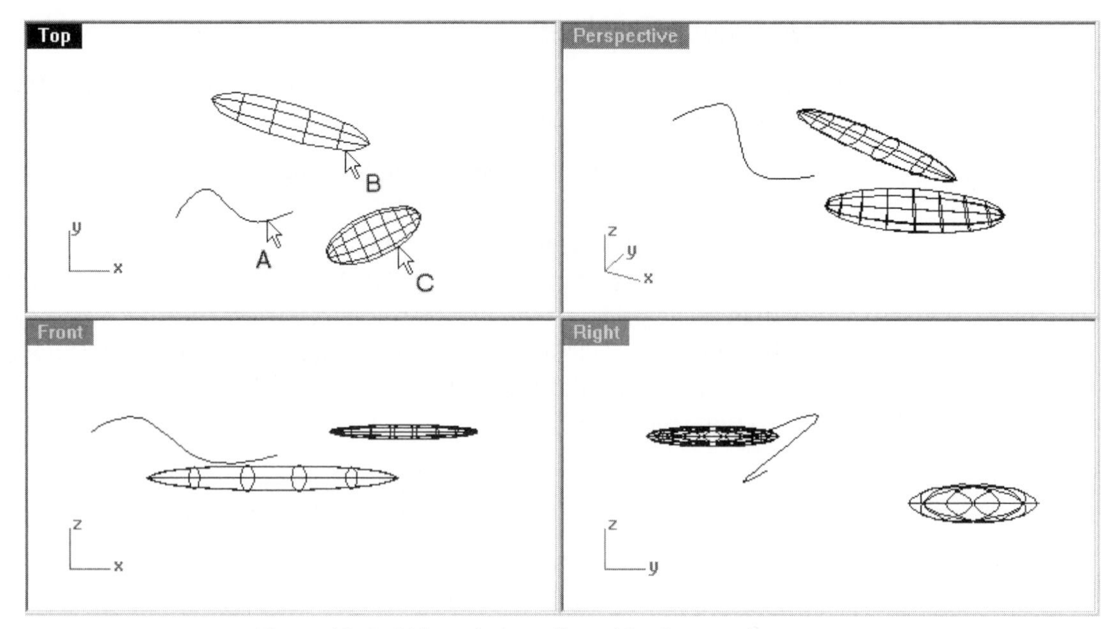

Figure 10–4. Objects being aligned in the top viewport

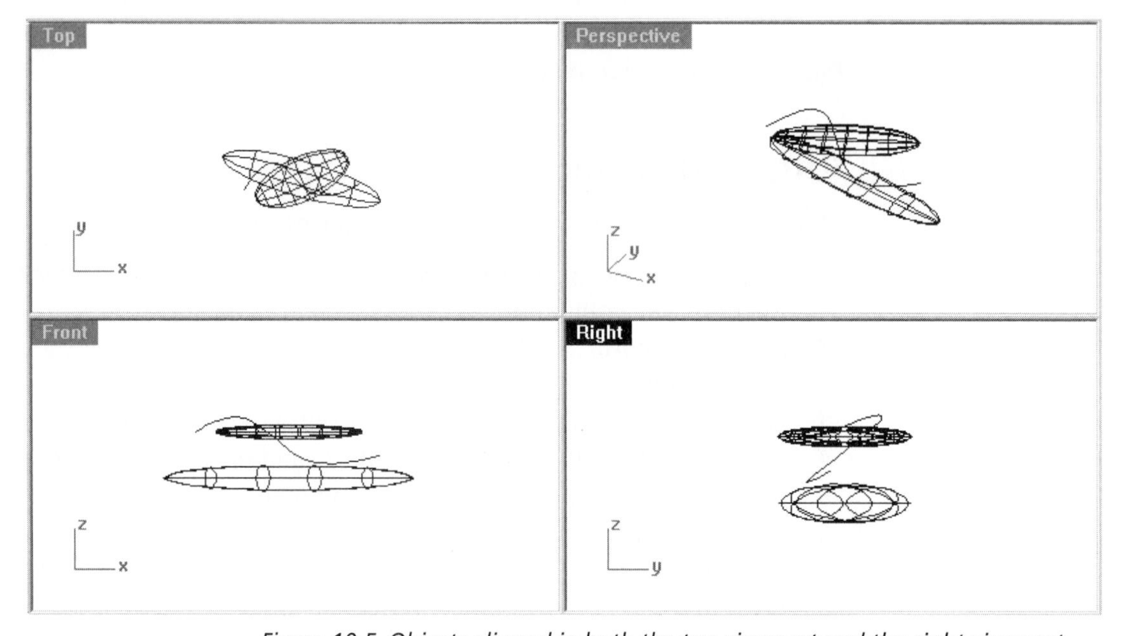

Figure 10-5. Objects aligned in both the top viewport and the right viewport

Orienting Objects

The Orient tool is used to change the orientation of selected objects and, where appropriate, change their scale.

Orient Two Points

A curve, surface, or polygon mesh can be repositioned and uniformly scaled by referencing two reference points and two target points. Basically, the object's first reference point will be relocated to the first target point, and the object's second reference point will be relocated to the second target point. As a result, the object's size is scaled by a ratio of the distances between the reference points and the target points. Optionally, you may choose not to scale the oriented object. In this case, the object is simply reoriented to position along an axis formed from the first target point and second target point. Perform the following steps to orient objects referencing two points:

1. Select File > Open and select the file Orient2Points.3dm from the Chapter 10 folder on the companion CD.
2. Check the End and Point boxes of the Osnap dialog box.
3. Select Transform > Orient > 2 Points, or click on the Orient: 2 Points/Orient: 3 Points button on the Transform toolbar.
4. Select curve A, shown in Figure 10–6, and press the ENTER key.
5. If Scale = No, select the option to change Scale = Yes. Otherwise, proceed to the next step.
6. Select end points B and C, shown in Figure 10–6, to specify the reference points.
7. Select end points D and E, shown in Figure 10–6, to specify the target points. The selected curve is relocated and scaled. If Scale = No, the curve will simply reposition using BC as the reference axis and DE as the target axis.

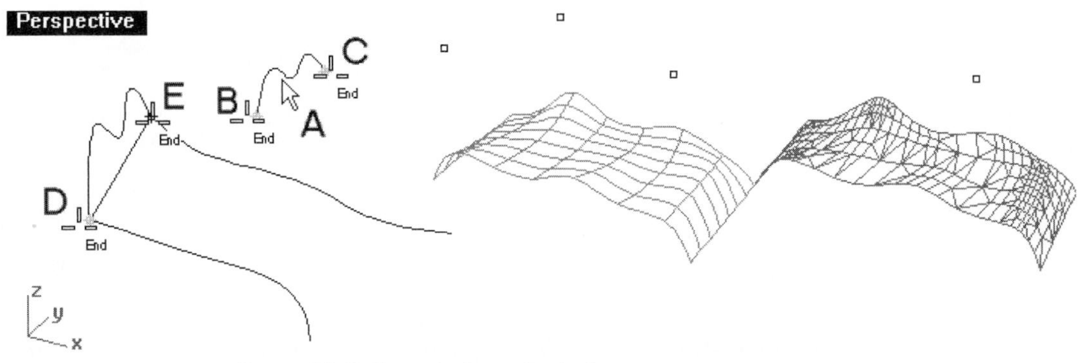

Figure 10–6. Curve being oriented

8. Select Transform > Orient > 2 Points, or click on the Orient: 2 Points/Orient: 3 Points button on the Transform toolbar.
9. Select surface A, shown in Figure 10–7, and press the ENTER key.
10. Select end points B and C, (Figure 10–7) to specify the reference points.
11. Select points D and E (Figure 10–7) to specify the target points. The selected surface is relocated and scaled.

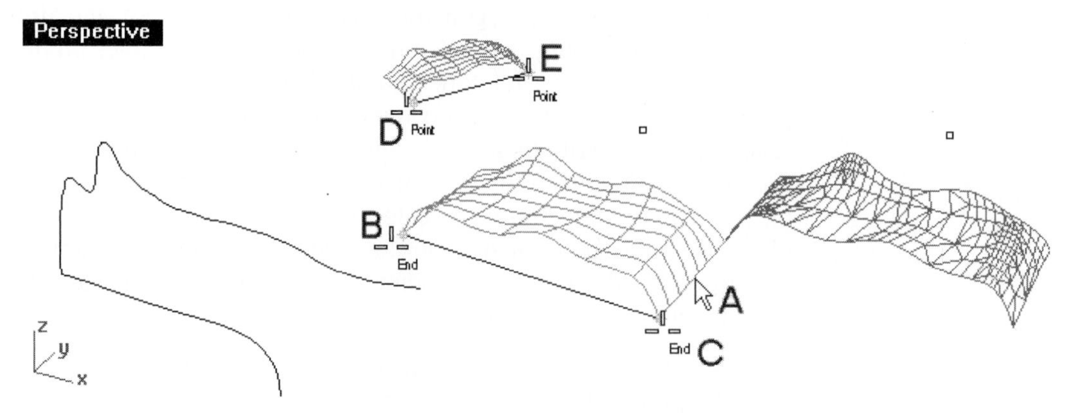

Figure 10–7. Surface being oriented

12. Select Transform > Orient > 2 Points, or click on the Orient: 2 Points/Orient: 3 Points button on the Transform toolbar.
13. Select polygon mesh A, shown in Figure 10–8, and press the ENTER key.
14. Select points B and C (Figure 10–8) to specify the reference points.
15. Select points D and E (Figure 10–8) to specify the target points. The selected polygon mesh is relocated and scaled.
16. Do not save your file.

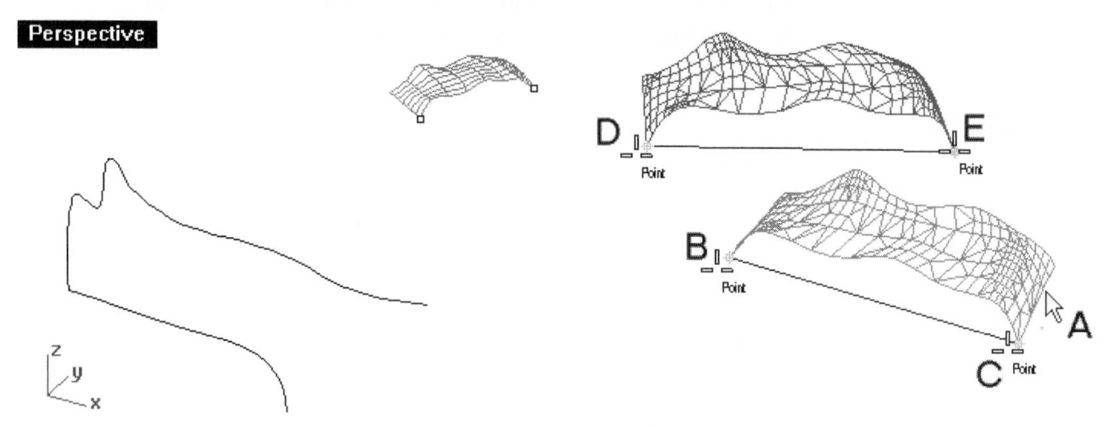

Figure 10–8. Polygon mesh being oriented (left) and polygon mesh oriented (right)

Orient Three Points

A curve, surface, or polygon mesh can be repositioned referencing an origin point, an axis, and a plane. The object is not scaled. (Note that you can also do this with the Orient 2 Points command using the No Scale option.)

During orientation, you need to input three reference points and three target points. The object's first reference point will be the origin of translation; it will relocate to the first target point. The object's first and second reference points will specify an axis, which will align with an axis formed by the first and second target points. The object's first, second, and third reference points will specify a plane, which will align with a plane formed by the first, second, and third target points. Perform the following steps:

1. Select File > Open and select the file Orient3Points.3dm from the Chapter 10 folder on the companion CD.
2. Check the End, Point, Cen, and Quad boxes, and clear all other boxes on the Osnap dialog box.
3. Select Transform > Orient > 3 Points, or right-click on the Orient: 2 Points/Orient: 3 Points button on the Transform toolbar.
4. Select circle A, shown in Figure 10–9, and press the ENTER key.
5. Select center point B, quadrant point C, and quadrant point D (Figure 10–9) to specify the reference points. (Point C is the source origin, CD is the source axis, and BCD is a source plane.)
6. Select end points E, F, and G (Figure 10–9) to specify the target points. The selected circle is relocated. (Point E is the target origin, EF is the target axis, and EFG is the target plane.)

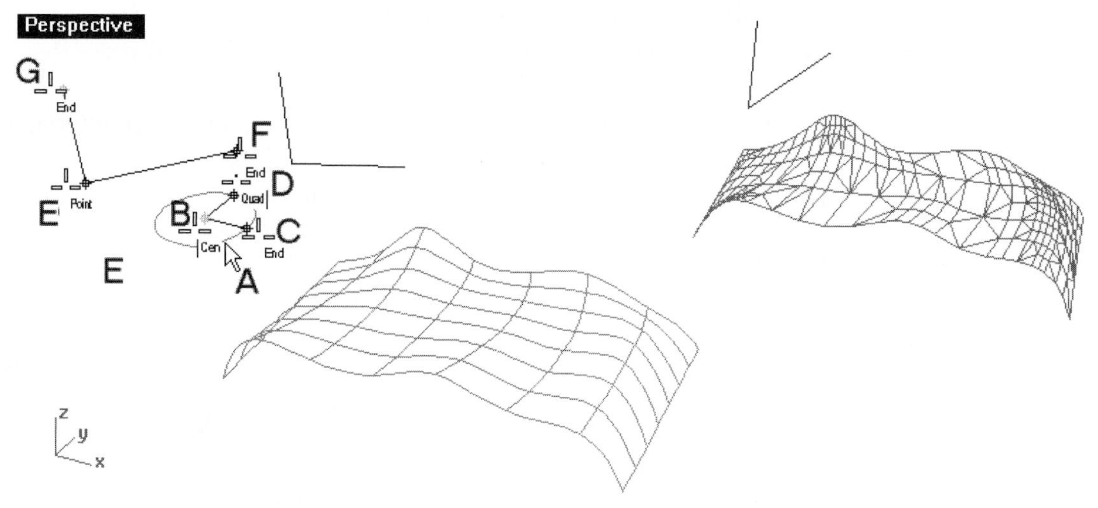

Figure 10–9. Circle being oriented with reference to three points

7. Select Transform > Orient > 3 Points, or right-click on the Orient: 2 Points/Orient: 3 Points button on the Transform toolbar.
8. Select surface A, shown in Figure 10–10, and press the ENTER key.
9. Select end points B, C, and D (Figure 10–10) to specify the reference points.
10. Select end points E, F, and G (Figure 10–10) to specify the target points. The selected surface is relocated.

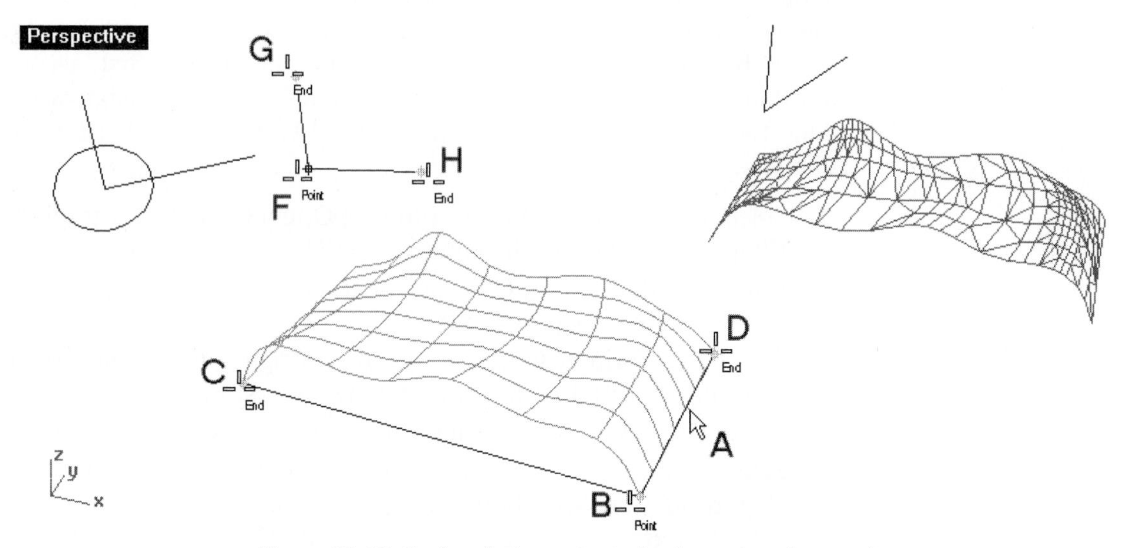

Figure 10–10. Surface being oriented referencing three points

11. Select Transform > Orient > 3 Points, or right-click on the Orient: 2 Points/Orient: 3 Points button on the Transform toolbar.
12. Select polygon mesh A, shown in Figure 10–11, and press the ENTER key.
13. Select points B, C, and D (Figure 10–11) to specify the reference points.
14. Select end points E, F, and G (Figure 10–11) to specify the target points. The selected polygon mesh is relocated.
15. Do not save your file.

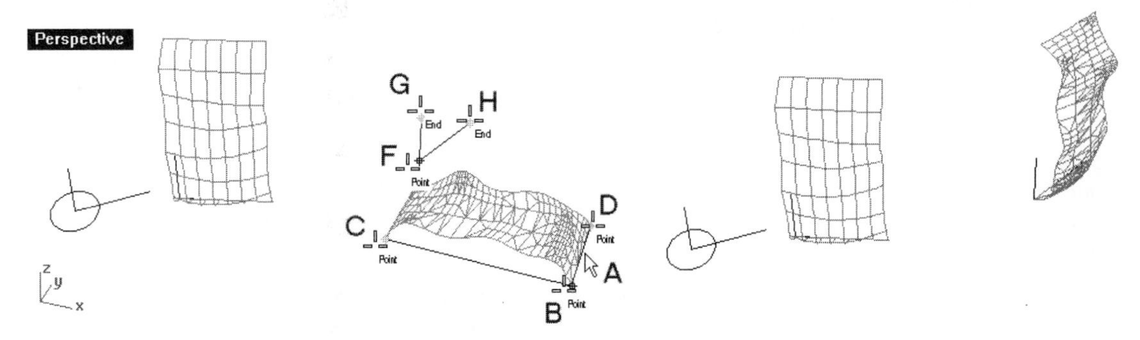

Figure 10–11. Polygon mesh being oriented with reference to three points (left) and polygon mesh oriented (right)

Orient on Surface

A simple way to construct an object (a curve, surface, or polygon mesh) on a surface is to first construct the object on any construction plane and then orient the object on the surface. After orientation, the reference plane (construction plane on which the objects are constructed) of the translated object will be tangent to the target surface. Perform the following steps:

1. Select File > Open and select the file OrientOnSurface.3dm from the Chapter 10 folder on the companion CD.
2. Check the Point and Cen boxes, and clear all other boxes on the Osnap dialog box.
3. Select Transform > Orient > On Surface, or click on the Orient on Surface button on the Transform toolbar.
4. Select circle A, shown in Figure 10–12 and press the ENTER key.
5. Select center B (Figure 10–12) of the circle as the point to be oriented.
6. Select surface C (Figure 10–12).
7. Select points D and E (Figure 10–12) and press the ENTER key. Two copies of the selected circle are oriented on the surface.

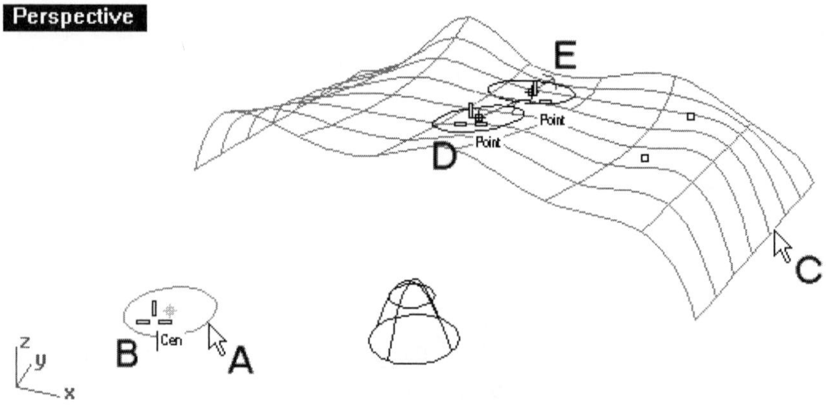

Figure 10–12. Circle being oriented to a surface in multiple copies

8. Select Transform > Orient > On Surface, or click on the Orient on Surface button on the Transform toolbar.
9. Select paraboloid A, shown in Figure 10–13, and press the ENTER key.
10. Select center B (Figure 10–13) of the circle as the point to be oriented.
11. Select surface C (Figure 10–13).
12. Select points D and E (Figure 10–13) and press the ENTER key. Two copies of the selected circle are oriented on the surface.
13. Do not save your file.

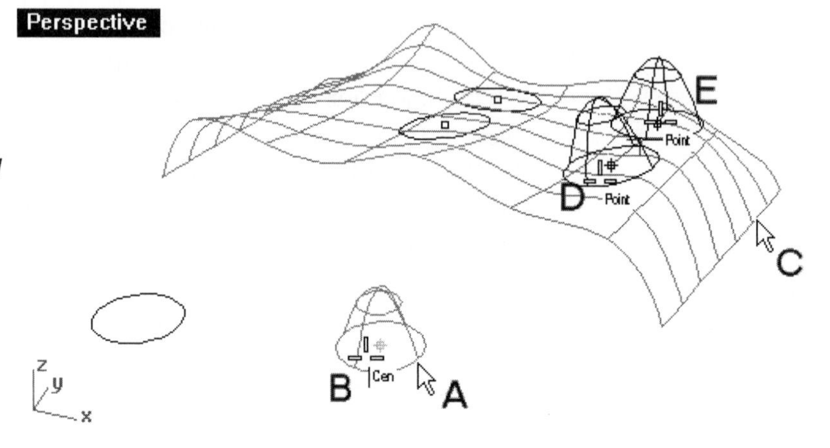

Figure 10–13. Paraboloid being oriented to a surface in multiple copies

Although a polygon mesh can also be oriented on a surface, it does not make much sense to mix polygon mesh and NURBS surface a file, except for the purpose of illustration.

Orient Perpendicular to Curve

Instead of setting up a construction plane perpendicular to a selected point on a curve for object construction, you may first construct objects on any construction plane and then orient them perpendicular to the target curve. This command is construction-plane dependent. The reference plane of the oriented object will become perpendicular to the target curve. Perform the following steps:

1. Select File > Open and select the file OrientPerpendicular-ToCurve.3dm from the Chapter 10 folder on the companion CD.
2. Check the End, Point, and Cen boxes, and clear all other boxes on the Osnap dialog box.
3. Select Transform > Orient > Perpendicular to Curve, or click on the Orient Perpendicular to Curve button on the Transform toolbar.
4. Select circle A, shown in Figure 10–14, and press the ENTER key.
5. Select center B as the base point.
6. Select curve C.
7. Select point D of the curve. The circle is oriented perpendicular to the curve.

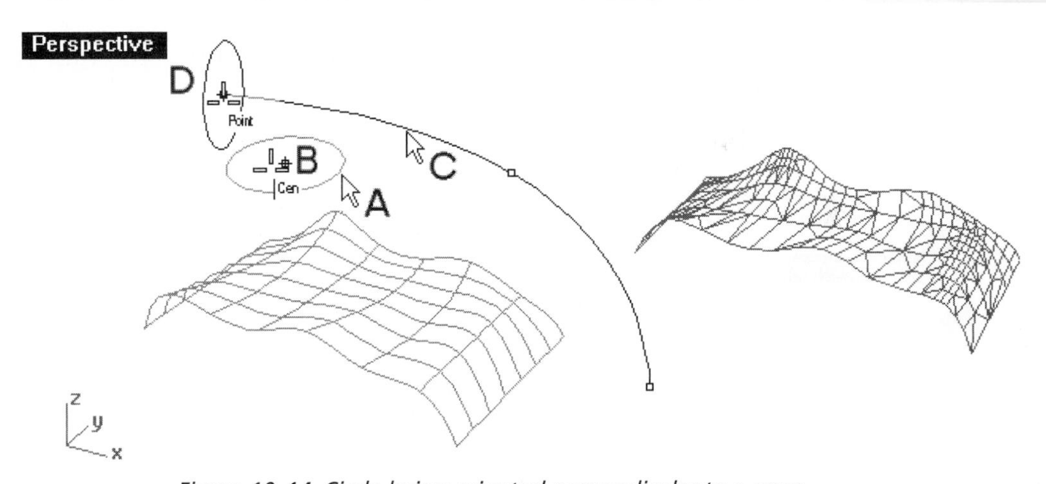

Figure 10–14. Circle being oriented perpendicular to a curve

8. Select Transform > Orient > Perpendicular to Curve, or click on the Orient Perpendicular to Curve button on the Transform toolbar.
9. Select surface A, shown in Figure 10–15, and press the ENTER key.
10. Select endpoint B, shown in Figure 10–15, as the base point.
11. Select the curve.
12. Select point C of the curve. The surface is oriented perpendicular to the curve.

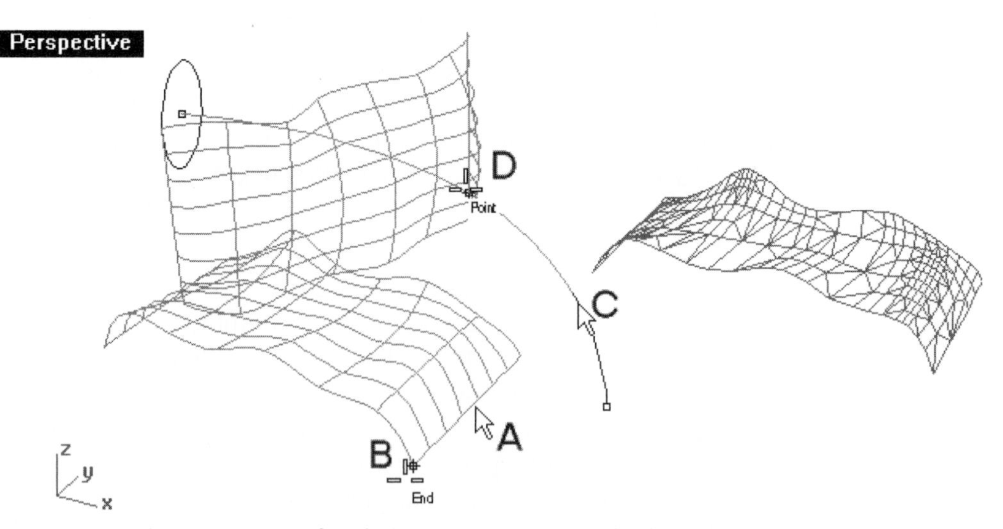

Figure 10–15. Surface being oriented perpendicular to a curve

13 Select Transform > Orient > Perpendicular to Curve, or click on the Orient Perpendicular to Curve button on the Transform toolbar.

14 Select polygon mesh A, shown in Figure 10–16, and press the ENTER key.

15 Select end point B, shown in Figure 10–16, as the base point.

16 Select the curve.

17 Select end point B of the curve. The polygon mesh is oriented perpendicular to the curve.

18 Do not save your file.

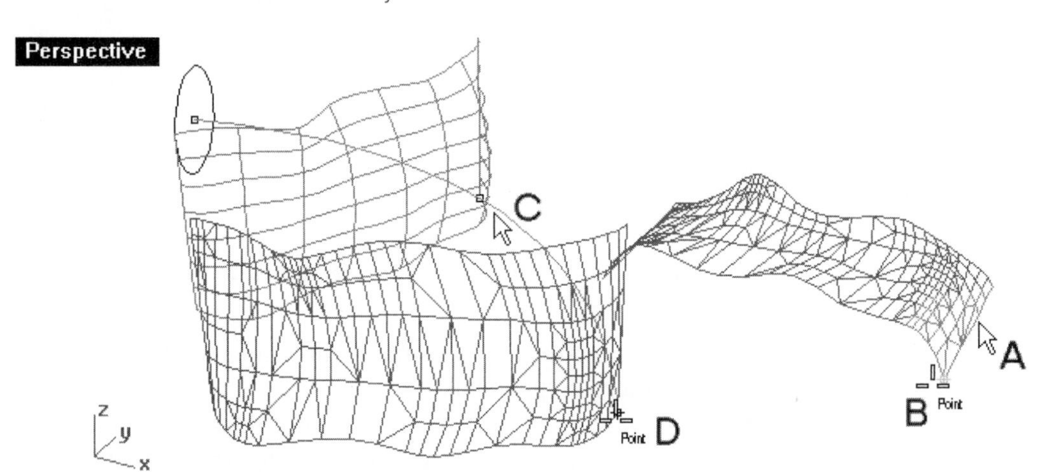

Figure 10–16. Polygon mesh being oriented perpendicular to a curve

Orient on Curve

You can orient selected curves, surfaces, and polygon meshes along curves, as follows:

1. Select File > Open and select the file OrientonCurve.3dm from the Chapter 10 folder on the companion CD.
2. Set object snap mode to Cen and End.
3. Select Transform > Orient > On Curve, click on the Orient Perperdicular to Curve button on the Transform toolbar.
4. Select circle A (Figure 10–17) and press the ENTER key.
5. Select center B.
6. Select curve C
7. If the Copy option is No, click on it to change it to Yes.
8. Select point D and press the ENTER key. The circle is oriented on the curve.
9. Repeat the command.
10. Select surface E and press the ENTER key.
11. Select center F.
12. Select curve C and point G. The surface is oriented.
13. Repeat the command.
14. Select polygon mesh H and press the ENTER key.
15. Select endpoint J.
16. Select curve C and point K. The polygon mesh is oriented.
17. Do not save your file.

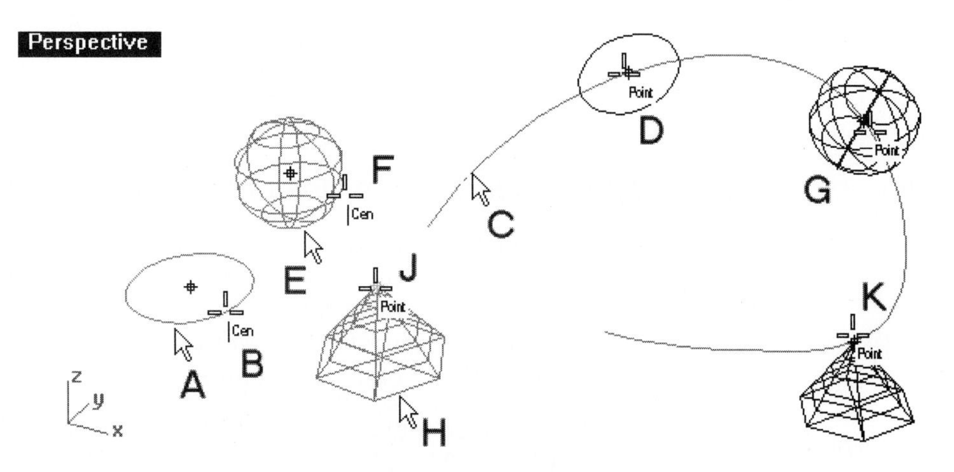

Figure 10–17. Curve, surface, and polygon mesh being oriented on the curve

Orient Curve to Edge

To construct curves tangent to an edge of a surface, you can first construct the curve and then orient it to the edge of the surface, as follows:

1. Select File > Open and select the file OrientCurveToEdge.3dm from the Chapter 10 folder on the companion CD.
2. Check the Point box and clear all other boxes on the Osnap dialog box.
3. Select Transform > Orient > Curve to Edge, or click on the Orient Curve to Edge button on the Transform toolbar.
4. Select curve A near its end point A (shown in Figure 10–18) and press the ENTER key. (Note that orientation of the curve is based on which end of the curve you select.)
5. Select surface edge B (Figure 10–18).
6. Select points C, D, E, and F (Figure 10–18) and press the ENTER key. Two copies of the curve are oriented to the edge of the surface.
7. Do not save your file.

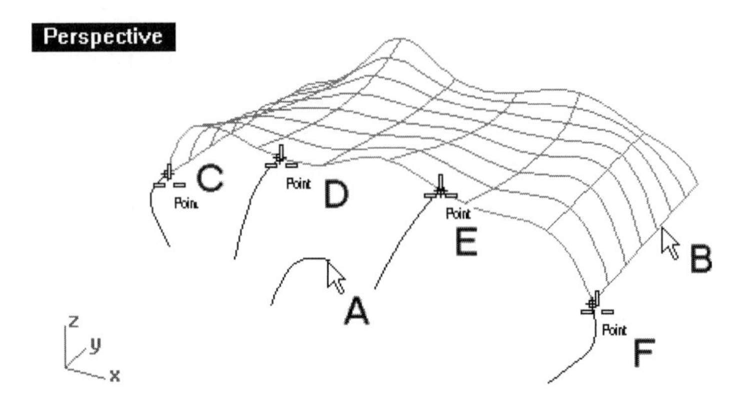

Figure 10–18. Curve being oriented to the edge of a surface

Remapping Objects from One CPlane to Another CPlane

Objects residing on a construction plane can be repositioned to another construction plane. This process is especially useful for orienting 2D drawings to 3D construction planes. You can map the front view to the Front viewport, the right view to the Right viewport, and so on. This is how most people use this command. Perform the following steps:

1. Select File > Open and select the file OrientRemapToCPlane.3dm from the Chapter 10 folder on the companion CD.
2. Click on the Front viewport to set it as the current viewport.
3. Select Transform > Orient > Remap to CPlane, or click on the Remap to CPlane button on the Transform toolbar.

4 Select curve A (shown in Figure 10–19) and press the ENTER key.
5 Select the CPlane option on the command area.
6 Type Right at the command area. The selected curve is remapped onto the Right viewport. (See Figure 10–19.)
7 Do not save your file.

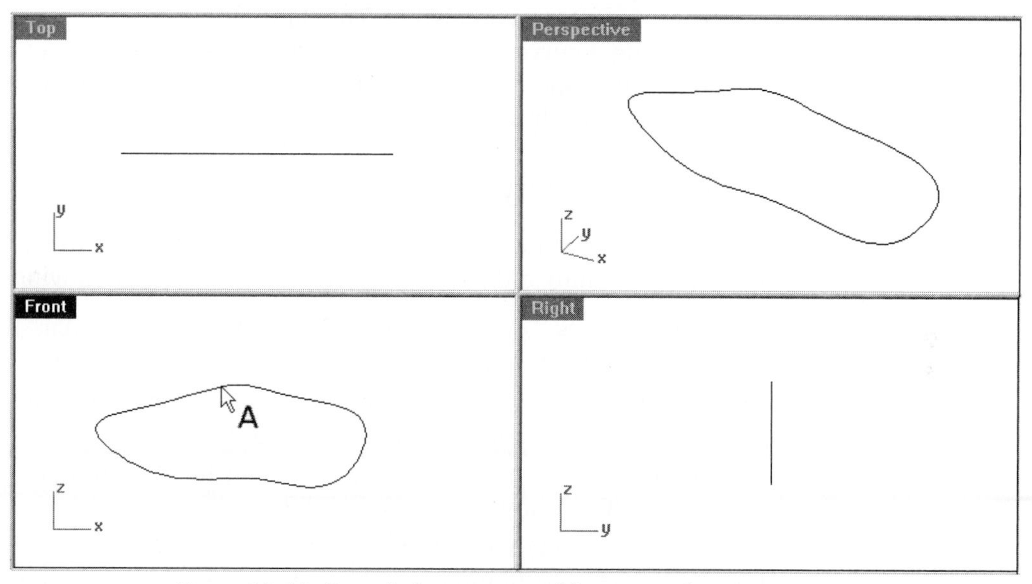

Figure 10–19. Curve being remapped from one viewport to another viewport

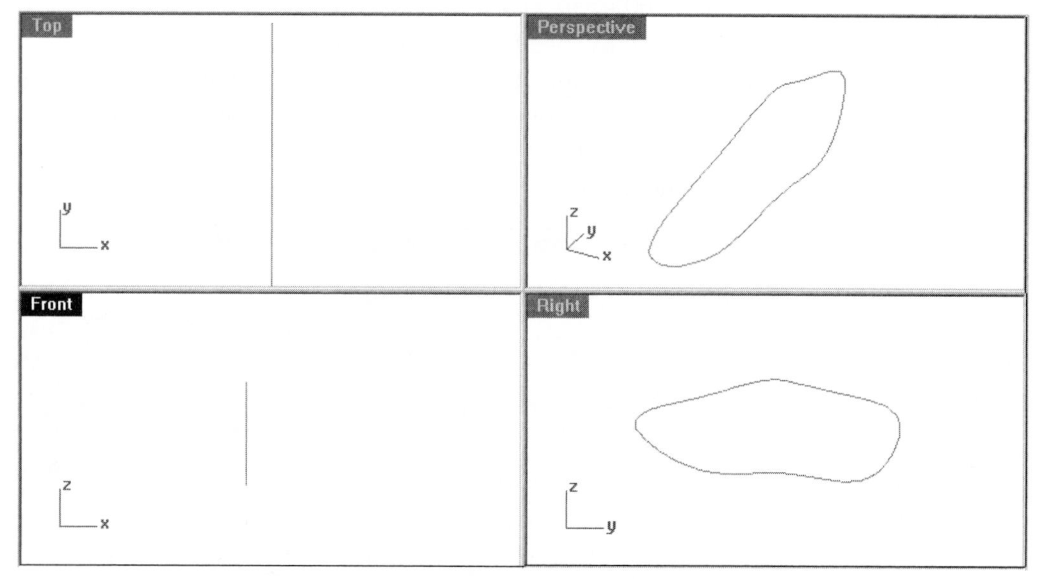

Figure 10–20. Curve remapped from one viewport to another viewport

Array

An array is a repetition of a selection of curves, surfaces, or polygon meshes. There are five types of arrays. The first is the rectangular array, in which objects are repeated in X, Y, and Z directions. The second type is a polar array, in which objects are repeated around a center of array on the active construction plane. The third type of array repeats selected objects along a curve. The fourth type repeats selected objects along the U and V directions of a surface. The fifth type of array repeats objects along a curve residing on a surface.

Rectangular Array

Curves, surfaces, and polygon meshes can be repeated in X, Y, and Z directions, forming a rectangular pattern. Perform the following steps:

1. Select File > Open and select the file ArrayRectangular.3dm from the Chapter 10 folder on the companion CD.
2. Select Transform > Array > Rectangular, or click on the Rectangular Array button on the Array toolbar.
3. Select polygon mesh A, curve B, and surface C (shown in Figure 10–21) and press the ENTER key.
4. Type 3 to specify the number of objects in the X direction.
5. Type 4 to specify the number of objects in the Y direction.
6. Type 2 to specify the number of objects in the Z direction.
7. Type 60 to specify the X spacing.
8. Type 40 to specify the Y spacing.
9. Type 50 to specify the Z spacing. Note that negative values for distances change the direction of an array. A rectangular array is constructed.
10. Do not save your file.

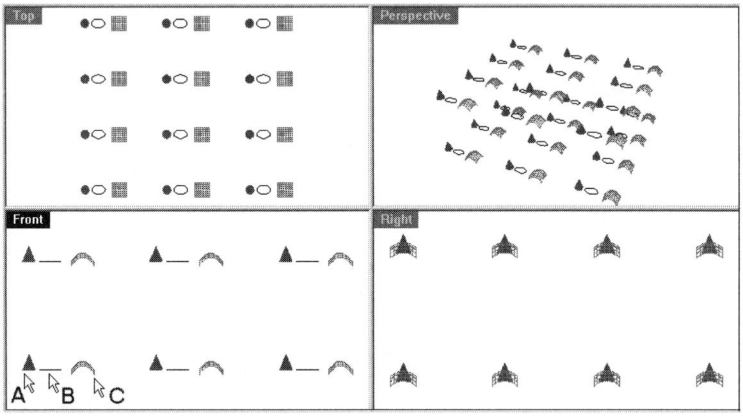

Figure 10–21. Rectangular array being constructed (left) and rectangular array constructed (right)

Polar Array

Curves, surfaces, and polygon meshes can be repeated in a circular pattern. You specify the center of the pattern, the number of repeated objects, and the angle to be filled by the repeated objects. Perform the following steps:

1. Select File > Open and select the file ArrayPolar.3dm from the Chapter 10 folder on the companion CD.
2. Select Transform > Array > Polar, or click on the Polar Array button on the Array toolbar.
3. Select polygon mesh A, curve B, and surface C (shown in Figure 10–22) and press the ENTER key.
4. Select point D (Figure 10–22) to specify the center of the array.
5. Type 4 to specify the number of elements in the array.
6. Type 90 (or press the ENTER key if the default is 360) to specify the angle to be filled. A polar array is constructed. *(Note: Step angle is the angle between contiguous copies.)*
7. Do not save your file.

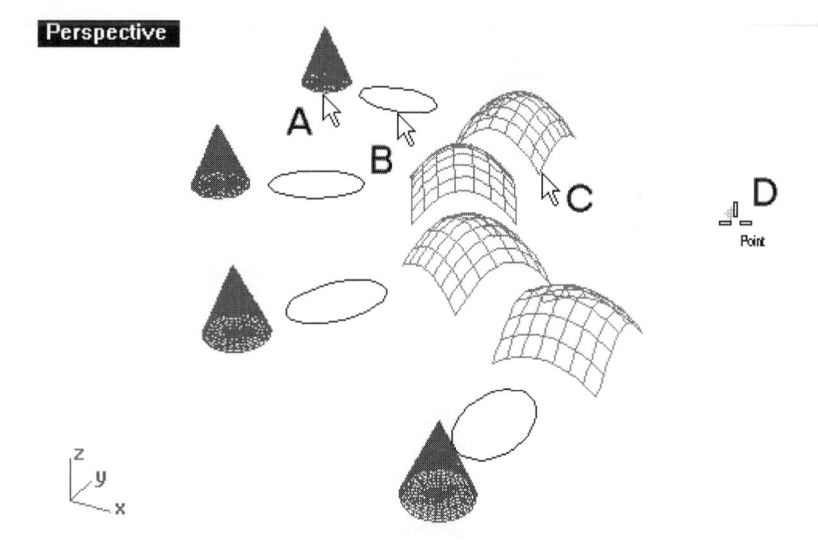

Figure 10–22. Polar array being constructed (left) and polar array constructed (right)

Array Along a Curve

Apart from repeating objects in a rectangular or circular pattern, you can array selected objects along a path curve. You specify the path curve and the number of repeated objects or the distance between contiguous objects. Perform the following steps:

1. Select File > Open and select the file ArrayAlongCurve.3dm from the Chapter 10 folder on the companion CD.
2. Select Transform > Array > Along Curve, or click on the Array Along Curve button on the Array tool bar.
3. Select polygon mesh A and surface B (shown in Figure 10–23) and press the ENTER key.
4. Select end point C to use it as the base point.
5. Select free-form curve D near end point C (Figure 10–23) to use it as the path curve. (Note that the direction of array depends on which end of the path curve you select.)
6. In the Array Along Curve Options dialog box, specify 3 items in the Number of Items box, select the Freeform twisting style, and then click on the OK button. The selected objects are arrayed along the curve.

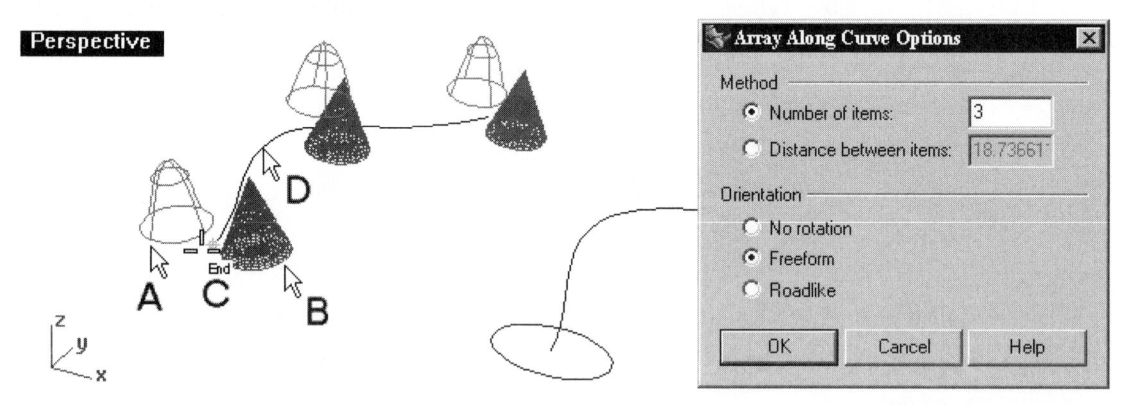

Figure 10–23. Polygon mesh and surface being arrayed along a curve

7. Select Transform > Array > Along Curve, or click on the Array along Curve button on the Array tool bar.
8. Select curve A (shown in Figure 10–24) and press the ENTER key.
9. Select end point B to use it as the base point.
10. Select free-form curve C (shown in Figure 10–24) to use it as the path curve.
11. In the Array Along Curve Options dialog box, specify distance between items to be 8 units, select the Road-like style, and then click on the OK button.

12 Click on the Perspective viewport to specify a construction plane. (The perspective viewport has the same construction plane as the Top viewport.) The selected objects are arrayed along the curve.

13 Do not save your file.

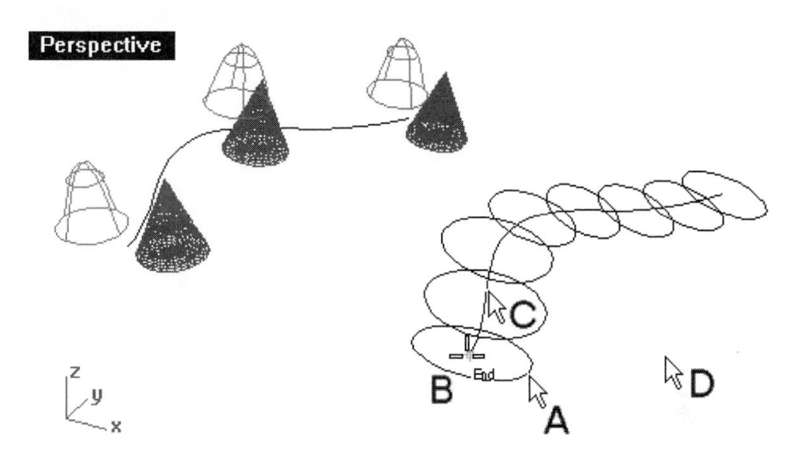

Figure 10–24. Curve being arrayed along a curve

Array Along Surface UV Directions

Objects can be arrayed along the U and V directions of a surface. Perform the following steps:

1 Select File > Open and select the file ArrayOnSurface.3dm from the Chapter 10 folder on the companion CD.

2 Check the Cen box on the Osnap dialog box.

3 Select Transform > Array > Along Surface, or click on the Array on Surface button on the Array toolbar.

4 Select surface A and curve B, shown in Figure 10–25, and press the ENTER key.

5 Select center C (Figure 10–25) to specify the base point.

6 Press the ENTER key to use the CPlane's Z axis as the reference normal.

7 Select surface D (Figure 10–25).

8 Type 3 to specify the number of elements in the U direction.

9 Type 4 to specify the number of elements in the V direction. An array is constructed.

10 Do not save your file.

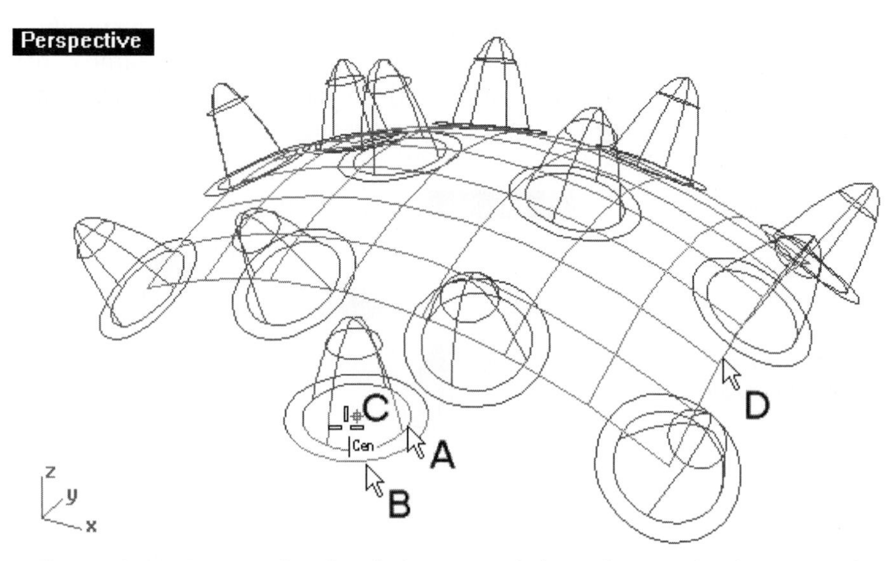

Figure 10–25. Curve and surface being arrayed along the U and V directions of a surface (left) and objects arrayed (right)

Array Along Curve on Surface

Apart from arraying along the U and V directions of a surface, you can also array objects on a curve residing on a surface. Perform the following steps:

1. Select File > Open and select the file ArrayCurveOnSurface.3dm from the Chapter 10 folder on the companion CD.
2. Select Transform > Array > Along Curve on Surface, or click on the Array along Curve on Surface button on the Transform toolbar.
3. Select surface A and curve B (shown in Figure 10–26) and press the ENTER key.
4. Select center point C (Figure 10–26) as the base point.
5. Select curve D (Figure 10–26).
6. Select surface E (Figure 10–26).
7. Select the Divide option on the command area.
8. Type 4 at the command area. The circle is arrayed along a curve on a surface.
9. Press the ENTER key to terminate the command.
10. Do not save your file.

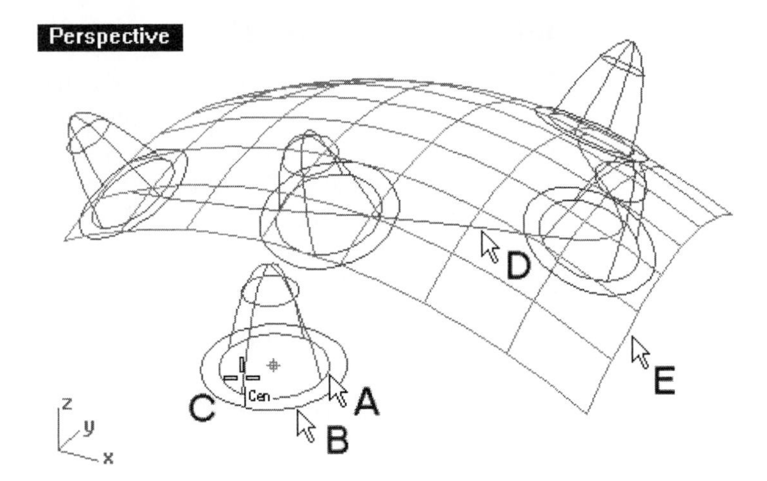

Figure 10–26. Circle being arrayed along a curve on the surface (left) and circles arrayed (right)

Making Symmetric Objects

You can construct symmetric curve elements or untrimmed surfaces by specifying a symmetry plane. Perform the following steps:

1. Select File > Open and select the file Symmetry.3dm from the Chapter 10 folder on the companion CD.
2. Select Transform > Symmetry, or click on the Symmetric button on the Surface Tools toolbar.
3. Select curve A, shown in Figure 10–27.
4. Select end points B and C. A symmetric curve is constructed.
5. Repeat the command.
6. Select surface D at edge D.
7. Select end points E and F. A symmetric surface is constructed.

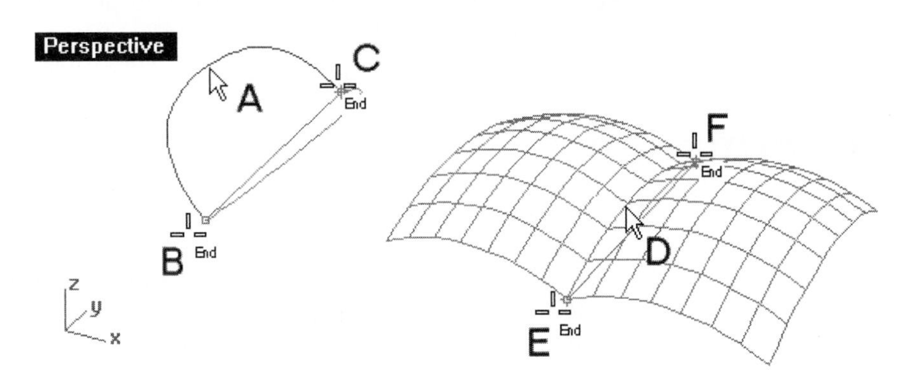

Figure 10–27. Symmetric objects being constructed

Updating History

If the History Manager is turned on beforehand, histories about symmetric and copied objects are kept in memory during the working session. If the source curve is modified, the symmetric or copied objects will update automatically. Copying includes the array of curves However, this history is only retained during the current working session. In other words, the next time you open a file, the history will be lost. Continue with the following steps:

8. Turn on the control points of curve A and surface B, shown in Figure 10–28.
9. Select Transform > Move, or click on the Move button on the Transform toolbar.
10. Select control points C and D.
11. Click anywhere on the graphics area.
12. Type r0,0,10 at the command area. The selected control points are moved 10 in the Z direction. Note that the symmetric curve and surface also change in shape.
13. Do not save your file.

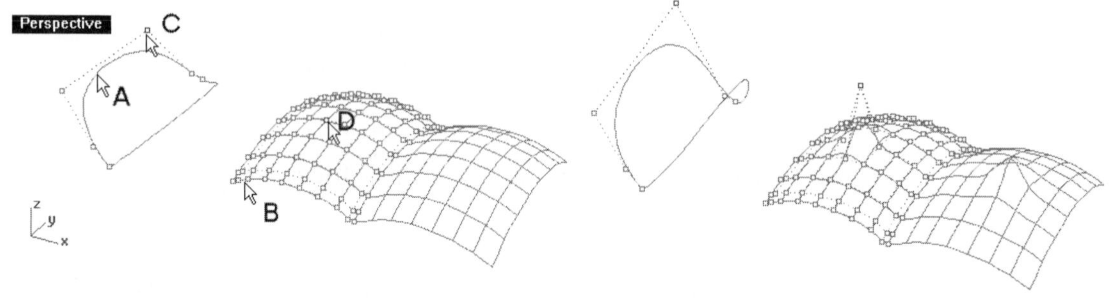

Figure 10–28. Control points turned on and selected (left) and symmetric objects change together (right)

Deformation

Deformation processes that are going to be described in this section change the shape of selected curves, surfaces, polysurfaces, and polygon meshes. Processes are setting points, soft editing, projecting, shearing, twisting, bending, tapering, smoothing, making uniform, attaching to another object, stretching, cage editing, and maelstrom editing.

(If a number of polygon meshes are to be deformed, welding them beforehand to eliminate duplicated edges will give a better result.)

Setting Points

Point-setting involves deforming the selected objects or their control points by setting selected control points of the objects to the X, Y, and Z coordinates of the world coordinate system or construction plane coordinate system. Perform the following steps:

1. Select File > Open and select the file SetPoint.3dm from the Chapter 10 folder on the companion CD.
2. Select Edit > Control Points > Control Points On, or click on the Control Points On/Points Off button of the Point Editing toolbar.
3. Select polygon mesh A, surface B, and curve C (shown in Figure 10–29) and press the ENTER key.
4. Select Transform > Set Points, or click on the Set XYZ Coordinates button on the Transform toolbar.
5. Click on location D and drag to location E, shown in Figure 10–29. (The exact location is unimportant here because this is only an example serving to illustrate the concept.)
6. Click on location F and drag to location G (Figure 10–29).
7. Select control points H and J, and press the ENTER key. A set of control points is selected.
8. In the Set Points dialog box, clear the Set X and Set Y boxes (if they are checked) and check the Set Z box.
9. Check the Align to World box, and click on the OK button.

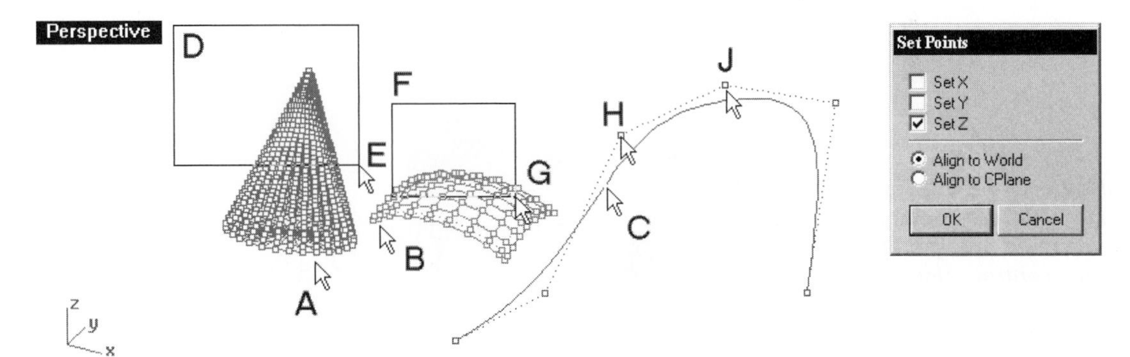

Figure 10–29. Control points of polygon mesh, surface, and curve turned on and being selected

10. Set the display to four-viewport configuration by double-clicking the Perspective viewport's label.
11. Click on location A, shown in Figure 10–30. The selected control points of the curve, surface, and polygon mesh are set to a specified Z value.

12 Maximize and shade the Perspective viewport to see the result. (See Figure 10–31.)

13 Do not save your file.

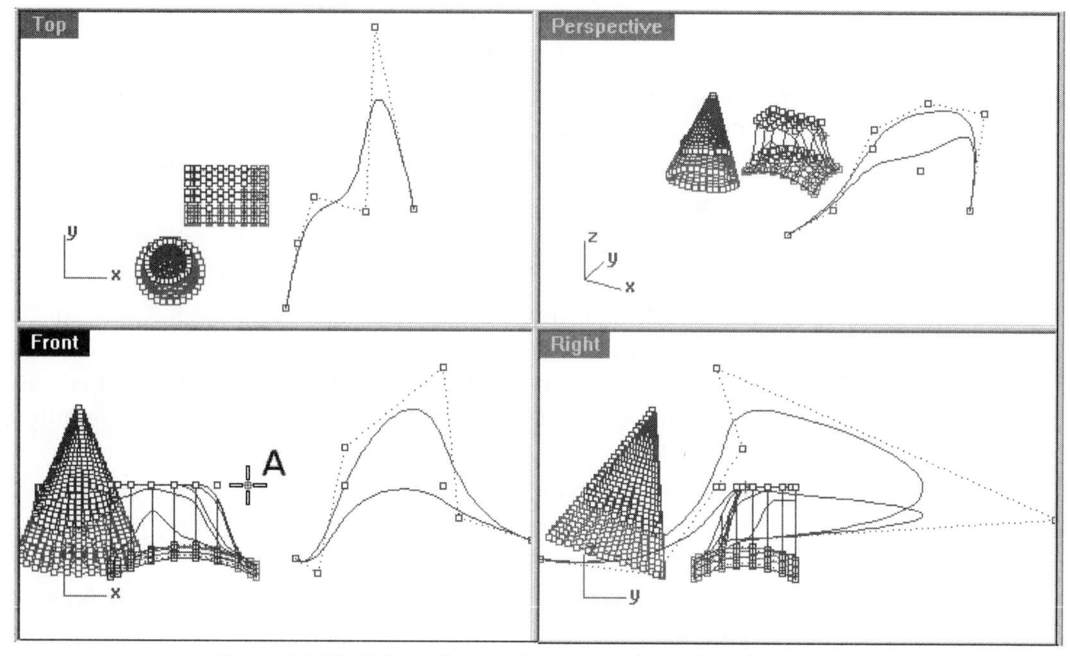

Figure 10–30. Selected control points being aligned

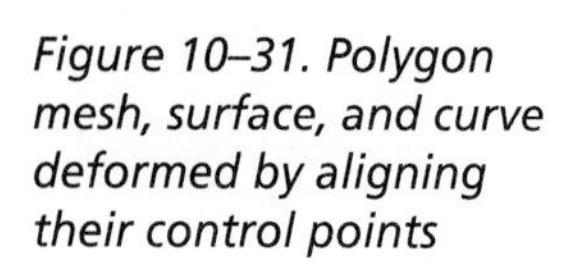

Figure 10–31. Polygon mesh, surface, and curve deformed by aligning their control points

Projecting to the Construction Plane

Using the set point operation described earlier, you set selected control points of a curve, surface, and polygon mesh. If you want to set the entire curve, surface, or polygon mesh to a construction plane, you project them. Perform the following steps:

1. Select File > Open and select the file ProjectToCplane.3dm from the Chapter 10 folder on the companion CD.
2. Select Transform > Project to CPlane, or click on the Project to CPlane button on the Transform toolbar.
3. Click on the Front viewport to make it the active construction plane. Note that this command is view dependent.
4. In the Front viewport, select polygon mesh A, surface B, and curve C (shown in Figure 10–32) and press the ENTER key.
5. Select the Yes option on the command area. Two of the three selected objects are projected onto the construction plane, corresponding to the Front viewport, and the original curve, surface, and polygon mesh are deleted. (See Figure 10–33.) The objects that cannot be projected are reported at the command area.
6. Do not save your file.

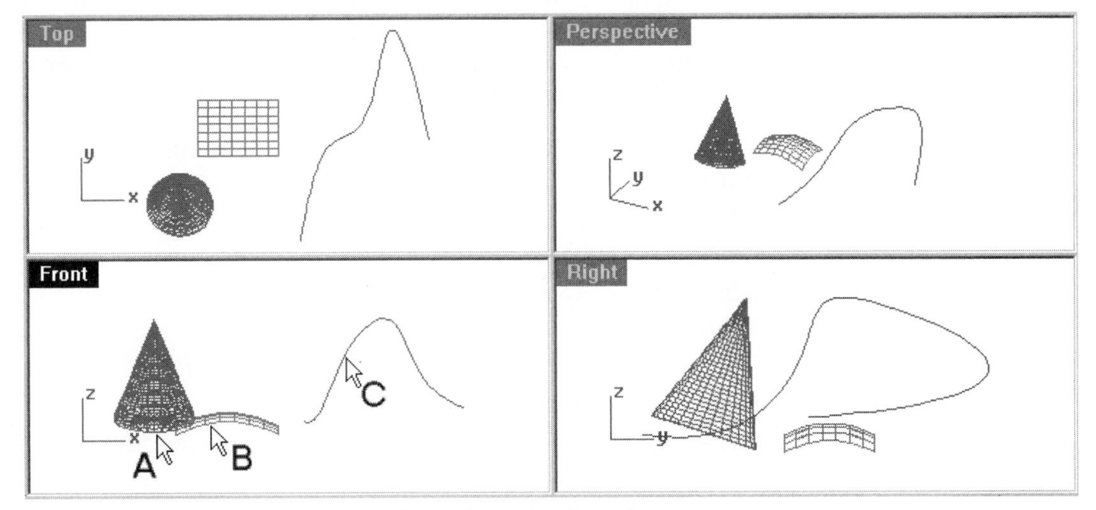

Figure 10–32. Objects being projected

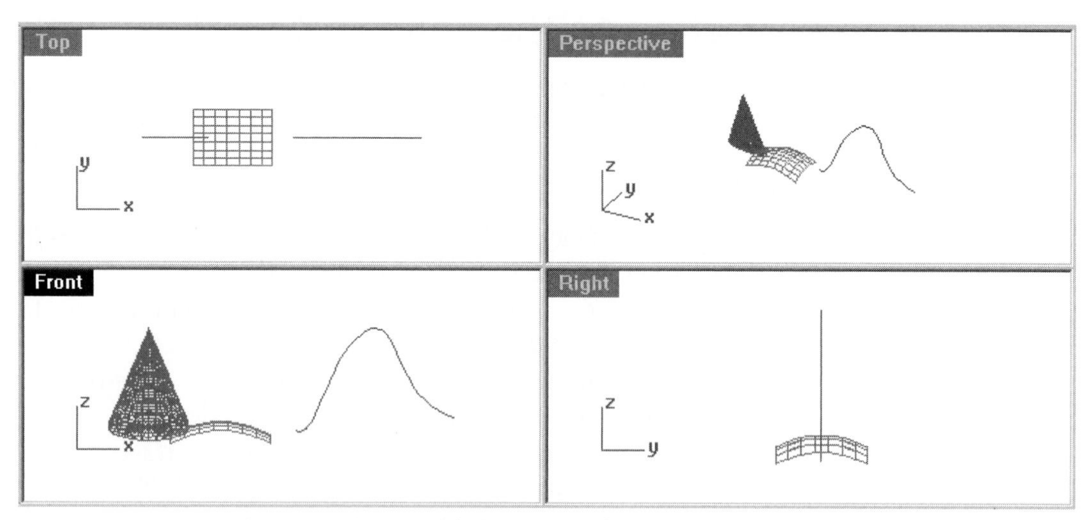

Figure 10–33. Two objects projected

Shearing Along a Plane

Shearing is a deformation process that changes the shape of selected objects by defining a shearing axis and specifying a shearing angle. To define the shearing axis, you specify an origin point and a reference point. Perform the following steps to shear a curve, surface, and a polygon mesh:

1. Select File > Open and select the file Shear.3dm from the Chapter 10 folder on the companion CD.
2. Select Transform > Shear, or click on the Shear button on the Transform toolbar.
3. Select surface A, polygon mesh B, and curve C (shown in Figure 10–34) and press the ENTER key.
4. Select point D (Figure 10–34) as the origin point.
5. Select point E (Figure 10–34) as the reference point.
6. Type -35 to specify a shear angle of -35 degrees. A negative angle measures in the clockwise direction. The selected objects are sheared.
7. Do not save your file.

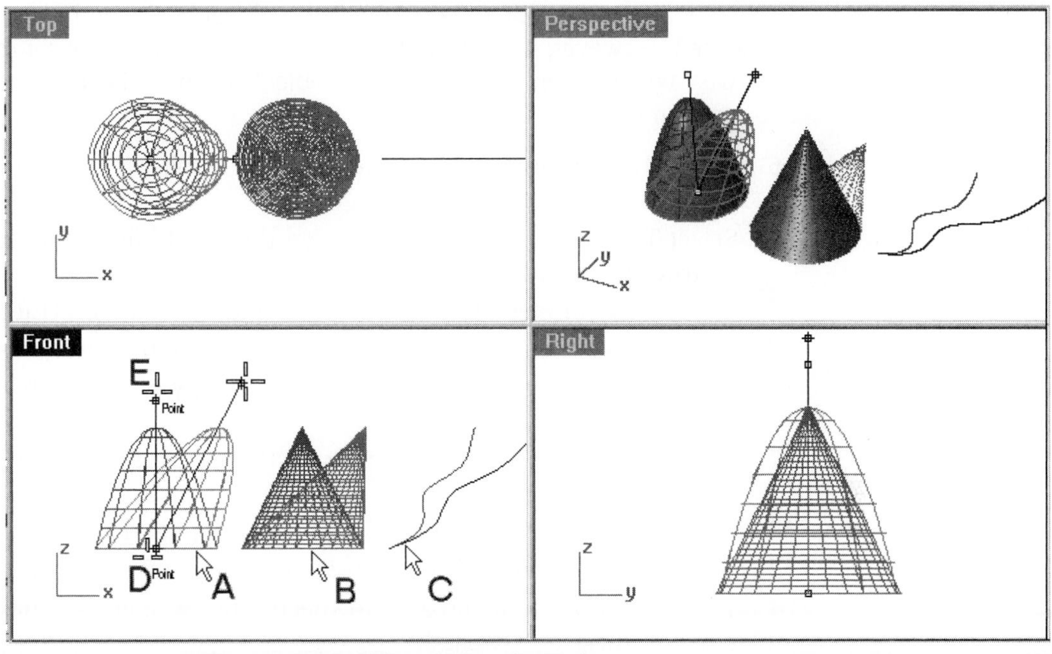

Figure 10–34. Objects being sheared

Twisting Around an Axis

The twisting operation twists selected objects (curve, surface, polygon mesh, or control points). Naturally, you need to define a twisting axis and specify a twisting angle. A twisting axis is formed by picking two reference points. Perform the following steps to twist a curve, surface, and a polygon mesh:

1. Select File > Open and select the file Twist.3dm from the Chapter 10 folder on the companion CD.
2. Check the Point box on the Osnap dialog box and clear all other boxes.
3. Select Transform > Twist, or click on the Twist button on the Transform toolbar.
4. Select curve A (shown in Figure 10–35) and press the ENTER key.
5. Select points B and C (Figure 10–35) to indicate the twist axis.

Points B and C not only define a twist axis through which the object is twisted, they also specify the portion of the object between the points to be twisted. However, if the Infinite option of the command is set to Yes, the twist will happen throughout the object, even if the twist axis is shorter than the object.

6. Type 90 at the command area to specify the twist angle. The selected curve is twisted.
7. Select Transform > Twist, or click on the Twist button on the Transform toolbar.
8. Select polygon mesh D (Figure 10–35) and press the ENTER key.
9. Select points E and F (Figure 10–35) to indicate the twist axis.
10. Type 90 at the command area to specify the twist angle. The selected polygon mesh is twisted.
11. Select Transform > Twist, or click on the Twist button on the Transform toolbar.
12. Select surface G (Figure 10–35) and press the ENTER key.
13. Select points H and J (Figure 10–35) to indicate the twist axis.
14. Type 90 at the command area to specify the twist angle. The selected surface is twisted.
15. Do not save your file.

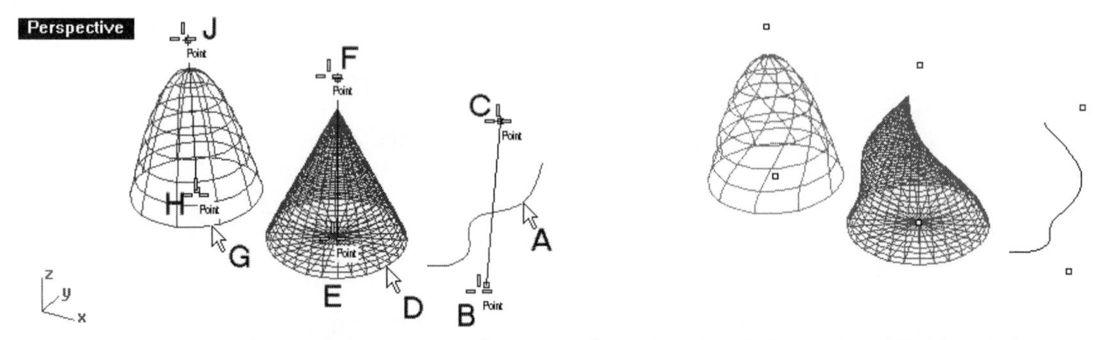

Figure 10–35. Curve, polygon mesh, and surface being twisted (left) and objects twisted (right)

Bending Along a Spline

Bending transforms objects (curve, surface, polygon mesh, or control points) by bending the object along an imaginary spline formed by two reference points. Perform the following steps to bend a curve, surface, and a polygon mesh:

1. Select File > Open and select the file Bend.3dm from the Chapter 10 folder on the companion CD.
2. Check the Point box on the Osnap dialog box and clear all other boxes.
3. Select Transform > Bend, or click on the Bend button on the Transform toolbar.
4. Select curve A (shown in Figure 10–36) and press the ENTER key.
5. Select points B and C (shown in Figure 10–36) to specify a reference spline.
6. Select point D (Figure 10–36) to specify the amount of bend. The selected curve is bent.
7. Select Transform > Bend, or click on the Bend button on the Transform toolbar.
8. Select polygon mesh E (Figure 10–36) and press the ENTER key.
9. Select points F and G (Figure 10–36) to specify a reference spline.
10. Select point H (Figure 10–36) to specify the amount of bend. The selected polygon mesh is bent.
11. Select Transform > Bend, or click on the Bend button on the Transform toolbar.
12. Select surface J (Figure 10–36) and press the ENTER key.
13. Select points K and L (Figure 10–36) to specify a reference spline.
14. Select point M (Figure 10–36) to specify the amount of bend. The selected surface is bent.
15. Do not save your file.

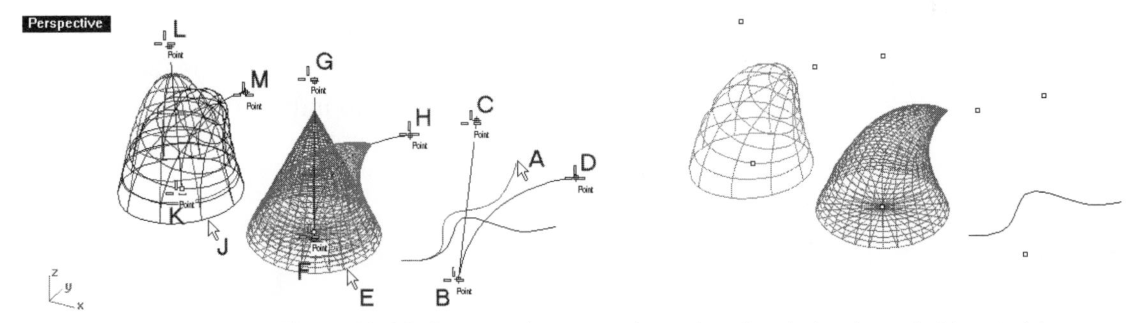

Figure 10–36. Curve, polygon mesh, and surface being bent (left) and objects bent (right)

Tapering along an Axis

The tapering operation deforms a selected object (curve, surface, polygon mesh, or control point) by bevelling the object with reference to a taper axis formed by two reference points. Perform the following steps to taper a curve, surface, and polygon mesh:

1. Select File > Open and select the file Taper.3dm from the Chapter 10 folder on the companion CD.
2. Check the Point box on the Osnap dialog box, and clear all other boxes.
3. Select Transform > Taper, or click on the Taper button on the Transform toolbar.
4. Select curve A (shown in Figure 10–37) and press the ENTER key.
5. Select points B and C (Figure 10–37) to indicate the taper axis.
6. Select points D and E (Figure 10–37) to indicate the start and end distances. The selected curve is tapered.
7. Select Transform > Taper, or click on the Taper button on the Transform toolbar.
8. Select polygon mesh F (Figure 10–37) and press the ENTER key.
9. Select points G and H (Figure 10–37) to indicate the taper axis.

Basically, points G and H define the taper axis and specify the portion within which the object is tapered. However, if the Infinite option is Yes, taper will happen throughout the object—even if the axis is shorter than the object.

10. Select points J and K (Figure 10–37) to indicate the start and end distances. The selected polygon mesh is tapered.
11. Select Transform > Taper, or click on the Taper button on the Transform toolbar.
12. Select surface L (Figure 10–37) and press the ENTER key.
13. Select points M and N (Figure 10–37) to indicate the taper axis.
14. Select points P and Q (Figure 10–37) to indicate the start and end distances. The selected surface is tapered.
15. Do not save your file.

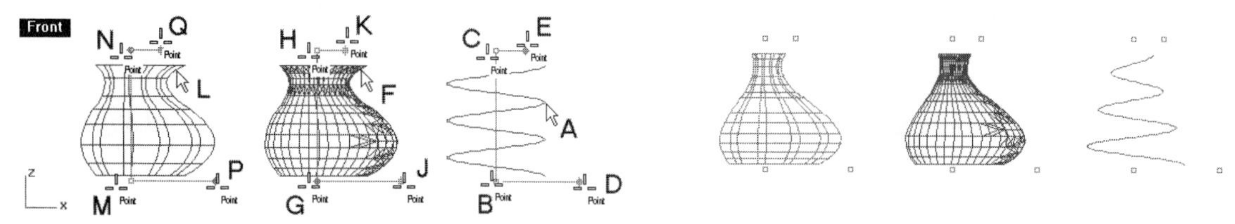

Figure 10–37. Curve, polygon mesh, and surface being tapered

Shearing, Bending, Twisting, and Tapering of Control Points

The following example demonstrates how deformation operations are applied to selected control points of objects:

1. Select File > Open and select the file ControlPointManipulation.3dm from the Chapter 10 folder on the companion CD.
2. Check the Point and End boxes on the Osnap dialog box, and clear all other boxes.
3. Select Transform > Bend, or click on the Bend button on the Transform toolbar.
4. Select surfaces A, B, and C, shown in Figure 10–38, and press the ENTER key.
5. Select end points D and E (Figure 10–38) to specify a reference spline.
6. Select point F (Figure 10–38) to specify the amount of bend. The surfaces are bent.

After bending, all three surfaces are deformed.

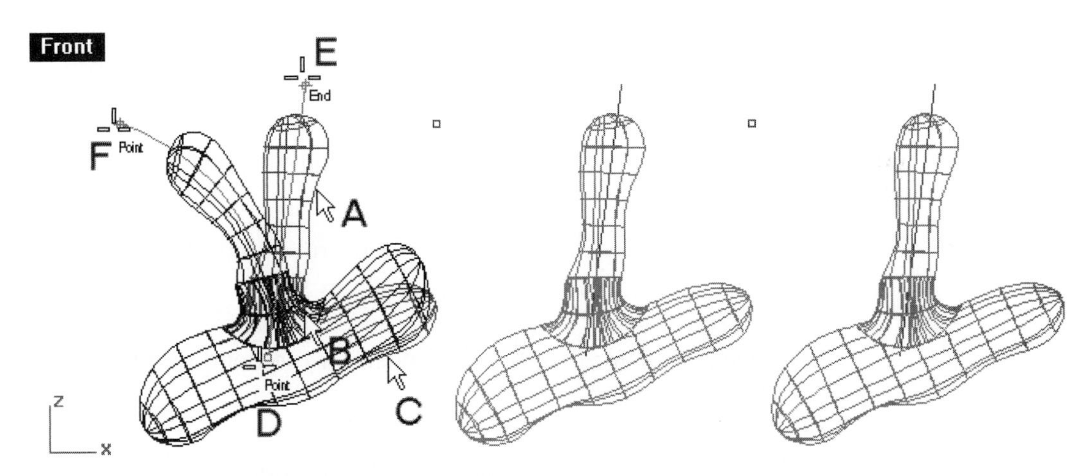

Figure 10–38. Three surfaces being bent

Continue with the following steps to bend two surfaces instead of all three surfaces.

7. Select Transform > Bend, or click on the Bend button on the Transform toolbar.

8. Select surfaces A and B, shown in Figure 10–39, and press the ENTER key.
9. Select end points C and D (Figure 10–39) to specify a reference spline.
10. Select point E (Figure 10–39) to specify the amount of bend. The surfaces are bent.

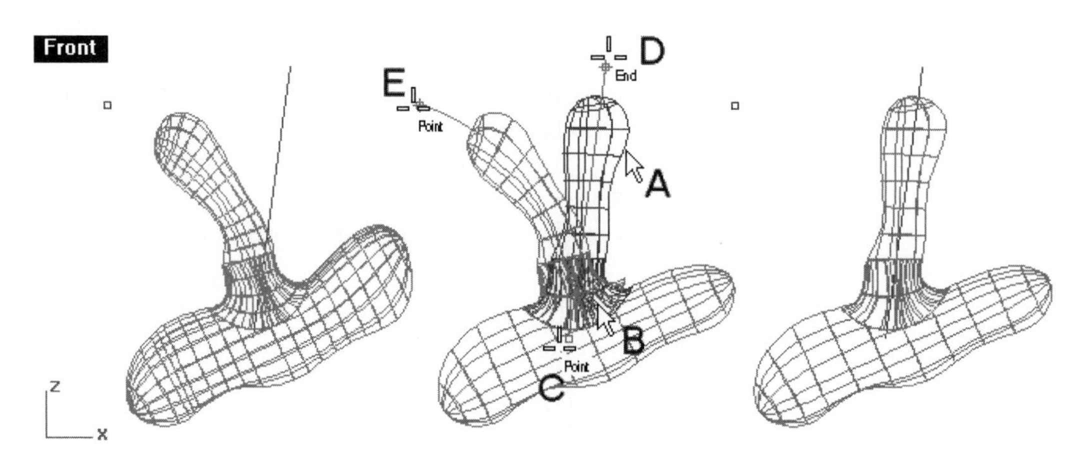

Figure 10–39. Two surfaces being bent

Continue with the following steps to bend selected control points of two surfaces.

11. Select Edit > Control Points > Control Points On.
12. Select surfaces A, B, and C, shown in Figure 10–40, and press the ENTER key.
13. Select Transform > Bend, or click on the Bend button on the Transform toolbar.
14. Type LASSO at the command area.
15. Click on D, shown in Figure 10–40, hold down the left mouse button, and drag a region. A set of control points is selected.
16. Press the ENTER key.

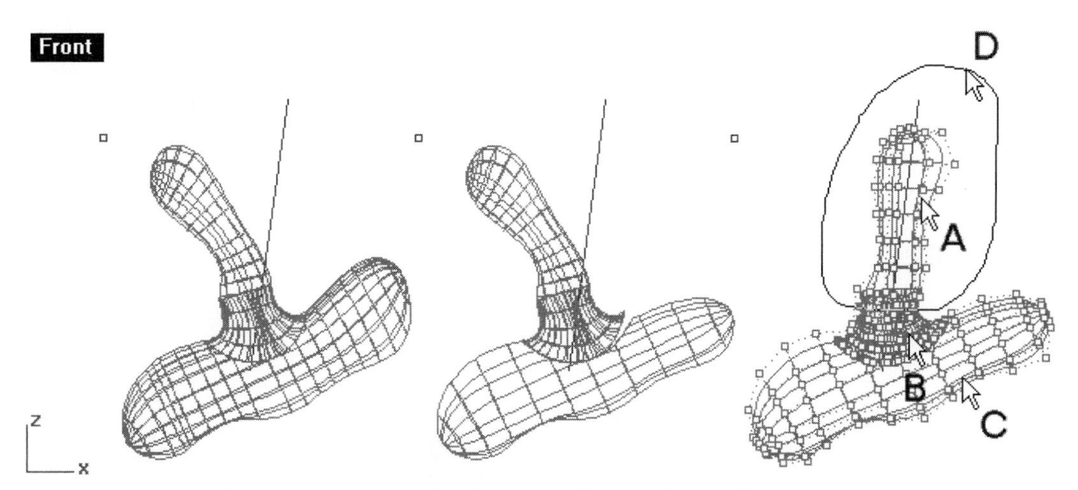

Figure 10–40. Control points turned on and control points being selected

17 Select end points A and B (shown in Figure 10–41) to specify a reference spline.

18 Select point C (Figure 10–41) to specify the amount of bend. The selected control points are relocated.

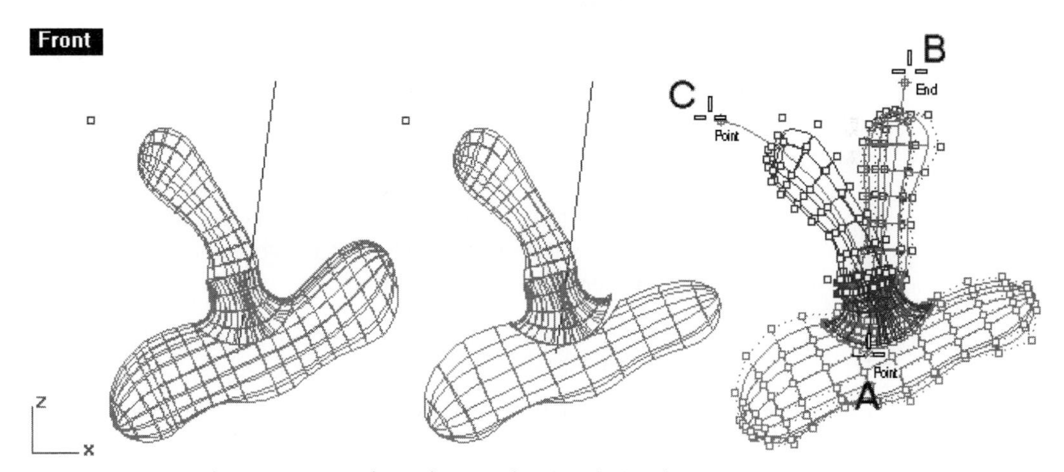

Figure 10–41. Selected control points being bent

As can be seen in Figure 10–42, portion A is deformed if all three surfaces are bent. If only two surfaces are bent, there is a gap in B. To bend two surfaces without creating a gap or opening, the solution is to bend selected control points.

19 Do not save your file.

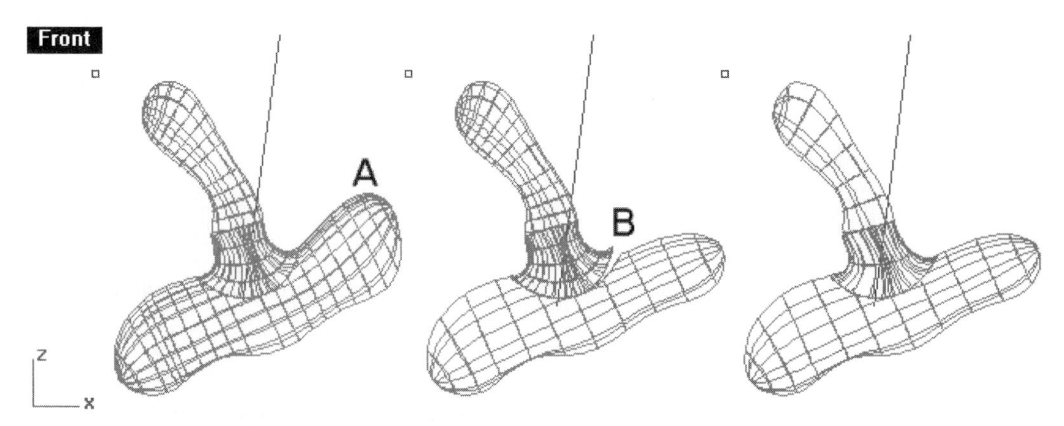

Figure 10–42. Comparing the results

Smoothing

Smoothing is a deformation process to remove irregularities of a curve, surface, or polygon mesh. In essence, smoothing averages the positions of the control points. While smoothing, you have to specify a smooth factor. If the smooth factor is between 0 and 1, the smoothing moves toward the average. If the smooth factor is greater than 1, the point moves past the average. If the smooth factor is negative, the point moves away from the average and causes the curve, surface, or polygon mesh to be rougher. Perform the following steps:

1 Select File > Open and select the file Smooth.3dm from the Chapter 10 folder on the companion CD.

2 Select Transform > Smooth, or click on the Smooth button on the Transform toolbar.

3 Select curve A, surface B, and polygon mesh C (shown in Figure 10–43) and press the ENTER key.

4 In the Smooth dialog box, select Smooth X, Smooth Y, and Smooth Z, set smooth factor to 0.5, and then click on the OK button. The selected objects are smoothed.

5 Do not save your file.

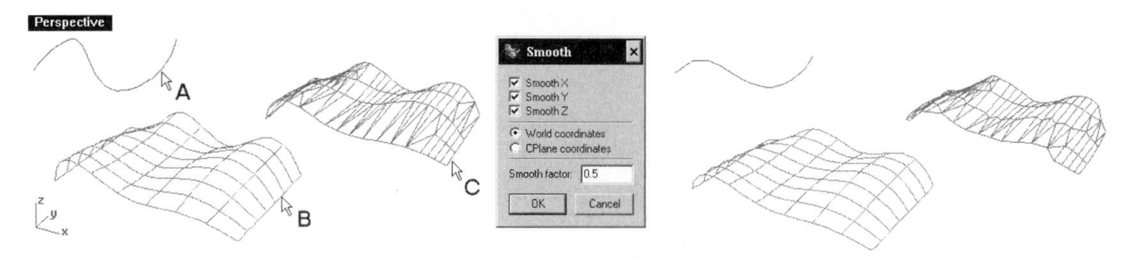

Figure 10–43. Objects being smoothed (left) and objects smoothed (right)

Making Objects Uniform

Another way to improve the smoothness of a curve or surface is to make it uniform in terms of the knot vectors. Unlike smoothing, which modifies the location of the control points, making an object uniform does not change the control point locations. To make uniform a curve and a surface, perform the following steps:

1. Select File > Open and select the file MakeUniform.3dm from the Chapter 10 folder on the companion CD.
2. Click on the Make Uniform button on the Curve Tools toolbar, or type MakeUniform at the command area.
3. Select curve A and surface B, shown in Figure 10–44, and press the ENTER key. The curve and surface are made uniform.
4. Do not save your file.

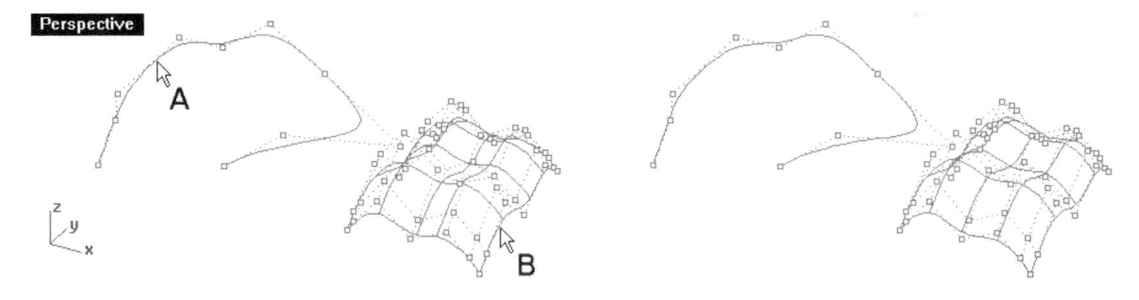

Figure 10–44. Original curve and surface (left) and treated curve and surface (right)

Attaching for Easy Manipulation

For easy manipulation of curves and surface, you can attach them to a backbone curve or backbone surface. After attaching, modification to the backbone curve or surface will cause corresponding changes to the original curves and surfaces. Perform the following steps to attach polysurfaces to a backbone surface and then deform the backbone surface.

1. Select File > Open and select the file Sped.3dm from the Chapter 10 folder on the companion CD.
2. Type Sped at the command area.
3. Select A, B, C, D, E, F, and G (shown in Figure 10–45) and press the ENTER key.
4. Select H (Figure 10–45) as the backbone.

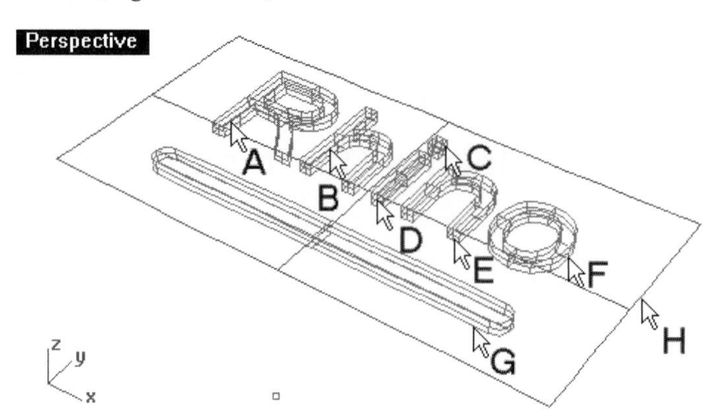

Figure 10–45. Polysurfaces being attached to a surface

Continue with the following steps to bend the backbone surface.

5. Set object snap mode to End and Point.
6. Select Transform > Bend, or click on the Bend button on the Transform toolbar.
7. Select surface A (shown in Figure 10–46) and press the ENTER key.
8. Select endpoints B and C (Figure 10–46) to specify a reference spline.
9. Select point D (Figure 10–46) to specify the amount of bend. The backbone surface, together with the attached polysurfaces, is bent.

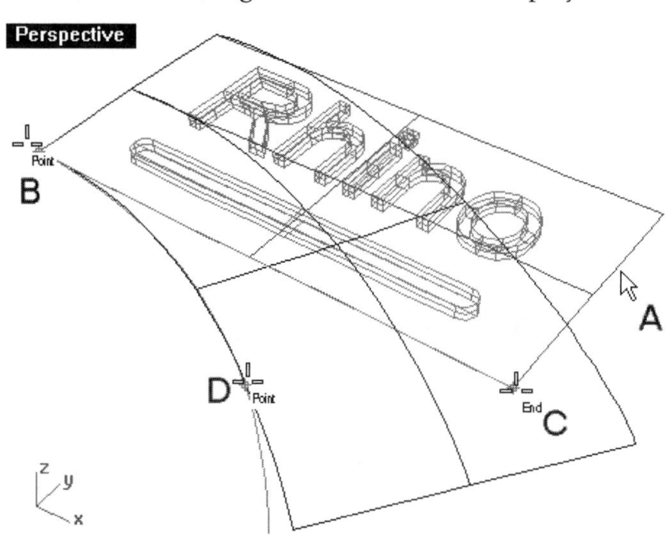

Figure 10–46. Backbone surface being bent

Continue with the following steps to manipulate the control points of the backbone surface.

10 Turn on the control points of the backbone surface A (Figure 10–47).
11 Select control point B, shown in Figure 10–47.
12 Select Transform > Move UVN, or click on the Move UVN/Turn Move UVN Off button of the Point Editing toolbar.
13 Select and drag the N sliding bar of the Move UVN dialog box. The selected control point is moved. Correspondingly, the attached surfaces are also deformed.
14 Do not save your file.

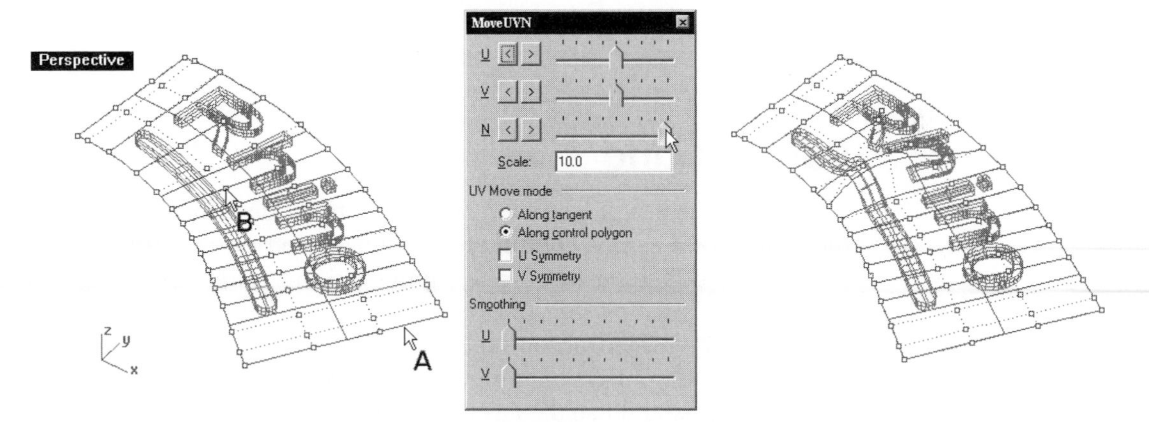

Figure 10–47. Control point being moved in N direction (left) and deformed surfaces (right)

Stretching

Stretching involves deforming a portion of selected objects (curve, surface, or polygon mesh) by lengthening or shortening them along a specified direction. You select objects to be stretched and pick two points to define the start of stretch axis and the end of stretch axis, then pick another point to define the location where the objects are stretched to, or specify a stretch factor. Perform the following steps:

1 Select File > Open and select the file Stretch.3dm from the Chapter 10 folder on the companion CD.
2 Click on the Stretch button on the UDT toolbar, or type Stretch at the command area.
3 Select curve A, polysurface B, and polygon mesh C (shown in Figure 10–48) and press the ENTER key.

4. Select points D, E, and F. (Point D is the start of the stretch axis, point E is the end of the stretch axis, and point F represents the ratio to be stretched.)
5. Do not save your file.

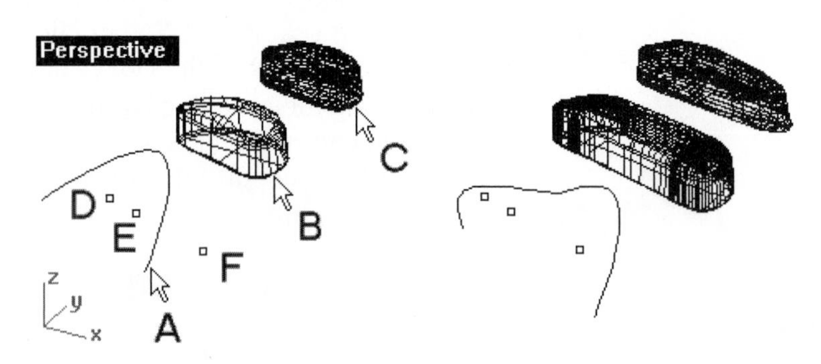

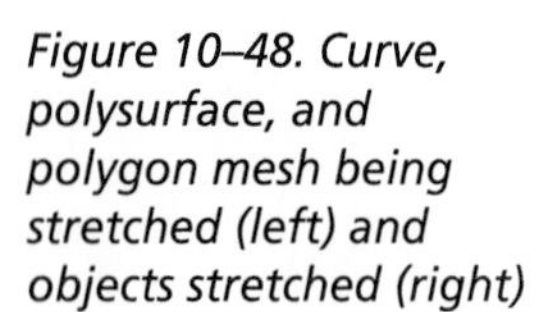
Figure 10–48. Curve, polysurface, and polygon mesh being stretched (left) and objects stretched (right)

Cage Editing

Cage editing is a process whereby objects to be deformed are treated as the captive objects of a control object, which can be any existing curve or surface, the bounding box of the captive objects, a line, a rectangle, or a box. By manipulating the control points of the control object, the captive objects are deformed accordingly. This deformation process is particularly suitable for objects of complex shape. Perform the following steps to construct a box and use it as a control object to deform a captive object:

1. Select File > Open and select the file CageEdit1.3dm from the Chapter 10 folder on the companion CD.
2. Select Transform > Cage Editing > Create Cage, or type Cage at the command area.
3. Click on points A, B, and C (shown in Figure 10–49) to define a cage box.
4. Select the XDegree option on the command area.
5. Type 3 to set the degree of polynomial of the cage box in X direction to 3.
6. Select the YDegree option on the command area.
7. Type 3 to set the degree of polynomial of the cage box in Y direction to 3
8. Select the ZDegree option on the command area.
9. Type 3 to set the degree of polynomial of the cage box in Z direction to 3.
10. Press the ENTER key. A cage box is constructed.

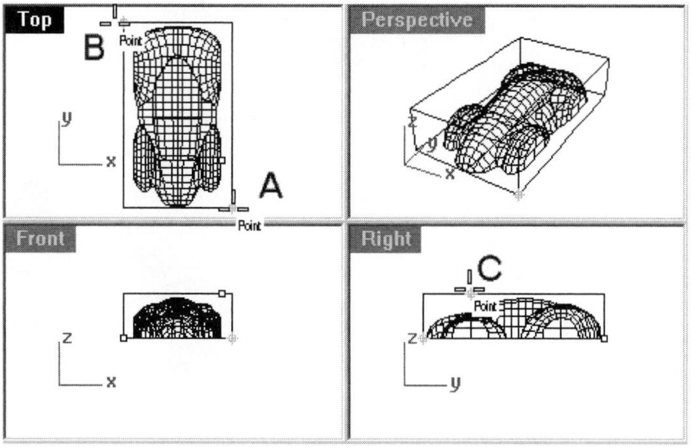

Figure 10–49. Cage box being constructed

11. Maximize the Perspective viewport.
12. Select Transform > Cage Editing > Cage Edit, or click on the Cage Edit Objects button on the UDT toolbar.
13. Select A(shown in Figure 10–50) as the captive object, and press the ENTER key.
14. Select B (Figure 10–50) as the control object.
15. If Region to edit is global, press the ENTER key to accept.

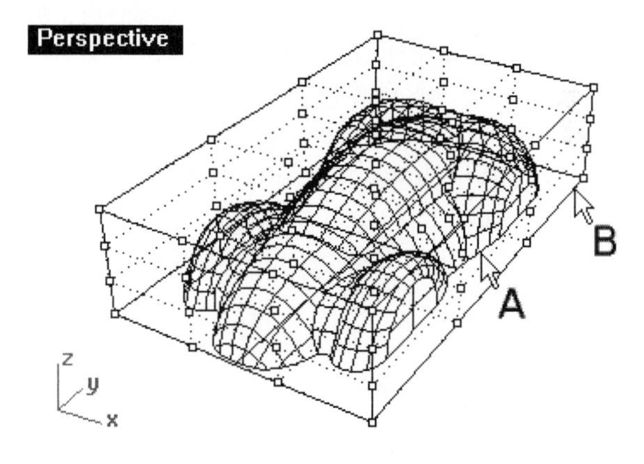

Figure 10–50. Captive object related to control cage box

16. Referencing Figure 10–51, click on the control points of the cage box and drag them to new locations. For the purpose of this tutorial, exact locations of the dragged points are unimportant.
17. Observe how the changes in the cage box are reflected in the caged object.
18. Do not save your file.

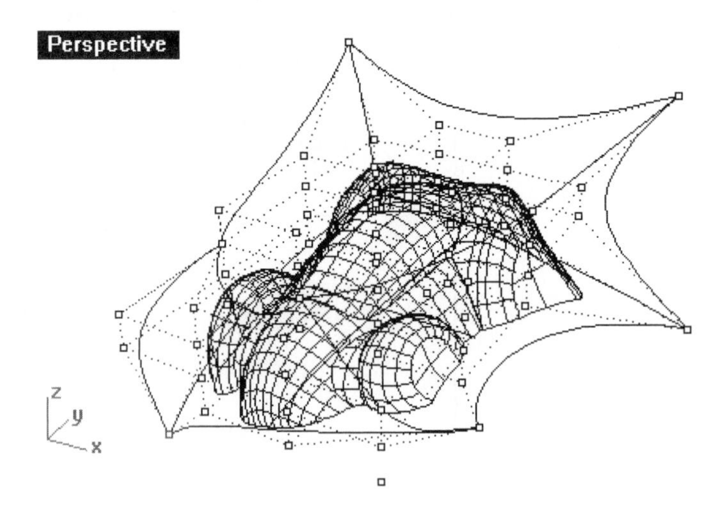

Figure 10–51. Caged object deformed by deforming the cage

Cage Type

Continue with the following steps to use an existing surface as a control object to deform selected captive objects:

1. Select File > Open and select the file CageEdit1.3dm from the Chapter 10 folder on the companion CD.
2. Select Transform > Cage Editing > Create Cage.
3. Click on A and drag to B, shown in Figure 10–52, and press the ENTER key.
4. Select rectangular surface C.
5. Drag the control points to cage edit.
6. The captive objects are deformed. Do not save your file.

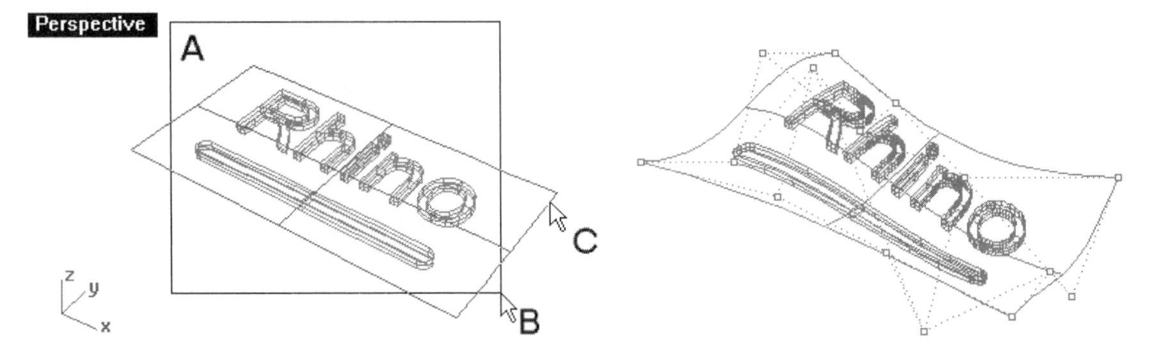

Figure 10–52. Rectangular surface used as control object (left) and cage edited (right)

Release from Cage

To release a captive object from its control object, you can click on the Release objects from control cage button on the Cage toolbar.

Maelstrom Editing

As the name implies, this deformation process changes the shape of an object as if there is a powerful and violent whirlpool sucking in the object within a given radius. Perform the following steps to appreciate how a polysurface is deformed by a maelstrom.

1. Select File > Open and select the file Maelstrom.3dm from the Chapter 10 folder on the companion CD.
2. Click on the Maelstrom button on the UDT toolbar, or type Maelstrom at the command area.
3. Select surface A, shown in Figure 10–53, and press the ENTER key. (*Note*: This command also works on polygon meshes.)
4. Select point B to specify the center of the maelstrom. (Because the maelstrom is defined by a circle, like constructing a circle, you can define the maelstrom in many ways.)
5. Select point C to specify the radius of the maelstrom.
6. Select point D to specify the second radius of the maelstrom.
7. Type 45 at the command area to specify the angle of the vortexification point. The selected object is deformed.
8. Do not save your file.

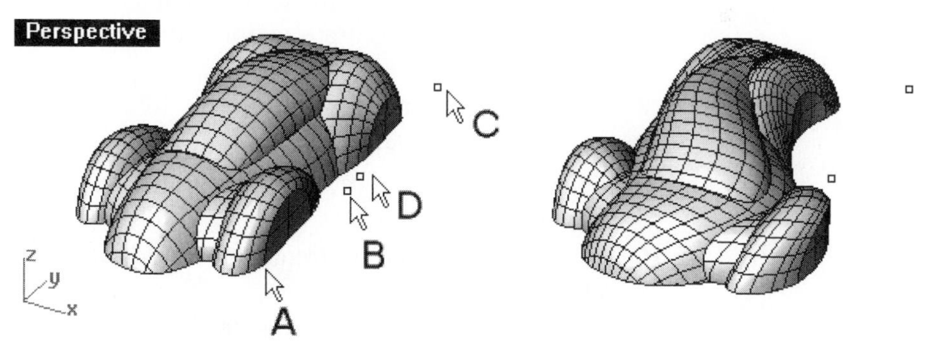

Figure 10–53. Maelstrom being defined (left) and object deformed (right)

Translate and Deform

A more advanced way of modeling is to translate and deform simple forms and shapes to achieve a specific outcome. These processes are as follows: flowing along a curve, flowing along a surface, applying to UVN of a surface, and dropping onto a surface.

Flowing Along a Curve

You can transform a curve, surface, or polygon mesh by flowing the object along a selected curve. This transformation process involves two curves: an original backbone curve and a new backbone curve. During the transformation process, the object is transformed to the new backbone curve with reference to the original backbone curve. Perform the following steps:

1. Select File > Open and select the file FlowAlongCurve.3dm from the Chapter 10 folder on the companion CD.
2. Select Transform > Flow along Curve, or click on the Flow Along Curve button on the UDT toolbar.
3. Select curves A, B, and C (shown in Figure 10–54) and press the ENTER key.
4. Select curve D (Figure 10–54) to use it as the original backbone curve.
5. If Stretch = No, select Stretch option on the command area to change to Yes.
6. Select curve E (Figure 10–54) to use it as the new backbone curve. The selected curves are transformed to flow along the new backbone curve. (See Figure 10–55.)
7. Repeat the command.
8. Select curve F (Figure 10–54) and press the ENTER key.
9. Select curve G (Figure 10–54) to use it as the original backbone curve.
10. If Stretch = No, select Stretch option on the command area to change to Yes.
11. Select curve H (Figure 10–54) to use it as the new backbone curve. The selected curve is transformed to flow along the new backbone curve. (See Figure 10–55.)

12 Repeat the command.

13 Select polygon mesh J (Figure 10–54) and press the ENTER key.

14 Select curve K (Figure 10–54) to use it as the original backbone curve.

15 If Stretch = No, select Stretch option on the command area to change to Yes.

16 Select curve L (Figure 10–54) to use it as the new backbone curve. The selected surface is transformed to flow along the new backbone curve. (See Figure 10–55.)

17 Repeat the command.

18 Select surface M (Figure 10–54) and press the ENTER key.

19 Select curve N (Figure 10–54) to use it as the original backbone curve.

20 If Stretch = No, select Stretch option on the command area to change to Yes.

21 Select curve P (Figure 10–54) to use it as the new backbone curve. The selected polygon mesh is transformed to flow along the new backbone curve. (See Figure 10–55.)

22 Do not save your file.

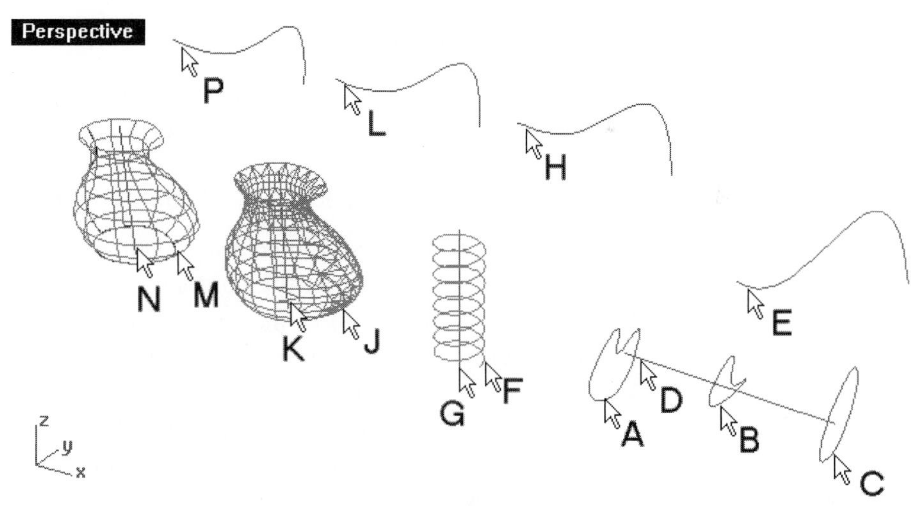

Figure 10–54. Objects being transformed by flowing from their original backbone curves to new backbone curve

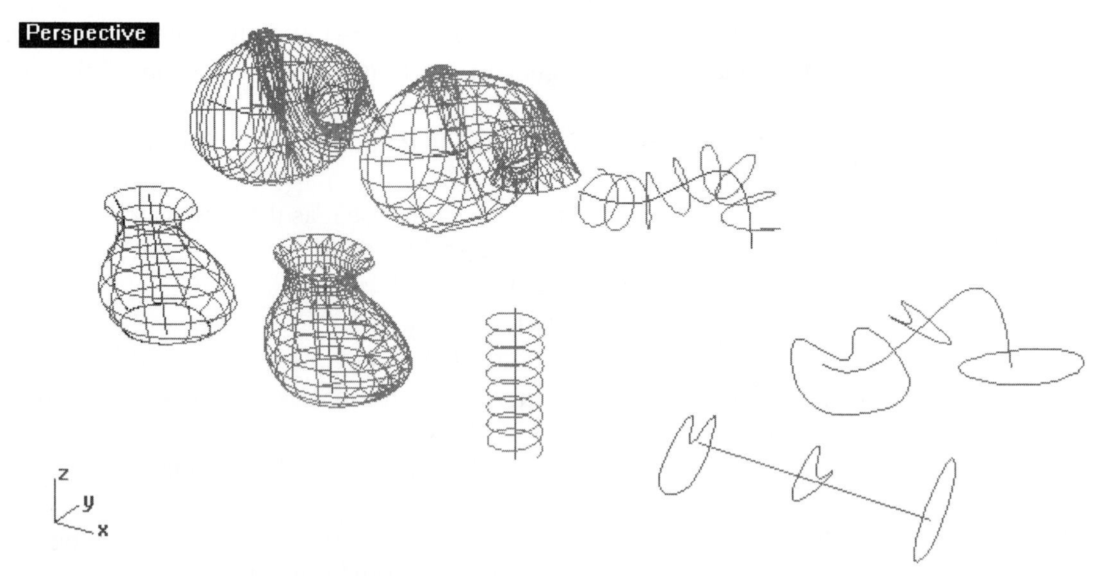

Figure 10–55. Objects transformed from their original backbone curves to new backbone curves

Flowing Along a Surface

One way to transform a set of surfaces onto a free-form surface is to perform space morphing. The set of surfaces is first constructed on a planar backbone surface, which is used as reference. The surfaces are then space morphed onto a target surface. Perform the following steps:

1. Select File > Open and select the file FlowAlongSurface.3dm from the Chapter 10 folder on the companion CD.
2. Select Transform > Flow along Surface, or click on the Flow along surface button on the UDT toolbar.
3. Referencing Figure 10–55, click on location A and drag to location B to select the text object "Rhinoceros 4."
4. Select ellipsoid surface C, shown in Figure 10–56, and press the ENTER key.
5. Select surface D near corner D. This is the reference backbone surface, and the corner is the reference corner.
6. Select surface E near corner E. This is the target backbone surface. The selected objects are flown onto the target surface.

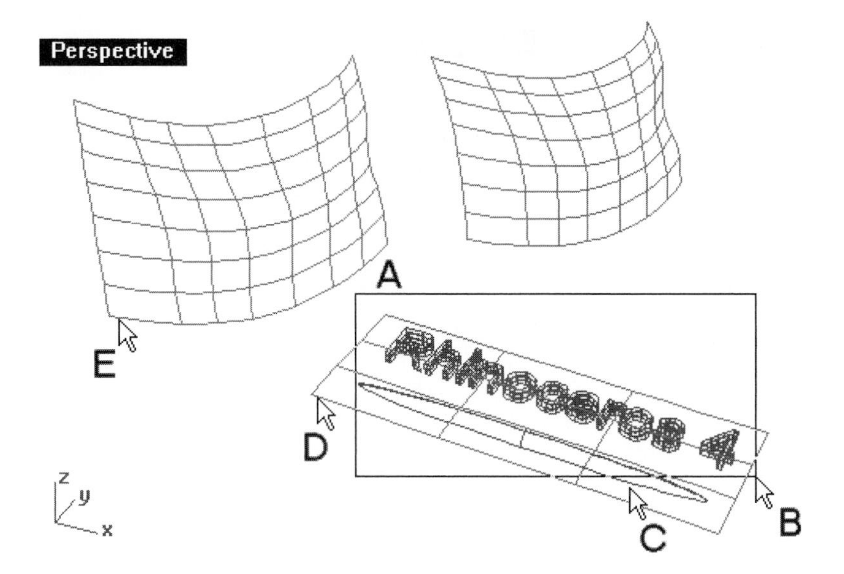

Figure 10–56. Objects being morphed

7 Repeat the command.

8 Click on location A, shown in Figure 10–57, and drag to location B.

9 Select C and press the ENTER key.

10 Select Rigid option on the command area.

11 Select surface D near corner D.

12 Select surface E near corner E. The selected objects are space morphed. (See Figure 10–58.)

13 Do not save your file.

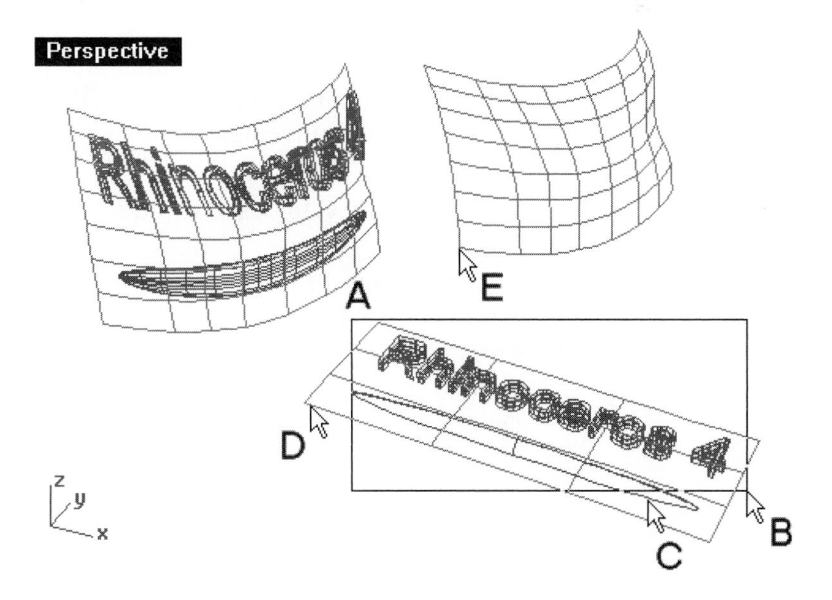

Figure 10–57. Objects for rigid morphing

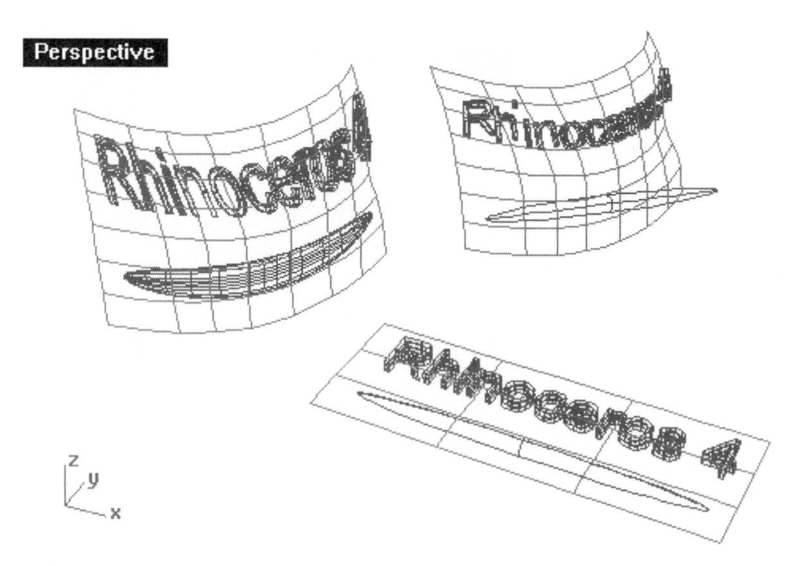

Figure 10–58. Objects rigidly morphed

Dropping onto a Surface

Perform the following steps to try another way of deforming objects and dropping onto a surface:

1. Select File > Open and select the file Splop.3dm from the Chapter 10 folder on the companion CD.
2. Click on the Splop button on the UDT toolbar, or type Splop at the command area.
3. Select surfaces A and B, shown in Figure 10–59, and press the ENTER key.
4. Click on points C and D in the front viewport to specify the reference sphere. Note that it is important to click on the construction plane parallel to the source objects.
5. Select surface E.
6. Click on F and G to describe a target sphere. A set of objects is dropped.
7. Click on H and J to describe another target sphere. Another set of objects is dropped.
8. Press the ENTER key.
9. Two copies of surfaces are dropped onto a surface. (See Figure 10–60.) Do not save your file.

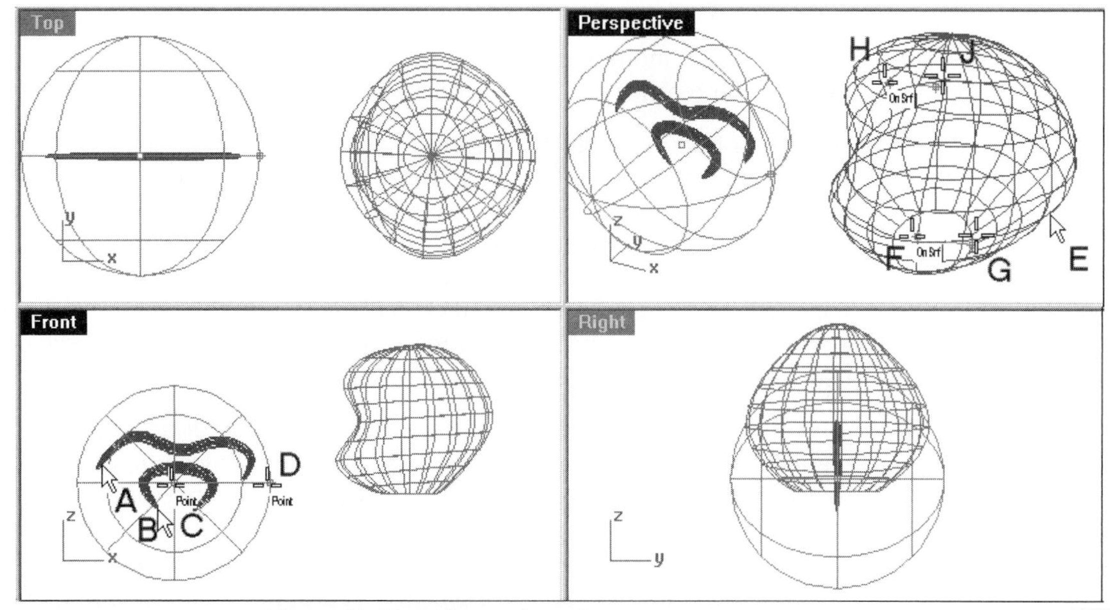

Figure 10–59. Objects selected

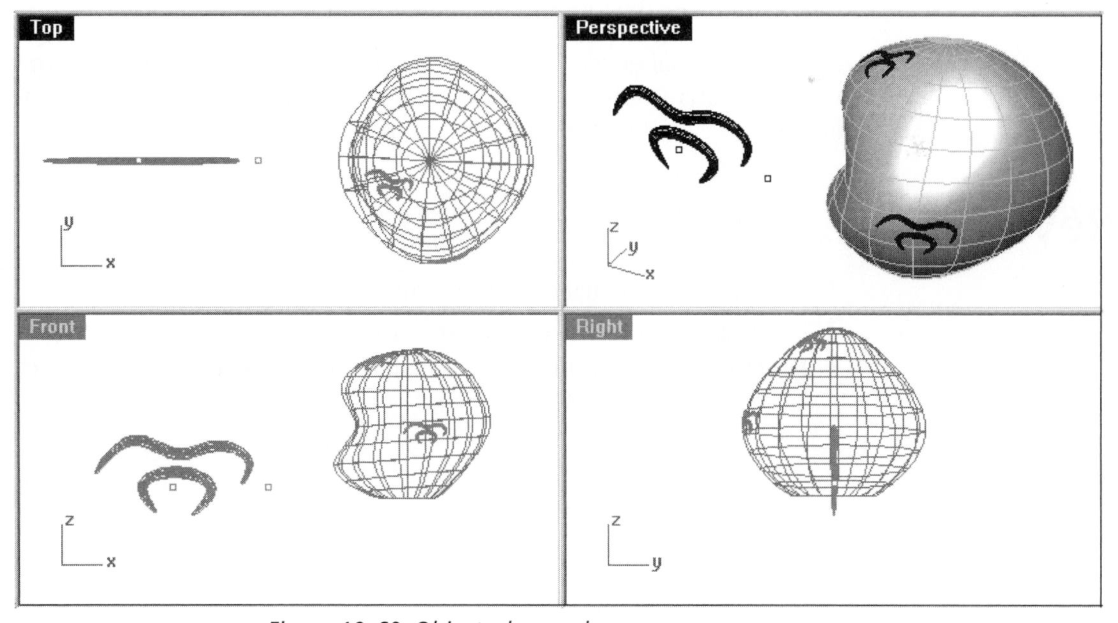

Figure 10–60. Objects dropped

Chapter Summary

To develop and modify your design, you can transform curves, surfaces, and polygon meshes and their control points as necessary.

In terms of translation, in addition to the basic translation tools described in Chapter 2, you can soft move a set of objects, align planar curves, and align objects with reference to their bounding boxes. You can orient objects referencing two or three points onto a surface, perpendicular to a curve, tangent to the edge of a surface, and to a selected construction plane. You can array objects in a rectangular pattern, polar pattern, along a curve, along the U and V directions of a surface, and along a curve on a surface. You can also construct symmetric objects.

You can soft edit a curve or a surface. Control points of objects can be aligned by setting points or flattened to a plane by projection. Shapes of objects can be manipulated by shearing, twisting, bending, and tapering. To manipulate individual surface elements of a polysurface or a portion of a surface, it is advisable to shear, twist, bend, or taper selected control points. To simplify a curve or a surface, you apply a smoothing operation or make them uniform. To handle a set of polysurfaces collectively, you attach them to a backbone surface, stretch them to extend or shorten a portion of them, use cage for editing, and deform it as if the objects are being acted upon by a maelstrom.

To achieve special effects, you can use combined translation and deformation operations—flowing along a curve, space morphing, applying to a surface's u, v, and n, and dropping onto a surface.

Review Questions

1. List the methods used to orient and array objects.
2. List the ways objects can be deformed.
3. Describe the operations that translate as well as deform an object.

CHAPTER 11

Rhinoceros Data Analysis

Introduction

This chapter explains various analysis tools to examine curves, surfaces, and polygon meshes.

Objectives

After studying this chapter, you should be able to

- ❒ Use Rhinoceros tools to analyze curves, surfaces, and meshes

Overview

To further improvise your design, you may need to examine the curves, surfaces, and polygon meshes that you already constructed. Tools for analysis can be divided into five major categories: general tools, dimensional analysis tools, surface profile analysis tools, mass properties tools, and diagnostics tools.

General Tools

The set of tools described below focuses on the form and shape of selected curves and surfaces.

What

To learn about properties of selected objects and display them in a separate window, you can use the What command, as follows:

1. Select File > Open and select the file What.3dm from the Chapter 11 folder on the companion CD.
2. Right-click on the List Object Database/What?! Button on the Diagnostics toolbar, or type What at the command area.

3. Select A and B, shown in Figure 11–1, and press the ENTER key. The Object Description dialog box is displayed.
4. If you want to save the information to a text file, click on the Save As button and then specify a file name.
5. Click on the Close button. Information about selected objects is retrieved.
6. Do not save your file.

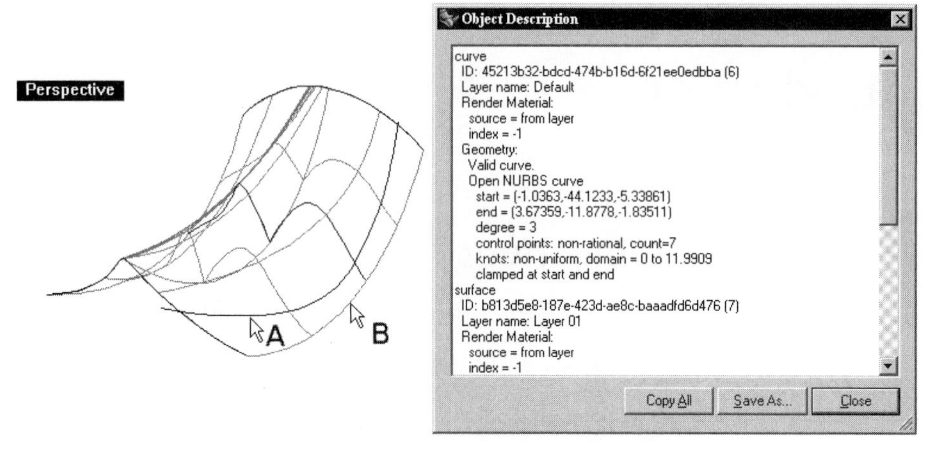

Figure 11–1. What dialog box

Curvature Graph

One way to determine the smoothness of a curve or a surface and discern any irregularities is to display the curvature graph, which tells the radius of curvature of a curve along the curve or radius of curvature of a surface along its U and V isocurves. Perform the following steps:

1. Select File > Open and select the file general01.3dm from the Chapter 11 folder on the companion CD.
2. Select Analyze > Curve > Curvature Graph On, or click on the Curvature Graph On/Curvature Graph Off button on the Analyze toolbar.
3. Select curve A and surface B (shown in Figure 11–2). The curvature graphs are displayed.
4. In the Curvature Graph dialog box, adjust the display scale and density as necessary.
5. Select Analyze > Curve > Curvature Graph Off, or right-click on the Curvature Graph On/Curvature Graph Off button on the Analyze toolbar. The curvature graph is turned off.

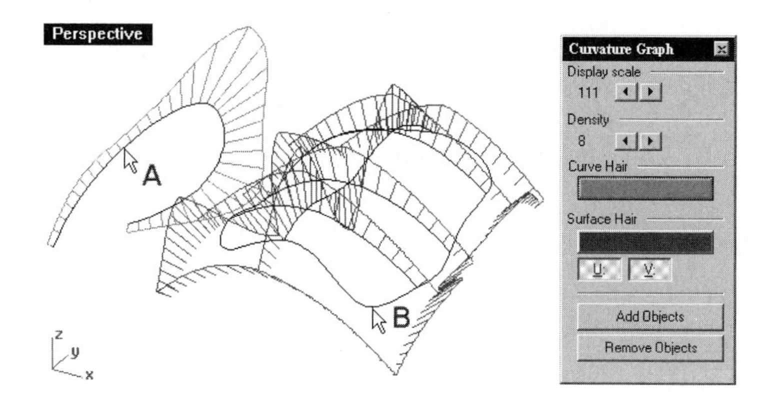

Figure 11–2. Curvature graph displayed

Curvature Circle

Another way to evaluate the profile of a curve or a surface is to display or construct a curvature circle at a selected location along the curve or two curvature arcs at a selected location on the surface. Continue with the following steps.

6 Select Analyze > Curvature Circle, or right-click on the Radius/Curvature button on the Analyze toolbar.
7 Select surface A, shown in Figure 11–3
8 If the MarkCurvature option is No, select the option on the command area to change it to Yes.
9 Select a point on the surface. The curvature arcs' radii are constructed.
10 Press the ENTER key to terminate the command.
11 Repeat the command.
12 Select curve B, shown in Figure 11–3
13 Select a point on the curve. The curvature circle is constructed.
14 Press the ENTER key to terminate the command.
15 Select Edit > Undo twice to undo the last two commands.

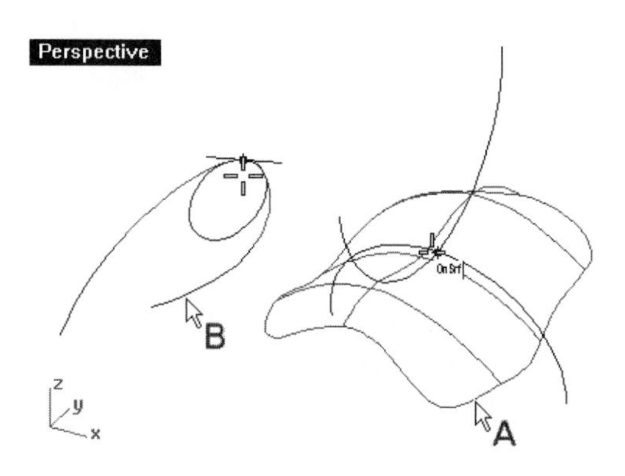

Figure 11–3. Curvature arcs and curvature circle constructed

Curvature Radius

You can find out the radius of the curvature at a selected location along a curve or along the edge of a surface. Continue with the following steps.

16 Check the Point option of the Osnap dialog box.

17 Select Analyze > Radius, or click on the Radius/Curvature button on the Analyze toolbar.

18 Select point A on the curve, shown in Figure 11–4. The radius and diameter of curvature of the selected location are displayed.

19 Repeat the command.

20 Select point B on the edge of the surface, shown in Figure 11–4. The radius of curvature of the point on the surface's edge is displayed.

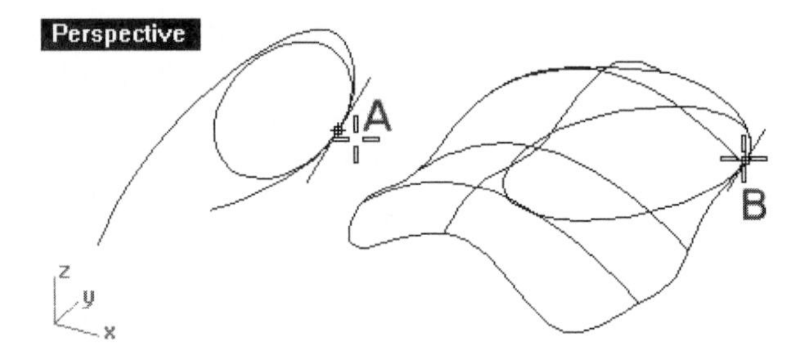

Figure 11–4. Radius of curvature being evaluated

Direction

Curves have a direction, from one end point to the other end point or vice versa. Direction of curve may have an effect on the final outcome of the constructed surfaces.

Surfaces have three directions: U, V, and N. The U and V directions are two orthogonal directions along the surface and N direction is the normal direction. You can change the U, V, and N direction and swap the U and V directions.

Although polygon mesh does not have U and V directions, it has N direction. Continue with the following steps.

21 Turn on Layer02.

22 Select Analyze > Direction, or click on the Analyze Direction/Flip Direction button on the Analyze toolbar.

23 Select curve A (shown in Figure 11–5). The direction of the curve is displayed.

24 Click to change the direction.

25. Press the ENTER key to terminate the command.
26. Select Analyze > Direction, or click on the Direction button on the Analyze toolbar.
27. Select surface B. The surface's U, V, and N direction arrows are displayed, as shown in Figure 11–5.
28. Click to change the normal direction.
29. Select the UReverse option to reverse the U direction, select the VReverse option to reverse the V direction, or select SwapUV option to swap U and V.
30. Press the ENTER key to terminate the command.
31. Select Analyze > Direction, or click on the Analyze Direction button on the Analyze toolbar.
32. Select polygon mesh C. The normal direction of the polygon mesh is displayed.
33. Click to change the normal direction.
34. Press the ENTER key to terminate the command.
35. Do not save your file.

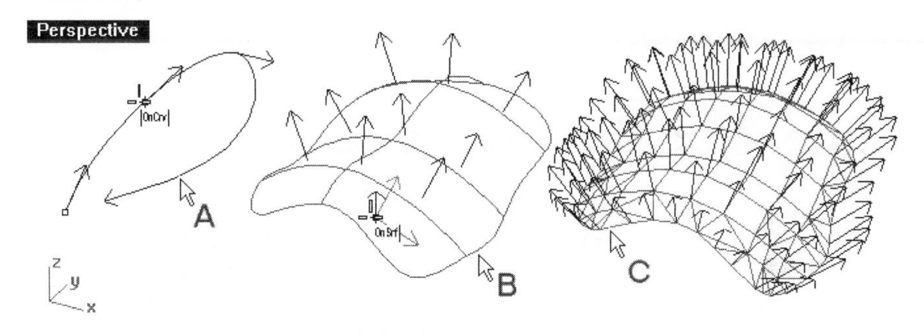

Figure 11–5. Directions being displayed and manipulated

Geometric Continuity

To discern the continuity of a joint in a curve, perform the following steps:

1. Select File > Open and select the file general02.3dm from the Chapter 11 folder on the companion CD.
2. Select Analyze > Curve > Geometric Continuity, or click on the Geometric Continuity of 2 Curves button on the Analyze toolbar.
3. Select curves A and B, shown in Figure 11–6. The geometric continuity is displayed.

Deviation Between Curves

If you have two curves that closely resemble each other, you can find out their deviation by performing the following steps.

4 Select Analyze > Curve > Deviation, or click on the Analyze Curve Deviation button on the Analyze toolbar.
5 Select curves A and C (shown in Figure 11–6). A report is displayed.
6 Do not save your file.

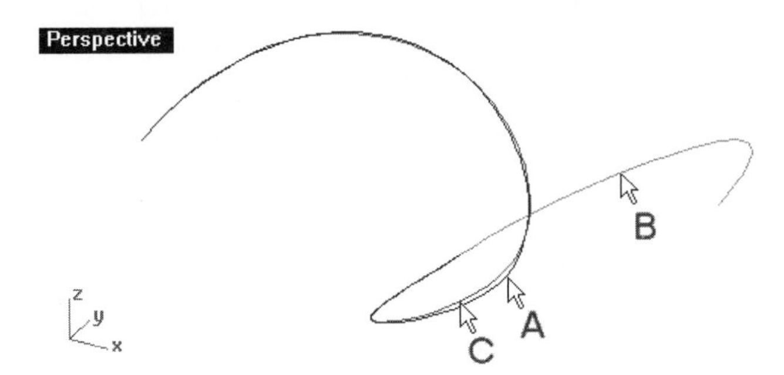

Figure 11–6. Geometric continuity of curves being analyzed

Dimensional Analysis Tools

You can find out the coordinates of a selected point, the bounding box of an object, the length of a curve or an edge of a surface, the distance between two selected points, and the angle between two selected lines.

Evaluate Point

It is sometimes necessary to find out the coordinates of a selected point. Perform the following steps:

1 Select File > Open and select the file Dimensional.3dm from the Chapter 11 folder on the companion CD.
2 Check the End box in the Osnap dialog box.
3 Select Analyze > Point, or click on the Evaluate Point button on the Analyze toolbar.
4 Select endpoint A, shown in Figure 11–7. The coordinates of the selected location are displayed in the command line area.

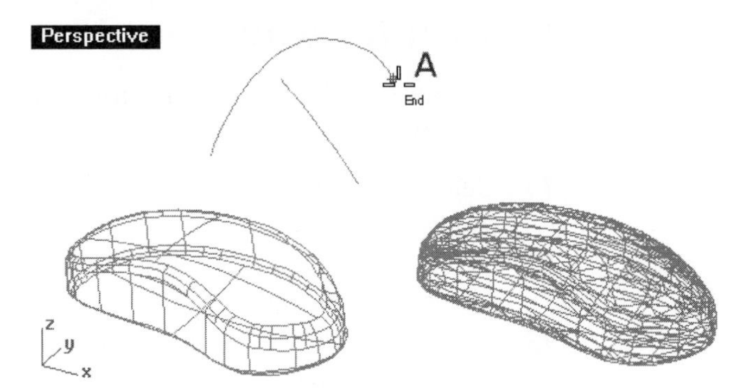

Figure 11–7. Coordinates of selected location being evaluated

Construction of a Bounding Box

A bounding box of a curve, surface, or polygon mesh is the minimum size of a rectangular box that encloses the entire curve, surface, or polygon mesh. Continue with the following steps.

5 Select Analyze > Bounding Box, or click on the Bounding Box/ Bounding Box, CPlane Orientation button on the Box toolbar.

6 Select curves A and B, shown in Figure 11–8, and press the Enter key twice. A bounding box enclosing the selected objects is constructed, and the box's dimensions are displayed at the command area.

7 Repeat the command.

8 Select surface C (Figure 11–8) and press the Enter key twice.

9 Repeat the command.

10 Select polygon mesh D (Figure 11–8) and press the Enter key twice.

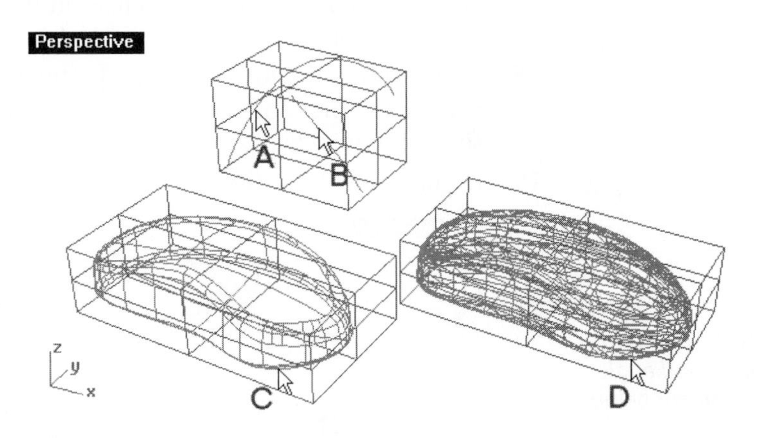

Figure 11–8. Bonding boxes constructed

Length Measurement

You can measure the length of a curve or the length of an edge of a surface. Continue with the following steps.

11 Select Analyze > Length, or click on the Length button on the Analyze toolbar.

12 Select curve A (shown in Figure 11–9). The length of the selected curve is displayed

13 Select Analyze > Length, or click on the Length button on the Analyze toolbar.

14 Select surface edge B (Figure 11–9). The length of the selected surface edge is displayed.

Distance Measurement

You can measure the distance between two selected points. Continue with the following steps.

15 Select Analyze > Distance, or click on the Distance button on the Analyze toolbar.

16 Select points C and D (Figure 11–9). The distance between the selected points is displayed.

Angle Measurement

You can measure the angle between two lines defined by four selected points. Continue with the following steps.

17 Select Analyze > Angle, or click on the Angle button on the Analyze toolbar.

18 Select points E and F and then points C and D (Figure 11–9) to specify the first and second lines. The angle between the first and second line is displayed.

19 Do not save your file.

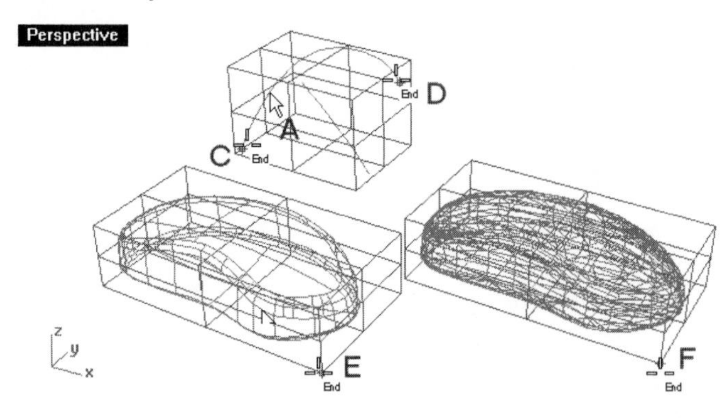

Figure 11–9. Length, distance, and angle measurement being taken

Surface Analysis Tools

Because a NURBS surface is a smooth surface with no facets, the profile and silhouette of a NURBS surface are displayed using U and V isocurves. To gain a visual picture of the surface, you use curvature rendering, draft angle rendering, environmental map rendering, and zebra stripes rendering and then construct curvature arcs.

In regard to individual points on a surface, you construct a point on a surface by specifying the U and V coordinates of the point. Given a point on the surface, you evaluate its U and V coordinates. On the other hand, you evaluate the deviation of any point from a surface. You can also evaluate a solid's thickness.

Curvature Rendering

You can display a color rendering to indicate the curvature values of a surface. In the rendering, different curvature values are represented by different colors. Perform the following steps:

1. Select File > Open and select the file Profile.3dm from the Chapter 11 folder on the companion CD.
2. Select Analyze > Surface > Curvature Analysis, or click on the Curvature Analysis/Curvature Analysis Off button on the Surface Analyze toolbar.
3. Select the surface and press the Enter key.
4. In the Curvature dialog box, select Mean in the Style pull-down list box. A color rendering showing the curvature values in various colors is displayed, as shown in Figure 11–10.
5. Close the Curvature dialog box.

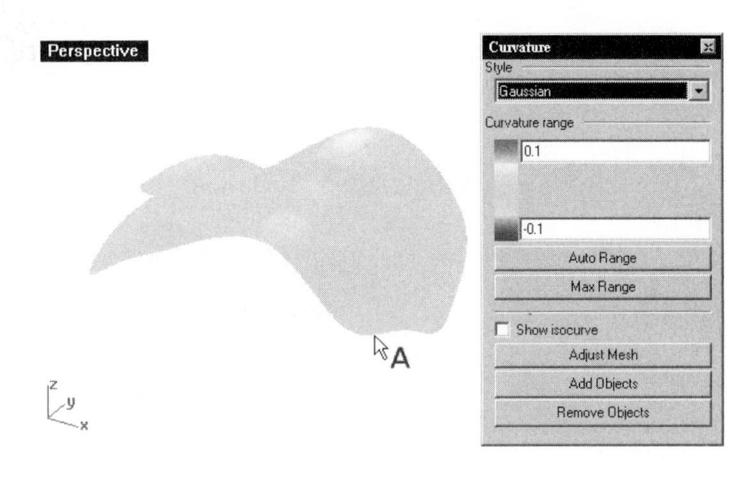

Figure 11–10. Curvature values displayed in various colors

Draft Angle Rendering

Draft angles are an essential consideration in mold and die design. You can display a color rendering to illustrate the draft angle value of a surface or a polygon mesh with respect to the construction plane. Continue with the following steps.

6 Select Analyze > Surface > Draft Angle Analysis, or click on the Draft Angle Analysis/Draft Angle Analysis Off button on the Surface Analyze toolbar.

7 Select the surface and press the Enter key. Draft angle values on the surface are displayed in various colors, as shown in Figure 11–11.

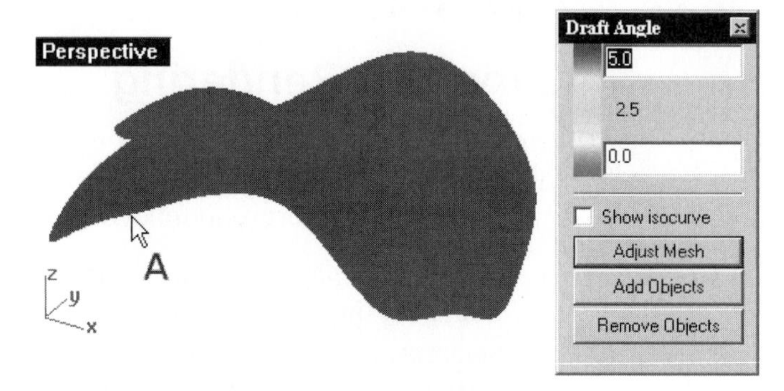

Figure 11–11. Draft angle values displayed in various colors

Environment Map Rendering

To easily inspect the smoothness of a surface or a polygon mesh, you place a bitmap image on the surface. Continue with the following steps.

8 Select Analyze > Surface > Environment Map, or click on the Environment Map/Environment Map Off button on the Surface Analyze toolbar.

9 Select the surface and press the Enter key.

10 In the Environment Map Options dialog box, select an image. The selected image is mapped onto the surface, as shown in Figure 11–12.

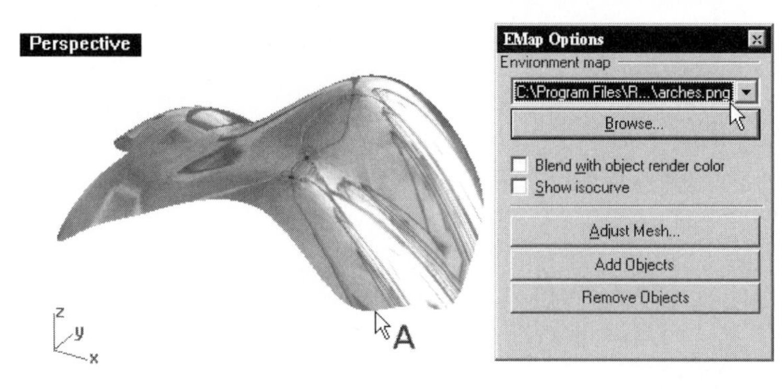

Figure 11–12. Bitmap mapped on surface

Zebra Stripes Rendering

To effectively view the smoothness of a surface or a polygon mesh, you use a set of zebra stripes instead of a bitmap. Continue with the following steps.

11 Select Analyze > Surface > Zebra, or click on the Zebra button/ Zebra Analysis Off button on the Surface Analysis toolbar.

12 Select the surface and press the Enter key.

13 In the Zebra Options dialog box, set the stripe direction and stripe size. Zebra stripes are placed on the surface, as shown in Figure 11–13. This is most valuable for checking the continuity between the surfaces of polysurfaces.

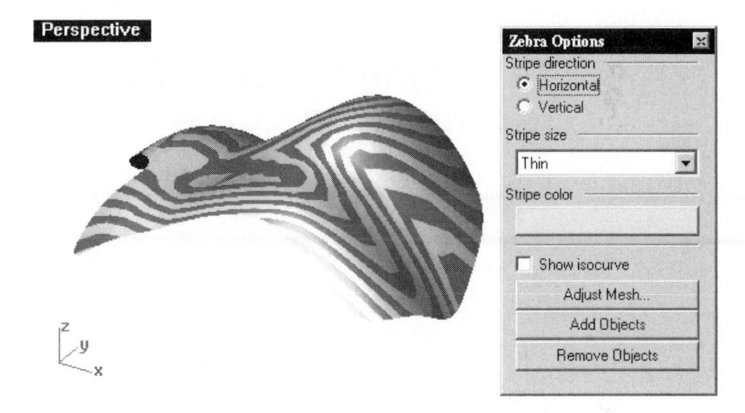

Figure 11–13. Zebra stripes

Points on a Surface

A surface has two sets of isocurves in two directions, U and V. Correspondingly, there are two axis directions, U and V. You can construct a point on a surface by specifying the point's U and V coordinate values. Continue with the following steps.

14 Select Analyze > Surface > Point from UV Coordinates, or click on the Point from UV Coordinates/UV Coordinates of a Point button on the Surface Analyze toolbar.

15 Select surface A and press the Enter key.

16 Type 20 to specify the U value.

17 Type 10 to specify the V value. A point is constructed on the surface, as shown in Figure 11–14.

Evaluating the U and V Coordinates of a Point

Contrary to constructing a point on a surface by specifying the U and V coordinates, you can evaluate the U and V coordinates of a point on a surface. Continue with the following steps.

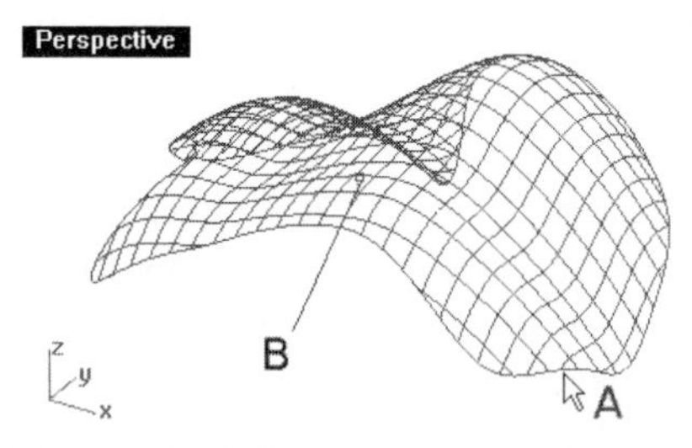

Figure 11–14. Point constructed on the surface

18 Select Analyze > Surface > UV Coordinates of a Point, or right-click on the Point from UV Coordinates/ UV Coordinates of a Point button on the Surface Analyze toolbar.

19 Select surface A.

20 Select point B (shown in Figure 11–14) and press the Enter key. The U and V coordinates of the selected point on the surface are displayed at the command area.

Checking the Deviation of a Point from a Surface or a Curve

You can find out the deviation of a point or set of points from a surface. Perform the following steps:

1. Open the file AnalyzeSurface.3dm from the Chapter 11 folder on the companion CD.
2. Select Analyze > Surface > Point Set Deviation, or click on the Point Set Deviation button on the Surface Analyze toolbar.
3. Select point A (shown in Figure 11–15) and press the Enter key.
4. Select surface B (shown in Figure 11–15) and press the Enter key.
5. Do not save your file.

A Point/Surface Deviation dialog box is displayed. The information is useful in identifying the deviation of a point or a set of point clouds and the surface constructed from the point objects.

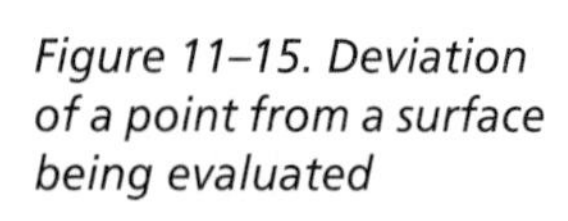
Figure 11–15. Deviation of a point from a surface being evaluated

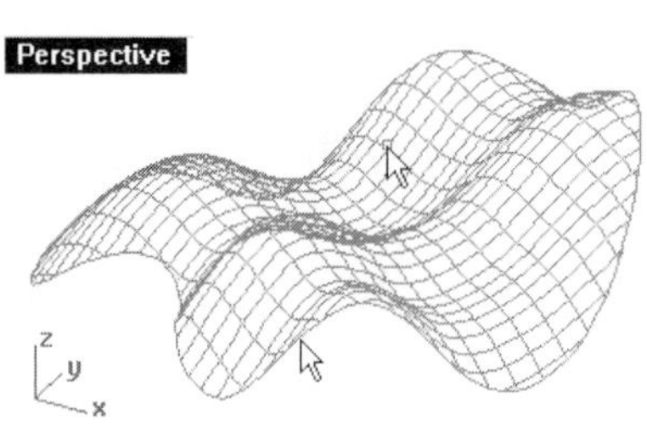

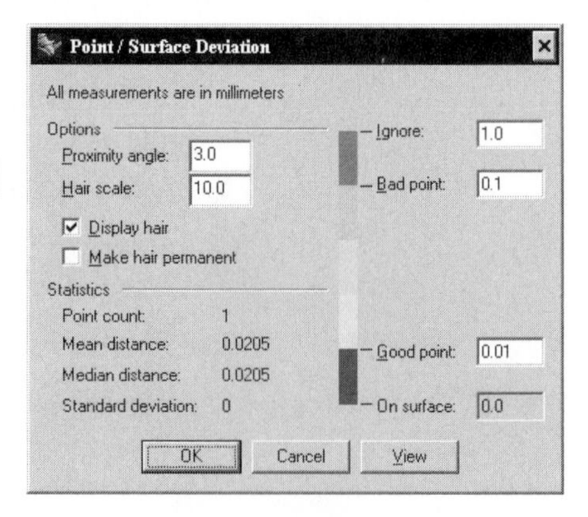

Thickness Evaluation

If you have a solid with thickness, you can use the thickness evaluation tool to find out the variation of thickness, as follows:

1. Open the file Thickeness.3dm from the Chapter 11 folder on the companion CD.
2. Click on the Thickness Analysis/Thickness Analysis Off button on the Surface Analysis toolbar.
3. Select object A, shown in Figure 11–16, and press the ENTER key. Variation in thickness is displayed, as shown.
4. Do not save your file.

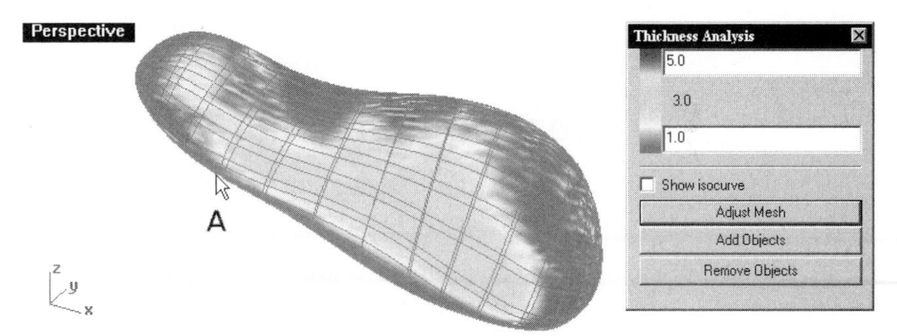

Figure 11–16. Thickness analysis

Mass Properties Tools

You can measure the area of a surface or a polygon mesh and the area centroid and area moments of a surface. Volume measure concerns objects that are closed, such as a solid or a polygon mesh that closes a volume without any openings. For such objects, you can measure their volumes. For solids, you can measure their volume centroid, volume moments, and hydrostatics.

Area Inquiry

You can determine the area of a surface or a polygon mesh. Perform the following steps:

1. Select File > Open and select the file Area.3dm from the Chapter 11 folder on the companion CD.
2. Select Analyze > Mass Properties > Area, or click on the Area button on the Mass Properties toolbar.

3. Select surface A (shown in Figure 11–17) and press the Enter key.
4. Select Analyze > Mass Properties > Area, or click on the Area button on the Mass Properties toolbar.
5. Select polygon mesh B (Figure 11–17) and press the Enter key.

Area Centroid Inquiry

You can determine the area centroid of a selected surface. Continue with the following steps.

6. Select Analyze > Mass Properties > Area Centroid, or click on the Area Centroid button on the Mass Properties toolbar.
7. Select surface A (Figure 11–17) and press the Enter key. The area centroid is displayed and a point (C) is constructed to indicate the location of the centroid.

Area Moments Inquiry

You can determine the area moments of a selected surface. Continue with the following steps.

8. Select Analyze > Mass Properties > Area Moments, or click on the Area Moments button on the Mass Properties toolbar.
9. Select surface A (Figure 11–17) and press the Enter key. Area moment information is displayed in the Area Moments dialog box, shown in Figure 11–17. You can save the information to a file and then click on the Close button to close the dialog box.
10. Do not save your file.

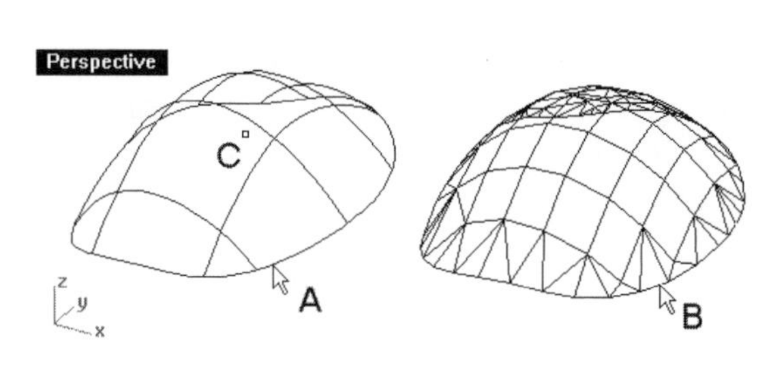

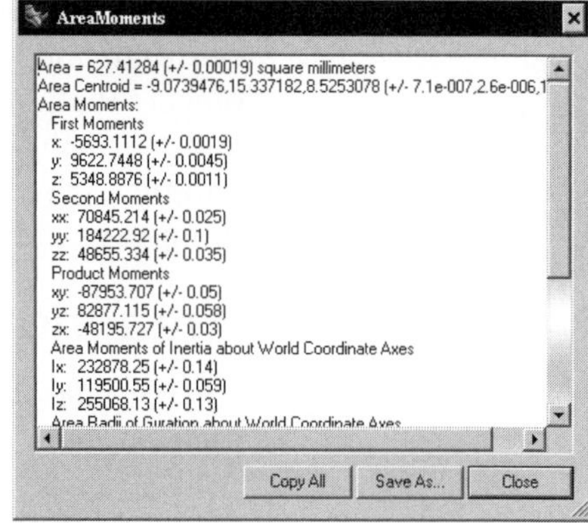

Figure 11–17. Area properties inquiry

Volume Inquiry

You can determine the volume of a solid (a closed polysurface) or a closed polygon mesh. Perform the following steps:

1. Select File > Open and select the file Volume.3dm from the Chapter 11 folder on the companion CD.
2. Select Analyze > Mass Properties > Volume, or click on the Volume button on the Mass Properties toolbar.
3. Select solid A (shown in Figure 11–18) and press the Enter key. Volume information of the solid is displayed in the command area.
4. Select Analyze > Mass Properties > Volume, or click on the Volume button on the Mass Properties toolbar.
5. Select polygon mesh B (Figure 11–18) and press the Enter key. Volume information of the polygon mesh is displayed in the command area.

Volume Centroid Inquiry

You can determine the volume centroid of a solid. Continue with the following steps.

6. Select Analyze > Mass Properties > Volume Centroid, or click on the Volume Centroid button on the Mass Properties toolbar.
7. Select solid A (Figure 11–18) and press the Enter key. Volume centroid (C) information is displayed at the command area.

Volume Moments Inquiry

You can determine the volume moments of a solid. Continue with the following steps.

8. Select Analyze > Mass Properties > Volume Moments, or click on the Volume Moments button on the Mass Properties toolbar.
9. Select solid A (Figure 11–18) and press the Enter key. Volume moment information is displayed in the dialog box shown in Figure 11–18.

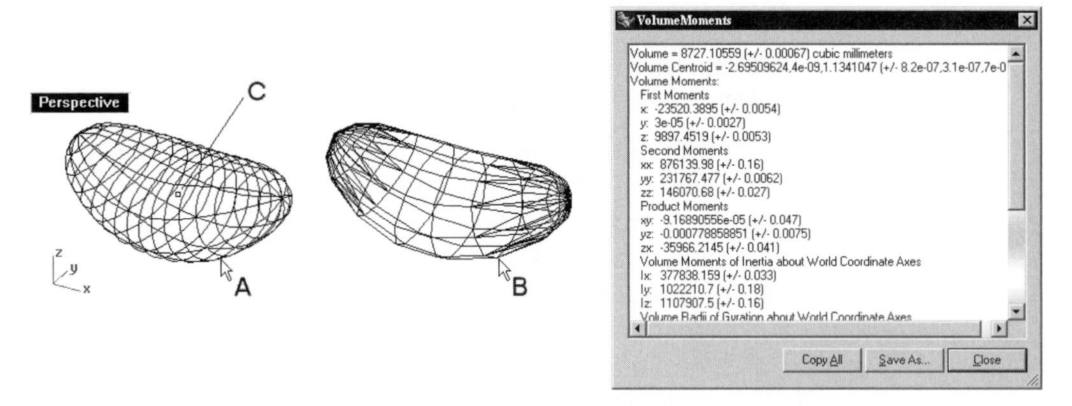

Figure 11–18. Volume analyze being taken

Cut Volume

You can evaluate the volume of intersection of a closed object and existing box as follows:

1. Select File > Open and select the file CutVolume.3dm from the Chapter 11 folder on the companion CD.
2. Type CutVolume at the command area.
3. Select object A (shown in Figure 11–19) and press the ENTER key.
4. Select box B. The cut volume is displayed at the command area.
5. Do not save your file.

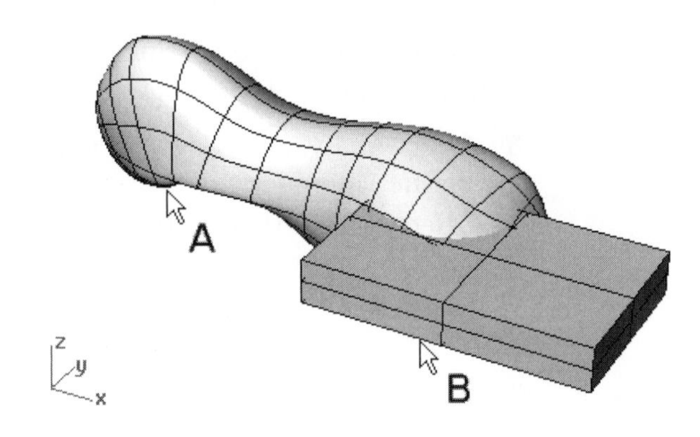

Figure 11–19. Closed object and a box

Hydrostatics

Assuming the waterline is set at the World coordinate plane, you can change the setting by typing *W* at the command area and specifying a new waterline. You can evaluate the hydrostatic value of a solid as follows.

6 Select Analyze > Mass Properties > Hydrostatic, or click on the Hydrostatics button on the Mass Properties toolbar.
7 Select the solid and press the Enter key. Hydrostatics information is displayed in the Hydrostatics dialog box, shown in Figure 11–20.
8 Do not save your file.

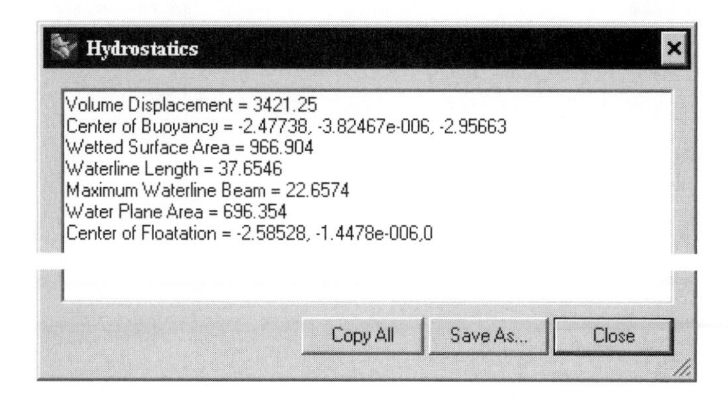

Figure 11–20. Hydrostatics dialog box

(***NOTE:*** *For professional marine application, it is advised to use RhinoMarine as a tool.*)

Diagnostics Tools

Diagnostics tools concern database listing, finding bad objects, and auditing.

Listing a Database

You can display the database of a selected object in list form. Perform the following steps:

1 Select File > Open and select the file Diagnostics.3dm from the Chapter 11 folder on the companion CD.

2. Select Analyze > Diagnostics > List, or click on the List Object Database button/What?! button on the Diagnostics toolbar.
3. Select surface A (shown in Figure 11–21) and press the Enter key. The database of the selected objected is displayed as a list, as shown in Figure 11–21.
4. Click on the Close button.

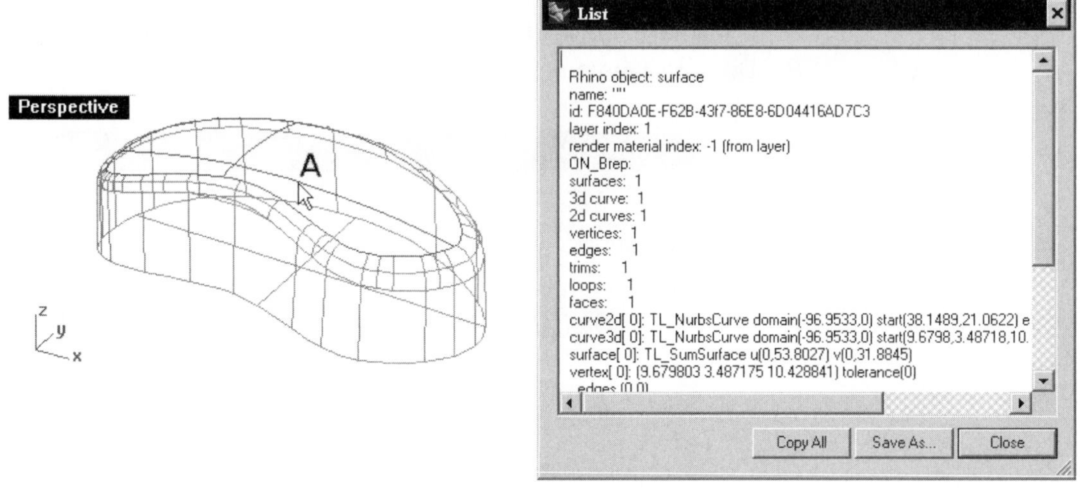

Figure 11–21. Curve data listed

Bad Object Report

You can check an object for errors and highlight erroneous objects. Continue with the following steps.

5. Select Analyze > Diagnostics > Check, or click on the Check Object button on the Diagnostics toolbar.
6. Select curve A (shown in Figure 11–21). The selected curve is checked.
7. Click on the Close button.

To check the entire file for bad objects, continue with the following.

8. Select Analyze > Diagnostics > Select Bad Objects, or click on the Bad Objects button on the Analyze toolbar. A report is displayed.

Auditing

You can audit the current Rhino file or other Rhino files.

- ❐ To audit the current file, select Analyze > Diagnostics > Audit, or click on the Audit button on the Diagnostics toolbar.
- ❐ To audit a different Rhino file, select Analyze > Diagnostics > Audit 3DM Files, or click on the Audit 3dm file button on the Diagnostics toolbar

Checking New Objects

Before inserting objects or files into a Rhino file, you can set Check-NewObjects to On in order to check the errors of any new objects.

- ❐ Right-click on the Check Objects/Check All New Objects button on the Diagnostic toolbar, or type CheckNewObjects at the command area and set it to Yes.

Chapter Summary

You can evaluate the profile of a curve or surface by displaying curvature graph, curvature circle, and curvature radius. You can manipulate the directions of a curve or a surface. You can also determine the continuity of contiguous curves or surfaces and the deviation between two curves. To perform dimensional analysis, you can evaluate a point's coordinates, an object's bounding box, the length of a curve or an edge of a surface, the distance between two points, and the angle between two lines.

To analyze a surface's profile, you can display a curvature rendering, draft angle rendering, environmental map rendering, and zebra stripe rendering. You can construct a point on a surface by specifying its U and V coordinates and determine the coordinates of a point on a surface in terms of the U and V coordinates. A point's deviation from a surface can also be determined.

In terms of mass properties, you can evaluate a surface's area, volume, and hydrostatics value. In addition, you can display an object's database list, report bad objects in a file, and perform file auditing.

Review Questions

1. What is the difference between displaying a database list and reporting bad objects in a file?
2. List the methods you might use to analyze a NURBS surface.

CHAPTER 12

Group, Block, and Work Session

Introduction

This chapter covers group, block, and work session and suggests how these tools can help enhance object management, enable assembly simulation, and facilitate design collaboration.

Objectives

After studying this chapter you should be able to:

- ❒ Organize Rhino objects into data groups
- ❒ Define and manipulate data blocks and use block definitions to help construct an assembly of objects
- ❒ Carry out design collaboration via the work session manager

Overview

In Rhino, a group is a collection of geometric objects put together so that selection on any part of the objects causes the entire set of objects to be selected. A block is a data definition for a set of objects residing within a file's database and optionally referencing to an external file. To simulate an assembly of components in the computer, you may consider using block definitions. To handle a large project, you can divide it into a number of smaller files and organize them in a work session. Work sessions can help facilitate design collaboration.

Group

To help select a set of objects (curves, surfaces, and/or polygon meshes) repeatedly in a file, you can think about organizing them into groups. After putting objects in a group, selecting any of them will automatically cause the entire group of objects to be selected. However, you may still select objects individually by holding down the Control and Shift keys while clicking on them.

Constructing Groups

Group construction is very simple. You use the group command and select objects that you want to group together. Perform the following steps:

1. Select File > Open and select the file Group.3dm from the Chapter 12 folder on the companion CD.
2. Select Edit > Groups > Group, or click on the Group button on the Grouping toolbar.
3. Select objects A, B, and C, shown in Figure 12–1, and press the ENTER key. Selected objects are put in a group.

Adding and Removing Group Elements

After a group is constructed, you may add new group elements to the group elements as well as remove existing group elements from the group. Continue with the following steps.

4. Click on the Add to Group button on the Grouping toolbar.
5. Select curve D, shown in Figure 12–1, and press the ENTER key.
6. Select object C (a member of a group already constructed). Curve D is added to the group.
7. Click on the Remove from Group button on the Grouping toolbar.
8. Select curve A, shown in Figure 12–1, and press the ENTER key. The selected object is removed from the group.

Assigning Group Names

Groups can be named for easy identification in case there are too many groups in a file. Continue with the following steps.

9. Select Edit > Groups > Set Group Name, or click on the Set Group Name button on the Grouping toolbar.
10. Select object A, shown in Figure 12–1, and press the ENTER key.
11. Type Group_A at the command area. The selected group is assigned a group name.

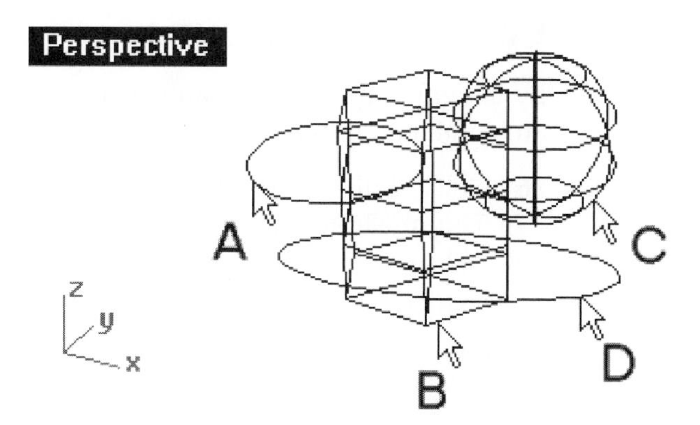

Figure 12–1. Objects being put in a group, added to a group, and removed from a group and group being assigned a group name

To appreciate how group names can help identify different groups in a file, continue with the following steps.

12 Turn on Layer 01. Objects constructed in this layer are already put in a group.

13 Select Transform > Move, or click on the Move button on the Transform toolbar.

14 Click on location A, shown in Figure 12–2. A pop-up menu is displayed. You may select one of the two groups.

15 Press the ESC key to cancel the command. (We do intend to move the objects.)

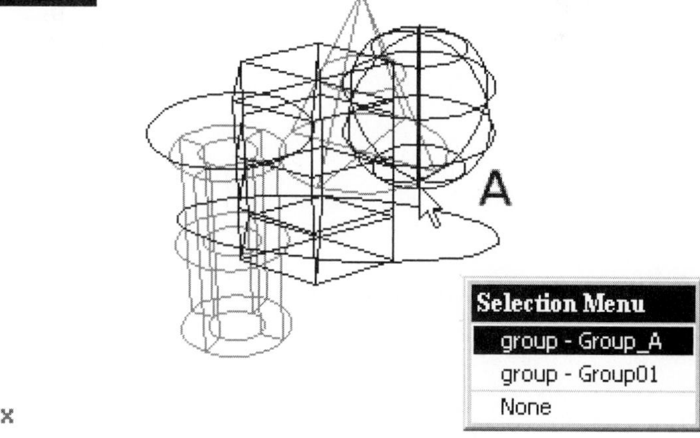

Figure 12–2. Group names displayed

Combining Groups

By assigning the name of an existing group to a new group, two groups are combined into one.

16 Select Edit > Groups > Set Group Name, or click on the Set Group Name button on the Grouping toolbar.

17 Select A, shown in Figure 12–3, and press the ENTER key.

18 Type Group_A at the command area. Because "Group_A" is already assigned to an existing group, the selected objects/groups are combined with Group_A.

Ungrouping

Grouped objects can be ungrouped, thus reverting back to individual objects.

19 Select Edit > Groups > Ungroup, or click on the Ungroup button on the Grouping toolbar.

20 Select A, shown in Figure 12–3, and press the ENTER key.

21 Do not save your file.

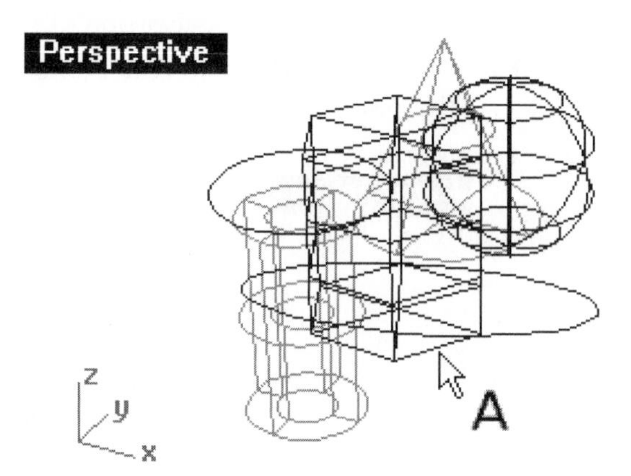

Figure 12–3. Groups being combined and ungrouped

Block

Think about a Rhino file withmany identical copies of geometric objects, including mirrored objects and scaled objects. Although each set of geometric objects may take up very little computer memory space, a hundred sets of such copies will increase the file size significantly. Now think about another scenario. You constructed a set of geometric objects in a Rhino file, and you have to repeat the same set of objects in another Rhino file. Both cases require proper management of the geometric objects.

Block and Block Instances

One way to save memory space in a Rhino file with many repeated objects is to define a block definition from the objects, give the block a name, and insert the block at designated locations. We call each insertion of the block an instance of the block definition. Even though there may be many instances of the blocks in your Rhino file, they take up very little memory space because each insertion instance only stores the information about the name of the block and the block's location, scale, and orientation. Using blocks, you can repeat a set of objects of a drawing in another drawing file as well.

Defining a Block from Existing Objects

A block is a set of objects residing in the Rhino file's database. In the block, you can include curves, surfaces, and/or polygon meshes. Blocks can be defined in two ways: by selecting a set of objects from within a Rhino file and by inserting a Rhino file. In the first method, you select a set of geometric objects from within a file, put them in a block, assign an insertion base point, and specify a block name.

1 Select File > Open and select the file Block.3dm from the Chapter 12 folder on the companion CD.

In the file, you will find two objects, a cup and a pyramid. You will now examine their properties.

2 Select Edit > Object Properties, or click on the Object Properties button on the Standard toolbar.

3 Select A, shown in Figure 12–4, and then click on the Details button of the Properties dialog box. This will display the Object Description dialog box. As can be seen, this is a polysurface.

4 Click on the Close button, select B, and click on the Details button of the Properties dialog box. As shown, this is a block definition. This block definition is predefined in this file.

5 Close the dialog boxes.

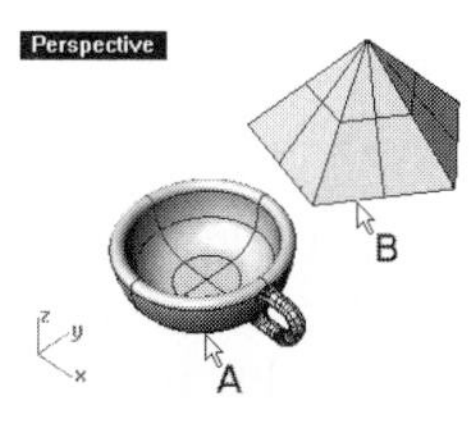

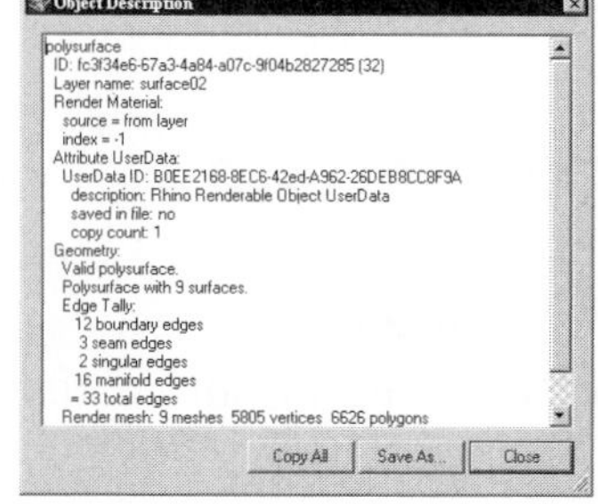

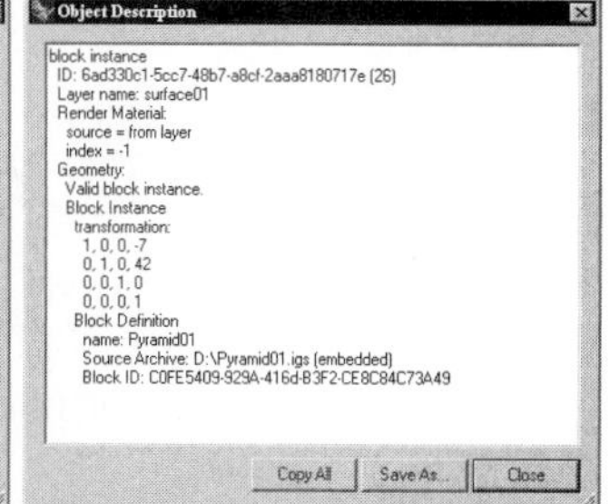

Figure 12–4. From left to right: Two objects in the file and Object Description of the objects.

6. Set object snap mode to End.
7. Select Edit > Blocks > Create Block Definition, or click on the Block Definition button on the Block toolbar.
8. Select object A, shown in Figure 12–5, and press the ENTER key.
9. Select end point B as the insertion base point.

   (***NOTE:*** *Insertion base point is the reference point when you insert the block.*)

10. In the Block Definition Properties dialog box, assign a block name "Cup."
11. Type www.rhino3d.com in the URL field of the dialog box.
12. Click on the OK button.

A block is constructed from the selected object. Although there is no visual change in the display, what you see now is an instance of the block. You can verify that this is a block definition by repeating steps 2 and 3 above.

As a hyperlink is constructed, you can open the hyperlink by selecting Tools > Hyperlink > Open Hyperlink and selecting the block object.

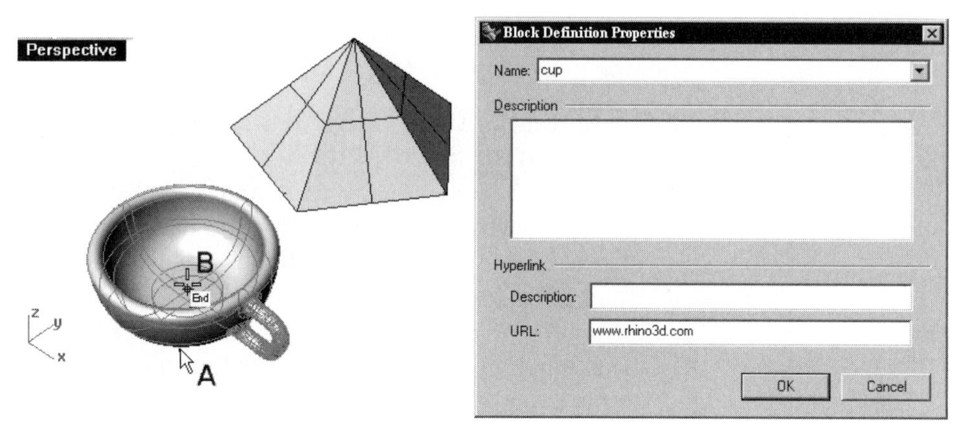

Figure 12–5. Block being constructed

Block Insertion

After a block definition is defined in a file, you can insert it as an instance. Optionally, you may specify a scale factor, which can be uniform or nonuniform, and a rotation angle. Being an instance of the block, objects will be referenced to the block definition. Changes in the block definition will cause automatic update of the block instance. Continue with the following steps to insert a block instance.

13 Select Edit > Blocks > Insert Block Instance, or click on the Insert/Export with Origin button on the Block toolbar.

14 Select the block "Cup" from the Name pull-down list box.

15 In the Insertion Point area of the dialog box, check the Prompt box.

16 In the Scale area, check the Uniform box and type 0.8 in the X box.

17 In the Rotation area, type 45 in the Angle box.

18 Select Block Instance in the Insert As area, if it is not already selected.

As shown in the dialog box, the block can be inserted as three kinds of objects: Block instance, Group, or Individual Objects. If you click on the Group button, the objects defined in the block definition will be inserted and then grouped. If you click on the Individual Objects button, the objects are simply copied and pasted at the inserted location.

19 Click on the OK button.

20 Select location A, shown in Figure 12–6 (exact location is unimportant for the purpose of this tutorial). The selected block is inserted as an instance, as shown in Figure 12–7.

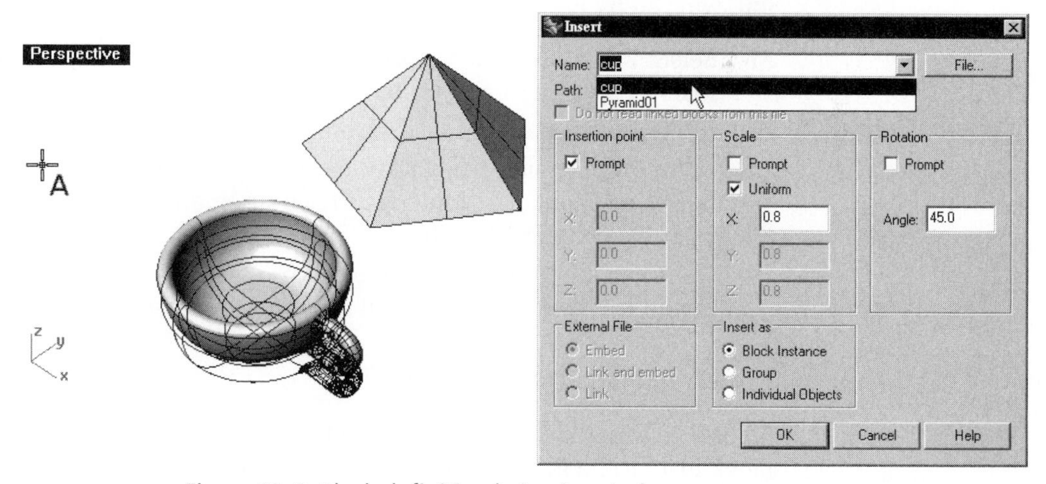

Figure 12–6. Block definition being inserted

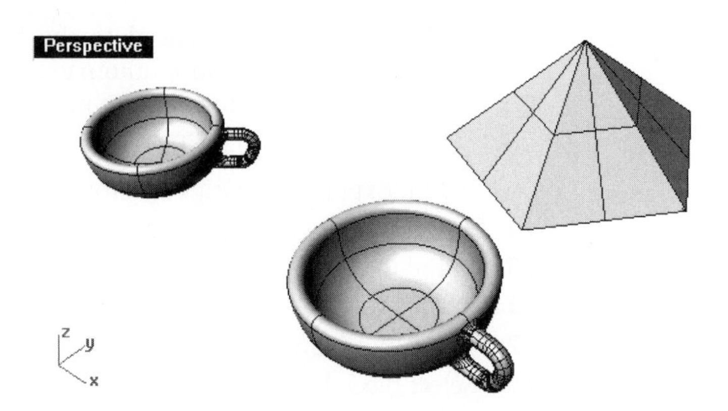

Figure 12–7. Block instance inserted

Inserting an External File

The second way to define a block definition is to insert an external file by using the same command for inserting a block instance from an internal block definition.

File Type

The type of file that you can insert is not restricted to Rhino files. You will learn more about file types in Chapter 13. Now continue with the following steps.

21 Select Edit > Blocks > Insert Block Instance, or click on the Insert/Export with Origin button on the Block toolbar.

22 In the Insert dialog box, click on the File button.

23 In the Import dialog box that opens, select the file Plate01.3dm from the Chapter 12 folder of the companion CD accompanying, and click on the OK button.

24 On returning to the Insert dialog box, check on the Block instance box in the Inserted As area, if it is not already checked.

Reference to the External File

Similar to inserting an internal block, the block can be inserted as a block instance, a group, or individual objects.

If the file is inserted as a block, there are three ways to treat the block definition:

- Embed
- Link and Embed
- Link

If the Embed option is selected, the file is first imported into the current file. The imported data is then used to construct a block definition, and an instance of the block is inserted. Although a block definition is already constructed in the active file from the external file, there is still a passive reference to the external file. You can always update the block via the Block Manager, described in the next section.

If the Link and Embed option is chosen, the inserted block data is always linked to the external file. Depending on the settings in the Block Manager, the inserted block can be set to update automatically if the external file changes.

Finally, if you choose the Link option, the data is only linked to the external file but not saved in the current file. Naturally, the current file size is much smaller because only the link information is saved. However, the update also depends on the settings in the Block Manager. The disadvantage of linking is that the external file must be in the location where it is linked every time you open the file with a linked block.

25 Click on the Link button, and then click the OK button. (See Figure 12–8.)

A block instance is inserted by linking to an external file. (See Figure 12–9.)

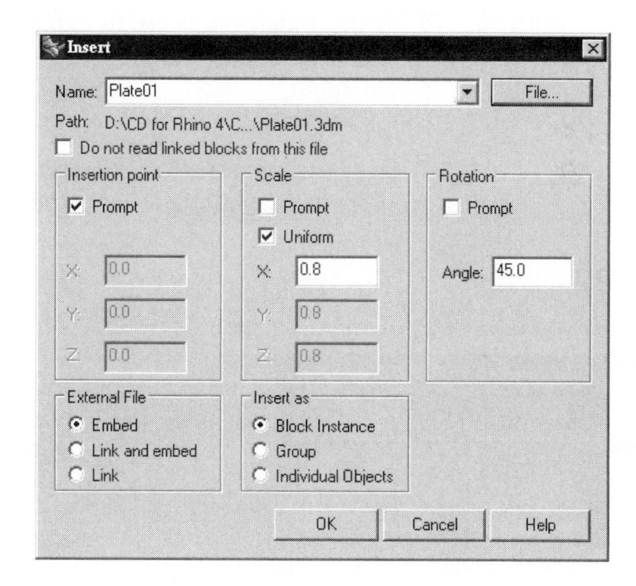

Figure 12–8. An external file being inserted

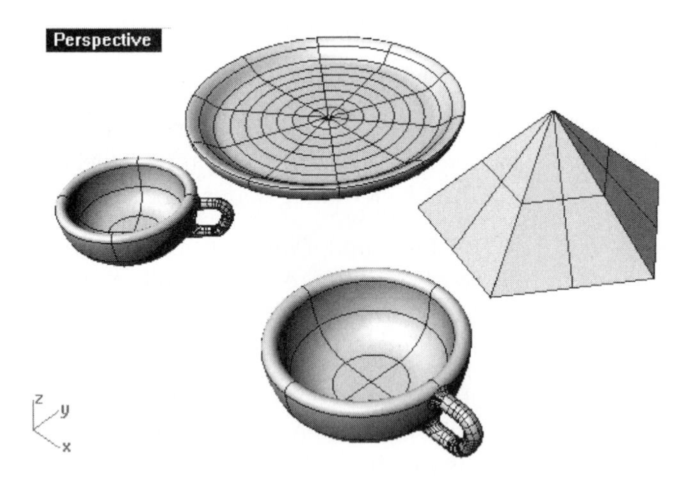

Figure 12–9. File imported and instance inserted

Block Manager

You can use the block manager to manage blocks in the following ways:

- You can modify a block's properties.
- You can export a block to form an individual file.
- You can delete a block from the computer's memory.
- If the block refers to an external file, you can update the link information so that the latest version of the imported file is reflected.
- If a block is nested in another block, you can determine the nesting information. (When a block is used as a drawing object in constructing another block, the block is said to be nested.)
- You can find out how many instances of the block are in the file.

Continue with the following steps.

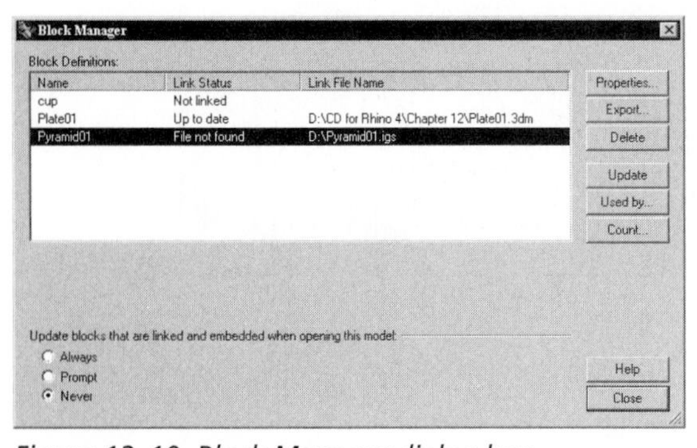

Figure 12–10. Block Manager dialog box

26 Select Edit > Blocks > Block Manager, or click on the Block Manager button on the Block toolbar. The Block Manager dialog box is displayed, as shown in Figure 12–10.

The lower-left sectionof the Block Manager dialog box includes three buttons:

- Always
- Prompt
- Never

If the Always button is checked, all linked external files will be updated automatically each time the file is opened. If the Prompt button is checked, you will be prompted to update when the external file is modified. If the Never button is checked, no action will be taken by the system.

As shown in the Block Manager dialog box, there are three block definitions in the file. The block definition "Cup" is not linked, because it is a block defined and saved in the file. The block definition "Plate01" is up to date, meaning that the block is defined from an external file and that the content of the block is referenced to the most up-to-date external file. Finally, the block definition "Pyramid01" has a status of "File Not Found," meaning that it was defined from an external file but the file does not exist anymore.

27 In the Block Manager dialog box, select the block "Pyramid01" and click on the Properties button.

28 In the Block Definition Properties dialog box, shown in Figure 12–11, click on the Browse button.

29 In the Open dialog box that follows, select the file Pyramid02.3dm from the Chapter 12 folder of the companion CD, and click on the OK button. The referenced file is changed.

30 Click on the OK button of the Block Definition Properties dialog box.

31 On returning to the Block Manager dialog box, click on the Update button.

32 Click on the OK button of the Rhinoceros 4 Update Block dialog box.

33 Click on the Close button of the Block Manager dialog box.

The unreferenced block definition is now referenced to the file Pyramid02.igs. (See Figure 12–12.)

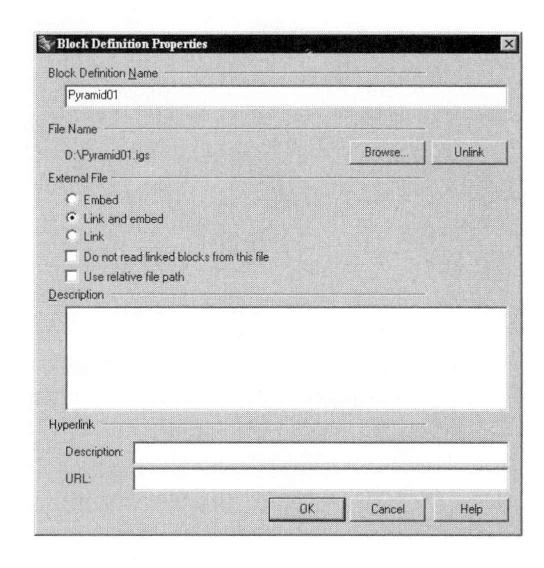

Figure 12–11. Block Definition Properties dialog box

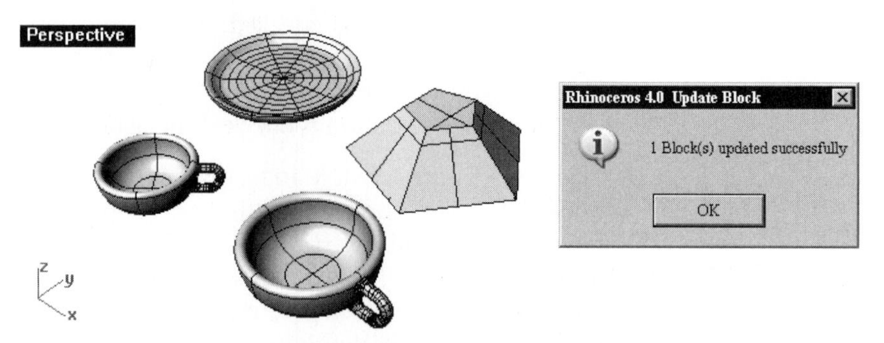

Figure 12–12. Unreference file changed and updated

Nested Block

When an instance of a block definition is used in the construction of another block definition, the block definition is said to be nested. Continue with the following steps:

34 Referencing Figure 12–13, drag an instance of the block definition "Cup" to a new location.

35 Select Edit > Blocks > Create Block Definition, or click on the Block Definition button on the Block toolbar.

36 Select A and B, shown in Figure 12–13, and press the ENTER key.

37 Select end point C.

38 In the Block Definition Properties dialog box, type TeaSet in the name box and click on the OK button. A block definition is constructed from two block instances.

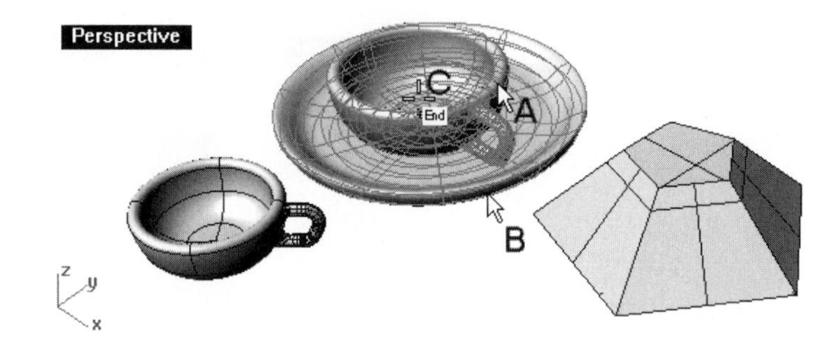

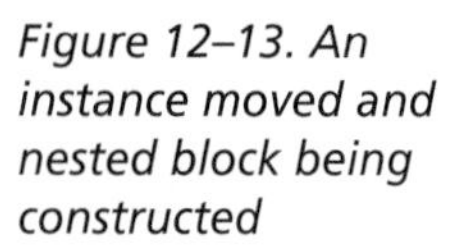

Figure 12–13. An instance moved and nested block being constructed

39 Select Edit > Blocks > Block Manager, or click on the Block Manager button on the Block toolbar.

40 Select the block definition "Cup" and click on the Used By button. The Nested Block Definition dialog box is displayed, showing that it is nested in the block definition "TeaSet."

41 Click on the Close button.

42 While the block definition "Cup" is still selected, click on the Count button. The Block Instance Count dialog box is displayed, indicating that there is one instance referenced to the block definition, as shown in Figure 12–14.

43 Click on the OK button.

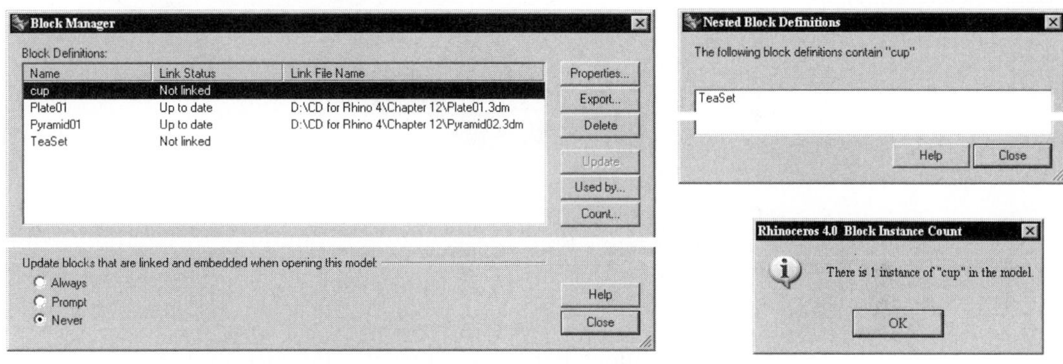

Figure 12–14. Nested block

Deleting Block Definition

If you delete a block definition via the Block Manager, all the instances and the block will be removed. However, if a block is nested, it cannot be deleted this way. Continue with the following steps:

44 Open the Block Manager dialog box, if it is closed.

45 Select the block definition "Pyrmid01" and click on the Delete button.

46 A warning dialog box is displayed. Click on the Yes button if you really want to delete the block definition and its instances.

47 Close the Block Manager dialog box. The block definition is removed.

Exploding a Block Instance

As mentioned, an internal block definition can only be inserted as a block instance, and an external file can be inserted as an instance as well as individual objects or grouped objects.

If you want to make a copy of the objects from an internal block definition so that the copied objects are independent of the block definition, you explode the block instance. After an instance is exploded, the instance becomes a set of objects without any reference to the block instance. Continue with the following steps:

48 Select Edit > Explode, or click on the Explode button on the Main2 toolbar.

49 Select A, shown in Figure 12–15, and press the ENTER key.

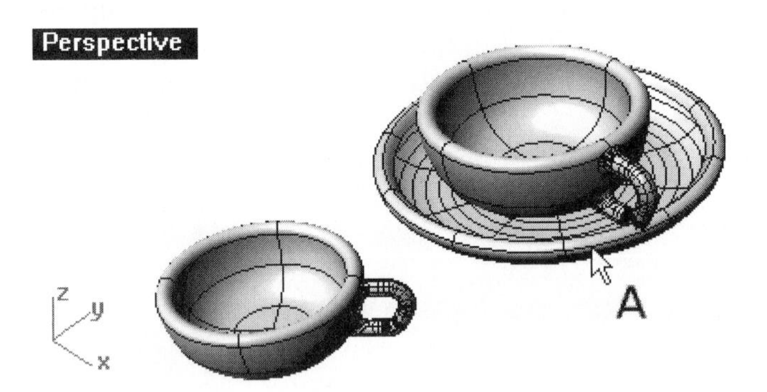

Figure 12–15. An instance being exploded

To appreciate what happens to the block definition, continue with the following steps:

50 Open the Block Manager dialog box.

51 Select the block definition "TeaSet" and click on the Count button. The Block Instance Count dialog box is displayed, indicating that there is 0 instance in the file.

52 Close the Block Instance Count dialog box and the Block Manager.

Purging Unreferenced Blocks

Because block definitions are residing in the memory space, even when the instances of the block are deleted or exploded, the block definitions are still in the file. If all the instances of a block definition are removed, either by deleting or exploding, the block is called an unreferenced block. You can remove unreferenced blocks by purging, as follows:

53 Select Tools > File Utilities > Purge Unused Blocks and Layers, or click on the Purge Used Block Definitions and layers Button on the Utilities toolbar.

54 Press the ENTER key. The unreferenced block is purged. If you open the Block Manager, you will find that the block definition "TeaSet" is removed.

Model Base Point

If a Rhino file is inserted into another file, the default base point is the origin. However, you can redefine the base point, as follows:

1 Open the Rhino file, which will be used as a block for insertion into another file.

2. Click on the Set Model Base Point button on the Block toolbar, or type ModelBasePoint at the command area.
3. Specify a point and save the file.

Although the objects in the file are changed, the specified base point will be used next time you insert this file.

Advantages of Insertion as Instances

The advantages of inserting as instances of the block rather than copying as individual objects are that file size is smaller (because block insertion does not actually duplicate the entities) and that change to a block can cause global change to all instances of the blocks (because instances refer to block definitions). You save a considerable amount of memory space when a block with numerous drawing objects is inserted many times in a drawing. In addition, redefining the block definition causes all the referenced instances to change as well.

Assembly Simulation

With the exception of very simple objects, such as a ruler, most objects have more than one component put together to form a useful whole. The set of components put together is called an assembly. There are three approaches to designing a product or a system. In the first approach, you start from making the individual components and work upward to build the assembly. In the second approach, you start from the assembly as a whole and work downward to construct individual components. In the third approach, you construct some components and work upward as well as work downward to construct individual components in the context of the assembly.

Components of an Assembly

For complex devices that have many parts, it is common practice to organize the parts into a number of smaller subassemblies such that each subassembly has fewer parts. Therefore, an assembly set may consist of a file depicting the assembly and a number of files to depict individual components, or an assembly file together with a number of subassembly files and component files. Collectively calling the individual parts or subassemblies components, you can define an assembly in the computer as a data set containing information about a collection of components linked to the assembly and how the components are assembled together. Figure 12–16 shows an assembly of a robot toy from a set of components residing on a number of individual files.

Figure 12–16. An assembly of components

The Bottom-Up Approach

The bottom-up approach is used when you already have a good idea on the size and shape of the components of an assembly or you are working as a team on an assembly. Using Rhino as a tool, you construct all the individual components in individual files. Then you start a new file to depict the assembly, insert the individual files into the assembly file as block references that linked to the original files, and orient the inserted instances as may be appropriate. Perform the following steps:

1. Select File > Open and select the file Robot.3dm from the Chapter 12 folder on the companion CD.
2. Select Edit > Blocks > Insert Block Instance, or click on the Insert/Export with Origin button on the Block toolbar.
3. Click on the File button on the Insert dialog box.
4. Select the file Robot-Body.3dm from the Chapter 12 folder of the companion CD.
5. Check the Prompt box in the Insertion point area.
6. In the Scale area, click on the Uniform button and set X to 1.
7. In the Rotation area, set the rotation angle to 0.

8 Click on the OK button and select location A (exact location is unimportant) shown in Figure 12–17.

A block definition is constructed from the external file and an instance of the block definition is inserted.

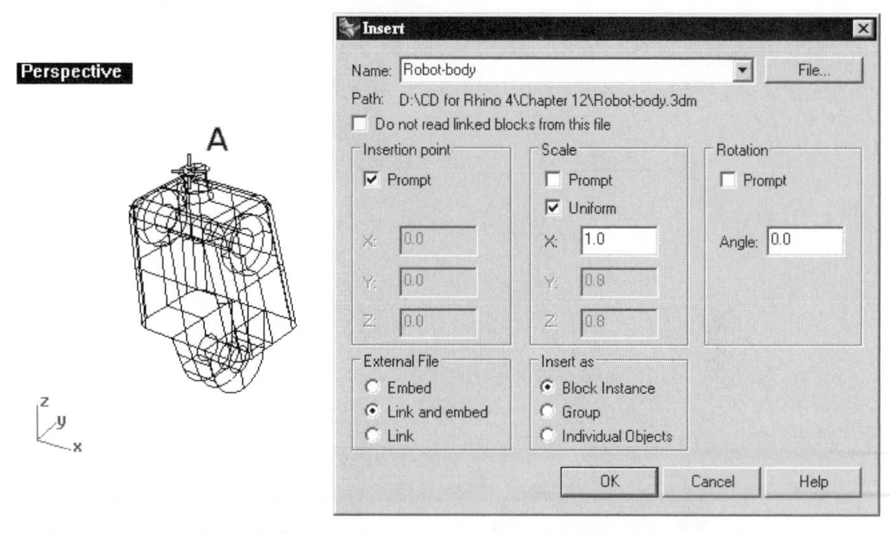

Figure 12–17. Block definition being constructed from an external file and an instance being inserted

9 Referencing Figure 12–18, insert the files Robot-Arm.3dm and Robot-Leg.3dm from the Chapter 12 folder of the accompanying CD, as shown in Figure 12–18.

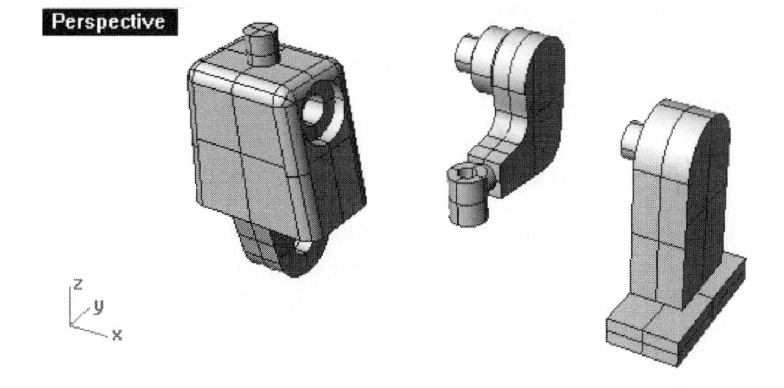

Figure 12–18. Robot-Arm and Robot-Leg inserted

10 Repeat the Insert command to insert the block definition Robot-Arm. While inserting, clear the Uniform box and set the X scale to -1, Y scale to 1, and Z scale to 1 in the Insert dialog box.

11 Repeat the command to insert the block definition Robot-Leg, with X scale = -1, Y scale = 1, and Z scale = 1.

Instances with X scale equal to -1 are inserted, as shown in Figure 12–19.

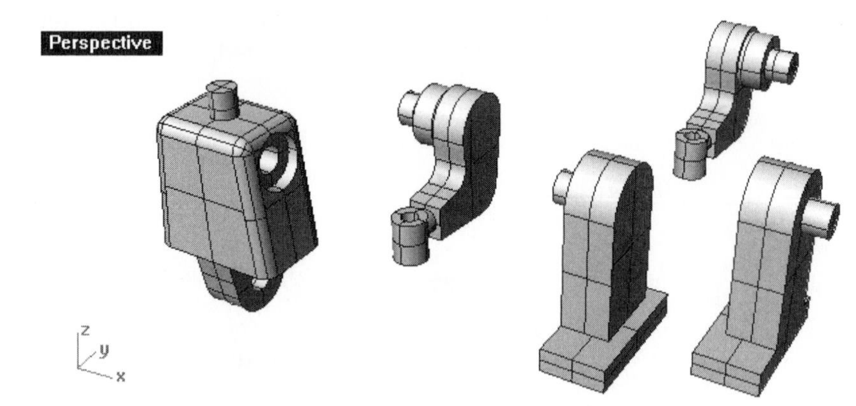

Figure 12–19. Instances with X scale equal to -1 inserted

Orient Objects

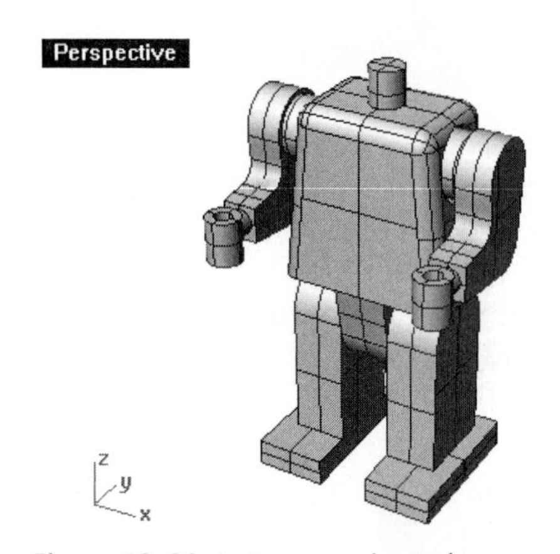

Figure 12–20. Instances oriented

After constructing three block definitions and five instances, refer to what you have learned in Chapter 10 to orient the instances, as shown in Figure 12–20.

Now you have constructed an assembly from three external files by inserting them into the current file as block definitions and instances of the block instances. Because the block definitions are constructed from external files, they are referenced to the external files in a passive way. Here passive means that the block definitions remember the source files, but an update has to be made explicitly and manually by clicking on the Update button of the Block Manager dialog box.

The Top-Down Approach

Sometimes you have a concept in your mind, but you have no concrete ideas about the component parts. You use the top-down approach—starting a Rhino file and designing some component parts there. From the preliminary component parts, you improvise. Upon finalizing the design, you construct block definitions for each of the components and export the block instances as external files. The main advantages in using this approach are that you see other parts while working on an individual part, and you can continuously switch from designing one part to another. Continue with the following steps to construct a component.

12 Set the current layer to Robot-Head. A curve is already constructed on this layer.

13 Referencing Figure 12–21, revolve curve A around an axis defined by BC. A revolved surface is constructed. This will become the head of the robot toy.

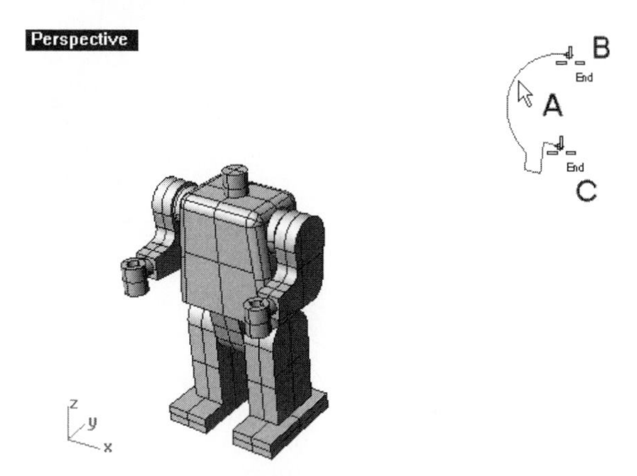

Figure 12–21. Robot-Head layer turned on and revolve surface being constructed

Defining a Block Definition

Now a component is constructed in the assembly file. If all the components are constructed this way, you are using the top-down approach. To put the component that you construct in a block definition, continue with the following steps.

14 Select Edit > Blocks > Create Block Definition, or click on the Block Definition button on the Block toolbar.

15 Select A (Figure 12–22) and press the ENTER key.

16 Select Center B.

17 Set the block definition name to Robot-Head.

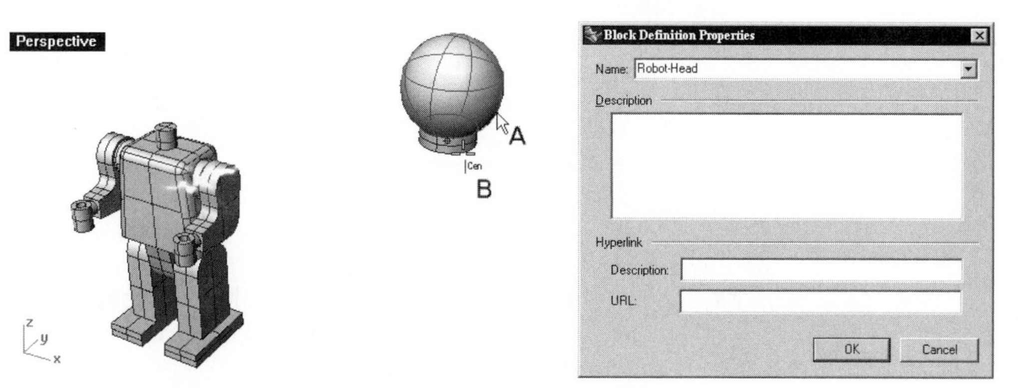

Figure 12–22. Block definition being constructed

18 Referencing Figure 12–23, orient the robot's head.

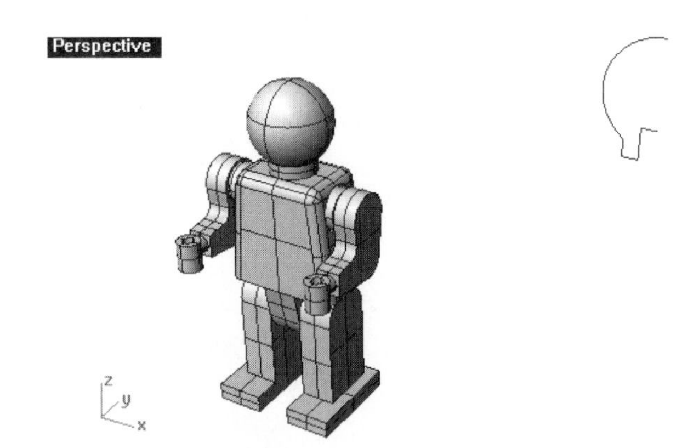

Figure 12–23. Robot's head oriented

Exporting and Referencing

To export the block definition to an external file and reference to the exported file, continue with the following steps:

19 Open the Block Manager dialog box.

20 Select the block definition "Robot-Head" and click on the Export button.

21 Save the file to a folder in your computer.

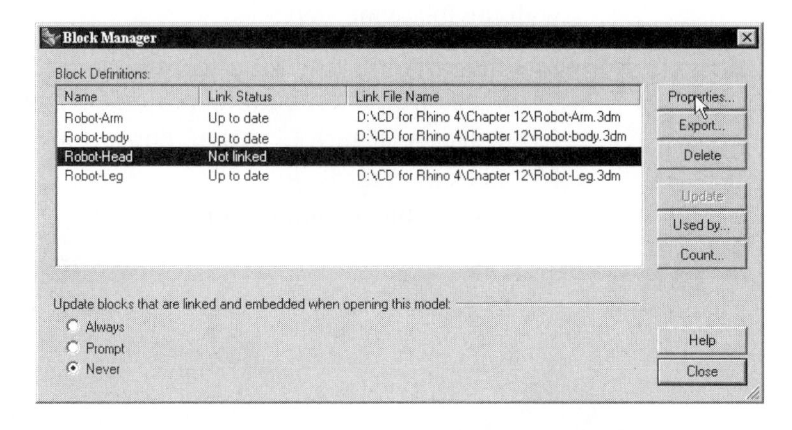

Figure 12–24. Internal block definition being exported

22 Select the block definition "Robot-Head" and click on the Properties button.

23 In the Block Definition Properties dialog box that follows, click on the Browse button and select the exported file to reference the block to the exported file.

24 Click on the OK button, and then click the Close button.

25 The assembly is complete. (Figure 12–16 shows the rendered image of the assembly.)

26 Do not save your file.

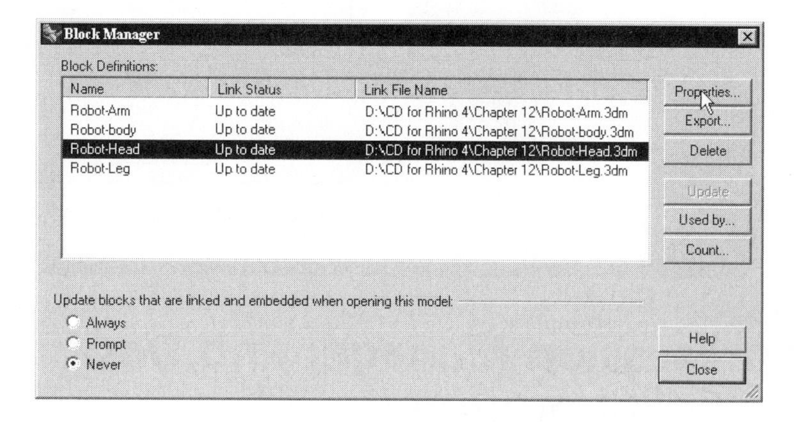

Figure 12–25. Block referenced to the external file

Hybrid Approach

Because the design you are now working on concerns making a component in the assembly as well as inserting components from external files, you are using a hybrid approach.

Named Position

In a file consisting of many objects, you can save the positions of selected objects and assign names to those arrangements, enabling you to store several arrangements of objects and recall them at will. Perform the following steps:

1 Select File > Open and select the file NamedPosition.3dm from the Chapter 12 folder on the companion CD.

2 Click on the Named Position button on the Move button, or type NamedPosition at the command area.

3 In the Named Position dialog box (Figure 12–26), two positions are already saved. Select Position 2, click on the Restore button and then click the OK button. The second saved position is restored.

4 Do not save your file.

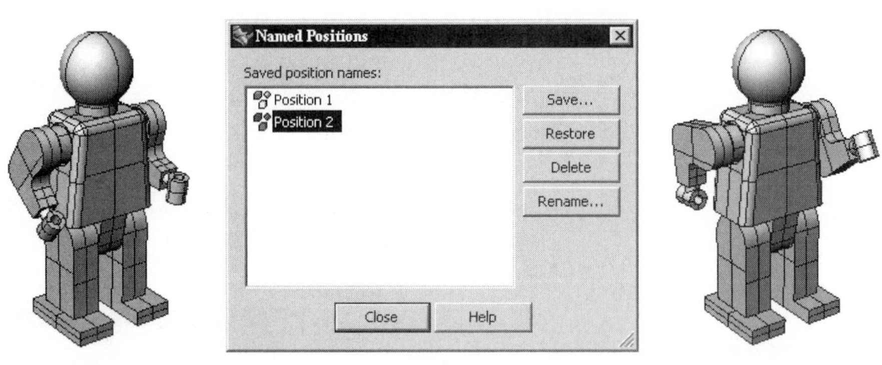

Figure 12–26. Two saved positions of objects and Named

Worksession Manager and Design Collaboration

In an industrial environment, it is a common practice to team up a group of designers to work collaboratively on a large project. In Rhino, a way to enable design work to be shared among the designers is to divide the project into a number of smaller projects. To work collaboratively, each designer works on and controls a file but can view other portions of the project.

To facilitate such design collaboration activity, you can use the Worksession Manager to perform four tasks: file attachment, file detachment, attached file activation, and refreshing attached files.

Working Directory

If a project involves a number of files, it is necessary to set the working directory as follows:

1. Click on the Set Working Directory button on the File toolbar.
2. Select a directory from the Select Directory dialog box.

File Attachment

You can use the Worksession Manager to attach external files of various formats that Rhino supports. (Supported files will be covered in the next chapter.) The same file can be attached concurrently by a number of designers working on different computers. Geometry in the attached file cannot be modified but can be used as input for constructing geometry.

Attached files are listed in the Worksession Manager dialog box. The attached file list will not be saved in the current file; you have to save the list in a worksession file. Perform the following steps:

1. Select File > Open and select the file carbody02.3dm from the Chapter 12 folder on the companion CD.
2. Select File > Worksession Manager, or right-click on the Attach/ Worksession button of the File toolbar. The Worksession Manager dialog box is displayed.

As shown in the Figure 12–27, the current file is listed in the Worksession Manager dialog box. There is a tick mark in the first column of the list, indicating that this is the active file.

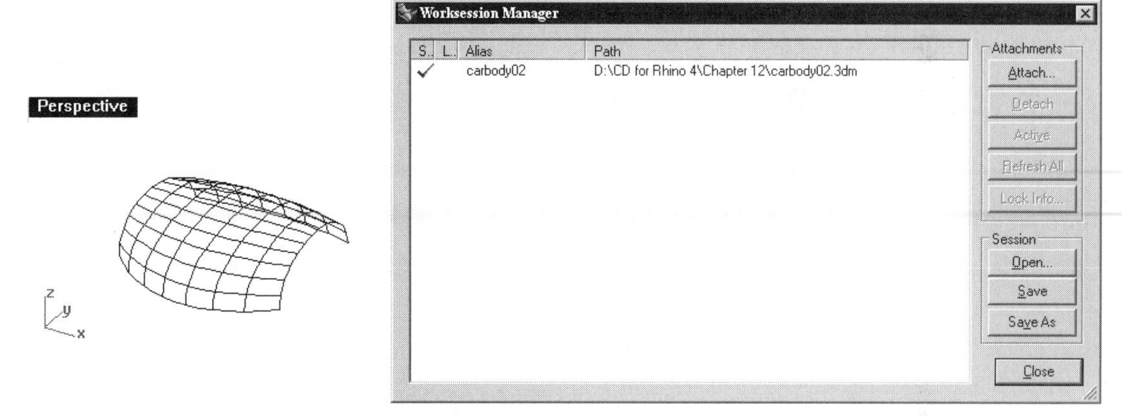

Figure 12–27. File opened and Worksession Manager dialog box displayed

3. In the Worksession Manager dialog box, click on the Attach button.
4. Select the file carbody01.3dm from the Chapter 12 folder of the companion CD. A file is attached.
5. Repeat steps 3 and 4 twice to attach the files carbody04.3dm and carbody05.3dm one by one. Now, the current working file has three external files attached. (See Figure 12–28.)

File Detachment

Contrary to file attachment, you can detach attached files, thus removing them from the Worksession Manager dialog box. It is important to note that if you close the Worksession Manager without saving it, closing the current file will detach the attached files automatically. Continue with the following steps.

6 Select the file carbody05.3dm from the list in the Worksession Manager dialog box.
7 Click on the Detach button. The selected file is detached.

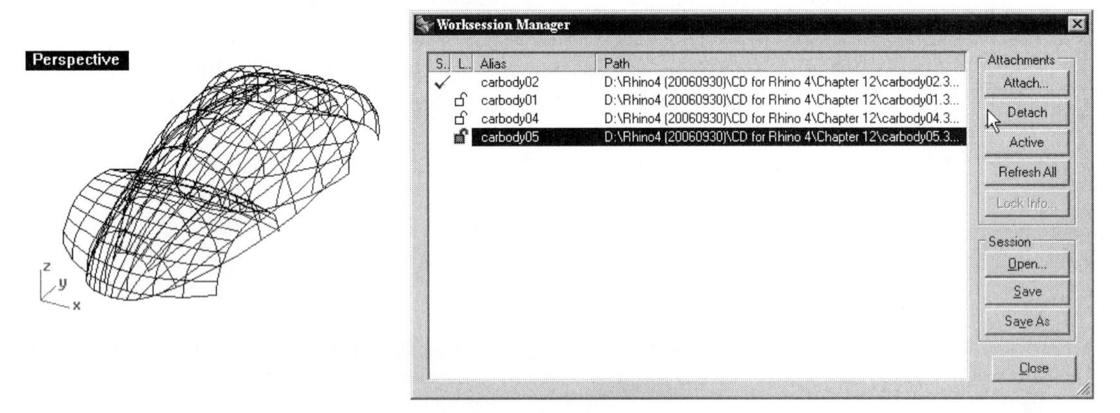

Figure 12–28. Three external files attached and one of them being detached

Attached File Activation

If you want to modify the geometry of an attached file, you can activate it via the Worksession Manager. Although a file can be attached by many persons for use as references, only one person can activate a file to modify it. Once an attached file is activated, the current file will be closed. Naturally, you will be prompted to save changes. Continue with the following steps.

8 Select the file carbody01 from the list in the Worksession Manager dialog box.
9 Click on the Active button. The selected file becomes the active file. If any changes are made to the previous active file, you will be prompted to save the changes. Because there is no modification made to the file carbody02.3dm, the active file simply changes to carbody01.3dm without any prompt.
10 Click on the Close button of the Worksession dialog box to close it. (See Figure 12–29.)

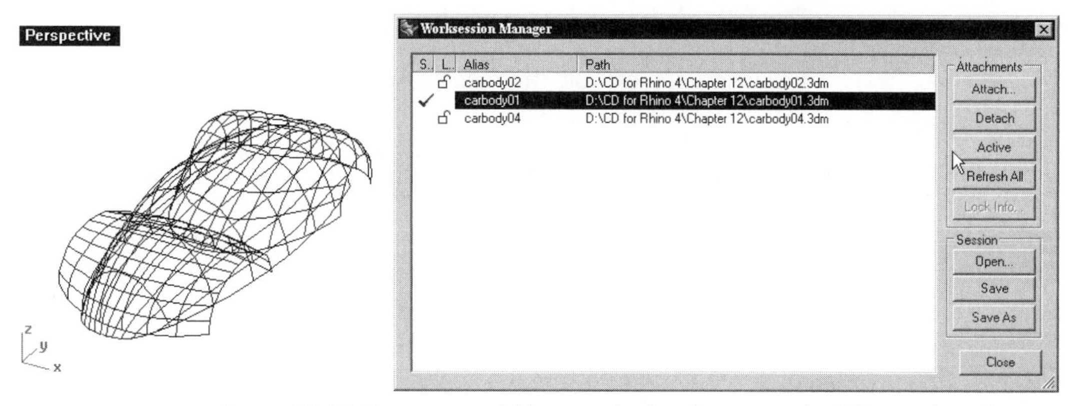

Figure 12–29. Two external files attached and one attached file made active

To appreciate how the attached file can be used to work on the current active file, continue with the following steps.

11 Referencing Figure 12–30, rotate the display.

12 Select Edit > Trim, or click on the Trim/Untrim Surface button on the Main 1 toolbar.

13 Select A and B, and press the ENTER key.

14 Select locations C, D, and E, and press the ENTER key. The selected portions of the surface in the active file are trimmed by surfaces from the attached file. (See Figure 12–31.)

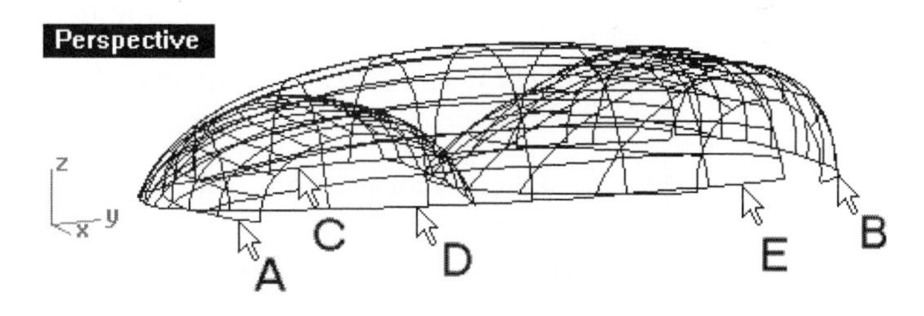

Figure 12–30. Display rotated and surface being trimmed

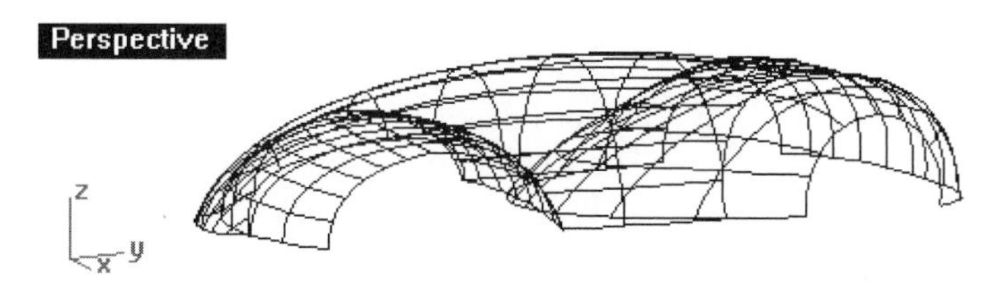

Figure 12–31. Surface in the current file trimmed by surfaces from the attached files

Refreshing Attached Files

Because a file can be attached by many designers working concurrently, it may have been activated and modified by one of the designers who attached the file. Therefore, you may need to refresh the attached files from time to time to display the most up-to-date version of the file. Continue with the following steps.

15 Select File > Worksession Manager, or right-click on the Attach/Worksession button of the File toolbar.

16 Click on the Refresh All button.

17 You will be prompted to save the file. If you click on the OK button, the attached files are refreshed. If these files are modified by another person, any change will be reflected.

18 Do not save your file.

Worksession File

The list of files, including the currently open file or activated file and all the attached files, together with their layer states, can be saved in a Worksession file.

Saving a Worksession File

To save a worksession file, open the Worksession Manager dialog box, click on the Save button, and specify a file name. If the worksession has already been saved and you want to save it to a new file, click on the Save As button and specify a new file name.

> *(**NOTE:** If you are using the evaluation version of Rhino 4, saving a worksession file will be counted as saving once.)*

The saved worksession file will have the extension rws. It remembers the list, layer states, and the window placement and size of the attached files. Now perform the following steps to open a worksession file:

1 Create a folder named CD for Rhino 4 in the D drive of your computer.

2 Copy the folder Chapter 12 from the accompanying CD to this folder.

3 Select File > Open and select the file worksession.rws from the Chapter 12 folder on the companion CD. You need to set the Files of Type to Rhino Worksession (*.rws).

When you open a Worksession file, the following operations will be performed:

- The current file will be closed. Naturally, you will be prompted to save changes.
- The current Worksession will be cleared.
- The files listed in the opened Worksession file will be opened, with the active file ready for editing and the other files attached for reference.
- The layer states of the listed files will be set.
- You can save the list of files (the currently open file and all attached files) and their layer states in a worksession file.

Layer States

Although you cannot edit the geometry of the attached files, you can modified their layer states, such as changing layer color, turning layers on and off, and locking layers. This way, you can change the color of the displayed geometry and control the display of some geometry to enhance visual effect.

If you want the computer to remember the attached file list in the Worksession Manager dialog box and the layer states of the attached file, you have to save the worksession. However, even if you save a worksession, layer states will not be saved in the attached files themselves. In other words, if a layer of an attached file is red and you change its color to blue in the worksession, the color of the layer will remain red if you open the file individually, and the color of the layer will become blue if the file is opened in the context of a worksession. Continue with the following steps.

4. Open the Edit Layer dialog box.
5. Use the Work Session manager to make active the attached files one by one, and modify the color settings of the layers as necessary. (See Figure 12–32.)

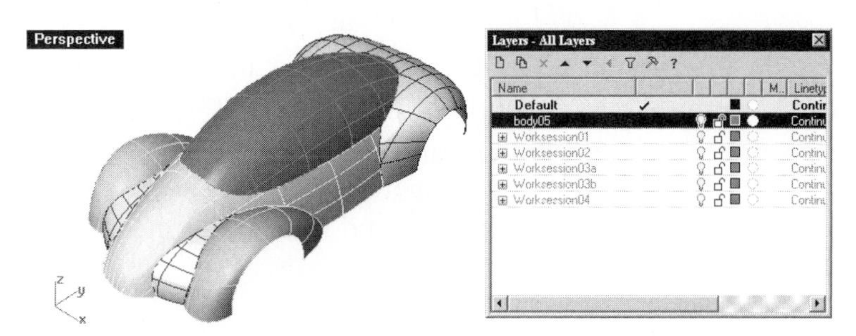

Figure 12–32. Layer state being modified

6 If you wish to save the layer states, open the Worksession Manager dialog box and click on the Save button. Otherwise, do not save your files.

Because the layer states of attached files saved in a worksession file do not affect the layer state of the attached files themselves, by attaching the same file to different worksession files and setting different layer states in the worksession files, you can save numerous layer states for a single file.

Reference Geometry Limit

Understanding that substantial computer resources are required if there are many attached files, you may limit the amount of reference geometry by constructing a sphere, outside which objects will be purged.

To set the limit, type LimitReferenceModel at the command area and then construct a sphere.

File Attachment and File Insertion

The major difference between file attachment and file insertion is that an attached file is not copied to the current file. If the worksession is not saved, the attachment information will be lost after the current file is closed. On the other hand, an inserted file is copied to the current file to become a set of objects or a block definition. If the file is inserted as a block definition, the external file is referenced. Change in the external file can be reflected by updating.

Design Collaboration

To sum up, the Worksession Manager can help enhance design collaboration as follows:

- Many designers can attach the same file to display it as reference.
- Designers can change the layer states of the attached files to enhance visual effect.
- One of the designers can activate an attached file to edit it.
- Other designers can refresh the attached files to display the most up-to-date version of the attached files.
- The attached file list and their layer states in the context of the worksession can be saved, thus saving the time to reattach the files and reset the layer states.

Drag and Drop

You can drag any file supported by Rhino from the window explorer to Rhinoceros windows to perform one of the following tasks: open, insert, import, and attach. After dragging, the File Options dialog box, shown in Figure 12–33, will display.

If the dragged file is to be opened, you will be prompted to save the current file before closing it.

If the dragged file is to be inserted, the Insert dialog box will be displayed. You can insert the file in one of three ways.

If the dragged file is to be imported, the content of the file will be imported and merged with the current file.

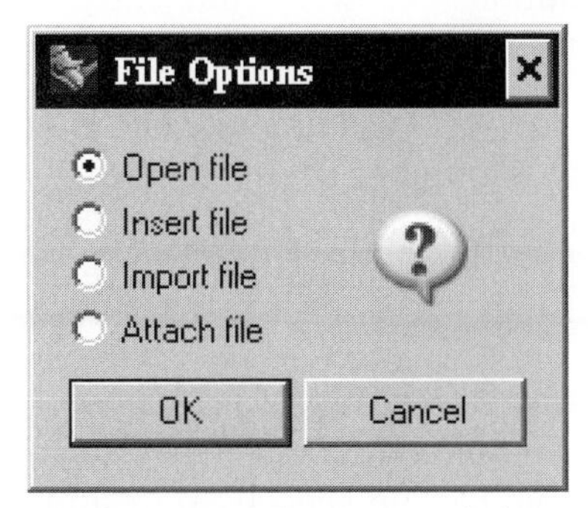

Figure 12–33. File Options dialog box

If the dragged file is to be attached, the file is attached and its file name will display in the list of the Worksession dialog box. You can reference to its geometry and change its layer state in context of the current file.

Textual Information

You can attach textual information to either object geometry or the attributes of an object by using the SetUserText command, and retrieve attached textual information by using the GetUserText command.

Chapter Summary

By putting a set of objects in a group, you can select the entire set of objects collectively by picking an element in the group. To select an individual object of the group, you hold down the Control and Shift keys simultaneously while picking the object. After a group is formed, you can add elements to it or remove elements from it. Groups can be named for easy identification, grouped objects can be ungrouped, and groups can be merged.

One way to manage repeated sets of geometric objects is to define block definitions and insert instances referenced to the block definitions. A block definition can be defined by selecting objects from the current file or inserting an external file. When an external file is inserted, it can be inserted in one of three ways: as an instance, as a group of objects, or as individual objects.

Because inserted instances are referenced to the block definition, instances will change automatically if the block definition is modified. Furthermore, if the block definition is defined from an external file and the external file is modified, you can update the change via the Block Manager.

There are three approaches in designing an assembly of components: the bottom-up approach, the top-down approach, and the hybrid approach. The term bottom refers to the individual components of an assembly and top refers to the assembly as a whole. Using the bottom-up approach, you construct a set of Rhino files to depict individual components of the assembly, start a new file for the assembly, and insert the component files into the assembly file. Using the top-down approach, you construct components in a Rhino file, construct block definitions for each component, export the block definitions as individual external files, and link the blocks to the exported files.

The Worksession Manager is a useful design collaboration tool. It enables you to divide a large project into a number of small files. By opening one of the files and attaching the remaining files via the Worksession Manager, you can work on a small part of the project and reference to the other parts of the project.

In a collaborative design environment in which a group of people work on a project, each people can open a file and attach other files. The latest version of the files done by other people can be seen by refreshing the attached files. By activating different files, design work can pass from one designer to another.

Review Questions

1. Differentiate between a group and a block.
2. What are the two ways to construct a block?
3. How can an external referenced block be updated?
4. What are the ways to remove block definitions in a file?
5. Briefly explain how bottom-up and top-down design approaches can be carried out using Rhino.
6. Outline how the Worksession Manager helps facilitate design collaboration.

CHAPTER 13

2D Drawing Output and Data Exchange

Introduction

This chapter explores the process of constructing 2D engineering drawings and delineates data exchange methods.

Objectives

After studying this chapter you should be able to:

- Construct 2D engineering drawings
- Perform data exchange

Overview

2D orthographic engineering drawings are the conventional means of communication among engineering personnel. Although the advent of computer-aided design applications replaced some of the uses of 2D drawings, there remain many situations in which 2D drawings and/or 2D drawing output is useful or necessary. In a drawing, you have to add dimensions and annotations to help delineate details of the object depicted by the drawing views. To address downstream and upstream operations, you export Rhino files to various data formats and read various data formats into Rhino.

Engineering Drawing

Apart from delineating the geometric shape of objects in two dimensions, one important function of engineering drawings is to specify precisely the dimensions of the objects they represent, along with annotations that convey other information about the object or objects represented. Engineering drawing construction consists of two major tasks:

- Constructing 2D orthographic projection drawings
- Dimensioning and annotating the 2D drawings

Methods to Construct 2D Drawings

Using Rhino as a tool, you can construct 2D engineering drawings in three ways, as follows:

- You can let the computer generate 2D drawings from 3D objects automatically.
- You can set out a page layout and construct detail viewports to delineate various orthographic projection views.
- You can construct 2D drawings from scratch, using the curve tools described in Chapters 4, 5, and 6. Construction of 2D drawings from scratches requires having a thorough understanding of orthographic projection principles and a good perception of how 3D objects would look when projected onto a planar face. Erroneous engineering drawings are quite often the result of human errors.

Generating 2D Engineering Drawing from 3D Objects

The process of outputting a 3D model as an engineering (2D) drawing is fairly simple using Rhino. Basically, you use an appropriate command and let the computer do all of the 2D drawing construction work. Because the 2D drawings are produced automatically, you can rely on Rhino and your computer in projecting 3D objects onto a plane, and sound knowledge in orthographic projection is not necessary. Perform the following steps:

1. Open the file 3Dto2D.3dm from the Chapter 13 folder on the companion CD.
2. Select Dimension > Make 2-D Drawings, or click on the Make 2-D Drawing button on the Dimension toolbar or the Curve from Object toolbar.
3. Referencing Figure 13–1, click on A and drag to B, and then press the ENTER key.
4. In the 2-D Drawing Options dialog box, select 4 View (USA) and then click on the OK button.

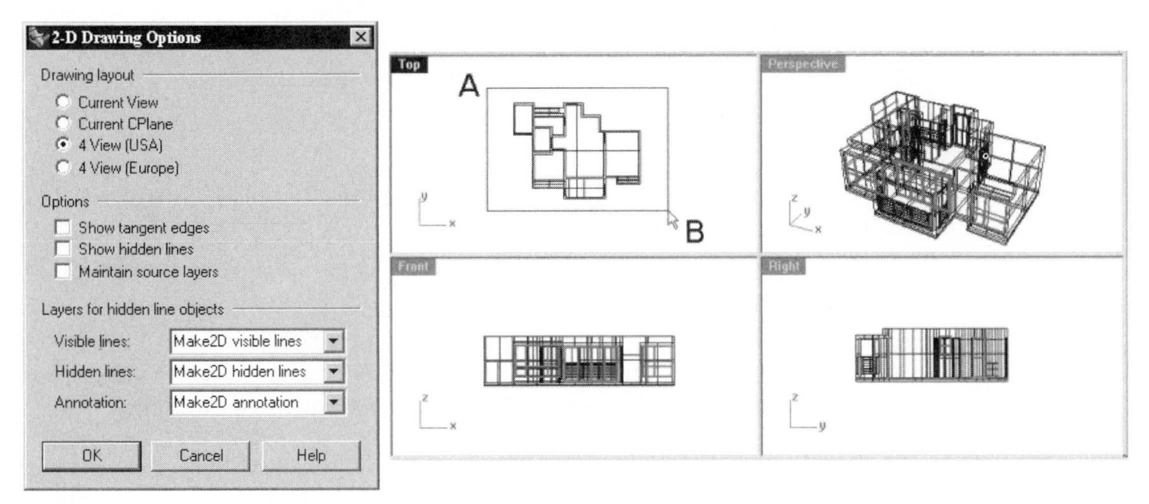

Figure 13–1. 3D object being selected and 2D Drawing Options dialog box

5 Set the current layer to Make2dvisiblelines and turn off all other Polysurface layers.

6 Maximize the Top viewport. A 2D drawing is constructed, as shown in Figure 13–2.

7 Do not save your file.

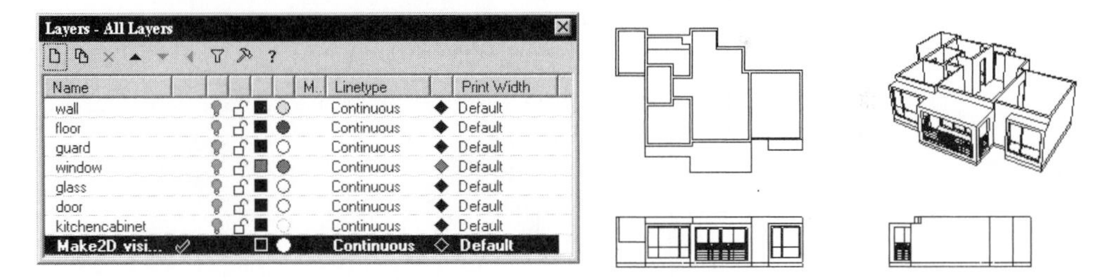

Figure 13–2. Front, side, top, and isometric views constructed

As you can see, generation of 2D drawing from 3D objects is very simple. However, this method of producing a drawing has a drawback, because the choice of viewing directions is restricted.

Constructing Page Layout and Detail Viewports

Another method to produce a 2D engineering drawing from a 3D model is to use page layout to depict a piece of drawing paper, and detail viewports to depict various orthographic drawing views.

Constructing a Page Layout

A page layout is analogous to a sheet of paper on which you construct 2D engineering drawings. Perform the following steps to construct a page layout for a Rhino file:

1. Open the file Locomotive.3dm from the Chapter 13 folder on the companion CD.
2. Select View > Page Layout > New Page Layout, or click on the New Layout button of the Sheet Layout toolbar.
3. In the New Page Layout dialog box, shown in Figure 13–3, select a printer (here the Acrobat PDFWriter is chosen), set the paper size to A4 Landscape, set the Initial Detail Count to 4, and click on the OK button.

Initial Detail Count refers to the number of detail viewports that will be constructed automatically by the system. If you choose 0, you will have a blank page layout.

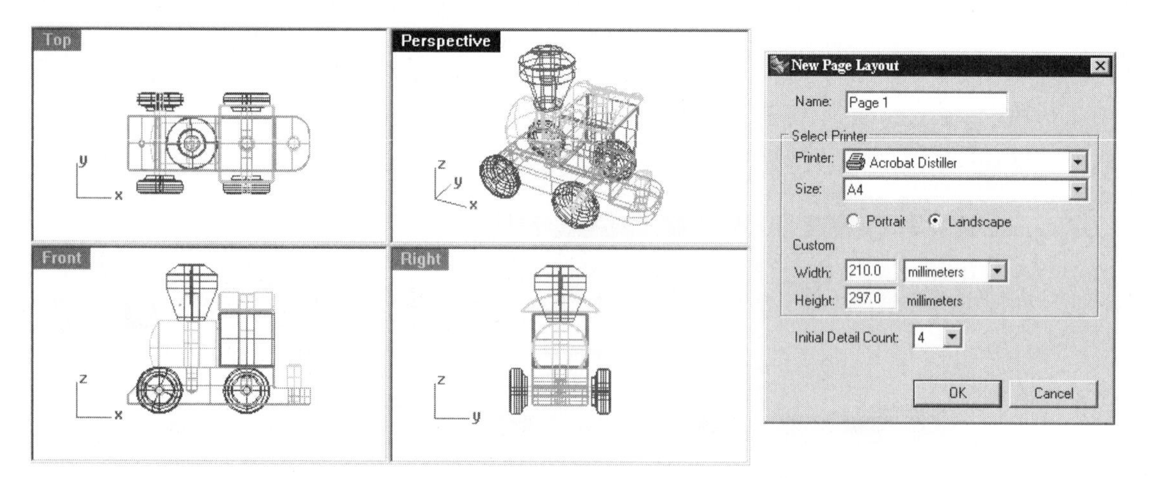

Figure 13–3. 3D model and New Page Layout dialog box

A page layout with four detail viewports is constructed. (See Figure 13–4.)

Manipulation of Detail Viewports

Detail viewports are viewports constructed on a page layout. By activating a detail viewport, you can perform pan, zoom, and rotate commands in order to obtain an appropriate view. At the lower-left corner of a detail viewport is a lock/unlock icon. Clicking on the icon locks/unlocks the viewport in terms of pan, zoom, and rotate. In a page layout, you can have many detail viewports, each depicting a different direction of viewing.

Detail viewports can be deleted and can be constructed anywhere on the page layout, as follows:

4. Select Edit > Delete.
5. Select detail viewport A, shown in Figure 13–4, and press the ENTER key.

A detail viewport is deleted.

6. Select View > Page Layout > Add Detail View, or click on the Add Detail View of the Sheet Layout toolbar.
7. Click on location A and B, shown in Figure 13–5. Exact location is unimportant.

A detail viewport is constructed.

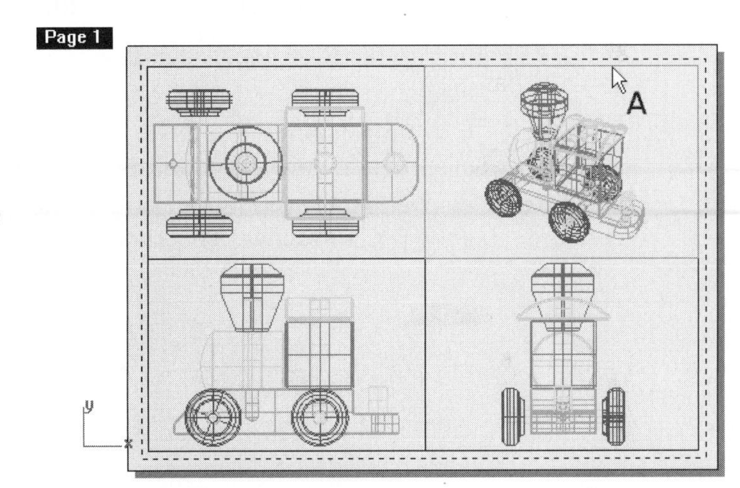

Figure 13–4. Page layout with four detail viewports and a viewport being deleted

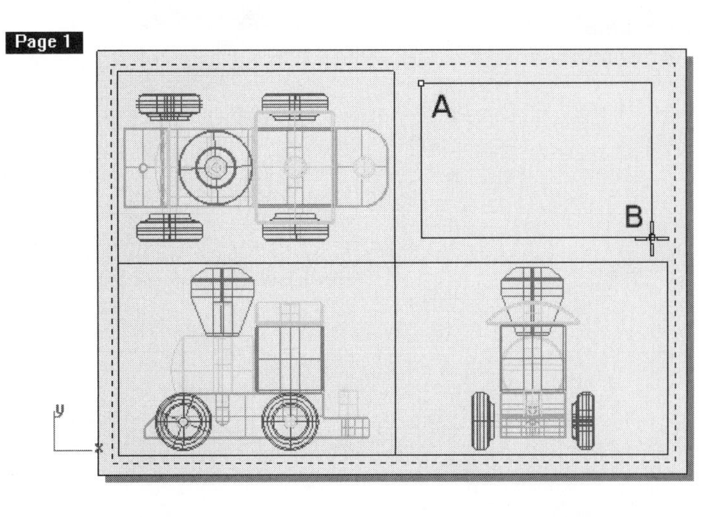

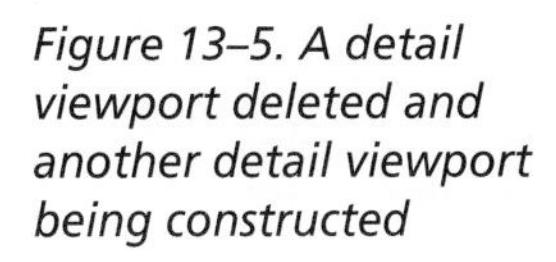

Figure 13–5. A detail viewport deleted and another detail viewport being constructed

To comply with engineering drawing standard, the display scale of the detail viewports has to be consistent and the detail viewports have to be aligned properly.

8. Select View > Page Layout > Scale Detail View.
9. Select detail viewport A, shown in Figure 13–6.
10. Set the distance on page to 1 unit.
11. Type 2 at the command area. This will set the display scale to "half" because one unit displayed in the detail viewport is equivalent to 2 units displayed in the page layout.
12. Repeat the command to set the scale of detail for viewports B, C, and D.

To help align the detail viewports, a dummy box is constructed and put in a layer that is turned off. Continue with the following steps:

13. Turn on the Reference layer, keeping the current layer unchanged.
14. Select Transform > Move, or click on the Move button on the Transform toolbar.
15. Select detail viewport A and press the ENTER key.
16. Select endpoint B (Figure 13–7) as the base point and end point C as the point to move. The viewport is moved.

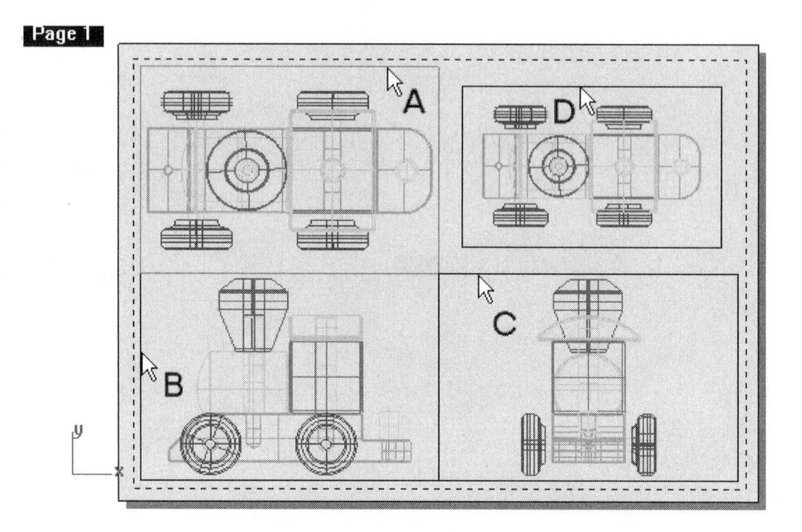

Figure 13–6. Detail viewports being scaled

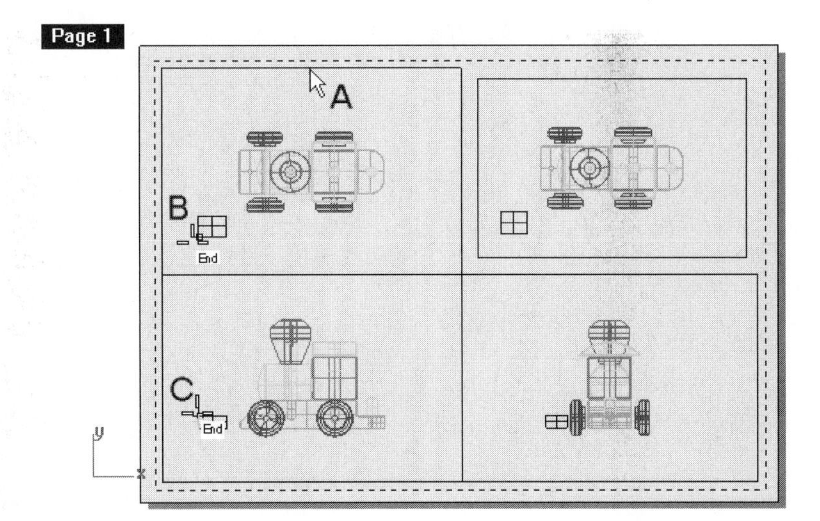

Figure 13–7. Detail viewport being moved

17 Turn on Ortho mode by checking the Ortho box at the status bar.

18 Referencing Figure 13–8, select and drag the moved detail viewport from A to B. The exact location is unimportant as long as the detail viewports are orthogonal to each other.

The top detail viewport is aligned properly with the front detail viewport.

19 Repeat steps 14 through 18 above to align the right side detail viewport, as shown in Figure 13–9.

20 Turn off the Reference layer.

21 Turn off Ortho mode.

To modify the display of the detail viewport, continue with the following steps:

22 Select View > Page Layout > Enable detail view and select detail viewport A, shown in Figure 13–9, or double-click on the interior of viewport A.

23 Right-click on B, the detail viewport's label, and select Shaded. The detail viewport is changed to shaded mode.

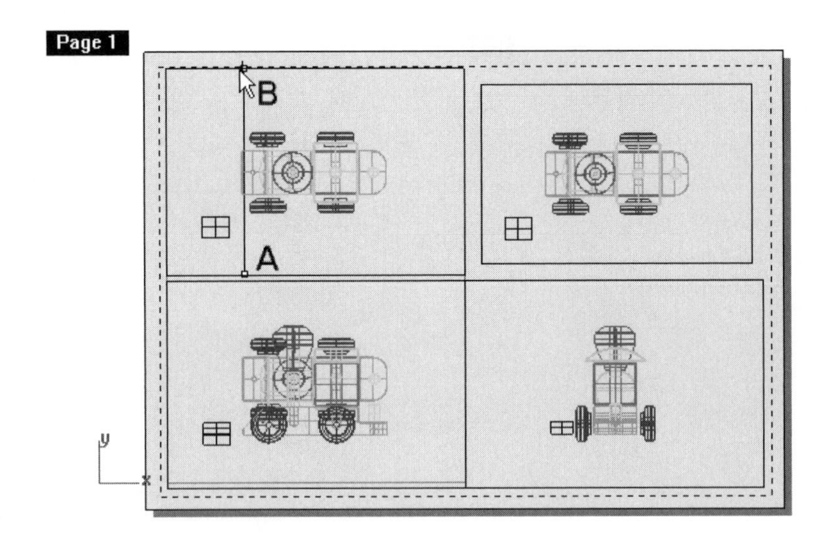

Figure 13–8. Top viewport moved and being dragged

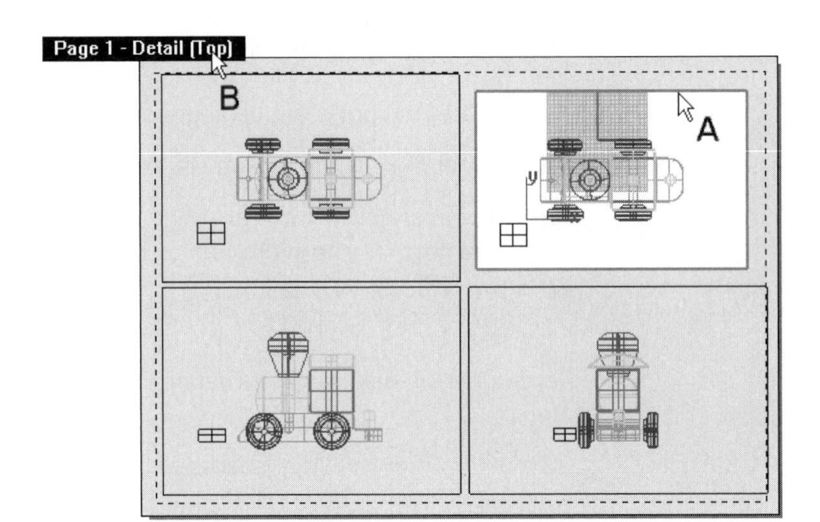

Figure 13–9. Detail viewport being enabled

24 Referencing Figure 13–10, rotate the viewport.

25 To exit the viewport, double-click anywhere outside the active detail viewport. The drawing is complete.

26 Do not save your file.

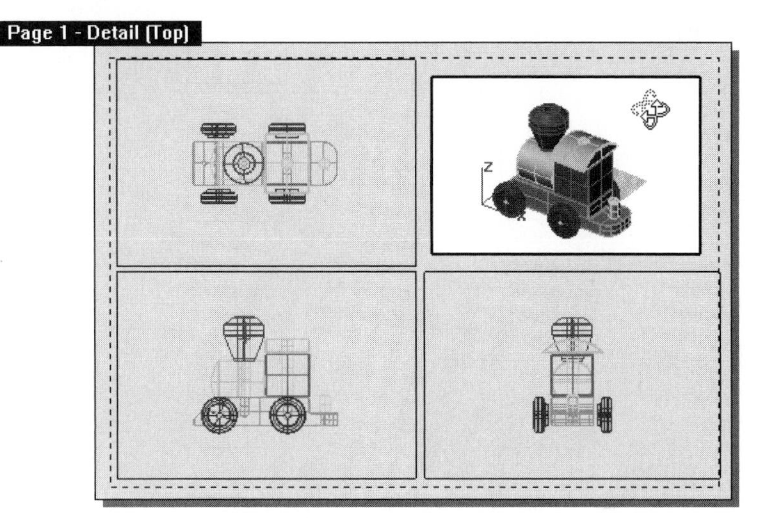

Figure 13–10. Detail viewport shaded and rotated

In order not to print the detail viewports, they are residing on a layer with print color set to white, the color of the paper on which the drawing is printed.

Copying Page Layout

In a Rhino file, you can have more than one page layout. Therefore, you can use a layout for a set of detail viewports showing a set of objects and use another layout for another set of drawing viewports.

To facilitate construction of a page layout from an existing page layout, you can use the CopyLayout command.

Constructing 2D Drawing from Scratch

To produce 2D a drawing from scratch, you are essentially using the computer as an electronic drawing board. Like constructing a drawing manually using pencil and paper, you need to have excellent knowledge of orthographic projection and good perception of 3D objects in terms of how they would look when they are projected on a plane. To appreciate a 2D drawing constructed from scratch, perform the following steps:

1. Open the file 2Ddrawing.3dm from the Chapter 13 folder on the companion CD. (See Figure 13–11.)

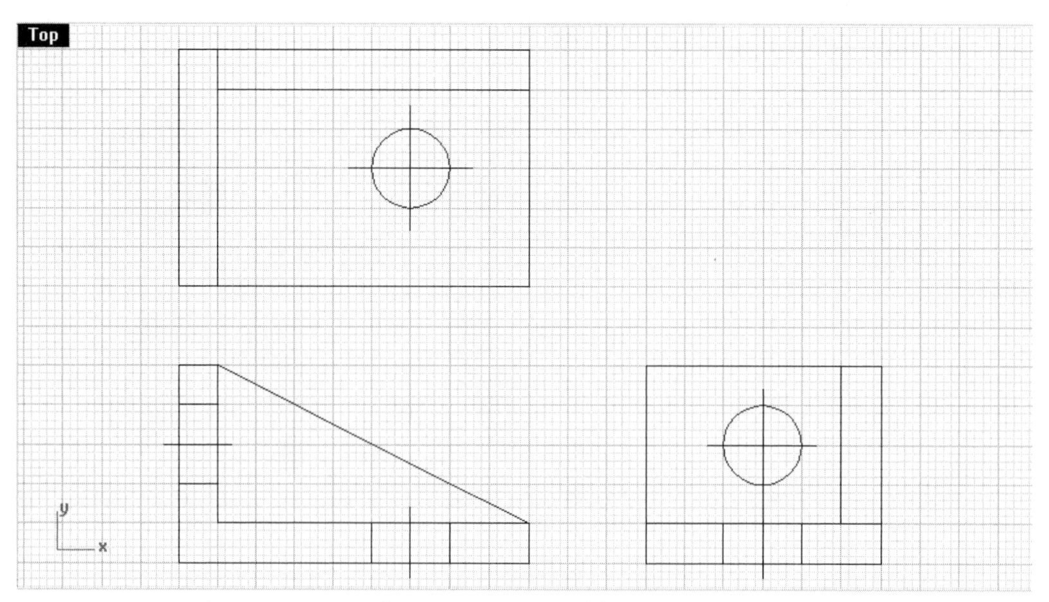

Figure 13–11. 2D drawing constructed from scratch.

2 Maximize the Perspective viewport.
3 Select View > pan, Zoom, and Rotate > Rotate View, or click on the Rotate View button on the Standard toolbar.
4 Rotate the view to appreciate how the 2D drawing looks.

Hatching

Hatching is a way to highlight a section face of an object cut by an imaginary plane. No matter which way you construct a drawing, you can construct a hatch pattern within a closed planar boundary curve. Continue with the following steps.

5 Maximize the Top viewport.
6 Select Dimension > Hatch, or right-click on the Hatch/Boundary Hatch button on the Dimension toolbar.
7 If you run the command from the pull-down menu, select the Boundary option at the command area. Otherwise, proceed to the next step.
8 Select curves A, B, C, D, E, F, G, H, J, and K (Figure 13–12) and press the ENTER key.
9 Click on locations L, M, and N (Figure 13–12) and press the ENTER key.

10 In the Hatch dialog box (Figure 13-12), select Hatch1, set Pattern Rotation to 45 deg and Pattern Scale to 20, and click on the OK button. The selected locations of the drawing are hatched. (See Figure 13–13.)

(***NOTE:*** *If solid hatch pattern is used and subsequently exploded, the solid hatch will become a surface.*)

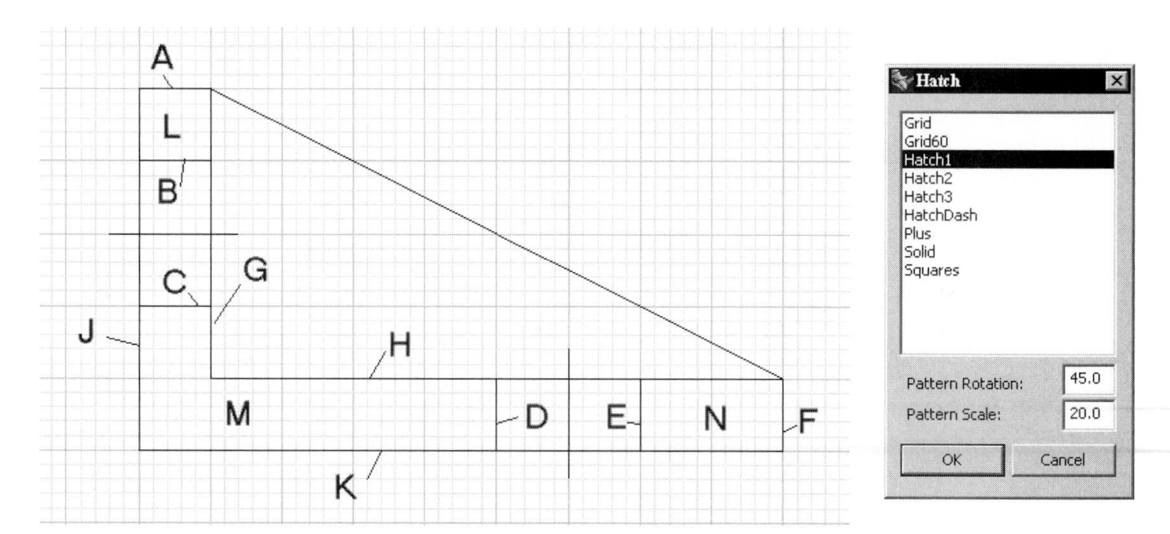

Figure 13–12. Boundaries selected and Hatch dialog box

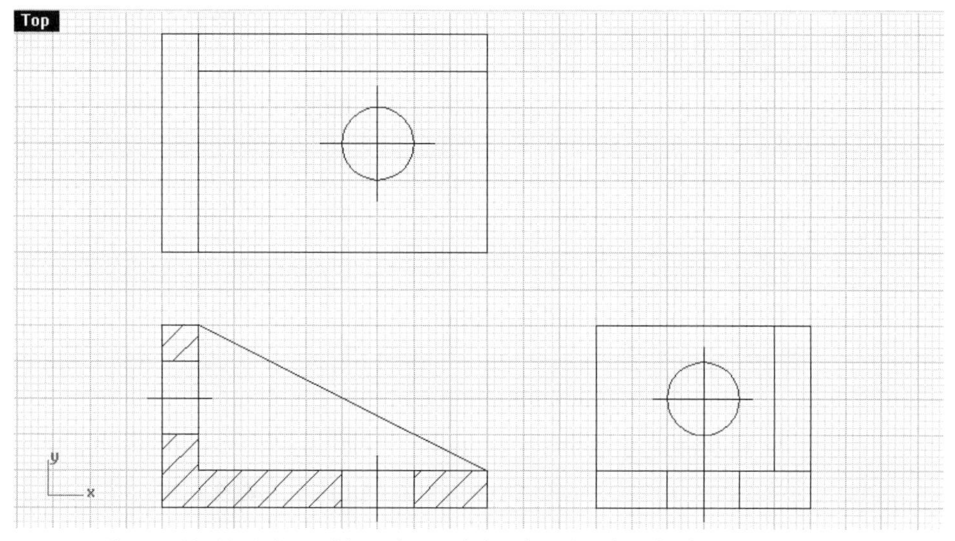

Figure 13–13. Selected locations of the drawing hatched

If you want to load or unload a hatch pattern, perform the following steps:

- Select Tools > Options.
- Select Hatch from the Document Properties folder in the Options dialog box.

If you want to modify a hatch pattern, perform the following steps:

- Select Edit > Properties, or click on the Object Properties button on the Properties toolbar.
- Select the hatch pattern that you want to modify.
- In the Properties dialog box, select Hatch from the pull-down list box.
- Select a pattern and specify pattern rotation angle and pattern scale.

Line Types

Because 2D orthographic projection drawings are line drawings, it is necessary to use various line types to depict different kinds of line objects, as follows:

- Thick continuous lines for visible outlines and edges
- Thin continuous lines for dimension lines, leaders, hatchings, outlines of adjacent parts, and outlines of revolved sections
- Thin dash lines for hidden outlines and hidden edges
- Thin chain lines for center lines and extreme positions of moving parts

To use these lines in your drawing, you might need to load them into your Rhino file, if such line types are not available in the template file that you use to construct the drawing, as follows:

- Select File > Properties.
- Select the Linetypes tab of the Document Properties dialog box.
- Click on the Load button.
- In the Load Linetype dialog box, select the linetype that is needed and click on the Add button.
- Click on the Close button to return to the Options dialog box.
- Click on the OK button of the Options dialog box.

By default, line type is ByLayer, meaning that objects will have a linetype assigned to the layer. Therefore, you can simply change the layer's linetype to set the linetype of objects residing on a layer. To set the linetype of an object individually, continue with the following steps.

11 Select Edit > Properties, or click on the Object Properties button on the Properties toolbar.

12 Hold down the SHIFT key and select lines A, B, C, D, E, and F (Figure 13–14).

13 In the Properties dialog box, select Object from the pull-down list box.

14 Select Center from the Linetype pull-down list box. The selected curves' linetype is changed.

15 Do not save your file.

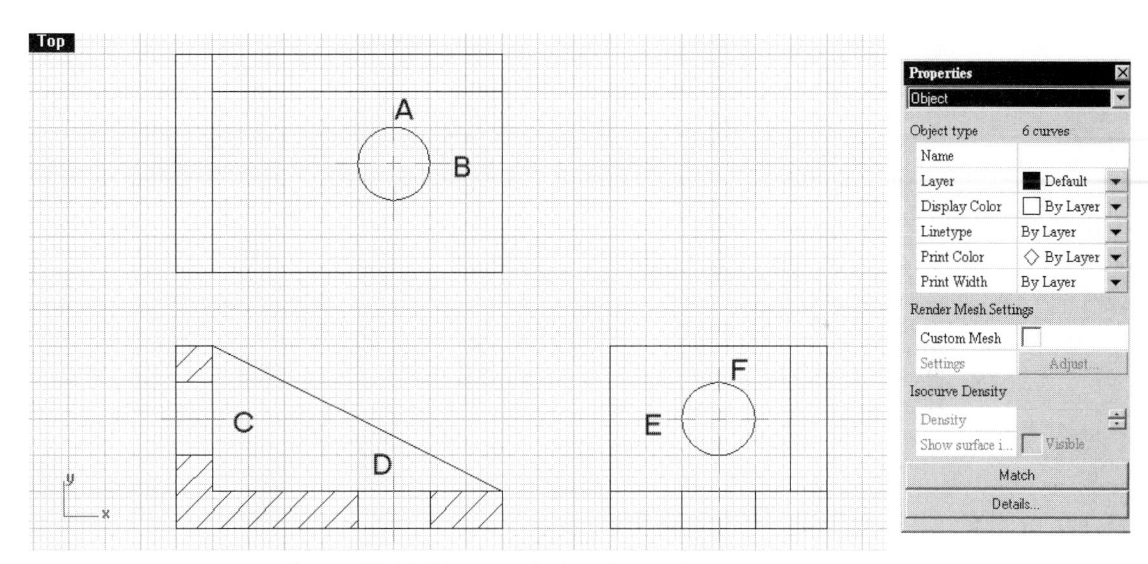

Figure 13–14. Line type being changed

Dimensioning and Annotation

To complement 2D drawing views that depict the shape of an object projected orthogonally, you add dimensions and annotations. Dimensions indicate the size of the object; annotations incorporate additional information other than dimensions. There are seven kinds of dimensions: linear, aligned, rotated, radial, diameter, angle, and ordinate. There are four kinds of annotations: leader, text box, dots, and annotation at curve's end points.

Options

Before adding dimensions and annotations to a drawing, it is recommended to check the Dimensions tab of the Document Properties dialog box and change the settings as necessary. Perform the following steps:

1. Open the file Dimension1.3dm from the Chapter 13 folder on the companion CD.
2. Select Dimension > Dimension Properties, or click on the Dimension Properties button on the Dimension toolbar.
3. In the Dimensions tab of the Document Properties dialog box, shown in Figure 13–15, set the text height to 3 units, the arrow length to 3 units, the extension line offset to 1 unit, and the extension line extension to 1 unit, and then click on the OK button.

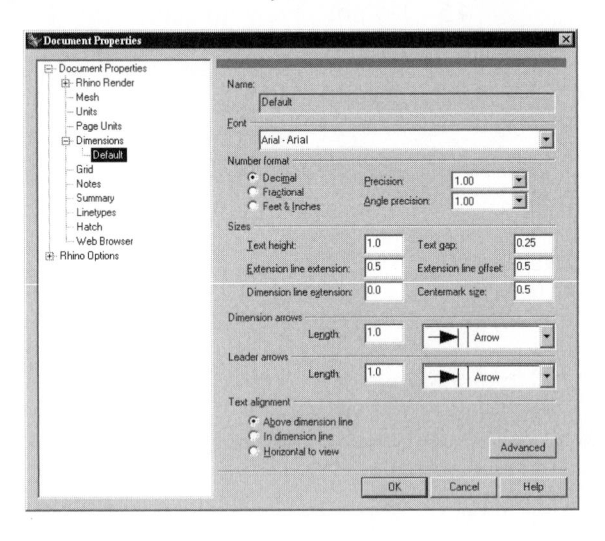

Figure 13–15. Dimension options

Linear Dimension

The most common way to specify the dimensions of an object is to state the vertical or horizontal distance between two selected points. Continue with the following steps.

4. Set object snap mode to End.
5. Select Dimension > Linear Dimension, or click on the Linear Dimension button on the Dimension toolbar.
6. Select end points A and B, and click on location C. A horizontal linear dimension is constructed. (See Figure 13–16.)
7. Repeat the command.
8. Select end points A and D, and click on location E. A vertical linear dimension is constructed.

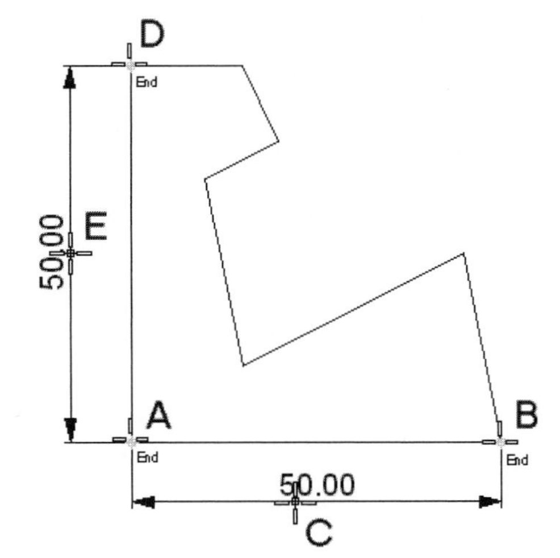

Figure 13–16. Horizontal and vertical linear dimension being constucted

Aligned Dimension

To specify the distance between two selected points, you use the aligned dimension. The dimension line is parallel to an imaginary line connecting the two selected points. Continue with the following steps.

9 Select Dimension > Aligned Dimension, or click on the Aligned Dimension button on the Dimension toolbar.

10 Select end points A and B, and click on location C. An aligned dimension is constructed. (See Figure 13–17.)

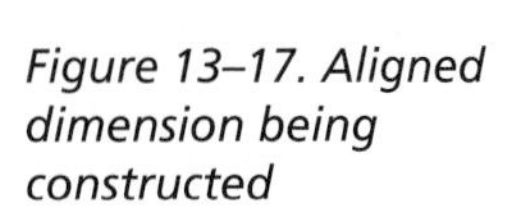
Figure 13–17. Aligned dimension being constructed

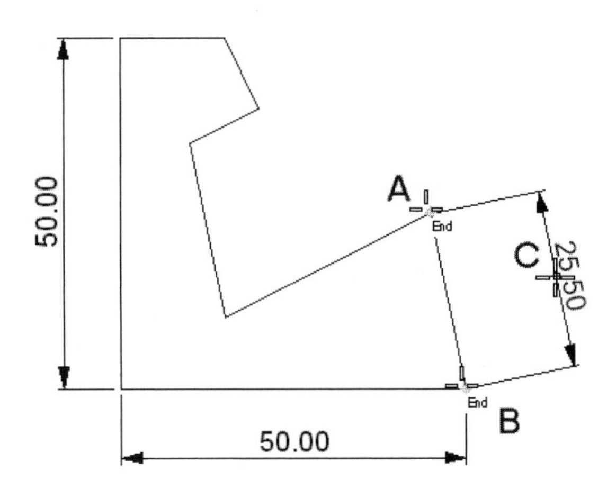

Rotated Dimension

To specify the distance between two select points measured in a direction defined by a rotation angle, you use the rotated dimension. Continue with the following steps.

11 Select Dimension > Rotated Dimension, or click on the Rotated Dimension button on the Dimension toolbar.

12 Check the Perp box on the Osnap dialog box.

13 Referencing Figure 13–18, select end point A and point B perpendicular to A. The rotation angle is defined. Alternatively, you can type a value at the command area to specify the rotation angle.

14 Select end points A and C, and click on location D. A rotated dimension is constructed.

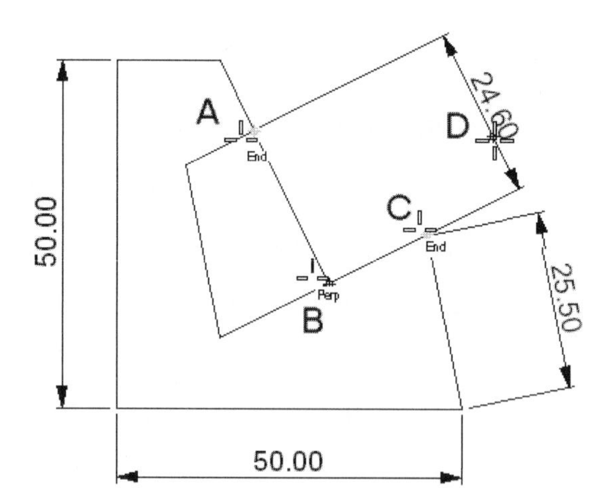

Figure 13–18. Rotated dimension being constructed

Angle Dimension

To specify the angle between two nonparallel lines, you use angle dimension. Continue with the following steps.

15 Select Dimension > Angle Dimension, or click on the Angle Dimension button on the Dimension toolbar.

16 Select lines A and B, and click on location C. An angle dimension is constructed. (See Figure 13–19.)

17 Do not save your file.

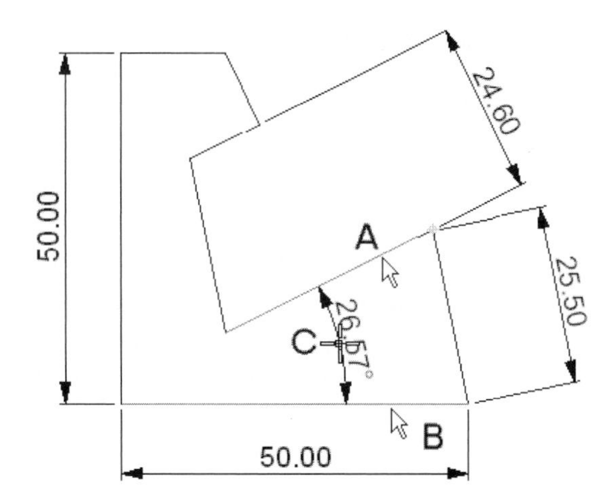

Figure 13–19. Angle dimension being constructed

Radial Dimension

To specify the radius of an arc or a circle, you use radial dimension. Perform the following steps:

1. Open the file Dimension2.3dm from the Chapter 13 folder on the companion CD.
2. Select Dimension > Radial Dimension, or click on the Radial Dimension button on the Dimension toolbar.
3. Select arc A and click on location B.
4. Repeat the command.
5. Select arc C and click on location D. Two radial dimensions are constructed. (See Figure 13–20.)

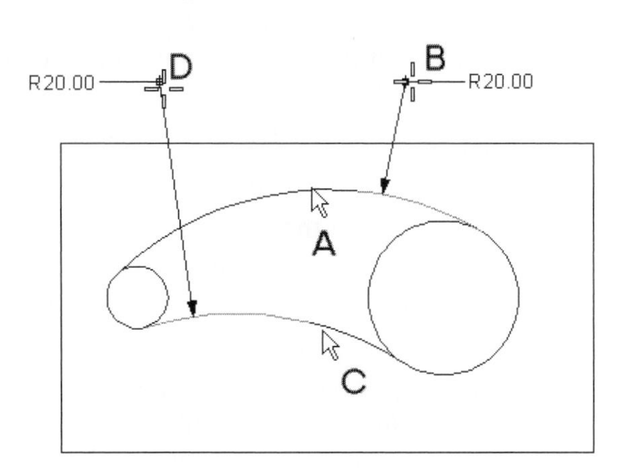

Figure 13–20. Radial dimensions being constructed

Diameter Dimension

To specify the diameter of an arc or circle, you use diameter dimension. Continue with the following steps.

6. Select Dimension > Diameter Dimension, or click on the Diameter Dimension button on the Dimension toolbar.
7. Select circle A and click on location B.
8. Repeat the command.
9. Select circle C and click on location D. Two diameter dimensions are constructed. (See Figure 13–21.)

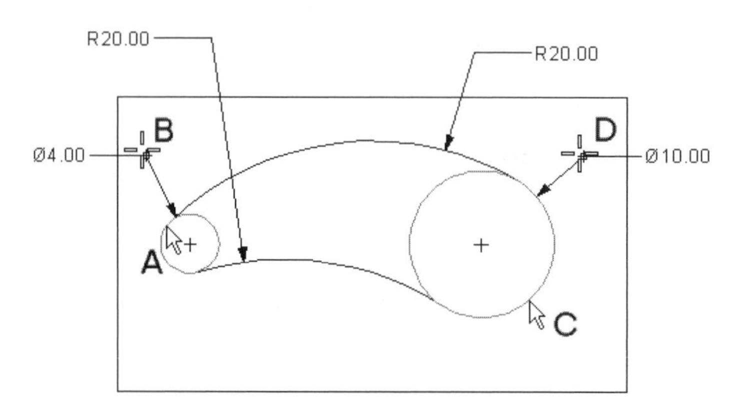

Figure 13–21. Diameter dimensions being constructed

Ordinate Dimension

You can construct ordinate dimensions, depicting the X coordinates and Y coordinates with reference to a specified base point. To facilitate the placement of ordinate dimensions, you can turn on ortho mode. Continue with the following steps.

10. Set object snap mode to End and Cen.
11. Check the Ortho box on the status line.
12. Click on the Ordinate Dimension button on the Dimension toolbar.
13. Select the Basepoint option on the command area.
14. Select end point A, shown in Figure 13–22.
15. Select center B and click on location C.
16. Repeat steps 12 through 15 two more times to construct a vertical ordinate dimension for center B at D and a horizontal ordinate dimension for center E at F.
17. Clear the check mark on the Ortho button of the status bar.

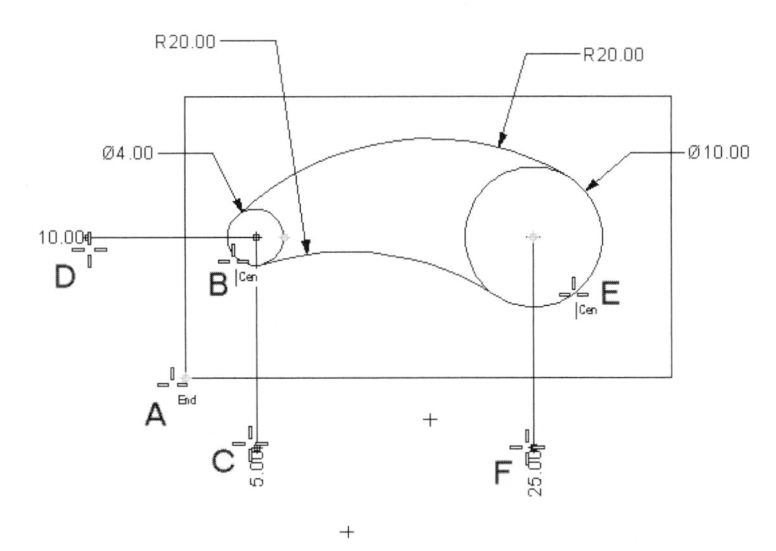

Figure 13–22. Ordinate dimensions being constructed

Leader

A leader is a set of lines with an arrowhead at one end of the connected lines and a text string at the other end. Continue with the following steps.

18 Select Dimension > Leader, or click on the Leader button on the Dimension toolbar.

19 Set object snap mode to Mid, and clear all other check boxes.

20 Referencing Figure 13–23, select midpoint A, click on locations B and C, and press the ENTER key.

21 Type the text string "RECTANGLE" in the Leader Text dialog box, and click on the OK button. A leader is constructed.

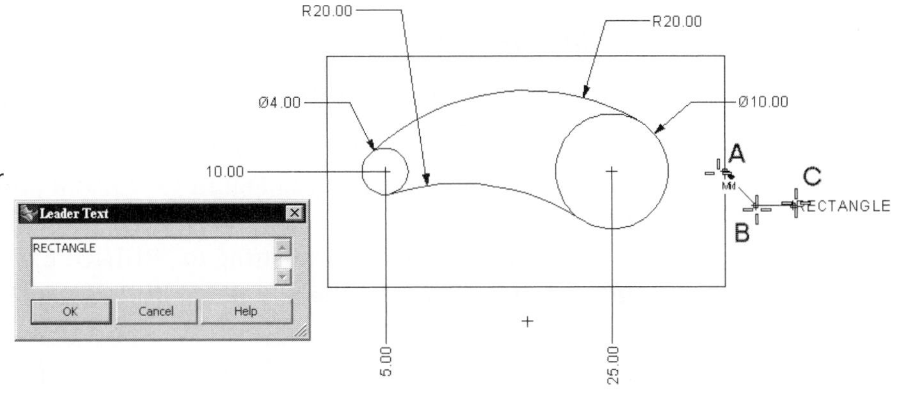

Figure 13–23. Leader being constructed

Text Box

A text box is a set of text string and is 2D; you must not confuse a text box with a 3D text object. If you explode a text box, a set of curves is obtained. On the other hand, if you explode a 3D text object, you obtain a set of surfaces. Continue with the following steps.

22 Select Dimension > Text Block, or click on the Text/Single Line of Text button on the Dimension toolbar.

23 Click on location A, shown in Figure 13–24.

24 In the Create Text dialog box, type the text string "RHINO 4" and click on the OK button. A text string is constructed.

Instead of inputting text string in the Create Text dialog box, you can click on the Import File button of the Create Text dialog box to import a text file (with a file extension of txt).

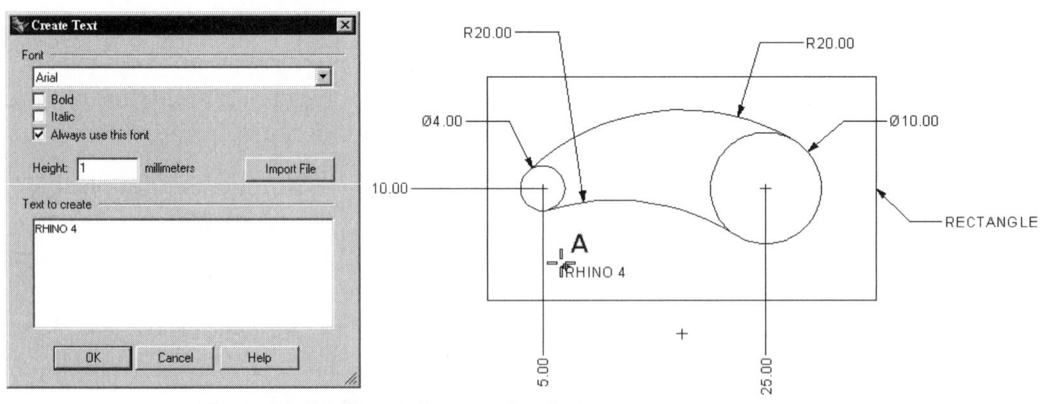

Figure 13–24. Text being constructed

Editing Text

Text that is already constructed can be modified, as follows.

25 Click on the Edit Text button on the Dimension toolbar.

26 Select A and B, shown in Figure 13–25, and press the ENTER key.

27 In the Properties dialog box, select Text from the pull-down list box and change the text string to "RHINOCEROS 4."

28 Select Leader from the pull-down list box, and change the leader text to "BASE."

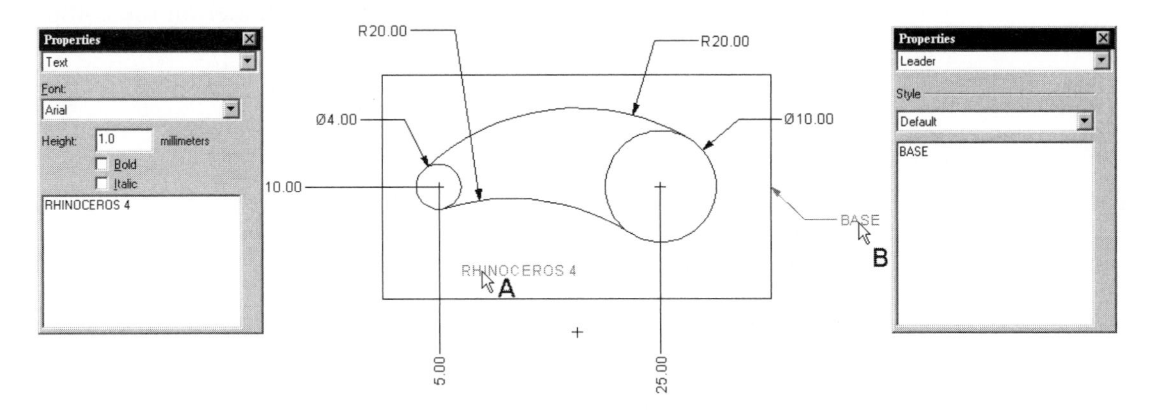

Figure 13–25. Text and leader being modified

Another method of editing the text and leader text is to double-click on the text or leader text to display the on-the-spot editing box, as shown in Figure 13–26.

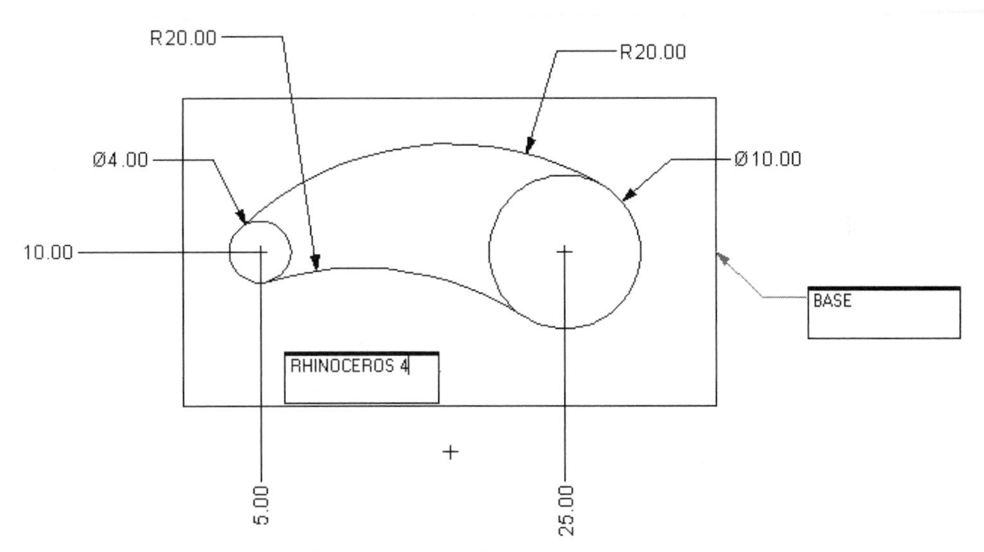

Figure 13–26. On-the-spot editing box

Dots

It is possible to place a dot with annotation that stays parallel to the viewport and sizes appropriately with the view.

1. Open the file Dot.3dm from the Chapter 13 folder on the companion CD.
2. Select Dimension > Annotate Dot, or click on the Annotate Dot button on the Annotate toolbar.
3. Type "CURVE" at the command area.
4. Select the midpoint of curve A, shown in Figure 13–27.

Arrow Head

You can add an arrow head at the end point of a curve.

5. Click on Add/Remove Arrowhead to curve on the Annotate toolbar, or type Arrowhead at the command area.
6. Select curve A, shown in Figure 13–27. An arrowhead is constructed.

To appreciate how the annotation dot, the curve end annotation, and arrowhead are parallel to the viewport's viewing direction, continue with the following steps.

7. Select View > Pan, Zoom, and Rotate > Rotate View, or click on the Rotate View button on the Standard toolbar.
8. Rotate the view.
9. Do not save your file.

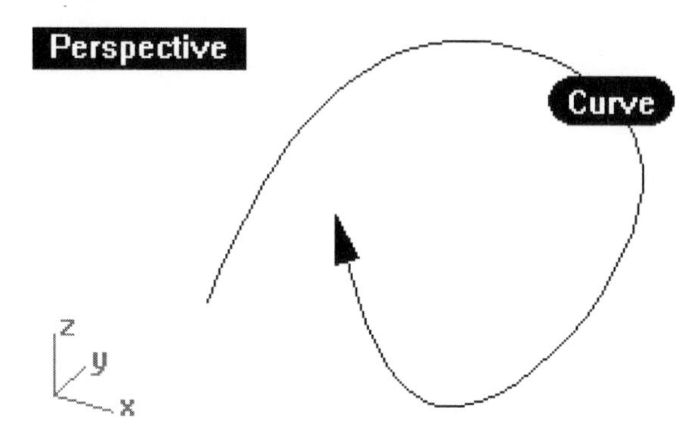

Figure 13–27. Annotation dots constructed

Exploding Dimensions

The explode command can be applied to dimensions, text, and hatch patterns, changing the dimensions to curves and text, the text to curves, and the hatch patterns to lines. However, this process is irreversible; exploded dimensions cannot be reconstructed as a single object.

Printing

To facilitate communication, you may need to output printed copies of your drawing. Perform the following steps:

1. Open the file Locomotive 1.3dm from the Chapter 13 folder on the companion CD.
2. To appreciate how the drawing looks printed, type PrintDisplay at the command area.
3. On the command area, set State = On and Color = Print.
4. Press the ENTER key. The display is set, as shown in Figure 13–28.

Figure 13–28. Display set to print

5. Select File > Print.

Figure 13–29 shows the Print Setup dialog box. At the left of the dialog box are five tabs: Destination, View and Scale, Margins and Position, Show, and Printer Details. Click on these tabs to make necessary changes.

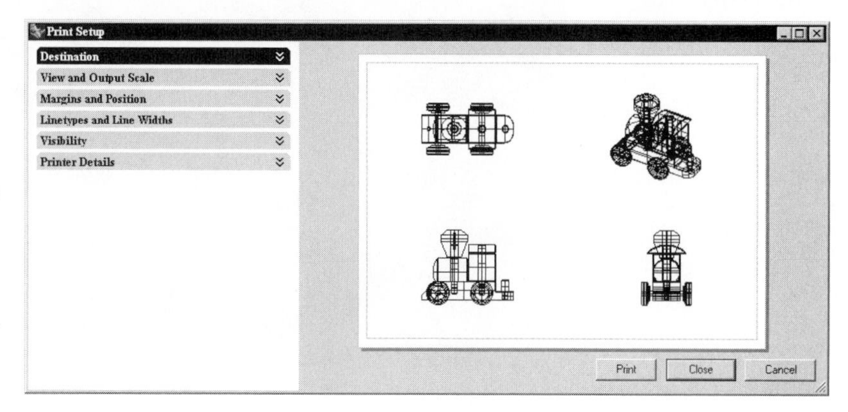

Figure 13–29. Print Setup dialog box

6 Click on the Print button to print the document, or click on the Close button to abort printing.
7 Do not save your file.

Data Exchange

The prime objective of exchanging data in various file formats is to facilitate downstream and upstream computerized operations.

Downstream Operations

Constructing 3D models and producing 2D drawings from 3D models is not necessarily the end of the digital modeling process. In a computerized manufacturing system, for example, the same digital model can (and often should be) designed to be used in all downstream operations, such as finite element analysis, rapid prototyping, CNC (computer numeric control) machining, and computerized assembly. Figure 13-30 shows a rapid prototyping machine making a 3D object, and Figure 13-31 shows the CNC machining of a free-form 3D object.

Figure 13–30. Rapid prototyping machine making a rapid prototype from a 3D model

Figure 13–31. CNC machining of a free-form 3D object

To enhance illustration of the 3D object, you construct renderings and animations. Because these operations may be done using different types of computer applications, and because each application may use a unique type of data format, the digital modeling system must enable the conversion of the 3D digital model into various file formats.

Rhino File Exporting

If your free-form models are constructed to facilitate downstream computerized operations, you can output them to various file formats for this purpose. You can save the entire Rhino file or export selected objects as various file formats, outlined in Table 13–1.

- ❒ To output the entire Rhino file in another format, select File > Save as.
- ❒ To export selected objects, select File > Export Selected.

Save as Template and Export with Origin

You must not confuse the Save as command with the Save as Template command and the Export selected command with the Export with Origin command.

- By selecting File > Save As Template, the entire drawing is saved as a Rhino template.
- By selecting File > Export with Origin, you export selected objects with origin point to Rhino format.

Upstream Operations

To take advantage of other computerized operation's strength, the digital modeling system must enable the opening of various file formats so that digital models constructed in other systems can be used for further elaboration of the design.

Other Files Importing

To reuse digital data from other applications, you can:

- Open a file by selecting File > Open
- Import a file by selecting File > Import
- Insert a file by selecting File > Insert

Naturally, opening a file converts the file to Rhino format, whereupon you can continue with the design. If you want to incorporate data into an existing Rhino file, you import or insert the file. Importing a file converts the imported data to individual Rhino curves or surfaces. Inserting a file also converts the data to Rhino format, but the file can be inserted as a block definition, a group of objects, or a set of individual objects.

Files that can be opened, imported, or inserted are listed in Table 13–1.

Table 13–1 File Formats Rhino Can Open and/or Save

Open File Format	Save As File Format
Rhino 3D model (*.3dm)	Rhino 4 3D model (*.3dm)
—	Rhino 3 3D model (*.3dm)
—	Rhino 2 3D model (*.3dm)
Rhino Worksession (*.rws)	—
Step (*.stp, *.step)	STEP (*.stp, *.step)

Open File Format	Save As File Format
DirectX (*.x)	DirectX (*.x)
SLC (*.slc)	SLC (*.slc)
Points File (*.asc, *.csv, *.txt, *.xyz, *.cgo_ascii, *.cc)	Points File (*.txt)
GHS Geometry File (*.gf; *.gft)	GHS Geometry File (*.gf)
—	GHS Part Maker file (*.pm)
—	XGL (*.xgl)
VDA (*.vda)	VDA (*.vda)
—	RenderMan (*.rib)
—	Object Properties (*.csv)
Adobe Illustrator (*.ai)	Adobe Illustrator (*.ai)
—	Windows Metafile (*.wmf)
Stereolithography (*.stl)	Stereolithography (*.stl)
VRML (*.vml, *.wrl)	VRML (*.wrl, *.vrml)
—	POV-Ray Mesh (*.pov)
—	Cult3D (*.cd)
—	Moray UDO (*.udo)
LightWave (*.lwo)	LightWave (*.lwo)
Raw Triangles (*.raw)	Raw Triangles (*.raw)
—	ACIS (*.sat)
IGES (*.igs, *.iges)	IGES (*.igs, *.iges)
—	Parasolid (*.x_t)
3D Studio (*.3ds)	3D Studio (*.3ds)
AutoCAD drawing file (*.dwg)	AutoCAD drawing file (*.dwg)
AutoCAD drawing exchange file (*.dxf)	AutoCAD drawing exchange file (*.dxf)
Wavefront (*.obj)	Wavefront (*.obj)
PLY – Polygon File Format (*.ply)	PLY – Polygon File Format (*.ply)
AutoCAD hatch pattern file (*.pat)	—
Recon M and PTS Files (*.m,*.pts)	—

Open File Format	Save As File Format
SolidWorks (*.sldprt; *.sldasm)	—
MotionBuilder (*.fbx)	—
SketchUp (*.skp)	—
—	ZCorp (*.xgl)
—	KML Google Earth (*.kml)
WAMIT (*.gdf)	—
MicroStation files (*.dgn)	—
Alias (*.fbx)	—
PDF Files (*.pdf; *.ai; *.eps)	—

Objects Import and Export

In essence, objects that you construct in a Rhino file are points, curves, surfaces, polysurfaces, and polygon meshes. In addition, you can incorporate dimensions, annotations, and rendering objects such as material properties, light objects, and environment information.

Objects to be Imported and Exported

Before saving the entire Rhino file or exporting selected Rhino objects into a particular file format, you need to know what kind of Rhino objects can be exported and whether such object types will be changed after exporting. For example, exporting to 3DS format will export rendering objects together with the 3D geometry, but this process converts all NURBS surfaces to polygon meshes. If you want to keep the NURBS surface data in the target file, you should export the file to IGES, SAT, or STP format. However, rendering objects may not be exported.

Another example is exporting to Adobe Illustrator format. The target file will become a 2D drawing projected in the active viewport. Importing the exported Adobe Illustrator file back into Rhino will convert the original 3D object into planar drawings of the 3D object.

When you work in the other direction, you have to understand what kinds of data are in the source file before importing. You cannot expect to have a set of NURBS surfaces by importing a file that contains only polygon meshes.

Conversion to Bézier Curves/Surfaces

By default, Rhino curves and surfaces are NURBS curves and surfaces. To convert such curves or surfaces to Bézier curves or surfaces, you use the ConvertToBeziers command.

File Formats

Because there are so many kinds of file formats that Rhino can export to and import from, you are advised to keep the source Rhino file before exporting and experiment with various kinds of exporting and importing before proceeding to real project work. The following delineation serves to provide some information about commonly used file formats that you may export to or import from.

STEP (.stp, .step) Files

STEP stands for Standard for the Exchange of Product model data. It is an international standard (ISO 10303) to provide a complete definition of the physical and functional characteristics of an object.

Direct X (.x) Files

Direct X (.x) is a set of technologies designed for multimedia elements. Rhino NURBS surfaces exported to Direct X will become a set of polygon meshes. Naturally, any surfaces that are imported via Direct X are polygon meshes.

Slice (.slc) Files

Slice files are StereoLithography Contour (.slc) files. It slices a 3D object to produce 2D contours. The contour lines are polylines. NURBS surfaces have to be translated into either.slc format or .stl format (described below) before the data can be read into the solid imaging machine software. Rhino solids exported will become a set of contour lines.

Point (*.asc, *.csv, *.txt, *.xyz, *.cgo_ascii, *.cc) Files

Point files are coordinates of point objects. Only point objects in a Rhino file are exported. If you open a point file, you get only points.

General Hydrostatic System (.gf, .pm) Files

GHS stands for General Hydrostatic System. These GHS files concern hydrostatic properties, stability, and other marine-related properties. If the Rhino file has such properties, you can export them to GHS formats. On the other hand, GHS files can be imported and attached to Rhino surfaces.

XGL Files

XGL files represent 3D objects for the purpose of visualization. They include all the 3D information related to SGI's OpenGL rendering library. NURBS surfaces are exported to polygon meshes.

Verband Der Automobileindustrie (.vda) Files

A VDA file is a neutral file format defined by the German Association of Automobile Industries for exchange of computer-aided design data across various systems.

RenderMan (.rib) Files

RIB stands for RenderMan Interface Bytestream. RenderMan is a technical specification for the interface between modeling application and rendering application. The file conveys model data, lights, camera, shaders, attributes, and options to rendering applications for it to produce a photorealistic image. Refer to *https://renderman.pixar.com/* for more information.

Object Properties (.csv) Files

CSV files stands for Comma-Separated Value. It is used to exchange data between applications. It has a tabular format with fields separated by a comma and quoted by a double-quote character. Rhino exports object properties to this format.

Adobe Illustrator (.ai) and Window Metafile (.wmf) Files

These two formats are viewport dependent. Exporting produces either a 2D Adobe Illustrator file or a 2D Window Metafile from the selected viewport. Therefore, you have to position the 3D object properly in the active viewport prior to exporting. Before importing from Adobe Illustrator, convert the text object in Illustrator to curves because Rhino does not read Adobe Illustrator text.

STL (.stl) Files

STL stands for Stereolithography. NURBS surfaces exported to an STL file will become triangular meshes, and imported STL files are triangular meshes. Although open objects in Rhino can also be exported to STL format, you should try to ensure the objects are water-tight by applying join, weld, and unifymeshnormals commands. To discover the location of open edges, use the SelNakedMeshEdgePt command.

VRML (.wrl, .vrml) Files

VRML stands for Virtual Reality Modeling Language. This is a standard for representing 3D scenes for delivery via the World Wide Web. Objects in VRML format are represented by polygon meshes. Naturally, NURBS surfaces will be exported as polygon meshes.

POV-Ray Mesh (.pov) Files

POV stands for Persistence of Vision. It is a raytracer that reads a source file describing the scene to be rendered and outputs a rendered image. A Rhino file exported to POV format becomes a set of data describing the rendering objects and geometry in polygon meshes. The file is written in ASCII format. In other words, you can open a POV file with the note pad or any word processor. Refer to *http://www.povray.org/* for more details about POV files.

Cult 3D (.cd) Files

Cult 3D is an application that enables you to apply interactivity to the geometry that is constructed in another 3D modeler and exported to .cd format. A Rhino file exported to Cult3D format is a polygon mesh. Refer to *http://www.cult3d.com/* to learn more about Cult 3D.

Moray UDO (.udo) Files

UDO stands for User-Defined Objects. It is a description on how objects should look in Moray and tells Moray how to render the wireframe of the object. For more information about Moray, refer to *http://www.stmuc.com/moray/*.

Light Wave (.lwo) Files

LWO stands for LightWave Object, a file format used by LightWave. The file stores information about points, NURBS surfaces, polygon meshes, splines, and surface attributes.

Raw Triangles (.raw) Files

This file format is used by Persistent of Vision applications for inputting triangular facet geometry. Naturally, surfaces are exported or imported as meshes.

ACIS (.sat, .sab) Files

ACIS is a 3D modeling kernel developed by Spatial Technology (*http://www.spatial.com/*). It stores modeling information (wireframe, surface, and solid) in two kinds of file formats: Standard ACIS Text (.sat) and Standard ACIS Binary (.sab).

IGES (.igs, .iges) Files

IGES stands for Initial Graphics Exchange Specifications. It was first introduced in 1979 as a standard platform to exchange computer-aided design data among various applications.

Parasolid (.x_t) Files

Parasolid is a kernel modeler used by many solid modeling applications. More information about it can be obtained from *http://www.ugs.com/products/open/parasolid/*.

3D Studio Files

Exporting Rhino surfaces to 3D Studio file format will convert the NURBS surfaces into polygon meshes. Importing a file back to Rhino will give a set of meshes.

Summary

2D orthographic engineering drawings are a standard means of communication among engineers. In Rhino, you create 2D drawings in two ways: by selecting a 3D object and letting the computer generate the drawing views, and by using the curve tools to construct a drawing from scratch. A Rhino 2D drawing can incorporate information on the dimensions of the object or objects the drawing represents as well as other annotation (textual information).

Rhino is a 3D digital modeling tool. You use it to construct points, curves, surfaces, polysurfaces, solids, and polygon meshes. In addition, you output photorealistic renderings, 2D drawings, and file formats of various types for downstream computerized operations. You also reuse upstream digital models by opening various file formats.

To export Rhino objects, you can use Save as command to export the entire file or use the Export selected command to export selected objects. To import other objects into Rhino, you can open, import, or insert a file.

Review Questions

1. Explain how 2D engineering drawings are produced.
2. In conjunction with the drawing views, what kinds of objects have to be incorporated in a drawing?
3. In what ways can Rhino objects be exported?
4. What are the ways to import other objects into Rhino?
5. Give a brief account of the file formats supported by Rhino.

CHAPTER 14

Rendering

Introduction

This chapter describes the use of Rhino as a tool in rendering and animation.

Objectives

After studying this chapter you should be able to:

- ❐ State the key concepts involved in rendering
- ❐ Use Rhino to produce rendered images and animations
- ❐ Manipulate the imaginary camera in a scene
- ❐ Add lighting effects to a scene
- ❐ Include environment elements in a scene
- ❐ Apply digital material properties to objects in a scene

Overview

Digital rendering is a method of producing a photorealistic image of an object in a 3D scene in the computer. Unlike shading, which simply applies a shaded color to the surface of an object, rendering takes into account the material properties (color and texture) assigned to the object, the effect of lighting in the scene, and any additional environment factors. To produce a photorealistic image, you need to:

1. Manipulate camera lens length, camera location, and camera's target location,
2. Use artificial lighting in the scene,
3. Include environment objects to enhance reality in the scene, and
4. Assign material properties to objects in a scene.

Rhino, together with Flamingo and Penguin, incorporate the capacity to produce output from four types of renderers: Rhino, Flamingo Photometric, Flamingo Raytrace, and Penguin. The Rhino renderer is the basic renderer. In producing a rendered image, Rhino takes into account the materials assigned to objects, lights added to a scene, and simple environment objects (ambient light and background color). Flamingo and Penguin are sold separately as plug-ins to the Rhino application.

Digital Rendering and Animation

Using Rhinoceros, rendering and animation can be done in any viewport, using the default camera setting, lights, environment material, and material properties. We will first focus on how to produce an image from the 3D computer models that are already constructed. Then we will try out several Rhinoceros animation tools.

Rendering

Rendering can be done in two basic ways: preview rendering and full rendering.

Render Preview

Because a fully rendered image takes into account all the digital data of a file, it may take a very long time to produce a good image, if the file includes substantial geometry and information on how the geometry is rendered. For the sake of previewing, you may want to speed up rendering time by applying minimal rendering settings. Perform the following steps:

1. Select File > Open and select the file BubbleCarRendering.3dm from the Chapter 14 folder on the companion CD.
2. Select Render > Render Preview, or click on the Render Preview button on the Render toolbar. A render preview is produced.

Note that the edges of the bubble car in the image shown in Figure 14–1 are jagged.

Figure 14–1. Render preview

To further reduce rendering time, you can preview a part of the viewport instead of rendering the entire viewport by describing a rectangular area in the screen.

3 Type RenderPreviewWindow at the command area.
4 Click on locations A and B, shown in Figure 14–2. A preview window is constructed.
5 Type RenderPreviewInWindow at the command area.
6 Click on locations A and B, shown in Figure 14–2. Another preview is constructed.

Note the difference that one command simply shows a preview window while the other enables you to save the preview as a bitmap image.

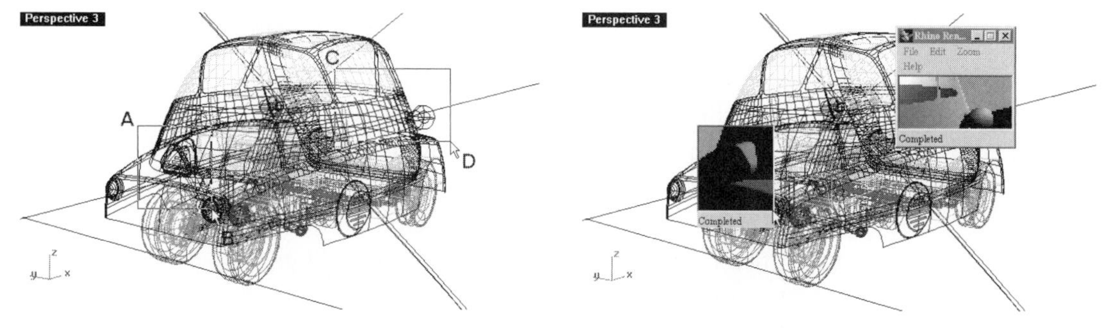

Figure 14–2. Preview window selected (left) and render preview window and render preview in window (right)

Full Render

A full render takes into consideration all the rendering information incorporated in the Rhino file.

7 Select Render > Render, or click on the Render button on the Render toolbar.

A rendered image is produced. (See Figure 14–3.) You will notice that it takes much longer to produce the rendered image than the render preview you did previously. However, if you compare the rendered image with the render preview image, you will find that the latter has a better quality. In particular, anti-aliasing is applied to remove the jagged edges. (Anti-aliasing will be explained later in this chapter.)

Figure 14–3. Full rendered image

In the process of fine tuning your image, you can render only a part of the viewport, as follows.

8 Type RenderWindow at the command area.

9 Click on locations A and B, shown in Figure 14–4. The selected rectangular area is rendered.

10 Type RenderInWindow at the command area.

11 Click on locations A and B, shown in Figure 14–4. The selected rectangular area is rendered.

12 Do not save the file.

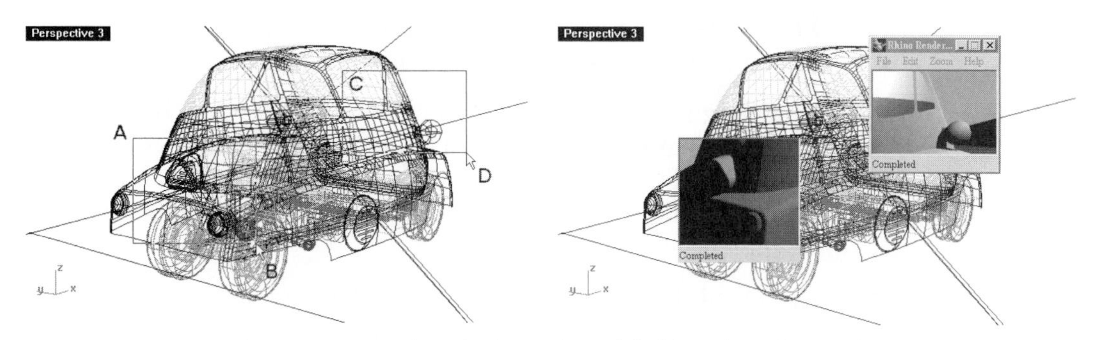

Figure 14–4. Render window selected (left) and render window and render in window (right)

Digital Animation

Simply speaking, an animation simulates motion by displaying a set of images sequentially at such a speed that the observer perceives movement of the objects in the images. Naturally, construction of an animation involves producing a set of sequential rendered images. In Rhino, the sequential rendered images can be done by manipulating the camera in several ways (turntable, path animation, and fly through animation) and by manipulating the sun's light source.

Turntable Animation

The turntable animation is the simplest way of animating the viewport. The imaginary camera simply rotates around the center of the viewport. Perform the following steps:

1. Select File > Open and select the file FoodGrinder.3dm from the Chapter 14 folder on the companion CD.
2. Click on the 360 Degree Turntable Animation button on the Setup Animation toolbar.
3. Type 36 at the command area to specify the number of frames.
4. Press the ENTER key to accept clockwise rotation.
5. Select jpg format on the command area.
6. Select the RenderFull option (or other options) on the command area.
7. Press the ENTER key to accept the Perspective viewport to render.
8. At the command area. specify a sequence name: turntable
9. Click on the Play Animation button on the Preview toolbar to observe the preview.
10. Click on the Record Animation button on the Animation toolbar to save the animation to file.

11 In the Rhino Animation dialog box, select a folder for the rendered animation to save, and click on the OK button.

Rendering will take place. The time to render depends on the number of frames, the resolution (specified in the Document Properties dialog box), and the kind of rendering. After rendering is completed, a set of rendered images, together with an html document, will be saved in the designated folder.

12 Open the folder where you save the rendered images, and click on the html file. (See Figure 14–5.)

13 Click on the Start button to run the animation, the > button to move to the next frame, the < button to go to the previous frame, the > > button to go to the last frame, and the < < to return to the first frame. To stop the animation while it is running, click on the Stop button.

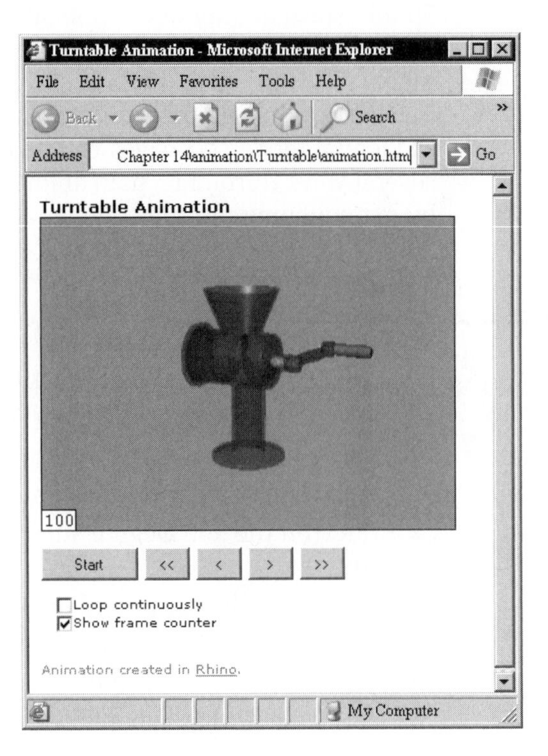

Figure 14–5. Turntable Animation html file opened

Path Animation

A path animation requires two separate paths, one for guiding the motion of the camera and one for the camera's target point. You can use a curve path for the camera and a point for the target. The target will be fixed and the camera will be moving. If you use a point for the camera and a curve for the target, the camera's location will be fixed and the target point will be moving. If you use curves for both the camera and the target, both will be moving. (If you use point objects for both the camera and the target, you will not get an animation effect.) Continue with the following steps.

14 Click on the Path Animation button on the Setup Animation toolbar.

15 Select curve A, shown in Figure 14–6, to specify the path for the camera.

16 Select point B, shown in Figure 14–6, to specify the target point.

Choosing a curve for camera path and a point for target path causes the camera to move along the path while targeting at a fixed point.

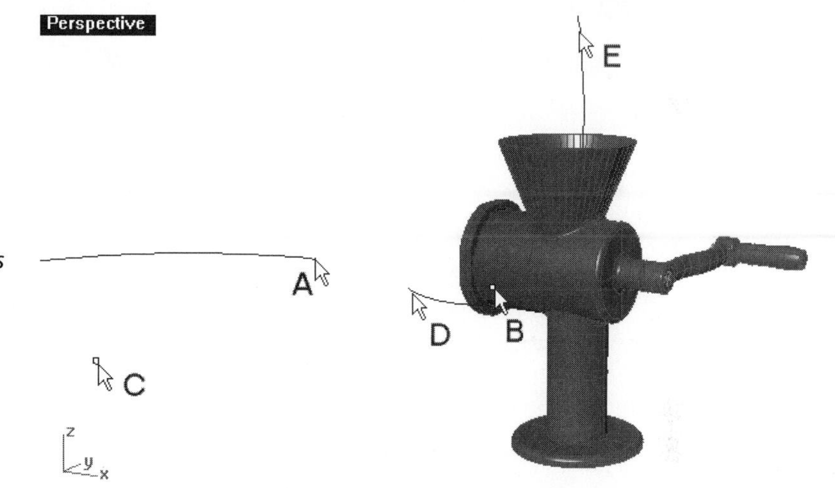

Figure 14–6. Specifying camera and target paths

17 Specify frame numbers, rendered image file type, capture method, viewport to render, and an animation sequence name.

18 Click on the Record Animation button on the Animation toolbar to save the rendered files. (Figure 14–7 shows a frame of the animation.)

Continue with the following steps to construct a path animation with the camera located at a fixed point and the target point moving along a path, and a path animation with the camera and target point both moving along path curves.

19 Repeat the procedure outlined in steps 14 through 16, but select point C as the camera path and curve D as the target path. (Figure 14–7 (left) shows a frame of the animation.)

20 Repeat the procedure outlined in steps 14 through 16, but select curve A as the camera path and curve D as the target path. (Figure 14–7 (right) shows a frame of the animation.)

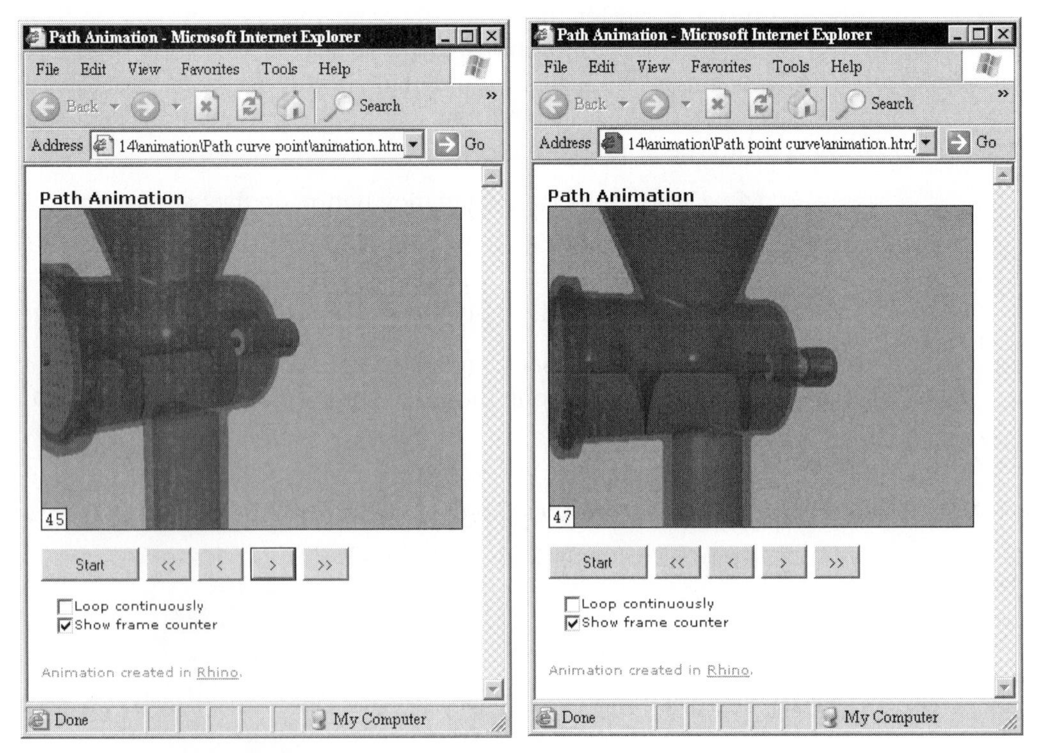

Figure 14–7. An image of the curve/point animation (left) and curve/curve animation (right)

Fly-Through Animation

A fly-through animation is a special kind of path animation. It requires only a path because the target is always tangent to the path. Continue with the following steps.

21 Click on the Fly-through Animation button on the Setup Animation toolbar.

22 Select curve ED, shown in Figure 14–6.

23 Specify the number of frames, file type, render method, viewport to render, and animation sequence.

24 Click on the Record Animation button on the Animation toolbar to save the animated images. (Figure 14–8 shows a frame of the animation.)

25 Do not save the file.

Five sets of rendered images are saved in the Chapter 14/Animation folder of the CD accompanying this book. You may open the folder to view the animations.

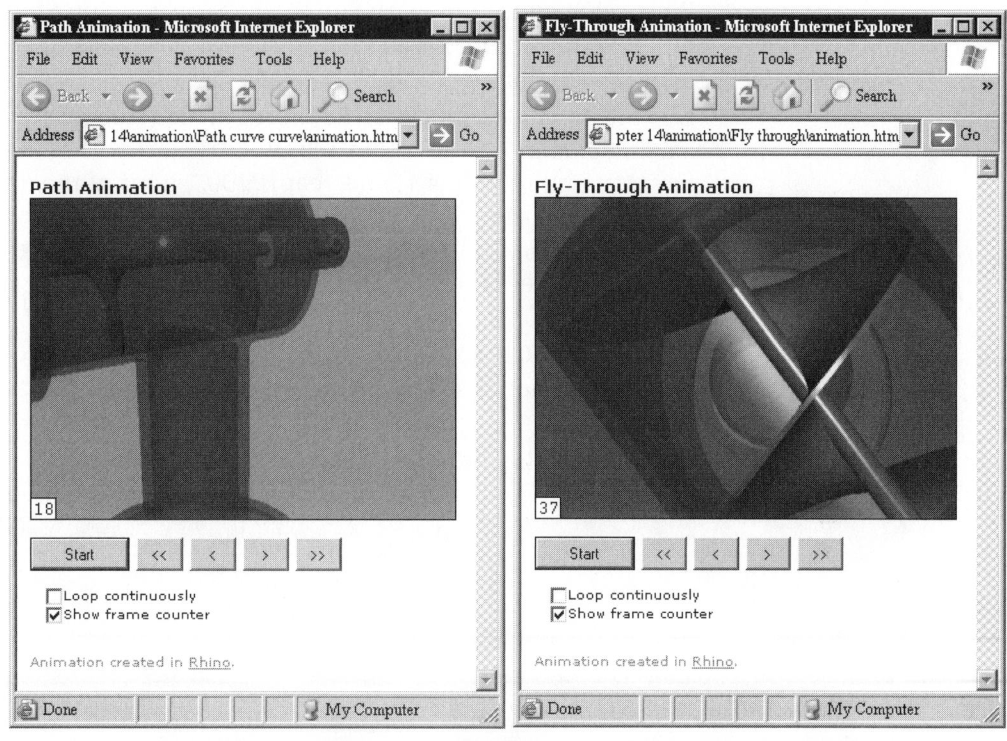

Figure 14–8. An image of the curve/curve animation (left) and fly-through animation (right)

Sun Study Animation

You can animate the sun shadow cast in a scene in a day or in a season, as follows:

1. Select File > Open and select the file LivingRoomStudy.3dm from the Chapter 14 folder on the companion CD.
2. Click on the One-Day Sun Study/Seasonal Sun Study button of the Setup Animation toolbar to set up one-day sun study animation, as follows:

   Latitude is 48 degrees, longitude is 122 degrees, north angle is 0, day of animation is 1, month of animation is 1, year of animation is 2006, start hour is 6, start minute is 0, end hour is 22, end minute is 0, minutes between frames is 15, file type is bmp, frame capture method is renderfull, viewport to render is perspective, and animation sequence is sunstudy,
3. Click on the Record button on the Animation toolbar. Sun study animation is saved. Figure 14–9 (left) shows a frame of the animation.

4 Right-click on the One-Day Sun Study/Seasonal Sun Study button of the Setup Animation toolbar to set up seasonal sun study animation, as follows:

Latitude is 48, longitude is 122, north angle (clockwise from Y axis) is 0, start day is 1, start month is 1, start year is 2006, end day is 1, end month is 1, end year is 2007, hour of the day for animation is 15, minute of the day is 0, days between frames is 10, file type is bmp, frame capture method is renderfull, viewport name to render is perspective, and animation sequence name is seasonstudy.

5 Click on the Record button on the Animation toolbar. Season study animation is saved. Figure 14–9 (right) shows a frame of the animation.

6 Do not save the file.

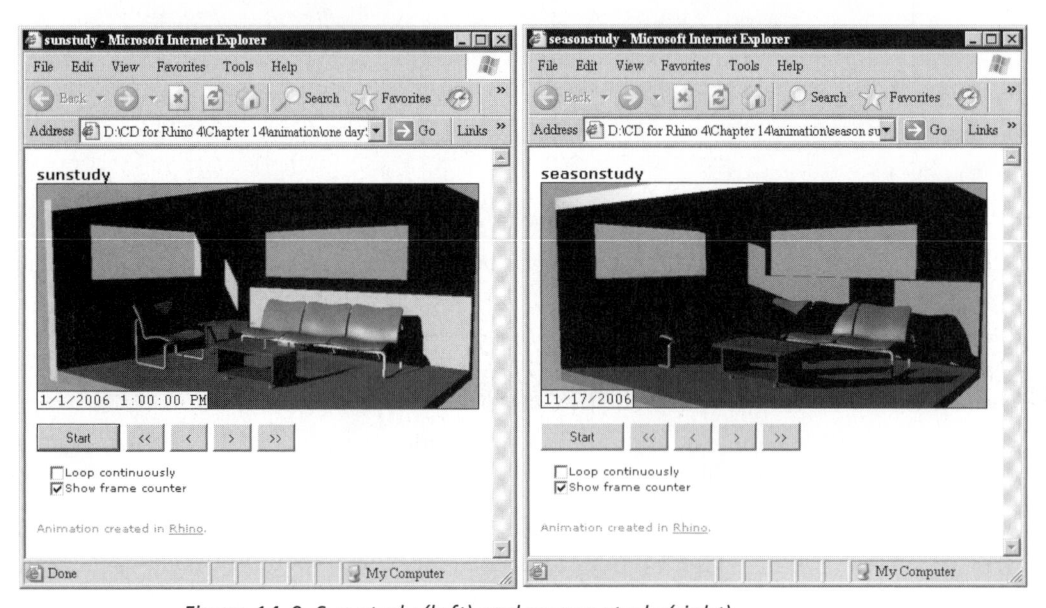

Figure 14–9. Sun study (left) and season study (right)

Camera Setting

In each viewport, there is an imaginary camera, which is defined in the Viewport Properties dialog box, delineating its lens size, location, and target point. You may open the Viewport Properties dialog box by right-clicking on a viewport's label and selecting Viewport Properties from the menu.

Lens Setting

Viewport can be classified as parallel projection viewport or perspective viewport. If you are going to produce a photorealistic rendered image, it is natural that you should set the viewport to be perspective, because this viewport allows you to set the lens size to simulate real camera photo-taking. To set lens size, you can input a value in the Lens length field in the Viewport Properties dialog box or use the Lens Length toolbar. Now perform the following steps:

1. Select File > Open and select the file SkateScooter.3dm from the Chapter 14 folder on the companion CD.
2. Click on the 17mm Camera button of the Lens Length toolbar, and then render the perspective viewport.
3. Click on other buttons on the Lens Length toolbar to appreciate the effect of having different lens lengths. (See Figure 14–10.)

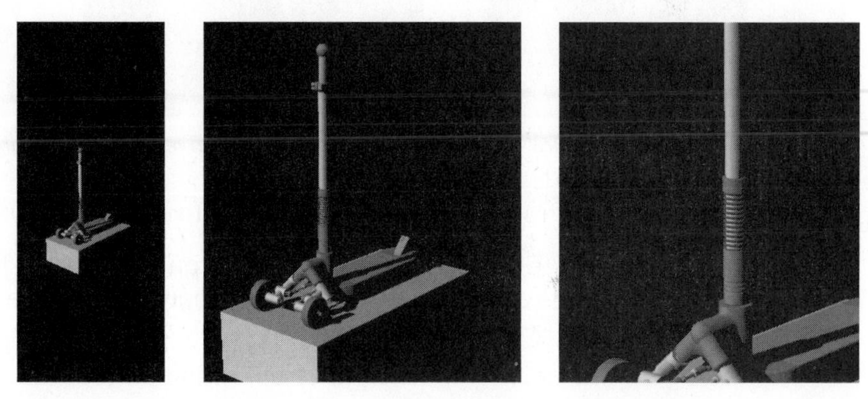

Figure 14–10. From left to right: 17mm lens, 50mm lens, and 100mm lens

Camera Orientation—Walk About

Imagining that you are holding a camera and watching the scene through the perspective viewport, you click on the buttons on the Walk About toolbar to simulate the effect of walking about in the scene. To control the step size, you can click on the buttons of the Step Size toolbar. Continue with the following steps.

4. Click on the 25mm lens button on the Lens Length toolbar. (See Figure 14–11.)
5. Click on the Large Steps button on the Step Size toolbar. The step size is now set.

6. Click on the Walk Forward button on the Walkabout toolbar. (See Figure 14–11.)
7. Click on the Walk Back button on the Walkabout toolbar.
8. Click on the Walk Right button on the Walkabout toolbar. (See Figure 14–11.)
9. Click on the Walk Left button on the Walkabout toolbar.

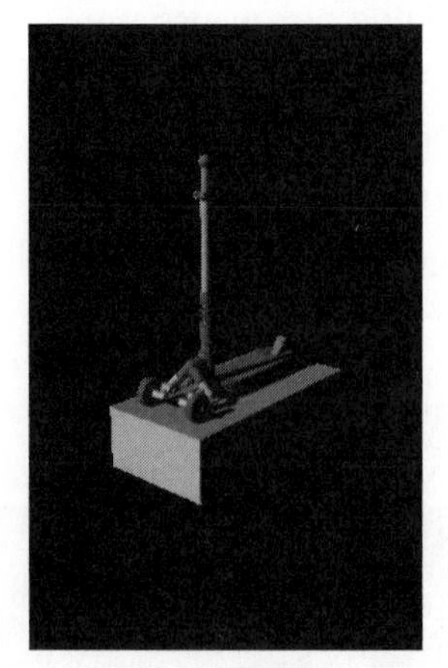
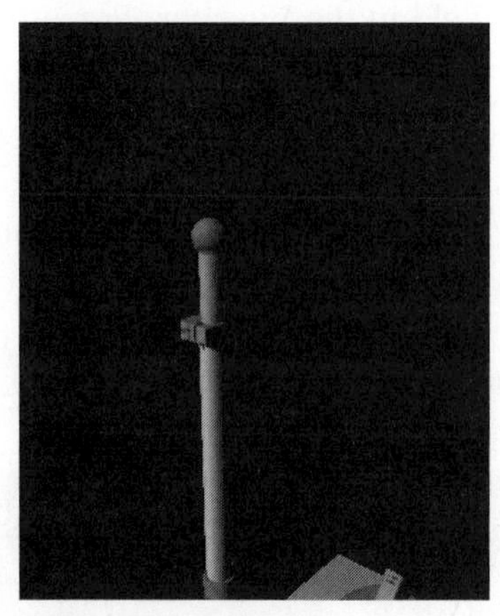
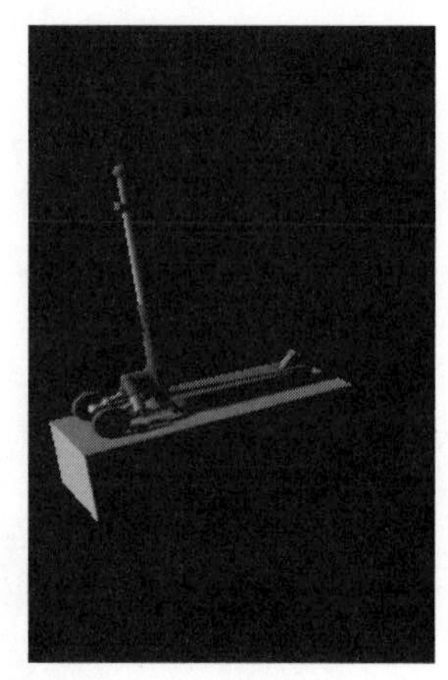

Figure 14–11. From left to right: 25mm lens, walk forward, and walk right

You can also elevate your perspective up and down, as follows.

10. Click on the 50mm lens button on the Lens Length toolbar and render the perspective viewport. (See Figure 14–12.)
11. Click on the Medium Large Steps button on the Step Size toolbar.
12. Click on the elevator up button on the Walkabout toolbar. (See Figure 14–12.)
13. Click on the elevator down button on the Walkabout toolbar twice. (See Figure 14–12.)

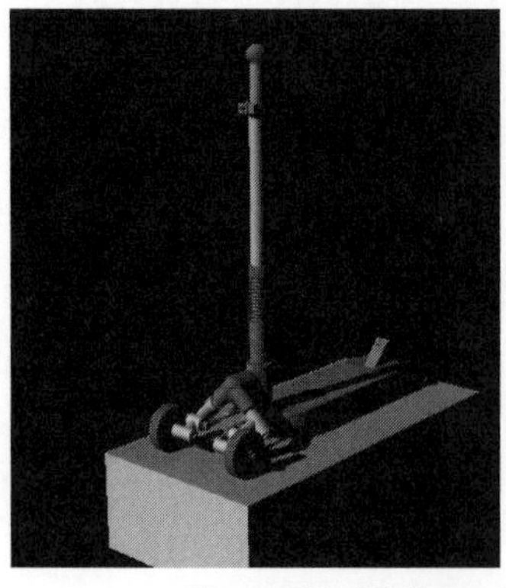

Figure 14–12. From left to right: 50mm lens, elevator up, and elevator down

Camera Orientation—Look About

In addition to simulating the effect of walking about in the scene, you can simulate the effect of looking about. That is, you stand still but turn your head around. The rotation increment of looking about has to be set beforehand. Continue with the following steps:

14 Click on the Rotate Increment button on the Lookabout toolbar.

15 Type 30 at the command area to set the rotation increment to 30 degree.

16 Click on the Look Right button on the Lookabout toolbar, and render the viewport. (See Figure 14–13.)

17 Click on the Look Left button on the Lookabout toolbar.

18 Click on the Look Up button on the Lookabout toolbar. (See Figure 14–13.)

19 Click on the Look Down button on the Lookabout toolbar twice. (See Figure 14–13.)

20 Click on the Look Up button on the Lookabout toolbar.

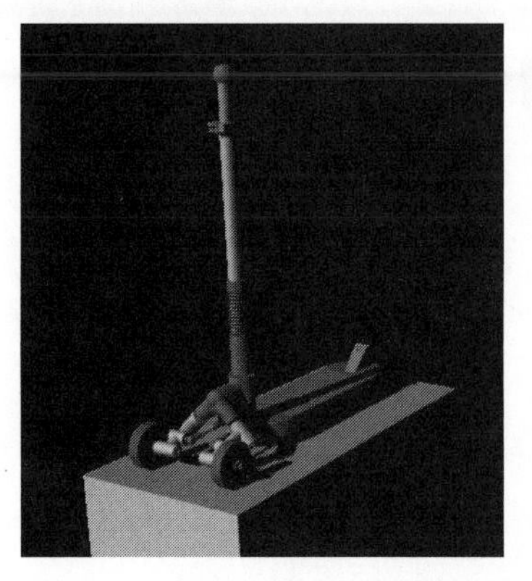

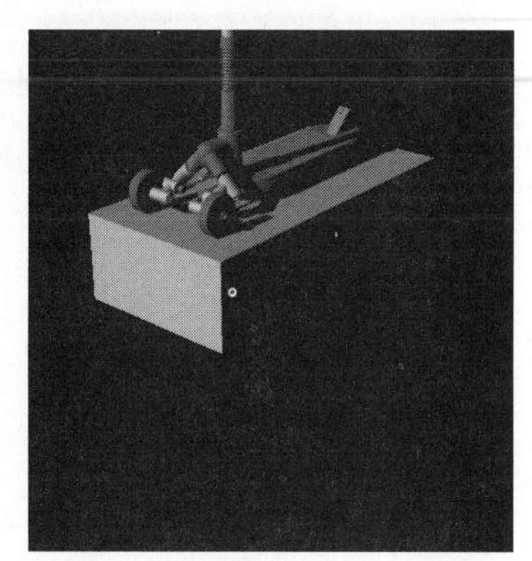

Figure 14–13. From left to right: look right, look up, and look down

Camera Orientation—Moving Target Point

To focus, you can move the target point of the camera towards the center of the bounding box of the target object, as follows:

21 Select View > Set Camera > Move Target to Object, or type Movetargetttoobjects at the command area.

22 Select polysurface A, shown in Figure 14–14, and press the ENTER key. (You may choose more than one object.) The target of the camera is set to the center of the bounding box of the selected object

23 Click on Undo View Change on the Standard toolbar to reset the viewport.

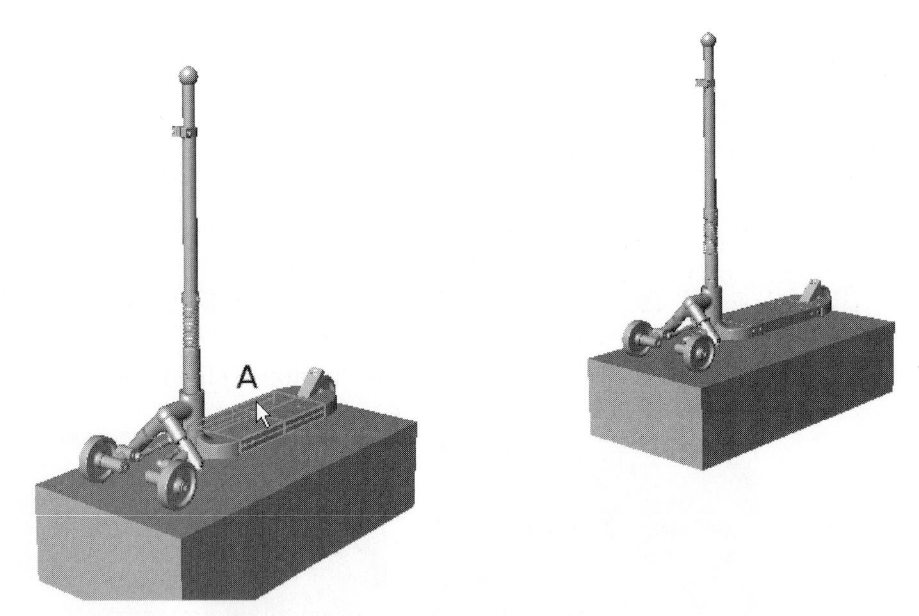

Figure 14–14. Original display (left) and camera target set to center of bounding box of object A (right)

Camera Orientation—To Surface

You can move a camera and align it with the normal direction of a selected surface, as follows:

24 Select View > Set Camera > Orient Camera to Surface, or type Orientcameratosrf at the command area.

25 Select surface A and location B on the surface (Figure 14–15). The camera and its target direction are aligned to the normal of the surface at designated location.

26 Click on Undo View Change on the Standard toolbar to reset the viewport.

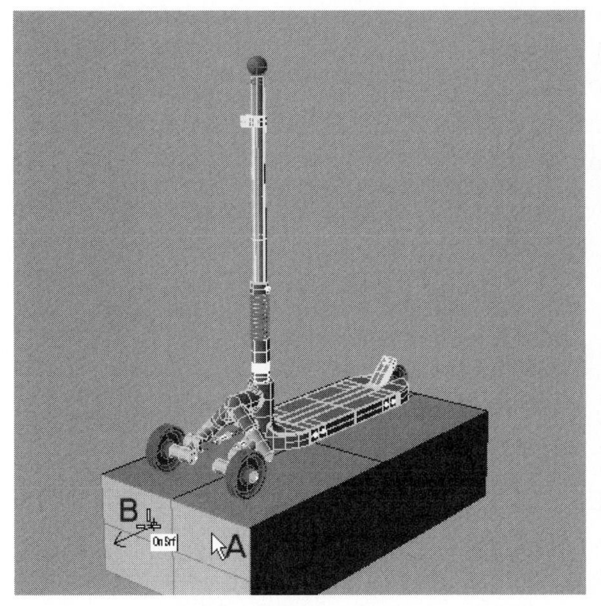

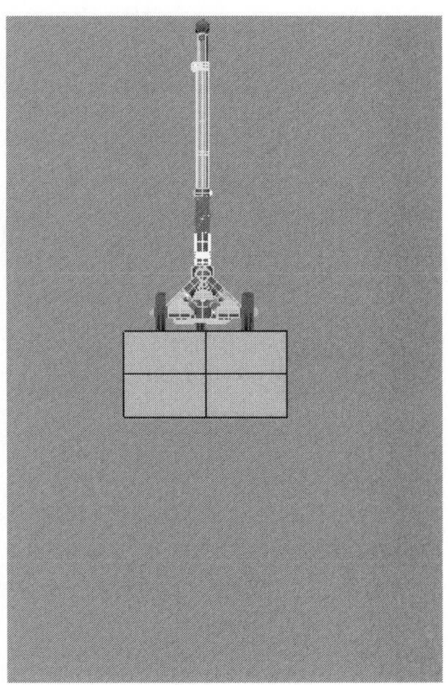

Figure 14–15. Face and location selected (left) and camera and target aligned (right)

Camera Manipulation—Camera Pyramid

A more interactive way of manipulating the camera of a selected viewport is to display its graphical representation in terms of a pyramid in other viewports. Continue with the following steps:

27 Double-click on the Perspective viewport's label to reset the display to 4 viewport display.

28 Check View > Show Camera, or right-click on any viewport's label and check Show Camera. The camera's graphical pyramid is displayed. This command toggles between showing and hiding the camera of a selected viewport.

The pyramid represents the camera's location, angle, and field of view. It has four control points. By dragging the control point at the tip of the pyramid, you adjust the camera's viewpoint; by dragging the control point at the opposite end of the straight line that runs through the axis of the pyramid, you adjust the target point location; by dragging the control point on the middle of one side of the pyramid base, you roll the camera; and by dragging the control point on one of the corners of the base of the pyramid, you adjust the field of view/lens angle. In addition, there are two rectangular planes in dotted lines that depict the far and near clipping planes.

29 Drag control point A, shown in Figure 14–16. The camera's viewpoint is changed.
30 Drag control point B. The camera's target point is changed.
31 Drag control point C. The camera's field of view is changed.
32 Drag control point D. The camera is rolled.
33 Do not save the file.

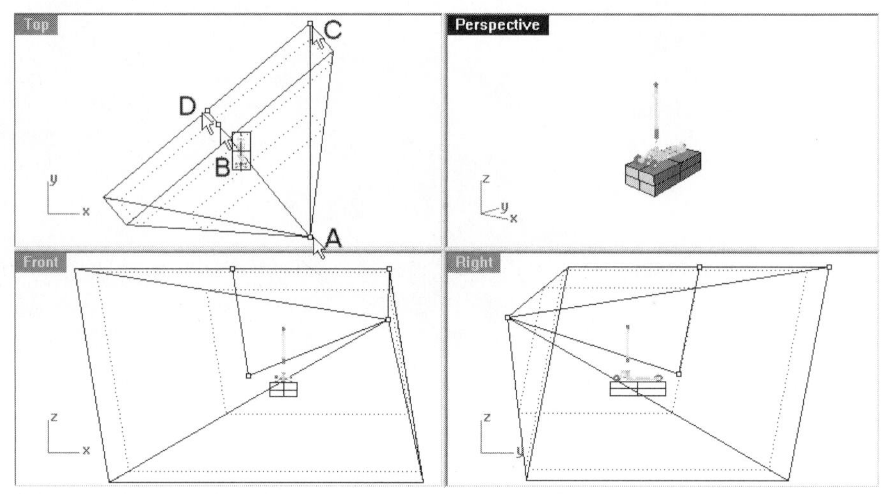

Figure 14–16. Camera's pyramid representation being manipulated

Camera Manipulation—Perspective Match

To add realism, you may place a wallpaper in a viewport, using it as background image during rendering. In order to match the 3D objects with the 2D image, you have to adjust the camera, as follows:

1 Select File > Open and select the file PerspectiveMatch.3dm from the Chapter 14 folder on the companion CD.

A bounding box is constructed, and lines are drawn on the wallpaper to help perspective match.

2 Select View > Set Camera > Match Wallpaper Image, or click on the Match Perspective Projection button on the Set View toolbar.
3 Click on location '1' and then select end point A. (See Figure 14–17.)
4 Click on location '2' and then select end point B.
5 Click on location '3' and then select end point C.
6 Click on location '4' and then select end point D.
7 Click on location '5' and then select end point E.

8 Click on location '6' and then select end point F. (A minimum of six points is required.)

9 Press the ENTER key. The camera is adjusted.

Figure 14–17. Perspective being matched

10 Turn off the BoundingBox layer. The bounding box is not invisible. Alternatively, you may delete the bounding box.

11 Right-click on the Viewport's label, and select Viewport Properties.

12 In the Viewport Properties dialog box, click on the Browse button of the Wallpaper options area.

13 Select the image "Toronto00.jpg" from the Chapter 14 folder on the companion CD, replacing the image with one without any reference lines.

14 Click on the OK button.

15 Matching is complete. (See Figure 14–18.) Do not save your file.

Figure 14–18. Bounding box removed and wallpaper changed (left) and rendered image (right)

Digital Lighting

The purpose of lighting is to illuminate the 3D scene and to achieve specific lighting effects. In a Rhino scene, you can add virtual lights and apply sunlight and environmental lighting effects. There are five categories of virtual lighting: spotlights, point light, directional light, rectangular light, and linear light.

Default Light

In a Rhino file, there is a default light. If you imagine the screen as a camera viewer through which you look into the 3D space in which you construct the model, the default light is located over your left shoulder. The intensity of this light is always good enough to illuminate the entire 3D space, no matter how large it is. This is why you always have a well-illuminated scene in any rendering processes, even though you have not added any light. If you are satisfied with the default lighting effect, you do not need to add any light.

Note that once a light of any kind is added, the default light is turned off automatically. It will turn on again if you delete all added lights. Therefore, to simulate a dark environment, you need to add a very dim light in the scene.

Spotlights

A spotlight is characterized by a light source emitting a conical beam of light toward a target. It illuminates a conical volume in the scene. In the scene, a spotlight is represented by two cones with a common vertex. They represent the direction of the light, not the range of the light. The inner cone represents the area of full brightness of the light, and brightness will decrease from the inner cone toward the outer cone. To adjust the size and location of the spotlight, you can turn on the control points and drag them to new locations. A spotlight that has narrower cones produces more detail than a spotlight with wider cones. Perform the following steps:

1. Select File > Open and select the file Modelcar.3dm from the Chapter 14 folder on the companion CD.
2. Select Render > Create Spotlight, or click on the Create Spotlight button on the Lights toolbar.
3. Referencing Figure 14–19, click on location A to specify the base of the cone of the spotlight and location B to specify the radius of the cone of the spotlight. Exact locations are unimportant.

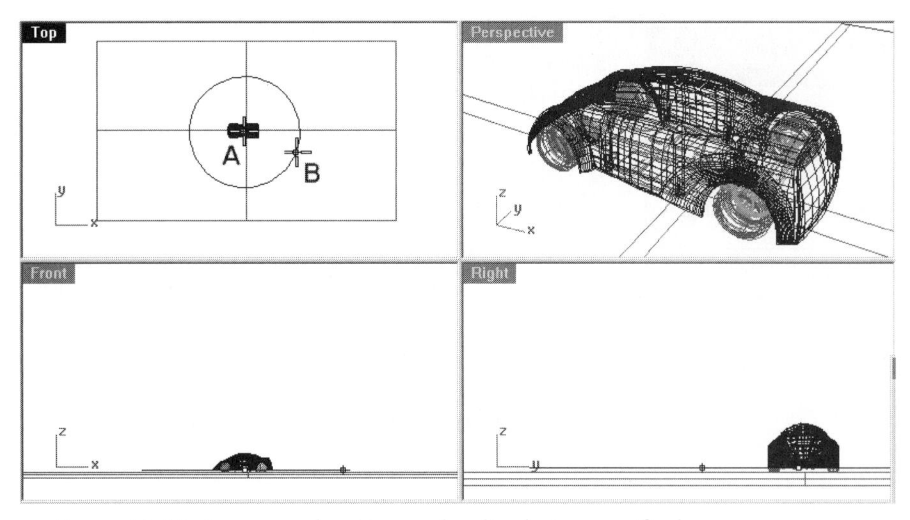

Figure 14–19. Base cone location and radius being specified

As explained earlier in this book, you can construct a 3D point by holding down the Control key while clicking on the screen. Continue with the following steps to specify a 3D point to locate the spotlight.

4 Hold down the control key, and then click on location C and then D, shown in Figure 14–20, to specify the end of the cone of the spotlight.

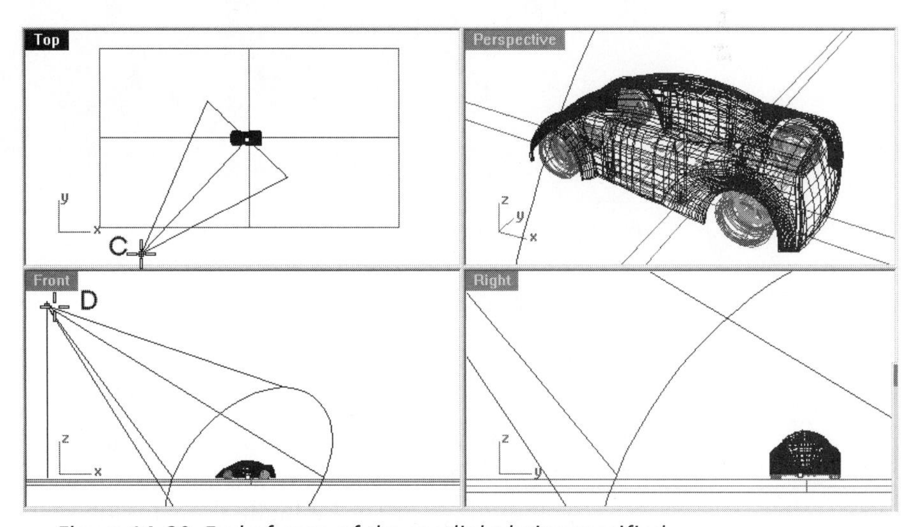

Figure 14–20. End of cone of the spotlight being specified

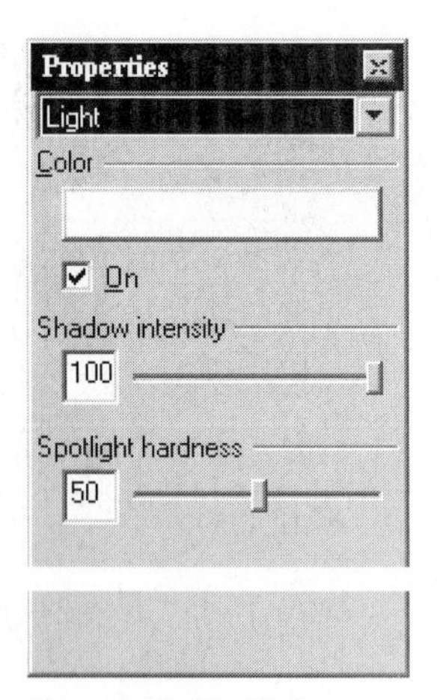

Figure 14–21. Light properties

As can be seen on your screen, the spotlight is displayed as two cones. The inner cone depicts the area of full intensity of the light. From the inner cone moving outward to the outer cone, the intensity of the light weakens. Continue with the following steps:

5. Click on the Edit Light Properties button of the Properties toolbar or the Lights toolbar.
6. As shown in Figure 14–21, the spotlight's properties include color, on/off, intensity, and hardness. Among them, the hardness slider bar determines size of the inner cone of the spotlight. The larger the value, the harder is the light's edge. Click on the X mark at the upper corner of the dialog box to close it.
7. Render the perspective viewport. (See Figure 14–22.)

Figure 14–22. Rendered with a spotlight added

Set View to Spotlight

To realize how the spotlight is located in relation to the objects to be illuminated, you can set the view to selected spotlight, as follows:

8. Click on the Top viewport to make it the current viewport.
9. Select Render > Set View to Spotlight, or click on the Set view to spotlight button on the Light Tools toolbar.

10 Select the spotlight. The viewport is set to the spotlight's direction, as shown in Figure 14–23.

11 Right-click on the Top viewport's label, and select Set View > Top to reset the Top viewport's orientation.

12 Zoom the viewport to its extents.

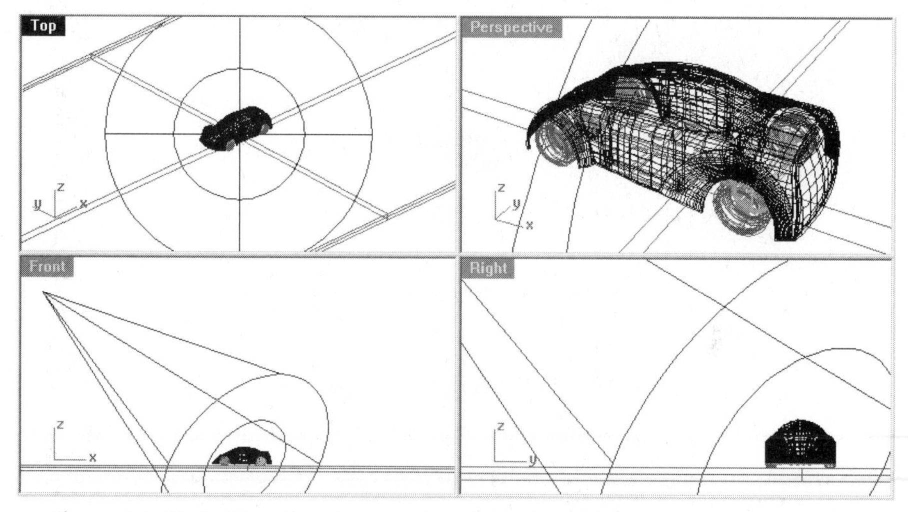

Figure 14–23. Setting the viewport to selected spotlight

Edit Light by Looking

Edit light by looking is a way to position a spotlight. You can adjust the spotlight's projection direction as if you are manipulating the spotlight at the spotlight location, as follows.

13 Select Render > Edit Light by Looking, or click on the Edit light by looking button on the Light Tools toolbar.

14 Select the spotlight in the Front viewport.

15 Click on the Rotate View button on the Standard toolbar.

16 Manipulate the Front viewport. (See Figure 14–24.)

17 Press the ENTER key. The spotlight direction is set and the display is returned to its previous orientation.

18 Undo the last command.

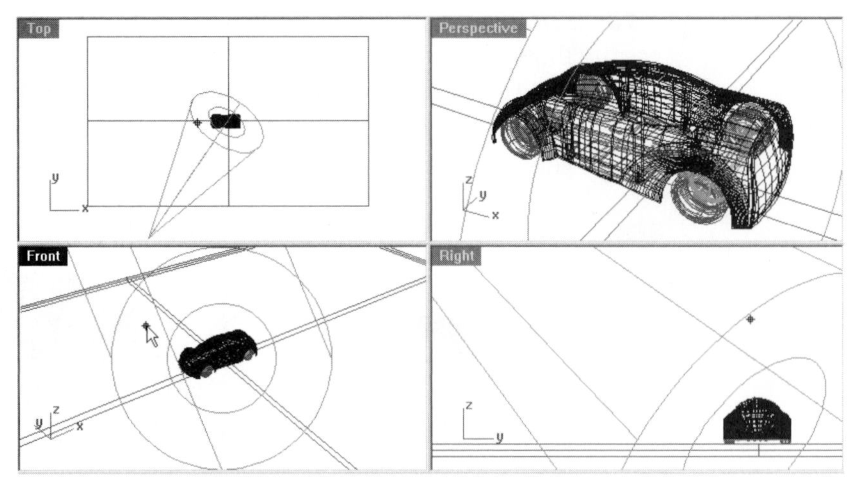

Figure 14–24. Manipulating spotlight

Set Spotlight to View

Another way to position a spotlight is to set the spotlight to a specified view, as follows:

19 Select Render > Set Spotlight to View, or click on the Set spotlight to view/Add a spotlight from current view button on the Light Tools toolbar.

20 Click on the Perspective viewport

21 Select spotlight A, shown in Figure 14–25. The spotlight is oriented in accordance with the orientation of the selected viewport.

22 Undo the last command.

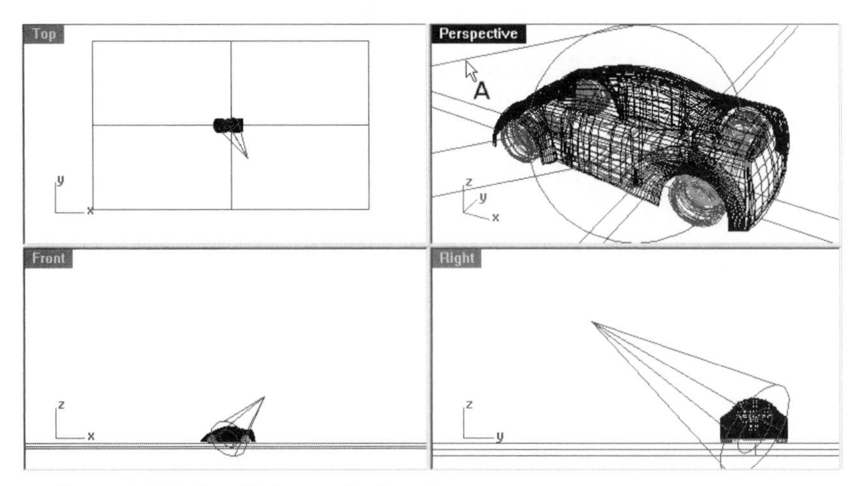

Figure 14–25. Spotlight manipulated

Contrary to setting a spotlight to the viewport's viewing direction, you can change the viewport's orientation to match the lighting direction by selecting Render > Set View to Spotlight, thus allowing you to see how the light is shining onto the scene.

Point Light

A point light is analogous to an ordinary electric light bulb that emits light in all directions. Point light illuminates an entire scene. To construct a point light, you specify a location of the light source, as follows:

23 Turn off the spotlight by clicking on the Edit Light Properties button of the Properties toolbar or the Lights toolbar, and clearing the On check box.

24 Select Render > Create Point Light, or click on the Create Point Light button on the Lights toolbar.

25 Hold down the control key, and click on location A and then B, shown in Figure 14–26. A point light is constructed.

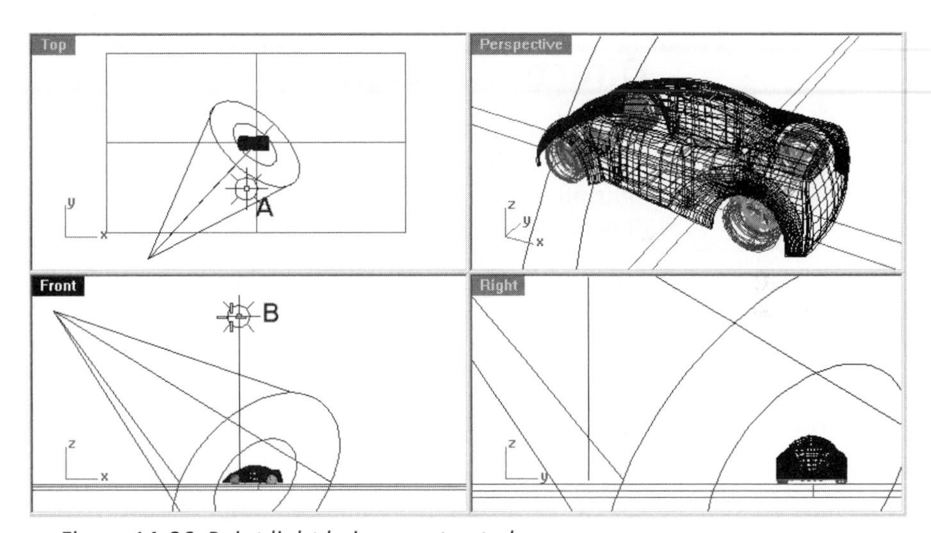

Figure 14–26. Point light being constructed

26 Render the perspective viewport. The result is shown in Figure 14–27.

27 Do not save the file.

Figure 14–27. Viewport rendered by using the point light

Directional Light

Directional lighting derives from a light source at a distance, such as the sun. This lighting, generally represented as a beam or parallel beams of light, illuminates an entire scene. You specify a direction vector for the directional light.

A directional light in a scene is a symbol. It depicts the direction. It does not matter where it is placed. Perform the following steps:

1. Select File > Open and select the file LivingRoom.3dm from the Chapter 14 folder on the companion CD.
2. Select Render > Create Directional Light, or click on the Create Directional Light button on the Lights toolbar.
3. Select location A, shown in Figure 14–28, to specify the end-of-light direction vector.
4. Hold down the control key, and click on location B and then location C, shown in Figure 14–28, to specify the start-of-light direction vector. A direction light is constructed.
5. Render the perspective viewport. (See Figure 14–29.)

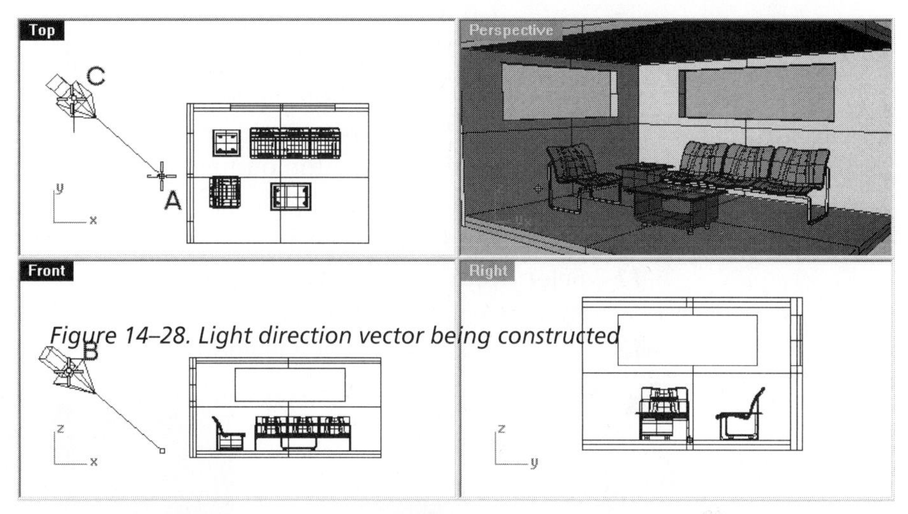

Figure 14–28. Light direction vector being constructed

Figure 14–28. Light direction vector being constructed

Figure 14–29. Direction light added and perspective viewport rendered

Rectangular Light

Rectangular lighting simulates the effect of a fluorescent light box. You specify the location, width, and length of the rectangular light. In a scene, a rectangular light is a rectangular symbol depicting a rectangular light box. Like other types of lights, a rectangular light is not shown in the rendered image. To add a rectangular light to a scene, continue with the following steps.

6 Click on the Edit Light Properties button of the Properties toolbar or the Lights toolbar.

7. Select the direction light and clear the On button of the Properties dialog box. The direction light is turned off.
8. Check the Ortho button on the Status bar.
9. Select Render > Create Rectangular Light, or click on the Create Rectangular Light button on the Lights toolbar.
10. Hold down the control key, select location A and then B, shown in Figure 14–30, to specify the location of the rectangular light.
11. Select locations C and D to specify the length and width of the light. A rectangular light is constructed.
12. Render the perspective viewport, as shown in Figure 14–31.

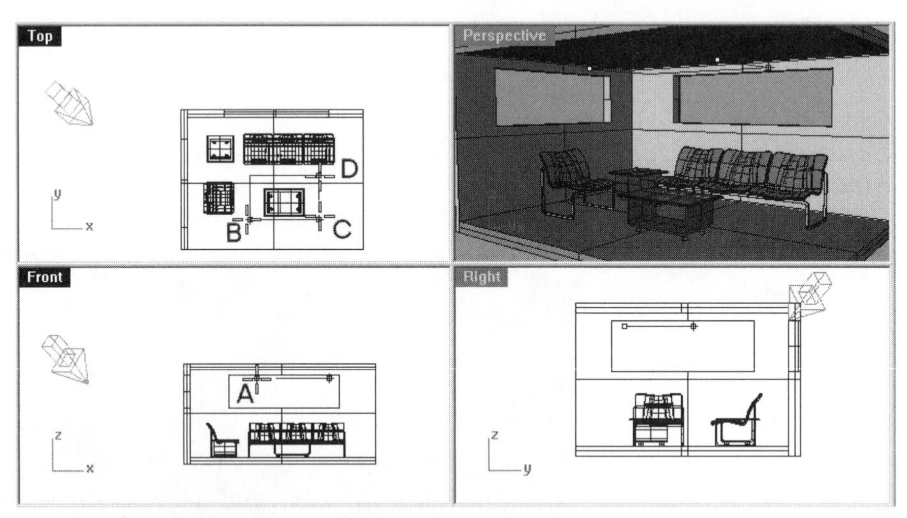

Figure 14–30. Rectangular light being constructed

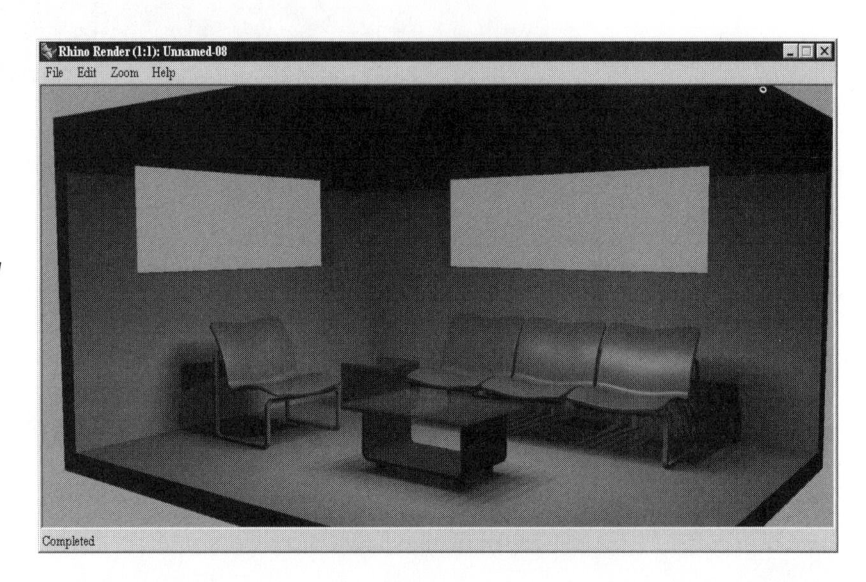

Figure 14–31. Rendered rectangular-light scene

Linear Light

Linear light simulates the effect of a fluorescent tube being added to a scene. You specify the start point and the end point of the linear light. In a scene, a linear light is a symbol in the shape of a cylinder. It is not displayed in the rendered image. To turn off the rectangular light and add a linear light to a scene, perform the following steps.

13 Click on the Edit Light Properties button of the Properties toolbar or the Lights toolbar.

14 Select the direction light and clear the On button of the Properties dialog box. The direction light is turned off.

15 Select Render > Create Linear Light, or click on the Create Linear Light button on the Lights toolbar.

16 Hold down the control key, select location A and then B, shown in Figure 14–33, to specify the origin of the linear light.

17 Select location C (shown in Figure 14–32) to specify the length and direction of the light. A linear light is constructed.

18 Render the scene, as shown in Figure 14–33.

19 Do not save your file.

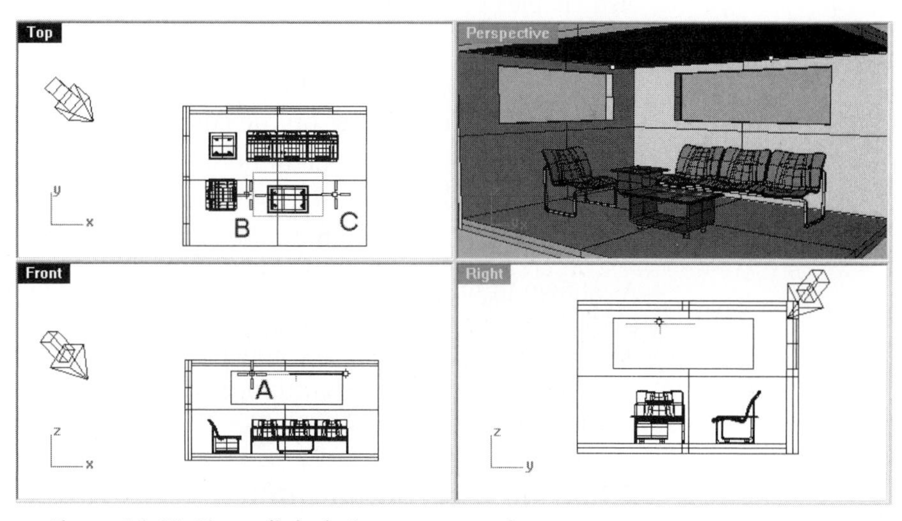

Figure 14–32. Linear light being constructed

Figure 14–33. Rendered linear-light scene

Bouncing Lights

To enhance visual effect, you may want to put a directional light, spotlight, or point light in such a location that produces highlight on designated positions of a surface. To construct new lights or reposition existing lights to produce such effect, you use the bouncing light tool. This tool also enables you to construct a helper line for placing objects that show in reflections. Perform the following steps:

1. Select File > Open and select the file ModelCarBounce.3dm from the Chapter 14 folder on the companion CD.
2. Click on the Bounce Light button on the Lights toolbar.
3. Press the ENTER key to construct a new bounce light.
4. Select the Type option on the command area.

As shown in the command line, there are four options: directional light, point light, spotlight, and line. Naturally, directional light, point light, and spotlight options enable you to place such lights as bouncing light into the active viewport. The fourth option, line, constructs a helper light for you to place a light precisely.

Bouncing Directional Light

To construct bouncing light from a directional light, continue with the following steps.

5. Select the DirectionalLight option on the command area.
6. Disable Osnap.
7. Select surface A, shown in Figure 14–34.

(***NOTE:*** *You have to click on the viewport that you want to render.*)

8 Click on location B, where you want to have specular highlight. A bouncing directional light is constructed.

9 Render the perspective viewport. (See Figure 14–35.)

Because the exact location where you place the bouncing light may not be the same, the rendered image shown in Figure 14-35 may be different from yours.

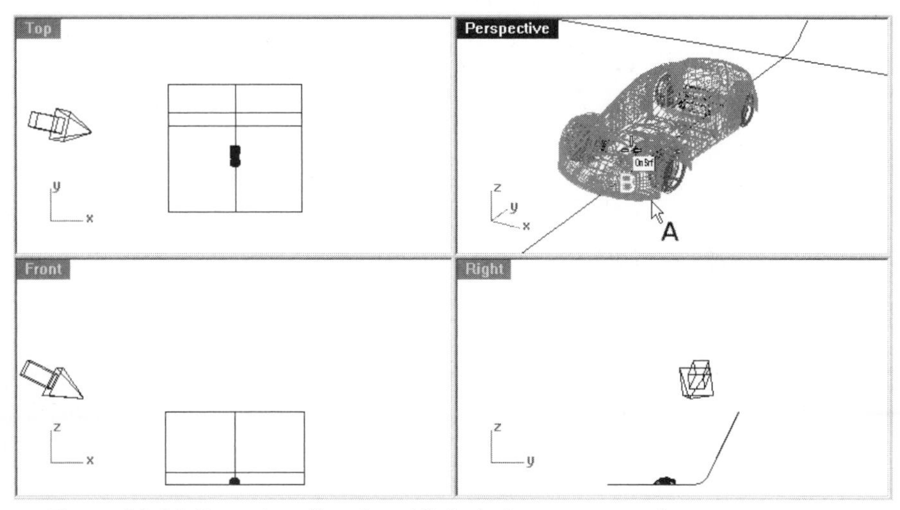

Figure 14–34. Bouncing directional light being constructed

Figure 14–35. Perspective viewport rendered

Bouncing Spotlight

To construct a spotlight that bounces at a designated location in the perspective viewport, continue with the following steps.

10 Set the current layer to Light02 and turn off the Light01 layer. The bouncing directional light is turned off. The default setting in the Render page of the Options dialog box does not use any light residing on layers that are turned off.

11 Click on the Bounce Light button on the Lights toolbar.

12 Press the ENTER key to construct a bounce light.

13 Select the Type option and then the Spotlight option on the command area.

14 Select surface A, shown in Figure 14–36.

15 Click on location B. A bouncing spotlight is constructed.

16 Render the perspective viewport. (See Figure 14–37.)

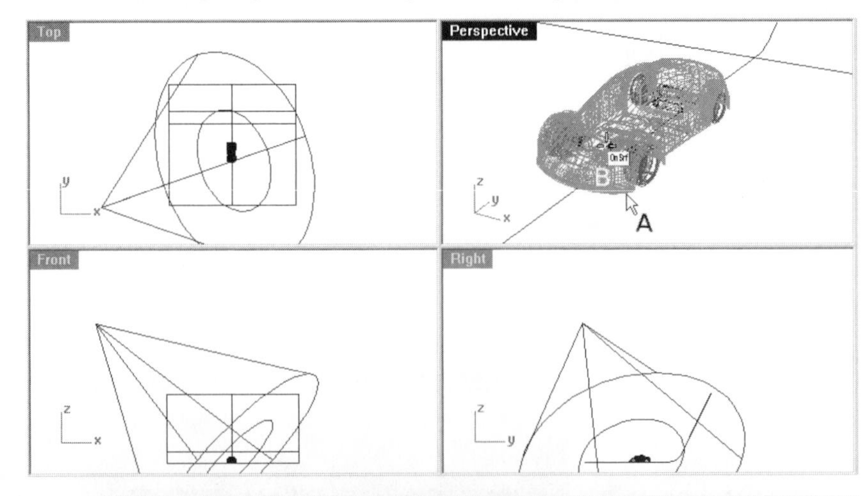

Figure 14–36. Bouncing spotlight being constructed

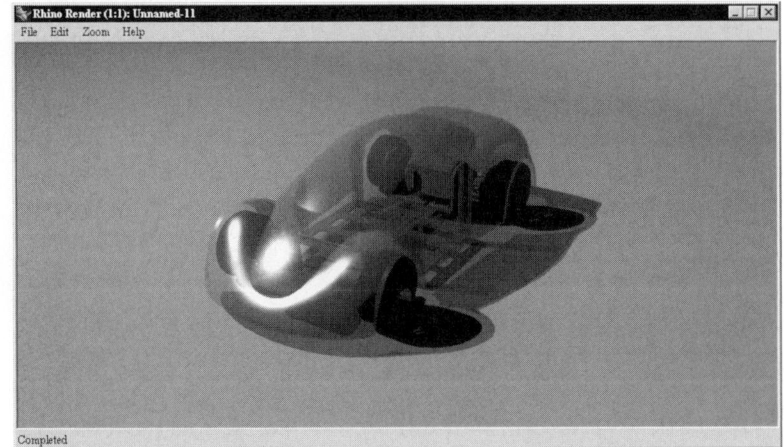

Figure 14–37. Viewport rendered

Bouncing Point Light

To place a point light that bounces light at designated location, continue with the following steps.

17 Set the current layer to Light03 and turn off the Light02 layer to turn off the bouncing spotlight.
18 Click on the Bounce Light button on the Lights toolbar.
19 Press the ENTER key to construct a bounce light.
20 Select the Type option and then the PointLight option on the command area.
21 Select surface A, shown in Figure 14–38.
22 Click on location B. A bouncing spotlight is constructed.
23 Render the perspective viewport. (See Figure 14–39.)

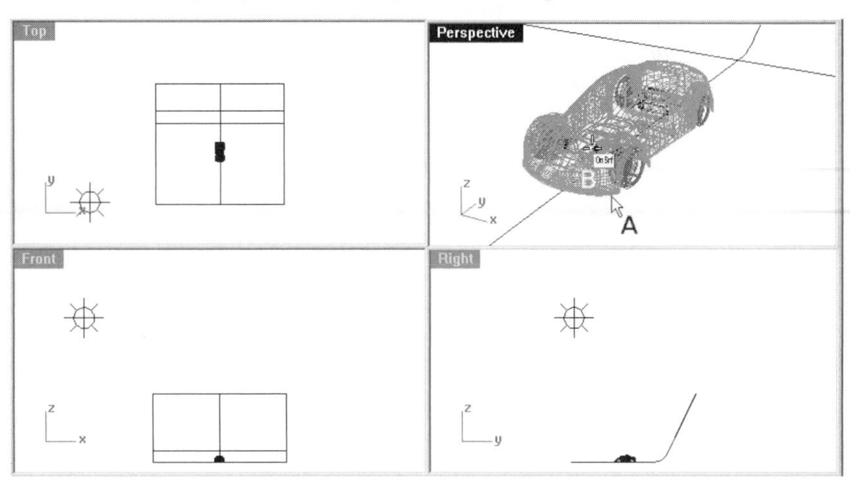

Figure 14–38. Bouncing point light being constructed

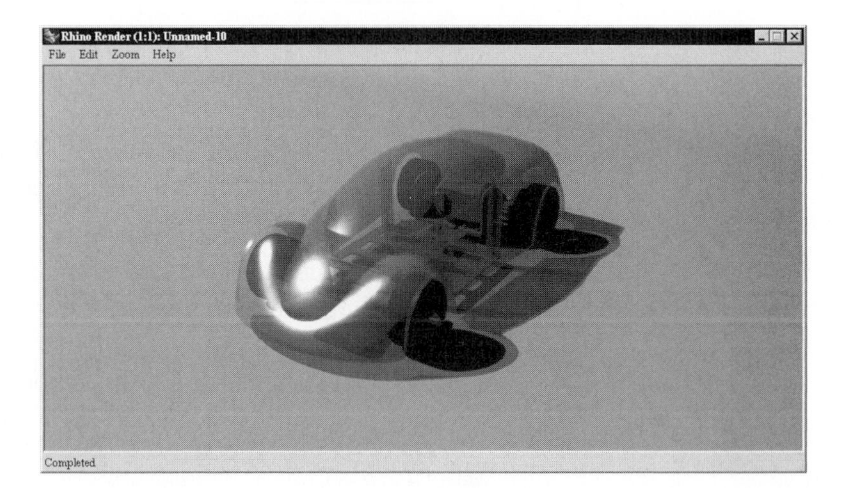

Figure 14–39. Viewport rendered

Helper Line

The bouncing line is a line that depicts where the light source should be directed onto the surface to give bouncing light effect. With this line, you can place a light or place an object to reflect under the light. To construct a bouncing line and then use the line for light placement, continue with the following steps.

24 Set the current layer to Light04 and turn off the Light03 layer.

25 Click on the Bounce Light button on the Lights toolbar.

26 Press the ENTER key to construct a bounce light.

27 Select the Line option on the command area.

28 Select surface A, shown in Figure 14–40.

29 Click on location B. A helper line for subsequent location of any light source is constructed.

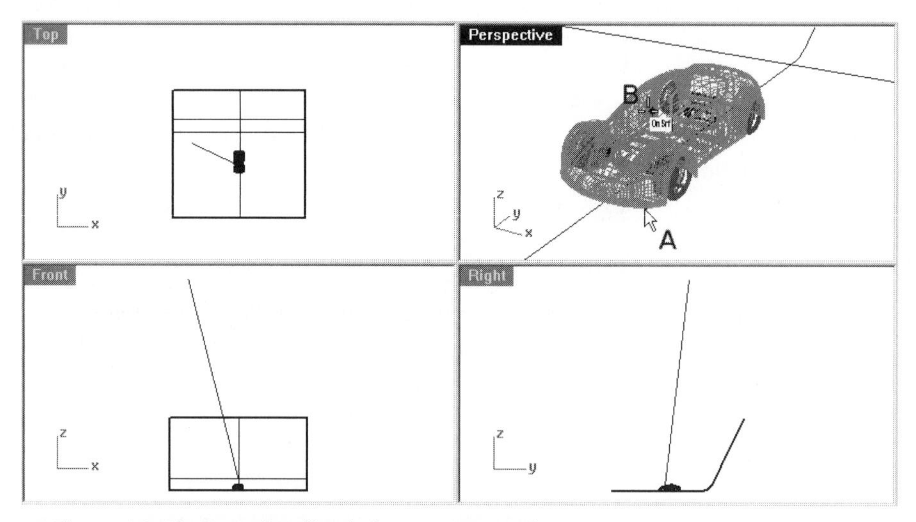

Figure 14–40. Bouncing line being constructed

Using the bouncing line, you can place a directional light, spotlight, or point light. Continue with the following steps to place a directional light along the bouncing line.

30 Select the DirectionalLight option on the command area.

31 Set object snap mode to nearest.

32 Select location A and then location B, shown in Figure 14–41. A directional light is placed along the bouncing line.

33 Render the perspective viewport. (See Figure 14–42.)

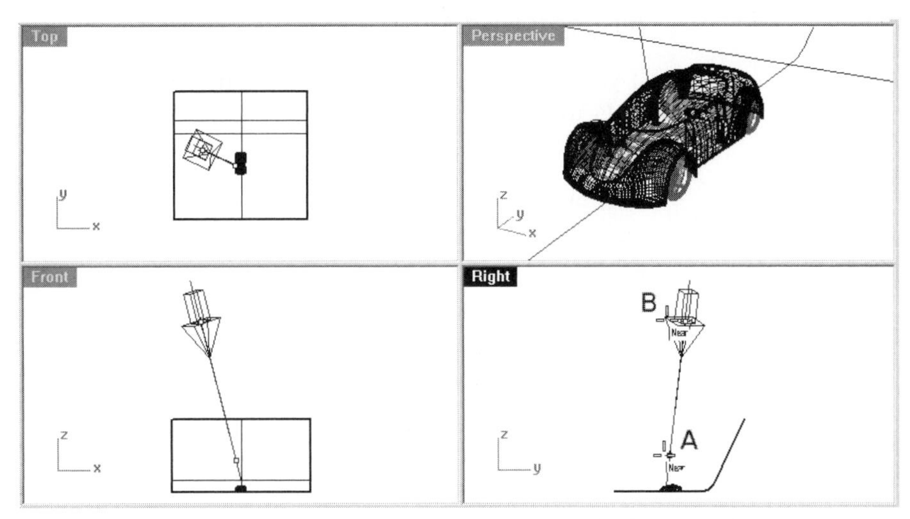

Figure 14–41. Directional light being placed along the bouncing line

Figure 14–42. Viewport rendered

(**NOTE:** *To not render the bouncing line in the rendered image, you can do one of the following:)*

- Clear the Render Curves and Isocurves check box in the Rhino Render pager of the Document Properties dialog box
- Hide the curve or put it in a layer that is turned off

Relocating a Light for It to Bounce

If you change the viewport orientation, the bouncing light effect will be lost. To relocate an existing light so that it bounces light, continue with the following steps.

34 Set the current layer to Light05 and turn off the Light04 layer. A spotlight is already constructed on this layer.

35 Click on the Bounce Light button on the Lights toolbar.

36 Select A, shown in Figure 14–44, the spotlight.

37 Select locations B (surface) and then location C (location of highlight) shown in Figure 14–43. The selected spotlight is relocated to bounce light in the perspective viewport.

38 Render the perspective viewport. (See Figure 14–44.)

39 Do not save the file.

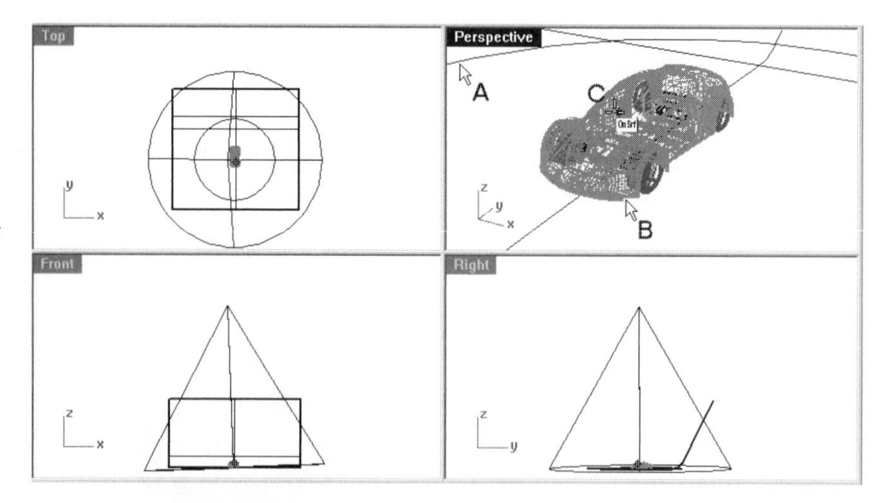

Figure 14–43. Spotlight being relocated

Figure 14–44. Rendered viewport

Turning Off Lights

To reiterate, there are two ways to turn off lights in a scene:

- ❐ Click on the Edit Light Properties button on the Properties or Lights toolbar, select the light, and clear the On check box.
- ❐ Clear the Use lights on layers that are off button of the Rhino Render page of the Document Properties dialog box, put the lights on separate layers, and turn off those layers.

Digital Material

Digital material assignment is a crucial element that contributes to the final digital rendered image. In Rhino, material properties can be assigned to an object in one of three ways, as follows:

- ❐ By default, material property is ByLayer, meaning that the material property of any object is in accordance with the material property assigned to the layer on which the object resides. Therefore, you can simply assign a material property to a layer and then have the object residing in that layer inherit the material properties of the layer.
- ❐ You can assign a material property to an object independent of the layer where it resides.
- ❐ You can have an object's material properties inheriting from its parent, which has its material properties assigned in one of the above two ways.

To set digital material to a layer, you can click on the material column of a layer in the Layer dialog box. To set digital material to an object, click on the Edit Material Properties button of the Properties toolbar and select an object. (See Figure 14–45.)

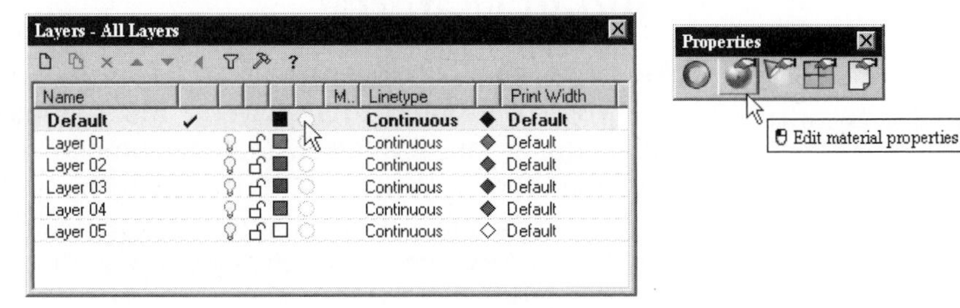

Figure 14–45. Clicking on the Material column of the Layer dialog box (left) and clicking on the Edit Material Properties dialog box (right)

If you click on the Material column of the Layer dialog box, the Material Editor dialog box is displayed. If you click on the Edit Material Properties button of the Properties toolbar and select an object, the Material page of the Properties dialog box will display. (See Figure 14–46.)

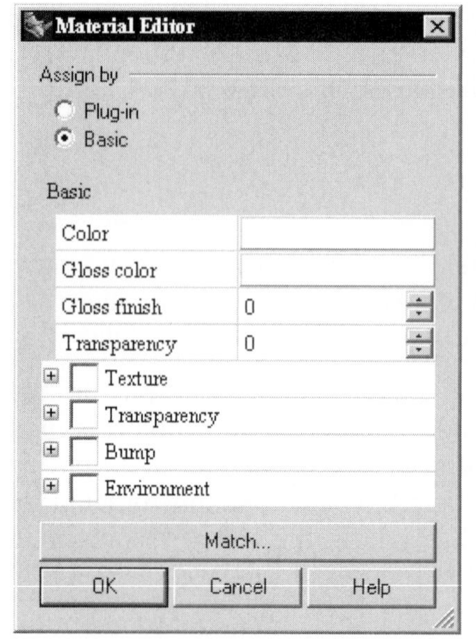

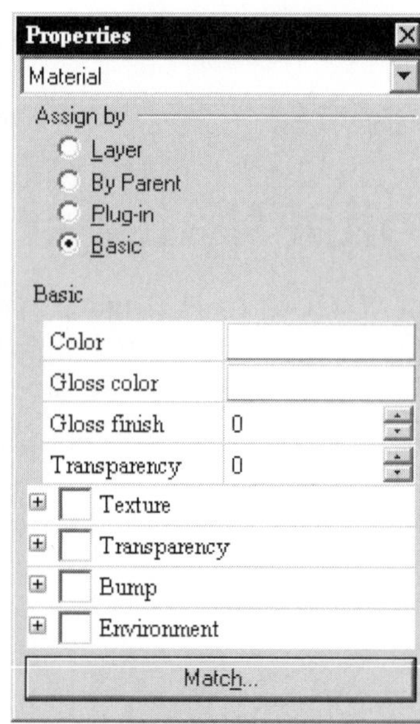

Figure 14–46. Material Editor dialog box (left) and Material page of the Properties dialog box

As shown in Figure 14–46, the Material Editor dialog box is similar to the Material page of the Properties dialog box. This is the case because basic material assignment method is the same for assigning to a layer or to individual objects. In any case, material properties can be defined in two ways:

- By using a plug-in
- By using Rhino's basic material properties

Plug-in Material

Assigning material properties via a plug-in means that you take a material properties source external to the Rhino program and bring it into a library (or create a library). Material properties can then be assigned from that library. (You need to use the TreeFrog, Flamingo Photometric, or Raytrace renderer to be able to use the default material library or to construct a new material library of your own.)

Basic Material

Basic material properties are defined by specifying a color, a gloss color, gloss color finish level, transparency percentage, and a number of mapping channels: texture, transparency, bump, and environment. Each of these channels is defined by specifying a bitmap and the bitmap's intensity, filter setting, and tiling.

Color

This is the basic color of the material; sometimes it is also called the diffuse color. It is the color of an object that we perceive when viewing under indirect light. You set the basic color by clicking on the color swatch to bring out the Select Color dialog box and selecting a color via the HSB color system or the RGB color system.

Gloss Color and Gloss Finish

These two parameters establish the appearance of the object when it is viewed under direct light. The gloss color is the color that the reflective area of the surface will elicit, and gloss finish concerns the strength of reflection.

Transparency

This parameter sets the overall opacity of the object. You can set the object to be 100 percent transparent, 100 percent opaque, or any value in between.

To appreciate the effects of setting color, gloss color, gloss finish, and transparency values and compare the results, perform the following steps:

1. Select File > Open and select the file Material01.3dm from the Chapter 14 folder on the companion CD.
2. Select Edit > Layers > Edit Layers, or click on the Edit Layers button on the Layer toolbar to open the Layer-All Layers dialog box.
3. Click on the material column of the Board layer, as shown in Figure 14–47.

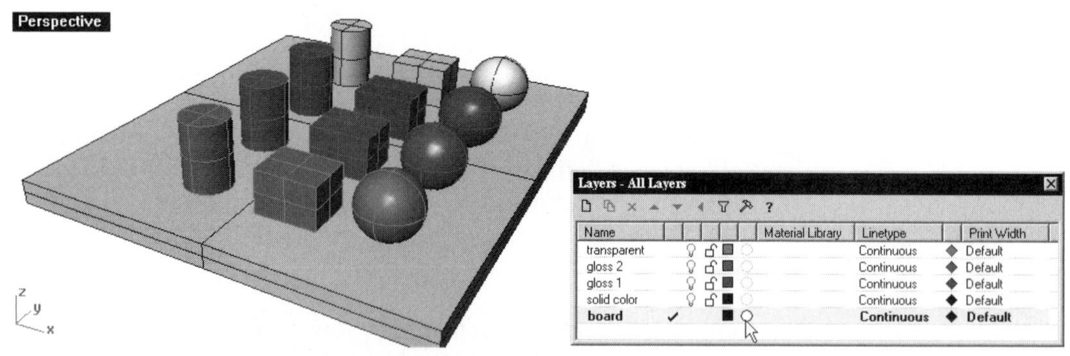

Figure 14–47. A layer being selected

4 In the Material Editor dialog box (Figure 14–46), click on A to check the Base box to use Base material, and click on B to display the Select Color dialog box.

5 In the Select Color dialog box, click on C to select a color (exact location is unimportant for this tutorial) and click on the OK button to close the Select Color dialog box.

6 Click on the OK button of the Material Editor dialog box. Color is set.

7 Repeat steps 3 through 6 above to set the color value for the Solid Color layer, setting its color to cyan.

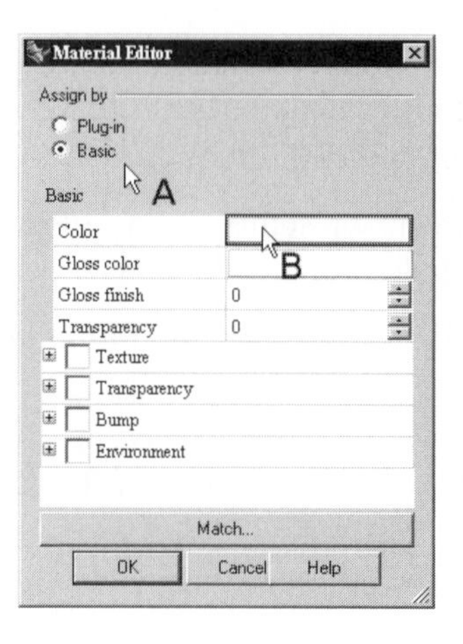

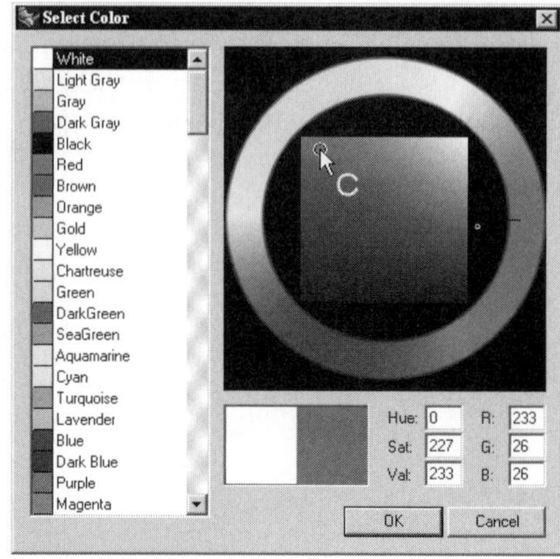

Figure 14–48. Color being set

8 Click on the Material column of the Gloss 1 layer.

9 In the Material Editor dialog box, check the Base button.

10 Set Color to Cyan, Gloss Color to Red, and Gloss finish to 25 percent.

11 Click on the Material column of the Gloss 2 layer.

12 Set Color to Cyan, Gloss Color to Red, and Gloss finish to 90 percent.
13 Click on the Material column of the Transparent layer.
14 Set Color to Cyan, Gloss Color to Red, Gloss finish to 80 percent, and Transparency to 60 percent.
15 Render the viewport by selecting Render > Render, or clicking on the Render button on the Render toolbar. (See Figure 14–49.)
16 Do not save your file.

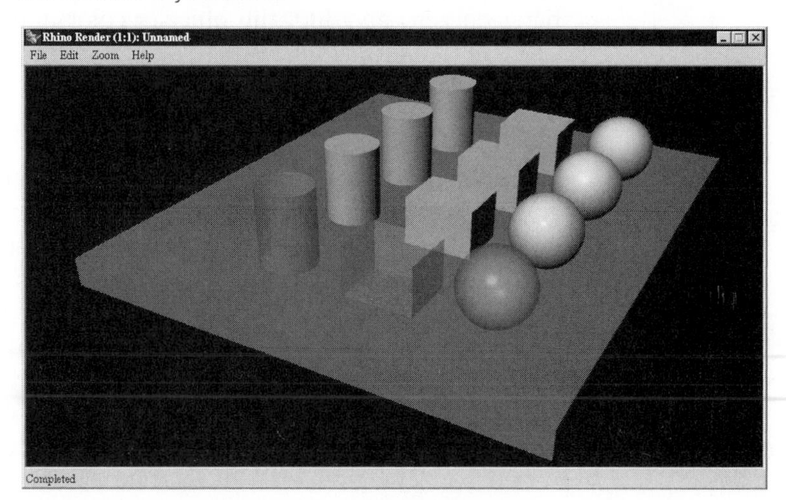

Figure 14–49. Rendered viewport

Using cylinders, cubes, and spheres to represent three different kinds of surfaces (single curve, flat, and double curve), it can be seen that glossiness of the cylinders and cubes are not readily perceivable, because there is no particular point of highlight. You can see that the more glossy the sphere is, the smaller is the highlight, so as to simulate the sharpness of reflection. Moreover, you should also notice that the transparent sphere has two highlights, one on the face nearer to the camera and one on the far side.

Basic Material's Mapping Methods

Apart from setting overall color, gloss color, glossiness, and transparency, you can wrap bitmaps onto selected objects or objects in a layer to simulate four kinds of visual effects: surface texture, partial transparency, surface bumpiness, and environment reflection.

Re-Parameterizing

To improve the effect of bitmap mapping, it may be necessary to re-parameterize the surface by typing Re-parameterize at the command area and selecting the objects to be re-parameterized.

Texture Map and Intensity

Applying texture map to an object is analogous to putting wallpaper on a wall. The intensity of the bitmap ranges from zero percent to 100 percent. In essence, a zero percent intensity is equivalent to having no texture mapping at all. At the other extreme, a 100 percent intensity setting will cause the object to possess the color and texture of the texture bitmap, with the object's original color and transparency totally masked by the texture map. If the intensity percentage is anywhere between zero and 100, the object's color and transparency setting will show off appropriately.

Transparency Map and Intensity

If you want only a portion of the object to be transparent instead of the entire object, you should wrap a bitmap onto the object, using the bitmap's color value to tell the computer what portion of the object is transparent and the degree of transparency. In essence, regardless of the color of the bitmap (red, green, or blue) only the color value is used to affect transparency. Therefore, it is more sensible to use a black and white or gray scale bitmap as the transparency map. After wrapping the transparent map onto the object, the area of the object covered by the darker color will become transparent. The degree of transparency depends on the gray scale of the bitmap and the intensity of the transparency map. For example, if you use a black and white bitmap and set the transparency intensity to 100 percent, areas covered by the black spots of the bitmap will be totally transparent, and areas covered by the white spots of the bitmap will be totally opaque.

Bump Map

If you want to have the objects look bumpy visually, you can wrap a bitmap onto the object, using the color value of a bitmap to adjust the apparent surface irregularity. The bitmap used is called bump map. Because the bump map is only used to affect bumpiness, the original color assignment to the object is not affected. Again, a black and white or gray scale bitmap is preferred because you can better foresee and estimate the effect of bumpiness. After applying a bump map, the object's surface will look bumpy, with the bumpiness degree dependent on the color value and the intensity of the bump map. It must be noted that a bump map only tells the computer to produce a bumpiness visual effect during rendering. In reality, the surface is still as it was. Therefore, if you really want to construct a bumpy surface, you should use appropriate surface manipulation tools.

Environment Map

To simulate reflective objects reflecting the scenes in the surrounding environment, you can use a color bitmap, delineating the imaginary environment in the scene and wrapping it around the objects. To control how reflective the object is, you set the intensity of the environment map. A value of 100 percent intensity will make the object look like a perfect mirror, reflecting the environment. Any value between 100 percent and zero percent will give the effect of varying degrees of reflectiveness.

Bitmap Map Filtering

Filtering is a technique in which the texture bitmap's individual pixel is reevaluated as the bitmap is stretched onto the surface where it is to be mapped, producing a better visual effect when the object being mapped is zoomed closely. Obviously, this takes a longer rendering time. Therefore, unless the camera is zoomed very close to the object, you may not need to check this box. To filter the bitmap, click on the Filter button.

Offset and Tiling

To better control how the bitmap (texture, transparency, bump, and environment) is wrapped onto objects in a layer or on selected objects, you manipulate two settings: offset and tiling.

By default, the bitmap is stretched onto the entire surface. To have the bitmap repeated as tiles on the surface, you set U and V tiling. If you set the U value to 2, the bitmap is repeated twice in the U direction.

Apart from repeating the bitmap, setting a negative value to U and V will mirror the bitmap about the U and V directions of the surface.

By default, the bitmap starts the origin of the surface, with U and V value both equal to zero. To adjust the starting point where the map is to apply on the surface, you adjust the U and V offsets.

Perform the following steps to appreciate how to apply bitmaps onto surfaces:

1. Select File > Open and select the file Material02.3dm from the Chapter 14 folder on the companion CD.

The objects are already assigned with color, gloss color, gloss finish, and transparency values.

2. Select Edit > Object Properties, or click on the Object Properties button on the Standard toolbar. The Properties dialog box displays.
3. Select A, shown in Figure 14–50.

4 In the Properties dialog box, select Material from the pull-down list box to display the Material page, if it is not already displayed.

5 Click on the Texture box or the browse button Next in the Map File field.

6 Select the bitmap file "Flowers.jpg" from the Chapter 14 folder of the accompanying CD.

7 Set the intensity value to 40%.

8 Clear the Filter button, if it is checked.

9 Click on the Modify button to display the Offset and Tiling button.

10 In the Offset and Tiling button, set Tiling U value to -1 and Tiling V value to -1, and click on the OK button. A texture map is applied.

11 Select Render > Render Preview to appreciate the effect and compare the rendered image with Figure 14–51.

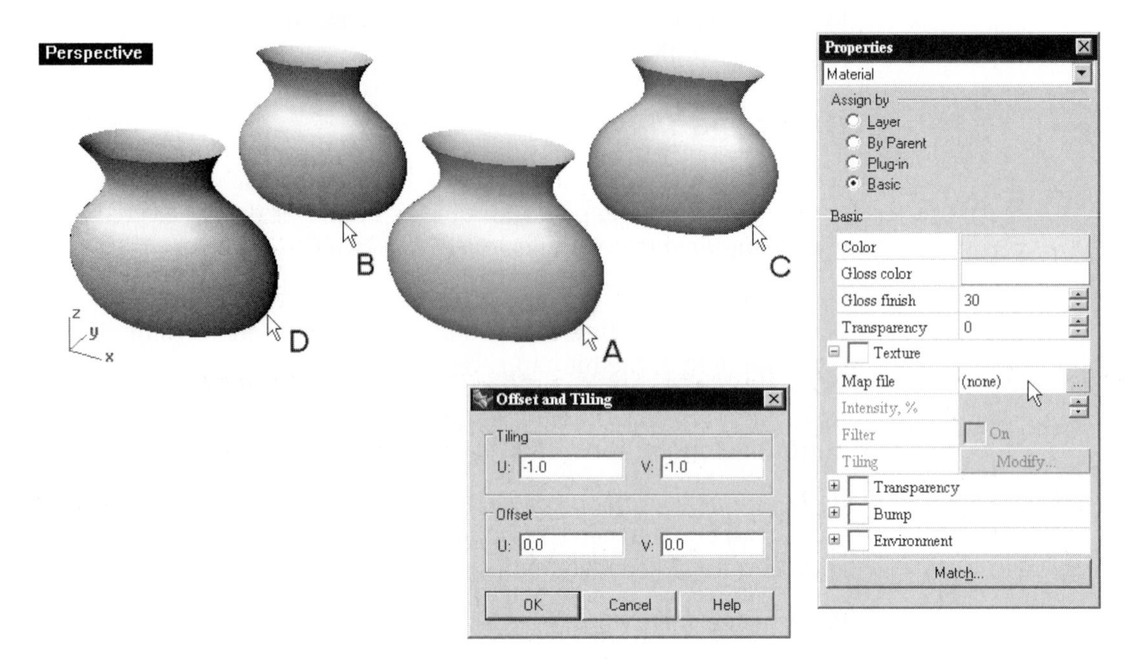

Figure 14–50. Bitmap being applied

12 Select B (Figure 14–50).

13 In the Properties dialog box, check the Transparency box and select the bitmap "Rhinoceros.jpg" from the Chapter 14 folder of the accompanying CD.

14 Set the intensity to 100 percent, if it is not 100 percent.

15 Clear the Filter check box.

16 Click on the Modify button.

17 In the Offset and Tiling dialog box, set Tiling U and Tiling V values to -1 and click on the OK button. A transparency bitmap is applied.

18 Select C (Figure 14–50).

19 In the Properties dialog box, check the Bump box and select the bitmap "Rhino.jpg" from the Chapter 14 folder of the accompanying CD.

20 Set the intensity to 25 percent.

21 Clear the Filter check box.

22 Click on the Modify button.

23 In the Offset and Tiling dialog box, set U tiling to -3 and V tiling to -3, and click on the OK button. A bump bitmap is applied.

24 Select D (Figure 14–50).

25 In the Properties dialog box, check the Environment box and select the bitmap "StreetScene.jpg" from the Chapter 14 folder of the accompanying CD.

26 Set the intensity to 50 percent.

27 Clear the Filter check box.

28 Click on the Modify button.

29 In the Offset and Tiling dialog box, set U tiling to -1, set V tiling to 1, and click on the OK button. An environment bitmap is applied.

30 Render the viewport. (See Figure 14–51.)

31 Do not save the file.

Figure 14–51. Bitmaps applied and rendered

Perform the following steps to learn about the effect of bitmap filtering:

1. Select File > Open and select the file Material02.3dm from the Chapter 14 folder on the companion CD.
2. Select Edit > Object Properties, or click on the Object Properties button on the Standard toolbar.
3. Select the surface.
4. In the Properties dialog, clear the Filter check box.
5. Render the viewport. (See Figure 14–52 left.)
6. Check the Filter check box.
7. Render the viewport. (See Figure 14–52 right.)
8. Compare the results.
9. Do not save your file.

Figure 14–52. Filter off (left) and filter on (right)

Texture Mapping Methods

Bitmaps are rectangular in shape, but surfaces on which bitmaps are mapped can be any shape: planar, spherical, or free-form. Therefore, if you simply put a bitmap on a surface that is not rectangular in shape, you may not get the desired outcome. To handle various shapes, you can apply the bitmap in a number of ways, including planar, box, cylindrical, and spherical. Perform the following steps to appreciate how bitmaps can be wrapped in different ways:

1. Select File > Open and select the file Mapping.3dm from the Chapter 14 folder on the companion CD.
2. Open the Edit Layers dialog box.
3. Click on the Material button to display the Material Editor dialog box.
4. In the Material Editor dialog box, check the Texture box, select the bitmap "Car Show.jpg" from the Chapter 14 folder on the companion CD, and click on the OK button. Texture is applied to the objects residing on this layer.

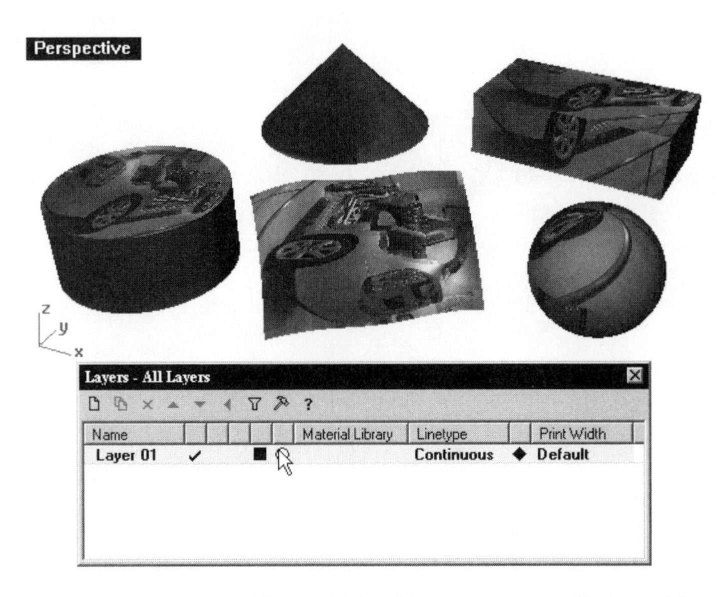

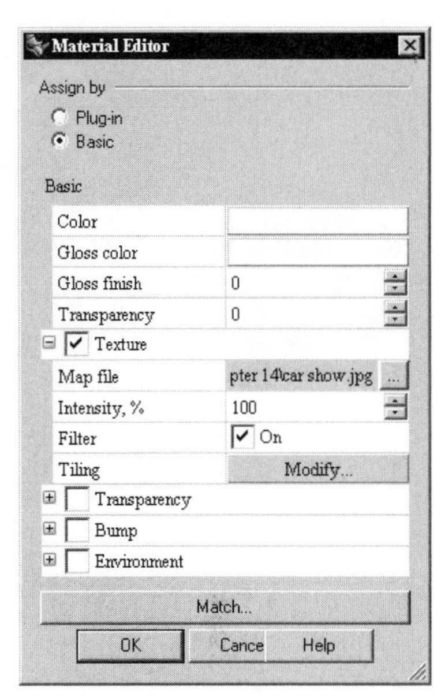

Figure 14–53. Texture map applied to objects

5. Select Edit > Object Properties, or click on the Object Properties button on the Standard toolbar to display the Properties dialog box.
6. Select A (Figure 14–54).
7. In the Properties dialog box, select the Texture Mapping page.
8. Check the Custom box.
9. Select Capped Cylindrical from the Project list box. The selected object's bitmap is projected from a capped cylinder.
10. Select B.
11. Select Cylindrical from the Project list box of the Properties dialog box. The cone's bitmap is projected conically.
12. Select C.
13. Select Box from the Project list box of the Properties dialog box. The box's bitmap is projected in a box form.
14. Select D.
15. Select Planar from the Project list box of the Properties dialog box. The surface's bitmap is projected in a planar direction.
16. Select E.
17. Select Spherical from the Project list box of the Properties dialog box. The sphere's bitmap is projected spherically.

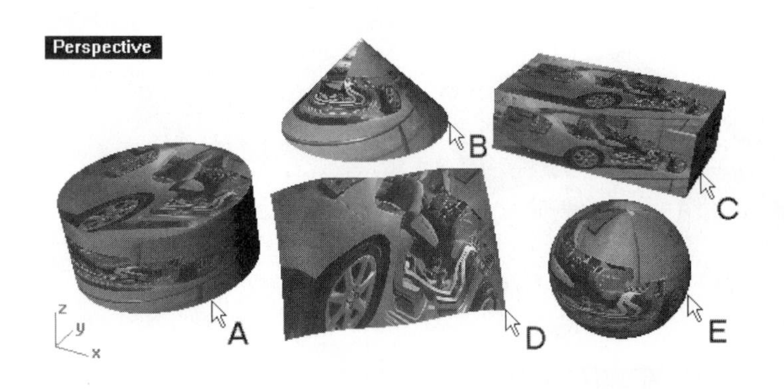

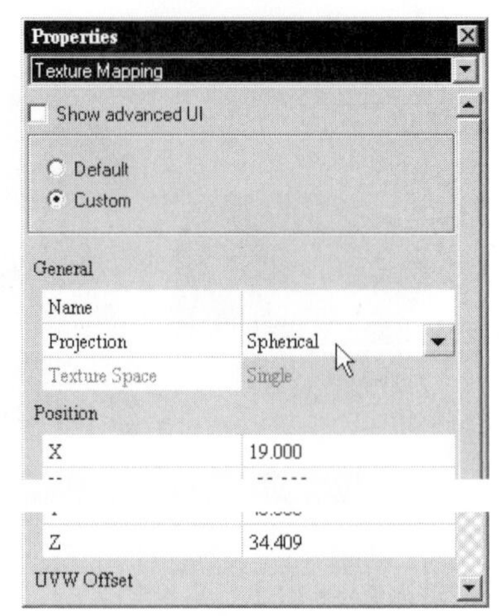

Figure 14–54. Mapping methods selected

To fine tune the bitmap's orientation, continue with the following steps:

18 Press the ESC key to clear any selection.

19 Click on the Apply Planar Mapping button on the Texture Mapping toolbar, or type ApplyPlanarMapping at the command area.

20 Select surface A (Figure 14–55) and press the ENTER key.

21 Type 1 at the command area to set the mapping method to channel 1, which is the default rendering channel.

22 Select end points B, C, and D.

23 Accept the default settings by pressing the ENTER key. The bitmap is applied in a planar direction as specified.

24 Do not save your file.

To fine tune other mapping methods, click on the Apply Box Mapping button (ApplyBoxMapping), Apply Spherical Mapping button (ApplySpherical-Mapping), Apply Cylindrical Mapping button (ApplyCylindricalMapping), and Apply Surface UV Mapping button (ApplySurfaceMapping) on the Texture Mapping toolbar.

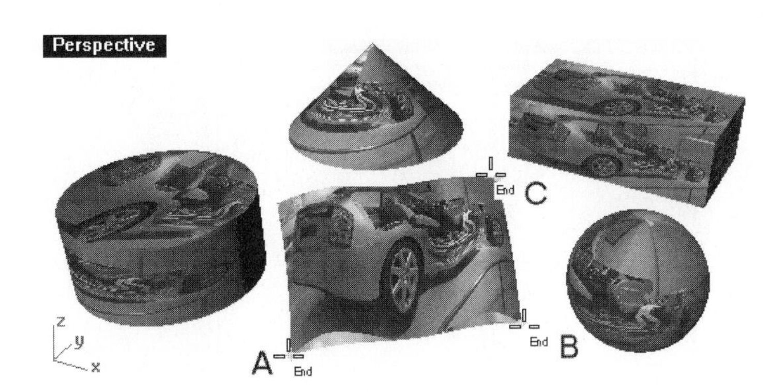

Figure 14–55. Mapping oriented

Environment Objects

Environment objects are objects you include in the scene to enhance its appearance. Using Rhino's basic renderer, you can adjust the ambient light and the background.

Setting Ambient Light

In reality, the combined effect of all lighting in a scene causes objects not directly under any light source to be illuminated. This is done by reflection of lights in the entire scene. The effect is commonly known as radiosity. The effect of radiosity depends on the strength of the light sources and how the lights are reflected by the objects in the scene. For example, an object placed inside a white box will be better illuminated than an object placed inside a box of dark color. The reason is that more light is reflected by the white box. Ambient light provides a means of approximating the radiosity effect. Using Rhino's basic renderer, you set the ambient light's color. To establish ambient lighting, perform the following steps:

1. Select File > Open and select the file Environment.3dm from the Chapter 14 folder on the companion CD.
2. Select Render > Render Properties.
3. In the Rhino Render tab of the Document Properties dialog box, select the Ambient Light color swatch.
4. In the Select Color dialog box, select Orange color, and then click on the OK button.
5. Close the Document Properties dialog box.
6. Render the scene, as shown in Figure 14–56.

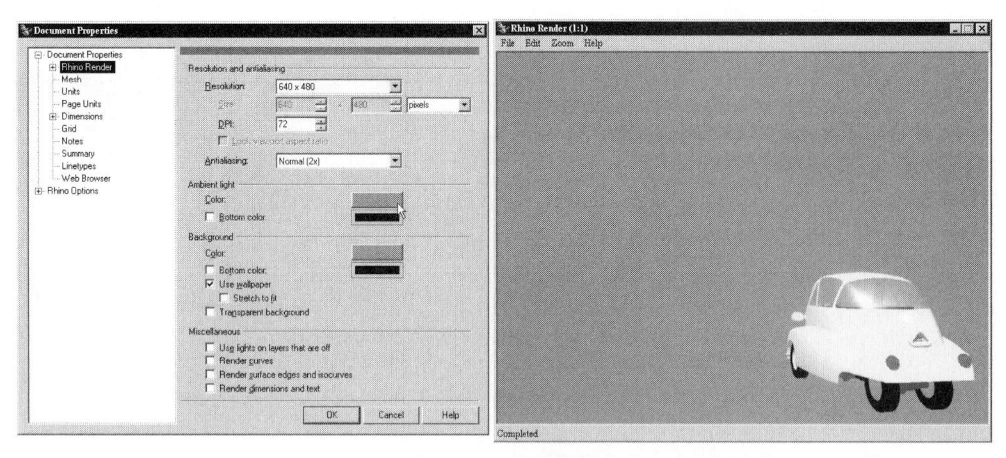

Figure 14–56. Ambient light changed and scene rendered

Setting Background Color

A simple way to enhance the visual effect of a rendered image is to change the background color, as follows.

7 Select Render > Render Properties.

8 In the Rhino Render tab of the Document Properties dialog box, select the Ambient Light color swatch and then change its color to black (the default color).

9 Select the Background color swatch.

10 In the Select Color dialog box, select red, and then click on the OK button.

11 Render the scene, as shown in Figure 14–57.

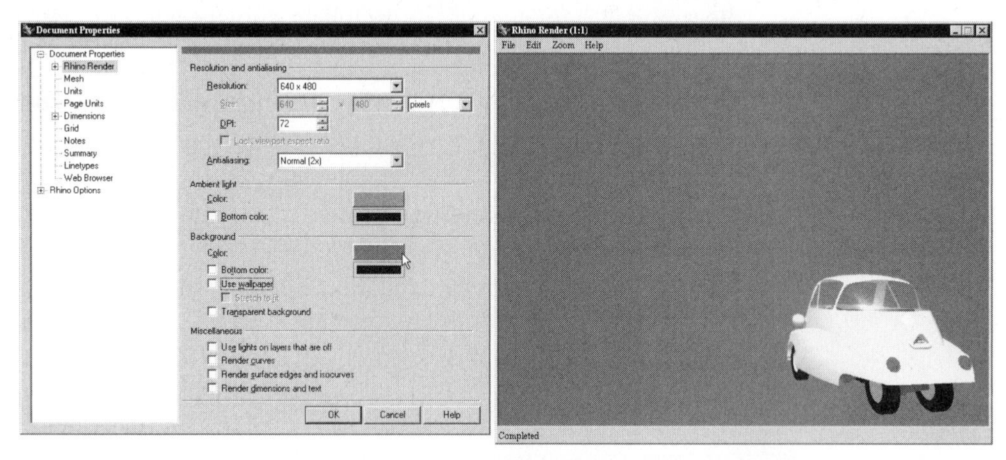

Figure 14–57. Background color changed and scene rendered

Using Wallpaper

To simulate a real environment, an image placed in the background may save a lot of effort in object construction and material assignment. Continue with the following steps.

12 Right-click on the Perspective viewport's label and select Viewport Properties, or click on the Viewport Properties button on the Properties toolbar.

13 In the Viewport Properties dialog box, shown in Figure 14–58, click on the Browse button of the Wallpaper options area.

14 Select the file "nightscene.jpg" from the Chapter 14 folder of the companion CD.

15 Click on the Show wallpaper check box, and clear the Show wallpaper as gray scale check box.

16 Click on the OK button to close the dialog box. A wallpaper is placed.

At this point, you can use the display tools to rotate, zoom, or pan the view to adjust the objects in the viewport to best suit the wallpaper.

17 Select Render > Render Properties.

18 Check the Use wallpaper and Stretch to fit check boxes, and click on the OK button.

19 Render the scene.

20 Do not save the file.

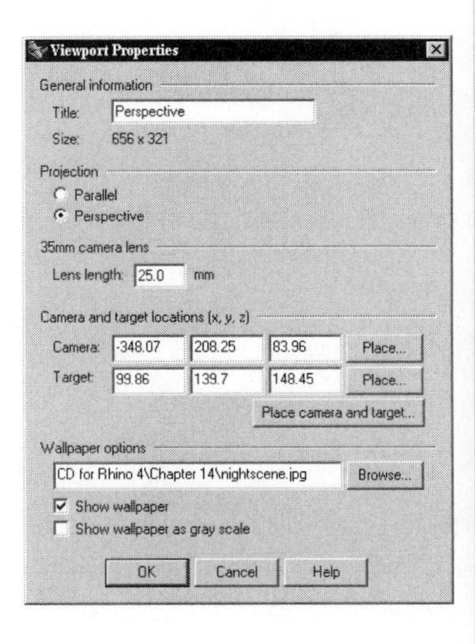

Figure 14–58. Viewport Properties dialog box and rendered perspective viewport

Render Properties

To produce a high quality rendered image, you may need to go through various render options that affect the outcome of rendering.

Resolution

To manage output requirements, you can set the resolution of the rendered image, as follows:

1. Select File > Open and select the file Environment.3dm from the Chapter 14 folder on the companion CD.
2. Select Render > Render Properties.
3. In the Rhino Render page of the Document Properties dialog box, select Custom from the Resolution pull-down list box of the Resolution and Anti-aliasing area.
4. Set the size to 100 x 80 pixels.
5. Click on the OK button.
6. Render the viewport.
7. Repeat steps 2 through 7 to set the resolution to 400 x 320 pixels, and render the viewport.
8. Compare the results, as shown in Figure 14-59.
9. Do not save your file.

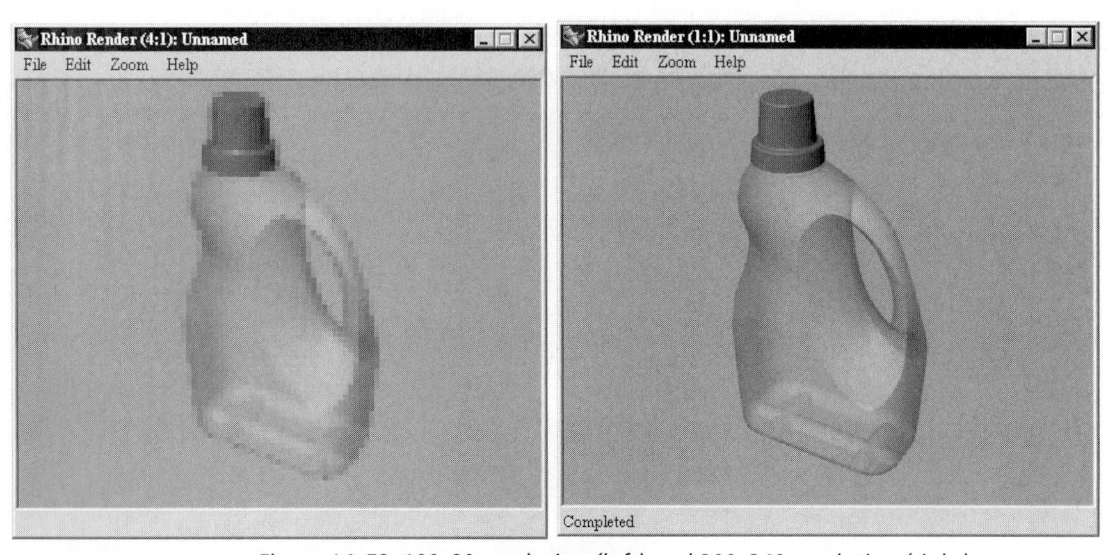

Figure 14–59. 100x80 resolution (left) and 300x240 resolution (right)

Anti-Aliasing

Because each discrete pixel in an image has a unique color value, an inclined edge in a rendered image may look jagged. To minimize this jagged effect, adjacent pixels in the inclined edges are averaged. The process is called anti-aliasing. Anti-aliasing produces a blurred edge to mask the jagged effect. For example, in a red box on a white background image, pixels are either red or white. Anti-aliasing produces a set of pinkish pixels between the red and white pixels. Continue with the following steps:

10 Select Render > Render Properties.

11 In the Rhino Render page of the Document Properties dialog box, set the resolution to 100 x 80 pixels.

12 Select None from the Antialiasing pull-down list box, and click on the OK button.

13 Render the viewport.

14 Repeat steps 11, 13, and 14, select High (10x) from the Antialiasing pull-down list box, and render the viewport.

15 Compare the results, as shown in Figure 14–60.

16 Do not save your file.

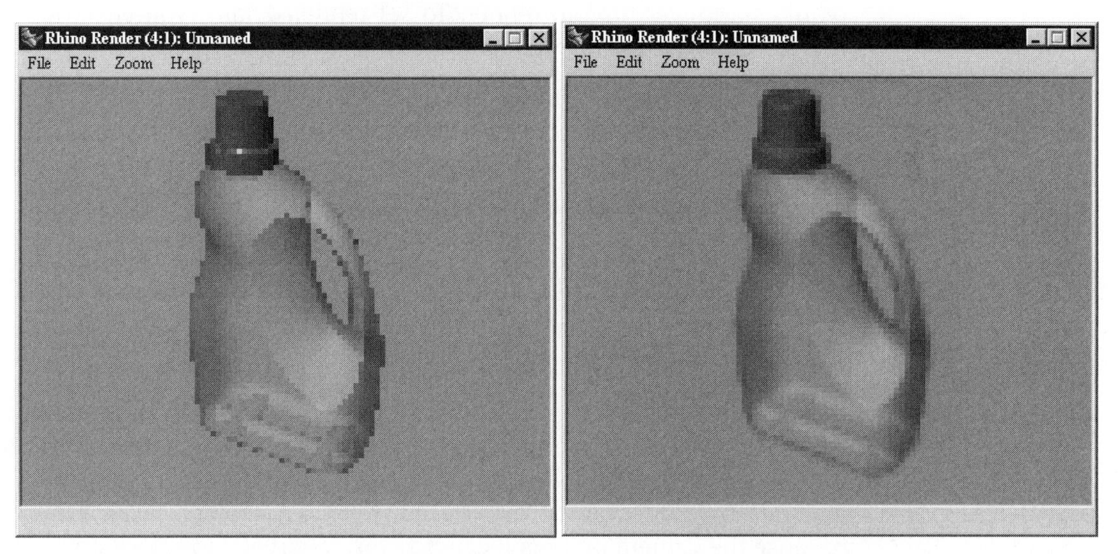

Figure 14–60. No anti-aliasing (left) and high anti-aliasing (right)

Miscellaneous

There are four check boxes at the bottom of the Rhino Render page of the Document Properties dialog box. Their functions are explained below:

Use lights on layers that are of Lights residing on layers that are off can be regarded as being turned off if this check box is cleared.

Render curve Curves are not rendered unless this box is checked.

Render surface edges and isocurves Edges and isocurves are not rendered unless this box is checked.

Render dimensions and text Dimensions and text are not rendered unless this box is checked.

Render Details

Several factors affect the quality of a rendered image and the amount of time to produce the image. In essence, a higher quality rendered image requires a longer time to render. To set render details, you click on the Render Details tab of the Document Properties dialog box.

Four major concerns are covered here: render acceleration grids, preventing self shadowing, object and polygon BSP tree, and transparency.

Render Acceleration Grids

Two settings, screen grid cell size and spotlight shadow grid size, have significant effect on rendering time.

Screen Grid Cell Size

When Rhino constructs a rendered image, rays are shot from the camera point to the scene to intersect with the geometry in the scene in order to determine which polygon is closer to the camera point and which polygons are hidden behind other polygons. These rays are called eye rays because the camera point is located at the observer's eye position.

To accelerate rendering, the render viewport is divided into a number of rectangular regions, called grid cells, building a list of polygons in these regions and sorting the polygons in each cell in accordance with their distance from the camera point.

In the process of rendering, computation is accelerated in such a way that eye rays are only needed to intersect with the polygons in the cell which is closest to the eye, and that some polygons farther away from the camera point are skipped after an intersection is found.

In Rhinoceros, the rectangular grid cells are square in shape, and you can set the size of these cells by specifying the width and height in terms of pixel number.

If we set a small grid cell size, there will be many grid cells in the scene. Naturally, it will require a significant amount of memory and take time to build the grids. However, subsequent rendering time will be faster. The default size is 16 pixels, and it works for most scenes.

Spotlight Shadow Grid Size

When an intersection between the eye ray and geometry in the scene is found, the ray is shot from the intersection to the light source. These rays are called shadow rays, for construction of shadows in the scene. In practice, lengthy rendering time is usually the result of tracing these shadow rays.

To accelerate the computation of the spotlight's shadow rays, the light cone of the spotlight is divided into rectangular regions, sorting the polygons within each region.

Shadow grid size is defined in terms of the number of grid cells in the spotlight's cone. The default value is 256, dividing the cone into 256 x 256 cells.

Preventing Self-Shadowing

Due to numerical fuzz in computing the shadow ray, it is possible that the shadow ray may hit the very same polygon, placing the shadow in the wrong place. This phenomenon is called self-shadowing. To prevent self-shadowing, the shadow ray is intentionally offset a small value. The default value is 0.001.

Object & Polygon BSP Tree

BSP stands for Binary Space Partitioning, a technique deployed for hidden face removal. It is a process in which the space is recursively subdivided until a stopping point is reached. In Rhino rendering, the stopping point is determined by the maximum tree depth value and the target node size.

Using BSP tree can accelerate rendering. However, building the BSP tree takes some time and memory. The bigger the depth and the smaller the node size, the more time and memory it will take to build.

Transparency

There may be a number of transparent objects in a scene. If the transparent objects are stacked, rendering time will increase significantly. In reality, if many transparent objects are stacked, it may be pointless to display objects at the very bottom. Therefore, you can, to save rendering time, set a limit on the number of stacked transparent objects for the rays to pass through. This value is called Max Bounces. For example, if you set Max Bounces = 5 and there are six transparent object stacked together, the sixth transparent object will not be regarded as transparent. The default maximum bounce is 12, which is adequate for most situations.

Render Mesh Quality

As explained earlier, a set of polygonal meshes is produced for the sake of rendering, even though the models are constructed by using NURBS surfaces. To have a higher quality rendered image, you may need to have a finer rendered mesh instead of a coarse rendered mesh. To set rendering mesh quality, you can click on the Mesh tab of the Document properties dialog box.

Render Options

By clicking on the Render Options of the Document Properties dialog box, the Render Options page is displayed. Here, you can click on Two Stage render to render the image in two stages, thus producing a more accurate image. If you have a multiple CPU computer, the rendering task can be divided among the CPUs by specifying the number of threads, such as 2 or 4, and also determining the priority of CPU usage.

Chapter Summary

To produce a photorealistic image from objects you construct in Rhino, you use one of the four renderers: Rhino's basic renderer, TreeFrog renderer, Flamingo's Raytrace renderer, or Flamingo's Photometric renderer. To produce nonphotorealistic renderings, you use the Penguin renderer. In this chapter, you learned how to use the basic renderer and the TreeFrog renderer.

There are five major steps in producing a photorealistic rendered image from the objects you construct in Rhino. You apply material properties to the objects, construct lights in the scene, include environment objects, set up a camera, and use a renderer to produce an image. Using the basic renderer, you add basic materials (basic color, reflective finish, transparency, texture, and bumpiness) to objects. If the TreeFrog renderer is used, more control of material properties is available through a set of plug-ins.

In a viewport, you can establish five kinds of lighting effects (spotlight, point light, directional light, rectangular light, and linear light). To enhance light bouncing effect, you can set up bouncing directional light, bouncing spotlight, and bouncing point light. If these lights are already constructed in a viewport, you can relocate them to produce bouncing light effect. In addition to material and lighting, you can add reality to the viewport by including three kinds of environment objects: ambient light, background color, and wallpaper.

Using Rhino's default animation scripts, you can create turntable animation, path animation, fly-through animation, and sun study animation. To produce more sophisticated animations, you can use the Bongo animator.

Review Questions

1. State the methods by which material properties are applied to an object.
2. Outline the ways that lights can be established in a viewport.
3. What environment objects can be included in a viewport?
4. In what ways can a camera be set?
5. Briefly describe the kind of animations that you can construct using Rhino.

Index

A

B

C

D

G

H

I

J

K

L

M

N

O

P

R

S

T

U

V

W

Z

IMPORTANT-READ CAREFULLY: This End User License Agreement ("Agreement") sets forth the conditions by which Delmar Learning, a division of Thomson Learning Inc. ("Thomson") will make electronic access to the Thomson Delmar Learning-owned licensed content and associated media, software, documentation, printed materials and electronic documentation contained in this package and/or made available to you via this product (the "Licensed Content"), available to you (the "End User"). BY CLICKING THE "I ACCEPT" BUTTON AND/OR OPENING THIS PACKAGE, YOU ACKNOWLEDGE THAT YOU HAVE READ ALL OF THE TERMS AND CONDITIONS, AND THAT YOU AGREE TO BE BOUND BY ITS TERMS CONDITIONS AND ALL APPLICABLE LAWS AND REGULATIONS GOVERNING THE USE OF THE LICENSED CONTENT.

1.0 SCOPE OF LICENSE

1.1 Licensed Content. The Licensed Content may contain portions of modifiable content ("Modifiable Content") and content which may not be modified or otherwise altered by the End User ("Non-Modifiable Content"). For purposes of this Agreement, Modifiable Content and Non-Modifiable Content may be collectively referred to herein as the "Licensed Content." All Licensed Content shall be considered Non-Modifiable Content, unless such Licensed Content is presented to the End User in a modifiable format and it is clearly indicated that modification of the Licensed Content is permitted.

1.2 Subject to the End User's compliance with the terms and conditions of this Agreement, Thomson Delmar Learning hereby grants the End User, a nontransferable, non-exclusive, limited right to access and view a single copy of the Licensed Content on a single personal computer system for non-commercial, internal, personal use only. The End User shall not (i) reproduce, copy, modify (except in the case of Modifiable Content), distribute, display, transfer, sublicense, prepare derivative work(s) based on, sell, exchange, barter or transfer, rent, lease, loan, resell, or in any other manner exploit the Licensed Content; (ii) remove, obscure or alter any notice of Thomson Delmar Learning's intellectual property rights present on or in the License Content, including, but not limited to, copyright, trademark and/or patent notices; or (iii) disassemble, decompile, translate, reverse engineer or otherwise reduce the Licensed Content.

2.0 TERMINATION

2.1 Thomson Delmar Learning may at any time (without prejudice to its other rights or remedies) immediately terminate this Agreement and/or suspend access to some or all of the Licensed Content, in the event that the End User does not comply with any of the terms and conditions of this Agreement. In the event of such termination by Thomson Delmar Learning, the End User shall immediately return any and all copies of the Licensed Content to Thomson Delmar Learning.

3.0 PROPRIETARY RIGHTS

3.1 The End User acknowledges that Thomson Delmar Learning owns all right, title and interest, including, but not limited to all copyright rights therein, in and to the Licensed Content, and that the End User shall not take any action inconsistent with such ownership. The Licensed Content is protected by U.S., Canadian and other applicable copyright laws and by international treaties, including the Berne Convention and the Universal Copyright Convention. Nothing contained in this Agreement shall be construed as granting the End User any ownership rights in or to the Licensed Content.

3.2 Thomson Delmar Learning reserves the right at any time to withdraw from the Licensed Content any item or part of an item for which it no longer retains the right to publish, or which it has reasonable grounds to believe infringes copyright or is defamatory, unlawful or otherwise objectionable.

4.0 PROTECTION AND SECURITY

4.1 The End User shall use its best efforts and take all reasonable steps to safeguard its copy of the Licensed Content to ensure that no unauthorized reproduction, publication, disclosure, modification or distribution of the Licensed Content, in whole or in part, is made. To the extent that the End User becomes aware of any such unauthorized use of the Licensed Content, the End User shall immediately notify Delmar Learning. Notification of such violations may be made by sending an Email to delmarhelp@thomson.com.

5.0 MISUSE OF THE LICENSED PRODUCT

5.1 In the event that the End User uses the Licensed Content in violation of this Agreement, Thomson Delmar Learning shall have the option of electing liquidated damages, which shall include all profits generated by the End User's use of the Licensed Content plus interest computed at the maximum rate permitted by law and all legal fees and other expenses incurred by Thomson Delmar Learning in enforcing its rights, plus penalties.

6.0 FEDERAL GOVERNMENT CLIENTS

6.1 Except as expressly authorized by Delmar Learning, Federal Government clients obtain only the rights specified in this Agreement and no other rights. The Government acknowledges that (i) all software and related documentation incorporated in the Licensed Content is existing commercial computer software within the meaning of FAR 27.405(b)(2); and (2) all other data delivered in whatever form, is limited rights data within the meaning of FAR 27.401. The restrictions in this section are acceptable as consistent with the Government's need for software and other data under this Agreement.

7.0 DISCLAIMER OF WARRANTIES AND LIABILITIES

7.1 Although Thomson Delmar Learning believes the Licensed Content to be reliable, Thomson Delmar Learning does not guarantee or warrant (i) any information or materials contained in or produced by the Licensed Content, (ii) the accuracy, completeness or reliability of the Licensed Content, or (iii) that the Licensed Content is free from errors or other material defects. THE LICENSED PRODUCT IS PROVIDED "AS IS," WITHOUT ANY WARRANTY OF ANY KIND AND THOMSON DELMAR LEARNING DISCLAIMS ANY AND ALL WARRANTIES, EXPRESSED OR IMPLIED, INCLUDING, WITHOUT LIMITATION, WARRANTIES OF MERCHANTABILITY OR FITNESS OR A PARTICULAR PURPOSE. IN NO EVENT SHALL THOMSON DELMAR LEARNING BE LIABLE FOR: INDIRECT, SPECIAL, PUNITIVE OR CONSEQUENTIAL DAMAGES INCLUDING FOR LOST PROFITS, LOST DATA, OR OTHERWISE. IN NO EVENT SHALL DELMAR LEARNING'S AGGREGATE LIABILITY HEREUNDER, WHETHER ARISING IN CONTRACT, TORT, STRICT LIABILITY OR OTHERWISE, EXCEED THE AMOUNT OF FEES PAID BY THE END USER HEREUNDER FOR THE LICENSE OF THE LICENSED CONTENT.

8.0 GENERAL

8.1 Entire Agreement. This Agreement shall constitute the entire Agreement between the Parties and supercedes all prior Agreements and understandings oral or written relating to the subject matter hereof.

8.2 Enhancements/Modifications of Licensed Content. From time to time, and in Delmar Learning's sole discretion, Thomson Thomson Delmar Learning may advise the End User of updates, upgrades, enhancements and/or improvements to the Licensed Content, and may permit the End User to access and use, subject to the terms and conditions of this Agreement, such modifications, upon payment of prices as may be established by Delmar Learning.

8.3 No Export. The End User shall use the Licensed Content solely in the United States and shall not transfer or export, directly or indirectly, the Licensed Content outside the United States.

8.4 Severability. If any provision of this Agreement is invalid, illegal, or unenforceable under any applicable statute or rule of law, the provision shall be deemed omitted to the extent that it is invalid, illegal, or unenforceable. In such a case, the remainder of the Agreement shall be construed in a manner as to give greatest effect to the original intention of the parties hereto.

8.5 Waiver. The waiver of any right or failure of either party to exercise in any respect any right provided in this Agreement in any instance shall not be deemed to be a waiver of such right in the future or a waiver of any other right under this Agreement.

8.6 Choice of Law/Venue. This Agreement shall be interpreted, construed, and governed by and in accordance with the laws of the State of New York, applicable to contracts executed and to be wholly preformed therein, without regard to its principles governing conflicts of law. Each party agrees that any proceeding arising out of or relating to this Agreement or the breach or threatened breach of this Agreement may be commenced and prosecuted in a court in the State and County of New York. Each party consents and submits to the non-exclusive personal jurisdiction of any court in the State and County of New York in respect of any such proceeding.

8.7 Acknowledgment. By opening this package and/or by accessing the Licensed Content on this Website, THE END USER ACKNOWLEDGES THAT IT HAS READ THIS AGREEMENT, UNDERSTANDS IT, AND AGREES TO BE BOUND BY ITS TERMS AND CONDITIONS. IF YOU DO NOT ACCEPT THESE TERMS AND CONDITIONS, YOU MUST NOT ACCESS THE LICENSED CONTENT AND RETURN THE LICENSED PRODUCT TO THOMSON DELMAR LEARNING (WITHIN 30 CALENDAR DAYS OF THE END USER'S PURCHASE) WITH PROOF OF PAYMENT ACCEPTABLE TO DELMAR LEARNING, FOR A CREDIT OR A REFUND. Should the End User have any questions/comments regarding this Agreement, please contact Thomson Delmar Learning at delmarhelp@thomson.com.